γ

# ENCYCLOPEDIA
## OF
# SEPARATION SCIENCE

# ENCYCLOPEDIA
## OF
# SEPARATION SCIENCE

Editor-in-Chief
## IAN D. WILSON

Managing Technical Editor
## EDWARD R. ADLARD

Editors
## MICHAEL COOKE
## COLIN F. POOLE

A Harcourt Science and Technology Company

San Diego   San Francisco   New York   Boston
London   Sydney   Tokyo

Copyright © 2000 by ACADEMIC PRESS

The following articles are US Government works in the
public domain and not subject to copyright:
III/FOOD TECHNOLOGY/Supercritical Fluid Chromatography
III/FORENSIC SCIENCES/Liquid Chromatography

III/INSECTICIDES/Gas Chromatography
Crown Copyright 1999

III/MECHANICAL TECHNIQUES: PARTICLE SIZE SEPARATION
Copyright © 1999 Minister of Natural Resources, Canada

II/CENTRIFUGATION/Large-Scale Centrifugation
Copyright © 2000 Minister of Public Works and Government Services, Canada

III/AIR LIQUEFACTION: DISTILLATION
Copyright © 2000 Air Products and Chemicals, Inc

Academic Press
*A Harcourt Science and Technology Company*
Harcourt Place, 32 Jamestown Road, London NW1 7BY, UK
http://www.academicpress.com

Academic Press
*A Harcourt Science and Technology Company*
525 B Street, Suite 1900, San Diego, California 92101-4495, USA
http://www.academicpress.com

ISBN 0-12-226770-2

Library of Congress Catalog Number: 99-61531

A catalogue record for this book is available from the British Library

Access for a limited period to an on-line version of the Encyclopedia of Separation Science is
included in the purchase price of the print edition.

This on-line version has been uniquely and persistently identified by the Digital Object Identifier
(DOI)

**10.1006/rwss.2000**

*By following the link*

**http://dx.doi.org/10.1006/rwss.2000**

from any Web Browser, buyers of the Encyclopedia of Separation Science
will find instructions on how to register for access.

If you have any problems with accessing the on-line version, e-mail:
**idealreferenceworks@harcourt.com**

Typeset by Macmillan India Limited, Bangalore, India
Printed and bound in Great Britain by The Bath Press, Bath, Somerset, UK
00 01 02 03 04 05 BP 9 8 7 6 5 4 3 2 1

# Editors

## EDITOR-IN-CHIEF

**Ian D. Wilson**
AstraZeneca Pharmaceuticals Limited
Mereside, Alderley Park
Macclesfield
Cheshire SK10 4TG, UK

## MANAGING TECHNICAL EDITOR

**Edward R. Adlard**
Formerly of Shell Research Limited
Thornton Research Centre
PO Box 1
Chester CH1 3SH, UK

## EDITORS

**Michael Cooke**
Royal Holloway, University of London
Centre for Chemical Sciences
Egham Hill, Egham
Surrey TW20 0EX, UK

**Colin F. Poole**
Wayne State University
Department of Chemistry
Detroit
MI 48202, USA

# Editorial Advisory Board

# Foreword

Separation science was first recognized as a distinct area of physical and analytical chemistry in the 1960s. The term was first coined, I believe, by the late J. Calvin Giddings, Research Professor at the University of Utah. Calvin Giddings recognized that the same basic physical principles governed a wide range of separation techniques, and that much could be learnt by applying our understanding of one such technique to others. This was especially true for his first loves, chromatography and electrophoresis and latterly field flow fractionation. Of course there are many separation techniques other than chromatography, many with a history at least as long, or indeed longer, than that of chromatography: distillation, crystallization, centrifugation, extraction, flotation and particle separation, spring to mind. Other separation techniques have emerged more recently: affinity separations, membrane separations and mass spectrometry. Most people, a few years ago, would not have classed mass spectrometry as a separation technique at all. However, with modern ionization methods, which minimize fragmentation, mixtures of compounds can first of all be separated and then each component identified through fragmentation by secondary ion-molecule collisions and further mass spectrometry. With the scale of mass spectrometry now matching that of microseparation methods such as capillary electrophoresis and capillary electrochromatography, combinations of orthogonal methods can now provide extremely powerful separation and identification platforms for characterizing complex mixtures.

Basically, all separation techniques rely on thermodynamic differences between components to discriminate one component from another, while kinetic factors determine the speed at which separation can be achieved. This applies most obviously to distillation, chromatography and electrophoresis, but is also obvious in most of the other techniques; even particle size separation by sieving can be classified in this way. The thermodynamic aspect is, of course, trivial being represented by the different sizes of the particles, as indeed it is for the size exclusion chromatography of polymers. However, the kinetic aspects are far from trivial. Anyone who has tried to sieve particles will have asked the question: is it better to fill the sieve nearly to the top and sieve for a long time, or is it better to dribble the material slowly into the sieve and just remove the heavies from time to time? One might further ask: how does one devise a continuous sieving process where large particles emerge from one port of the equipment, and small ones emerge from the other port? And how does one optimize throughput and minimize unit cost?

The publication of this *Encyclopedia of Separation Science* is a landmark for this area of science at the start of the third millennium. It will undoubtedly be of enormous value to practitioners of separation science looking for an overview and for guidance as to which method to select for a new problem, as well as to those who are at an early stage, simply dipping their toes into the waters, and trying to find out just what it is all about. Most important of all, by providing a comprehensive picture, it advances the whole field of separation science and stimulates further work on its development and application. The publishers, their editors and their authors are to be congratulated on a splendid effort.

John H. Knox
Edinburgh
8 March 2000

# Preface

The ability to perform separations for the analysis, concentration or isolation of substances present in mixtures (of varying degrees of complexity) is arguably fundamental to the maintenance of our technological civilization. Separations can also be rather difficult to define, and over the course of a number of debates where we tried to 'separate the wheat from the chaff', we defined separations for the purposes of this work as '*processes of any scale by which the components of a mixture are separated from each other without substantial chemical modification*'. Of course some of the processes that are used in separations have a very long history, and the terms used to describe them are in widespread general use. So, we talk about how it is possible to distil wisdom, precipitate an argument, extract meaning and crystallize an idea and who can predict the future uses of the word chromatography? Whilst separations have been practised as an *art* for millennia, the last hundred years or so has seen the elucidation of the fundamentals that lie behind many of these processes. Thus, although separations are most widely used for achieving some practical objective, a firm theoretical understanding has been put into place that does allow the use of the term *separation science*. We have tried here to reflect the theoretical and practical aspects of the topics in this encyclopedia, and have attempted to achieve a blend of theory, practice and applications that will enable someone knowledgeable in a field to go directly to a relevant article; whilst the novice can begin with an overview and gradually iterate towards the practical application.

One thing is clear, separations cover such a wide range of topics that no single individual can be knowledgeable, let alone expert, in them all. It is against this background that we decided that an encyclopedia designed to cover this science would be of value as a single source of reference that would provide access to the whole field of separations.

For the purposes of defining the scope and coverage of the encyclopedia we have divided the area of separations into 12 families, or topic areas, using separation principles based on affinity, centrifugation, chromatography, crystallization, distillation, electrophoresis, extraction, flotation, ion exchange, mass spectrometry, membranes and particle size. Whilst there is no doubt that different editors might have grouped these slightly differently, they did not seem to us to be capable of further reduction. Taken together, we believe that they provide coverage of the whole field.

In preparing this multi-author and multi-volume work, all of the editors have been conscious of the gaps in their own expertise and the debt which they and the publishers owe to the Editorial Advisory Board, who to a large degree have compensated for the deficiencies in our own knowledge. Their help has been invaluable, as without them we would not have been able to achieve the necessary balance and it was, therefore, a source of particular sadness that one of them, Ted Woodburn, died prior to publication. Without Ted it is quite clear that the topic area of flotation would not have been so well covered, and we would like to think that he would have been well pleased with the finished work. We also hope that the masterly overview which he contributed to the encyclopedia, will be a lasting memorial to him. We are grateful to Jan Cilliers for stepping into the role of Editorial Board Advisor on flotation at short notice. We would also like to acknowledge the valuable input of G. J. Arkenbout at the beginning of this project, who was able to work with us for a short period of time before his death.

Assembling this knowledgeable and enthusiastic group of experts was a difficult task, and the editors would also wish to acknowledge the role of the Major Reference Works development team at Academic Press in this whole area, as well as thanking them for their assistance and patience throughout the project, from the initial planning to its final publication.

Finally, of course, we must acknowledge the contributions of the authors whose expertise constitutes this encyclopedia. Some of them have become firm friends in the period between the inception of this project and its completion. After all, separation science separates things but brings people together.

Ian D. Wilson, Edward R. Adlard, Michael Cooke, Colin F. Poole
Editors

# Introduction

The need for separations as a means for performing the isolation, purification or analysis of substances, at scales ranging from tonnage quantities down to picograms or less, is an important feature of modern life. Such separations underpin virtually all aspects of research and commerce and indeed a vast industry has arisen to provide the equipment and instrumentation to perform and control these essential processes; indeed it is impossible to envisage our world without ready access to separations. To provide these capabilities a whole family of techniques has evolved to exploit differences in the physical or chemical properties of the compounds of interest, and to accommodate the scales on which the separations are performed. After all, a separation that works on the picogram scale based perhaps on capillary electrophoresis, may not easily be transferred to the gram scale and will be utterly impossible on the kilogram scale. In such instances an alternative type of separation, based on a totally different principle must be sought. And therein lies the problem–most scientists are specialists and while having an excellent knowledge of their own, often narrow, sphere of expertise are generally possessed of a much more hazy view of the capabilities and attributes of techniques outside that area. Even worse, such ignorance may persuade them to adopt an approach that is quite unsuited to the solution of the problem in hand.

The large number of articles in this encyclopedia, and the wide variety of the subject matter described in them, represents an attempt to provide separation scientists with a single authoritative source covering the broad range of separation methods currently available. Inevitably there is some overlap in places between articles dealing with closely related topics. However, in a work such as this where each article is meant to be a self-contained source of information, some overlap is unavoidable and not wholly undesirable in the context of ensuring full coverage of a topic.

The articles in the *Encyclopedia of Separation Science* fall into three categories as follows: 'Level I', which provides overviews of a particular separation area, e.g. flotation, distillation, crystallization, etc., written by acknowledged experts in the particular fields. These articles are presented with the aim of providing a wide ranging introduction to the topic from which the reader can then, if it seems appropriate, move on to 'Level II' articles.

Level II articles cover the theory, development, instrumentation and practice of the various techniques contained within each broad classification. For example the Level I article on chromatography is supported in Level II by descriptions of gas, liquid and supercritical fluid chromatography, together with information on instrumentation. Separations are, however, often of interest to practitioners because of their applications and Level II serves as an introduction to 'Level III'.

Level III provides detailed descriptions of the use of the various methods described in Levels I and II for solving real problems. These might include articles on the various methods for extraction of pesticides or drugs from a matrix, with other articles on the chromatographic or electrophoretic techniques that could be used for their subsequent analysis. The extensive cross-referencing and exhaustive indexing for all of the articles in the work should enable the reader to obtain easily and rapidly all the relevant information in the encyclopedia. In addition, each article at whatever level contains a brief but carefully selected bibliography of key books, review articles and important papers. In this way the encyclopedia provides the reader with an invaluable 'gateway' into the separation science literature on a topic.

Lastly, because the importance of separation science lies in its value as an applied technique, we have commissioned a number of 'essential guides' to method development, in a limited number of key areas such as the isolation and purification of proteins and enzymes, etc., or the development of chromatographic separations.

While the techniques described in these volumes can in many cases be used as the basis of analytical methods, the emphasis of this work is on the methods of separation of mixtures, rather than their determination. For a treatise devoted to Analytical Science the reader is referred to the *Encyclopedia of Analytical Science*, also published by Academic Press, which can be considered to be complementary to the present work, in that it deals in depth with analysis rather than the application of separations.

The reader faced with the need to perform a separation, requiring information on a type of detector, or the use of a technique for a particular class of compound, etc., should therefore be able to look into the encyclopedia for information on that topic. Even where there is no information that is directly relevant to the problem, it should still be possible to begin at Level I in order to determine the potential of a technique to solve the problem and then progress down through the levels until a solution begins to emerge.

The editors believe that, taken as a whole, this encyclopedia and its electronic version should provide a valuable source of knowledge and expertise for both those already skilled in the art of some aspect of separations and also for the novice. That, at least, is our hope.

This encyclopedia is a guide providing general information concerning its subject matter; it is not a procedure manual. The readers should consult current procedural manuals for state-of-the-art instructions and applicable government safety regulations. The publisher and the authors do not accept responsibility for any misuse of this encyclopedia, including its use as a procedural manual or as a source of specific instructions.

# Guide to Use of the Encyclopedia

## Structure of the Encyclopedia

The material in the encyclopedia comprises a series of entries on three levels, as follows:

- Level I entries, 'Overviews', provide broad overviews of the separation techniques covered by the encyclopedia, and comprise 12 single articles, arranged alphabetically.
- Level II entries, 'Methods and Instrumentation', provide detailed theoretical and technical descriptions of separation techniques. Entries for each separation technique are listed under the broad heading for each, with the latter following, in alphabetical sequence. Entries may consist of either single articles or several articles that deal with various aspects of the topic. In the latter case the articles are arranged in a logical sequence within their technique headings, for example:

> ION EXCHANGE:
>   Catalysis: Organic Ion Exchangers
>   Historical Development
>   Inorganic Ion Exchangers
>   Multispecies Ion Exchange Equilibria       *See* Ion Exchange: Surface Complexation Theory: Multispecies
>       Ion Exchange Equilibria.
>   Non-Phosphates: Novel Layered Materials       *See* Ion Exchange: Novel Layered Materials: Non-Phosphates.
>   Novel Layered Materials: Phosphates
>   Novel Layered Materials: Non-Phosphates
>   Organic Ion Exchangers
>   Organic Membranes
>   Phosphates: Novel Layered Materials       *See* Ion Exchange: Novel Layered Materials: Phosphates.
>   Surface Complexation Theory: Multispecies Ion Exchange Equilibria
>   Theory of Ion Exchange

- Level III entries, 'Practical Applications', (comprising single or multi-article entries) describe applications of these separation techniques to particular separation problems, and are arranged in alphabetical sequence.

To help you realize the full potential of the material in the encyclopedia we have provided the following features to help you find the topic of your choice.

## 1. Contents List

Your first point of reference will probably be the contents list. The complete contents list appearing in each volume will provide you with both the volume number and page number of the entry. On the opening page of an entry a contents list is provided so that the full details of the articles within the entry are immediately available.

Alternatively you may choose to browse through a volume; to assist you in identifying your location within the encyclopedia a running headline indicates the current level, the current entry and the current article within that entry.

You will find 'dummy entries' where obvious synonyms exist for entries or where we have grouped together related topics. Dummy entries appear in both the contents list and the body of the text. For example a dummy entry appears within Level II for Membrane Separations: Kidney Dialysis which directs you to **Membrane Separations: Dialysis in Medical Separations,** where the material on this subject is located.

## Level II Example

If you were attempting to locate material on Polyacrylamide Gel Electrophoresis via the Electrophoresis section, you could do this either via **A** the contents list, or **B** browsing through the text.

### A The contents list

ELECTROPHORESIS (*Continued*)

One-dimensional Sodium Dodecyl Sulfate Polyacrylamide Gel Electrophoresis

Polyacrylamide Gel Electrophoresis      *See* ELECTROPHORESIS: Two-dimensional Polyacrylamide Gel Electrophoresis, ELECTROPHORESIS: One-dimensional Sodium Dodecyl Sulfate Polyacrylamide Gel Electrophoresis, ELECTROPHORESIS: One-dimensional Polyacrylamide Gel Electrophoresis.

Porosity Gradient Gels
Proteins, Detection of
Staining      *See* ELECTROPHORESIS: Detection Techniques: Staining, Autoradiography and Blotting
Theory of Electrophoresis
Two-dimensional Electrophoresis
Two-dimensional Polyacrylamide Gel Electrophoresis

At the appropriate location in the contents list, the page numbers for these articles are given.

### B Browsing through the text

If you were trying to locate the material by browsing through the text and you looked up Electrophoresis: Polyacrylamide Gel Electrophoresis then the following information would be provided.

## ELECTROPHORESIS: POLYACRYLAMIDE GEL ELECTROPHORESIS

*See* **II / ELECTROPHORESIS: One-dimensional Polyacrylamide Gel Electrophoresis; One-dimensional Sodium Dodecyl Sulfate Polyacrylamide Gel Electrophoresis; Two-dimensional Polyacrylamide Gel Electrophoresis**

Alternatively, if you were looking up Electrophoresis: One-dimensional Polyacrylamide Gel Electrophoresis the following information would be provided. The icon that appears alongside the article title indicates which of the 12 main areas of separation the article falls within.

**ELECTROPHORESIS/One-Dimensional Polyacrylamide Gel Electrophoresis**

## ELECTROPHORESIS

## One-Dimensional Polyacrylamide Gel Electrophoresis

### Level III Example

If you were attempting to locate material on Aromas, you could do this either via **A** the contents list, or **B** browsing through the text.

#### A The contents list

Aromas: Gas Chromatography* *See* Fragrances: Gas Chromatography

*Note articles are arranged alphabetically within each entry.*

At the appropriate location in the contents list, the page numbers for these articles are given.

#### B Browsing through the text

If you were trying to locate the material by browsing through the text and you looked up Aromas then the following information would be provided.

---

# AROMAS: GAS CHROMATOGRAPHY

*See*   **III / Fragrances: Gas Chromatography**

---

Alternatively, if you were looking up Fragrances the following information would be provided. The icon that appears alongside the article title indicates which of the 12 main areas of separation the article falls within.

---

**FRAGRANCES: GAS CHROMATOGRAPHY**

# FRAGRANCES: GAS CHROMATOGRAPHY

---

## 2. Cross References

All of the articles in the encyclopedia have been extensively cross-referenced. These include 'see' cross references that are located within the text of each article that direct the reader to articles which are of immediate relevance to the topic. '*See also*' cross-references appear at the end of the article, and point to related topics. These take the following form:

i. To indicate where a topic is discussed in greater detail elsewhere.

---

**III / AMINO ACIDS AND DERIVATIVES: CHIRAL SEPARATIONS**
*See also:* **II / Chromatography: Liquid:** Derivatization. **III / Chiral Separations:** Capillary Electrophoresis; Cellulose and Cellulose Derived Phases; Chiral Derivatization; Countercurrent Chromatography; Crystallization; Cyclodextrins and Other Inclusion Complexation Approaches; Gas Chromatography; Ion-Pair Chromatography; Ligand Exchange Chromatography; Liquid Chromatography; Molecular Imprints as Stationary Phases; Protein Stationary Phases; Synthetic Multiple Interaction ('Pirkle') Stationary Phases; Supercritical Fluid Chromatography; Thin-Layer (Planar) Chromatography.

---

ii. To draw the reader's attention to parallel discussions in other articles.

---

**III / AMINO ACIDS AND DERIVATIVES: CHIRAL SEPARATIONS**
*See also:* **II/Chromatography: Liquid:** Derivatization. **III/Chiral Separations:** Capillary Electrophoresis; Cellulose and Cellulose Derived Phases; Chiral Derivatization; Countercurrent Chromatography; Crystallization; Cyclodextrins and Other Inclusion Complexation Approaches; Gas Chromatography; Ion-Pair Chromatography; Ligand Exchange Chromatography; Liquid Chromatography; Molecular Imprints as Stationary Phases; Protein Stationary Phases; Synthetic Multiple Interaction ('Pirkle') Stationary Phases; Supercritical Fluid Chromatography; Thin-Layer (Planar) Chromatography.

---

iii. To indicate material that broadens the discussion.

---

**III / AMINO ACIDS AND DERIVATIVES: CHIRAL SEPARATIONS**
*See also:* **II/Chromatography: Liquid:** Derivatization. **III/Chiral Separations:** Capillary Electrophoresis; Cellulose and Cellulose Derived Phases; Chiral Derivatization; Countercurrent Chromatography; Crystallization; Cyclodextrins and Other Inclusion Complexation Approaches; Gas Chromatography; Ion-Pair Chromatography; Ligand Exchange Chromatography; Liquid Chromatography; Molecular Imprints as Stationary Phases; Protein Stationary Phases; Synthetic Multiple Interaction ('Pirkle') Stationary Phases; Supercritical Fluid Chromatography; Thin-Layer (Planar) Chromatography.

---

## 3. Index

The index in Volume 10 will provide you with the volume number and page number showing where the material is located. The index entries differentiate between material that is a whole article, is part of an article or contains data presented in a table. On the opening page of the index detailed notes are provided.

## 4. Colour Plates

The colour figures for each volume have been grouped together in a plate section. The location of this section is cited both in the contents list and before the *See also* list of the pertinent articles.

## 5. Appendices

In addition to the articles that form the main body of the encyclopedia, you will find material in the appendices that provides 'Essential Guides' to topics such as method development in a particular separation technique, IUPAC definitions of chromatographic terms and conventions, and a range of tabulated material providing physical constants.

The Appendices are located in Volume 10.

## 6. Contributors

A full list of contributors appears in Volume 10.

# Contents

## Volume 1

## Level I

| | | |
|---|---|---|
| AFFINITY SEPARATION | *K Jones* | 3 |
| CENTRIFUGATION | *DN Taulbee, MM Maroto-Valer* | 17 |
| CHROMATOGRAPHY | *CF Poole* | 40 |
| CRYSTALLIZATION | *HJM Kramer, GM van Rosmalen* | 64 |
| DISTILLATION | *R Smith, M Jobson* | 84 |
| ELECTROPHORESIS | *D Perrett* | 103 |
| EXTRACTION | *DE Raynie* | 118 |
| FLOTATION | *T Woodburn* | 128 |
| ION EXCHANGE | *A Dyer* | 156 |
| MASS SPECTROMETRY | *KL Busch* | 174 |
| MEMBRANE SEPARATIONS | *RW Baker* | 189 |
| PARTICLE SIZE SEPARATION | *J Janca* | 210 |

## Level II

AFFINITY SEPARATION

| | | |
|---|---|---|
| Affinity Membranes | *K Haupt, SMA Bueno* | 229 |
| Affinity Partitioning in Aqueous Two-Phase Systems | *G Johansson* | 235 |
| Aqueous Two-Phase Systems | *See* AFFINITY SEPARATION: Affinity Partitioning in Aqueous Two-Phase Systems. | 246 |
| Biochemical Engineering Aspects | *HA Chase* | 246 |
| Covalent Chromatography | *K Brocklehurst* | 252 |
| Dye Ligands | *YD Clonis* | 259 |
| Hydrophobic Interaction Chromatography | *HP Jennissen* | 265 |
| Immobilized Boronates and Lectins | *WH Scouten* | 273 |
| Immobilized Metal Ion Chromatography | *DP Blowers* | 277 |
| Immunoaffinity Chromatography | *ID Wilson, D Stevenson* | 283 |
| Imprint Polymers | *PAG Cormack, K Haupt, K Mosbach* | 288 |
| Molecular Imprint Polymers | *See* AFFINITY SEPARATION: Imprint Polymers. | 296 |
| Rational Design, Synthesis and Evaluation: Affinity Ligands | *G Gupta, CR Lowe* | 297 |
| Theory and Development of Affinity Chromatography | *R Scopes* | 306 |

CENTRIFUGATION

| | | |
|---|---|---|
| Analytical Centrifugation | *JL Cole* | 313 |
| Large-Scale Centrifugation | *T Beveridge* | 320 |
| Macromolecular Interactions: Characterization by Analytical Ultracentrifugation *DJ Winzor* | | 329 |

CENTRIFUGATION (*Continued*)

    Preparative Centrifugation    *See* CENTRIFUGATION: Large-Scale Centrifugation.    336

    Theory of Centrifugation    *AG Letki*    336

CHROMATOGRAPHY

    Automation    *E Verette, SA Gilson*    343

    Convective Transport in Chromatographic Media    *AE Rodrigues*    352

    Correlation Chromatography    *HC Smit*    358

    Countercurrent Chromatography and High-Speed Countercurrent Chromatography: Instrumentation    *WD Conway*    365

    Detectors: Laser Light Scattering    *RPW Scott*    374

    Hydrodynamic Chromatography    *A Revillon*    379

    Laser Light Scattering Detectors    *See* CHROMATOGRAPHY: Detectors: Laser Light Scattering.    393

    Liquid Chromatography-Gas Chromatography    *K Grob*    393

    Paper Chromatography    *ID Wilson*    397

    Polymer Separation by Size Exclusion Chromatography    *See* CHROMATOGRAPHY: Size Exclusion Chromatography of Polymers.    405

    Protein Separation    *RK Scopes*    405

    Size Exclusion Chromatography of Polymers    *B Trathnigg*    411

    Universal Chromatography    *D Ishii, T Takeuchi*    419

## Volume 2

CHROMATOGRAPHY: GAS

    Column Technology    *W Jennings*    427

    Derivatization    *P Husek*    434

    Detectors: General (Flame Ionization Detectors and Thermal Conductivity Detectors)    *D McMinn*    443

    Detectors: Mass Spectrometry    *MR Clench, LW Tetler*    448

    Detectors: Selective    *ER Adlard*    456

    Gas Chromatography-Infrared    *PR Griffiths*    464

    Gas Chromatography-Mass Spectrometry    *See* CHROMATOGRAPHY: GAS: Detectors: Mass Spectrometry.    470

    Gas Chromatography-Ultraviolet    *VL Lagesson, L Lagesson-Andrasko*    470

    Gas-Solid Gas Chromatography    *J de Zeeuw*    481

    Headspace Gas Chromatography    *B Kolb*    489

    High Temperature Gas Chromatography    *P Sandra, F David*    497

CHROMATOGRAPHY: GAS (*Continued*)

  High-Speed Gas Chromatography    *A Andrews*                                          505

  Historical Development    *ER Adlard, CF Poole*                                       513

  Ion Mobility Mass Spectrometry    *D Young, P Thomas*                                 520

  Large-Scale Gas Chromatography    *P Jusforgues*                                      529

  Multidimensional Gas Chromatography    *P Marriott*                                   536

  Pyrolysis Gas Chromatography    *CER Jones*                                           544

  Sampling Systems    *IW Davies*                                                       550

  Theory of Gas Chromatography    *PA Sewell*                                           558

CHROMATOGRAPHY: LIQUID

  Electron Spin Resonance Detectors in Liquid Chromatography    *See* CHROMATOGRAPHY:
    LIQUID: Detectors: Electron Spin Resonance.                                         566

  Chiral Separations in Liquid Chromatography: Mechanisms    *See* CHROMATOGRAPHY:
    LIQUID: Mechanisms: Chiral.                                                         566

  Column Technology    *P Myers*                                                        567

  Countercurrent Liquid Chromatography    *Y Ito*                                       573

  Derivatization    *IS Krull, R Strong*                                               583

  Detectors: Electron Spin Resonance    *K Osterloh, HH Borchert, C Kroll*             591

  Detectors: Evaporative Light Scattering    *RPW Scott*                               597

  Detectors: Fluorescence Detection    *RPW Scott*                                     602

  Detectors: Infrared    *RPW Scott*                                                   608

  Detectors: Mass Spectrometry    *MR Clench, LW Tetler*                               616

  Detectors: Refractive Index Detectors    *RPW Scott*                                 623

  Detectors: Ultraviolet and Visible Detection    *RPW Scott*                          630

  Electrochemically Modulated Liquid Chromatography    *MD Porter, H Takano*           636

  Electrochromatography    *N Smith*                                                   646

  Enhanced Fluidity Liquid Chromatography    *SV Olesik*                               654

  Evaporative Light Scattering Detectors in Liquid Chromatography    *See* CHROMATOGRAPHY:
    LIQUID: Detectors: Evaporative Light Scattering.                                    662

  Fluorescence Detectors in Liquid Chromatography    *See* CHROMATOGRAPHY: LIQUID:
    Detectors: Fluorescence Detection.                                                 663

  Historical Development    *VR Meyer*                                                  663

  Infrared Detectors in Liquid Chromatography    *See* CHROMATOGRAPHY: LIQUID:
    Detectors: Infrared.                                                               670

  Instrumentation    *WR LaCourse*                                                     670

  Ion Chromatography: Mechanisms    *See* CHROMATOGRAPHY: LIQUID: Mechanisms:
    Ion Chromatography.                                                                676

  Ion Pair Liquid Chromatography    *J Ståhlberg*                                       670

  Large-Scale Liquid Chromatography    *H Colin, GB Cox*                               685

CHROMATOGRAPHY: LIQUID *(Continued)*

Mass Spectrometry Detection in Liquid Chromatography    *See* CHROMATOGRAPHY: LIQUID: Detectors: Mass Spectrometry.    690

Mechanisms: Chiral    *IW Wainer*    691

Mechanisms: Ion Chromatography    *PR Haddad*    696

Mechanisms: Normal Phase    *RPW Scott*    706

Mechanisms: Reversed Phases    *RPW Scott*    711

Mechanisms: Size Exclusion Chromatography    *SR Holding*    718

Micellar Liquid Chromatography    *ML Marina, MA García*    726

Multidimensional Chromatography    *P Campins-Falco, R Herraez-Hernandez*    738

Normal Phase Chromatography: Mechanisms    *See* CHROMATOGRAPHY: LIQUID: Mechanisms: Normal Phase.    747

Nuclear Magnetic Resonance Detectors    *M Dachtler, T Glaser, H Händel, T Lacker, L-H Tseng, K Albert*    747

Partition Chromatography (Liquid–Liquid)    *C Wingren*    760

Physico-Chemical Measurements    *RPW Scott*    770

Proteins    *See* CHROMATOGRAPHY: Protein Separation.    779

Refractive Index Detectors in Liquid Chromatography    *See* CHROMATOGRAPHY: LIQUID: Detectors: Refractive Index Detectors.    779

Size Exclusion Chromatography: Mechanisms    *See* CHROMATOGRAPHY: LIQUID: Mechanisms: Size Exclusion Chromatography.    779

Theory of Liquid Chromatography    *PA Sewell*    779

Ultraviolet and Visible Detection in Liquid Chromatography    *See* CHROMATOGRAPHY: LIQUID: Detectors: Ultraviolet and Visible Detection.    787

CHROMATOGRAPHY: SUPERCRITICAL FLUID

Fourier Transform Infrared Spectrometry Detection    *MW Raynor, KD Bartle*    788

Historical Development    *KD Bartle*    798

Instrumentation    *TA Berger*    802

Large-Scale Supercritical Fluid Chromatography    *P Jusforgues, M Shaimi*    809

Theory of Supercritical Fluid Chromatography    *KD Bartle*    819

CHROMATOGRAPHY: THIN-LAYER (PLANAR)

Densitometry and Image Analysis    *PE Wall*    824

Historical Development    *E Reich, DE Jaenchen*    834

Instrumentation    *DE Jaenchen, E Reich*    839

Ion Pair Thin-Layer (Planar) Chromatography    *ID Wilson*    847

Layers    *F Rabel*    853

Mass Spectrometry    *WE Morden*    860

Modes of Development: Conventional    *T-H Dzido*    866

Modes of Development: Forced Flow, Overpressured Layer Chromatography and Centrifugation    *S Nyiredy*    876

Preparative Thin-Layer (Planar) Chromatography    *S Nyiredy*    888

CHROMATOGRAPHY: THIN-LAYER (PLANAR) (*Continued*)

Radioactivity Detection *T Clark* 899

Spray Reagents *PE Wall* 907

Theory of Thin-Layer (Planar) Chromatography *AM Siouffi* 915

# Volume 3

CRYSTALLIZATION

Additives: Molecular Design *JH ter Horst, RM van Rosmalen, RM Geertman* 931

Biomineralization *D Volkmer* 940

Control of Crystallizers and Dynamic Behaviour *HJM Kramer* 950

Dynamic Behaviour *See* CRYSTALLIZATION: Control of Crystallizers and Dynamic Behaviour 961

Geocrystallization *JA Gamble* 961

Melt Crystallization *PJ Jansens, M Matsuoka* 966

Polymorphism *MR Caira* 975

Zone Refining *C-D Ho, H-M Yeh* 985

DISTILLATION

Azeotropic Distillation *F-M Lee, RW Wytcherley, MC Hunt* 990

Batch Distillation *M Barolo* 995

Control Systems *See* DISTILLATION: Instrumentation and Control Systems. 1000

Energy Management *TP Ognisty* 1005

Extractive Distillation *F-M Lee, MC Hunt, RW Wytcherley* 1013

Freeze-Drying *G-W Oetjen* 1023

High and Low Pressure Distillation *JA Rocha Uribe, J Lopez-Toledo* 1035

Historical Development *MS Ray* 1045

Instrumentation and Control Systems *B Roffel* 1050

Laboratory Scale Distillation *RC Gillman* 1059

Modelling and Simulation *JR Haas* 1062

Multicomponent Distillation *V Rico-Ramirez UM Diwekar* 1071

Packed Columns: Design and Performance *L Klemas, JA Bonilla* 1081

Pilot Plant Batch Distillation *WR Curtis* 1098

Sublimation *JD Green* 1113

Theory of Distillation *IJ Halvorsen, S Skogestad* 1117

Tray Columns: Design *KT Chuang, K Nandakumar* 1135

Tray Columns: Performance *K Nandakumar, KT Chuang* 1140

Vapour–Liquid Equilibrium: Correlation and Prediction *BC-Y Lu, D-Y Peng* 1145

Vapour–Liquid Equilibrium: Theory *AS Teja, LJ Holm* 1159

ELECTROPHORESIS

Agarose Gels    *JR Shainoff*                                                                 1169

Autoradiography Electrophoresis    *See* ELECTROPHORESIS: Detection Techniques:
    Staining.                                                                                 1175

Blotting    *See* ELECTROPHORESIS: Detection Techniques: Staining.                           1175

Capillary Electrophoresis    *SFY Li, YS Wu*                                                 1176

Capillary Electrophoresis Detection    *See* ELECTROPHORESIS: Detectors for Capillary
    Electrophoresis.                                                                         1188

Capillary Electrophoresis-Mass Spectrometry    *M Hamdan, PG Righetti*                       1188

Capillary Electrophoresis-Nuclear Magnetic Resonance    *K Pusecker, J Schewitz*            1194

Capillary Gel Electrophoresis    *R Freitag*                                                 1201

Capillary Isoelectric Focusing    *PG Righetti, C. Gelfi*                                    1208

Capillary Isotachophoresis    *J Sádecká, J Polonsky*                                        1215

Cellulose Acetate    *G Destro-Bisol, M Dobosz, VL Pascali*                                  1222

Deoxyribonucleic Acid, Theory of Techniques for Separation    *J Noolandi*                   1227

Detection of Proteins in Electrophoresis    *See* ELECTROPHORESIS: Proteins, Detection of    1233

Detection Techniques: Staining, Autoradiography and Blotting    *PJ Wirth*                   1233

Detectors for Capillary Electrophoresis    *T Kappes, PC Hauser*                             1239

Discontinuous Electrophoresis    *MJ Doktycz*                                                1245

Electrochromatography: Thin Layer    *T Shafik, AG Howard*                                   1250

Electrochromatography in Thin-Layer Electrophoresis    *See* ELECTROPHORESIS:
    Electrochromatography: Thin Layer.                                                       1256

Electrophoresis Using Cellulose Acetate    *See* ELECTROPHORESIS: Cellulose Acetate.         1256

Electrophoresis: Discontinuous    *See* ELECTROPHORESIS: Discontinuous Electrophoresis.      1256

Gel Electrophoresis in Capillary Electrophoresis    *See* ELECTROPHORESIS: Capillary Gel
    Electrophoresis.                                                                         1257

Immunoelectrophoresis    *F Lampreave, M Piñeiro, S Carmona, MA Alava*                       1257

Isoelectric Focusing    *PG Righetti, C Gelfi*                                               1263

Isoelectric Focusing in Capillary Electrophoresis    *See* ELECTROPHORESIS:
    Capillary Isoelectric Focusing.                                                          1271

Isotachophoresis    *T Hirokawa*                                                             1272

Isotachophoresis in Capillary Electrophoresis    *See* ELECTROPHORESIS:
    Capillary Isotachophoresis.                                                              1280

Mass Spectrometry Detection in Capillary Electrophoresis    *See* ELECTROPHORESIS:
    Capillary Electrophoresis-Mass Spectrometry.                                             1280

Micellar Electrokinetic Chromatography    *M-L Riekkola*                                     1280

Microtechnology    *T McCreedy*                                                              1286

Nonaqueous Capillary Electrophoresis    *SH Hansen, I Bjørnsdottir,
    J Tjørnelund*                                                                            1293

Nuclear Magnetic Resonance Detection in Capillary Electrophoresis    *See* ELECTROPHORESIS:
    Capillary Electrophoresis-Nuclear Magnetic Resonance.                                    1301

One-dimensional Polyacrylamide Gel Electrophoresis    *PG Righetti*                          1301

ELECTROPHORESIS (*Continued*)

One-dimensional Sodium Dodecyl Sulfate Polyacrylamide Gel Electrophoresis
*GL Jones*                                                                                                    1309

Polyacrylamide Gel Electrophoresis    *See* ELECTROPHORESIS: Two-dimensional
Polyacrylamide Gel Electrophoresis, ELECTROPHORESIS: One-dimensional Sodium
Dodecyl Sulfate Polyacrylamide Gel Electrophoresis, ELECTROPHORESIS:
One-dimensional Polyacrylamide Gel Electrophoresis.                                                           1315

Porosity Gradient Gels    *GM Rothe*                                                                          1315

Proteins, Detection of    *MJ Dunn*                                                                           1342

Staining    *See* ELECTROPHORESIS: Detection Techniques: Staining, Autoradiography and
Blotting.                                                                                                    1348

Theory of Electrophoresis    *KS Pitre*                                                                       1348

Two-dimensional Electrophoresis    *M Fountoulakis*                                                           1356

Two-dimensional Polyacrylamide Gel Electrophoresis    *J-D Tissot, P Schneider,
MA Duchosal*                                                                                                  1364

EXTRACTION

Analytical Extractions    *MKL Bicking*                                                                       1371

Analytical Inorganic Extractions    *KA Anderson*                                                             1383

Extraction With Supercritical Fluid    *See* EXTRACTION: Supercritical Fluid Extraction.                      1389

Inorganic Extractions    *See* EXTRACTION: Analytical Inorganic Extractions.                                  1389

Microwave-Assisted Extraction    *V Lopez-Avila*                                                              1389

Multistage Countercurrent Distribution    *G Johansson*                                                       1398

Solid-Phase Extraction    *CF Poole*                                                                          1405

Solid-Phase Microextraction    *JB Pawliszyn*                                                                 1416

Solvent Based Separation    *R Gani, PM Harper, M Hostrup*                                                    1424

Steam Distillation    *L Ramos*                                                                               1434

Supercritical Fluid Extraction    *AA Clifford*                                                               1442

Ultrasound Extractions    *C Bendicho, I Lavilla*                                                             1448

## Volume 4

FLOTATION

Bubble-Particle Adherence: Synergistic Effect of Reagents    *BJ Bradshaw, CT O'Connor*                       1455

Bubble-Particle Capture    *J Ralston*                                                                        1464

Column Cells    *IM Flint, MA Burstein*                                                                       1471

Column Flotation Cells    *See* FLOTATION: Froth Processes and the Design of Column
Flotation Cells.                                                                                              1480

Cyclones for Oil/Water Separations    *MT Thew*                                                               1480

Dissolved Air    *D Shekhawat, P Srivastava*                                                                  1490

FLOTATION (*Continued*)

Electrochemistry: Contaminant Ions and Sulfide Mineral Interactions    *JT Smit, RF Sandenbergh, J. Gnoinsky* — 1494

Flotation Cell Design: Application of Fundamental Principles    *BK Gorain, JP Franzidis, EV Manlapig* — 1502

Foam Fractionation    *G Narsimhan* — 1513

Froth Processes and the Design of Column Flotation Cells    *I Flint, MA Burstein* — 1521

Historical Development    *Z Xu* — 1527

Hydrophobic Surface State Flotation    *JD Miller* — 1537

Intensive Cells: Design    *GJ Jameson* — 1541

Oil and Water Separation    *B Knox-Holmes* — 1548

Pre-aeration of Feed    *M Xu, Z Zhou, Z Xu* — 1556

Reagent Adsorption on Phosphates    *P Somasundaran, L Zhang* — 1562

ION EXCHANGE

Catalysis: Organic Ion Exchangers    *RL Albright* — 1572

Historical Development    *I Grafova* — 1577

Inorganic Ion Exchangers    *EN Coker* — 1584

Multispecies Ion Exchange Equilibria    *See* ION EXCHANGE: Surface Complexation Theory: Multispecies Ion Exchange Equilibria. — 1595

Non-Phosphates: Novel Layered Materials    *See* ION EXCHANGE: Novel Layered Materials: Non-Phosphates. — 1595

Novel Layered Materials: Phosphates    *U Costantino* — 1595

Novel Layered Materials: Non-Phosphates    *R Mokaya* — 1610

Organic Ion Exchangers    *C Luca* — 1617

Organic Membranes    *R Wodzki* — 1632

Phosphates: Novel Layered Materials    *See* ION EXCHANGE: Novel Layered Materials: Phosphates. — 1639

Surface Complexation Theory: Multispecies Ion Exchange Equilibria    *WH Höll, J Horst* — 1640

Theory of Ion Exchange    *R Harjula* — 1651

MASS SPECTROMETRY

Spectrometry-Mass Spectrometry Ion Mobility    *HR Bollan* — 1661

MEMBRANE SEPARATIONS

Bipolar Membranes and Membrane Processes    *H Strathmann* — 1667

Catalytic Membrane Reactors    *ME Rezac* — 1676

Concentration Polarization    *H Wijmans* — 1682

Dialysis in Medical Separations    *WR Clark, MJ Lysaght* — 1687

Diffusion Dialysis    *TA Davis* — 1693

Donnan Dialysis    *TA Davis* — 1701

Electrodialysis    *H Strathmann* — 1707

Filtration    *R Sahai, R Lombardi* — 1717

Gas Separations With Polymer Membranes    *DV Laciak, M Langsam* — 1725

MEMBRANE SEPARATIONS (*Continued*)

Haemodialysis    *See* MEMBRANE SEPARATIONS: Dialysis in Medical Separations.    1739

Kidney Dialysis    *See* MEMBRANE SEPARATIONS: Dialysis in Medical Separations.    1739

Liquid Membranes    *L Boyadzhiev*    1739

Membrane Bioseparations    *AL Zydney*    1748

Membrane Preparation    *I Pinnau*    1755

Microfiltration    *IH Huisman*    1764

Pervaporation    *HEA Brüschke, NP Wynn*    1777

Polymer Membranes    *See* MEMBRANE SEPARATIONS: Gas Separations With Polymer
Membranes.    1787

Reverse Osmosis    *V Spohn*    1787

Ultrafiltration    *M Cheryan*    1797

PARTICLE SIZE SEPARATION

Electric Fields in Field Flow Fractionation    *See* PARTICLE SIZE SEPARATION: Field Flow
Fractionation: Electric Fields.    1802

Electrostatic Precipitation    *JJ Harwood*    1802

Field Flow Fractionation: Electric Fields    *SN Semenov*    1811

Field Flow Fractionation: Thermal    *SN Semenov*    1815

Hydrocyclones for PARTICLE SIZE SEPARATION    *JJ Cilliers*    1819

Instrumentation of Field Flow Fractionation    *See* PARTICLE SIZE SEPARATION: Theory
and Instrumentation of Field Flow Fractionation.    1825

Sedimentation    *See* PARTICLE SIZE SEPARATION: Split Flow Thin Cell (SPLITT)
Separation.    1825

Sieving/Screening    *J Skopp*    1826

Split Flow Thin Cell (SPLITT) Separation    *C Contado*    1831

Theory and Instrumentation of Field Flow Fractionation    *J Janca*    1837

## Volume 5

## Level III

ACIDS

Gas Chromatography    *G Gutnikov, N Scott*    1847

Liquid Chromatography    *PR Haddad, KI Ng*    1854

Thin-Layer (Planar) Chromatography    *JHP Tyman*    1863

AFLATOXINS AND MYCOTOXINS

Chromatography    *RD Coker*    1873

Thin-Layer (Planar) Chromatography    *ME Stack*    1888

AIR LIQUEFACTION: DISTILLATION    *R Agrawal, DM Herron*    1895

AIRBORNE SAMPLES: SOLID PHASE EXTRACTION    *DJ Eatough*    1910

ALCOHOL AND BIOLOGICAL MARKERS OF ALCOHOL ABUSE:
GAS CHROMATOGRAPHY    *F Musshoff*    1921

ALCOHOLIC BEVERAGES: DISTILLATION     *See* Whisky: Distillation.     1931

ALDEHYDES AND KETONES: GAS CHROMATOGRAPHY     *H Nishikawa*     1931

ALKALOIDS

    Gas Chromatography     *M Muzquiz*     1938

    High Speed Countercurrent Chromatography     *See* Medicinal Herb Compounds: High-Speed
    Countercurrent Chromatography.     1949

    Liquid Chromatography     *R Verpoorte*     1949

    Thin-Layer (Planar) Chromatography     *J Flieger*     1956

ALLERGENS IN PERFUMES: GAS CHROMATOGRAPHY-MASS SPECTROMETRY
    *SC Rastogi*     1974

AMINES: GAS CHROMATOGRAPHY     *H Kataoka, S Yamamoto, S Narimatsu*     1982

AMINO ACIDS

    Gas Chromatography     *SL Mackenzie*     1990

    Liquid Chromatography     *I Molnár-Perl*     1999

    Thin-Layer (Planar) Chromatography     *R Bhushan, J Martens*     2012

AMINO ACIDS AND DERIVATIVES: CHIRAL SEPARATIONS     *ID Wilson, RPW Scott*     2033

AMINO ACIDS AND PEPTIDES: CAPILLARY ELECTROPHORESIS     *P Bohn*     2038

ANAESTHETIC MIXTURES: GAS CHROMATOGRAPHY     *A Uyanik*     2047

ANALYTICAL APPLICATIONS: DISTILLATION     *JD Green*     2052

ANION EXCHANGERS FOR WATER TREATMENT: ION EXCHANGE
    *See* WATER TREATMENT: Anion Exchangers: Ion Exchange.     2058

ANTIBIOTICS

    High-Speed Countercurrent Chromatography     *H Oka, Y Ito*     2058

    Liquid Chromatography     *T Itoh*     2067

    Supercritical Fluid Chromatography     *FJ Señoráns, KE Markides*     2077

ARCHAEOLOGY: USES OF CHROMATOGRAPHY IN     *C Heron, R Stacey*     2083

AROMAS: GAS CHROMATOGRAPHY     *See* Fragrances: Gas Chromatography.     2089

ART CONSERVATION: USE OF CHROMATOGRAPHY IN     *SL Vallance*     2089

ATMOSPHERIC ANALYSIS: GAS CHROMATOGRAPHY     *AC Lewis*     2096

BACTERIOPHAGES: SEPARATION OF     *P Serwer*     2102

BALSAMS AND RESINS: THIN-LAYER (PLANAR) CHROMATOGRAPHY
    *See* ESSENTIAL OILS: Thin-Layer (Planar) Chromatography.     2109

BASES: THIN-LAYER (PLANAR) CHROMATOGRAPHY     *L Lepri, A Cincinelli*     2109

BILE ACIDS

    Gas Chromatography     *AK Batta, G Salen*     2124

    Liquid Chromatography     *K Saar, S Müllner*     2130

BILE COMPOUNDS: THIN-LAYER (PLANAR) CHROMATOGRAPHY     *K Saar, S Müllner*     2135

BIOANALYTICAL APPLICATIONS: SOLID-PHASE EXTRACTION     *DA Wells*     2142

BIOGENIC AMINES: GAS CHROMATOGRAPHY     *R Draisci, PL Buldini, S Cavalli*     2146

BIOLOGICAL SYSTEMS: ION EXCHANGE     *RJP Williams*     2156

BIOLOGICALLY ACTIVE COMPOUNDS AND XENOBIOTICS: MAGNETIC AFFINITY
        SEPARATIONS     *I Safarik, M Safarikova*     2163

BIOMEDICAL APPLICATIONS

    Gas Chromatography-Mass Spectrometry     *V Garner*     2170

    Thin-Layer (Planar) Chromatography     *See* Clinical Chemistry: Thin-Layer (Planar)
        Chromatography.     2180

BITUMENS: LIQUID CHROMATOGRAPHY     *K Dunn, GV Chilingarian, TF Yen*     2180

CARBAMATE INSECTICIDES IN FOODSTUFFS: CHROMATOGRAPHY AND
        IMMUNOASSAY     *GS Nunes, D Barcelo*     2191

CARBOHYDRATES

    Electrophoresis     *O Grosche*     2201

    Gas Chromatography and Gas Chromatography-Mass Spectrometry     *A Fox, MP Kozar,*
        *PA Steinberg*     2211

    Liquid Chromatography     *C Corradini*     2224

    Thin-Layer (Planar) Chromatography     *JF Robyt*     2235

CAROTENOID PIGMENTS: SUPERCRITICAL FLUID CHROMATOGRAPHY     *V Sewram,*
        *MW Raynor*     2245

CATALYST STUDIES: CHROMATOGRAPHY     *B Mile*     2252

CELLS: ISOLATION: MAGNETIC TECHNIQUES     *I Safarik, M Safarikova*     2260

CELLS AND CELL ORGANELLES: FIELD FLOW FRACTIONATION     *PJP Cardot, S Battu,*
        *T Chianea, S Rasouli*     2267

CHELATING ION EXCHANGE RESINS     *RJ Eldridge*     2271

CHEMICAL WARFARE AGENTS: CHROMATOGRAPHY     *PA D'Agostino*     2279

CHIRAL SEPARATIONS

    Amino Acids and Derivatives     *See* Amino Acids and Derivatives: Chiral Separations.     2287

    Capillary Electrophoresis     *BJ Clark*     2287

    Cellulose and Cellulose Derived Phases     *J Dingenen*     2294

    Chiral Derivatization     *S Görög*     2310

    Countercurrent Chromatography     *Y Ma, Y Ito*     2321

    Crystallization     *A Collet*     2326

    Cyclodextrins and Other Inclusion Complexation Approaches     *J Dingenen*     2335

Gas Chromatography      *V Schurig*                                                                2349

Ion-Pair Chromatography      *E Heldin*                                                            2358

Ligand Exchange Chromatography      *V Davankov*                                              2369

Liquid Chromatography      *J Haginaka*                                                          2381

Molecular Imprints as Stationary Phases      *M Kempe*                                        2387

Protein Stationary Phases      *J Haginaka*                                                      2397

Supercritical Fluid Chromatography      *N Bargmann-Leyder, M Caude, A Tambute*            2406

Synthetic Multiple Interaction ('Pirkle') Stationary Phases      *CJ Welch*                    2418

Thin-Layer (Planar) Chromatography      *L Lepri*                                                2426

# Volume 6

CITRUS OILS: LIQUID CHROMATOGRAPHY      *P Dugo*                                              2441

CLINICAL APPLICATIONS

Capillary Electrophoresis      *PG Righetti*                                                      2454

Electrophoresis      *J-D Tissot, A Layer, P Schneider, H Henry*                                2461

Gel Electrophoresis      *J-D Tissot, P Hohlfeld, A Layer, F Forestier, P Schneider, H Henry*   2468

CLINICAL CHEMISTRY: THIN-LAYER (PLANAR) CHROMATOGRAPHY      *J Bladek,*
*A Zdrojewski*                                                                                    2475

CLINICAL DIAGNOSIS: CHROMATOGRAPHY      *ID Watson*                                          2484

COAL: FLOTATION      *BK Parekh*                                                                2490

COBALT ORES: FLOTATION      *See* Nickel and Cobalt Ores: Flotation.                            2496

COLLOIDS: FIELD FLOW FRACTIONATION      *G Karaiskakis*                                        2496

COMPUTER DATABASES FOR TWO-DIMENSIONAL ELECTROPHORESIS      *T Toda*        2503

CONTINUOUS ION EXCHANGE USING POWDERED RESINS      *See* Powdered Resins:
Continuous Ion Exchange.                                                                          2510

COPRECIPITATION: TRACE ELEMENTS: EXTRACTION      *See* Trace Elements by
Coprecipitation: Extraction.                                                                      2511

COSMETICS AND TOILETRIES: CHROMATOGRAPHY      *M Carini, R Maffei Facino*        2511

CRUDE OIL: LIQUID CHROMATOGRAPHY      *BN Barman*                                            2526

DECANTER CENTRIFUGES IN PHARMACEUTICAL APPLICATIONS      *JV McKenna*        2532

DE-INKING OF WASTE PAPER: FLOTATION      *C Jiang, J Ma*                                    2537

DEOXYRIBONUCLEIC ACID PROFILING

Overview      *R Coquoz*                                                                          2545

Capillary Electrophoresis      *M Chiari, L Ceriotti*                                            2552

DETERGENT FORMULATIONS: ION EXCHANGE    *SP Chopade, K Nagarajan*    2560

DNA    *See* Capillary Electrophoresis, Overview.    2567

DRUGS AND METABOLITES

    Liquid Chromatography-Mass Spectrometry    *RPW Scott*    2567

    Liquid Chromatography-Nuclear Magnetic Resonance-Mass Spectrometry    *R Plumb, G Dear, I Ismail, B Sweatman*    2573

DRUGS OF ABUSE: SOLID-PHASE EXTRACTION    *F Musshoff*    2580

DYES

    High-Speed Countercurrent Chromatography    *A Weisz, Y Ito*    2588

    Liquid Chromatography    *W Nowik*    2602

    Thin-Layer (Planar) Chromatography    *PE Wall*    2619

ECDYSTEROIDS: CHROMATOGRAPHY    *R Lafont, C Blais, J Harmatha*    2631

ECOLOGICALLY SAFE ION EXCHANGE TECHNOLOGIES    *D Muraviev*    2644

ELECTROCHEMICAL ION EXCHANGE    *JPH Sukamto, SD Rassat, RJ Orth, MA Lilga*    2655

ELECTRODIALYSIS: ION EXCHANGE    *G Pourcelly*    2665

ENVIRONMENTAL APPLICATIONS

    Flotation    *MA Burstein, IM Flint*    2675

    Gas Chromatography-Mass Spectrometry    *N Scott, G Gutnikov*    2678

    Pressurized Fluid Extraction    *SR Sumpter*    2687

    Solid-Phase Microextraction    *T Nilsson*    2695

    Soxhlet Extraction    *MD Luque de Castro, LE Garcia Ayuso*    2701

    Supercritical Fluid Extraction    *V Camel*    2709

ENZYMES

    Chromatography    *S Nilsson, S Santesson*    2721

    Liquid Chromatography    *D Shekhawat, N Kirthivasan*    2732

ESSENTIAL OILS

    Distillation    *E Hernandez*    2739

    Gas Chromatography    *C Bicchi*    2744

    Thin-Layer (Planar) Chromatography    *P Dugo, L Mondello, G Dugo*    2755

EXPLOSIVES

    Gas Chromatography    *J Yinon*    2762

    Liquid Chromatography    *U Lewin-Kretzschmar, J Efer, W Engewald*    2767

    Thin-Layer (Planar) Chromatography    *J Bladek*    2782

EXTRACTION: PRESSURIZED FLUID EXTRACTION    *See* ENVIRONMENTAL APPLICATIONS: Pressurized Fluid Extraction    2789

FATS

Crystallization    *K Sato*    2789

Extraction by Solvent Based Methods    *EJ Birch*    2794

Supercritical Fluid Chromatography    *See* Oils, Fats and Waxes: Supercritical Fluid
Chromatography.    2801

FATTY ACIDS: GAS CHROMATOGRAPHY    *See* LIPIDS: Gas Chromatography.    2801

FLAME IONIZATION DETECTION: THIN-LAYER (PLANAR) CHROMATOGRAPHY
*RG Ackman*    2801

FLASH CHROMATOGRAPHY    *CF Poole*    2808

FLAVOURS: GAS CHROMATOGRAPHY    *FP Scanlan*    2814

FOAM COUNTERCURRENT CHROMATOGRAPHY    *H Oka, Y Ito*    2822

FOOD ADDITIVES

Liquid Chromatography    *VD Sattigeri, BK Lonsane, MN Krishnamurthy, LR Gouda,
PR Ramasarma, V Prakash*    2829

Thin-Layer (Planar) Chromatography    *M Vega Herrera*    2838

FOOD MICROORGANISMS: BUOYANT DENSITY CENTRIFUGATION    *R Lindqvist*    2843

FOOD TECHNOLOGY

Membrane Separations    *M Cheryan*    2849

Supercritical Fluid Chromatography    *JW King*    2855

Supercritical Fluid Extraction    *SSH Rizvi*    2860

FORENSIC SCIENCES

Capillary Electrophoresis    *J Sádecká*    2862

Liquid Chromatography    *LA Kaine, CL Flurer, K-A Wolnik*    2870

FORENSIC TOXICOLOGY: THIN-LAYER (PLANAR) CHROMATOGRAPHY    *I Ojanperä*    2879

FRAGRANCES: GAS CHROMATOGRAPHY    *ER Adlard, M Cooke*    2885

FUELS AND LUBRICANTS: SUPERCRITICAL FLUID CHROMATOGRAPHY    *M Robson*    2894

FULLERENES: LIQUID CHROMATOGRAPHY    *VL Cebolla, L Membrado, J Vela*    2901

FUNGICIDES

Gas Chromatography    *JL Bernal*    2908

Liquid Chromatography    *M Jesús del Nozal Nalda*    2915

FUSED SALTS: ELECTROPHORESIS    *M Lederer*    2921

GAS ANALYSIS: GAS CHROMATOGRAPHY    *CJ Cowper*    2925

GAS CENTRIFUGE: ISOTOPES SEPARATION    *See* ISOTOPE SEPARATIONS:
Gas Centrifugation.    2932

GAS CHROMATOGRAPHY-MASS SPECTROMETRY IN MEDICINE
*See* BIOMEDICAL APPLICATIONS: Gas Chromatography-Mass Spectrometry.    2932

GAS SEPARATION BY METAL COMPLEXES: MEMBRANE SEPARATIONS    *N Toshima,
S Hara*    2933

# Volume 7

GENE TYPING: TWO-DIMENSIONAL ELECTROPHORESIS    NJ van Orsouw, SB McGrath, J Vijg, RK Dhanda, CB Scott    2939

GEOCHEMICAL ANALYSIS: GAS CHROMATOGRAPHY    RP Philp    2948

GLYCOPROTEINS: LIQUID CHROMATOGRAPHY    K Miyazaki    2960

GOLD RECOVERY: FLOTATION    SM Bulatovic, DM Wyslouzil    2965

GRADIENT POLYMER CHROMATOGRAPHY: LIQUID CHROMATOGRAPHY
G Glöckner    2975

HERBICIDES

Gas Chromatography    JL Tadeo, C Sanchez-Brunete    2984

Solid-Phase Extraction    Y Picó    2991

Thin-Layer (Planar) Chromatography    V Pacáková    3006

HEROIN: LIQUID CHROMATOGRAPHY AND CAPILLARY ELECTROPHORESIS
RB Taylor, AS Low, RG Reid    3010

HOT-PRESSURIZED WATER: EXTRACTION    See Superheated Water Mobile Phases: Liquid Chromatography    3017

HUMIC SUBSTANCES

Capillary Zone Electrophoresis    J Havel, D Fetsch    3018

Gas Chromatography    J Pörschmann    3026

Liquid Chromatography    DK Ryan    3032

HYDRODYNAMIC CHROMATOGRAPHY: PRACTICAL APPLICATIONS    A Revillon    3039

HYDROGEN RECOVERY USING INORGANIC MEMBRANES    R Hughes    3046

IMMOBILIZED BORONIC ACIDS: EXTRACTION    P Martin, ID Wilson    3051

IMMUNOAFFINITY EXTRACTION    D Stevenson    3060

IMPREGNATION TECHNIQUES: THIN-LAYER (PLANAR) CHROMATOGRAPHY
ID Wilson    3065

IMPRINTED POLYMERS: AFFINITY SEPARATION    See Selectivity of Imprinted Polymers: Affinity Separation.    3071

IN-BORN METABOLIC DISORDERS: DETECTION: THIN-LAYER (PLANAR) CHROMATOGRAPHY    E Marklová    3072

INCLUSION COMPLEXATION: LIQUID CHROMATOGRAPHY    SR Gratz, BM Gamble, AM Stalcup    3079

IN-DEPTH DISTRIBUTION IN QUANTITATIVE THIN-LAYER CHROMATOGRAPHY
I Vovk, M Prosek    3087

INDUSTRIAL ANALYTICAL APPLICATIONS: SUPERCRITICAL FLUID EXTRACTION    MEP McNally    3094

INKS: FORENSIC ANALYSIS BY THIN-LAYER (PLANAR) CHROMATOGRAPHY
LW Pagano, MJ Surrency, AA Cantu    3101

INORGANIC EXTRACTION: MOLECULAR RECOGNITION TECHNOLOGY
See Molecular Recognition Technology in Inorganic Extraction.    3110

INSECTICIDES

    Gas Chromatography    *P Brown*    3110

    Solid-Phase Extraction    *A Przyjazny*    3118

INSECTICIDES IN FOODSTUFFS    *See* Carbamate Insecticides in Foodstuffs:
    Chromatography and Immunoassay.    3128

ION ANALYSIS

    Capillary Electrophoresis    *M Macka, PR Haddad*    3128

    Electrophoresis    *See* ION ANALYSIS: Capillary Electrophoresis.    3141

    High-Speed Countercurrent Chromatography    *E Kitazume*    3141

    Liquid Chromatography    *CA Lucy*    3149

    Thin-Layer (Planar) Chromatography    *A Mohammad*    3156

ION EXCHANGE RESINS: CHARACTERIZATION OF    *LS Golden*    3172

ION EXCHANGE: ZEOLITES    *See* Zeolites: Ion Exchangers.    3179

ION FLOTATION    *LO Filippov*    3179

ION-CONDUCTING MEMBRANES: MEMBRANE SEPARATIONS    *JA Kilner*    3187

ION-EXCLUSION CHROMATOGRAPHY: LIQUID CHROMATOGRAPHY
    *K Tanaka, PR Haddad*    3193

ISOTOPE SEPARATIONS

    Gas Centrifugation    *VD Borisevich, HG Wood*    3202

    Liquid Chromatography    *L Leseticky*    3207

LEAD AND ZINC ORES: FLOTATION    *M Barbaro*

LIPIDS    3215

    Gas Chromatography    *A Kuksis*    3219

    Liquid Chromatography    *A Kuksis*    3237

    Thin-Layer (Planar) Chromatography    *B Fried*    3253

LIQUID CHROMATOGRAPHY-GAS CHROMATOGRAPHY    *L Mondello, P Dugo, G Dugo,*
    *KD Bartle, AC Lewis*    3261

MARINE TOXINS: CHROMATOGRAPHY    *A Gago-Martínez, JA Rodríguez-Vázquez*    3269

MECHANICAL TECHNIQUES: PARTICLE SIZE SEPARATION    *AIA Salama*    3277

MEDICINAL HERB COMPOUNDS: HIGH-SPEED COUNTERCURRENT
    CHROMATOGRAPHY    *T-Y Zhang*    3289

MEDIUM-PRESSURE LIQUID CHROMATOGRAPHY    *K Hostettmann, C Terreaux*    3296

MEMBRANE CONTACTORS: MEMBRANE SEPARATIONS    *JGSG Crespo, IM Coelhoso,*
    *RMC Viegas*    3303

MEMBRANE PREPARATION

    Hollow Fibre Membranes    *M van Bruijnsvoort, PJ Schoenmakers*    3312

    Interfacial Composite Membranes    *JE Tomaschke*    3319

    Phase Inversion Membranes    *MHV Mulder*    3331

METABLITES    *See* DRUGS AND METABOLITES: Liquid Chromatography-Mass
Spectrometry, DRUGS AND METABOLITES: Liquid Chromatography-Nuclear Magnetic
Resonance-Mass Spectrometry.    3346

METAL ANALYSIS: GAS AND LIQUID CHROMATOGRAPHY    *PC Uden*    3347

METAL COMPLEXES

Ion Chromatography    *DJ Pietrzyk*    3354

Use in Gas Separation    *See* Gas Separation by Metal Complexes: Membrane Separations.    3365

METAL MEMBRANES: MEMBRANE SEPARATIONS    *YS Lin, R Buxbaum*    3365

METAL UPTAKE ON MICROORGANISMS AND BIOMATERIALS: ION EXCHANGE
*H Eccles*    3372

METALLOPROTEINS: CHROMATOGRAPHY    *E Parisi*    3380

MICROORGANISMS: BUOYANT DENSITY CENTRIFUGATION    *See* Food Microorganisms:
Buoyant Density Centrifugation.    3387

MICROWAVE-ASSISTED EXTRACTION: ENVIRONMENTAL APPLICATIONS
*GN LeBlanc*    3387

MOLECULAR IMPRINTS FOR SOLID-PHASE EXTRACTION    *PAG Cormack, K Haupt*    3395

MOLECULAR RECOGNITION TECHNOLOGY IN INORGANIC EXTRACTION
*JD Glennon*    3400

MULTIRESIDUE METHODS: EXTRACTION    *SJ Lehotay, FJ Schenck*    3409

NATURAL PRODUCTS

High-Speed Countercurrent Chromatography    *A Marston, K Hostettmann*    3415

Liquid Chromatography    *K Hostettmann, J-L Wolfender*    3424

Liquid Chromatography-Nuclear Magnetic Resonance    *B Schneider*    3434

Supercritical Fluid Chromatography    *ED Morgan*    3445

Supercritical Fluid Extraction    *ED Morgan*    3451

Thin-Layer (Planar) Chromatography    *J Pothier*    3459

NEUROTOXINS: CHROMATOGRAPHY    *KJ James, A Furey*    3482

## Volume 8

NICKEL AND COBALT ORES: FLOTATION    *GV Rao*    3491

NOVEL INORGANIC MATERIALS: ION EXCHANGE    *DJ Jones*    3501

NUCLEAR INDUSTRY: ION EXCHANGE    *J Lehto*    3509

NUCLEIC ACIDS    3517

Centrifugation    *A Marziali*    3517

Extraction    *SJ Walker, KE Vrana*    3524

NUCLEIC ACIDS (*Continued*)

    Liquid Chromatography    *CW Gehrke, K Kuo*    3528

    Thin-Layer (Planar) Chromatography    *JJ Steinberg*    3543

OCCUPATIONAL HYGIENE: GAS CHROMATOGRAPHY    *M Harper*    3554

OILS, FATS AND WAXES: SUPERCRITICAL FLUID CHROMATOGRAPHY    *F David, A Medvedovici, P Sandra*    3567

OILS: EXTRACTION BY SOLVENT BASED METHODS    *See* FATS: Extraction by Solvent Based Methods.    3575

OLIGOMERS: THIN-LAYER (PLANAR) CHROMATOGRAPHY    *See* SYNTHETIC POLYMERS: Thin-Layer (Planar) Chromatography.    3575

ON-LINE SAMPLE PREPARATION: SUPERCRITICAL FLUID EXTRACTION    *JM Levy*    3575

OPIATES    *See* Heroin: Liquid Chromatography and Capillary Electrophoresis.    3585

ORGANELLES

    Centrifugation    *JA Garner*    3586

    Field Flow Fractionation    *See* Cells and Cell Organelles: Field Flow Fractionation.    3596

PAINTS AND COATINGS: PYROLYSIS: GAS CHROMATOGRAPHY    *TP Wampler*    3596

PARTICULATE CHARACTERIZATION: INVERSE GAS CHROMATOGRAPHY    *DR Williams, D Butler*    3609

PEPTIDES AND AMINO ACIDS: CAPILLARY ELECTROPHORESIS    *See* Amino Acids and Peptides: Capillary Electrophoresis.    3614

PEPTIDES AND PROTEINS

    Liquid Chromatography    *CT Mant, RS Hodges*    3615

    Thin-Layer (Planar) Chromatography    *R Bhushan, J Martens*    3626

PERVAPORATION: MEMBRANE SEPARATIONS    *SP Chopade, K Nagarajan*    3636

PESTICIDES

    Extraction from Water    *MC Hennion, V Pichon*    3642

    Gas Chromatography    *M-R Lee, B-H Hwang*    3652

    Supercritical Fluid Chromatography    *MEP McNally*    3657

    Thin-Layer (Planar) Chromatography    *J Blådek, A Rostkowski*    3669

PETROLEUM PRODUCTS

    Gas Chromatography    *JP Durand*    3678

    Liquid Chromatography    *VL Cebolla, L Membrado, J Vela*    3683

    Thin-Layer (Planar) Chromatography    *AA Herod, M-J Lazaro*    3690

PHARMACEUTICALS

    Basic Drugs: Liquid Chromatography    *B Law*    3701

    Capillary Electrophoresis    *KD Altria, SM Bryant*    3708

    Chiral Separations: Liquid Chromatography    *WJ Lough*    3714

    Chromatographic Separations    *J Vessman*    3719

PHARMACEUTICALS (*Continued*)

    Crystallization    *W Beckmann, U Budde*    3729

    Neutral and Acidic Drugs: Liquid Chromatography    *RK Gilpin, CS Gilpin*    3738

    Supercritical Fluid Chromatography    *WH Wilson*    3749

    Thin-Layer (Planar) Chromatography    *B Renger*    3754

PHENOLS

    Gas Chromatography    *M-R Lee*    3760

    Liquid Chromatography    *RM Marcé, F Borrull*    3766

    Solid-Phase Extraction    *J Blådek, M Sliwakowski*    3776

    Thin-Layer (Planar) Chromatography    *JHP Tyman*    3783

PHEROMONES

    Gas Chromatography    *NG Agelopoulos, LJ Wadhams*    3796

    Thin-Layer (Planar) Chromatography    *ED Morgan*    3803

PHYSICO-CHEMICAL MEASUREMENTS: GAS CHROMATOGRAPHY    *JR Conder*    3808

pH-ZONE REFINING COUNTERCURRENT CHROMATOGRAPHY    *Y Ito*    3815

PIGMENTS

    Liquid Chromatography    *S Roy*    3832

    Thin-Layer (Planar) Chromatography    *GW Francis*    3839

POLYCHLORINATED BIPHENYLS: GAS CHROMATOGRAPHY    *DE Wells*    3851

POLYCYCLIC AROMATIC HYDROCARBONS

    Gas Chromatography    *HK Lee*    3860

    Solid-Phase Extraction    *F Borrull, RM Marcé*    3867

    Supercritical Fluid Chromatography    *KD Bartle*    3877

    Thin-Layer (Planar) Chromatography    *G Donnevert*    3881

POLYETHERS: LIQUID CHROMATOGRAPHY    *K Rissler*    3889

POLYMER ADDITIVES: SUPERCRITICAL FLUID CHROMATOGRAPHY    *TP Hunt*    3901

POLYMERS

    Field Flow Fractionation    *ME Schimpf*    3906

    Supercritical Fluid Extraction    *HJ Vandenburg*    3915

POLYSACCHARIDES

    Centrifugation    *SE Harding*    3921

    Liquid Chromatography    *JM Mates, C Pérez-Gómez*    3929

POROUS GRAPHITIC CARBON: LIQUID CHROMATOGRAPHY    *M-C Hennion*    3937

POROUS POLYMER COMPLEXES FOR GAS SEPARATIONS: GAS ADSORBENTS
    *H Asanuma, N Toshima*    3947

POROUS POLYMERS: LIQUID CHROMATOGRAPHY    *A Coffey*    3954

PORPHYRINS: LIQUID CHROMATOGRAPHY    *CK Lim*    3960

POWDERED RESINS: CONTINUOUS ION EXCHANGE  *PA Yarnell*  3973

PREPARATIVE ELECTROPHORESIS  *RMC Sutton, AM Stalcup*  3981

PREPARATIVE SUPERCRITICAL FLUID CHROMATOGRAPHY  *JR Williams, R Dmoch*  3988

PRESSURIZED FLUID EXTRACTION: NON-ENVIRONMENTAL APPLICATIONS
    *JL Ezzell*  3993

PROSTAGLANDINS: GAS CHROMATOGRAPHY  *J Nourooz-Zadeh, CCT Smith*  4000

# Volume 9

PROTEINS

    Capillary Electrophoresis  *SP Radko*  4009

    Centrifugation  *A Yamazaki*  4014

    Crystallization  *MY Gamarnik*  4020

    Electrophoresis  *MJ Schmerr*  4026

    Field Flow Fractionation  *R Hecker, H Cölfen*  4031

    Glycoproteins: Liquid Chromatography  *See* Glycoproteins: Liquid Chromatography.  4038

    High-Speed Countercurrent Chromatography  *Y Shibuswa*  4039

    Ion Exchange  *PR Levison*  4047

    Metalloproteins: Chromatography  *See* Metalloproteins: Chromatography.  4052

    Thin-Layer (Planar) Chromatography  *See* PEPTIDES AND PROTEINS: Thin-Layer (Planar)
        Chromatography.  4052

PROTEOMICS: ELECTROPHORESIS  *MJ Dunn*  4052

PURGE-AND-TRAP: GAS CHROMATOGRAPHY  *See* Volatile Organic Compounds in Water:
    Gas Chromatography.  4062

PYROLYSIS: GAS CHROMATOGRAPHY  *See* CHROMATOGRAPHY: GAS: Pyrolysis: Gas
    Chromatography.  4062

QUANTITATIVE STRUCTURE-RETENTION RELATIONSHIPS IN
    CHROMATOGRAPHY  *R Kaliszan*  4063

REACTIVE DISTILLATION  *SM Mahajani, S Chopade*  4075

RESINS AS BIOSORBENTS: ION EXCHANGE  *S Belfer*  4082

RESTRICTED-ACCESS MEDIA: SOLID-PHASE EXTRACTION  *J Haginaka*  4087

REVERSE-FLOW GAS CHROMATOGRAPHY  *NA Katsanos, F Roubani-Kalatzopoulou*  4091

RIBONUCLEIC ACIDS: CAPILLARY ELECTROPHORESIS  *J Skeidsvoll*  4098

RNA  *See* DEOXYRIBONUCLEIC ACID PROFILING: Capillary Electrophoresis.  4104

SELECTIVITY OF IMPRINTED POLYMERS: AFFINITY SEPARATION  *O Ramström*  4104

SILVER ION

    Liquid Chromatography  *WW Christie*  4112

    Thin-Layer (Planar) Chromatography  *B Nikolova-Damyanova*  4117

SODIUM CHLORIDE: CRYSTALLIZATION    *R Geertman*    4127

SOLID-PHASE EXTRACTION OF DRUGS    *See* Bioanalytical Applications: Solid-Phase
Extraction.    4134

SOLID-PHASE EXTRACTION WITH CARTRIDGES    *DA Wells*    4135

SOLID-PHASE EXTRACTION WITH DISCS    *CF Poole*    4141

SOLID-PHASE EXTRACTION: SORBENT SELECTION    *See* Sorbent Selection for Solid-Phase
Extraction.    4148

SOLID-PHASE MATRIX DISPERSION: EXTRACTION    *SA Barker*    4148

SOLID-PHASE MICROEXTRACTION

Biomedical Applications    *H Kataoka, HL Lord, JB Pawliszyn*    4153

Environmental Applications    *A Andrews*    4170

Food Technology Applications    *R Marsili*    4178

Overview    *JR Dean*    4190

SOLVENTS: DISTILLATION    *B Buszewski*    4199

SORBENT SELECTION FOR SOLID-PHASE EXTRACTION    *EM Thurman*    4204

SPACE EXPLORATION: GAS CHROMATOGRAPHY    *R Sternberg, F Raulin, C Vidal-Madjar*    4212

STEROIDS

Gas Chromatography    *HLJ Makin*    4220

Liquid Chromatography and Thin-Layer (Planar) Chromatography    *HLJ Makin*    4230

Supercritical Fluid Chromatography    *K Yaku, K Aoe, N Nishimura,
T Sato*    4243

STEROLS

Supercritical Fluid Chromatography    *FJ Señoráns, KE Markides*    4249

Thin-Layer (Planar) Chromatography    *J Novakovic, K Nesmerak*    4255

STRONTIUM FROM NUCLEAR WASTES: ION EXCHANGE    *P Sylvestor*    4261

SUB-CRITICAL WATER: EXTRACTION    *See* Superheated Water Mobile Phases:
Liquid Chromatography.    4267

SUGAR DERIVATIVES: CHROMATOGRAPHY    *SC Churms*    4267

SULFUR COMPOUNDS: GAS CHROMATOGRAPHY    *W Wardencki*    4285

SUPERCRITICAL FLUID CRYSTALLIZATION    *AS Teja, T Furuya*    4301

SUPERCRITICAL FLUID EXTRACTION-SUPERCRITICAL FLUID CHROMATOGRAPHY
*HJ Vandenburg, KD Bartle*    4307

SUPERHEATED WATER MOBILE PHASES: LIQUID CHROMATOGRAPHY    *RM Smith*    4313

SURFACTANTS

Chromatography    *JG Lawrence*    4318

Liquid Chromatography    *TM Schmitt*    4327

SYNTHETIC POLYMERS

    Gas Chromatography    *JK Haken*    4334

    Liquid Chromatography    *CH Lochmuller*    4343

    Thin-Layer (Planar) Chromatography    *LS Litvinova*    4348

TERPENOIDS: LIQUID CHROMATOGRAPHY    *PK Inamdar, S Chatterjee*    4354

THERMALLY-COUPLED COLUMNS: DISTILLATION    *R Smith*    4363

THIN-LAYER CHROMATOGRAPHY-VIBRATION SPECTROSCOPY    *E Koglin*    4371

TOBACCO VOLATILES: GAS CHROMATOGRAPHY    *WM Coleman III*    4380

TOXICOLOGICAL ANALYSIS: LIQUID CHROMATOGRAPHY    *AP de Leenheer, W Lambert, J van Bocxlaer*    4388

TOXINS: CHROMATOGRAPHY    *See* Marine Toxins: Chromatography, Neurotoxins: Chromatography.    4394

TRACE ELEMENTS BY COPRECIPITATION: EXTRACTION    *K Terada*    4394

TRIGLYCERIDES

    Liquid Chromatography    *V Ruiz-Gutierrez, JS Perona*    4402

    Thin-Layer (Planar) Chromatography    *PE Wall*    4412

ULTRASOUND-ASSISTED METAL EXTRACTIONS    *Carlos Bendicho, Isela Lavilla*    4421

VENOMS: CHROMATOGRAPHY    *See* Neurotoxins: Chromatography.    4426

VETERINARY DRUGS: LIQUID CHROMATOGRAPHY    *HF De Brabander, K De Wasch*    4426

VIRUSES: CENTRIFUGATION    *Larry L Bondoc Jr*    4433

VITAMINS

    Fat-Soluble: Thin-Layer (Planar) Chromatography    *WE Lambert, AP De Leenheer*    4437

    Liquid Chromatography    *MH Bui*    4443

    Water-Soluble: Thin-Layer (Planar) Chromatography    *JC Linnell*    4454

VOLATILE ORGANIC COMPOUNDS IN WATER: GAS CHROMATOGRAPHY    *MC Tombs*    4460

WATER TREATMENT

    Overview: Ion exchange    *J Irving*    4469

    Anion Exchangers: Ion Exchange    *W Höll*    4477

WATER-SOLUBLE VITAMINS: THIN-LAYER (PLANAR) CHROMATOGRAPHY    *See* VITAMINS: Water-Soluble: Thin-Layer (Planar) Chromatography.    4484

WAXES: SUPERCRITICAL FLUID CHROMATOGRAPHY    *See* Oils, Fats and Waxes: Supercritical Fluid Chromatography.    4484

WHISKY: DISTILLATION    *DS Pickerell*    4485

WINE: GAS AND LIQUID CHROMATOGRAPHY    *J Guasch, O Busto*    4490

XENOBIOTICS: MAGNETIC AFFINITY SEPARATIONS    *See* Biologically Active Compounds and Xenobiotics: Magnetic Affinity Separations.    4498

ZEOLITES: ION EXCHANGERS    *CD Williams*    4498

ZINC ORES: FLOTATION    *See* Lead and Zinc Ores: Flotation.    4502

ZONE REFINING COUNTERCURRENT CHROMATOGRAPHY    *See* pH-Zone Refining
Countercurrent Chromatography.    4502

## Volume 10

## Contributors

4503

## Appendices

4531

1.  ESSENTIAL GUIDES FOR ISOLATION/PURIFICATION    4533
2.  ESSENTIAL GUIDES FOR METHOD DEVELOPMENT    4575
3.  ABBREVIATIONS    4673
4.  ANALYTICAL CHIRAL SEPARATION METHODS    4681
5.  BIOLOGICAL BUFFERS    4685
6.  CLASSIFICATION AND CHARACTERIZATION OF STATIONARY PHASES FOR
    LIQUID CHROMATOGRAPHY    4685
7.  CONVERSION OF UNITS    4695
8.  DEFINITIONS AND SYMBOLS FOR UNITS    4701
9.  FUNDAMENTAL PHYSICAL CONSTANTS    4706
10. IMPORTANT PEAKS IN MASS SPECTRA OF COMMON SOLVENTS    4708
11. NOMENCLATURE AND TERMINOLOGY FOR ANALYTICAL PYROLYSIS    4709
12. NOMENCLATURE    4712
13. pH SCALE FOR AQUEOUS SOLUTIONS    4781
14. PROPERTIES OF PARTICLES, ELEMENTS AND NUCLIDES    4783
15. SOLVENTS FOR ULTRAVIOLET SPECTROPHOTOMETRY    4797
16. STATISTICAL TABLES    4798
17. THIN-LAYER (PLANAR) CHROMATOGRAPHY: Detection    4802
18. WAVELENGTH SCALE    4808

## Index

4809

## Colour Plate Sections

Volume 1    between pages    282–283
Volume 2    between pages    644–645
Volume 3    between pages 1244–1245
Volume 4    between pages 1576–1577
Volume 5    between pages 2030–2031
Volume 6    between pages 2754–2755
Volume 7    between pages 3252–3253
Volume 8    between pages 3676–3677
Volume 9    between pages 4386–4387

# GENE TYPING: TWO-DIMENSIONAL ELECTROPHORESIS

**N. J. van Orsouw, S. B. McGrath and J. Vijg,**
Institute for Drug Development, Cancer Therapy and
Research Center, San Antonio, TX, USA
**R. K. Dhanda,** Mosaic Technologies, Boston, MA, USA
**C. B. Scott,** CBS Scientific Company, Del Mar, CA, USA

## Introduction

With the human genome program drawing to a close, attention is now rapidly shifting from obtaining consensus sequences of all human genes to the detection of individual DNA sequence variations. Based on complete sequence information for all human genes, it is theoretically possible to generate a catalogue of all gene mutations and polymorphisms in the human genome and test them directly for association to relevant phenotypes, e.g. of health and disease. Unfortunately, current methods for detecting DNA sequence variants are not optimized for generating data on multiple genes in large numbers of individuals, e.g. in population-based studies or in the clinical setting. The most reliable system for comprehensive gene sequence analysis is still nucleotide sequencing itself, which is not compatible with cost-effective large scale population-based genetic screening.

Recently, various systems have been proposed to analyse gene-coding and regulatory sequences more effectively for all possible variations. Here we review the development and application of one such system, two-dimensional gene scanning (TDGS). This method is based on the two-dimensional separation of polymerase chain reaction (PCR)-amplified gene fragments on the basis of both size and base pair sequence in polyacrylamide gels. Attention will be focused on most recent developments in automation and miniaturization of the two-dimensional electrophoresis procedure. Future developments towards a dedicated fully automated high-throughput system for gene analysis will be discussed.

## Two-Dimensional Gene Scanning: Background and Principles

### Denaturing Gradient Gel Electrophoresis

TDGS is based on denaturing gradient gel electrophoresis (DGGE) as the mutation detection principle, in combination with PCR amplification to prepare the target sequences. In DGGE, DNA fragments are subjected to electrophoresis in a polyacrylamide gel against a gradient of ever higher temperature or chemical denaturants (i.e. a mixture of urea and formamide). Unlike nucleotide sequencing, DGGE detects mutations, including base pair substitutions and small insertions and deletions, on the basis of differences in the melting temperature of the target fragments. A given DNA fragment comprises one or more domains, each representing a stretch of between 50 and 300 base pairs with equal melting temperature (the temperature at which each base pair has a 50% probability of being in either the helical or the denatured state). Since the stability of each domain depends on its sequence, mutational differences among different fragments are revealed as migrational differences in the gel (**Figure 1A**).

In order to obtain virtually 100% accuracy in mutation detection, fragments to be subjected to DGGE can be clamped to a GC-rich sequence (a stretch of 30–50 G and C bases). A convenient way of attaching a GC-clamp to the target fragment is by making it part of one of the primers in a PCR. Without GC-clamping, a DNA fragment consisting of one melting domain will become completely single-stranded upon denaturation and run off the gel. By adding a GC-clamp, a single high-melting domain is artificially created at one end of the target fragment. As the GC-clamped target fragment migrates through the gradient of denaturants, melting of the target domain causes partial branching and halting of the fragment in the gel (Figure 1B). Thus, one function of the GC-clamp is to ensure branch formation after melting of the target fragment. However, when the target DNA fragment consists of multiple melting domains (Figure 1C), only mutations in the lowest melting domain are readily detected. To facilitate detection of all possible mutations, it is imperative that the target fragment represents only one melting domain. Fortunately, since the addition of a GC-clamp allows for stacking interactions with neighbouring bases, the entire fragment will often behave as one melting domain (Figure 1C). However, this is not always the case, and in practice, the target fragment needs to be designed, e.g. through the strategic positioning of PCR primers to achieve the ideal single

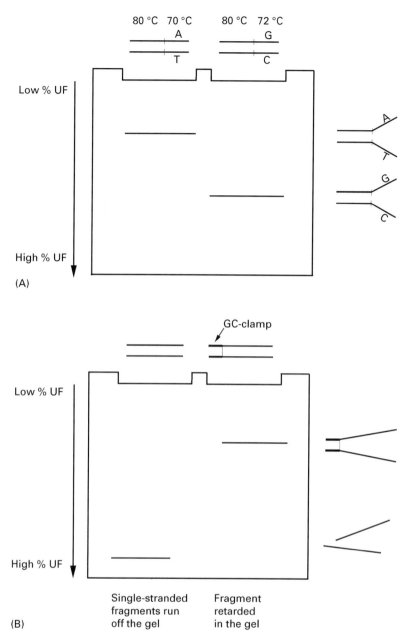

**Figure 1** Principles of denaturing gradient gel electrophoresis. (A) Single base changes affect the melting temperature of a fragment, which results in a gel shift. (B) After complete denaturation, single-stranded fragments will run off the gel; the addition of a GC-clamp to the target fragment prevents complete denaturation and therefore fragments will be retarded in the gel. (C) The addition of a GC-clamp to a multiple-domain fragment can make the fragment behave as a single-domain fragment. Continuous line: target fragment without a GC-clamp. Dashed line: target fragment, including a 40 bp GC-clamp. (D) The introduction of a heteroduplex cycle at the end of PCR amplification of target fragments facilitates detection of heterozygous mutations as four molecules: two homoduplexes and two (early-melting) heteroduplexes.

melting domain. In general, target fragments in DGGE have an average size of 275 bp, including a GC-clamp and PCR primer sequences.

The sensitivity of DGGE for detecting variants is further enhanced by the introduction of a heteroduplexing step using one round of denaturation/renaturation, usually at the end of PCR amplification of the target fragment. In this manner, a heterozygous mutation is revealed as four different double-stranded fragments: two homoduplex molecules (one wild-type homoduplex and one mutant homoduplex) and two heteroduplex molecules (each comprising one wild-type and one mutant strand). Since the stability of heteroduplexes is so much lower, they always melt earlier than the homoduplex molecules (Figure 1D).

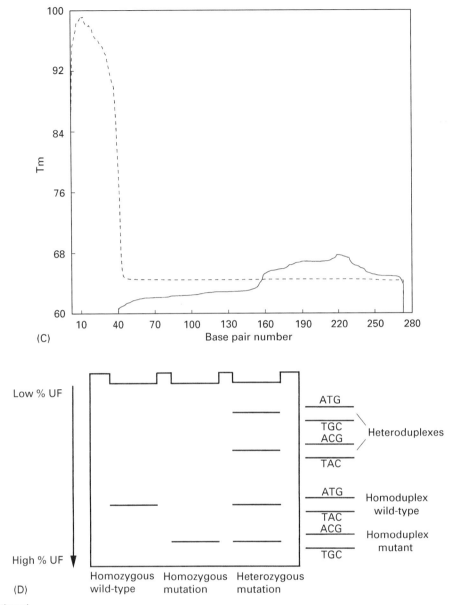

**Figure 1** *Continued*

Although DGGE has the crucial advantage of having virtually 100% sensitivity in detecting mutations, it has typically been applied in a serial fashion, e.g. on a fragment-by-fragment basis. For analysing large genes or multiple genes this is not practical. A solution for this problem, which we adopted, is to apply the DGGE principle in the format as it was originally described, i.e. a two-dimensional system of separation by size followed by DGGE. Successful implementation of such a two-dimensional DNA electrophoresis system in mutation scanning of large genes requires an efficient multiplex PCR protocol. Indeed, without the possibility to PCR-amplify multiple target fragments (i.e. typically 10 or more) in one single reaction, the application of a parallel analysis system offers only a limited advantage. Multiplex PCR systems for genes and genetic markers are now becoming available and it has been demonstrated that as many as 26 fragments can be co-amplified in one single tube under the same reaction conditions.

## Two-Dimensional DNA Electrophoresis

The major advantage of two-dimensional electrophoresis is that it provides a high resolution system to screen multiple fragments under the same conditions. It has been demonstrated that DGGE provides virtually 100% mutation detection sensitivity even when applied with a broad range gradient of denaturants. This opens up the possibility to analyse multiple fragments for all possible mutations under the same set of experimental conditions. The total number of target fragments that can be analysed simultaneously depends on the resolution of the gel system used. Although high resolution can be obtained by using one-dimensional denaturing gradient gels, two-dimensional separation allows characterization of each fragment on the basis of two independent criteria, size and melting temperature. In practice, a fragment mixture corresponding to all exons of a gene is electrophoresed in a nondenaturing size gel. Fragments are further sorted out in a denaturing gradient gel as the second dimension (**Figure 2**). By using the two-dimensional system, it is possible to visualize completely all fragments corresponding to

an entire gene for a particular DNA sample and immediately recognize each exon and variants therein. This has been demonstrated for several large human disease genes, including *CFTR*, *RB1*, *MLH1*, *TP53*, *TSC1*, *BRCA1*, as well as for a part of the mitochondrial genome.

## Design Software and Instrumentation for TDGS Tests

### Computer-Automated Design of Target Fragments for PCR and DGGE

A potential hindrance to the widespread application of TDGS to multiple novel genes involves the difficulties in the design of PCR primers generating

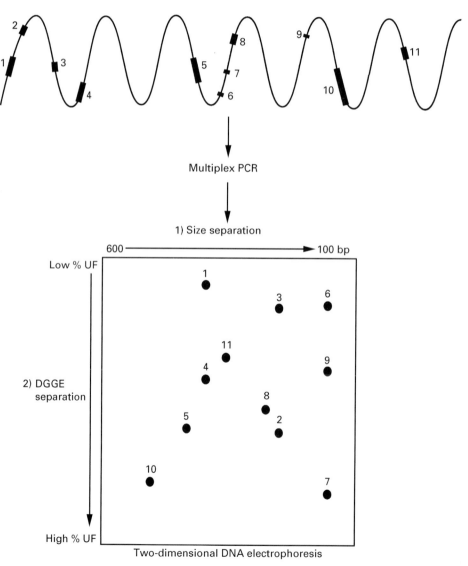

**Figure 2**  Schematic depiction of a TDGS test. All exons are amplified in an extensive multiplex reaction, and the fragments are resolved by size separation, followed by separation in a gradient of denaturants. Heterozygous mutations would show up as four spots instead of one.

single-domain fragments which can be resolved under one set of electrophoretic conditions. To design complete gene tests for mutational analysis by TDGS, an automated generally applicable computer program was developed, which was based on a commercially available primer design program (Primer Designer 3; Scientific and Educational Software, State Line, PA), the melting routine MELT87 and a newly generated spot distribution routine. After entering a gene's coding sequence as exons with their flanking intronic sequences, a rank of suitable PCR primers for each exon is designed by the PCR design subroutine. Next, the best primer pair is used in the melt subroutine to check for a one-domain target fragment. The program uses different GC-clamps at either the 5' or 3' end of the target fragment and, if necessary, additional small GC- or AT-clamps at either side of the target fragment. If it is impossible to design a one-domain fragment, the next optimal primer pair is tested, and so on. If a primer pair suitable to create a one-domain fragment cannot be found, the exon is split.

As soon as primers fulfil PCR and melting criteria, the fragment is positioned according to its size ($x$) and melting ($y$) coordinates. The spot distribution routine then checks for possible overlap. The output file of the program is a complete list of primers to be used in TDGS. (Questions regarding the use of the TDGS software should be directed to Accelerated Genomics, Concord, NH, http://www.accelerated genomics.com Tel.: (210) 616-5910; fax: (210) 692-7502.)

### Electrophoresis

For two-dimensional DNA electrophoresis, originally two different gels were used for the first-dimension (separation according to size) and the subsequent second-dimension separation of these fragments by DGGE on the basis of their melting temperature. The first-dimension separation was carried out in polyacrylamide slab gels, which required staining of the gel to visualize the one-dimension separation pattern before this could be excised and transferred to the second-dimension denaturing gradient gel. Alternatively, tube gels have been employed for size separation, which obviated the need for gel staining and lane excision. However, routine application of TDGS requires standardization and automation, which is incompatible with the labour-intensive step of manual interference between the first- and second-dimension separation.

Recently, we developed a simple automated two-dimension instrument, which is based on an existing vertical electrophoresis system with an isolated horizontal unit on top (**Figure 3**). This top unit consists of two outer chambers and one middle chamber. The necessary contacts between the outer buffer chambers and the gel are provided by two strategically located openings in the inner glass plate.

In this system only one gel (a denaturing gradient gel) is used with the top part nondenaturing. This nondenaturing part functions as the lane for the first-dimension size separation. A slot former is placed in the top left part (**Figure 4**). In the current configuration, a gel is attached to each side of a gel holder, which can be placed in a buffer tank. Buffer tanks can hold multiple units so that multiple gels can be run simultaneously (**Figure 5**). The sample is electrophoresed on the basis of size horizontally in the nondenaturing top gel, and the second-dimension

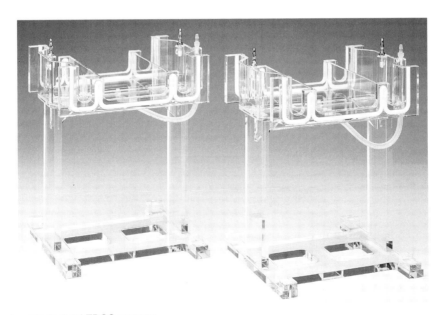

**Figure 3**   Two automatic dual-gel TDGS systems.

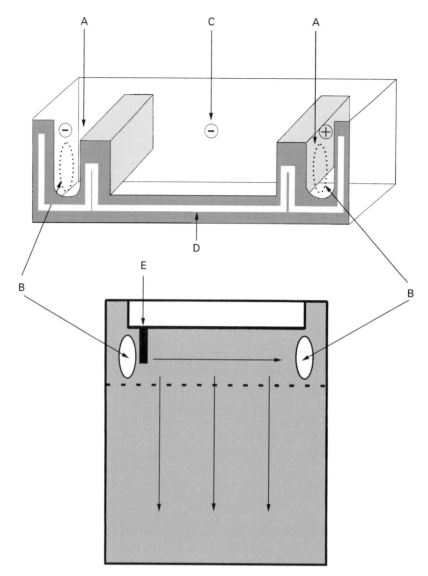

**Figure 4** Schematic depiction of the automated two-dimensional electrophoresis unit. Buffer chambers for the first-dimension separation (A) are connected with the gel through openings in the (inner) glass plate (B). During the first-dimension electrophoresis, the middle buffer chamber (C) is isolated from the outer chambers. For the second-dimension run, buffer chamber C is flooded with buffer and the upper electrode is turned on in conjunction with a positive electrode in the lower reservoir (not shown in this figure). The gel cassette is sealed to the top unit with a serpentine silicone gasket (D), and sample is loaded in the single slot (E). The dashed line indicates the beginning of the gradient of urea/formamide.

electrophoresis is carried out vertically in the denaturing gradient gel. All components of the automatic TDGS electrophoresis system are depicted in **Figure 6**.

Gradient gels can be poured, up to nine at a time, using a simple linear gradient maker in combination with a multiple gel caster. The exact gradient that is to be applied is dependent on the GC-content of the gene(s) of interest and is determined by the TDGS primer designer software.

**Miniaturization**

Miniaturization of gene analysis systems, such as the TDGS system described here, offers two major advantages: increased speed and lower cost.

**Speed** The duration of electrophoresis depends on the voltage applied. For example, the optimal electrophoresis conditions for the retinoblastoma susceptibility gene *RB1* using standard 1.0 mm thick gels,

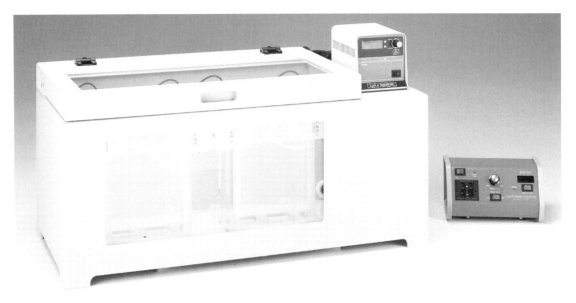

**Figure 5**  The entire automatic TDGS system. In this version of the system, four gels can be run simultaneously submerged in a buffer tank, which is equipped with a heater/stirrer to provide for a constant temperature. For more information, see http://www.cbssci.com

are 100 V, 5 h for the size separation and 100 V, 16 h for the second-dimension separation. Increasing the voltage increases the heat production, which negatively affects gel resolution. An obvious strategy is the use of thinner gels, which facilitate rapid heat dissipation into the surrounding buffer and thereby allow increasing the voltage while maintaining a good resolution. Currently, gels as thin as 0.35 mm are now run at 500 V, 0.8 h for the size separation and 500 V, 3.5 h for the second-dimension separation.

Cost   The cost factor is of major importance for the large scale implementation of genetic testing. Since this is determined to a major extent by reagent and material cost, as well as space, miniaturization of analytical systems is of crucial importance. Miniaturization of TDGS results in thinner and smaller gels, which require less sample (smaller PCR volumes can now be applied) and lower gel and buffer amounts. Moreover, they take up less space. Instead of the current $17 \times 22$ cm format, two-dimensional patterns have already been produced on $10 \times 10$ cm mini-gels, and it is not unreasonable to expect that ultimately electrophoresis will be carried out on glass slides.

### Detection of TDGS Patterns

After electrophoresis, the two-dimensional DNA fragment patterns can be visualized by incubating the gels with DNA staining dyes. Examples are ethidium bromide or the more sensitive dye Sybr-green. Patterns are photographed under UV light and evaluated

for the occurrence of variations (in the form of four spot patterns; see Figure 1D). An example of a TDGS pattern is shown in **Figure 7**, depicting the *RB1* gene, containing a mutation in exon 2.

However, for large scale application of TDGS, dye primer technology for the in-gel detection of two-dimensional spot patterns is an obvious strategy. Test results indicate similar two-dimensional patterns and sensitivity for fluorescein-labelled primers compared to Sybr-green-stained gels.

Introduction of fluorescent detection offers two advantages over gel staining. First, the reduction in labour is considerable and loss of gels due to breakage is prevented. Second, since there is no need to release the gel from between the glass plates it has become possible to use thinner gels, which will allow shorter electrophoresis times (see above). To increase the efficiency even further it is possible to label different samples with different fluorophores. Current fluorescence imagers have the option to analyse multiples fluorophores in the same gel.

### Future Developments

Routine application of TDGS requires standardization and further streamlining of the procedure. Ultimately, one could envisage a fully automated system of PCR amplification, sample loading, electrophoresis, scanning of gels by a fluorescent imager, followed by online interpretation of gels by image analysis systems.

Much of the labour that is involved in PCR amplification, as well as the error rate, can be greatly

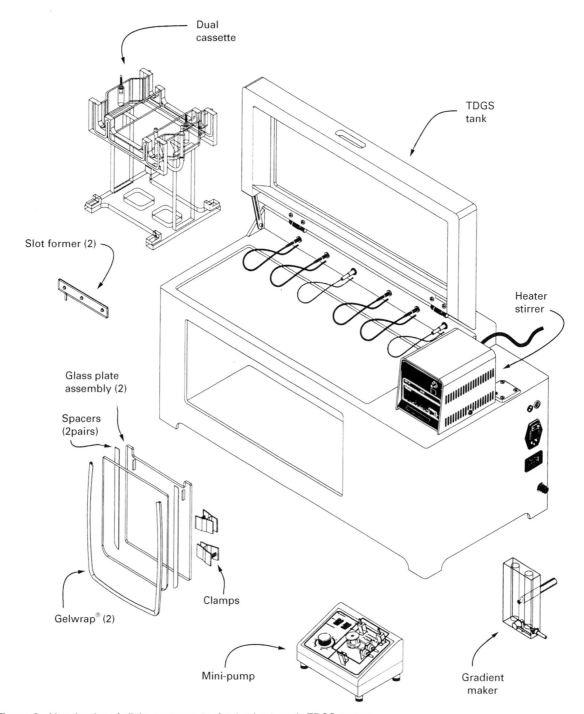

**Figure 6**  Line drawing of all the components of a 4-gel automatic TDGS system.

diminished by PCR robotics. Such instruments have now become widely available and, in combination with an ongoing effort to increase multiplex groups, are expected to increase greatly the front-end throughput of genetic testing. Multiple two-dimensional gels can be stacked for simultaneous electrophoresis of manifold samples. A simple robot arm could load the gel sandwiches into the fluorescent imager for quick gel scanning. Finally, while the actual interpretation of spot patterns is currently most conveniently done by eye, automated image analysis software is commercially available. The use of such software may, for example, facilitate the detection of subtle positional changes in the context of other spot variations. In this respect, one could envisage a programme with information on all possible spot positional variants to identify quickly recurrent mutations and polymorphisms on the basis of their unique

400 bp                                        100 bp

0% UF

65% UF

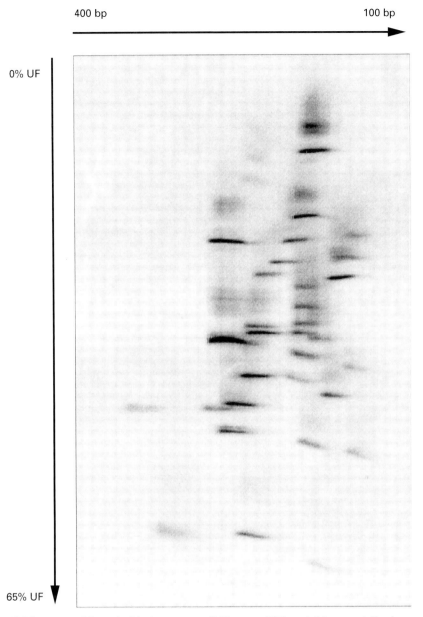

**Figure 7** Empirical TDGS pattern of the retinoblastoma susceptibility gene *RB1*, containing a mutation in exon 2.

configuration. Such software should also be capable of storing two-dimensional patterns and link subsets of them in particular experiments requiring comparisons of large numbers of individuals. It could also provide for a sample tracking system.

## Further Reading

Cotton RGH (1997) *Mutation Detection*. Oxford: Oxford University Press.

Dhanda RK, Smith WM, Scott CB *et al.* (1998) A simple system for automated two-dimensional electrophoresis: applications to genetic testing. *Genetic Testing* 2: 67–70.

Eng C and Vijg J (1997) Genetic testing: the problems and the promise. *Nature Biotechnology* 15: 422–426.

Fischer SG and Lerman LS (1979) Length-independent separation of DNA restriction fragments in two-dimensional gel electrophoresis. *Cell* 16: 191–200.

Lerman LS and Silverstein K (1987) Computational simulation of DNA melting and its application to denaturing gradient gel electrophoresis. *Methods in Enzymology* 155: 501–527.

Sheffield VC, Cox DR, Lerman LS and Myers RM (1989) Attachment of a 40-base pair G + C rich sequence (GC-clamp) to genomic DNA fragments by the polymerase chain reaction results in improved detection of single-base changes. *Proceedings of the National Academy of Science of the USA* 86: 232–236.

van Orsouw NJ, Li D, van der Vlies P *et al.* (1996) Mutational scanning of large genes by extensive PCR multiplexing and two-dimensional electrophoresis: application to the RB1 gene. *Human Molecular Genetics* 5: 755–761.

van Orsouw NJ, Dhanda RK, Rines DR *et al.* (1998) Rapid design of denaturing gradient-based two-dimensional

electrophoretic gene mutational scanning tests. *Nucleic Acids Research* 10: 2398–2406.

Vijg J and van Orsouw NJ (1999) Two-dimensional gene scanning: exploring human genetic variability. *Electrophoresis* 20: 1239–1249.

# GEOCHEMICAL ANALYSIS: GAS CHROMATOGRAPHY AND GC-MS

**R. P. Philp**, University of Oklahoma, Norman, OK, USA

## Introduction

Geochemical analysis, and more specifically chromatography, is concerned with samples derived from two different sources: those of relatively recent origin, related to environmental problems; and those of a much greater geological age, related to fossil fuel exploration and exploitation. The chromatographic techniques utilized to analyse and characterize such samples are virtually identical regardless of the age and origin of the sample. The extracts from geochemical samples, whether they are rocks, soils, crude oil spills, contaminated wildlife or spills of refined products, are very complex mixtures of a wide variety of organic compounds. Compounds derived from fossil fuels typically will be complex mixtures of hydrocarbons, and the environmental samples from more recent sediments probably will contain a variety of other compounds such as chlorinated compounds, pesticides or herbicides. In view of the similarities of the techniques used for analysing the samples from these different sources, the majority of examples used in this article to illustrate the techniques will be based on the characterization of fossil fuel samples.

The major goal of any geochemical analysis is to take a sample and, through a variety of fractionations and analytical techniques, reach a point where either the presence or absence of specific target compounds can be determined, or fingerprints for specific classes of compounds can be obtained and used for correlation purposes. Applications related to petroleum exploration might use such fingerprints for oil–source rock or oil–oil correlation studies, whereas in envir-

onmental studies one is more concerned with correlating a spilled product with its original source material or trying to evaluate the extent of removal during clean-up procedures.

Geochemical samples are extremely complex mixtures of a wide variety of compound classes. The analytical techniques commonly used to characterize such mixtures involve some form of chromatography, such as gas chromatography (GC), gas chromatography–mass spectrometry (GC-MS), gas chromatography–mass spectrometry/mass spectrometry (GC-MS/MS), and more recently gas chromatography-isotope ratio mass spectrometry (GC-IRMS). Liquid chromatography (LC) and combined liquid chromatography–mass spectrometry (LC-MS) are also used in certain applications, but not to the same extent as GC and GC-MS. In addition to the analytical chromatographic separations, most geochemical analyses require some sort of fractionation into compounds classes prior to the actual analysis. There are certain cases where total sediment extracts or whole crude oils are analysed directly but generally the mixtures are so complex that an initial fractionation(s) is required to simplify the extracts for subsequent analyses. For example gas chromatograms of many crude oils (**Figure 1**) are dominated by *n*-alkanes but, for the most part, compounds that are of much greater geochemical importance are not readily observable in these chromatograms but are hidden in the baseline of the chromatogram. It should be noted that there are also many naphthenic crudes not dominated by *n*-alkanes, e.g. Venezuelan and Russian crudes. Most of these naphthenic crudes are either severely biodegraded or have been generated at relatively low levels of maturity from sulfur-rich kerogens. A fractionation step involving thin-layer chromatography, column chromatography or liquid chromatography, all of which involve partitioning of components between a liquid and solid phase, leads to the separation of

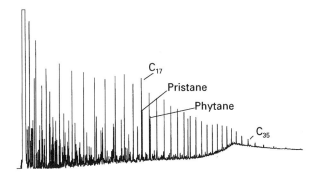

**Figure 1** Gas chromatograms of crude oils, rock extracts, or refined petroleum products are typically dominated by *n*-alkanes and isoprenoids. While GC alone does not permit their identification, the fact that the isoprenoids pristane and phytane have very similar elution times to the $C_{17}$ and $C_{18}$ *n*-alkanes, respectively, generally make it relatively easy to identify the other members of the homologous series with a reasonably high degree of confidence.

compounds on the basis of factors such as polarity, shape and size. For hydrocarbon-containing samples, the fractionation step typically involves separation into three fractions – saturate and aromatic hydrocarbons and a polar fraction containing nitrogen, sulfur and oxygen compounds. In general it is the saturate and aromatic fractions that receive most attention in terms of additional analyses. Although the nitrogen, sulfur and oxygen fractions contain many compounds that possess useful information, the complexity of these fractions has precluded their detailed analyses. It is not proposed to go into the experimental details of such chromatographic fractionations since these are very basic techniques and descriptions of specific methods for particular classes of compounds are readily available in the literature.

Fractionation of the crude oil used for Figure 1 into various fractions produces saturate and aromatic fractions as shown in **Figure 2**. It should be noted when comparing Figures 1 and 2 that the result of the fractionation and evaporation of the solvents used in the fractionation process will lead to the loss of some of the more volatile compounds in the $C_1$–$C_{15}$ range of the saturate and aromatic fractions. GC analyses of the saturate fraction produces a chromatogram dominated by *n*-alkanes, typically in the carbon number range from $C_{15}$ to around $C_{40}$ when the analyses are performed using conventional GC. The isoprenoids pristane (Pr) and phytane (Ph) can also be clearly discerned on these chromatograms. Once again, it should be emphasized that although the *n*-alkanes are the predominant compounds in the chromatograms, there are many more minor compounds in the fraction that generally provide a great deal more information than the *n*-alkanes. These

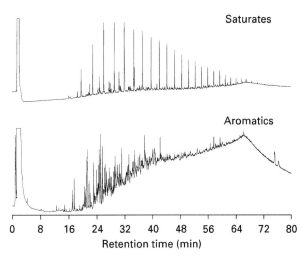

**Figure 2** The whole oil chromatogram shown in Figure 1 does not give a true impression of the complexity of the mixture of compounds in a crude oil. While the *n*-alkanes are the dominant components in the chromatogram, a vast array of branched, cyclic, aromatic and polar compounds are also present. This figure shows the chromatograms for a saturate and aromatic fraction separated from a crude oil by thin-layer chromatography.

compounds can be further concentrated by such processes as molecular sieving or urea adduction, both of which will separate the *n*-alkanes from the branched and cyclic alkanes as shown in **Figure 3**. At this point fractionation and sieving of the original extract or crude oil will have produced a fraction that is readily

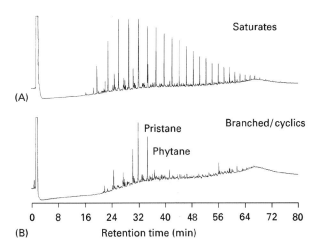

**Figure 3** The saturate fraction shown in Figure 2 is again dominated by *n*-alkanes, which tend to mask the presence of a very complex mixture of branched and cyclic compounds also present in this fraction. The *n*-alkanes can be separated from these branched and cyclic compounds by processes such as molecular sieving or urea adduction to produce the branched and cyclic fraction shown in the bottom chromatogram of this figure. The top chromatogram (A), shown for comparison purposes, is the total saturate fraction from which the branched and cyclic compounds were isolated.

amenable to analysis by the techniques mentioned above, such as GC, GC-MS, GC-MS/MS or GC-IRMS. In the following sections a brief description of each technique and typical applications will be given. Again it should be reiterated that for the most part this article uses hydrocarbons for illustration purposes, but the majority of the techniques are equally applicable to the analysis of other samples of geochemical interest such as environmental samples, possibly with some slight modifications in the operating conditions.

## Gas Chromatography

As can be seen from the chromatograms used in the Introduction, GC provides a great deal of information on the composition of geochemical samples. With the inclusion of an internal standard, this information can be both quantitative and qualitative in nature. However, it is important to remember that, for the most part, GC does not provide any information on the identification of individual components. In the saturate hydrocarbon fractions, individual compounds such as n-alkanes in the chromatograms can be readily recognized and their identification confirmed either by the use of co-injected standards or analysis of the sample by GC-MS as described below. In other cases where it may be necessary to detect certain classes of chlorinated compounds, or organosulfur compounds, additional information on the presence or absence of these compounds can be obtained by using detectors that are specific for these classes of compounds, such as electron capture, flame photometric or Hall detectors. It should also be noted that a number of recent studies have shown that atomic emission detection (AED) is a particularly sensitive and specific method of detection for sulfur-containing compounds. Although the flame photometric detector (FPD) is probably the most widely used detector for sulfur-containing compounds, it has some drawbacks including nonlinear compound-dependent response, and the quenching of sulfur signals by co-eluting hydrocarbons. The AED has a linear response and is compound-independent, permitting easy calibration using any sulfur-containing compounds. This feature is unique because other detectors require the construction of curves using target analytes as standards, which becomes a very time-consuming exercise.

GC has been utilized widely in geochemistry since the 1950s. As column technology and instrumentation have improved, so has the quality of the analytical data. There are many similarities between the gas chromatographs of today and the systems that were developed in the 1950s and 1960s. Detectors and

injectors have improved and temperature control of the ovens has improved, but probably the greatest advances have occurred in the field of column technology. Early columns were short, large-diameter packed columns made of stainless steel or copper. Over the years narrow-bore capillary columns were developed, initially made of stainless steel, then glass and more recently fused silica. At the same time as the evolution of capillary columns, the variety of liquid phases for different applications has also greatly improved. The most recent advance, and one that is quite significant for petroleum exploration and production, has been the development of high temperature GC phases. Traditionally most crude oil analyses have characterized hydrocarbons in the $C_1$–$C_{40}$ range but the advent of high temperature GC (HTGC) significantly changed the way in which we look at hydrocarbon distributions of crude oils. The use of HTGC has demonstrated that many crude oils contain a wide range of hydrocarbons significantly above $C_{40}$, extending to as far as $C_{100}$ and possibly higher. This in turn has also led to changes in one of the very basic premises of geochemistry – that oils with a high wax content were thought to be derived only from higher plant sources. It is now clear that such oils may also be derived from lacustrine and marine source rocks as a result of analysing a number of samples from such source rocks using HTGC. Figure 4 provides an excellent illustration of the additional information obtained from the use of HTGC. The upper chromatogram shows the analysis of an ozocerite extract by conventional GC and the bottom chromatogram shows the same sample analysed by HTGC. Clearly the distribution of hydrocarbons is quite different in the lower chromatogram. The significance of this is related to the fact that the greater the concentration of the higher molecular weight alkanes, the greater the production problems associated with oils that contain such compounds. In other words, if the oils were only characterized by conventional GC, high molecular weight hydrocarbons would remain undetected. Once a production programme was initiated it would not be long before the wellhead facilities and pipelines would become blocked with paraffin deposits, which require costly measures to remove. While HTGC analyses do not eliminate the problem, production engineers would be aware of the potential for such a problem and steps could be introduced to minimize its occurrence.

With the availability of HTGC an increasing number of samples have been analysed using this approach and steps have been taken to develop methods that will quantitatively separate high molecular weight alkanes from the asphaltene fraction. Analyses

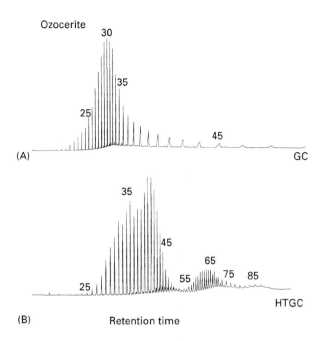

**Figure 4** One of the most significant recent developments within GC has been the development of high temperature phases for the columns. Before this development it was generally only possible to analyse compounds with up to approximately 40 carbon atoms. The newer HTGC columns permit samples containing up to approximately 120 carbon atoms to be analysed. This figure illustrates the comparison between the analyses of the hydrocarbons in the fossil bitumen ozocerite by (A) conventional GC and (B) HTGC. The difference in distributions is very clear and also demonstrates how the composition of a fraction obtained by conventional GC may not necessarily reflect the true composition of the sample being investigated.

of a wide range of such samples have shown that, in addition to the *n*-alkanes, there is also a wide range of additional compounds in the higher molecular weight fraction including branched alkanes and alkylcyclohexanes. The distribution patterns for these compounds has provided an additional powerful tool for determination of the type of environment from which a sample has been generated. For example **Figure 5**A and **B** show the distributions of high molecular weight fractions from oils whose source materials were known to be deposited in lacustrine and marine depositional environments, respectively. Note the difference in the distributions of these monocyclic hydrocarbons, which are characteristic of the different environments. At present relatively little information is available concerning the origin of these compounds, although their widespread distribution suggests that they are probably related to an algal/phytoplanktonic source, possibly with additional contributions from higher plant waxes. It is known that many marine organisms contain abundant quantities of higher molecular weight esters, alcohols and fatty acids. Relatively simple transformations

could readily convert these compounds to the corresponding hydrocarbon and could easily represent a viable source for such hydrocarbons.

Another area of geochemistry that has become particularly important in the past few years is reservoir geochemistry, and GC has played an extremely important role in its development. Oil and gas reservoirs are very complex geological features with many compartments. A knowledge of the relative position of these compartments is extremely important for reservoir management, determination of where additional production wells should be drilled, and evaluating how a specific reservoir may have been filled. There are a number of ways in which the reservoir compartments may be delineated but one particularly interesting, innovative and relatively cheap method involves the utilization of high resolution GC. As indicated above, crude oils are very complex mixtures of hydrocarbons, but when the chromatograms are expanded the complexity of the mixtures becomes far more apparent and the presence of a large number of minor components is clearly visible. Reservoir geochemistry utilizes these minor components to assist in the delineation of the reservoir compartments. In brief it is first necessary to determine whether all of the oils in the reservoir are derived from the same source materials. Once this has been established, all of the oils need to be analysed by high resolution GC, the early eluting region of the chromatogram expanded and a number of pairs of minor peaks selected as shown in **Figure 6**. Ratios based on pairs of selected peaks are measured and subsequently plotted on a star or polar diagram. This process is then repeated for all the oils to be examined from the reservoir. It is important to ensure that the same pairs of peaks are selected for each oil, even if the identity of these peaks in unknown. Since the differences between the pairs of peaks in individual samples are often quite small it is extremely important to ensure that the GC analyses are highly reproducible for this particular application. However, once all of the oils have been analysed and the data plotted on the star diagram, it will be found that oils that are in the same compartment or in communication will appear virtually on top of each other, whereas those oils in different compartments will be slightly separated (**Figure 7**). These small difference may result from slight differences in oil–rock interactions, slight maturity differences or generation from slightly different sources.

## Gas Chromatography– Mass Spectrometry

While GC can provide a great deal of information that is of interest and useful from a geochemical

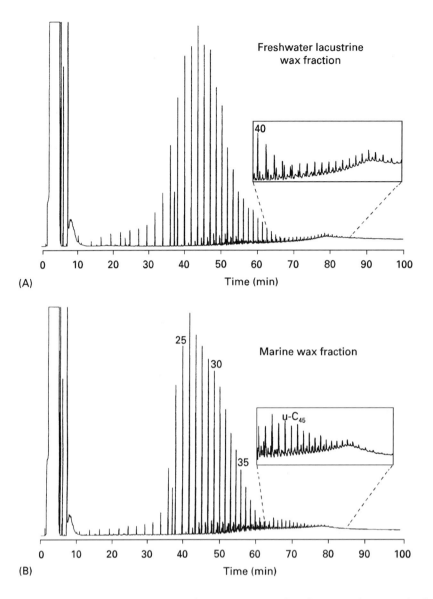

**Figure 5** The availability of HTGC has led to the discovery of numerous new series of compounds present in oils and source rocks above $C_{40}$. Several of these series, and in particular a series of alkylcyclohexanes, have been shown to be useful in discriminating between oils derived from source materials deposited in lacustrine (A) versus marine settings (B). More subtle variations in these distributions allow the salinity levels of the depositional environments to be distinguished.

perspective, it should also be noted that in most cases GC only provides information on the distribution of the major components in the sample, and for the most part these are generally dominated by the *n*-alkanes. The more useful compounds are the more complex molecules, or biomarkers, which are typically present in relatively low concentrations and which require the use of GC-MS and more specifically single ion or multiple ion detection (MID) in order to determine their distributions. While there are many classes of biomarkers that are commonly used for correlation and other purposes, compounds such as the steranes and terpanes will typically provide the greatest

amounts of useful information for both an environmental and exploration context.

To illustrate the utility of the biomaker fingerprints, the gas chromatograms of three oils are shown in **Figure 8**. From the gas chromatograms alone it is virtually impossible to determine what relationship, if any, exists between these samples. In other words, are they from the same source rock or can they be correlated with each other? The effects of biodegradation are clearly evident in sample B since all of the *n*-alkanes have been removed, making it appear even more significantly different from the other two samples. Detailed analyses of the same samples by

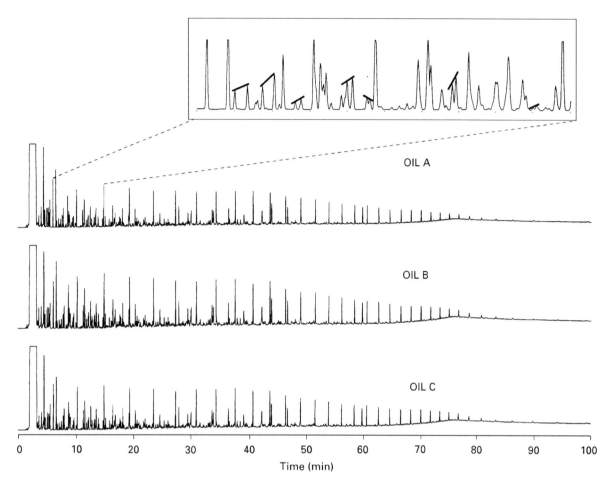

**Figure 6** Reservoir geochemistry has provided an important means of determining continuity and communication within reservoir compartments. Once it has been established that the oil in a reservoir is from a common source, high resolution gas chromatograms are obtained for individual samples and ratios of various pairs of peaks are determined, as shown in the figure. The identity of the components does not have to be known; the important point is that the same pairs of peaks are used for all the samples examined in any particular study. Although the early studies typically used peaks in the early part of the chromatogram, it has been shown that the minor components in the higher regions of the chromatogram can also be used for the same purpose.

GC-MS and MID using the characteristic ions for the sterane and terpane biomarkers at $m/z$ 217 and 191, respectively, produces the additional data shown in **Figure 9**. On the basis of the chromatograms shown in Figure 9, samples B and C are in all probability related to each other. It is not necessary to identify each component, rather one should think of the mass chromatograms as fingerprints. If two samples are derived from the same source, then their fingerprints should be the same, or at least very similar; samples from different sources will be different from each other. Hence when the fingerprint for sample A is compared with those for B and C in Figure 9, there are a number of significant differences between these samples that permit one to conclude that A is from a different source than B or C. The biomarker fingerprints obtained in this way are very

specific for a variety of applications, in addition to this type of correlation. The presence of individual compounds, for example oleanane and gammacerane, can provide information on the presence of specific types of source materials or the nature of the depositional environment.

To illustrate this type of application, **Figure 10** shows the $m/z$ 191 and 217 mass chromatograms of an oil that is derived from source material dominated by higher plant or terrestrial source material. This evidence is contained in the fact that the predominant component in the $m/z$ 191 mass chromatogram is the terpane called 18α(H)-oleanane. It has been established that this compound has its precursor in higher plant material and hence the presence of this compound in an oil will indicate that the sample is derived from such material. In support of such evidence is the

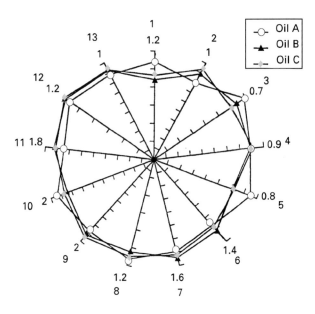

**Figure 7** The ratios of the pairs of peaks measured in Figure 6 are plotted on a star or polar diagram. If two oils are in the same compartment, or in communication with each other, then on such a star plot the two oils will have identical plots (i.e. B and C). If they are not in communication with each other, then their plots will show some subtle differences (i.e. oil A).

fact that the sterane chromatogram is dominated by the $C_{29}$ steranes. For oils of this nature it has been clearly established that the $C_{29}$ steranes are also associated with higher plant source materials. In this

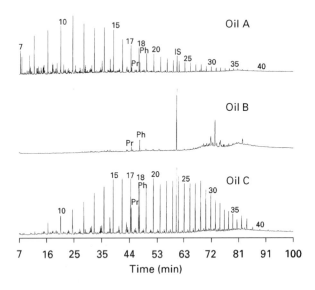

**Figure 8** Gas chromatograms of three different oils provides limited information concerning the relationship between the samples. In this figure oil B clearly appears to be quite different from oils A and C, but all that can really be said is that oil B has been heavily biodegraded and the *n*-alkanes have been removed. It is impossible to say what, if any, similarities there may have been between the alkane distribution for this sample and the other two samples.

manner, pieces of evidence can be put together that in many cases will provide a very clear indication as to the origin of the material being analysed.

In the second example, shown in **Figure 11**, the presence of another very specific compound, gammacerane, can also be very clearly seen in the $m/z$ 191 chromatogram. This compound is a very specific indicator of depositional environments of enhanced salinity. Recent attempts have been made to relate the presence of certain compounds, for example, dinosterane to the age of the source rock from which the sample was generated. Specific ratios of different sterane isomers or terpane isomers are also used extensively for determining the relative maturity of oils or source rocks.

The sterane and methylsterane distributions in crude oils are far more complex than the terpanes and no matter how good the GC resolution, it is impossible to obtain complete separation of all co-eluting isomers, epimers and homologues (**Figure 12**). In order to optimize this separation it is necessary to utilize GC combined with tandem mass spectrometry, or MS/MS, which provides an additional degree of separation based on the utilization of the MS/MS capability.

## Gas Chromatography–Mass Spectrometry/Mass Spectrometry

To demonstrate the utility of the GC-MS/MS approach to the characterization and determination of biomarkers in geochemical samples, the resolution of a complex mixture of sterane isomers and homologues will be described. While this example utilizes the steranes, it should be borne in mind that the same approach can be used to resolve any very complex mixture of organic compounds from geochemical samples.

The mixture of steranes commonly analysed by MS/MS is in the $C_{27}$–$C_{30}$ carbon number range and each homologue has a molecular mass at $m/z$ 372, 386, 400 and 414, respectively. For each of the steranes the parent ions will undergo a collision-activated decomposition to produce a daughter ion at $m/z$ 217. Hence a series of MS/MS parent–daughter experiments are performed utilizing these parent–daughter transitions in combination with the GC separation. The GC-MS analysis and single ion monitoring of $m/z$ 217 produces the mass chromatogram shown in Figure 12 but with the GC-MS/MS analyses, the results shown in **Figure 13** are obtained. It can be seen in Figure 13 that by using the $C_{27}$ parent–daughter ion pair at $m/z$ 372/217, respectively, the result of analysing the sample by MS/MS is to

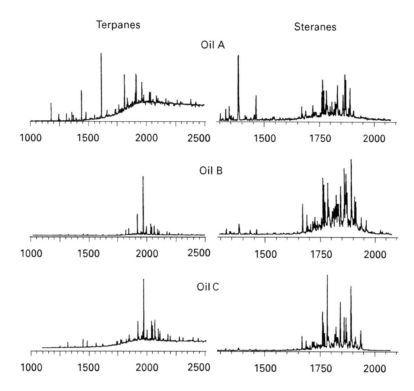

**Figure 9**  The terpane and sterane distributions for the same three oils as used in Figure 8 provide a more specific indication of the relationship between different samples. It is not necessary to know the identification of all the individual peaks but rather to think of the total chromatogram as a fingerprint. In this case it can be seen that oil A has quite a different set of fingerprints than oils B and C, suggesting it is not related to these other two samples. While the fingerprints of oils B and C may show some differences, these are actually due to small maturity differences in the samples.

totally resolve the $C_{27}$ components from the rest of the complex mixture.

Similar results would be obtained if the parent–daughter pairs for the other members of the series were also illustrated. A similar approach could be applied to the methylsterane mixture using the parent ions and the daughter ion at mass 231 and a similar simplification of the mixture would be obtained. In this particular application the MS/MS serves to introduce an additional element of separation following the initial separation by GC.

## Pyrolysis–Gas Chromatography–Mass Spectrometry

While a large proportion of the geochemical samples analysed are soluble in organic solvents and readily amenable to direct analysis by GC or GC-MS, there is another aspect to geochemical samples that is often overlooked. Samples of geochemical interest such as soils, source rocks or coals also have a significant insoluble organic component such as the humic fraction of soils or the kerogen fraction of a source rock. Characterization of these insoluble fractions requries

some type of degradation step prior to analysis. At present for geochemical purposes this degradation step typically consists of some type of pyrolysis reaction with the pyrolyser interfaced to the gas chromatograph or GC-MS system. There are also reports of the use of various NMR techniques to characterize this insoluble fraction, although this is a little less specific than the pyrolysis approach.

An example of the pyrolysis of the insoluble fraction of an organic-rich source rock in shown in **Figure 14**. This was produced by pyrolysing the sample at a temperature of 600°C for a short period of time and allowing the pyrolysis products to be transferred directly to the GC column. (There are of course a wide variety of pyrolysis conditions that could be used, but those cited here give a general idea of the typical conditions used.) The products of a sample pyrolysed in this manner produce a chromatogram dominated by alkane/alkene doublets plus a wide variety of minor components. From these distributions it is often possible to gain information about the nature of the source materials originally responsible for the formation of the kerogen plus the type of products it will subsequently produce if buried to

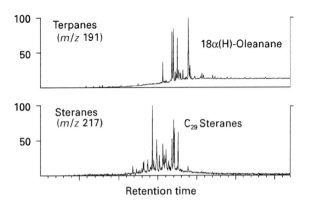

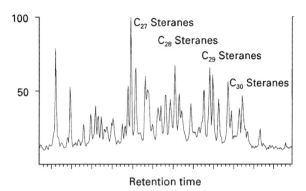

**Figure 10** GC-MS analysis of crude oils reveals complex fingerprints of biomarkers. In many cases these compounds may be very specific indicators of particular types of source materials responsible for sourcing the oil. In this example the predominant component in the terpane chromatogram is 18α(H)-oleanane. Not only is this compound very specific in terms of being derived from higher plant source materials, but it is also more specifically related to the flowering plants or angiosperms that have only evolved since the Late Cretaceous–Early Tertiary periods. The presence of this compound can therefore be used to constrain the age of the source rock from which the oil was generated.

**Figure 12** Sterane distributions in crude oils as typically determined by single ion monitoring of $m/z$ 217 reflect the complex nature of the mixture of these components in a crude oil. It is virtually impossible to separate all of the co-eluting isomers, epimers and homologues by GC-MS, no matter how good the chromatography.

appropriate depths and subjected to thermal degradation.

Another useful application is the pyrolysis of asphaltenes, particularly those isolated from biodegraded crude oil samples. It is often difficult to determine the origin of a biodegraded oil sample. However, if there are a number of possible nondegraded samples with which it can be compared, then the asphaltenes can be isolated from all the samples

by pentane precipitation and pyrolysed. In this way it will be observed that the $n$-alkane/alkene fingerprints generated from the degraded and nondegraded samples will be virtually identical if the samples are derived from the same source, but quite different if the samples are unrelated.

## Gas Chromatography–High Resolution Mass Spectrometry

The majority of geochemical analyses reported in the literature are concerned with the detection and identification of hydrocarbons. However, many

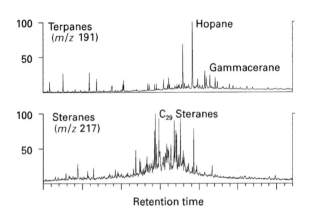

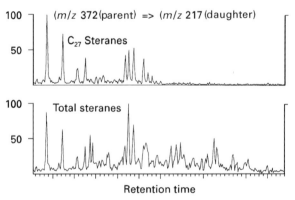

**Figure 11** Gammacerane is another very specific biomarker that can be readily observed in the $m/z$ 191 chromatogram. The relative concentration of this compound varies with the salinity of the original depositional environment. Hence samples deposited in a very saline lacustrine setting will generally have high gammacerane contents whereas those from freshwater lacustrine environments are generally depleted in gammacerane.

**Figure 13** To simplify the sterane fingerprint to some degree GC-MS/MS becomes an important tool. In this diagram the same sterane mixture used for Figure 12 has been analysed in the GC-MS/MS mode. By filtering out the parent–daughter ions for the $C_{27}$ steranes, the top chromatogram shows only the $C_{27}$ compounds. Distributions for the $C_{28}$–$C_{30}$ steranes could also be obtained using the appropriate parent–daughter ions.

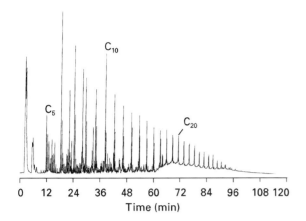

**Figure 14** One of the most useful ways for characterizing the insoluble organic matter in a source rock, coal, shale or soil sample is by pyrolysis-GC. This figure illustrates the results obtained from pyrolysis-GC of Messel shale and shows that the major components obtained in the approach are a series of alkane/alkene doublets. Variations in these distributions can be used to characterize the organic matter in terms of whether it is algal or terrestrial as well as provide information on the relative maturity of the samples.

geochemical mixtures contain complex mixtures of compounds in which heteroatoms are mixed with the hydrocarbons in varying amounts. In certain applications knowledge of these components, particularly sulfur-containing compounds, may be extremely important. There are two approaches by which such distributions may be obtained. The first is by the use of element-selective GC detectors, such as the FPD, Hall detector, or one of the more recent types of AED that are selective for sulfur-containing compounds at the exclusion of non-sulfur-containing compounds. The second method combines GC with high resolution MS. Although this approach is not used routinely, it is a very powerful and specific technique for this type of application, as discussed by Tibbetts and Large in 1988. While the use of low resolution GC-MS and ancillary techniques such as single ion monitoring and multiple ion detection have been discussed elsewhere, it needs to be recognized that utilization of nominal masses in MID may lead to ambiguous results. As demonstrated by Tibbets and Large, while the ion at nominal mass 184 may be used for the determination of dibenzothiophenes (DBT), it is also the nominal mass for the $C_4$-substituted naphthalenes, leading to possible misinterpretation of the resulting chromatograms. However, use of the accurate mass at $m/z$ 184.0347 permits detection of only the DBT and no substituted naphthalenes. Several examples have been given by Tibbetts and Large on the use of this approach for the correlation of crude oils or to distinguish oils from different sources or reservoirs.

Used in conjunction with the conventional detection of biomarkers, this method provides a very powerful and additional tool for geochemical analyses.

## Gas Chromatography–Isotope Ratio Mass Spectrometry

Carbon naturally exists as a mixture of its two stable isotopes, $^{12}C$ and $^{13}C$, in an approximate $^{12}C/^{13}C$ ratio of 99 : 1. The carbon isotopic composition of living organic matter in part depends on the species but is also determined by a number of environmental properties. For example, atmospheric carbon dioxide is assimilated by living plants during photosynthesis and the nature of the plants and whether they assimilate $CO_2$ via a C3 or C4 photosynthetic cycle will determine the extent of preferential assimilation of the lighter $^{12}C$ isotope. C3 plants are typically associated with warmer and more arid climates and in general have isotopic values in the $-10$ to $-18‰$ range. C4 plants are more typically associated with colder climates and have lighter isotopic values in the $-22$ to $-30‰$ range. To determine the $^{13}C$ composition, the sample is combusted to convert all of the carbon to $CO_2$, which is analysed in a stable isotope ratio mass spectrometer and compared with the isotopic composition of a standard material (Pee Dee Belimnite, PDB), whose isotopic composition has been assigned a value of 0.

## GC-IRMS and Isotopic Composition of Individual Components

GC-IRMS permits acquisition of $\delta^{13}C$ values for individual components in complex mixtures. The important part of the system is the interface between the GC and the isotope ratio mass spectrometer. This consists of a reactor tube, generally a narrow-bore ceramic tube, containing a bundle of wires, typically copper, nickel or platinum where complete combustion to $CO_2$ must occur. After combustion the water and $CO_2$ pass through a membrane separator to remove the water before the $CO_2$ enters the mass spectrometer, where the isotopic composition of the gas is determined relative to the standard. The isotopic values of individual components can be interpreted to obtain information on the diagenetic history of an individual component and the nature of the microbial community during deposition.

The isotopic composition of individual compounds is also of importance from an environmental viewpoint. For example, analysis of a whole oil, or the saturate fraction of a whole oil, allows the ready determination of $\delta^{13}C$ values of the $n$-alkanes and the major isoprenoids, pristane and phytane. These values can be used for correlation purposes, to distinguish oils from different sources, to correlate oil spills with their suspected sources, or to determine the source of hydrocarbons that have contaminated wildlife. Examples of this approach are shown in **Figure 15**. Gas chromatograms of oil extracted from the feathers of birds that had been exposed to a crude oil spill and a suspected source are compared with each other and show certain differences, particularly at the lighter end of the chromatograms. Such differences could lead to a dispute as to whether or not this oil was actually the one that was responsible for contaminating the birds. The lower part of the figure shows the carbon isotope data for individual compounds in the two samples. It can be seen

that, despite the loss of some of the light ends through evaporation, a good relationship between the two samples can be established. These data could also be used in support of the biomarker and other properties normally used to establish relationships between samples thought to be related. In an additional example, **Figure 16** shows the results from the GC-IRMS analyses of 20 oils from a region in SE Asia. It can be seen that on the basis of these analyses two distinct families of oils are present in the region. One family has isotopic values of around $-20‰$ for each compound whereas the other family has values around $-28‰$. This information can be used in conjunction with the biomarker data to determine the significance of these differences.

While these are just two examples, we have also shown that GC-IRMS can be used in an environmental context to correlate weathered and unweathered oils and refined products and their weathered counterparts. If there are small amounts of the *n*-alkanes remaining it should be possible to obtain their isotopic composition and subsequently use these values to make the correlation. Alternatively if the samples have been so extensively biodegraded that all of the *n*-alkanes have been removed, it is possible to isolate the asphaltenes, pyrolyse them and analyse the pyrolysates by GC-IRMS. Correlations can then be made using these data. This is a particularly valuable approach for the correlation of refined products that only contain lower carbon number compounds and none of the more reliable biomarker compounds that are typically used for correlation purposes.

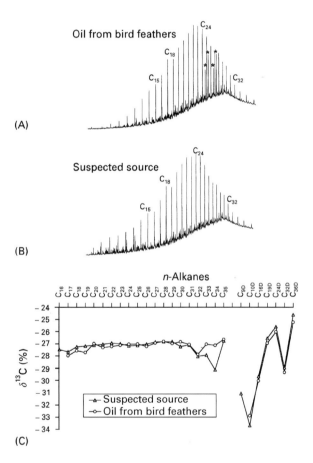

**Figure 15** (A) Chromatogram of an extract from bird feathers contaminated by a crude oil; (B) shows the chromatogram for the suspected source of the contamination. The two chromatograms show some significant differences, particularly in the early part of the chromatogram, mainly as a result of weathering. However, in (C) we see that there is a strong relationship between the isotopic compositions for each individual *n*-alkane in the two samples. This isotopic information can be used in conjunction with the biomarker data and other parameters to establish a relationship between the two samples.

## Summary

This article has attempted to illustrate the importance of GC to geochemical analyses. Geochemical samples from all sources, whether recent or ancient, oils or synthetic chemicals, refined or crude, are incredibly complex mixtures of organic compounds in most cases. To try and analyse such samples, whether simply for correlation purposes or to detect and identify unknown compounds, almost inevitably requires some level of chromatography to facilitate the analytical process. The most common forms of chromatography generally involve some form of liquid chromatography in the initial steps to simplify the mixture into compound classes, followed by GC to separate and resolve as many compounds as possible in the resulting fractions. Chromatography alone simply separates the components, hence it is very common in most geochemical analyses for the chromatographic step to be combined with an identification technique such as MS. The results of such

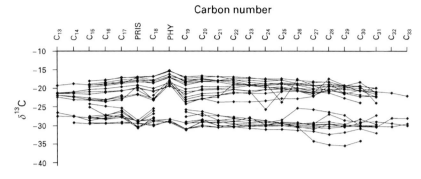

**Figure 16** Analysis of 20 oils by GC-IRMS provides isotopic data for each major compound present in the oils. The oils plotted on this chart can be divided into isotopically heavy and isotopically light groups. Supporting evidence for these groupings can be obtained from the GC-MS data.

analyses generally provide the information necessary to determine the origin of a particular sample and, in the context or crude oil exploration, relate it to possible source rocks and such information as age, maturity, and migration pathways. For environmental samples the information obtained is generally for the purpose of determining the source of a spill and hence a great deal of use is made of these distributions in terms of their fingerprinting capability. As chromatographic and spectroscopic techniques continue to improve, clearly the degree of separation achievable for these complex mixtures will also greatly improve, but mixtures of even greater complexities will always be available to provide that next level of challenge.

*See also:* **II/Chromatography: Gas:** Column Technology; Detectors: General (Flame Ionization Detectors and Thermal Conductivity Detectors); Detectors: Mass Spectrometry; Detectors: Selective; Historical Development; Pyrolysis Gas Chromatography; Theory of Gas Chromatography.

## Further Reading

Baskin DK, Hwang RJ and Purdy RK (1995) Prediction of gas, oil, and water intervals in Niger Delta reservoirs using gas chromatography. *American Association of Petroleum Geologists Bulletin* 79: 337–350.

Brooks J and Fleet AJ (eds) (1987) *Marine Petroleum Source Rocks*, Geological Society Special Publication, vol. 26, 444 pp. Oxford: Blackwell Scientific Publications.

Carlson RMK, Teerman SC, Moldowan JM, Jacobson SR, Chan EI, Dorrough KS, Seetoo WC and Mertani B (1993) High temperature gas chromatography of high wax oils. In: *Proceedings of the 20th Annual Convention of the Indonesian Petroleum Association*, Jakarta, pp. 483–507.

Connan J (1984) Biodegradation of crude oils in reservoirs. In: Brooks J and Welte DH (eds) *Advances in Petroleum Geochemistry*, vol. 1, pp. 299–335. London: Academic Press.

England WA (1989) The organic geochemistry of petroleum reservoirs. *Organic Geochemistry* 16(1–3): 415–425.

Hayes JM, Freeman KH, Popp BN and Hoham CH (1990) Compound-specific isotopic analyses: a novel tool for reconstruction of ancient biogeochemical processes. *Organic Geochemistry* 16 (1–3): 1115–1128.

Johnson D, Quimby BD and Sullivan JJ (1995) An atomic emission detector for gas chromatography. *American Laboratory*, pp. 13–20.

Killops SD and Readman JW (1985) HPLC fractionation and GC-MS determination of aromatic hydrocarbons from oils and sediments. *Organic Geochemistry* 8: 247–257.

Levy JM (1994) Fossil fuel applications of SFC and SFE: a review. *Journal of High Resolution Chromatography* 17: 212–216.

Peters KE and Moldowan JM (1992) *The Biomarker Guide: Interpreting Molecular Fossils in Petroleum and Ancient Sediments*. Englewood Cliffs, NJ: Prentice Hall.

Philp RP and Oung J-N (1992) Biomarkers, occurrence, utility and detection. In: Voress L (ed.) *Instrumentation in Analytical Chemistry*, 1988–1991, pp. 368–376. Washington, DC: American Chemical Society.

Del Rio JC and Philp RP (1992) High molecular weight hydrocarbons: a new frontier in organic geochemistry. *Trends in Analytical Chemistry* 11 (15): 187–193.

Tibbetts PJC and Large R (1988) Improvements in oil fingerprinting: GC/HRMS of sulphur heterocycles. In: Crump GB (ed.) *Petroanalysis 87*, pp. 45–57. London: John Wiley.

# GLYCOPROTEINS: LIQUID CHROMATOGRAPHY

**K. Miyazaki**, Hokkaido University Hospital, Hokkaido, Japan

## Introduction

Many proteins in cells and biological fluids are glycosylated, and these glycoproteins are present in animals, plants, microorganisms and viruses. The most commonly occurring monosaccharides found in oligosaccharide attachments to mammalian proteins are D-mannose (Man), D-galactose (Gal), D-glucose (Glu), L-fucose (Fuc), N-acetylglucosamine (GlcNAc), N-acetylgalactosamine (GalNAc), and N-acetylneuraminic acid (sialic acid or NeuAc).

The primary structure of glycoprotein glycans and their biological functions have been gradually unravelled by the improvements of methods for isolation and structure determination. High performance liquid chromatography (HPLC) is one of the most commonly used methods for the isolation and analysis of both glycoproteins and their derived carbohydrates, mainly due to the excellent resolution, ease of use, the generally high recoveries, the excellent reproducibility of repetitive separations, and the high productivity in terms of cost parameters.

## Chemistry and Importance of the Glycan Chain of Glycoproteins

### Basic Structure of Glycoprotein

In glycoproteins, glycans are conjugated to peptide chains by two types of primary covalent linkage, N-glycosyl and O-glycosyl. The former is called an N-linked sugar chain and contains a GlcNAc residue that is linked to the amide group of asparagine residues of a polypeptide. As shown in **Figure 1**, almost all N-linked glycoproteins have a common core of two GlcNAc and three Man residues. N-linked glycoproteins have three types of carbohydrate moieties: complex type (Figure 1), high mannose type and hybrid type. The hybrid type is a mixture of the complex and high mannose types. Complex type structures usually have from two to four branches attached to the two outer core Man residues. The branches are distributed over the two terminating core Man residues. The complex structures are termed diantennary, triantennary and tetraantennary, according to the number of antennae. The basic branch structures are composed in most instances of one GlcNAc and one Gal residue (Figure 1).

O-Glycosyl glycoproteins contain at their reducing end a GalNAc residue that is linked to the hydroxyl group of either serine or threonine residues of a polypeptide. This linkage is called an O-linked or mucin-type sugar chain. In general, O-linked structures appear to be less complex than N-linkages in terms of the number of antennae and monosaccharides. However, they can be fucosylated and sialylated. Some glycoproteins have both the N-linked and O-linked forms in their molecules (N, O-glycoproteins).

The addition of carbohydrate to a peptide chain changes the shape and size of the protein structure. Several important discoveries have revealed the following biological roles of glycans: (i) protection of polypeptide chains against proteolytic enzymes; (ii) influence on heat stability, solubility, and many physicochemical properties; and (iii) interaction with other proteins or nonprotein components of the cell, including control of the lifetime of circulating glycoproteins and cells.

### Microheterogeneity of Glycans

In addition to genetically determined variants expressed as variations in their polypeptide chains, almost all glycoproteins exhibit polymorphism associated with their glycan moieties. This type of diversity is termed microheterogeneity, and these different forms have recently been called glycoforms. These variants were first characterized in the $\alpha_1$-acid glycoprotein (AAG) from human serum using electrophoresis. As shown in the structure of major oligosaccharides of AAG (Figure 1), microheterogeneity was found to be due to the occurrence of di-, tri-, and tetraantennary glycans of the N-acetyllactosamine type at the five glycosylation sites.

This feature is widespread and has been observed in natural as well as in recombinant DNA glycoproteins. The existence of microheterogeneity gives rise to many interesting questions regarding the origin of this phenomenon and its relevance to the biological functioning of the glycoproteins that can be distinguished.

Recent interest has been shown in glycoproteins in the industrial field of genetic engineering of human glycoproteins of therapeutic interest. This gives rise to

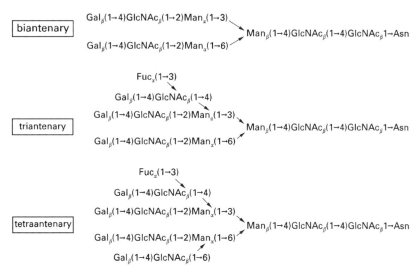

**Figure 1** Structure of the major oligosaccharides of $\alpha_1$-acid glycoprotein (a complex type of the N-linked form). Several NeuAc are linked to Gal residues.

an enormous problem, because the production of recombinant human glycoproteins in nonhuman eukaryotic cells or in prokaryotic cells devoid of glycan biosynthesis machinery leads to the production of incorrectly glycosylated proteins. Incorrectly glycosylated glycoproteins may have an undesirable effect on therapeutic effectiveness and safety due to changes in the properties of the products, including a decrease in the stability against heat or protease, shortening of the *in vivo* life span of the molecules by an increase in clearance, a decrease in the affinity for specific receptors, and an increase in antigenicity.

## Isolation and Quantitation of Glycoprotein Molecules and Analysis of Glycan Chains

Determination of the primary structure of glycoproteins necessitates analysis of the protein sequence, identification of the glycosylation sites, unravelling of the glycan structures and determination of the microheterogeneity of the glycans at each glycosylation. For these studies, it is essential that adequate starting materials are available.

For the isolation or purification of glycoproteins, a combination of several complementary separation methods such as gel permeation chromatography, affinity chromatography (lectin or others), anion- or cation-column chromatography, high performance capillary electrophoresis, and HPLC using several sorbents is generally used.

Determination of the carbohydrate composition, type and branching pattern is an important step for understanding the biological function of the native

glycoprotein molecules as well as for the development of a recombinant DNA-derived glycoprotein as a pharmaceutical agent. However, the composition, type and branching pattern of carbohydrates are complex due to the diversity of monosaccharides and the variety of possible linkages. Unlike amino acids, which are linked through an amido bond, monosaccharides are joined through a variety of hydroxyl groups present on the sugar to form glycosidic linkages. For example, two different amino acids can form only two dipeptides, while two different monosaccharides can lead to more than 60 disaccharides. However, the availability of improved and sophisticated methods for the isolation and characterization of glycoproteins and their derived glycans has paved the way for the analysis and characterization of the carbohydrate chains of glycoprotein. These analytical methods include mass spectrometry, enzymatic microsequencing, nuclear magnetic resonance (NMR), capillary electrophoresis, reversed-phase HPLC (RP-HPLC), and high pH anion exchange chromatography with pulsed amperometric detection (HPAEC-PAD), and generally a combination of several complementary analytical methods is needed to determine the carbohydrate structure.

In this section, we give an outline of the recently developed HPLC procedures for the purification, separation, and determination of glycoproteins and their glycoforms.

### Examples of Isolation or Purification of Glycoproteins by Using HPLC

$\alpha_1$-Acid glycoprotein (AAG)    AAG is a characteristic and dominant fraction of human serum sialoglycoproteins with a molecular mass approximately 44 000

Da, an unusually high carbohydrate content (45%) and a large number of sialyl residues. Although its exact biological function is still unknown, AAG is an acute-phase reactant that increases following cancer, myocardial infarction, and congestive heart failure and has also been reported to play an important role in immunoregulation.

Ion exchange chromatography and gel permeation chromatography previously used for purification are time-consuming and require a large volume of plasma or serum because of the low quantities recovered. These methods also lead to a strong possibility of denaturation and desialylation. Moreover, separation of AAG and $\alpha_1$-antitrypsin has been difficult, because the chromatographic behaviour of these compounds during anion exchange chromatography is similar. Recently, these problems have been overcome by the introduction of an HPLC system equipped with a hydroxyapatite column as the last step after the clean-up procedures with commercially available cartridge ion exchange columns.

**Ceruloplasmin (CP)**    CP is a serum $\alpha_2$-glycoprotein that carries more than 95% of the copper present in plasma and is believed to have an active role in the regulation of copper and iron homeostasis. It has been pointed out that fragmentation of CP during purification and storage has hampered the study of its structure. The rapid degradation of purified CP reported by many laboratories may be largely due to the presence of one or more copurifying or contaminating proteases, at least one of which is a metalloproteinase.

Recently, a highly purified and nonlabile CP has been obtained from human plasma by combining the previously reported chromatographic steps with additional gel permeation and fast protein liquid chromatography (FPLC) steps. In the latter steps, further purification of CP by Sephadex G-50 chromatography and Mono Q FPLC were essential for the removal of plasma metalloproteinase, and this purification procedure yielded a protein that was completely stable even after incubation at 37°C for 4 weeks.

**Erythropoietin (EPO)**    EPO, an acidic glycoprotein hormone, is synthesized in the kidney and circulates in the blood to stimulate red cell proliferation and differentiation in bone marrow. Native human EPO was first purified from the urine of patients suffering from severe aplastic anaemia. Since then, several methods for the purification of urinary human EPO (uHuEPO) have been developed. RP-HPLC has recently been used for the purification of uHuEPO with high *in vivo* activity. This purification procedure involves two membrane filtration steps, Sephadex G-25, two DEAE-agarose steps, Sephadex G-75, wheat germ agglutinin (WGA)-agarose, and RP-HPLC. The final HPLC step is essential for the removal of nucleic acids.

**Chromatographic Determination of AAG**

There have been very few reports on chromatographic methods for determining concentrations of glycoproteins other than AAG in biological samples. Radial immunodiffusion (RID) utilizing the antibody against AAG has been widely used to determine AAG in serum because of its strong specificity. This method, however, is time-consuming and is not easily applicable to experimental animals. To overcome these problems, some simple and rapid HPLC methods have been developed.

After pretreatment of human serum with a chloroform/methanol mixture (2 : 1, v/v), 500 μL of the upper phase was applied to the anion exchange FPLC system (Mono Q HR 5/5 column), and AAG was eluted with a pH/NaCl gradient elution programme. To measure the serum AAG content, the Mono Q HR column was calibrated with commercial AAG in the range 100–200 μg/500 μL of sample volume.

A rapid and sensitive determination method starting from the diluted human serum itself has also been reported. This procedure involves the anion exchange step for cleaning up serum (commercially available cartridge column, DEAE-M) and a hydroxyapatite HPLC system. A linear relationship between standard AAG concentration and peak height was observed over the concentration range 0.5–2.5 mg mL$^{-1}$ serum. The coefficient of variation at 0.5 mg mL$^{-1}$ AAG was 3.7% ($n = 8$). A good correlation was observed between this HPLC method ($y$) and the conventional RID ($x$) ($y = 1.009x + 0.004$, $r = 0.996$).

**Determining the Carbohydrate Composition, Type and Branching Pattern**

RP-HPLC has become a commonly used method for the analysis and purification of peptides, proteins and glycoproteins. The RP-HPLC experimental system usually comprises an n-alkylsilica-based sorbent. By using modern instrumentation and columns, complex mixtures of peptides and proteins can be separated and low picomolar amounts of resolved components can be collected. Separation can be easily performed by changing the gradient slope of solvents such as acetonitrile containing an ionic modifier (e.g. trifluoroacetic acid (TFA)); column temperature; or the organic solvent composition. The technique is equally applicable to the analysis of enzymatically derived mixtures of peptides from proteins and glycoproteins. Separated fractions can be subsequently subjected to further analysis of carbohydrates and amino acids.

A new HPLC method, HPAEC-PAD, which bypasses the derivatization steps by using pulsed electrochemical detection on gold electrodes, has been developed. Monosaccharides and oligosaccharides can be directly resolved by anion exchange chromatography, because the hydroxyl groups of carbohydrates are weakly acidic and reveal anionic forms at pH values greater than pH 12. In addition to high sensitivity in the low picomole range of PAD, a major advantage of HPAEC-PAD is its usefulness in analysing both monosaccharides and all classes of oligosaccharides without derivatization. HPAEC-PAD has therefore been used successfully for resolving and quantitating the constituent monosaccharides released by acidic hydrolysis (e.g. TFA) of glycan chains and for resolving N-linked oligosaccharides separated by enzyme digestion (e.g. PNGase F).

**Figure 2** shows the HPAEC-PAD chromatograms of fractions 23–27 from the RP-HPLC separation of a tryptic digest of recombinant tissue plasminogen activator (tPA). The peaks from RP-HPLC separation were collected manually, and aliquots of all 62 fractions were analysed for neutral and amino monosaccharides after acid hydrolysis. The chromatograms in

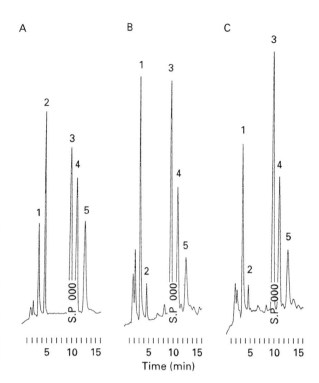

**Figure 3** Representative chromatograms of monosaccharides after treatment of standard and plasma samples. (A) Standard sample containing 5.0 μg mL⁻¹ of each monosaccharide; (B) healthy subject; (C) patient with renal insufficiency. Peaks: 1 = mannitol (internal standard); 2 = Fuc; 3 = GlcNAc; 4 = Gal; 5 = Man. (Reproduced with permission from Kishino *et al.*, 1995.)

Figure 2 indicate that fractions 24–26 contain glycopeptides, and the ratio of constituent monosaccharides suggests that their oligosaccharide structures are those of fucosylated *N*-acetyllactosamine-type oligosaccharides. Another *N*-acetyllactosamine-type chain and an oligomannose-type chain were identified similarly by the same analytical procedure.

The HPAEC-PAD method is also applicable to the quantitation of concentrations of monosaccharides after release by acid hydrolysis and following clean-up procedures with commercially available cartridge columns. **Figure 3** shows the chromatograms of four monosaccharides in purified serum AAG from healthy subjects and from patients with renal insufficiency. The concentration of NeuAc can also be determined under different solvent conditions. This method enables composition analysis of the carbohydrate moiety of AAG with only 1.0 mL of plasma. Linear relations between the amount of NeuAc or monosaccharides and the peak-height ratio of NeuAc or monosaccharides to the internal standards are observed over the concentration range of 5.0 to 100 μg mL⁻¹. *N*-Glycolylneuraminic acid and mannitol are used as the internal standard for NeuAc and the four monosaccharides, respectively.

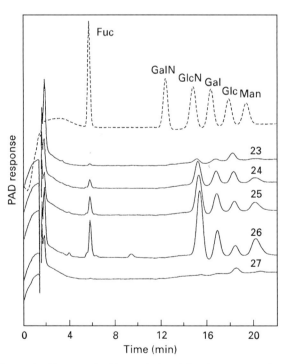

**Figure 2** Monosaccharide analysis of fractions 23–27 from RP-HPLC separation of a tryptic digest of recombinant tissue plasminogen activator. Elution positions of monosaccharide standards are indicated on the upper trace. (Reproduced with permission from Townsend *et al.*, 1996.)

**Table 1**  Analysis of NeuAc and monosaccharide levels in purified $\alpha_1$-acid glycoprotein (AAG) from plasma of healthy subjects, patients with renal insufficiency and patients with myocardial infarction

|  | Healthy subjects (n = 8) | Renal insufficiency (n = 6) | Myocardial infarction (n = 4) |
|---|---|---|---|
| Age (years) | 65.8 ± 12.5 | 60.2 ± 7.2 | 69.7 ± 11.7 |
| AAG conc. (mg mL$^{-1}$) | 0.79 ± 0.14 | 1.01 ± 0.19[a] | 1.66 ± 0.34[b] |
| NeuAc (mg g$^{-1}$·AAG) | 82.77 ± 12.55 | 89.29 ± 15.55 | 132.70 ± 6.89[b] |
| Fuc (mg g$^{-1}$·AAG) | 9.84 ± 3.08 | 12.79 ± 3.37 | 10.40 ± 3.42 |
| GluNAc (mg g$^{-1}$·AAG) | 113.01 ± 10.07 | 135.44 ± 10.51[b] | 142.68 ± 12.96[b] |
| Gal (mg g$^{-1}$·AAG) | 76.97 ± 4.23 | 88.78 ± 5.61[b] | 86.92 ± 11.73[a] |
| Man (mg g$^{-1}$·AAG) | 49.14 ± 2.57 | 57.26 ± 3.80[b] | 50.16 ± 1.67 |
| GluNAc/Man | 2.18 ± 0.37 | 2.50 ± 0.35 | 2.84 ± 0.21[a] |

Source: Kishino et al. (1995).
AAG concentration was determined by the HPLC method with a hydroxyapatite column.
NeuAc and each monosaccharide concentration were determined by HPAEC-PAD.
Values in the table are means ± SD.
[a]Significantly different ($p < 0.05$) from healthy subjects.
[b]Significantly different ($p < 0.01$) from healthy subjects.

The resultant quantitation data (**Table 1**) for healthy subjects, patients with renal insufficiency and patients with myocardial infarction show that not only AAG levels but also the concentrations of several monosaccharides in patients increased significantly compared to those of healthy subjects, suggesting a change in the carbohydrate branching pattern in such pathologic conditions.

It is well known that the microheterogeneity of AAG is due to the occurrence of di-, tri-, and tetra-antennary glycans of the N-acetyllactosamine type at the five glycosylation sites. Moreover, the Man content is constant among the antennary glycans, and the number of branches increases with the addition of GlcNAc to Man residues. A highly branched glycan chain of AAG is constructed by the linkage of Gal to GlcNAc (Figure 1), which results in the formation of an antennary structure (N-acetyllactosamine). Therefore, in the case of AAG, determination of the concentration ratio of GlcNAc to Man (GlcNAc/Man) is important for estimating whether the carbohydrate moiety of glycoforms has a highly or less-branched glycan chain. The significantly higher GlcNAc/Man ratio in the patients with myocardial infarction suggests that a highly branched glycan chain was synthesized. Changes in the carbohydrate moiety in the glycoproteins have been reported in patients with various types of disease.

As shown in **Figure 4**, at least six fractions, which are possibly based on carbohydrate-mediated microheterogeneity, have been obtained from healthy human (Japanese) serum AAG by HPLC using a hydroxyapatite column under a gradient elution programme. From the determination of five monosac-charides (NeuAc, Fuc, GlcNAc, Gal, Man) in each fraction by HPAEC-PAD, it was found that glycoforms rich in carbohydrates were eluted later (fractions 4, 5, 6) and that NeuAc was relatively abundant in these highly adsorbed glycoforms, especially in fraction 6. Furthermore, the ratio of GlcNAc/Man in fraction 2 was significantly higher than those in the other fractions, suggesting the presence of a highly branched glycan chain. Interestingly, it has been also shown that fractions 1 and 2, both relatively rich in highly branched glycan chains, showed a significantly lower binding capacity to disopyramide, a drug for the treatment of arrhythmia, than did the other fractions. This result suggests that the binding sites of AAG to disopyramide are hindered by relatively large carbohydrate moieties, such as a tetraantennary structure. These results are consistent with the findings that the binding capacity of purified AAG isolated from patients (with renal insufficiency or myocardial infarction) to disopyramide is significantly lower than that of healthy subjects and that the AAGs of these patients revealed a higher concentration ratio of GlcNAc/Man, an index of the abundance of highly branched glycan chains.

In conclusion, in order to gain further insight into the structure–function relations of the carbohydrate moiety, it is essential that sufficient quantities of glycoprotein variants are available. An effective combination of the sophisticated separation methods for glycoforms, such as the HPLC system using a hydroxyapatite column, and the qualitative and quantitative analytical methods for monosaccharides/oligosaccharides, such as HPAEC-PAD, must be established.

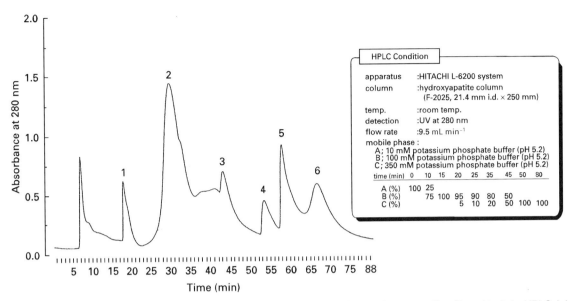

**Figure 4**  Typical chromatograms of the glycoforms of $\alpha_1$-acid glycoprotein (AAG) from the serum of healthy subjects by HPLC. Inlet is the gradient programme for the fractionation of the glycoforms of AAG. (Sampling time of each fraction: fraction 1, 17–22 min; 2, 27–36 min; 3, 43–50 min; 4, 53–57 min; 5, 58–62 min and 6, 65–72 min.) (Reproduced with permission from Kishino *et al.*, 1997.)

*See also:* **III/Carbohydrates:** Liquid Chromatography. **Peptides and Proteins:** Liquid Chromatography. **Polysaccharides:** Liquid Chromatography.

## Further Reading

Clemetson KJ (1997) In: Montreuil J, Vliegenthart JFG and Schachter H (eds) *Glycoproteins II*, pp. 173–201. Amsterdam: Elsevier.

Hancock WS, Chakel AAJ, Souders C, M'Timkulu T, Pungor E Jr and Guzzetta AW (1996) In: Karger BL and Hancock WS (eds) *Methods in Enzymology*, vol. 271, pp. 403–427. San Diego: Academic Press.

Hardy MR and Townsend RR (1994) In: Lennarz WJ and Hart GW (eds) *Methods in Enzymology*, vol. 230, pp. 208–225. San Diego: Academic Press.

Kishino S, Nomura A, Sugawara M, Iseki K, Kakinoki S, Kitabatake A and Miyazaki K (1995) *Journal of Chromatography* 672: 199–205.

Kishino S and Miyazaki K (1997) *Journal of Chromatography* 699: 371–381.

Kishino S, Nomura A, Saitoh M, Sugawara M, Iseki K, Kitabatake A and Miyazaki K (1997) *Journal of Chromatography* 703: 1–6.

Montreuil J (1995) In: Montreuil J, Schachter H and Vliegenthart JFG (eds) *Glycoproteins*, pp. 1–12. Amsterdam: Elsevier.

Schmid K (1989) In: Bauman P, Eap CB, Muler WE and Tillement J-P (eds) *Alpha₁-Acid Glycoprotein*, pp. 7–22. New York: Alan R. Liss.

Townsend RR, Basa LJ and Spellman MW (1996) In: Karger BL and Hancock WS (eds) *Methods in Enzymology*, vol. 271, pp. 135–147. San Diego: Academic Press.

Vliegenthart JFG and Montreuil J (1995) In: Montreuil J, Schachter H and Vliegenthart JFG (eds) *Glycoproteins*, pp. 13–28. Amsterdam: Elsevier.

# GOLD RECOVERY: FLOTATION

**S. Bulatovic and D. M. Wyslouzil**, Lakefield Research, Lakefield, Ontario, Canada

## Introduction

The recovery of gold from gold-bearing ores depends largely on the nature of the deposit, the mineralogy of the ore and the distribution of gold in the ore. The methods used for the recovery of gold consist of the following unit operations:

- The gravity preconcentration method, which is mainly used for recovery of gold from placer deposits that contain coarse native gold. Gravity is often used in combination with flotation and/or cyanidation.

- Hydrometallurgical methods are normally employed for recovery of gold from oxidized deposits (heap leach), low grade sulfide ores (cyanidation, carbon-in-pulp (CIP), carbon-in-leach (CIL)) and refractory gold ores (autoclave, biological decomposition followed by cyanidation).
- A combination of pyrometallurgical (roasting) and hydrometallurgical route is used for highly refractory gold ores (carbonaceous sulfides, arsenical gold ores) and the ores that contain impurities that result in a high consumption of cyanide, which have to be removed before cyanidation.
- The flotation method is a widely used technique for the recovery of gold from gold-containing copper ores, base metal ores, copper–nickel ores, platinum group ores and many other ores where other processes are not applicable. Flotation is also used for the removal of interfering impurities before hydrometallurgical treatment (i.e. carbon prefloat), for upgrading of low sulfide and refractory ores for further treatment. Flotation is considered to be the most cost-effective method for concentrating gold.

Significant progress has been made over the past several decades in the recovery of gold using hydrometallurgical methods, including cyanidation (CIL, resin-in-pulp) and bio-oxidation. All of these processes are well documented in the literature and abundantly described. However, very little is known about the flotation properties of gold contained in various ores and the sulfides that carry gold. The sparse distribution of discrete gold minerals, as well as their exceedingly low concentrations in the ore, is one of the principal reasons for the lack of fundamental work on the flotation of gold-bearing ores.

In spite of the lack of basic research on flotation of gold-bearing ores, the flotation technique is used, not only for upgrading of low grade gold ore for further treatment, but also for beneficiation and separation of difficult-to-treat (refractory) gold ores. Flotation is also the best method for recovery of gold from base metal ores and gold-containing platinum group metals (PGM) ores. Excluding gravity preconcentration, flotation remains the most cost-effective beneficiation method.

Gold itself is a rare metal and the average grades for low grade deposits vary between 3 and 6 p.p.m. Gold occurs predominantly in its native form in silicate veins, alluvial and placer deposits or encapsulated in sulfides. Other common occurrences of gold are alloys with copper, tellurium, antimony, selenium, PGMs and silver. In massive sulfide ores,

gold may occur in several of the above forms, which affects flotation recovery.

During flotation of gold-bearing massive sulfide ores, the emphasis is generally placed on the production of base metal concentrates and gold recovery becomes a secondary consideration. In some cases, where significant quantities of gold are contained in base metal ores, the gold is floated from the base metal tailings.

The flotation of gold-bearing ores is classified according to ore type (i.e. gold ore, gold–copper ore, gold–antimony ores), because the flotation methods used for the recovery of gold from different ores is vastly different.

## Geology and General Mineralogy of Gold-bearing Ores

The geology of the deposit and the mineralogy of the ore play a decisive role in the selection of the best treatment method for a particular gold ore. Geology of the gold deposits varies considerably, not only from deposit to deposit, but also within the deposit. Table 1 shows major genetic types of gold ores and their mineral composition. More than 50% of the total world gold production comes from clastic sedimentary deposits.

In many geological ore types, several subtypes can be found, including primary ores, secondary ores and

**Table 1** Common genetic types of gold deposits

| Ore type | Description |
|---|---|
| Magmatic | Gold occurs as an alloy with copper, nickel and platinum group metals Typically contains low amount of gold |
| Ores in clastic sedimentary rock | Placer deposits, in general conglomerates, which contain quartz, sericite, chlorite, tourmaline and sometimes rutile and graphite. Gold can be coarse. Some deposits contain up to 3% pyrite. Size of the gold contained in pyrite ranges from 0.01 to 0.07 μm |
| Hydrothermal | This type contains a variety of ores, including: Gold–pyrite ores Gold–copper ores Gold–polymetallic ores Gold–oxide ore, usually upper zone of sulfide zones The pyrite content of the ore varies from 3% to 90%. Other common waste minerals are quartz, aluminosilicates, dolomite |
| Metasomatic or scarn ores | Sometimes very complex and refractory gold ores. Normally the ores are composed of quartz, sericite, chlorites, calcite, magnetite. Sometimes the ore contains wolframite and sheelite |

**Table 2**  Major gold minerals

| Group | Mineral | Chemical formula | Impurity content |
|---|---|---|---|
| Native gold and its alloys | Native gold | Au | 0–15% Ag |
| | Electrum | Au/Ag | 15–50% Ag |
| | Cuproauride | Au/Cu | 5–10% Cu |
| | Amalgam | Hg/Au | 10–34% Au |
| | Bismuthauride | Au/Bi | 2–4% Bi |
| Tellurides | Calaverite | $AuTe_2$ | |
| | Sylvanite | $(Au, Ag)Te_2$ | |
| | Petzite | $(Au, Ag)Te$ | |
| | Magyazite | $Au(Pb, Sb, Fe)(S, Te_{II})$ | Unstable |
| Gold associated with platinum group metals | Krennerite | $AuTe_2(Pt, Pl)$ | |
| | Platinum gold | AuPt | Up to 10% Pt |
| | Rhodite | AuRh | 30–40% Rh |
| | Rhodian gold | AuRh | 5–11% Rh |
| | Aurosmiride | Au, Ir, Os | 5% Os + 5–7% Ir |

oxide ores. Some of the secondary ores belong to a group of highly refractory ores, such as those from Nevada (USA), and El Indio (Chile). The number of gold minerals and their associations are relatively small and can be divided into the following three groups: native gold and its alloys, tellurides and gold associated with PGMs.

Table 2 lists major gold minerals and their associations.

## Flotation Properties of Gold Minerals and Factors Affecting Floatability

Native gold and its alloys, which are free from surface contaminants, are readily floatable with xanthate collectors. Very often, however, gold surfaces are contaminated or covered with varieties of impurities. The impurities present on gold surfaces may be argentite, iron oxides, galena, arsenopyrite or copper oxides. The thickness of the layer may be in the order of 1–5 µm. Because of this, the flotation properties of native gold and its alloys vary widely. Gold covered with iron oxides or oxide copper is very difficult to float and requires special treatment to remove the contaminants.

Tellurides on the other hand are readily floatable in the presence of small quantities of collector, and it is believed that tellurides are naturally hydrophobic. Tellurides from Minnesota (USA) were floated using dithiophosphate collectors, with over 95% gold recovery.

Flotation behaviour of gold associated in the platinum group metals is apparently the same as that for the PGMs or other minerals associated with the PGMs (i.e. nickel, pyrrhotite, copper and pyrite). Therefore, the reagent scheme developed for PGMs

also recovers gold. Normally, for the flotation of PGMs and associated gold, a combination of xanthate and dithiophosphate is used, along with gangue depressants guar gum, dextrin or modified cellulose. In the South African PGM operations, gold recovery into the PGM concentrate ranges from 75% to 80%.

Perhaps the most difficult problem in flotation of native gold and its alloys is the tendency of gold to plate, vein, flake and assume many shapes during grinding. Particles with sharp edges tend to detach from the air bubbles, resulting in gold losses. This shape factor also affects gold recovery using a gravity method.

In flotation of gold-containing base metal ores, a number of modifiers normally used for selective flotation of copper–lead, lead–zinc and copper–lead–zinc have a negative effect on the floatability of gold. Such modifiers include $ZnSO_4 \times 7H_2O$, $SO_2$, $Na_2S_2O_5$, and cyanide when added in excessive amounts.

The adsorption of collector on gold and its floatability are considerably improved by the presence of oxygen. **Figure 1** shows the relationship between collector adsorption, oxygen concentration in the pulp and conditioning time. The type of modifier and the pH are also important parameters in flotation of gold.

### Flotation of Low Sulfide-containing Gold Ores

The beneficiation of this ore type usually involves a combination of gravity concentration, cyanidation and flotation. For an ore with coarse gold, gold is often recovered by gravity and flotation, followed by cyanidation of the reground flotation concentrate. In some cases, flotation is also conducted on the cyanidation tailing. The reagent combination used in flotation depends on the nature of gangue present in the ore. The usual collectors are xanthates,

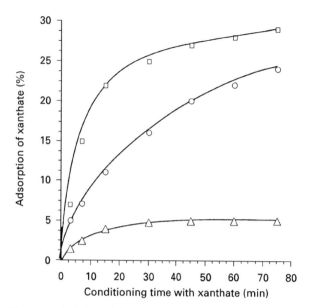

**Figure 1**  Relationship between adsorption of xanthate on gold and conditioning time in the presence of various concentrations of xanthate. Triangles, $O_2$ 2 mg $L^{-1}$, circles; $O_2$ 9 mg $L^{-1}$; squares, $O_2$ 45 mg $L^{-1}$.

dithiophosphates and mercaptans. In the scavenging section of the flotation circuit, two types of collector are used as secondary collectors. In the case of a partially oxidized ore, auxiliary collectors, such as hydrocarbon oils with sulfidizer, often yield improved results. The preferred pH regulator is soda ash, which acts as a dispersant and also as a complexing reagent for some heavy metal cations that have a negative effect on gold flotation. Use of lime often results in the depression of native gold and gold-bearing sulfides. The optimum flotation pH ranges between 8.5 and 10.0. The type of frother also plays an important role in the flotation of native gold and gold-bearing sulfides. Glycol esters and cyclic alcohols (pine oil) can improve gold recovery significantly.

Amongst the modifying reagents (depressant), sodium silicate starch dextrins and low molecular weight polyacrylamides are often selected as gangue depressants. Fluorosilicic acid and its salts can also have a positive effect on the floatability of gold. The presence of soluble iron in a pulp is highly detrimental to gold flotation. The use of small quantities of iron-complexing agents, such as polyphosphates and organic acids, can eliminate the harmful effect of iron.

### Flotation of Gold-containing Mercury/Antimony Ores

In general, these ores belong to a group of difficult-to-treat ores, where cyanidation usually produces poor extraction. Mercury is partially soluble in cyanide, which increases cyanide consumption and reduces extraction. A successful flotation method has been

developed using the flow sheet shown in **Figure 2**, where the best metallurgical results were obtained using a three-stage grinding and flotation approach. The metallurgical results obtained with different grinding configurations are shown in **Table 3**.

Flotation was carried out at an alkaline pH, controlled by lime. A xanthate collector with cyclic alcohol frother (pine oil, cresylic acid) was shown to be the most effective. The use of small quantities of a dithiophosphate-type collector, together with xanthate, was beneficial.

### Flotation of Carbonaceous Clay-containing Gold Ores

These ores belong to a group of refractory gold ores, where flotation techniques can be used to remove interfering impurities before the hydrometallurgical treatment process of the ore for gold recovery and to preconcentrate the ore for further pyrometallurgical or hydrometallurgical treatment. There are several flotation methods used for beneficiation of this ore type. Some of the most important methods are described as follows:

- Preflotation of carbonaceous gangue and carbon. In this case, only carbonaceous gangue and carbon are recovered by flotation, in preparation for further hydrometallurgical treatment of the float tails for gold recovery. Carbonaceous gangue and carbon are naturally floatable using only a frother, or a combination of a frother and a light hydrocarbon oil (fuel oil, kerosene). When the ore contains clay, regulators for clay dispersion are used. Some of the more effective regulating reagents include sodium silicates and oxidized starch.
- Two-stage flotation method. In this case, carbonaceous gangue is prefloated using the method described above, followed by flotation of gold-containing sulfides using activator–collector combinations. In extensive studies conducted on carbonaceous gold-containing ores, it was established that primary amine-treated copper sulfate improved gold recovery considerably. Ammonium salts and sodium sulfide ($Na_2S \times 9H_2O$) also have a positive effect on gold-bearing sulfide flotation, at a pH between 7.5 and 9.0. The metallurgical results obtained with and without modified copper sulfate are shown in **Table 4**.
- Nitrogen atmosphere flotation method. This technique uses a nitrogen atmosphere in grinding and flotation to retard oxidation of reactive sulfides, and has been successfully applied on carbonaceous ores from Nevada (USA). The effectiveness of the method depends on the amount of carbonaceous gangue present in the ore, and the amount and type

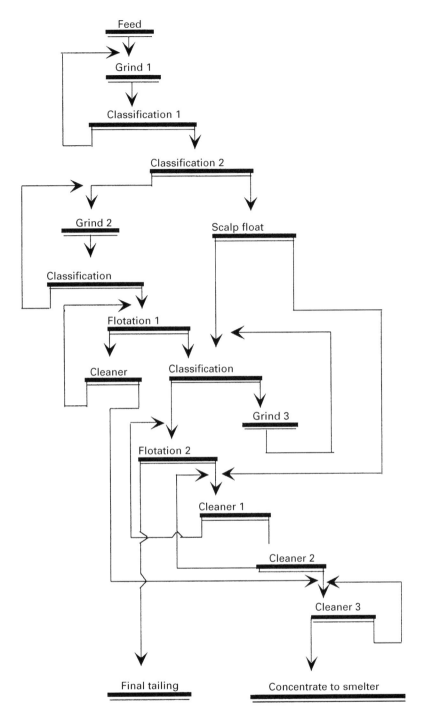

**Figure 2**  Flotation flow sheet developed for the treatment of gold-containing mercury–antimony ore.

of clay. Ores that are high in carbon or contain high clay content (or both) are not amenable for nitrogen atmosphere flotation.

### Flotation of Gold-containing Copper Ores

The floatability of gold from gold-containing copper–gold ores depends on the nature and occurrence of gold in these ores, and its association with iron sulfides.

Gold in the porphyry copper ore may appear as native gold, electrum, cuproaurid and sulfosalts associated with silver. During the flotation of porphyry copper–gold ores, emphasis is usually placed on the production of a marketable copper–gold concentrate and optimization of gold recovery is usually constrained by the marketability of its concentrate.

**Table 3**  Gold recovery obtained using different flow sheets

| Flow sheet | Recovery in concentrate (%) | | | | | Tailing Assays (%, g t$^{-1}$) | | | | |
|---|---|---|---|---|---|---|---|---|---|---|
| | Au | Ag | Sb | As | S | Au | Ag | Sb | As | S |
| Single-stage grind flotation | 88.1 | 89.2 | 72.9 | 68.4 | 70.1 | 1.7 | 5.0 | 0.04 | 0.035 | 0.38 |
| Two-stage grind flotation | 92.2 | 91.8 | 93.4 | 78.7 | 81.2 | 1.0 | 4.1 | 0.015 | 0.022 | 0.27 |
| Three-stage grind flotation | 95.3 | 95.2 | 95.7 | 81.2 | 85.7 | 0.7 | 2.2 | 0.005 | 0.015 | 0.19 |

Reproduced from Sristinov (1964) with permission.

The minerals that influence gold recovery in these ores are iron sulfides (i.e. pyrite, marcasite), in which gold is usually associated as minute inclusions. Thus, the iron sulfide content of the ore determines gold recovery in the final concentrate. **Figure 3** shows the relationship between pyrite content of the ore and gold recovery in the copper concentrate for two different ore types. Most of the gold losses occur in the pyrite.

The reagent schemes used in commercial operations treating porphyry copper–gold ores vary considerably. Some operations, where pyrite rejection is a problem, use a dithiophosphate collector at an alkaline pH between 9.0 and 11.8 (e.g. OK Tedi, PNG Grasberg, Indonesia). When the pyrite content in the ore is low, xanthate and dithiophosphates are used in a lime or soda ash environment.

In more recent years, in the development of commercial processes for the recovery of gold from porphyry copper–gold ores, bulk flotation of all the sulfides has been emphasized, followed by regrinding of the bulk concentrate and sequential flotation of copper–gold from pyrite. Such a flow sheet (**Figure 4**) can also incorporate high intensity conditioning in the cleaner–scavenger stage. Comparison of metallurgical results using the standard sequential flotation flow sheet and the bulk flotation flow sheet is shown in **Table 5**. A considerable improvement in gold recovery was achieved using the bulk flotation flow sheet.

During beneficiation of clay-containing copper–gold ores, the use of small quantities of $Na_2S$ (at natural pH) improves both copper and gold metallurgy considerably.

In the presence of soluble cations (e.g. Fe, Cu), additions of small quantities of organic acid (e.g. oxalic, tartaric) improve gold recovery in the copper concentrate.

Some porphyry copper ores contain naturally floatable gangue minerals, such as chlorites and aluminosilicates, as well as preactivated quartz. Sodium silicate, carboxymethylcellulose and dextrins are common depressants used to control gangue flotation.

Gold recovery from massive sulfide copper–gold ores is usually much lower than that of porphyry copper–gold ores, because very often a large portion of the gold is associated with pyrite. Normally, gold recovery from these ores does not exceed 60%. During the treatment of copper–gold ores containing pyrrhotite and marcasite, the use of $Na_2H_2PO_4$ at alkaline pHs depresses pyrrhotite and marcasite, and also improves copper and gold metallurgy.

**Flotation of Oxide Copper–Gold Ores**

Oxide copper–gold ores are usually accompanied by iron hydroxide slimes and various clay minerals. There are several deposits of this ore type around the world, some of which are located in Australia (Red Dome), Brazil (Igarape Bahia) and the Soviet Union (Kalima). Treatment of these ores is difficult, and even more complicated in the presence of clay minerals.

Recently, a new class of collectors, based on ester-modified xanthates, has been successfully used to treat gold-containing oxide copper ores, using a sulfidization method. **Table 6** compares the metallurgical results obtained on the Igarape Bahia ore using

**Table 4**  Effect of amine-modified CuSO$_4$ on gold-bearing sulfide flotation from carbonaceous refractory ore

| Reagent used | Product | Weight (%) | Assays (%, g t$^{-1}$) | | Distribution (%) | |
|---|---|---|---|---|---|---|
| | | | Au | S | Au | S |
| CuSO$_4$ + xanthate | Gold sulfide conc. | 30.11 | 9.63 | 4.50 | 69.1 | 79.7 |
| | Gold sulfide tail | 69.89 | 1.86 | 0.49 | 30.9 | 20.3 |
| | Head | 100.00 | 4.20 | 1.70 | 100.0 | 100.0 |
| Amine modified CuSO$_4$ + xanthate | Gold sulfide conc. | 26.30 | 13.2 | 5.80 | 84.7 | 90.8 |
| | Gold sulfide tail | 73.70 | 0.85 | 0.21 | 15.3 | 9.2 |
| | Head | 100.00 | 4.10 | 1.68 | 100.0 | 100.0 |

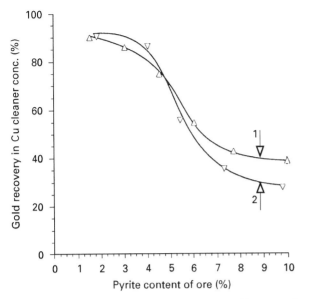

**Figure 3** Effect of pyrite content of the ore on gold recovery in the copper–gold concentrate at 30% Cu concentrate grade. 1, Ore from Peru; 2, ore from Indonesia.

xanthate and a new collector (PM230, supplied by Senmin in South Africa).

The modifier used in the flotation of these ores included a mixture of sodium silicate and Calgon. Good selectivity was also achieved using boiled starch.

### Flotation of Gold–Antimony Ores

Gold–antimony ores usually contain stibnite (1.5–4.0% Sb), pyrite, arsenopyrite gold (1.5–3.0 g t$^{-1}$) and silver (40–150 g t$^{-1}$). Several plants in the USA (Stibnite, Minnesota and Bradly) and Russia have been in operation for some time. There are two commercial processes available for treatment of these ores.

- Selective flotation of gold-containing sulfides followed by flotation of stibnite with pH change.

Stibnite floats well in neutral and weak acid pH, while in an alkaline pH (i.e. > 8), it is reduced. Utilizing this phenomenon, gold-bearing sulfides are floated with xanthate and alcohol frother in alkaline medium (pH > 9.3) followed by stibnite flotation at about pH 6.0, after activation with lead nitrate. Typical metallurgical results using this method are shown in **Table 7**.

- Bulk flotation followed by sequential flotation of gold-bearing sulfides, and depression of stibnite. This method was practised at the Bradly concentrator (USA) and consisted of the following steps: first, bulk flotation of stibnite and gold bearing sulfides at pH 6.5 using lead nitrate (Sb activator) and xanthate; second, the bulk concentrate is reground in the presence of NaOH (pH 10.5) and $CuSO_4$, and the gold-bearing sulfides are refloated with additions of small quantities of xanthate; third, cleaning of the gold concentrate in the presence of NaOH and NaHS. The plant metallurgical results employing this method are shown in **Table 8**.

Recent studies conducted on ore from Kazakhstan have shown that sequential flotation using thionocarbamate collector gave better metallurgical results than those obtained with xanthate.

### Flotation of Arsenical Gold Ores

There are two major groups of arsenical gold ores of economic value. These are the massive base metal sulfides with arsenical gold (e.g. the lead–zinc Olympias deposit, Greece) and arsenical gold ores without the presence of base metals. Massive, base metal arsenical gold ores are rare, and there are only a few deposits in the world. A typical arsenical gold ore contains arsenopyrite as the major arsenic mineral. However, some arsenical gold ores, such as those from Nevada in the USA (Getchel deposit), contain realgar and orpiment as the major arsenic-bearing

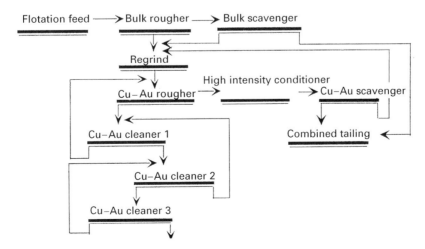

**Figure 4** Bulk flow sheet used in the treatment of pyritic copper–gold ores. Reproduced with permission from Bulatovic (1997).

**Table 5** Comparison of metallurgical results using conventional and bulk flotation flow sheets on ore from Peru

| Flow sheet used | Product | Weight (%) | Assays (%, g t$^{-1}$) | | Distribution (%) | |
|---|---|---|---|---|---|---|
| | | | Cu | Au | Cu | Au |
| Conventional | Cu/Au conc. | 2.28 | 27.6 | 32.97 | 95.4 | 76.7 |
| (sequential Cu/Au) | Cu/Au tail | 97.72 | 0.031 | 0.23 | 4.6 | 23.3 |
| | Head | 100.00 | 0.66 | 0.98 | 100.0 | 100.0 |
| Bulk | Cu/Au conc. | 2.32 | 27.1 | 36.94 | 95.2 | 85.8 |
| (Figure 4) | Cu/Au tail | 97.68 | 0.032 | 0.14 | 4.8 | 14.2 |
| | Head | 100.00 | 0.66 | 0.96 | 100.0 | 100.0 |

minerals. Pyrite, if present in an arsenical gold ore, may contain some gold as minute inclusions.

Flotation of arsenical gold ores associated with base metals is accomplished using a sequential flotation technique, with flotation of base metals followed by flotation of gold-containing pyrite–arsenopyrite. The pyrite–arsenopyrite is floated at a weakly acid pH with a xanthate collector.

Arsenical gold ores that do not contain significant base metals are treated using a bulk flotation method, where all the sulfides are first recovered into a bulk concentrate. In case the gold is contained either in pyrite or arsenopyrite, separation of pyrite and arsenopyrite is practised. There are two commercial methods available. The first method utilizes arsenopyrite depression and pyrite flotation, and consists of the following steps:

- Heat the bulk concentrate to 75°C at a pH of 4.5 (controlled by $H_2SO_4$) in the presence of small quantities of potassium permanganate or disodium phosphate. The temperature is maintained for about 10 min.
- Flotation of pyrite using either ethylxanthate or potassium butylxanthate as collector. Glycol frother is also usually employed in this separation.

This method is highly sensitive to temperature. **Figure 5** shows the effect of temperature on pyrite–

arsenopyrite separation. In this particular case, most of the gold was associated with pyrite. Successful pyrite–arsenopyrite separation can also be achieved with the use of potassium peroxydisulfide as the arsenopyrite depressant.

The second method involves depression of pyrite and flotation of arsenopyrite. In this method, the bulk concentrate is treated with high dosages of lime (pH > 12), followed by a conditioning step with $CuSO_4$ to activate arsenopyrite. The arsenopyrite is then floated using a thionocarbamate collector.

Separation of arsenopyrite and pyrite is important from the point of view of reducing downstream processing costs. Normally, roasting or pressure oxidation followed by cyanidation is used to recover gold.

### Flotation of Gold from Base Metal Sulfide Ores

Very often lead–zinc, copper–zinc, copper–lead–zinc and copper–nickel ores contain significant quantities of gold (i.e. between 1 and 9 g/t). The gold in these ore types is usually found as elemental gold. A large portion of the gold in these ores is finely disseminated in pyrite, which is considered nonrecoverable. Because of the importance of producing commercial-grade copper, lead and zinc concentrates, little or no consideration is given to improvement in gold recovery, although the possibility exists to optimize gold

**Table 6** Effect of collector PM230 on copper–gold recovery from Igarape Bahia oxide copper–gold ore

| Reagent used | Product | Weight (%) | Assays (%, g t$^{-1}$) | | Distribution (%) | |
|---|---|---|---|---|---|---|
| | | | Cu | Au | Cu | Au |
| $Na_2S$ = 2500 g t$^{-1}$ | Copper Cl conc. | 9.36 | 33.3 | 14.15 | 67.0 | 50.0 |
| PAX = 200 g t$^{-1}$ | Copper tail | 90.64 | 1.61 | 1.46 | 33.0 | 50.0 |
| | Feed | 100.00 | 4.65 | 2.65 | 100.0 | 100.0 |
| $Na_2S$ = 2500 g t$^{-1}$ | Copper Cl conc. | 10.20 | 39.5 | 21.79 | 88.0 | 85.5 |
| PAX/PM230 | Copper tail | 89.80 | 0.61 | 0.42 | 12.0 | 14.5 |
| (1 : 1) = 200 g t$^{-1}$ | Feed | 100.00 | 0.61 | 0.42 | 12.0 | 14.5 |

PAX, potassium amylxanthate.
Reproduced from Bulatovic (1997) with permission.

**Table 7** Metallurgical results obtained using a sequential flotation method

| Product | Weight (%) | Assays (%, g t$^{-1}$) | | | Distribution (%) | | |
|---|---|---|---|---|---|---|---|
| | | Au | Ag | Sb | Au | Ag | Sb |
| Gold concentrate | 2.34 | 42.3 | 269.3 | 20.0 | 53 | 13 | 15 |
| Stibnite concentrate | 4.04 | 6.2 | 559.8 | 51.0 | 13 | 51 | 64 |
| Tailing | 93.62 | 0.65 | 18.7 | 0.7 | 34 | 36 | 21 |
| Feed | 100.00 | 1.86 | 46.4 | 3.2 | 100 | 100 | 100 |

recovery in many cases. Normally, gold recovery from base metal ores ranges from 30 to 75%.

In the case of a copper–zinc and copper–lead–zinc ore, gold collects in the copper concentrate. During the treatment of lead–zinc ores, the gold tends to report to the lead concentrate. Information regarding gold recovery from base metal ores is sparse.

The most recent studies conducted on various base metal ores revealed some important features of flotation behaviour of gold from these ores. It has been demonstrated that gold recovery to the base metal concentrate can be substantially improved with the proper selection of reagent schemes. Some of these studies are discussed below.

**Gold-containing lead–zinc ores**  Some of these ores contain significant quantities of gold, ranging from 0.9 to 6.0 g t$^{-1}$ (e.g. Grum, Yukon, Canada; Greens Creek, Alaska and Milpo, Peru). The gold recovery from these ores ranges from 35 to 75%. Laboratory studies have shown that the use of high dosages of zinc sulfate, which is a common zinc depressant used in lead flotation, reduces gold floatability significantly. The effect of $ZnSO_4 \times 7H_2O$ addition on gold recovery in the lead concentrate is illustrated in **Figure 6**.

In order to improve gold recovery in the lead concentrate, an alternative depressant to $ZnSO_4 \times 7H_2O$ can be used. Depressant combinations such as $Na_2S + NaCN$, or $Na_2SO_3 + NaCN$, may be used. The type of collector also plays an important role in gold flotation of lead–zinc ores. A phosphine-based collector, in combination with xanthate, gave better gold recovery than dithiophosphates.

**Copper–zinc gold-containing ores**  Gold recovery from copper–zinc ores is usually higher than that obtained from either a lead–zinc or copper–lead–zinc ore. This is attributed to two main factors. When selecting a reagent scheme for treatment of copper–zinc ores, there are more choices than for the other ore types, which can lead to the selection of a reagent scheme which is more favourable for gold flotation. In addition, a noncyanide depressant system can be used for the treatment of these ores, which in turn results in improved gold recovery. This option is not available during treatment of lead–zinc ores. **Table 9** shows the effect of different depressant combinations on gold recovery from a copper–zinc ore.

The use of a noncyanide depressant system resulted in a substantial improvement in gold recovery in the copper concentrate.

**Gold-containing copper–lead–zinc ores**  Because of the complex nature of these ores, and the requirement for a relatively complex reagent scheme for treatment of this ore, the gold recovery is generally lower than that achieved from a lead–zinc or copper–zinc ore. One of the major problems associated with the flotation of gold from these ores is related to gold mineralogy. A large portion of the gold is usually contained in pyrite, at sub-micron size. If coarser elemental gold and electrum are present, the gold surfaces are often coated with iron or lead, which can result in a substantial reduction in floatability.

The type of collector and flow sheet configuration play an important role in gold recovery from these ores. With a flow sheet that uses bulk copper–lead flotation followed by copper–lead separation, the gold recovery is higher than that achieved with a

**Table 8**  Plant metallurgical results obtained using a bulk flotation method

| Product | Weight (%) | Assays (%, g t$^{-1}$) | | | Distribution (%) | | |
|---|---|---|---|---|---|---|---|
| | | Au | Ag | Sb | Au | Ag | Sb |
| Gold concentrate | 1.80 | 91.1 | 248.8 | 1.5 | 61.0 | 31.3 | 2.0 |
| Antimony concentrate | 1.80 | 13.0 | 684.2 | 51.3 | 9.0 | 58.6 | 75.0 |
| Middlings | 0.50 | 46.6 | 248.8 | 20.0 | 8.6 | 6.0 | 8.0 |
| Bulk concentrate | 4.10 | 51.7 | 440.0 | 29.0 | 78.6 | 85.9 | 85.0 |
| Tailing | 95.90 | 0.6 | 3.1 | 0.2 | 21.4 | 14.1 | 15.0 |
| Feed | 100.00 | 2.7 | 21.0 | 1.3 | 100.0 | 100.0 | 100.0 |

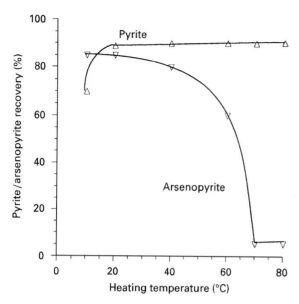

**Figure 5** Effect of temperature on separation of pyrite and arsenopyrite from a bulk pyrite–arsenopyrite concentrate.

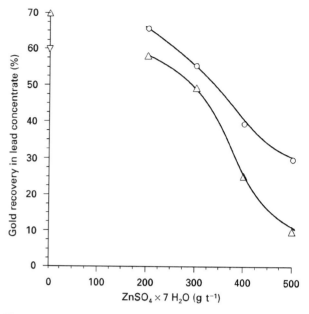

**Figure 6** Effect of $ZnSO_4$ additions on gold recovery from lead–zinc ores. Circles, Greens Creek ore (Alaska); triangles, Grum ore Yukon (Canada).

sequential copper–lead flotation flow sheet. In laboratory tests, an aerophine collector type, in combination with xanthate, had a positive effect on gold recovery as compared to either dithiophosphate or thionocarbamate collectors. **Table 10** compares the metallurgical results obtained with an aerophine collector to those obtained with a dithiophosphate collector.

Because of the complex nature of gold-containing copper–lead–zinc ores, the reagent schemes used are also complex. Reagent modifiers such as $ZnSO_4$, NaCN and lime have to be used, all of which have a negative effect on gold flotation.

## Conclusions

- The flotation of gold-bearing ores, whether for production of bulk concentrates for further gold

recovery processes (i.e. pyrometallurgy, hydrometallurgy) or for recovery of gold to base metal concentrates, is a very important method for concentrating the gold and reducing downstream costs.

- The flotation of elemental gold, electrum and tellurides is usually very efficient, except when these minerals are floated from base metal massive sulfides.

- Flotation of gold-bearing sulfides from ores containing base metal sulfides presents many challenges and should be viewed as flotation of the particular mineral that contains gold (i.e. pyrite, arsenopyrite, copper), because gold is usually associated with these minerals at micron size.

- Selection of a flotation technique for gold preconcentration depends very much on the ore

**Table 9** Effect of different depressant combinations on gold recovery to the copper concentrate from lower zone Kutcho Creek Ore

| Depressant system | Product | Weight (%) | Assays (%, $g\,t^{-1}$) | | | Distribution (%) | | |
|---|---|---|---|---|---|---|---|---|
| | | | Au | Cu | Zn | Au | Cu | Zn |
| $ZnSO_4$, NaCN, CaO | Cu concentrate | 3.10 | 20.4 | 26.2 | 3.30 | 45.1 | 85.6 | 2.8 |
| pH 8.5 Cu, 10.5 Zn | Zn concentrate | 5.34 | 1.20 | 0.61 | 55.4 | 4.6 | 3.4 | 82.2 |
| | Tailings | 91.56 | 0.77 | 0.11 | 0.58 | 50.3 | 11.0 | 15.0 |
| | Feed | 100.00 | 1.4 | 0.95 | 3.60 | 100.0 | 100.0 | 100.0 |
| $Na_2SO_3$, NaHS, CaO | Cu concentrate | 3.05 | 32.5 | 28.1 | 2.80 | 68.3 | 87.4 | 2.3 |
| pH 8.5 Cu, 10.5 Zn | Zn concentrate | 5.65 | 1.20 | 0.55 | 54.8 | 4.7 | 3.2 | 84.6 |
| | Tailings | 91.30 | 0.43 | 0.10 | 0.52 | 27.0 | 9.4 | 13.1 |
| | Feed | 100.00 | 1.45 | 0.98 | 3.66 | 100.0 | 100.0 | 100.0 |

**Table 10** Effect of collector type on Cu-Pb-Zn-Au metallurgical results from a high lead ore

| Collector used | Product | Weight (%) | Assays (%, g t⁻¹) | | | | Distribution (%) | | | |
| --- | --- | --- | --- | --- | --- | --- | --- | --- | --- | --- |
| | | | Au | Cu | Pb | Zn | Au | Cu | Pb | Zn |
| Xanthate = 30 g/t | Cu concentrate | 2.47 | 22.4 | 25.5 | 1.20 | 4.50 | 41.6 | 78.6 | 2.3 | 1.3 |
| Dithiophosphate | Pb concentrate | 1.80 | 2.50 | 0.80 | 51.5 | 8.30 | 3.4 | 1.8 | 71.3 | 1.7 |
| 3477 = 20 g/t | Zn concentrate | 13.94 | 1.10 | 0.60 | 0.80 | 58.2 | 11.5 | 10.4 | 8.6 | 92.2 |
| | Tailing | 81.79 | 0.71 | 0.089 | 0.28 | 0.52 | 43.5 | 9.2 | 17.8 | 4.8 |
| | Feed | 100.00 | 1.33 | 0.80 | 1.30 | 8.80 | 100.0 | 100.0 | 100.0 | 100.0 |
| Xanthate = 30 g/t | Cu concentrate | 2.52 | 31.3 | 26.1 | 1.10 | 5.00 | 60.6 | 80.1 | 2.1 | 1.4 |
| Aerophine 3418A = 20 g/t | Pb concentrate | 1.92 | 2.80 | 0.90 | 51.1 | 9.20 | 4.1 | 2.1 | 72.5 | 2.0 |
| | Zn concentrate | 13.91 | 0.90 | 0.50 | 0.72 | 58.5 | 9.6 | 8.5 | 7.4 | 92.5 |
| | Tailing | 81.65 | 0.41 | 0.093 | 0.30 | 0.44 | 25.7 | 9.3 | 18.0 | 4.1 |
| | Feed | 100.00 | 1.30 | 0.82 | 1.35 | 8.80 | 100.0 | 100.0 | 100.0 | 100.0 |

mineralogy, gangue composition and gold particle size. There is no universal method for flotation of the gold-bearing minerals, and the process is tailored to the ore characteristics. A specific reagent scheme and flow sheet are required for each ore.

- There are opportunities on most operating plants for improving gold metallurgy. Most of these improvements come from selection of more effective reagent schemes, including collectors and modifiers.

- Perhaps the most difficult ores to treat are the clay-containing carbonaceous sulfides. Significant progress has been made in treatment options for these ores. New sulfide activators (e.g. amine-treated $CuSO_4$, ammonium salts) and nitrogen gas flotation are amongst the new methods available.

## Further Reading

Baum W (1990) *Mineralogy as a Metallurgical Tool in Refractory Ore, Progress Selection and Optimization.* Squaw Valley, Salt Lake City: Randol Gold Forum.

Bulatovic SM (1993) Evaluation of new HD collectors in flotation of pyritic copper–gold ores from BC Canada. LR-029. Interim R&D report.

Bulatovic SM (1996) An investigation of gold flotation from base metal lead–zinc and copper–zinc ores. *Interim Report* LR-049.

Bulatovic SM (1997) An investigation of the recovery of copper and gold from Igarape Bahia oxide copper–gold ores. *Report of Investigation* LR-4533.

Bulatovic SM and Wyslouzil DM (1996) Flotation behaviour of gold during processing of porphyry copper–gold ores and refractory gold-bearing sulphides. *Second International Gold Symposium.* Lima, Peru.

Fishman MA and Zelenov BI (1967) Practice in treatment of sulphides and precious metal ores. *Izdatelstro Nedra* (in Russian) 5: 22–101.

Kudryk V, Carigan DA and Liang WW (1982) *Precious Metals.* Mining Extraction and Processing, AIME.

Martins V, Dunne RC and Gelfi P (1991) *Treatment of Partially Refractory Gold Ores.* Perth, Australia: Randol Gold Forum.

Sristinov NB (1964) The effect of the use of stage grinding in processing of refractory clay-containing gold ore. *Kolima* 1: 34–40.

# GRADIENT POLYMER CHROMATOGRAPHY: LIQUID CHROMATOGRAPHY

**G. Glöckner**, Dresden University of Technology, Dresden, Germany

## Classical Precipitation Chromatography

### Polymer Solubility and Precipitation

Solubility is governed by the general requirement that the change in Gibbs' free energy must be negative.

With low molecular weight substances this condition is easily fulfilled, because the entropy contribution is large owing to the large number of particles involved. But with polymer compounds, the entropy of dissolution is comparatively small and the enthalpy contribution gains in importance. The precept that '*similia similibus solventur*' becomes a stringent requirement; in terms of Hildebrand's solubility concept, this means that a polymer can dissolve only in fluids whose solubility parameters are very closely related to those of the polymer. Therefore, most liquids are non-solvents and the number of solvents available for a given polymer is far fewer than the number of solvents available for a low molecular weight substance of comparable structure.

The solubility of polymers decreases with increasing molecular weight (MW) and can be measured easily by the controlled addition of a non-solvent to the solution of a polymer. The volume fraction $\phi_{NS}$ of non-solvent at the cloud point is related to the square root, $M^{0.5}$, of molecular weight by

$$100 \, \phi_{NS} = C_1 + C_2/M^{0.5}$$

where $C_1$ and $C_2$ are constants for the particular system.

This dependence can be used to separate polymers by either fractional precipitation or dissolution. The latter method can also be performed in packed columns by gradients whose solvent power increases in the course of the elution.

The solubility of polymers also depends on temperature. Usually, the temperature coefficient is positive, i.e. fractional dissolution can be carried out with a given solvent (or a non-solvent/solvent mixture at constant composition) by raising the temperature. This procedure can also be performed in columns.

### Baker–Williams Fractionation

In 1956, Baker and Williams described 'a new chromatographic procedure and its application to high polymers'. This was column elution combining the effects of solvent strength and temperature. The important innovation was a temperature gradient along the column. The top of the column was heated to a temperature about 50 K higher than that of the cooled bottom. An aluminium jacket ensured a linear temperature profile. The polymer to be investigated was coated onto the part of the inert packing that subsequently was put into the uniformly heated uppermost section of the column. The temperature gradient enabled multistage separation to be performed. Any component dissolved from the sample bed was reprecipitated in a cooler zone of the column. Here it was redissolved later by a non-solvent/solvent mixture of higher solvent strength and transported to the next cooler zone for another reprecipitation. Thus, Baker–Williams fractionation was described as 'a chromatographic method based upon the equilibration of substances between a stationary precipitated phase and a moving solution'. Baker and Williams investigated polystyrene in a glass tube, 350 mm long and 24 mm wide, packed with glass beads of average

size 0.1 mm diameter. The sample size was 300 mg, the gradient ran from ethanol (non-solvent) to methyl ethyl ketone, and the temperature gradient spanned $10–60°C$. The multistage mechanism ensured a high separation power, which was confirmed by both theory and experiment. The method became popular in polymer characterization; the second citation in the Bibliography provides a survey of application of the method to about 30 different polymers.

The development of size exclusion chromatography (SEC) made separation according to MW feasible and convenient. SEC allows the investigation of different polymers in a common eluent with very little preliminary work. Dissolved samples can be injected into a running eluent, e.g. tetrahydrofuran, which has sufficient solvent strength for a great many polymers. The elution curve can be monitored with a suitable detector and provides at least a first guess at the MW distribution (MWD). Using MW-sensitive detectors and sophisticated software, reliable MWD curves can be measured within minutes. Interest in the demanding Baker–Williams technique therefore faded away. Although this technique is no longer a competitor in analytical separations according to MW, it should still be considered a powerful tool for separations according to chemical differences and for preparative fractionation. The chemical composition distribution of copolymers, blends or modified polymers can be measured by SEC only in rare cases (if coupled with MWD in a known ratio). This was realized some years ago (see Bibliography).

## High Performance Precipitation Liquid Chromatography (HPPLC)

### Principle and Instrumentation

The renaissance of precipitation chromatography requires modern equipment, e.g. detectors and programmable gradient devices. The samples to be investigated should be applied in solution and injected into the eluent stream ahead of the column.

About 80% of all high performance liquid chromatography (HPLC) investigations are performed in the reversed-phase mode. Reversed-phase packing materials have a nonpolar surface. They usually consist of particles with a silica core and a bonded layer of alkane chains. Reversed-phase gradients run from a highly polar initial eluent to a final eluent of low polarity. The polar eluent forces nonpolar solutes to be retained by the stationary phase. Retention increases with decreasing polarity of the sample components. The mechanism is understood to be a solvophobic interaction that requires the mobile phase to be an unfavourable environment for the solute.

The measures taken to force the polymer towards the stationary phase may easily reach or even transgress the limits of solubility. The latter effect has been observed occasionally in reversed-phase chromatography of low molecular weight compounds, but is almost the rule with polymers whose solubility is more restricted.

In normal-phase chromatography, the column is polar and gradient elution is performed with a nonpolar starting component A and a polar component B is added during the run. Retention increases with increasing polarity of the sample constituents.

In order to achieve proper retention of a polymer, the starting eluent A must usually be a non-solvent. This means that sample solutions cannot be prepared in a portion of the starting eluent and that the polymer is precipitated at the top of the column. Since proper retention is required, the separation is by this step classified as precipitation chromatography. The precipitation at the top of the column yields preconcentration of the sample. Thus, HPPLC can cope with samples differing widely in concentration, e.g. SEC fractions. The column permeability is not affected. If the sample solvent is a portion of eluent B, the amount of solvent injected will not cause difficulties. The use of another solvent is not recommended because it could overload the column with an additional substance.

The mechanism of separation is, in general, a combination of precipitation and adsorption. The detector must be capable of measuring the eluting sample components without being affected by the solvent gradient. Suitable equipment became available in the late 1970s.

## High Performance Precipitation Liquid Chromatography of Styrene–Acrylonitrile Copolymers

Styrene is a polymerizable substance of formula $CH_2=CH(C_6H_5)$, whose homopolymerization yields polystyrene (PS). It can be polymerized with numerous other monomers to yield copolymers. Styrene units have a strong UV absorption, which means that polystyrene and styrene-containing copolymers can be monitored by UV detectors. Copolymers of styrene and acrylonitrile are of commercial interest. Well-characterized samples graded in composition are available together with a considerable knowledge of styrene–acrylonitrile dissolution/precipitation behaviour. The polarity of acrylonitrile is higher than that of styrene units. Therefore a separation of styrene–acrylonitrile copolymers according to composition is also a separation into constituents dif-

fering in polarity, which is of basic interest in the framework of chromatography. Styrene–acrylonitrile copolymers therefore seemed to be well suited to early studies of high performance precipitation chromatography.

Preliminary studies published in 1982 showed that tetrahydrofuran (THF) has the capacity to separate a mixed styrene–acrylonitrile sample into its constituents, provided that the starting gradient component A enables proper retention of the injected samples. This was achieved by using at least 80% $n$-hexane in THF, i.e. with a non-solvent. The injected polymer was therefore precipitated at the top of the column. The elution characteristic (percentage THF in the eluent versus acrylonitrile content of the sample) was similar to the solubility borderline determined by turbidimetric titration. It was found that equivalent separations could be achieved on a silica column as well as on a nonpolar $C_8$ column. This surprising result was confirmed in systematic studies performed by Glöckner and van den Berg in 1987 using other polar and nonpolar columns including silica CN bonded phase, small-pore $C_{18}$, wide-pore $C_{18}$ and $\mu$-Bondagel E1000-10. Thus, the surface of the packings did not actively participate in the separation. It was found that the peak shapes obtained could not be improved further by a temperature gradient along the column (which had been so essential in Baker–Williams fractionation).

Multistage separation without the use of a temperature gradient or interaction with the surface can be achieved on porous packings where the polymer solute is excluded from the pores. The polymer solute then has a higher linear velocity than the eluent, which fills the interstitial volume as well as the pore volume of the column. The polymer bypasses the pores and thus overtakes the eluent which has sufficient solvent strength. The polymer is precipitated and retained until a more powerful eluent reaches its position.

In chromatographic terms, the gradient hexane → THF is a normal-phase gradient, i.e. increasing in polarity. In combination with a polar column, e.g. silica or a CN bonded phase, it forms a standard normal-phase system, which should elute more polar sample constituents after less-polar ones. Thus, the observed efficiency of irregular combinations with nonpolar $C_8$ or $C_{18}$ columns shows that the separation was not governed by the common polarity rules of chromatography. The separation of styrene–acrylonitrile copolymers was, under the conditions of these studies, dominated by a precipitation mechanism. Another example of precipitation mechanism in styrene–acrylonitrile gradient chromatography is given in the next section. However, it should be firmly stated that styrene–acrylonitrile is an exception rather than the rule. In general, gradient chromatography of synthetic polymers is governed by the *combination of precipitation and adsorption*. Irregular phase combination will not often work, but they do with styrene–acrylonitrile.

Normal- and reversed-phase chromatography are like mirror images. It was a challenge to find out whether or not a given synthetic copolymer could be separated by both mechanisms. The first positive report appeared in 1987 when copolymers of styrene and ethyl methacrylate were measured by both modes of chromatography. All previous related work was by normal-phase separation. As expected, the elution order achieved by reversed-phase chromatography was the opposite of that in normal-phase chromatography. Since then, several polymer systems have been separated by normal-phase and reversed-phase chromatography with inversion of elution order, e.g. styrene–methyl methacrylate copolymers, styrene–methyl acrylate copolymers or methacrylate homopolymers graded in polarity of the ester group.

## High Performance Precipitation Liquid Chromatography of Styrene–Acrylonitrile Copolymers with Inversion of Elution Order

The separation of styrene–acrylonitrile copolymers with an elution order as in reversed-phase chromatography was achieved on a column packed with polystyrene gel. Eluent A was methanol (MeOH), a polar non-solvent for styrene–acrylonitrile samples; the less polar eluent, B, was either THF or dichloromethane. In both cases, the gradient rate was 0–100% B in 25 min. Copolymers with acrylonitrile content between 2.3% and 27.3% were retained longer the less acrylonitrile they contained (see **Figure 1**). Although the phase system and the elution order conformed to the rules of reversed-phase chromatography, the solubility mechanism prevailed.

The samples were prepared by copolymerization to only about 5% conversion, but they still consisted of macromolecules differing in composition. The chemical composition distributions of the samples are essentially responsible for the shape of the elution curves. The chemical composition distributions of samples with, say, 8.6% or 17.6% average acrylonitrile content are obviously narrower than that of a sample with 12.5% acrylonitrile.

The shape of the elution curve for 36.2% acrylonitrile in Figure 1 looks rather odd. In addition, the position of its maximum is not where it might be expected. According to its high acrylonitrile content, the sample is the most polar of the series investigated. It should therefore be eluted before the copolymer

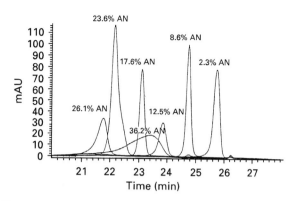

**Figure 1** Merged plot of elution curves of seven styrene–acrylonitrile copolymers on a column (250 mm × 7.1 mm i.d.) packed with polystyrene gel. Gradient: methanol → tetrahydrofuran, 0–100% B in 25 min: UV signal detected at 254 nm. The acrylonitrile (AN) content of the samples is indicated on the curves; the amount of each injected was 30 μg. (Reproduced from Glöckner *et al.*, 1991, by courtesy of Vieweg-Verlag.)

labelled 26.1% acrylonitrile, but is was eluted between the samples 17.6% and 12.5% acrylonitrile. This puzzling observation can be understood with the help of the solubility diagram for styrene–acrylonitrile in THF/MeOH, which is shown in **Figure 2**. The solubility boundary has a maximum at 20–25% acrylonitrile content, where samples require only about 45 vol% THF in MeOH for dissolution, whereas copolymers with more or less acrylonitrile need up to 12% more THF.

Along the left-hand branch of the solubility boundary (0–20% acrylonitrile), both polarity and solubility decrease with decreasing acrylonitrile content. The sequence of the five late-eluting peaks in Figure 1 is supported by polarity and solubility. Beyond the point of inflection, polarity increases but solubility decreases with increasing acrylonitrile content. A sample with 36.2% acrylonitrile requires about 50% THF for dissolution but should, according to polarity, already be released from the column in a mixture of 35% THF in MeOH. The measured peak position between the peaks for 17.6% and 12.5% acrylonitrile is determined by solubility. This is another indication of precipitation prevailing in the chromatography of styrene–acrylonitrile copolymers.

Along the left-hand branch of the solubility boundary, polarity supports the effect of solubility but beyond the turning point the two effects counteract each other. This is the reason why the elution curve for 36.2% acrylonitrile is much broader than the others. With normal-phase gradients, the 36.2% sample yielded an elution curve of the usual narrow shape, even in irregular phase systems (see **Figure 3**).

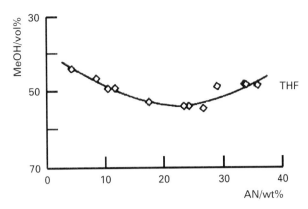

**Figure 2** Solubility boundary for styrene-acrylonitrile copolymers in THF/MeOH as measured by turbidimetric titration at 20°C using THF as a sample solvent and methanol as the precipitating non-solvent. Phase separation (precipitation) occurs on crossing the curve from the upper part of the diagram (homogeneous solutions) to the lower part. (Reproduced from Glöckner *et al.*, 1991, by courtesy of Vieweg-Verlag.)

# Sudden-Transition Gradient Chromatography of Synthetic Polymers

## Interaction of Precipitation and Adsorption in Polymer Gradient Chromatography

Chromatographic retention and elution of synthetic polymers is generally governed by precipitation/dissolution and adsorption/desorption. The contribution of adsorption can be judged by comparing the solubility and elution characteristics of the sample

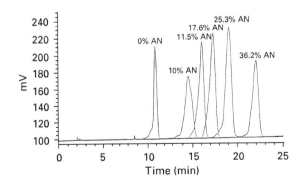

**Figure 3** Merged plot of elution curves of six styrene–acrylonitrile copolymers on a reversed-phase column (250 mm × 4.3 mm i.d.) packed with $C_{18}$ bonded phase; in irregular combination with a normal-phase gradient *n*-heptane → (THF + 20% methanol), 0–100% B in 25 min. Detection by signal from an evaporative light-scattering detector. The acrylonitrile (AN) content of the samples is indicated on the curves; the amount of each injected was 30 µg. (Reproduced from Glöckner *et al.*, 1991, by courtesy of Vieweg-Verlag.)

system. If solubility prevails (and both temperature and concentration are suitable), both curves coincide. Noticeable adsorption shifts the elution characteristic above the solubility boundary, i.e. a higher concentration of solvent B is necessary for *eluting* a given sample than for *dissolving* it. The least adsorption was observed in reversed-phase systems with polystyrene samples. The predominance of an adsorption mechanism causes a retention behaviour different from (or even opposite to) that observed with a precipitation mechanism (see **Table 1**).

Baker and Williams reported that classical precipitation chromatography can be performed with 'a column of inert material … providing that the polymer gel does not flow through the column'. This type of flow can also occur in high performance precipitation chromatography in which case an optical detector may register the strong signal characteristic of a turbid liquid. Such a signal is affected by many parameters including time and is therefore poorly reproducible. Gel breakthrough can be avoided if there is some contribution of adsorption to retention. Separation according to composition with the least superimposition of molecular weight effects requires adsorption to dominate, i.e. elution by changing polarity of the eluent rather than by solvent strength. On the other hand, the unfavourable effect of sample size owing to strong adsorption can be compensated by increasing the contribution of precipitation to retention.

## Independent Control of Adsorption and Precipitation

In common binary gradients, the solvent power and polarity of the mobile phase change simultaneously in the course of the run. An optimum can be sought by using a variety of different eluents A and B and their combinations. However, this is a cumbersome procedure requiring an adequate supply of chemicals and a prolonged time. In addition, it may not be successful because thermodynamic reasons restrict the number of possible solvents for a given polymer, and several of these may be further ruled out by physical, physiological, or financial reasons. More promising and efficient is the use of ternary systems consisting of two non-solvents (A and B) and one solvent (C) for the polymers under investigation. A and B must be opposite in polarity, i.e. if A is a polar non-solvent, B is a nonpolar one. The polarity of solvent C is in between those of A and B. Solvent C must be miscible with A and B and must have sufficient strength to dissolve samples in the whole range of molecular weight and composition under investigation.

**Table 1** Features of polymer gradient chromatography with predominance of either precipitation/dissolution or adsorption/desorption

| Dominating mechanism | Elution characteristic | Irregular phase combinations | Increasing temperature | Increasing sample size (overload) | Increasing molecular weight[a] |
|---|---|---|---|---|---|
| Percipitation | Coincident with the solubility boundary | Separating like standard ones | Decreases retention | Increases retention | Increases, retention, $C_2 = 2\text{--}4 \times 10^3$ |
| Adsorption | Above the solubility boundary | Ineffective in separation | Increases the retention of polymers | Decreases retention | Increases retention slightly, $C_2 = 2\text{--}5 \times 10^2$ |

[a]For $C_2$ factors see the equation in the text. Values of $C_2$ are compiled in Glöckner G (1991) *Gradient HPLC of Copolymers and Chromatographic Cross-Fractionation*, p. 107. New York: Springer-Verlag.

Since with gradients of this kind the chromatographically significant process is the result of interactions between non-solvents there is, owing to the large variety of the latter, more freedom in adjusting optimum conditions than with binary non-solvent/solvent gradients.

The samples to be investigated are dissolved in solvent C and injected into a starting eluent (e.g. A), whose polarity and precipitating power ensure proper retention at the top of the column. Solvent C is then added to the eluent at a concentration that in itself does not suffice for elution. In order to achieve short chromatograms, the concentration of C is changed as rapidly as the apparatus allows. No unfavourable side effects of the shock caused by the sudden transition from injection to elution conditions have ever been observed. The disturbance is visible with the help of optical detectors. With cyanopropyl or $C_{18}$ packings, it is swept through by the approximately three-fold volume of mobile phase in the column. The elution of the sample is then triggered by a gradient A → B at a constant level of solvent C.

The first results of gradient elution with sudden transition of solvent concentration were achieved in the normal-phase mode of chromatography. The column packing was polar (CN-modified silica), A was iso-octane (with addition of 2% MeOH to the starting eluent), B was MeOH and C was THF. The gradient A → B was performed at 5% min$^{-1}$ and applied to copolymers of styrene and ethyl methacrylate (EMA), methyl methacrylate (MMA), or 2-methoxyethyl methacrylate.

**Figure 4** is the merged plot of UV signals measured on the elution of a mixture of five styrene–EMA copolymers through a gradient iso-octane → MeOH after sudden transition to 20, 25, 30 or 35% THF solvent. Both iso-octane and MeOH are non-solvents for styrene–EMA. The addition of 35% THF yielded too high a solubility: the sample with 4.7% EMA was swept through the column by the sample solvent. A proportion of 20 or 25% THF did not suffice

for baseline separation. The best result was obtained after addition of 30% THF.

The advantage of this technique in comparison with binary gradient elution is obvious (see **Figure 5**). The chromatogram in Figure 5 was obtained in the same laboratory as those of Figure 4 with the same instrument and identical solvents. The baseline shift in Figure 5 is due to the UV absorption of THF which, at 259 nm, is slightly higher than that of iso-octane. This causes the baseline rise with

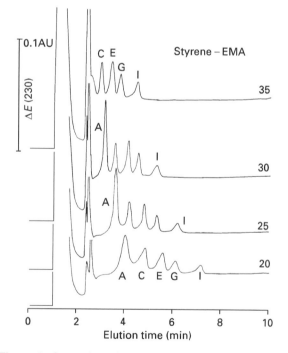

**Figure 4** Separation of the mixture of five styrene–EMA copolymers at 50°C on a column (60 mm × 4 mm i.d.) packed with cyanopropyl bonded phase. Gradient: *iso*-octane → methanol (5% min$^{-1}$) after increase of THF concentration from zero to the percentage indicated at the curves; flow rate 0.5 mL min$^{-1}$. Sample: 1.8 µg copolymer A (4.7% EMA) + 1.2 µg C (32.2% EMA) + 2.0 µg E (54.6% EMA) + 1.2 µg G (68.0% EMA) + 2.0 µg I (92.5% EMA); UV signal detected at 230 nm. (Reproduced from Glöckner, 1991, by courtesy of Springer-Verlag.)

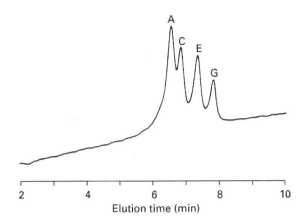

**Figure 5** Separation of the mixture of four styrene–EMA copolymers at 50°C on a column (60 mm × 4 mm i.d.) packed with silica. Gradient: iso-octane → methanol (5% min⁻¹); flow rate 0.5 mL min⁻¹. Samples A to G as in Figure 4, 2.5 μg each; UV signal detected at 259 nm. (Reproduced from Glöckner, 1987a, by courtesy of Elsevier Science Publishers.)

increasing THF content of the eluent. The effect would be still more dramatic at a shorter wavelength, e.g. at 230 nm. Figure 4 presents horizontal baselines although the chromatograms were monitored at 230 nm. This is due to the constant concentration of THF throughout the elution, which disturbs the traces much less than a changing amount of THF does. The higher the THF addition, the higher the level of the baseline at the end of the chromatogram in comparison to the starting position in Figure 4.

The poor separation in Figure 5 is explained by the comparatively low molecular weight of these samples $(50–80 \times 10^3)$ and the superimposition of separation by molecular weight and by composition. The peaks are indeed quite well separated when SEC fractions of the copolymer mixture are injected. Figure 4 indicates that the molecular weight effect in the investigation of the raw copolymers can be suppressed by the sudden-transition technique. **Table 2**

**Table 2** Characteristics of polymer separation with separate control of solubility and adsorption[a]

| Factor | Details |
|---|---|
| Sample | 20–100 μg polymer per injection, dissolved in about 50 μL solvent |
| Solvent | C, capable of dissolving samples of the system under investigation in the whole range of composition and molecular weight, used also for sample solutions (recommended: tetrahydrofuran, dichloromethane) |
| Non-solvents | A and B, opposite in polarity, both miscible with solvent C, e.g. A, acetonitrile, methanol; B, n-heptane.<br>In general, the variety of non-solvents for a given polymer system is much broader than the list of suitable solvents |
| Interactions of eluents and detector | Eluents must not impede the monitoring of the eluting sample components, they must be transparent if optical detection is employed. This demand is more stringent for the gradient components A and B than for solvent C, whose concentration is not changed during the elution of sample components. For instance, separations at constant concentration of THF can be monitored at 230 nm or at constant DCM concentration with an evaporative light scattering detector without disturbance |
| Reversed-phase separation | Non-polar column, e.g. reversed-phase $C_{18}$ bonded phase, injection into polar non-solvent A, gradient A → B after adjusting the solvent concentration to a suitable constant value |
| Normal-phase separation | Polar column, e.g. cyanopropyl bonded phase, injection into non-polar non-solvent B, gradient B → A after adjusting the solvent concentration to a suitable constant value |
| Reversed-phase and normal-phase separations | Can be performed with a common set of three eluents |
| Automated search for optimum separation method | Possible with programmable apparatus equipped with three storage bottles and a device for column switching |
| Balance between solubility and adsorption | Can be adjusted by the solvent concentration, which remains constant during the elution |
| Length of chromatograms | Can be optimized by sudden transition[a] of solvent concentration from zero to the selected level |

[a]Information on how to perform sudden-transition gradients is available in Glöckner G, Wolf D and Engelhardt H (1994) *Chromatographia* 39: 557–563.

summarizes the characteristics of sudden-transition gradient elution.

## Chromatography in Normal-Phase and Reversed-Phase Modes Using a Solvent and Two Non-Solvents

Independent control of adsorption and solubility enables normal-phase *and* reversed-phase separations to be performed with a common set of three liquids. This was first demonstrated with styrene–MMA copolymers in the system A (acetonitrile), B (*n*-heptane) and C (dichloromethane, DCM) on either CN or $C_{18}$ bonded phases.

**Figure 6** shows chromatograms measured under reversed-phase and normal-phase conditions. Both modes yielded good separations. The elution order is inverted in the reversed-phase mode, as expected. The elution of styrene–MMA copolymers by the strong precipitant heptane (Figure 6A) is rather surprising.

**Figure 7** shows the composition triangle of the eluent system used in Figure 6 with dichloromethane at the top, the polar non-solvent acetonitrile at the bottom left and the non-polar precipitant heptane at the bottom right. Acetonitrile and heptane have a miscibility gap that diminishes as dichloromethane is added. Eluent mixtures containing 25% or more dichloromethane are homogeneous. The elution characteristics of the styrene–MMA copolymers investigated in reversed-phase mode with 25–50% DCM or

in normal-phase mode with 25–40% dichloromethane are indicated. The characteristics of reversed-phase elutions form a group in the left-hand area of the triangle. The proportion of acetonitrile present means that eluent systems in this region have a higher polarity than those in the right-hand region. Reversed-phase chromatography starts with retention in a strongly polar medium. Sample components are released when the polarity of the eluent is no longer sufficient for retention. Thus, the characteristics of reversed-phase elution are to be expected on the polar side of the composition diagram. On the other hand, normal-phase elution characteristics are located in the right-hand part of the triangle. This can be understood by complementary reasoning because normal-phase chromatography starts with retention in a nonpolar medium.

The characteristics in Figure 7 are due to samples containing methyl methacrylate in the proportions (from left to right) 83.7%, 62.2%, 48.1%, 34.1%, 14.1% or 0% (polystyrene homopolymer). This sequence holds true with reversed-phase as well as with normal-phase elutions. In both modes, the copolymer with the highest content in polar methyl methacrylate units yields characteristics nearer to the polar (left) side of the diagram than the other samples. As expected, the least polar sample (polystyrene) marks the right border of the elution area in each mode.

All polymers considered here are soluble in the region beneath the solvent apex. The addition of

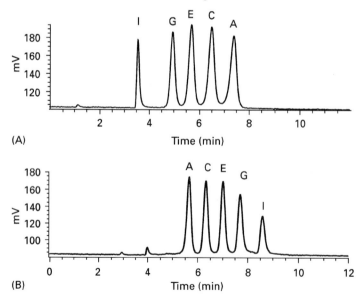

**Figure 6** Separation of the mixture of five styrene–MMA copolymers at 35°C and flow rate 1 mL min⁻¹ by gradient elution in reversed-phase (A) or normal-phase mode (B) after a sudden increase of dichloromethane concentration from zero to 30%, monitored by an evaporative light-scattering detector. Sample in each mode: 6.76 μg copolymer A (14.1% MMA) + 5.54 μg C (34.1% MMA) + 5.28 μg E (48.1% MMA) + 5.48 μg G (62.2% MMA) + 5.02 μg I (83.7% MMA), dissolved in 10 μL DCM. (A) Column (250 mm × 4.1 mm i.d.) packed with reversed-phase $C_{18}$ bonded phase. Gradient: acetonitrile → *n*-heptane (4.99% min⁻¹). (B) Column (250 mm × 4.1 mm i.d.) packed with cyanopropyl bonded phase. Gradient: *n*-heptane → acetonitrile (4.99% min⁻¹). (Reproduced from Glöckner *et al.*, 1994, by courtesy of Vieweg-Verlag.)

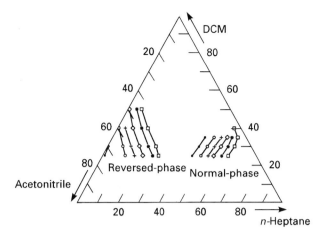

**Figure 7** Composition triangle for acetonitrile/n-heptane/dichloromethane with elution characteristics of styrene–MMA copolymers in normal-phase and reversed-phase sudden-transition gradients. Samples: ●, 83.7% MMA; ○, 62.2% MMA; +, 48,1% MMA; ◇, 34.1% MMA; *, 14.1% MMA; □, polystyrene. (Reproduced from Glöckner, 1996, by courtesy of Gordon & Breach.)

small quantities of heptane or acetonitrile to dichloromethane will impair solubility but will not immediately cause precipitation. The polymers are still soluble in mixtures of dichloromethane with about 40% acetonitrile or 40% heptane. Thus, the upper sections of the solubility boundary follow the left and right sides of the eluent triangle. With increasing concentration of nonsolvent, a precipitation threshold is reached on each side. From these points, both branches of the solubility boundary bend towards each other. These sections may be determined experimentally by turbidimetric titration. For example, the elution characteristics of the copolymer containing 48.1% MMA run almost parallel to the corresponding sections of the solubility boundary. In both reversed-phase and normal-phase modes, the elution characteristics are shifted from the solubility boundary towards the centre of the solubility window. This shift indicates the contribution of adsorption to retention, which is well known in gradient HPLC of styrene–methyl methacrylate copolymers. Finally, both branches of the boundary will merge inside the triangle (above the miscibility gap). For details, see Glöckner G (1996).

Solubility windows of similar shape can be expected with many polymers in mixtures of a solvent with two non-solvents differing in polarity. Hence, HPLC separation generally should be possible in normal-phase as well as in reversed-phase mode with a suitable ternary eluent system. These separations should be achievable near the respective side of the solubility boundary. Thus, the use of ternary gradients consisting of a solvent and two non-sol-

vents and control of solubility by a sudden increase of solvent concentration to a constant level will not only offer the opportunity to improve separations with small additional effort, but will also contribute to a better understanding of the mechanisms of polymer chromatography.

*See also:* **II/Chromatography: Liquid:** Mechanisms: Normal Phase; Mechanisms: Reversed Phases; Mechanisms: Size Exclusion Chromatography. **III/Polyethers: Liquid Chromatography. Synthetic Polymers:** Liquid Chromatography.

## Further Reading

Baker CA and Williams RJP (1956) A new chromatographic procedure and its application to high polymers. *Journal of Chemistry Society (London)* 1956: 2352–2362.

Glöckner G (1987a) Normal- and reversed-phase separation of copolymers prepared from styrene and ethyl methacrylate. *Journal of Chromatography* 403: 280–284.

Glöckner G (1987b) *Polymer Characterization by Liquid Chromatography.* Amsterdam: Elsevier Science Publishers.

Glöckner G (1991) *Gradient HPLC of Copolymers and Chromatographic Cross-Fractionation.* New York: Heidelberg, Tokyo: Springer-Verlag.

Glöckner G (1996) Solubility and chromatographic separation of styrene/methacrylate copolymers in ternary eluent systems. *International Journal of Polymer Analysis and Characterization* 2: 237–251.

Glöckner G and van den Berg JHM (1987) Copolymer fractionation by gradient high-performance liquid chromatography. *Journal of Chromatography* 384: 135–144.

Glöckner G, Wolf D and Engelhardt H (1991) Separation of copoly(styrene/acrylonitrile) samples according to composition under reversed phase conditions. *Chromatographia* 32: 107–112.

Glöckner G, Wolf D and Engelhardt H (1994) Control of adsorption and solubility in gradient high performance liquid chromatography 5: separation of styrene/methyl methacrylate copolymers by sudden-transition gradients in normal-phase and reversed phase mode. *Chromatographia* 39: 557–563.

Mourey TH and Schunk TC (1992) In E. Heftmann (ed.) *Chromatography – Fundamentals and Application of Chromatography and Related Differential Migration Methods.* Chapter 22. Amsterdam: Elsevier Science Publishers.

Pasch H and Trathnigg B (1997) *HPLC of Polymers.* New York: Heidelberg, Tokyo: Springer-Verlag.

Quarry MA, Stadalius MA, Mourey TH and Snyder LR (1986) General model for the separation of large molecules by gradient elution: sorption versus precipitation. *Journal of Chromatography* 358: 1–16.

Schultz R and Engelhardt H (1990) HPLC of synthetic polymers: characterization of polystyrenes by high performance precipitation liquid chromatography (HPPLC). *Chromatographia* 29: 205–213.

Schunk TC (1993) Chemical composition separation of synthetic polymers by reversed-phase liquid chromatography (review). *Journal of Chromatography A* 656: 591–615.

# HERBICIDES

## Gas Chromatography

**J. L. Tadeo and C. Sanchez-Brunete**, Department of Sustainable Use of Natural Resources, INIA, Madrid, Spain

### Herbicide Formulations

Weeds have been controlled by humans since the beginning of agriculture by means of mechanical tools or by hand. It was early in the 20th century that some inorganic compounds were first used with this aim. The discovery of the herbicidal properties of 2,4-D (2,4-dichlorophenoxyacetic acid) in 1945 can be considered the initiation of use of organic herbicides in agriculture. Since then, more than 130 different active compounds have been synthesized for their application as herbicides. These compounds can be grouped, according to their chemical structures, into different herbicide classes (**Table 1**).

Compounds belonging to the principal herbicide groups will be considered in this study. These compounds control weeds in a variety of ways, showing different modes of action, selectivity and application characteristics. Soil-applied herbicides are absorbed by roots or emerging shoots and foliage-applied herbicides are absorbed into the leaves, where they may be translocated to other parts of the plant.

The active ingredient of a herbicide is a compound, usually obtained by synthesis, which is formulated by a manufacturer in soil particles or liquid concentrates. These commercial formulations of herbicides are diluted with water before application in agriculture at the recommended doses. Herbicide formulations generally contain other materials to improve the efficiency of application.

Analysis of herbicide formulations was initially carried out by wet chemical procedures, such as determination of total chlorine, nitrogen or phosphorus, or by spectrometric procedures like ultraviolet absorption. The development of gas chromatography (GC) allowed the analysis of these compounds in commercial formulations with high selectivity and sensitivity. The analytical procedure is commonly based on the dissolution of a known amount of the formulation in an organic solvent, which often contains an internal standard to improve the precision and accuracy of the determination. An aliquot of this solution is analysed by GC. Packed columns were used initially, but have now been replaced by capillary columns of low or medium polarity and flame ionization is the detection technique more widely used. When herbicides are not volatile or thermally stable, high performance liquid chromatography (HPLC) is the preferred technique for their determination in commercial formulations. **Figure 1** shows the gas chromatographic separation of a mixture of phenoxy esters.

### Herbicide Residue Analysis

Residues of herbicides will persist in the plant or in the soil for a variable time, depending on their physicochemical properties and on the environmental conditions. Analysis of herbicide residues in these matrices is important, not only from the point of view of the efficacy of application, but also to know the distribution and persistence of these compounds in food and in the environment. Therefore, herbicides of a wide range of polarities have to be determined in complex environmental matrices at very low levels.

Initially, herbicide residues were analysed by colorimetric methods. These procedures were generally based on acidic or basic hydrolysis followed by formation of derivatives. These methods are time-consuming and do not usually distinguish between the parent herbicide and metabolites.

Since the development of GC, this technique has been widely used in the analysis of these compounds. **Table 2** summarizes the preparation of different types of samples for residue determination. These samples are generally analysed by a procedure with the following main steps: sample extraction, clean-up of extracts, then GC determination and identification. Some compounds are not volatile or thermally stable

**Table 1**  Chemical structures of herbicides

*Benzonitriles*
Bromoxynil; ioxynil

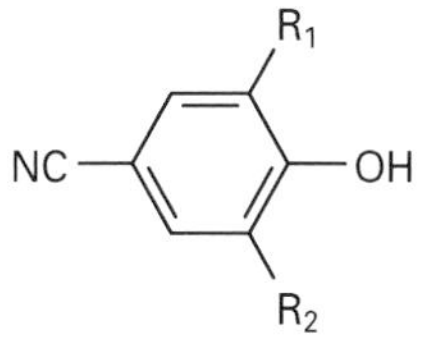

*Phenoxyacids*
2,4-D; MCPA; MCPP; dichlorprop; diclofop; fenoxaprop

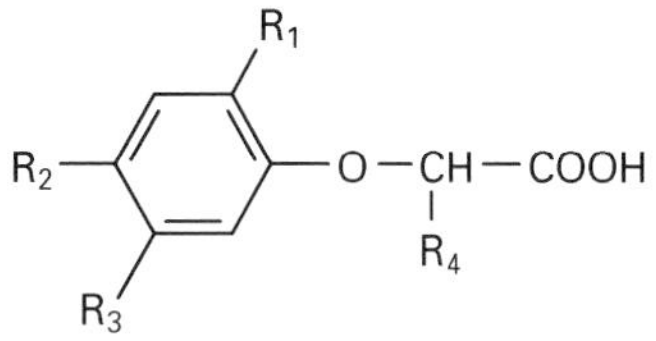

*Carbamates*
EPTC; triallate

$R_1 - O - CO - NH - R_2$

*Chloroacetamides*
Alachlor; metolachlor

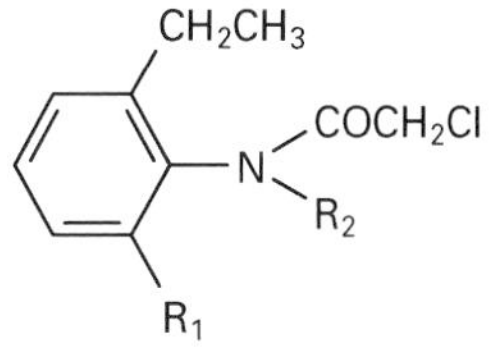

*Dinitroanilines*
Butralin; ethalfluralin; pendimethalin; trifluralin

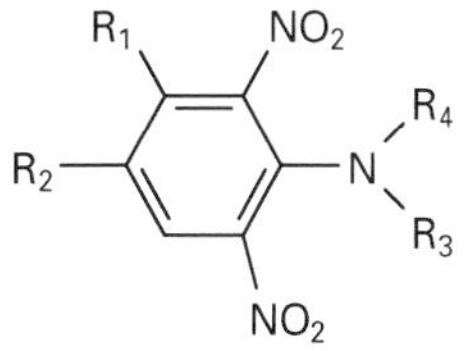

*Triazines*
Ametryn; atrazine; cyanazine; simazine; terbutryn

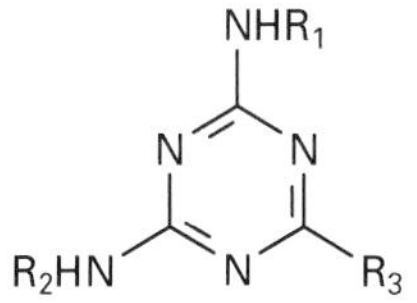

*Uracils*
Bromacil; lenacil; terbacil

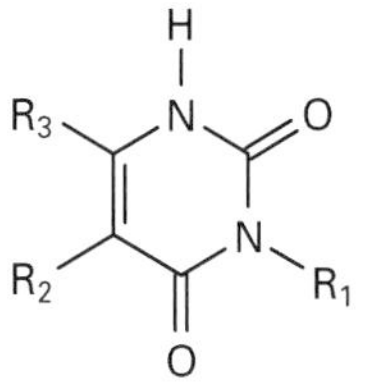

*Ureas*
Chlorotoluron; isoproturon; linuron; chlorsulfuron; metsulfuron; triasulfuron

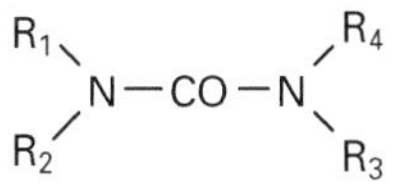

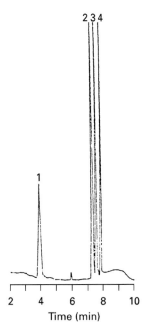

**Figure 1** Analysis of a mixture of phenoxy esters by gas chromatography on a BP-5 fused silica column, 12 m × 0.53 mm i.d., with helium as carrier gas at 10 mL min$^{-1}$ and flame ionization detection. Oven temperature was held at 180°C for 5 min, increased at 25°C min$^{-1}$ to 250°C and held for 10 min. 1, 2,4-D isobutyl ester; 2, MCPA 2-butoxyethyl ester, 3, 2,4-DP 2-butoxyethyl ester; 4, 2,4-D 2-butoxyethyl ester. Adapted from Sánchez-Brunete C, Pérez S and Tadeo JL (1991) Determination of phenoxy ester herbicides by gas and high-performance liquid chromatography. *Journal of Chromatography* 552: 235 with permission from Elsevier Science.

and need to be derivatized before being analysed by GC.

Currently capillary columns are most commonly used for residue analysis. **Table 3** shows some representative examples of different packed and capillary columns used in the determination of herbicides.

In trace analysis, blank samples are commonly processed through the analytical method to identify the possibility of interferences in the herbicide determination from other sample or reagent components. In addition, recoveries of the analysed compound are also carried out through the extraction and clean-up procedures. The accepted normal range of these re-

coveries is 70–120%, with a standard deviation of 20%.

Several injection techniques are used in herbicide residue analysis. The techniques most often employed are splitless, on-column and programmed-temperature vaporizer injection and the volume normally injected is 1–2 μL.

Various selective detectors are used in trace analysis of herbicides. The electron-capture detector (ECD) was first used for the determination of halogen-containing compounds or halogenated derivatives, due to its high sensitivity for these compounds. The nitrogen–phosphorus detector (NPD), also known as thermionic or alkali flame detector, is commonly used in the analysis of nitrogen-containing herbicides. Both detectors are highly sensitive and can detect herbicide concentrations lower than 1 pg. The flame photometric detector (FPD) has sometimes been used in the determination of sulfur-containing herbicides. The limit of detection (LOD) reflects the sensitivity of a detector for a given compound and it is defined as the amount producing a signal-to-noise ratio equal to 3. When this ratio is determined with extracts of real samples, processed through the whole analytical procedure, this parameter is known as the limit of quantification (LOQ), which depends on the efficacy of the extraction and clean-up procedures and on the selectivity and sensitivity of the detector. The coupling of mass spectrometry (MS) with GC allows the determination of herbicide residues with high selectivity and sensitivity and, in addition, the identification of residues by means of their mass spectra obtained at trace levels. The atomic emission detector allows monitoring characteristic wavelengths for carbon and hydrogen atoms, as well as the more specific emission lines for phosphorus, nitrogen and sulfur. **Table 4** summarizes the detection techniques for herbicide residue determination by GC.

The analysis of herbicides, grouped into various chemical classes, is considered in more detail below.

### Benzonitriles and Phenoxy Acids

Benzonitriles and phenoxy acids are widely applied as salts or esters, but they are hydrolysed to their

**Table 2** Procedures used in sample preparation for herbicide residue determination

| Matrix | Amount sampled | Sample preparation | Amount extracted | Extraction procedure |
| --- | --- | --- | --- | --- |
| Soil | 1 kg | Sieving ( < 2 mm) | 10–20 g | Shaking, Soxhlet, SFE |
| Water | 1 L | Filtration | 0.1–1 L | SPE, LLE |
| Plants | 1 kg | Blending | 20–50 g | Homogenization |
| Air | 25–250 L | Adsorption or trapping | | Thermal or solvent desorption |

SFE, Supercritical fluid extraction; SPE, solid-phase extraction; LLE, liquid–liquid extraction.

**Table 3**  Chromatographic columns used in herbicide residue analysis

| Column (length/diameter) | Stationary phase | Applications |
|---|---|---|
| Packed (1–3 m/2–4 mm) | Dimethylpolysiloxane (SE-30, DC-200) | Carbamates, ureas, dinitroanilines, triazines |
| | Phenylmethylpolysiloxane (OV-17, OV-25) | Benzonitriles, phenoxy acids |
| | Trifluoropropylpolysiloxane (QF-1, OV-210) | Carbamates, chloroacetamides, phenoxy acids |
| | Polyethyleneglycol (Carbowax) | Triazines |
| | Cyanoethylpolysiloxane (XE-60) | Triazines |
| | Cyanopropylpolysiloxane (OV-225) | Ureas |
| Capillary (10–30 m/0.2–0.5 mm) | Dimethylpolysiloxane | Nitrogen-containing herbicides |
| | Phenylmethylpolysiloxane | Triazines |
| | Polyethylene glycol | Triazines, phenoxy acids, benzonitriles |
| | Cyanopropylphenylmethylpolysiloxane | Multiresidue |

respective phenols or acids in the matrix. Extraction of residues from soil and water is commonly performed at acidic pH with organic solvents of medium polarity. The extraction of these herbicides from vegetable matter is often done with aqueous solutions at basic pH, followed by extraction with organic solvents.

Purification of extracts is required in most cases and this step is accomplished by liquid–liquid partition at basic pH or by chromatography on silica columns.

Analysis of these compounds in air is carried out by trapping herbicides in ethylene glycol or in various adsorbents, like polyurethane or amberlite resins.

Derivatization of phenoxy acids, before GC determination, is necessary to make them volatile. Various alkyl, silyl or pentafluorobenzyl derivatives are obtained with this aim. Methyl esters have been commonly prepared for the determination of phenoxy acids and the reagents most often used are diazomethane and boron trifluoride–methanol. Benzo-

nitriles can be determined directly by GC, but the sensitivity and reproducibility achieved are poor. Various derivatives overcome these problems and diazomethane and heptafluorobutyric anhydride are the reagents most often used.

The determination of herbicides is widely carried out by GC with ECD, if the compound has halogen substituents or halogenated derivatives are obtained. MS detection has the advantage of being more selective and requiring less clean-up of extracts.

## Carbamates

Carbamates are a wide group of pesticides and some of them have herbicide properties, like the thiocarbamates S-ethyl dipropylthiocarbamate (EPTC) and triallate. These compounds are extracted from soil with methanol or acetone and from water by means of hexane or dichloromethane. The extraction from plants is commonly accomplished with acetonitrile or by steam distillation. Clean-up of extracts is often

**Table 4**  Detection of herbicides in environmental samples

| Herbicides | Detectors | LOD ($\mu g\,g^{-1}$) | Derivatives |
|---|---|---|---|
| Benzonitriles | ECD, MS | 0.05–0.0003 | Methyl ethers |
| | MS | 0.001 | Heptafluorobutyryl |
| Phenoxyacids | ECD, MS | 0.05–0.001 | Methyl esters |
| Carbamates | ECD, NPD, FPD, MS | 0.1–0.001 | |
| Chloroacetamides | NPD, MS | 0.05–0.001 | |
| Dinitroanilines | ECD, NPD, MS | 0.05–0.0001 | |
| Triazines | NPD, ECD, MS | 0.01–0.0001 | |
| Uracils | NPD, ECD, MS | 0.04–0.001 | |
| Ureas | | | |
|   Phenylureas | NPD, MS | 0.1–0.01 | Methyl or ethyl |
| | ECD | 0.01–0.001 | Heptafluorobutyryl |
|   Sulfonylureas | ECD | 0.1–0.002 | Methyl or PFB |
| Multiresidue | NPD, MS | 0.02–0.001 | |

ECD, Electron capture detector; MS, mass spectrometry; NPD, nitrogen–phosphorus detector; FPD, flame photometric detector; PFB, pentafluorobenzyl.

done by chromatography on silica columns. Most carbamates are thermally unstable and are usually analysed by HPLC. Some carbamates, such as the thiocarbamates considered above, can be determined by GC and their residues are determined by this technique using different detectors, like ECD, NPD, FPD and MS.

## Chloroacetamides

These compounds, also known as anilides, are very often used for weed control in maize, in combination with triazines. Chloroacetamides are extracted from soil by polar and medium polarity solvents and their determination is generally carried out without further purification. Extraction from water is performed by reversed-phase solid-phase extraction (SPE) or by liquid–liquid extraction with low polarity solvents and clean-up of extracts on silica columns is necessary in some cases.

Analysis of these herbicides in plants is done by extraction with polar solvents, followed by purification by liquid–liquid partition or column chromatography on silica or alumina adsorbents.

Analytical methods for the determination of these herbicides in air have been reported, using several adsorbents or an ethylene glycol phase for their extraction from air. NPD and MS are the detectors most widely used in the determination of chloroacetamides; the ECD is also sometimes used.

## Dinitroanilines

Dinitroaniline herbicides are highly lipophilic with a low solubility in water and some compounds have a remarkable volatility. Extraction of these compounds from soil is carried out by polar as well as by low polarity solvents. Dinitroanilines are extracted from water by SPE or by liquid–liquid extraction with low polarity solvents, such as dichloromethane. Clean-up of water and soil extracts on silica columns is sometimes needed.

Extraction with polar solvents, like methanol, is the method often used for the determination of dinitroanilines in plants, followed by liquid–liquid partition or Florisil column clean-up of extracts.

Analysis of these herbicides in air is performed by means of several adsorbents or organic solvents as trapping phases to remove these compounds from air.

ECD is often used in the determination of these compounds due to the large response obtained, particularly with some halogen-containing dinitroanilines. NPD and MS are also used in the gas chromatographic analysis of these herbicides.

## Triazines

These compounds form a wide group of herbicides often employed in fruit trees and cereals; in particular, simazine and atrazine are the most widely used triazines.

Triazine herbicides are extracted from soil by polar or medium polarity organic solvents, followed by liquid–liquid partition or column chromatography clean-up if necessary. Supercritical fluid extraction is also used in the analysis of triazines in soil.

These herbicides are extracted from water by reversed-phase SPE, by anion exchange columns or by liquid–liquid extraction with low polarity solvents, followed in some cases by Florisil column clean-up.

Analysis of these compounds in plants is performed by sample homogenization with polar organic solvents and column chromatography or liquid–liquid partition clean-up. Triazines are sometimes analysed in air and plugs of polyurethane foam have been used to extract these compounds from air.

The gas chromatographic determination of triazines is commonly performed with NPD, due to the high response obtained with this detector because of the number of nitrogen atoms in their molecules. ECD is also employed but its sensitivity and selectivity for these compounds are lower. GC-MS is used for confirmation of residues as well as for routine determination, due to the good sensitivity obtained with selected ion monitoring. **Figure 2** shows some chromatograms of the determination of various triazines, together with other herbicides, by GC with NPD and MS detection.

## Uracils

These compounds, also named pyrimidines, are used to control weeds in some fruit trees, vegetables and sugar beet. Extraction of these herbicides from soil is carried out with polar organic solvents or with basic aqueous solutions. Extraction from water is performed by SPE or by liquid–liquid extraction with low polarity solvents. Uracils are extracted from plants by homogenization with basic aqueous solutions or with mixtures of polar solvents with water. Clean-up of extracts is commonly carried out by liquid–liquid partition at different pHs. The gas chromatographic determination of these herbicides is performed using ECD and NPD detectors, and also by MS and atomic emission detection.

## Ureas

Substituted ureas constitute one important group of herbicides formed by two different classes of compounds: phenylureas, one of the first used herbicides, and sulfonylureas, introduced later and applied

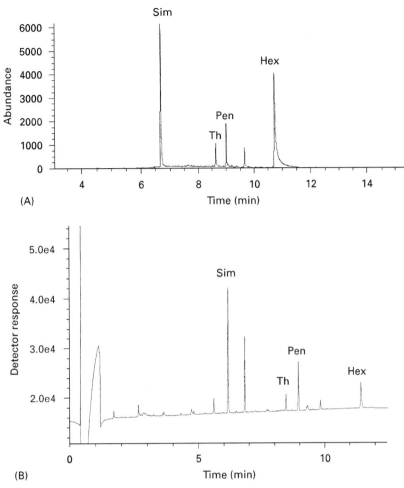

**Figure 2** Gas chromatograms of herbicide residues in soil, fortified at 0.01 μg g$^{-1}$, separated on an HP-1 capillary column, 12.5 m × 0.20 mm i.d., with helium as carrier gas at 1 mL min$^{-1}$ and detected (A) by GC-MS with selected ion monitoring or (B) by GC-NPD. Oven temperature was maintained at 100°C for 1 min, programmed at 15°C min$^{-1}$ to 250°C and held for 1 min. Sim, simazine; Th, thiazopyr; Pen, pendimethalin; Hex, hexazinone. Adapted with permission from Pérez RA, Sanchez-Brunet C, Miguel E and Tadeo JL (1998) Analytical methods for the determination in soil of herbicides used in forestry by GC-NPD and GC-MS. *Journal of Agriculture and Food Chemistry* 46: 1864.

at lower doses due to their high herbicidal activity. Both types of compounds suffer from thermal instability and their decomposition products, rather than the parent compounds, have been determined in some cases. To overcome this problem, various derivatives amenable to GC have been obtained, mainly their methyl, heptafluorobutyryl or perfluorobenzyl derivatives.

Substituted ureas are extracted from soil by shaking with organic solvents of medium or high polarity, like methanol or acetone, sometimes followed by silica column or liquid–liquid partitioning clean-up. These herbicides are extracted from water by reversed-phase SPE or by liquid–liquid extraction with dichloromethane.

Urea herbicides are generally extracted from plants by homogenization with polar organic solvents. Supercritical fluid extraction of these compounds

from plants has also been reported. Clean-up of extracts through column chromatography or liquid–liquid partition is necessary before GC determination.

Detection of these herbicides is performed by NPD or by ECD, mainly when halogenated derivatives are obtained. MS is also used in the determination of substituted ureas in environmental matrices.

## Multiresidue Analysis

The wide range of polarities and physico-chemical properties of herbicides does not allow the determination of the more than 130 available herbicides in one analytical method. Nevertheless, it is advisable to use analytical procedures that allow the determination of as many herbicides as possible in one method. These multiresidue methods permit reduction in time and cost of analysis as well as detection of the possible presence of herbicide residues in samples with unknown origin or contamination. A multiresidue method is able to determine all the herbicides that can be extracted, cleaned up, separated and detected in the conditions used in the analytical procedure. In some cases, two detectors – generally ECD and NPD – are connected at the end of the same chromatographic column to allow the detection of a wider type of compounds. MS is used as a universal detector and also for residue identification purposes. Atomic emission detection is a more recent detection technique which is increasingly used in trace analysis.

Herbicide residues are extracted from soil and plants with medium or high polarity solvents, like methanol, ethyl acetate and acetone. Vegetable samples are generally homogenized with organic solvent and soil samples are normally shaken and filtered. Matrix solid-phase dispersion (MSPD) is a technique which has recently been employed for pesticide residue determination in food samples: it performs sample extraction and purification at the same time. A simple multiresidue method, based on soil extraction in small columns, has been reported. **Figure 3** shows representative chromatograms of the analysis of various nitrogen-containing herbicides by this method.

Analysis of herbicides in water is generally based on liquid–liquid extraction with dichloromethane or on SPE using reversed phase columns, mainly $C_{18}$. Determination of herbicide residues in air is

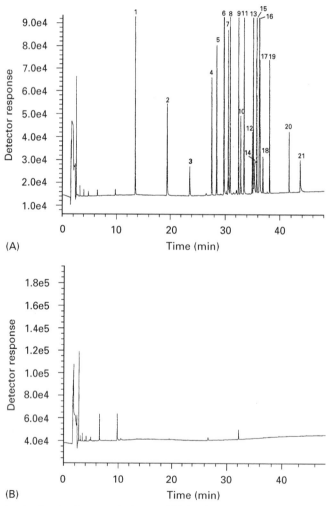

**Figure 3** Multiresidue analysis of soil extracts separated on an HP-1 column, 30 m × 0.25 mm i.d., with helium as carrier gas at 1 mL min$^{-1}$ and detected by GC-NPD. Oven temperature was kept at 80°C for 1 min, programmed at 5°C min$^{-1}$ to 140°C, held for 10 min and programmed at 5°C min$^{-1}$ to 250°C, held 15 min. (A) Soil fortified with nitrogen-containing herbicides at 0.5 μg g$^{-1}$. (B) Blank soil. 1, EPTC; 2, molinate; 3, propachlor; 4, ethalfluralin; 5, trifluralin; 6, atrazine; 7, terbumeton; 8, terbuthylazin; 9, dinitramine, 10, triallate; 11, prometryn; 12, alachlor; 13, metribuzin; 14, bromacil; 15, terbutryn; 16, cyanazine; 17, thiobencarb; 18, metolachlor; 19, butralin; 20, oxadiazon; 21, lenacil. Reproduced from Sánchez-Brunete C, Pérez RA, Miguel E and Tadeo JL (1998) Multiresidue herbicide analysis in soil samples by means of extraction in small columns and GC with NPD and MS detection. *Journal of Chromatography A* 823: 17, with permission from Elsevier Science.

accomplished by trapping these compounds on adsorbents, followed by extraction with organic solvents.

## Future Developments

GC will continue to be the main chromatographic technique used in herbicide residue analysis in the near future, due to the high sensitivity and selectivity given by the detectors that can be coupled with this technique. In particular, the use of less expensive and more robust and sensitive GC-MS equipment will keep growing in the routine determination and confirmation of herbicide residues.

The time needed for sample processing is expected to be reduced as a consequence of the continuation in the development of automatic processes for sample preparation, extraction and clean-up. These processes will use less sample and lower volumes of organic solvents in the analytical procedure.

New improvements in the gas chromatographic equipment to allow higher injection volumes of less purified extracts can also be expected.

**See Colour Plate 85.**

*See also:* **II/Chromatography: Gas:** Detectors: Mass Spectrometry; Detectors: Selective. **Insecticides:** Gas Chromatography. **Pesticides:** Supercritical Fluid Chromatography; Gas Chromatography. **Solid-Phase Matrix Dispersion: Extraction. III/Sorbent Selection for Solid-Phase Extraction.**

## Further Reading

Barceló D and Henion MC (1997) *Trace Determination of Pesticides and their Degradation Products in Water.* Amsterdam: Elsevier.

Blau K and King GS (eds) (1978) *Handbook of Derivatives for Chromatography.* London: Heyden.

Dobrat W and Martijn A (eds) (1998) *CIPAC Handbook: Vol. H Analysis of Technical and Formulated Pesticides.* Cambridge: Black Bear Press.

Hutson DH and Roberts TR (eds) (1987) *Progress in Pesticide Biochemistry and Toxicology,* vol. 6 *Herbicides.* New York: Wiley.

Milne GWA (1995) *CRC Handbook of Pesticides.* Boca Raton: CRC Press.

Nollet LML (ed.) (1996) *Handbook of Food Analysis.* New York: Marcel Dekker.

Sherma J (ed.) (1989) *Analytical Methods for Pesticides and Plant Growth Regulators:* vol. XVII. *Advanced Analytical Techniques.* San Diego: Academic Press.

Tadeo JL, Sánchez-Brunete C, García-Valcarcel AI *et al.* (1996) Review: determination of cereal herbicide residues in environmental samples by gas chromatography. *Journal of Chromatography A* 754: 347.

Tekel' J and Kovačičová J (1993) Review: chromatographic methods in the determination of herbicide residues in crops, food and environmental samples. *Journal of Chromatography* 643: 291

Zweig G (ed.) (1972) *Analytical Methods for Pesticides and Plant Growth Regulators:* vol. VI. *Gas Chromatographic Analysis.* New York: Academic Press.

# Solid-Phase Extraction

**Y. Picó,** Universitat de València, València, Spain

Solid-phase extraction (SPE) methods, using bonded-silicas, were first introduced in 1971 as an alternative to liquid partitioning. The method combines extraction and preconcentration of organic compounds in water by adsorption on proper solid material followed by desorption with a small quantity of an organic solvent. In comparison with liquid–liquid extraction, the following advantages are offered: the amount of solvent required for the clean up is greatly reduced, thus saving time for the evaporative concentration step and minimizing exposure of the analyst to the toxic solvent; the final eluate has less interfering material, and it could be analysed using any of a variety of detection, separation and identification techniques, including high performance liquid chromatography (HPLC) or gas chromatography (GC); accuracy and precision are improved; and it is rapid and easily automated.

Another impressive feature of SPE is the commercial availability of sorbents in small and inexpensive cartridges. $C_{18}$-bonded silica cartridges, styrene-divinylbenzene Empore® extraction discs and Carbopack® cartridges have been extensively used for the extraction of organic molecules from water samples. Automated column switching systems and on-line SPE coupled to determination devices have also been often reported for determination of pollutants in drinking and surface water.

Because of the reasons given above, in recent years much analysis of herbicides in fruit, vegetable and water has been conducted using SPE. Phenoxy acids, phenylureas, aryloxyphenoxypropionic acids, triazines, sulfonylureas, imidazolinones, glyphosate, phenoxyacetic acids, bipyridynium compounds, chloroacetamides, dinitroanilines and substituted phenols are examples of herbicides usually extracted and isolated by this technique.

It is undeniable that SPE is gaining in importance and, today, is a well-established and validated method, since the Environmental Protection Agency (EPA) in the United States currently offers one SPE procedure for the analysis of organic compounds (including neutral herbicides) and two for the analysis of acid herbicides.

## Solid-Phase Extraction

### Analyte Characteristics

The determination of herbicide residues is an intricate problem because of the large number of chemicals involved. As a general rule, to classify them into a wide variety of classes depending on their chemical structure results in a lot of groups that barely provide enough information in order to select the best SPE procedure.

In this way, the most practical approach is to organize the herbicides according to their acid/base character or other properties that condition the protocol following by SPE. **Table 1** shows these characteristics for the major classes of herbicides and some examples of the structures included in each group.

### Disposable Solid Phases

The modern SPE technique began in 1978 with the introduction of Sep-Pack cartridges, the first compact silica-based solid-phase extraction device for sample preparation on the market. Present-day, disposable prepacked columns or cartridges are available from more than 30 manufacturers, who offer phases such as $C_{18}$, $C_8$, cyano and amino. The containers are generally made of polypropylene. The sorbent bed varies from 100 to 1000 mg and is retained between two porous frits.

The use of Empore® discs are described in more recent studies. These devices include flat discs with large cross-sectional areas that provide advantages for preconcentration and clean-up methods with respect to the sorption, capacity, back pressure and stability after repeated use.

Reversed-phase silica-based sorbents, especially $C_8$ and $C_{18}$ bonded-silicas, are the most widely used packings for SPE. A typical SPE requires a previous sorbent activation step (wetting), usually with methanol, and removal of activation solvent excess (conditioning), usually with water.

Neutral herbicides can be extracted from 1 L samples with an average amount of sorbent (500 mg).

The sample is extracted under neutral or slightly alkaline conditions, and the pH is adjusted before the extraction to between 6 and 8. Under these conditions, salts of humic acids, which generally cause considerable interference in herbicide determination, are unlikely to be adsorbed during enrichment. As acid herbicides are highly polar, they are soluble in water and in aqueous solutions and are less soluble (in their dissociated form) in apolar sorbents. To overcome this difficulty, the aqueous phase has to be acidified before extraction to suppress the dissociation of this class of herbicides and to facilitate the transfer of the undissociated molecular species to the solid phase.

The recoveries and the relative standard deviation of the performance of different devices and solid phases are compared in **Tables 2** and **3** for basic/neutral and acid/phenolic herbicides, respectively. The $C_{18}$ cartridges showed good recoveries with most of the basic/neutral and acidic/phenolic herbicides. Compounds having a small (deisopropylatrazine, tribensulfuron-methyl) or a very high (beta-cyfluthrine) affinity to the $C_{18}$ material gave the worst recoveries. In comparison with Empore® discs a lower breakthrough of polar metabolites of atrazine was reported, possibly due to the fact that Empore® discs contain only half the quantity of $C_{18}$ material. However, lower recoveries were achieved for medium polar and non-polar pesticides except trifluralin and trialate.

As reported in the literature on the subject, the bonded-silicas, in cartridge or in disc configuration, are the most commonly used supports, but they also have some limitations:

- For polar analytes, the retention is weak and often results in breakthrough during the loading step.
- Basic analytes interact strongly with the residual silanols, which in turn cause low recovery.
- The sorbent must remain wet prior to sample loading. (If one accidentally lets the cartridges run dry, the recovery is low and variable.)
- Poor stability in very acidic and basic media, which limits their use to the pH range of between 2 and 8.

These limitations have led to a search for new materials with improved characteristics. For example, the styrene-divinylbenzene resins have been extensively checked for their use in the extraction of pesticides. These polymers show higher retention of analytes and a wider pH range than $C_{18}$ silicas. The LC-grade polymers used as stationary phases have more commonly been used in precolumns (mainly PRP-1 and PLRP-S) for on-line purposes, because

**Table 1** Chemical structure of major classes of herbicides according to the character that determines the SPE procedure used

| Character | Class | Typical herbicide | Chemical structure |
|---|---|---|---|
| Basic/neutral | Triazine | Atrazine | |
| | Chloroacetamide | Metolachlor | |
| | Urea | Monuron | |
| | Carbamate | Desmedifam | |
| | Dinitroaniline | Pendimethalin | |
| Acid phenolic | Phenoxy acid | 2,4,5-T | |
| | Substituted phenol | Bromoxynil | |
| | Aryloxyphenoxy propanoic acid | Fluazifop | |
| Cationics | Bipyridilium compound | Diquat | |
| Very soluble | Organophosphate | Glyphosate | |

**Table 2** Relative standard deviation (RSD), recoveries (Rec.) and determination limits (DL, P 95%) for the preconcentration of basic/neutral pesticides from 1000-mL Milli-Q-water

| | Concentration range (ng L⁻¹) | Preconcentration on | | | | | | | | | | | |
|---|---|---|---|---|---|---|---|---|---|---|---|---|---|
| | | Bond Elut C₁₈ cartridges | | | Empore® C₁₈ discs | | | Empore® SDB discs | | | ENVI-Carb cartridges | | |
| | | RSD (%) | Rec. (%) | DL (ng L⁻¹) | RSD (%) | Rec. (%) | DL (ng L⁻¹) | RSD (%) | Rec. (%) | DL (ng L⁻¹) | RSD (%) | Rec. (%) | DL (ng L⁻¹) |
| Desisopropylatrazine | 54–270 | 9.9 | 33 | 133 | 10.6 | 21 | 143 | 15.9 | 16 | 221 | 6.9 | 99 | 96 |
| Desethylatrazine | 50–248 | 8.7 | 82 | 113 | 8.8 | 49 | 110 | 3.6 | 72 | 46 | 7.3 | 101 | 94 |
| Metoxuron | 108–538 | 1.9 | 95 | 58 | 3.4 | 106 | 95 | 4.8 | 94 | 134 | 5.7 | 106 | 159 |
| Hexazinon | 107–535 | 8.1 | 96 | 224 | 6.9 | 102 | 186 | 6.7 | 104 | 182 | 7.4 | 108 | 200 |
| Simazine | 49–245 | 1.5 | 102 | 20 | 4.5 | 102 | 57 | 5.2 | 100 | 66 | 9.3 | 95 | 117 |
| Metribuzin | 94–470 | 4.7 | 94 | 120 | 4.0 | 104 | 99 | 4.1 | 109 | 99 | 10.2 | 97 | 249 |
| Cyanazine | 57–285 | 7.9 | 81 | 119 | 5.0 | 110 | 73 | 6.3 | 93 | 91 | 5.6 | 82 | 81 |
| Carbofuran | 179–895 | 6.4 | 92 | 309 | 4.4 | 107 | 211 | 6.8 | 95 | 311 | 23.1 | 80 | 1057 |
| Methabenzthiazuron | 100–498 | 5.2 | 89 | 138 | 5.0 | 98 | 127 | 3.9 | 103 | 101 | 5.2 | 86 | 135 |
| Chlortoluron | 101–503 | 4.9 | 96 | 133 | 4.5 | 102 | 118 | 2.3 | 98 | 61 | 5.9 | 97 | 158 |
| Atrazine | 50–250 | 5.6 | 96 | 75 | 5.6 | 100 | 72 | 3.6 | 93 | 47 | 7.1 | 90 | 92 |
| Monolinuron | 149–745 | 4.0 | 96 | 166 | 13.7 | 105 | 473 | 5.7 | 114 | 221 | 7.7 | 89 | 296 |
| Diuron | 102–510 | 7.3 | 89 | 194 | 5.1 | 101 | 134 | 4.9 | 94 | 130 | 6.5 | 92 | 171 |
| Isoproturon | 109–543 | 6.7 | 90 | 191 | 5.7 | 102 | 158 | 7.3 | 95 | 199 | 6.0 | 100 | 164 |
| Metobromuron | 102–508 | 13.4 | 88 | 334 | 9.2 | 93 | 230 | 4.5 | 96 | 119 | 9.4 | 96 | 248 |
| Metazachlor | 123–165 | 7.3 | 85 | 235 | 6.5 | 106 | 204 | 2.2 | 100 | 72 | 5.3 | 89 | 177 |
| Sebutylatrazine | 53–265 | 4.9 | 97 | 69 | 4.8 | 103 | 66 | 4.3 | 102 | 58 | 4.1 | 97 | 56 |
| Terbutylatrazine | 53–263 | 6.3 | 104 | 87 | 6.6 | 104 | 89 | 5.7 | 98 | 77 | 6.2 | 92 | 83 |
| Linuron | 102–508 | 2.3 | 101 | 66 | 3.6 | 96 | 97 | 3.2 | 106 | 87 | 3.5 | 98 | 94 |
| Napropamide | 45–225 | 5.8 | 98 | 69 | 2.1 | 100 | 24 | 4.9 | 100 | 57 | 3.1 | 98 | 36 |
| Terbuconazol | 164–820 | 6.5 | 90 | 282 | 2.1 | 101 | 98 | 5.2 | 101 | 224 | 3.9 | 98 | 168 |
| Metolachlor | 127–633 | 3.3 | 103 | 115 | 6.3 | 100 | 204 | 7.1 | 104 | 226 | 5.2 | 89 | 166 |
| Propiconazol | 225–1125 | 4.9 | 96 | 316 | 2.3 | 99 | 156 | 4.7 | 107 | 290 | 6.3 | 97 | 391 |
| Dinosebacetate | 138–688 | 6.6 | 41 | 239 | 16.2 | 52 | 509 | 7.6 | 70 | 262 | 32.2 | 33 | 1108 |
| Parathion-ethyl | 148–740 | 4.6 | 90 | 189 | 8.0 | 101 | 294 | 7.1 | 97 | 264 | 6.0 | 88 | 224 |
| Pyrazophos | 125–623 | 7.1 | 82 | 230 | 3.6 | 93 | 120 | 5.1 | 107 | 165 | 32.8 | 38 | 1058 |
| Bifenox | 67–333 | 3.9 | 98 | 70 | 3.5 | 93 | 61 | 8.8 | 92 | 147 | 4.2 | 89 | 70 |
| Prosulfocarb | 140–698 | 4.8 | 91 | 182 | 11.6 | 94 | 385 | 5.5 | 72 | 198 | 6.2 | 66 | 224 |
| Pendimethalin | 121–603 | 8.0 | 84 | 248 | 9.9 | 93 | 290 | 4.5 | 107 | 142 | 10.4 | 62 | 327 |
| Trifluoralin | 125–625 | 8.3 | 63 | 268 | 14.2 | 21 | 415 | 0.0 | 0 | 0 | 10.9 | 60 | 458 |
| Triallate | 128–638 | 4.1 | 96 | 144 | 14.7 | 35 | 434 | 0.0 | 0 | 0 | 8.5 | 52 | 513 |
| Fluoroxypyrester | 81–403 | 3.2 | 86 | 71 | 5.1 | 89 | 105 | 6.0 | 98 | 123 | 8.3 | 52 | 170 |
| Beta-cyfluthrine | 190–950 | 3.7 | 61 | 202 | 8.1 | 82 | 383 | 2.2 | 89 | 122 | 15.0 | 42 | 818 |

(Reproduced with permission from Schülein J, Martens D, Spizauer P and Kertrup A (1995) Comparison of different solid phase extraction materials and techniques by application of multiresidue methods for the determination of pesticides in water. *Fresenius Journal of Analytical Chemistry* 352: 565–571.)

**Table 3** Relative standard deviation (RSD), recoveries (Rec.) and determination limits (DL, P 95%) for the preconcentration of acidic phenolic pesticides from 1000-mL Milli-Q-water

| | Concentration range (ng L⁻¹) | Preconcentration on | | | | | | | | | | |
| | | Bond Elut C$_{18}$ cartridges | | | Empore® C$_{18}$ discs | | | Empore-SDB discs | | | ENVI-Carb cartridges | | |
| | | RSD (%) | Rec. (%) | DL (ng L⁻¹) | RSD (%) | Rec. (%) | DL (ng L⁻¹) | RSD (%) | Rec. (%) | DL (ng L⁻¹) | RSD (%) | Rec. (%) | DL (ng L⁻¹) |
|---|---|---|---|---|---|---|---|---|---|---|---|---|---|
| Trifensulfuron-methyl | 52–258 | 5.3 | 38 | 71 | 7.4 | 93 | 99 | 5.2 | 68 | 67 | 5.3 | 76 | 69 |
| Metsulfuron-methyl | 53–263 | 6.7 | 43 | 92 | 4.8 | 85 | 66 | 5.0 | 71 | 66 | 9.5 | 77 | 126 |
| Dicamba | 91–464 | 3.9 | 63 | 96 | 3.8 | 46 | 93 | 4.9 | 115 | 116 | 6.5 | 97 | 153 |
| MCPA | 96–487 | 8.5 | 98 | 208 | 6.5 | 99 | 164 | 12.8 | 59 | 300 | 10.7 | 124 | 252 |
| Bromoxynil | 110–561 | 3.9 | 76 | 119 | 6.1 | 65 | 177 | 3.8 | 67 | 111 | 7.1 | 89 | 206 |
| Dichlorprop | 105–533 | 4.1 | 99 | 118 | 4.2 | 94 | 119 | 3.0 | 83 | 84 | 2.4 | 104 | 67 |
| Ioxynil | 88–446 | 6.6 | 100 | 151 | 4.7 | 95 | 110 | 6.3 | 76 | 140 | 7.0 | 91 | 156 |
| MCPB | 91–464 | 8.7 | 88 | 202 | 5.0 | 94 | 121 | 9.9 | 106 | 220 | 3.6 | 88 | 80 |
| Bifenox acid | 70–357 | 5.6 | 88 | 104 | 4.5 | 103 | 85 | 4.4 | 97 | 80 | 3.8 | 96 | 69 |
| Haloxyfop | 104–528 | 6.1 | 100 | 166 | 4.3 | 99 | 121 | 4.7 | 104 | 127 | 4.2 | 91 | 113 |

(Reproduced with permission from Schülein J, Martens D, Spizauer P and Kertrup A (1995) Comparison of different solid phase extraction materials and techniques by application of multiresidue methods for the determination of pesticides in water. *Fresenius Journal of Analytical Chemistry* 352: 565–571.)

they are too expensive for use in disposable SPE cartridges. Empore® extraction discs have recently become available, with styrene divinylbenzene (SDB) copolymer sorbents enmeshed in the matrix.

Recoveries of acid herbicides from water samples have been compared by using C$_{18}$ and SDB discs; the results of this comparison are controversial. Some authors documented an improvement in the recoveries of phenoxycarboxylic acid and phenols on SDB discs for the enrichment of samples down to 500 mL (see **Table 4**). The addition of salt considerably enhances the recovery and decreases the differences between the extraction efficiency of C$_{18}$ and resin discs. However, other authors reported that SDB discs showed worse recovery rates under acidic conditions, in comparison with C$_{18}$ when preconcentrations were carried out with 1-L samples (see Table 3). In any case, salting out the water sample enhances the retention of substances on both materials; this increases recovery rates for hydrophilic substances. However, salting out is avoided as it may introduce impurities into the samples. The addition of a small quantity of methanol or other organic solvent also enhances the recovery by the so-called 'dynamic solvation'. However, it is not recommended, as it produces a relatively early breakthrough of hydrophilic substances.

Graphitized carbon black (GCB) has been confirmed to be a valuable adsorbing material for SPE of pesticides in aqueous environmental samples. GBC cartridges proved to be more efficient than the more commonly used C$_{18}$ bonded-silica cartridges for the SPE of polar herbicides, whereas the extraction of non-polar compounds showed inferior results (see Table 3). Although GCB is known to behave as a natural reversed phase, it contains chemical heterogeneities on its surface, which are able to bind anions via electrostatic forces. GBC can behave as both a reversed-phase sorbent and an anion exchanger, retaining the acidic pesticides in their ionic form under acidic conditions. In this situation, the base–neutral/acid fractionation can be achieved by using solvent mixtures at different pHs.

Silica-based ion exchangers are found in disposable SPE cartridges. They are not widely used for the preconcentration of environmental samples owing to their low capacity. Strong anion exchanger discs have been used for the analysis of chlorinated acid and phenoxy acid herbicides. The main problem with these comes from the fact that environmental waters contain high amounts of inorganic ions, which overload the capacity of the sorbent.

Selective SPE from environmental waters has been accomplished by using different sorbents coupled in the same or in different cartridges.

**Table 4**  $C_{18}$ and resin recoveries and effect of salting water

| Analyte | Recoveries $\pm$ RSD (%; n = 3) | | | |
| --- | --- | --- | --- | --- |
| | $C_{18}{}^1$ | Resin[1] | $C_{18}{}^2$ | Resin[2] |
| Acifluorfen | 77 ± 20 | 82 ± 5 | 104 ± 5 | 121 ± 1 |
| Bentazon | 0 | ND | 90 ± 13 | 71 ± 5 |
| Chloramben | 8 ± 11 | 3 ± 15 | 72 ± 14 | 77 ± 7 |
| 2,4-D | 86 ± 12 | 83 ± 6 | 81 ± 8 | 94 ± 15 |
| Dalapon | 0 | 42 ± 25 | 12 ± 75 | 31 ± 30 |
| 2,4-DB | 81 ± 13 | 80 ± 14 | 118 ± 10 | 130 ± 8 |
| Dacthal | 53 ± 17 | 99 ± 8 | 67 ± 16 | 97 ± 5 |
| Dicamba | 73 ± 13 | 71 ± 14 | 83 ± 3 | 94 ± 15 |
| 3,5-Dichlorobenzoic acid | 70 ± 17 | 76 ± 2 | 86 ± 25 | 107 ± 20 |
| Dichloroprop | 77 ± 11 | 78 ± 3 | 85 ± 9 | 94 ± 10 |
| Dinoseb | 72 ± 16 | 75 ± 5 | 92 ± 26 | 85 ± 6 |
| Pentachlorophenol | 69 ± 14 | 70 ± 2 | 65 ± 15 | 73 ± 8 |
| Pichloram | 49 ± 19 | 74 ± 7 | 96 ± 24 | 99 ± 21 |
| 2,4,5-T | 76 ± 11 | 75 ± 14 | 93 ± 10 | 89 ± 5 |
| Silvex | 73 ± 14 | 74 ± 14 | 82 ± 9 | 80 ± 5 |

[1]Fortified, unsalted reagent water.
[2]Fortified reagent water with 20% (w/w) $Na_2SO_4$.
ND, no data.
(Reproduced with permission from Hodgeson J, Collins J and Bashe W (1994) Determination of acid herbicides in aqueous samples by liquid–solid disk extraction and capillary gas chromatography. *Journal of Chromatography A* 659: 395–401.)

One of the sorbents is non-specific, such as GCB, which traps the analytes of interest and many other compounds, while the other is more specific, such as a cation or anion exchanger, which retains and reconcentrates the analytes of interest. **Table 5** shows the recoveries of nine phenoxy acid herbicides extracted by one miniaturized cartridge containing 50 mg of GCB at the top and 70 mg of a silica-based strong anion exchanger (SAX) at the bottom, compared with $C_{18}$ and anion exchanger extraction. A large loss of dicamba and incomplete recovery of phenoxyacids were obtained by using the resin-based exchanger material.

Ammonium quaternary compounds and glyphosate constitute a special and complicated case. Their determination is very important because they are among the top herbicides used in the world. An important drawback in the preconcentration of such compounds from water is their high polarity. Several efforts have been made to analyse them in environmental samples.

SPE of ammonium quaternary herbicides has been mainly performed with silica, which is a well-known example of the solid phase using adsorption and ionic interaction mechanisms with the silanol groups. Recoveries are rather acceptable in the pH range 7.5–9. Taking into account that at pH values higher than 7 the silanol groups of the stationary phase are ionized, at these pH values the cation-exchange capacity of the solid-phase will be increased.

Glyphosate, due to its ionic form, can be preconcentrated using anionic and cationic resins. Derivatization of the analyte, prior to SPE of the water sample, seems to help the concentration from water samples.

**Table 5**  Recovery of herbicides from 200-mL groundwater samples by using one cartridge containing GCB and anion exchanger compared with that from two other extraction methods

| | % Recovery[1] | | | |
| --- | --- | --- | --- | --- |
| | $C_{18}$ | | Anion exchanger | GCB + anion exchanger |
| | pH 2 | pH 7.9 | | |
| Dicamba | 94.2 | < 10 | 23.0 | 97.3 |
| 2,4-D | 96.1 | < 10 | 78.3 | 98.4 |
| MCPA | 96.4 | < 10 | 77.7 | 97.6 |
| 2,4-DP | 97.7 | < 10 | 76.8 | 96.4 |
| MCPP | 94.4 | < 10 | 82.0 | 97.3 |
| 2,4,5-T | 93.5 | < 10 | 81.4 | 95.4 |
| 2,4-DB | 96.0 | < 10 | 82.3 | 98.9 |
| MCPB | 95.8 | < 10 | 81.5 | 96.5 |
| 2,4,5-T | 93.1 | < 10 | 77.3 | 95.2 |

[1]Mean values were calculated for two determinations.
(Reproduced with permision from Di Corcia A, Marchetti M and Sampieri R (1989) Extraction and isolation of phenoxyacid herbicides in environmental waters using two adsorbents in one minicartridge. *Analytical Chemistry* 61: 1363–1367.)

**Table 6** Desorption efficiencies from the solid phase and overall recoveries of herbicides

| Herbicide | Desorption efficiency (%) | | | | | | | | | | Recovery from filtered river water (%)[5] | |
| | Methanol | | Ethyl acetate | | Acetone (3 mL) | | Acetone | | Acetone (6 mL) | | | |
| | 3 mL | 6 mL | 3 mL | 6 mL | Mean[1] | RSD[2] | 3 mL + HX[3] | 3 mL + DCM[4] | Mean[1] | RSD[2] | Mean[1] | RSD[2] |
|---|---|---|---|---|---|---|---|---|---|---|---|---|
| ACN | 68.9 | 68.9 | 78.3 | 78.3 | 74.6 | 8.4 | 74.6 | 74.6 | 81.9 | 12 | 128 | 14 |
| Alachlor | 67.7 | 67.7 | 80.5 | 80.5 | 80.1 | 7.4 | 80.1 | 80.1 | 101 | 6.1 | 105 | 3.4 |
| Benfluralin | 48.1 | 48.1 | 73.4 | 73.4 | 55.0 | 13 | 69.6 | 70.3 | 67.9 | 1.8 | 83.6 | 2.2 |
| Bifenox | <5 | <5 | 78.3 | 78.3 | 69.2 | 9.1 | 69.2 | 79.0 | 92.9 | 8.2 | 98.1 | 5.7 |
| Bromobutide | 77.1 | 77.1 | 78.2 | 78.2 | 80.1 | 8.5 | 80.1 | 80.1 | 103 | 2.7 | 108 | 3.9 |
| Bromobutide-debromo | 76.4 | 76.4 | 76.6 | 76.6 | 79.5 | 7.8 | 79.5 | 79.5 | 92.7 | 7.8 | 101 | 7.1 |
| Butachlor | 50.6 | 64.0 | 79.6 | 79.6 | 73.4 | 9.3 | 83.5 | 83.1 | 100 | 5.2 | 112 | 1.1 |
| Butamifos | 73.9 | 73.9 | 81.4 | 81.4 | 81.2 | 9.2 | 81.2 | 81.2 | 96.5 | 3.9 | 104 | 2.6 |
| Chlomethoxyfen | <5 | <5 | 74.9 | 74.9 | 70.9 | 5.9 | 70.9 | 83.2 | 91.3 | 8.7 | 97.9 | 6.7 |
| Chlornitrofen | <5 | <5 | 77.8 | 77.8 | 63.2 | 13 | 70.6 | 80.5 | 71.9 | 3.2 | 81.6 | 4.4 |
| Chlorprofam | 57.5 | 57.5 | 81.1 | 81.1 | 68.2 | 5.1 | 68.2 | 79.1 | 95.7 | 5.4 | 103 | 4.6 |
| Dimepiperate | <5 | 42.2 | 80.1 | 80.1 | 71.7 | 7.8 | 84.0 | 85.8 | 104 | 8.9 | 110 | 4.8 |
| Dimethametryn | 64.5 | 64.5 | 83.1 | 83.1 | 79.4 | 6.9 | 79.4 | 79.4 | 96.4 | 3.2 | 105 | 5.8 |
| Dithiopyr | 73.5 | 73.5 | 79.5 | 79.5 | 74.4 | 6.6 | 74.4 | 82.4 | 84.1 | 8.6 | 85.9 | 5.0 |
| Esprocarb | <5 | 35.9 | 76.2 | 76.2 | 63.4 | 13 | 77.5 | 78.3 | 87.3 | 3.6 | 94.1 | 0.23 |
| MCPA-ethyl | 22.5 | 47.3 | 72.6 | 72.6 | 57.9 | 12 | 70.3 | 74.8 | 89.7 | 5.8 | 89.3 | 5.3 |
| MCPA-thioethyl | <5 | <5 | 76.6 | 76.6 | 57.9 | 14 | 74.8 | 82.6 | 98.1 | 4.0 | 98.9 | 6.5 |
| Mefenacet | <5 | 55.4 | 78.2 | 78.2 | 82.5 | 2.5 | 82.5 | 82.5 | 90.7 | 2.9 | 103 | 12 |
| Molinate | 25.8 | 45.9 | 76.8 | 76.8 | 64.1 | 9.7 | 75.9 | 80.5 | 90.8 | 7.0 | 98.7 | 6.2 |
| Naproanilide | <5 | 60.0 | 76.8 | 76.8 | 79.8 | 2.4 | 79.8 | 79.8 | 96.0 | 3.1 | 103 | 9.5 |
| Oxadiazon | 38.0 | 59.4 | 76.1 | 76.1 | 70.8 | 6.7 | 83.2 | 84.0 | 93.2 | 1.8 | 96.8 | 1.6 |
| Pendimethalin | <5 | 42.9 | 92.6 | 92.6 | 71.1 | 8.9 | 86.4 | 86.5 | 73.2 | 7.3 | 86.4 | 2.8 |
| Piperophos | 54.1 | 71.8 | 79.9 | 79.9 | 85.9 | 5.7 | 85.9 | 85.9 | 93.5 | 5.3 | 100 | 9.1 |
| Pretilachlor | 64.1 | 64.1 | 78.5 | 78.5 | 79.8 | 8.6 | 79.8 | 79.8 | 108 | 6.5 | 112 | 2.7 |
| Prometryn | 70.9 | 70.9 | 79.4 | 79.4 | 82.8 | 8.2 | 82.8 | 82.8 | 93.6 | 4.5 | 101 | 4.1 |
| Simazine | 84.6 | 84.6 | 78.1 | 78.1 | 78.9 | 11 | 78.9 | 78.9 | 102 | 9.6 | 112 | 7.3 |
| Simetryn | 78.8 | 78.8 | 77.9 | 77.9 | 80.4 | 9.2 | 80.4 | 80.4 | 93.5 | 9.4 | 120 | 11 |
| Thiobencarb | <5 | 30.7 | 80.3 | 80.3 | 64.4 | 13 | 77.6 | 80.6 | 97.6 | 5.1 | 110 | 0.63 |
| Trifluralin | 50.3 | 50.3 | 76.8 | 76.8 | 57.3 | 13 | 71.1 | 72.3 | 72.1 | 1.2 | 87.1 | 5.8 |

[1]Percentage mean recovery ($n = 3$).
[2]Percentage relative standard deviation.
[3]A 3-mL volume of hexane.
[4]A 3-mL volume of dichloromethane.
[5]The herbicides collected on the cartridges were eluted with 6 mL of acetone.
(Reproduced with permission from Tanabe A, Mitobe H, Kawata K and Sakai M (1996) Monitoring of herbicides in river water by gas chromatography–mass spectrometry and solid phase extraction. *Journal of Chromatography A* 754: 159–168.)

It can be concluded that $C_{18}$ material is inappropriate for some herbicides, especially more polar and very non-polar herbicides. In these cases, the SDB polymers and GCB offer a valuable alternative. The appropriate choice of solid phase for application to a separation problem will vary from case to case and must be adapted accordingly.

## Elution of the Target Analytes

Desorption of the compounds from the concentration columns is mainly performed with a small volume of liquid. The partition coefficient in a given solid-phase eluent system should favour the shift of the studied herbicides. On the other hand, SPE is not a separate step, but it is part of a process that includes subsequent determination and so it should be taken into account that some determination systems, such as GC, are incompatible with the presence of water. In this way, the selection of the eluting solvent depends on the selected sorbent, the analytes and the detection method. Air-drying is often applied before analyte elution in order to remove residual water.

Methanol and acetonitrile are recommended solvents for the elution of herbicides adsorbed to $C_8$ or $C_{18}$ silicas. Dichloromethane and ethyl acetate have also been extensively used, especially when the presence of water is undesirable. **Table 6** presents the results obtained when herbicides were eluted from a cartridge using different solvents, such as methanol, ethyl acetate, acetone, hexane and dichloromethane following acetone.

Desorption of acid herbicides from the sorbents can also be performed using a solution adjusted to a pH where the analytes are in their ionic form (two units below or above the p$K_a$). The uniqueness of GCB is that acid compounds are retained in their ionic forms and neutral compounds are adsorbed by unspecific mechanisms. In this situation, base–neutral/acid fractionation can be easily achieved by first eluting base–neutral species with a neutral organic solvent mixture and then passing a basified or acidified solvent system to desorb acidic compounds. **Table 7** reports the results obtained using base–neutral/acid fractionation in three kinds of GCB. In all cases, there was some carryover of 2,4-DB, which is the weakest compound included in this table.

With ion exchange sorbents, the analytes can be eluted from the SPE column by either adjusting the pH in order to neutralize the charge on the analyte or by using a buffer of high ionic strength.

## Sample Requirements

Samples undergoing SPE need to be filtered to separate suspended matter. Filtration is especially necessary before extraction of surface water, but is also

**Table 7** Base–neutral/acid fractionation by differential elution of selected compounds with cartridges containing three different types of GCB at two eluents

| Compound | Sorbent material | | | | | |
|---|---|---|---|---|---|---|
| | Carbograph 1 | | Carbograph 4 | | Carbograph 5 | |
| | Eluent A[1] | Eluent B[3] | Eluent A | Eluent B | Eluent A | Eluent B |
| *Base/neutral* | | | | | | |
| Atrazine | 97 | — | 95 | — | 94 | — |
| Linuron | 99 | — | 98 | — | 95 | — |
| Aldicarb | 92 | — | 92 | — | 92 | — |
| *Acidic* | | | | | | |
| Dichlorprop (3.5)[3] | — | 95 | — | 97 | 30 | 73 |
| 2,4,5-T (2.2) | — | 97 | — | 102 | — | 99 |
| Ioxynil (3.9) | — | 101 | — | 102 | — | 93 |
| 2,4-D (2.6) | — | 99 | — | 100 | — | 93 |
| 2,4-DB (4.8) | 40 | 63 | 18 | 81 | 50 | 49 |
| Mecoprop (3.7) | — | 99 | — | 99 | — | 96 |

Extraction from 1 L of Aldrich humic acid-spiked drinking water (spiked level, 10 µg L$^{-1}$). Mean recovery values obtained from three measurements.
[1]Eluent phase: $CH_2Cl_2$–$CH_3OH$ (80 : 20).
[2]Eluent phase: $CH_2Cl_2$–$CH_3OH$ (80 : 20) + 10 mmol L$^{-1}$ tetrabutylammonium chloride (TBACl).
[3]Reported p$K_a$ values of the acidic compounds are given in parentheses.
(Reproduced with permission from Crescenzi C, Di Corcia A, Passariello G, Samperi R and Turnes MI (1996) Evaluation of two new examples of graphitized carbon blacks for use in solid-phase extraction cartridges. *Journal of Chromatography A* 733: 41–55.)

often advisable for extraction of ground water to avoid blocking up the cartridge material. However, waters from different sources are very different in chemical composition. Matrix effects from the water itself can cause errors in quantitation and determination. The presence in waters of common contaminants (natural or xenobiotics), such as humic acids, surfactants, inorganic salts, phenols, polycyclic aromatic hydrocarbons (PAH), other pesticides and related compounds, can negatively affect the analysis, significantly diminishing the recovery efficacy or interfering with the posterior determination. **Figure 1** shows that when natural samples are acidified, humic and fulvic acids are

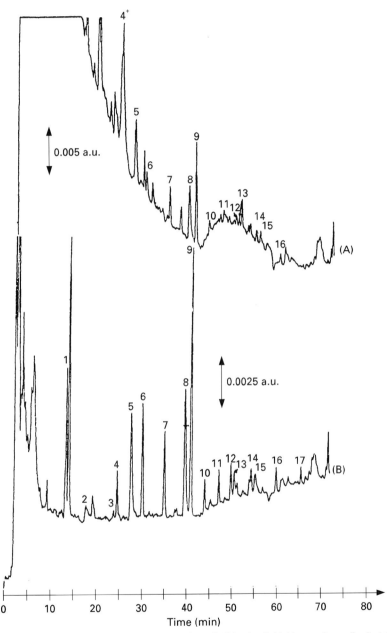

**Figure 1**  Effect of the pH of the sample on the preconcentration of 500 mL of drinking water spiked at 0.1 µg L$^{-1}$. Sample (A) adjusted to pH 3 with perchloric acid and (B) not adjusted (pH 7). Analytical conditions: flow-rate, 1 mL min$^{-1}$, loop, 50 µL; mobile phase, acetonitrile gradient with 0.005 M phosphate buffer acidified to pH 3 with HClO$_4$, gradient from 10 to 30% acetonitrile from 0 to 10 min, and from 30 to 77% from 10 to 80 min; UV detection at 220 nm. Peaks: 1, chloridazon; 2, dicamba; 3, aldicarb; 4, methoxuron; 5, simazine; 6, cyanazine; 7, bentazone; 8, atrazine; 9, carbaryl; 10, isoproturon; 11, ioxynil; 12, MCPP; 13, difenoxuron; 14, 2,4-DB; 15, 2,4,5-T; 16, metolaclor; 17, dinoterb. (Reproduced with permission from Pichon V, Cau Dit Coumes C, Chen L, Guenu S and Henion MC (1996) Simple removal of humic and fulvic acid interferences using polymeric sorbents for the simultaneous solid-phase extraction of polar acidic, neutral and basic pesticides. *Journal of Chromatography A* 737: 25–33.)

**Table 8**   Recoveries of cationic herbicides (4 µg L$^{-1}$) from 0.25 L of water samples containing various concentrations of different surfactants

| Surfactants | Concentration (µg L$^{-1}$) | Recovery (%)[1] | | |
|---|---|---|---|---|
| | | Diquat | Paraquat | Difenzoquat |
| Cetrimide | 5 | 98 | 99 | 90 |
| | 50 | 93 | 92 | 92 |
| | 300 | 95 | 91 | 93 |
| | 3000 | 87 | 89 | 92 |
| Benzalkonium chloride | 5 | 114 | 114 | 94 |
| | 50 | 107 | 107 | 93 |
| | 300 | 102 | 102 | 95 |
| | 3000 | 105 | 105 | 103 |
| Sodium tetradecyl sulfate | 5 | 99 | 102 | 87 |
| | 50 | 98 | 95 | 93 |
| | 300 | 41 | 47 | 41 |
| | 3000 | 34 | 30 | 37 |
| Lauryl sulfate | 5 | 84 | 85 | 92 |
| | 50 | 109 | 100 | 93 |
| | 300 | 42 | 35 | 41 |
| | 3000 | 34 | 30 | 39 |
| Laurylsulfobetaine | 5 | 89 | 83 | 79 |
| | 50 | 91 | 94 | 84 |
| | 300 | 47 | 57 | 42 |
| | 3000 | 54 | 55 | 48 |
| Brij-35 | 5 | 85 | 89 | 92 |
| | 50 | 83 | 101 | 91 |
| | 300 | 45 | 33 | 37 |
| | 3000 | 36 | 30 | 42 |
| Triton X-100 | 5 | 86 | 89 | 92 |
| | 50 | 102 | 101 | 91 |
| | 300 | 45 | 33 | 37 |
| | 3000 | 36 | 30 | 42 |

[1]Average recovery calculated from four determinations.
(Reproduced with permission from Ibáñez M, Picó Y and Mañes J (1996) Influence of organic matter and surfactants on solid-phase extraction of diqua, paraquat and difenzoquat from waters. *Journal of Chromatography A* 727: 245–252.)

co-extracted and co-eluted, which generates a large, unresolved peak in the chromatogram when HPLC with UV detection is used (chromatogram A). At pH 7, humic and fulvic acids are not co-extracted, as can be seen by the flat baseline from the beginning to the end of chromatogram (B).

Organic matter and anionic or non-ionic surfactants have demonstrated negative effect on the recovery of any class of herbicides. Although these undesirable effects are well known, only a few analytical studies have focused on ways in which to avoid them. The proposed methods for removing interferences are based on the use of chemical reagents, such as sulfite or cationic surfactants. In these cases, the recovery values after chemical treatment were similar to those when a Milli-Q-quality water standard was analysed. The recoveries reported in **Table 8** show that the quantitative SPE of diquat, paraquat and difenzoquat is affected by the presence of anionic, zwitterionic and non-ionic surfactants when they are present in water at a level of up to 50 µg L$^{-1}$.

Although some common contaminants of natural waters have a negative effect on the recoveries, SPE is useful for analysing herbicides in drinking and surface water because only in very extreme conditions does the concentration of these contaminants reach levels at which recoveries are significantly decreased.

The application of SPE to the isolation of herbicide residues from other matrices presents difficulties that must be overcome, which have, up to now, discouraged investigation into the use of other matrices. For liquid matrices (plasma, urine, blood or milk), acceptable recoveries have been obtained using protein precipitation prior to SPE but the impurities present can accumulate in the analytical columns and affect the chromatogram. The recoveries obtained by SPE for determining triazines from milk are compared with those obtained by liquid–liquid extraction (Hajšlova *et al.*) in **Table 9**. SPE was performed using a double trap: first, a non-specific adsorbent (GCB), and then a cation exchanger. The liquid–liquid extraction method, after an initial double protein precipitation using methanol in

**Table 9** Recovery ($n = 6$) of triazines from fortified (50 ng mL$^{-1}$) skimmed milk using the proposed method and that of Hajšlová et al.

| Compounds | Recovery % (mean ± RSD) | |
|---|---|---|
| | SPE method | Hajšlová et al. |
| Simazine | 89.7 ± 4.1 | 86.1 ± 5.2 |
| Atrazine | 89.3 ± 3.9 | 84.3 ± 4.7 |
| Prometon | 90.4 ± 4.0 | 92.4 ± 4.3 |
| Ametryn | 89.5 ± 3.5 | 90.0 ± 4.7 |
| Propazine | 93.4 ± 3.6 | 88.6 ± 4.2 |
| Terbutylazine | 91.6 ± 3.4 | 87.3 ± 4.2 |
| Prometryn | 87.2 ± 3.8 | 85.5 ± 4.0 |
| Terbutryn | 77.8 ± 3.2 | 80.9 ± 3.9 |

(Reproduced with permission from Lagana A, Marino A and Fago C (1995) Evaluation of double solid-phase extraction system for determining triazine herbicides in milk. *Chromatographia* 41: 178–182.)

basic and acid environments, used a partition with chloroform followed by a sample clean up using a silica cartridge. There were no significant differences in the triazine recovery using the two methods.

Solid matrices can also be extracted by SPE with cartridge or disc devices but require a separate homogenization step and other laborious processes. The reported recoveries are lower than those obtained with water, and the addition of methanol or acetonitrile as organic modifier is necessary. However, these recoveries are comparable to those obtained by other well known extraction methods for solid matrices. **Table 10** gives a comparison of the features of three extraction procedures for tribenuron methyl analysis in soil. Solid-phase and supercritical fluid extractions are the most adequate in terms of recovery percentage and precision, with acceptable detection limits; nevertheless, the recovery is affected by the amount of herbicide present in soil. SPE can also be performed by blending directly a homogenized sample with C$_{18}$ sorbent, transferring the mixture to a glass chromatography column and eluting the analytes with appropriate solvent.

The SPE of matrices other than water requires further investigation.

## On-line and Off-line Procedures

Nowadays, SPE methods using off-line procedures can be converted into on-line SPE methods by direct connection of the precolumn to the analytical column via switching valves. The concentrated analytes are then directly desorbed and transferred to the analytical system. Such systems often involve microprocessor control of the stages for sample switching and flushing of solvents and eluents through the concentration and chromatographic columns.

On-line procedures have gained popularity since European Union (EU) guidelines were introduced which limited the maximum amount allowed for a single pesticide in drinking water to 0.1 µg L$^{-1}$ and for several pesticides to 0.5 µg L$^{-1}$, including toxic transformation products. Very sensitive methods are required for monitoring herbicide residues in drinking water at such low concentrations. Furthermore the recent commercialization of automatic devices has certainly helped in the development of on-line trace enrichment methods in environmental analysis, because the sequence can be totally automated using systems such as the Prospect module (Spark Holland) or the OSP-2 system (Merck).

On-line SPE-LC is the most common procedure used because it is easily performed in any laboratory. The extracted compounds are eluted directly from the precolumn to the analytical column by a suitable mobile phase, which permits the separation of the trapped compounds. It is well established that on-line procedures enable lower concentrations of pesticides to be determined, and most compounds can be kept within EU limits. **Table 11** illustrates the improvement in detection limits obtained for triazine and phenylurea herbicides using on-line procedures when compared with off-line ones.

Breakthrough is the key parameter in on-line SPE because it indicates the sample volume and the amount of analyte that can be preconcentrated. Two factors can be responsible for breakthrough: insufficient retention of the analytes by the sorbent and overloading of the sorbent. One important factor of the concentration procedure is the selection of the

**Table 10** Comparison of the extraction procedures for tribenuron methyl analysis

| Extraction | Efficacy | Precision | Selectivity | Operation time | Affecting factors | Detection limit |
|---|---|---|---|---|---|---|
| Solvent | + | + + | + + + | + + | No data | No data |
| Solid-phase | + + | + + + | + + + | + + | Concentration | + + + |
| Supercritical fluid | + + | + + + | + + + | + + + | Concentration | + + + |

+ , Bad; + + , regular; + + + , good.
(Reproduced with permission from Berna JL, Jiménez JJ, Herguedas A and Atienza J (1997) Determination of chlorsulfuron and tribenuron-methyl residues in agricultural soils. *Journal of Chromatography A* 778: 119–125.)

**Table 11**  Range of linearity, $r^2$ and detection limit (LOD) for the on-line method

| Pesticide | Off-line method | | | On-line method | | |
|---|---|---|---|---|---|---|
| | Range of linearity ($\mu g\,L^{-1}$) | $r^2$ | LOD ($\mu g\,L^{-1}$) | Range of linearity ($\mu g\,L^{-1}$) | $r^2$ | LOD ($\mu g\,L^{-1}$) |
| Simazine | 0.5–50 | 0.9985 | 0.1 | 0.1–8 | 0.9990 | 0.03 |
| Cyanazine | 0.5–50 | 0.9973 | 0.1 | 0.1–8 | 0.9987 | 0.03 |
| Chlortoluron | 0.5–50 | 0.9960 | 0.1 | 0.2–8 | 0.9956 | 0.05 |
| Atrazine | 0.5–50 | 0.9980 | 0.1 | 0.1–8 | 0.9999 | 0.03 |
| Isoproturon | 1.0–50 | 0.9990 | 0.1 | 0.2–8 | 0.9993 | 0.05 |
| Ametryn | 0.5–50 | 0.9985 | 0.05 | 0.1–8 | 0.9993 | 0.03 |
| Prometryn | 0.5–50 | 0.9989 | 0.05 | 0.1–8 | 0.9995 | 0.03 |
| Terbutryn | 0.5–50 | 0.9980 | 0.05 | 0.1–8 | 0.9985 | 0.03 |
| Chlorpyriphos-methyl | 2.0–50 | 0.9962 | 0.5 | 0.5–8 | 0.9980 | 0.20 |
| Fenitrothion | 2.0–50 | 0.9983 | 0.5 | 0.5–8 | 0.9993 | 0.20 |
| Fenchlorphos | 5.0–50 | 0.9950 | 1.0 | 1.0–8 | 0.9927 | 0.30 |
| Parathion-ethyl | 5.0–50 | 0.9944 | 1.0 | 1.0–8 | 0.9995 | 0.30 |

(Reproduced with permission from Aguilar C, Borrull F and Marcé RM (1996) On-line and off-line solid-phase extraction with styrene-divenylbenzene-membrane extraction disks for determining pesticides in water by reversed-phase liquid-chromatography-diode array detection. *Journal of Chromatography A* 754: 77–84)

sorbent, which must allow a convenient break-through of the analytes. **Table 12** shows a comparison of SPE sorbents for analysis of phenyl carbamate herbicides. The results were unsatisfactory with some herbicides. GCB is not used much in online SPE because it is not sufficiently pressureresistant.

Another factor in the procedure is to evaluate the maximum sample volume that can be preconcentrated without breakthrough of analytes, thus avoiding peak broadening. Generally, 50 mL was considered as optimum, but it could be increased for a particular kind of herbicide.

It should be taken into account that sorbents used in on-line SPE are not selective and numerous compounds from the matrix of natural samples are preconcentrated and can be eluted with the analytes of interest. Interferences depend on the nature of the water. They have an effect on both detection limits and quantification. **Figure 2** shows some chromatograms obtained with different waters. In spite of the presence of interference peaks, it can be seen that making a good choice of preconcentration parameter and analytical conditions, allows low levels of many pesticides to be determined, even in highly contaminated surface waters.

In this way, the EPA in the United States currently offers an on-line SPE procedure followed by HPLC for the analysis of acidic herbicides in drinking water. The sample is first adjusted to pH 12 to hydrolyse esterified analytes, then it is acidified to a pH of 1 and a 20-mL aliquot is pumped through a reversed-phase concentration column. By use of a switching valve, the concentration column is then pumped in line with the analytical column and the sample constituents are then passed to the analytical column for separation and detection.

**Table 12**  Average recoveries and RSDs (%) of the analytes by the proposed on-line SPE-LC-DAD procedures in environmental water samples spiked at different levels

| Compound | $C_{18}$ pre-column | | | | PRP-1 pre-column | | | |
|---|---|---|---|---|---|---|---|---|
| | Drinking water | | Surface water | | Drinking water | | Surface water | |
| | 0.5 $\mu g\,L^{-1}$ | 4 $\mu g\,L^{-1}$ | 0.5 $\mu g\,L^{-1}$ | 4 $\mu g\,L^{-1}$ | 0.2 $\mu g\,L^{-1}$ | 1 $\mu g\,L^{-1}$ | 0.2 $\mu g\,L^{-1}$ | 1 $\mu g\,L^{-1}$ |
| Carbetamide | — | 105 (3) | — | 105 (4) | 84 (12) | 101 (3) | 102 (8) | 101 (3) |
| Propham | 101 (2) | 98 (3) | 99 (3) | 97 (3) | 90 (5) | 102 (5) | 97 (6) | 102 (3) |
| Desmedipham | 84 (9) | 86 (8) | 94 (7) | 98 (7) | — | | | |
| Phenmedipham | 87 (2) | 97 (6) | 98 (10) | 108 (7) | 87 (3) | 101 (2) | 93 (6) | 104 (3) |
| Chlorbufam | 105 (5) | 99 (2) | 102 (4) | 97 (1) | 106 (7) | 101 (3) | 99 (5) | 105 (2) |
| Chlorpropham | 103 (2) | 99 (1) | 108 (3) | 106 (2) | 105 (5) | 101 (4) | 99 (5) | 108 (2) |

(Reproduced with permission from Hidalgo C, Sancho JV, López FJ, and Hernández F (1998) Automated determination of phenylcarbamate herbicides in environmental water by on-line trace enrichment and reversed-phase liquid chromatography-diode array detection. *Journal of Chromatography A* 823: 121–128.)

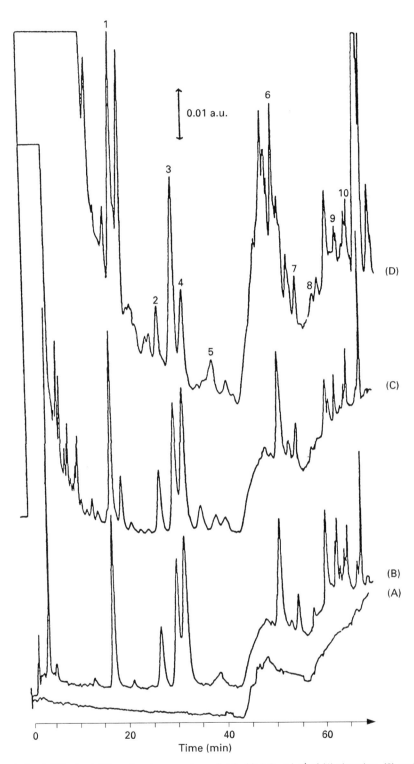

**Figure 2**   On-line analysis of 150 mL of different water samples spiked with 0.3 µg L$^{-1}$ of (1) simazine, (2) methabenzthiazuron, (3) atrazine, (4) carbaryl, (5) isoproturon, (6) propanil, (7) linuron, (8) fenamiphos, (9) fenitrothion and (10) parathion. Precolumn, PRLP-S. (A) Blank gradient; (B) Milli-Q-purified water; (C) drinking water; (D) surface water from the Seine (28 June 1993). (Reproduced with permission from Pichon V and Henion MC (1994) Determination of pesticides in environmental water by automated on-line trace-enrichment and liquid chromatography. *Journal of Chromatography A* 665: 269–281.)

On-line SPE-GC is another interesting approach that has gained popularity over the last few years. The SPE-GC coupled techniques generally use an uncoated, deactivated capillary precolumn, also known as a retention gap, which accommodates the liquid SPE eluent while it vaporizes, thereby providing

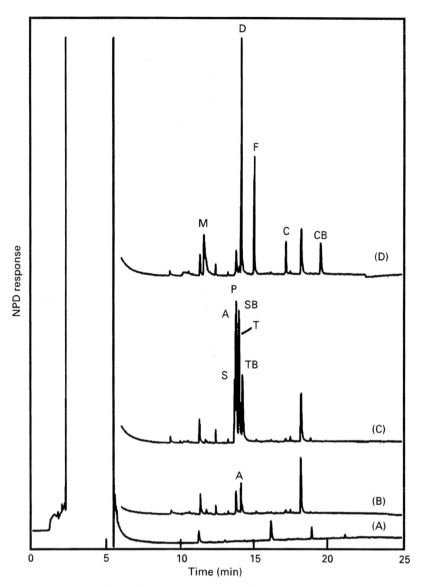

**Figure 3** SPE-GC-NPD chromatograms obtained after preconcentration of 10 mL of (A) HPLC grade water, (B) Amsterdam drinking water, and drinking water spiked with (C) triazines (0.1 μg L$^{-1}$) and (D) OPPs (0.03 μg L$^{-1}$). Peak assignment for the herbicides: S, simazine; A, atrazine; P, propazine; SB, secbumeton; T, trietazine; and TB, terbutylazine. GC programme: 75°C during sample introduction, then to 300°C at 15°C min$^{-1}$; held at 300°C for 5 min. (Reproduced with permission from Picó Y, Louter AJH, Vreuls JJ and Brinkman UATh (1994) On-line trace-level enrichment gas chromatography of triazine herbicides, organophosphorus pesticides, and organosulfur compounds from drinking and surface waters. *Analyst* 119: 2025–2031.)

solute preconcentration. **Figure 3** shows typical results for the SPE-GC-nitrogen phosphorus detector (NPD) analysis of triazines. The most striking observation is the good baseline stability, because NPD is a very selective detector. The drawback of this technique is the high cost involved, which makes it unaffordable by most of the laboratories involved in herbicide analysis.

The analysis of herbicides, using an automated on-line solid-phase extraction device results in:

- a reduction in error
- a more efficient use of time
- savings in amount of solvent used
- an improved chromatographic separation
- a reduction of sample volume needed to achieve good results (up to 200 mL)
- a ten-fold improvement detection limit over that required by EPA and EU regulations (limit values)

The advantages cited for on-line procedures are convenient for some analysts, but many prefer the off-line approach, which gives a convenient extract in an organic solvent suitable for multiple analyses. Moreover, such an extract is generally much more stable than the aqueous sample from which it was derived, and is therefore more suitable for long-term storage. Also, the off-line approach allows the processing of many samples at one time, an approach which is generally more productive in laboratories that are not fully automated.

Both off-line and on-line techniques are not mutually exclusive. The possibility of employing both methods gives the analyst more tools at his/her disposal for performing the analysis adequately.

## Future Developments

Today, SPE has become generally accepted as the analytical method of choice for determination of all major herbicide groups in water. It is suitable for detecting approximately 300 pesticides and pesticide-related compounds and has undergone rigorous multi-laboratory calibration studies. SPE is also the backbone of residue analysis protocols for government agencies such as the EPA in the US. However, there is still much to be done. The development of new, more selective supports for SPE, its coupling with high separation power techniques, such as capillary electrophoresis (CE), and its application to extract herbicide residues from solid samples, may further reduce the detection limit and will represent an exciting challenge for researchers working in the area of herbicide residue analysis.

Looking to the future, it is interesting to note that new SPE sorbents involving antigen–antibody interaction, so-called immunosorbents, have been described. Due to their high affinity and high selectivity for these interactions, extraction and clean up of complex aqueous environmental samples is achieved in the same step. Their application to extracts from solid samples is solvent-free and simpler than any other clean-up procedure. Two class-selective immunosorbents have been optimized up to now that enable the trapping of two groups of widely used herbicides, phenyl urea and triazines.

Experiments have been recently designed to explore the possibility of recovering herbicide residues from food, soil, biological liquid and tissue samples by SPE. For liquid matrices, such as plasma, urine, fruit juice or milk, acceptable residue recovery may be obtained almost without clean up. Before SPE can be used with solid matrices (e.g. muscle, vegetables or soil) a separate homogenization step and often multiple filtration, sonication and centrifugation are required. Despite these drawbacks, SPE has been used a few times to extract residues of triazines, carbamates, ureas and other herbicides. More work is needed to further develop SPE for use with the many different types of matrices that may contain herbicide residues.

Capillary electrophoresis (CE) is very much suited for those analytes that are not amenable to GC or when existing LC methods do not offer sufficient separation power. Many impressive CE separations, including the separation of triazines and sulfonylureas, have been demonstrated in the last few years. The main disadvantages of these techniques are its inadequate detection limits and lack of selective detectors for the determination of residues in environmental matrices. As a result of coupling with SPE, use of CE has become competitive in trace analysis, and the door has been opened to environmental applications in real matrices. Thus, the potential of CE is very good indeed.

*See also:* **II/Extraction:** Solid-Phase Extraction. **III/Immunoaffinity Extraction. Porous Graphitic Carbon: Liquid Chromatography. Solid-Phase Extraction with Cartridges. Sorbent Selection for Solid-Phase Extraction.**

## Further Reading

Barceló D (ed) (1993) *Environmental Analysis. Techniques, Applications and Quality Assurance.* Amsterdam: Elsevier.

Barceló D and Hennion MC (eds) (1997) *Trace Determination of Pesticides and Their Degradation Products in Water.* Amsterdam: Elsevier.

Berrueta LA, Gallo B and Vicente F (1995) A review of solid-phase extraction: basic principles and new developments. *Chromatographia* 40: 474–483.

Csarhárti T and Forgács E (1998) Phenoxyacetic acids: separation and quantitative determination. *Journal of Chromatography A* 717: 157–178.

Dean JR, Wade G and Barnabas IJ (1996) Determination of triazine herbicides in environmental samples. *Journal of Chromatography A* 733: 295–335.

Font G, Mañes J, Moltó JC and Picó Y (1993) Solid phase extraction in multi-residue pesticide analysis of water. *Journal of Chromatography A* 642: 135–161.

International Union of Pure and Applied Chemistry (1994) Analyte isolation by solid-phase extraction (SPE) on silica-bonded phases. Classification and recommended practices. *Pure and Applied Chemistry* 62: 277–304.

Liska I, Kupcik J and Leclercq PA (1989) The use of solid sorbents for direct accumulation of organic compounds from water matrices. A review of solid-phase extraction techniques. *Journal of High Resolution Chromatography* 12: 577–590.

Nollet LML (ed.) (1996) *Handbook of Food Analysis.* New York: Marcel Dekker.

Pichon V (1998) Multiresidues solid-phase extraction for trace analysis of pesticides and their metabolites in environmental waters. *Analusis* 26: M91–M98.

Picó Y, Moltó JC, Mañes J and Font G (1994) Solid phase techniques in the extraction of pesticides and related compounds from food and soils. *Journal of Microcolumn Separations* 6: 331–359.

Tekel, J and Kovačičová (1993) Chromatographic methods in the determination of herbicide residues in crops, food and environmental samples. *Journal of Chromatography A* 643: 291–303.

Zweig G and Sherma J (Eds) (1986) *Analytical Methods for Pesticides and Plant Growth Regulations*. New York: Academic Press.

# Thin-Layer (Planar) Chromatography

**V. Pacáková**, Charles University, Prague, Czech Republic

## Introduction

Modern agricultural production in major agriculture countries depends heavily on the use of pesticides. Herbicides represent more than 50% of all pesticides used (in the USA and Germany the figure is about 60%) and are found in soil, ground and surface water and food. Triazines are the most common herbicides, and they represent c. 30% of all herbicides used. As they are relatively stable they are also the ones most commonly found in the environment. Triazine toxicity is low and they do not usually present a risk to humans. However, as they are extensively used, their presence in the environment and in food must be monitored (triazine residues, especially those of atrazine, can be found, in milk, butter and sugar). Phenylureas are replacing triazines as they are easily degraded, but this fact makes their analysis more difficult. Phenoxyacids are the oldest synthetic herbicides. Other herbicides include triazones, carbamates, uracils, pyridazines, substituted ureas and anilines (**Figure 1**).

Methods used to analyse herbicides and their residua are similar to those used for other pesticides. The methods applied should be able to determine multicomponent pesticide mixtures simultaneously, have a good reproducibility and a high recovery and a low limit of determination (the maximum permissible concentration in drinking water is as low as $0.1\ \mu g\ L^{-1}$).

## Principles of Thin Layer Chromatography and High Performance TLC (HPTLC)

Thin-layer chromatography (TLC) remains an important practical analytical method for the analysis of herbicides with well-developed standard procedures. Its main advantages for this type of analysis are simple equipment, the possibility of varying a large number of the experimental parameters, high throughput (up to 36 samples can be analysed simultaneously), fast analyses, economy and low solvent consumption. Sample clean-up is either simple or not required at all.

TLC thus provides a simple and inexpensive screening method for the analysis of herbicides.

In classical TLC the samples are applied to the thin layer and then the layer is developed by a mobile phase (a solvent or a solvent mixture). After the mobile phase is evaporated, the separated zones are evaluated. As capillary forces govern the migration of the solute through the stationary phase, the mobile-phase velocity is less than optimal. Higher velocity can be obtained by forced-flow development.

High performance TLC (HPTLC), which has developed from classical TLC, offers greater separation efficiency, greater sensitivity and reproducibility, accurate quantification and automation.

Modern instrumental HPTLC is thus a complementary technique to high performance liquid chromatography (HPLC) in the analysis of herbicides and is increasingly used in this application.

## Chromatographic Systems

Silica gel is the most common stationary phase in TLC and HPTLC of herbicides but reversed-phases (silica gel modified with $C_8$, $C_{18}$, e.g., RP-18 W, Nano-Sil $C_{18}$-100, silica gel impregnated with paraffin oil) can also be used. Silica gel impregnated with diethylene glycol is suitable for triazine herbicides. Good separation of herbicides can also be obtained on alumina. TLC layers covered with a transparent polymer film have been recommended to prevent evaporation of the mobile phase and volatile herbicide samples and to suppress the adsorption of environmental impurities.

Mobile phases frequently used in the TLC of herbicides in combination with silica gel include dichloromethane, hexane–acetone and hexane–butyl acetate mixtures. Methanol or acetonitrile with water is recommended when reversed-phase stationary phases are employed. Hexane–dioxane mixtures are used with alumina layers. The hydrophobicity of herbicides can be modified by the addition of cyclodextrins.

For the TLC systems commonly used in the analysis of herbicides, see **Table 1**.

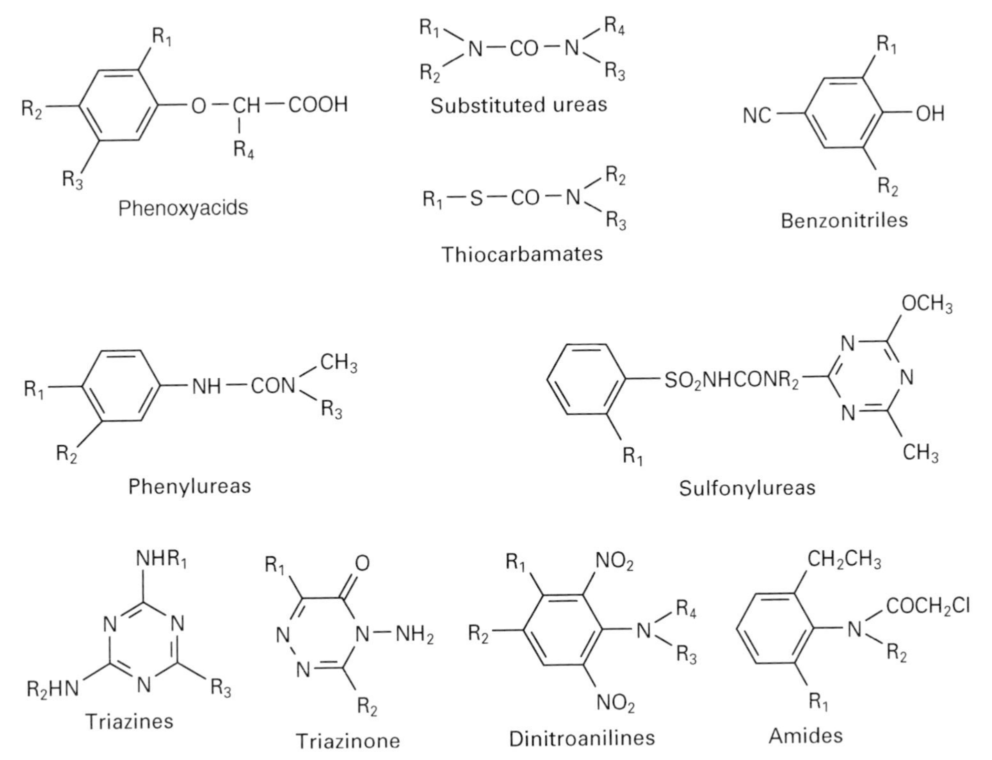

**Figure 1**  Structure of common herbicides.

## Multiple Development

In multiple development the sample spots are re-concentrated whenever the solvent front contacts the chromatographic zone and the spots are re-focused to narrow bands. This results in increased separation efficiency and improved detection limits.

Automated multiple development (AMD) HPTLC has been applied to the screening of pesticides (including triazine, phenylurea, carbamate, phenoxycarboxylic acids and other herbicides) in environmental samples. Herbicides from different classes can be resolved by a universal gradient, based on dichloromethane. Positive results are confirmed by a second analysis using special gradients optimized for indi-

vidual classes. Separated pesticides can be evaluated by densitometric detection and characterized by their UV spectra and migration distance. Most pesticides have detection limits in the range from 5 to 60 ng and their analysis does not require preconcentration methods. Only a few pesticides have higher detection limits and require preconcentration, e.g. by solid-phase extraction. An AMD-HPTLC method, DIN 38407, part 11, has been included in Germany's official methods for water analysis. An example of the AMD-HPTLC analysis of drinking water spiked with herbicides is shown in **Figure 2**; multi-wavelength detection was used. The optimized 20-step gradient for this separation is shown in **Figure 3**. Analysis takes 90 min (12 samples are analysed on one plate)

**Table 1**  Examples of TLC systems for the analysis of herbicides

| Stationary phase | Mobile phase | Detection |
|---|---|---|
| Silica gel (60 or G) | Dichloromethane<br>Ethyl acetate<br>Hexane–acetone (8 : 2)<br>Toluene–acetone (85 : 15)<br>Benzene–chloroform–methanol (9 : 3 : 2)<br>Chloroform–methanol (3 : 1) | Hill's reaction, sprayed with silver nitrate, o-toluidine 4,4′-tetramethyldiaminodiphenylmethane, 1% ferric chloride in butanol |
| PR-18W, PR-18, Nano-Sil C$_{18}$-100 | Methanol–water (7 : 3)<br>Acetonitrile–water (7 : 3) | Sprayed with silver nitrate or o-toluidine |

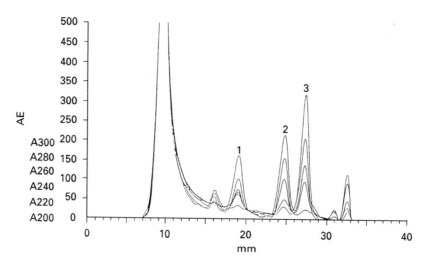

**Figure 2** Multi-wavelength scan of drinking water spiked with 1, 2,4-dichlorophenoxyacetic acid; 2, mecoprop; 3, simazine. The optimized gradient is given in Figure 3. Reproduced with permission from Morlock (1996).

and peak capacity (separation number) is more than 40. This method was shown to be suitable for 283 pesticides, including herbicides.

## Detection

Various agents have been used to visualize herbicide spots. They include silver nitrate, Gibbs reagent, diphenylamine and ninhydrin. The commonest detection method in TLC of herbicides is densitometry. Pre- and post-column derivatization can be carried out if the substances do not absorb visible or UV light. Simultaneous use of densitometry and fluorescence quenching is advantageous in the analysis of herbicides in multicomponent formulation, e.g. bromacil and diuron, chlotoluron and terbutryn.

Biochemical detection based on Hill's reaction is a very sensitive and selective method for herbicide residues that inhibit the enzyme system of isolated chloroplast. The concentration of the herbicide residues can also be determined from the lifetime of the spot. This method can be used for triazines, phenylureas, carbamates, uracils and pyridazone. Detection limits for herbicide residues in water, agriculture crops, foods and soil are in the range $1-10\ \mu g\ kg^{-1}$.

TLC with radioactive detection is useful in environmental studies, for example, in studies on the degradation of herbicides, formation of residues and their uptake and metabolism by plants.

Combination of TLC with mass spectrometry, recommended for the identification of herbicide metabolites, is less frequently used because of its high price.

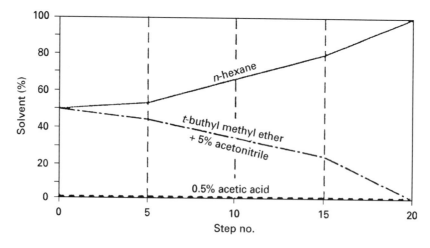

**Figure 3** Optimized 20-step gradient for the determination of 2,4-dichlorophenoxyacetic acid, 4-chloro-2-methylphenoxyacetic acid (isooctyl ester) and atrazine. Reproduced with permission from Morlock (1996).

## Selected Applications

TLC analysis of herbicides present in various matrices is usually preceded by isolation and/or preconcentration methods, e.g. solid-phase extraction (SPE) for water samples and supercritical fluid extraction (SFE) for solid samples, e.g. soils and plants; sample clean-up is usually not required. Some typical examples of the application of TLC in herbicide analysis are given below.

TLC can be used to determine triazine herbicides and their metabolites in drinking and surface water. Preconcentration by SPE is followed by chromatography on $C_{18}$ plates with hexane–ethyl acetate–acetone (4 : 4 : 1 v/v) or silica gel with nitromethane–tetrachloromethane (1 : 1 v/v) and densitometric detection. The same chromatographic system can be used to determine triazine and triazone herbicides in soil after SFE. Detection limits range from 30 to 60 ng $L^{-1}$. A simple TLC method to determine atrazine in drinking and ground water is based on its extraction with chloroform, followed by separation on silica gel with toluene–acetone (85 : 15) as mobile phase and detection by spraying with 4,4′-tetramethyldiaminodiphenylmethane. The detection limit is 20 ng $L^{-1}$.

Phenoxyacetic acid herbicides can be analysed on silica gel with acidified mobile phases, e.g. ethyl acetate–acetic acid (49 : 1 v/v) or toluene–acetone–acetic acid (2 : 2 : 1 v/v) and detected by spraying with a solution of 3,5-dichloro-p-benzoquinonechlorimine.

Residue analysis of phenylurea herbicides in water, potatoes and soil is based on extraction with acetone or dichloromethane, clean-up on silica gel, hydrolysis to anilines and derivatization to fluorescent dansyl derivatives in situ on plate. This is followed by separation on silica gel with dichloromethane–methanol (99 : 1 v/v) as mobile phase and fluorescence detection. Detection limits of ca. 1, 20 and 200 mg $kg^{-1}$ can be attained for water, potato and soil samples.

Pre-coated silica gel plates impregnated either with a 20% solution of formamide (A) or diethylene glycol (B), with hexane–chloroform–diethyl ether (2 : 1 : 1 v/v) as a mobile phase for system A and hexane-benzene–acetone (1 : 1 : 1) for system B gave satisfactory separation of 12 substituted urea herbicides (**Table 2**).

The screening of drinking water (about 300 compounds, including herbicides) is based on SPE followed by AMD. Identification is by analysing the sample under different separation conditions, together with reference compounds, measurement of the UV spectra in situ and by the use of post-chromatographic detection with various reagents. Both SPE and HPTLC can be automated.

**Table 2**  $R_F$ values of substituted urea herbicides

| Herbicide | A | B |
|---|---|---|
| N′-(3-chloro-4-methoxyphenyl)-N,N-dimethylurea | 0.16 | 0.36 |
| N′-Phenyl-N,N-dimethylurea | 0.20 | 0.40 |
| N′-(4-Chlorophenyl)-N,N-dimethylurea | 0.26 | 0.47 |
| N′-(3,4-Dichlorophenyl)-N,N-dimethylurea | 0.35 | 0.54 |
| N′-(3-Chloro-4-methylphenyl)-N,N-dimethylurea | 0.39 | 0.58 |
| N′-[4-(4′-Chlorophenoxyphenyl)]-N,N-dimethylurea | 0.45 | 0.62 |
| N′-(4-Isopropylphenyl)-N,N-dimethylurea | 0.46 | 0.64 |
| N′-(4-Chlorophenyl)-N,N-methoxyphenylurea | 0.70 | 0.77 |
| N′-(4-Bromohenyl)-N,N-methoxyphenylurea | 0.74 | 0.80 |
| N′-(3,4-Dichlorophenyl)-N,N-methoxyphenylurea | 0.78 | 0.83 |
| N′-(4-Chlorophenyl)-N,N-methylisobutyl-2-urea | 0.56 | 0.70 |
| N′-(3,4-Dichlorophenyl)-N,N-methylbutylurea | 0.72 | 0.79 |

Reproduced with permission from Ogierman L and Brysz G (1981) Partition thin-layer chromatography of some phenylurea herbicides. *Fresenius Zeitschrift für Analytische Chemie* 308: 464.

HPTLC can be used for the optimization of HPLC conditions where direct method development is more time-consuming and expensive. The herbicide migration data obtained in HPTLC on silica gel, RP-18, RP-8 and CN chemically bonded phases correlate well with HPLC.

## Conclusions and Future Trends

TLC is useful in herbicide analysis when many samples have to be analysed. It requires minimal sample pretreatment and reduces the number of separation steps. The reason for using HPTLC for herbicide residue is that it can analyse samples in complex matrices with minimal matrix modification. The ability to separate a number of samples simultaneously is important in screening studies. HPLC should be applied in cases of complicated samples when TLC fails and when full automation is required.

It is expected that TLC will keep its place as a simple and rapid method for qualitative and semiquantitative analysis of herbicides. HPTLC will find increasing use in herbicide residue determinations due to its main advantages of increased separation efficiency, decreased detection limit, more accurate quantitative analysis and automation. There will be developments in mobile-phase optimization. Greater use of combined TLC-spectroscopic methods can be expected.

A growing demand for method validation can be expected.

*See also:* **II/Chromatography: Thin-Layer (Planar):** Densitometry and Image Analysis; Layers; Modes of Development: Conventional; Modes of Development: Forced Flow, Over Pressured; Spray Reagents. **Extraction:** Analytical Extractions; Solid-Phase Extraction; **III/Herbicides:** Gas Chromatography; Solid-Phase Extraction. **Impregnation Techniques: Thin-Layer (Planar) Chromatography. Pesticides:** Gas Chromatography; Thin-Layer (Planar) Chromatography.

## Further Reading

Frey HP and Zieloff K (1992) *Qualitative und quantitative Dünschicht-Chromatographie.* Weinheim: VCH.

Fried B and Sherma J (eds) (1996) *Practical Thin Layer Chromatography: A Multidisciplinary Approach.* Boca Raton, FL: CRC Press.

Jork H, Funk W, Fischer W and Wimmer H (1990) *Thin Layer Chromatography,* vol. 1a. *Physical and Chemical Detection Methods.* Weinheim: VCH.

Jork H, Funk W, Fischer W and Wimmer H (1994) *Thin Layer Chromatography,* vol. 1b. *Reagents and Detection Methods.* Weinheim: VCH.

Kaiser RE, Günther W, Gunz H and Wulf G (eds) (1996) *Thin Layer Chromatography.* Düsseldorf: InCom.

Morlock GM (1996) Analysis of pesticide residues in drinking water by planar chromatography. *Journal of Chromatography A* 754: 423–430.

Pacáková V, Štulík K and Jiskra J (1996) High performance separations in the determination of triazine herbicides and their residues. *Journal of Chromatography A* 754: 17–31.

Rathore H-S and Begum T (1993) TLC methods for use in pesticide residue analysis. *Journal of Chromatography* 643: 271–290.

Sherma J (1994) Determination of pesticides by thin-layer chromatography. *Journal of Planar Chromatography* 7: 265–272.

Sherma J and Fried B (eds) (1996) *Handbook of Thin Layer Chromatography,* 2nd edn. New York: Marcel Dekker.

Touchstone J (1992) *Practice of Thin Layer Chromatography,* 3rd edn. New York: Wiley.

# HEROIN: LIQUID CHROMATOGRAPHY AND CAPILLARY ELECTROPHORESIS

**R. B. Taylor, A. S. Low and R. G. Reid,**
The Robert Gordon University, Aberdeen, UK

## Introduction

The separation and quantitative determination of opiates is required for a wide variety of purposes and applications. These include therapeutic drug monitoring, metabolism and pharmacokinetic studies and forensic investigations, as well as the detection and control of drug abuse. The determination of opiates in human urine is of considerable analytical interest, particularly in the context of detecting the consumption of heroin; it is in this context that the present article is written.

The first step in the establishment of such presumptive consumption of heroin is usually by enzyme immunoassay. This has the merit of low detection limit and large sample throughput, making it very suitable for large screening programmes. The immunoassay technique, however, is nonselective with respect to individual opiates and a positive result in such a screen for legal purposes must be followed by identification of individual opiates present. The purpose of this is usually to confirm or refute the hypothesis that heroin has been consumed. The short metabolic half-life of heroin complicates its confirmation, so that its consumption is inferred by detection of its metabolites. One generally sought metabolite is morphine owing to its relatively long half-life. This approach has the disadvantage that morphine is also produced as a metabolite of codeine, so that heroin consumption is presumed or not on the basis of notional codeine-to-morphine ratios. The detection of the first metabolite of morphine, 6-monoacetyl-morphine, is now taken as an unequivocal indicator of heroin consumption.

The detection of heroin consumption is further complicated by the quite general inclusion of legal opiates such as codeine, pholcodine and dihydrocodeine in commonly available medicines. This results in numbers of subjects being screened as positive for opiate consumption who are not confirmed by alternative methods as having consumed heroin or morphine.

To confirm the presence of individual opiates, techniques are required that are more selective than immunoassay. These are usually based on established chromatographic techniques. Several thin-layer systems have been used but in general these have insufficiently low limits of detection to confirm adequately results obtained by immunoassay. This is a particular problem since immunoassay detects total opiates and, following separation, individual concentrations may be much lower. The most generally accepted confirmation is by combined gas chromatography–mass spectrometry (GC-MS), owing to the selectivity of the capillary gas chromatography separation coupled with the unequivocal identification by electron impact ionization mass spectrometry. The technique of high performance liquid chromatography (HPLC) has been extensively applied to heroin and some of its metabolites as well as individual related opiates for a variety of purposes associated with drug abuse. No liquid chromatography (LC) method has specifically addressed the problem of confirmation of results obtained from immunoassay. Work in this area has concentrated on developing LC methods capable of quantifying morphine and codeine individually since, as indicated above, the codeine-to-morphine ratio has been a major criterion in establishing heroin consumption. The emerging technique of capillary electrophoresis (CE) is potentially attractive as a separation technique because of its capability for high resolution and peak capacity. As yet little has appeared in the literature concerned with the capability of this technique to determine low concentrations of opiates in biological matrices.

The literature on the determination of drugs of abuse and screening procedures using GC-MS is extensive and will not be discussed here. The present article describes both normal and reversed phase LC systems for the separation and quantification of a set of opiates comprising heroin, 6-monoacetylmorphine, morphine, codeine, dihydrocodeine and pholcodine, and discusses their potential in acting as an intermediate test in determining opiate abuse. Such a test may eliminate so-called false positives arising from urine samples screened as containing opiate that is subsequently found not to originate from consumption of heroin. A method based on CE is also described and its potential in relation to the LC methods discussed. Heroin is rapidly metabolized and cannot be detected in urine. This drug is included in this set, however, as the separation with respect to legal opiates may be of relevance in forensic applications.

## Reversed-Phase LC System

Several reversed-phase separations of selected groups of opiates have been reported in the literature using column switching, gradient elution and ion pairing techniques. From our experience in the analysis of

basic drugs in biological fluids, the most general approach to ion pairing is the use of mobile-phase additives. An anionic surfactant such as sodium dodecyl sulfate (SDS) or alkylsulfonic acid, which is adsorbed onto a $C_{18}$ surface, is incorporated in the mobile phase to increase retention and thus separation. The inclusion, in addition, of a less hydrophobic species of similar charge to the analyte, such as a tetraalkylammonium bromide, has been found to reduce tailing of such basic compounds and also in some cases to alter selectivity. In using such systems the pH of the mobile phase is made acid, thus protonating the basic analytes.

For the opiates morphine, pholcodine, dihydrocodeine and codeine the best anionic mobile phase additive is pentanesulfonic acid used in conjunction with tetraethylammonium bromide. At concentrations of 1 mmol $L^{-1}$ and 100 mmol $L^{-1}$, respectively, these additives in methanol/aqueous buffer (10 : 90 by vol.) containing disodium hydrogen phosphate at pH 2.5 give excellent separation of these four opiates in 15 min when dissolved in water. This system allows the inclusion of nalorphine as an internal standard, which elutes between pholcodine and dihydrocodeine. This optimum separation is shown in **Figure 1**. Both heroin and 6-monoacetylmorphine elute much later than codeine and therefore cannot be determined at sensitivities comparable with the other compounds. The separation of these four opiates is accompanied by excellent individual linearity of response of peak height ratios versus concentration over the range 1–5 $\mu$g cm$^{-3}$ using UV detection at 285 nm.

A standard solid-phase extraction procedure was applied to urine spiked with the four opiates and the internal standard. Bond-Elut Certify extraction cartridges containing a mixed stationary phase of nonpolar C8 and strong cation exchanger (SCX) were used. The extraction consisted of cartridge conditioning with methanol, water and acetate buffer followed by sample addition and washing with acetate buffer, water and methanol. After drying, the analytes were eluted with dichloromethane/isopropyl alcohol (80 : 20 by vol.) mixture containing 2% (v/v) ammonia. The resultant solution was dried at 40°C under nitrogen, reconstituted in 250 $\mu$L of ethyl acetate, and again dried under nitrogen. Immediately prior to injection, the sample was redissolved in 500 $\mu$L of mobile phase. This procedure gave percentage recoveries in excess of 80% for all the compounds other than morphine, for which the recovery was 71%, presumably as a result of its known propensity for being adsorbed onto glass surfaces.

This extraction procedure gives a large solvent peak in this reversed-phase solvent system. At the

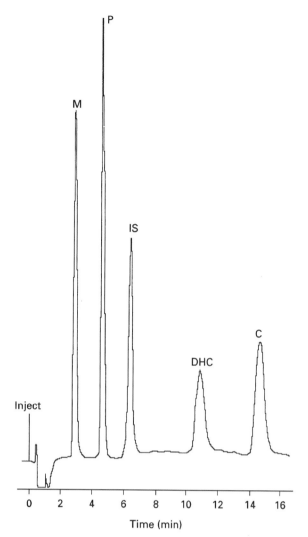

**Figure 1** Representative reversed-phase separation of morphine (M), pholcodine (P), dihydrocodeine (DHC), codeine (C) and nalorphine (IS) using 10% methanol/90% water, containing 50 mmol L$^{-1}$ disodium hydrogen orthophosphate, 1 mmol L$^{-1}$ pentanesulfonic acid and 100 mmol L$^{-1}$ tetraethylammonium bromide as mobile phase at a flow rate of 0.4 mL min$^{-1}$, column 100 × 2 mm i.d. packed with 3 μm microporous octadecylsilica, detection by UV at 280 nm.

detector sensitivity required for 500 ng cm$^{-3}$ opiate concentrations, it is not possible to quantify morphine or pholcodine since they are not separated from the solvent front. A specimen chromatogram of urine spiked (500 ng cm$^{-3}$) with the four opiates and nalorphine is shown in **Figure 2**.

It appears that for such a set of opiates, which vary widely in hydrophobicity, the reversed-phase system is of minimal use for the purpose of detecting heroin consumption by urine analysis when coupled with UV detection. Neither heroin nor 6-monoacetylmorphine interfere with the quantification of the legal opiates. However, the retention time of the 6-monoacetyl metabolite is excessive for quantification and

the codeine-to-morphine ratio cannot be established because of the masking of the morphine peak at levels below 500 ng cm$^{-3}$ by endogenous compounds in the solvent front when using UV detection. The inherent high resolution of reversed-phase chromatography is not helpful in the quantification of this widely differing range of analytes. The obvious choice of gradient elution, by analogy with GC, would result in compression of the timescale for elution of the complete range of compounds. However, the potential improvement in detection limits may not be wholly achievable owing to increased baseline noise. The use

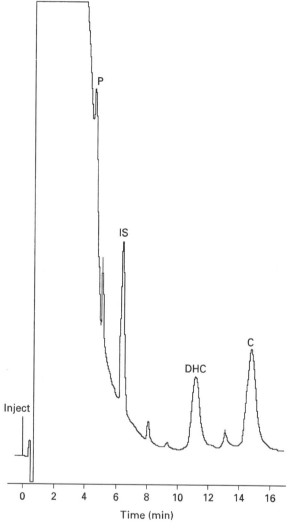

**Figure 2** Representative chromatogram of a urine sample spiked with morphine (M), pholcodine (P), dihydrocodeine (DHC) and codeine (C) at a concentration of 500 ng cm$^{-3}$ and incorporating nalorphine (IS) as internal standard after solid-phase extraction using 10% methanol/90% water containing 50 mmol L$^{-1}$ disodium hydrogen orthophosphate, 1 mmol L$^{-1}$ pentanesulfonic acid and 100 mmol L$^{-1}$ tetraethylammonium bromide as mobile phase at a flow rate of 0.4 mL min$^{-1}$, column 100 × 2 mm i.d. packed with 3 μm microporous octadecylsilica, detection by UV at 280 nm.

of electrochemical detection, while it has been shown to be highly sensitive in the detection of opiates, would not overcome the long retention time required to elute 6-monoacetylmorphine. Its characteristics were not examined using this separation system.

## Normal-Phase LC System

In common with most drug separations by LC, reversed-phase methods predominate in the literature for the separation and quantification of opiates. Normal-phase solvent systems have been reported for the quantitative determination of selected groups of opiates. None of these, however, is suitable for the purpose of determining heroin consumption, either by identification of 6-monoacetylmorphine or by quantitative estimation of the concentration of morphine relation to other legal opiates present.

Using a 200 mm × 2.1 mm column packed with 3 μm Hypersil it has been found possible to resolve completely morphine, dihydrocodeine, pholcodine, codeine, 6-monoacetylmorphine, nalorphine (as internal standard) and heroin. The optimized mobile phase consists of dichloromethane/pentane/methanol (29.8 : 65 : 5.2 by vol.) containing 0.026% (v/v) diethylamine. **Figure 3** shows a representative chromatogram of this set of opiates. The resolutions between individual pairs of opiates are less than in the reversed-phase system, although still in excess of one. This system results in complete elution of all components of interest in approximately 16 min. This separation appears to be more advantageous for the detection of heroin consumption than the reversed-phase system in that the order of elution is reversed and heroin and 6-monoacetylmorphine are eluted fairly early in the chromatogram, thus allowing maximum sensitivity of detection of the latter.

When a solid-phase extraction procedure identical to that described above is used, it is found that the choice of solvent is critical to redissolve the dried analytes completely while preserving peak shape on injection. The final injection solvent consists of dichloromethane/pentane (10 : 90 by vol.). This ensures complete solution of the extracted analytes and allows injection in a solvent chromatographically weaker than the mobile phase, which maintains peak sharpness. **Figure 4** shows a chromatogram of the six opiates together with the internal standard, nalorphine, after extraction from a spiked urine sample. In contrast to the reversed-phase system, all the analytes of interest are completely resolved from the solvent front and can thus be readily quantified using UV absorption at 280 nm. The main quantitative characteristics of analysis for these six opiates are shown in **Table 1**. The separation is adequately rugged with

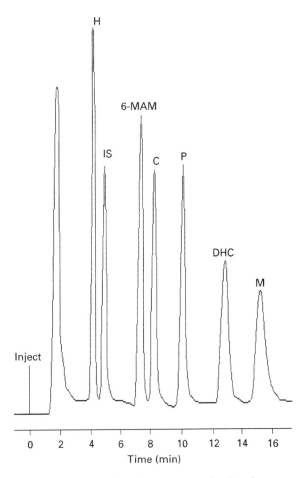

**Figure 3** Representative chromatogram showing the separation of a pentane solution of morphine (M), dihydrocodeine (DHC), pholcodine (P), codeine (C), 6-monoacetylmorphine (6-MAM), nalorphine (IS) and heroin (H) (column 200 × 2 mm i.d. packed with 3 μm microporous siilca mobile phase–dichloromethane/pentane/methanol 29.8 : 65 : 5.2 by volume containing 0.026% v/v diethylamine at a volumetric flow rate of 0.4 mL min$^{-1}$, detection by UV at 280 nm).

respect to the retention times as a result of the inclusion of nalorphine as an internal standard, even allowing for the relatively volatile solvents used in the mobile phase. The selectivity aspect of the separation is excellent with respect to both endogenous materials and the opiates of interest. The retention times of commonly ingested drugs, opiate metabolites and other compounds relative to nalorphine are listed in **Table 2**.

Table 2 shows that the selectivity with respect to the more commonly ingested drugs such as aspirin and caffeine, and also to several of the established opiate metabolites, is good in that retention times are appreciably different from the compounds of interest. Paracetamol, however, elutes between codeine and pholcodine and is incompletely resolved from the latter. At the wavelength of 280 nm used, the sensitivity for paracetamol is very low and

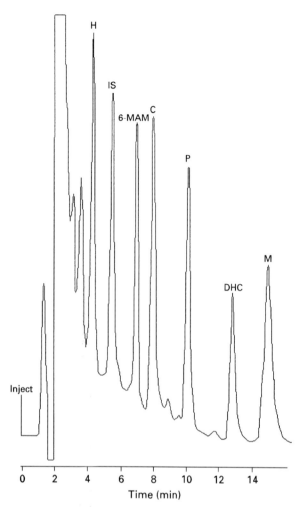

**Figure 4** Representative chromatogram of a urine sample spiked with 500 ng cm$^{-3}$ morphine (M), dihydrocodeine (DHC), pholcodine (P), codeine (C), 6-monoacetylmorphine (6-MAM), nalorphine (IS) and heroin (H) following solid-phase extraction on subsequent reconstitution in 10% dichloromethane/90% pentane and injected (column 200 × 2 mm i.d. packed with 3 μm microporous silica mobile phase dichloromethane/pentane/methanol 29.8 : 65 : 5.2 by volume containing 0.026% v/v diethylamine at a volumetric flow rate of 0.4 mL min$^{-1}$, detection by UV at 280 nm).

there would be minimal interference from paracetamol in the determination of any pholcodine present. The linearity and limits of quantification show that quantification after extraction and separation is adequate to confirm the presence of 6-monoacetylmorphine. The relative amounts of legal opiates as well as morphine detectable in urine are at levels corresponding to those that would result in the detection of opiate by immunoassay; the cutoff value generally used is 300 ng mL$^{-1}$.

An advantage of the normal-phase separation is the facility of coupling this with MS detection via an appropriate interface. The system described has been successfully linked via a particle beam interface with

some loss of resolution. Electron impact ionization allows unequivocal identification of all of the opiates.

## Capillary Electrophoresis

Separation of charged analytes by utilizing their differential migration rates in an electric field has been achievable for many decades. The high theoretical plate numbers readily realized by using narrow capillaries, and the relatively recent (compared to LC) availability of reliable microprocessor-controlled equipment, have resulted in increased interest in the theory and application of this technique. The high efficiency can potentially be exploited to achieve resolution and also to accommodate higher peak capacity than is readily or conveniently achieved in LC. Both of these aspects are relevant to the identification and quantification of this group of opiates. However, because only a small sample volume can be injected onto a capillary and only a short pathlength is available when direct UV detection is used, the concentration sensitivity of capillary electrophoresis with UV detection suffers in comparison with LC methods when hydrodynamic injection is used. Within the extensive literature that now exists on CE methods, there are relatively few reports on applications concerning the determination of drugs in biological fluids such as plasma or urine, in which the analyte is generally at low concentration levels. Most applications in the literature reporting quantification of drugs at the ng mL$^{-1}$ concentration level in biological fluids have used sample pretreatment methods that involve a preconcentration step in order to reach the required limits of detection, and thus exploit the undoubted separational advantages of the technique.

Samples can also be introduced into the capillary electrokinetically, i.e. by application of a relatively low voltage with the capillary end immersed in the solution of analyte. It has been established both theoretically and practically that the amount of analyte of a particular charge introduced electrokinetically can be dramatically increased if the solution from which the sample is injected is of low ionic strength compared with that of the buffer used for the electrophoretic separation. While this injection method is subject to bias (not all analytes will be introduced into the capillary in this way), this can be a positive advantage when quantifying particular analytes in a given matrix.

An optimized separation of the six opiates and levallorphan (as internal standard) by CE is shown in **Figure 5**. The parameters affecting separation are buffer concentration and pH. As is general, increase of buffer ionic strength over the range 20 to 140 mmol L$^{-1}$ increases retention and resolu-

**Table 1** Analytical characteristics of opiate determination in urine following normal-phase separation

| | Heroin | 6-MAM | Codeine | Pholcodine | Dihydrocodeine | Morphine |
|---|---|---|---|---|---|---|
| Relative retention drug/IS (%RSD) | 0.785 (6.1) | 1.25 (6.1) | 1.51 (5.2) | 1.99 (5.6) | 2.57 (3.2) | 3.04 (3.2) |
| Recovery (%) | 89.1 | 82.7 | 82 | 88.4 | 79.7 | 79.3 |
| Calibration correlation coefficient | 0.9987 | 0.9901 | 0.9900 | 0.9980 | 0.9909 | 0.9964 |
| Slope of calibration line $1 \times 10^3$ ($\pm$ SD) | 12.6 (0.26) | 9.5 (0.51) | 3.8 (0.32) | 3.1 (0.11) | 2.2 (0.17) | 7.8 (0.04) |
| Limit of quantification at $S/N = 4$ (ng cm$^{-3}$) | 3.0 | 5.4 | 4.2 | 6.3 | 6.0 | 15 |
| Within day precision as %RSD at 200 ng cm$^{-3}$ | 5.03 | 4.45 | 4.34 | 6.38 | 4.29 | 5.85 |
| Day to day precision as %RSD at 200 ng cm$^{-3}$ | 5.43 | 4.92 | 5.11 | 7.22 | 6.1 | 8.07 |

IS, internal standard; 6-MAM, 6-monoacetylmorphine; RSD, relative standard deviation; SD, standard deviation; S/N, signal-to-noise ratio. Reprinted in part from Low and Taylor (1995) with permission from Elsevier Scientific.

tion. Increase of buffer pH from 4 to 8 decreases the retention as a consequence of increased electro-osmotic flow. Figure 5 shows that all the opiates of interest are eluted in 12 min with resolution between individual pairs well in excess of that achieved by normal-phase LC. What is equally significant is that all seven compounds are eluted within a narrow time-scale of 9–12 min, so that there is much less disparity between the relative amounts of peak broadening than is the case in either of the LC methods described earlier.

A urine extraction procedure based on the method developed for LC has been developed in which a 0.5 cm$^3$ urine sample is extracted as previously but reconstituted in 1 cm$^3$ of a solvent consisting of water/methanol (9 : 1). The resultant electrophero-gram obtained from urine spiked at approximately 300 ng cm$^{-3}$ subsequent to extraction and elec-

**Table 2** Retention times relative to nalorphine of some drugs that may interfere with normal-phase LC determination of opiates

| Drug | Relative retention time |
|---|---|
| Chlodiazepoxide | < 0.39 |
| Papaverine | 0.43 |
| Diazepam | 0.48 |
| Lignocaine | 0.54 |
| Methadone | 0.58 |
| Naloxone | 0.62 |
| Theophylline | 0.65 |
| Diphenylhydramine | 0.74 |
| Ephedrine | 1.33 |
| Hydrocodone | 1.76 |
| Paracetamol | 1.84 |
| Dextropropoxyphene | 2.09 |
| Quinine | 2.28 |
| Norcodeine | 3.29 |
| Caffeine | 4.84 |
| Normorphine | 5.62 |
| Acetylsalicylic acid | > 8 |
| Procaine | > 8 |
| Theobromine | > 8 |

trokinetic injection is shown in **Figure 6A**. Figure 6B shows the corresponding electropherogram of a blank urine sample. In contrast to the LC methods there are very few endogenous compounds brought through the extraction and electrokinetic injection procedures. Only one of these is a potential interference as it elutes close to pholcodine.

The separational abilities of the CE method are clearly superior to those of either of the liquid LC methods, taking into account both the selectivity with respect to individual opiates and also the separation from endogenous urine components. The quantitative aspects of the CE method are also significant. It has been found that reconstitution of extracted and dried

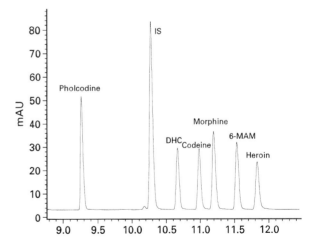

**Figure 5** Representative separation of morphine, pholcodine, heroin, codeine, 6-monoacetylmorphine (6-MAM), dihydrocodeine (DHC) and levallorphan (IS) as internal standard in aqueous solutions using capillary electrophoresis (capillary 50 μm i.d. × 500 mm effective length, running buffer 100 mmol L$^{-1}$ disodium hydrogen phosphate at pH 6, applied voltage 20 kV, electrokinetic injection for 10 s at 5 kV, UV detection at 200 nm). (Reprinted in modified form from Taylor et al., 1996, with permission from Elsevier Scientific.)

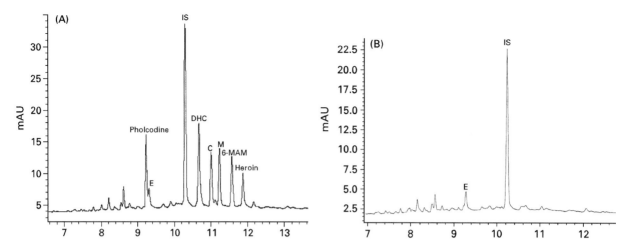

**Figure 6** (A) Representative electropherogram of a urine sample spiked with 300.8 ng cm$^{-3}$ pholcodine, 805 ng cm$^{-3}$ levallorphan (IS), 241 ng cm$^{-3}$ dihydrocodeine (DHC), 274 ng cm$^{-3}$ codeine (C), 304.8 ng cm$^{-3}$ morphine (M), 288.3 ng cm$^{-3}$ 6-monoacetylmorphine (6-MAM) and 284.4 ng cm$^{-3}$ heroin after solid-phase extraction. (E) designates an endogenous component in urine. (B) Representative electropherogram of a blank urine sample spiked with 805 ng cm$^{-3}$ levallorphan after solid-phase extraction (capillary 50 μm i.d. × 500 mm effective length, running buffer 100 mmol L$^{-1}$ disodium hydrogen phosphate at pH6, applied voltage 20 kV, electrokinetic injection for 10 s at 5 kV, UV detection at 200 nm). (Reprinted in modified form from Taylor *et al.*, 1996, with permission from Elsevier Science.)

analytes in aqueous methanol allows the field-amplified sample injection technique to be applied extremely effectively. Indeed, in the method development it was found that the increase of peak areas when comparing electrokinetic injection from water with hydrodynamic injection ranged form 90 in the case of heroin to 160 for 6-monoacetylmorphine and pholcodine. This, coupled with the use of 200 nm as the wavelength of detection, allows limits of detection well below the cutoff level for immunoassay to be achieved. The electrokinetic injection technique for the enhancement of concentration sensitivity applied to this assay procedure is capable of overcoming the perceived limitation of CE as technique for determining these compounds is urine. **Table 3** summarizes the main validation parameters of the assay of these drugs based on the CE procedure described.

The validation parameters in Table 3 show that the quantitative aspects are very comparable with those quoted for LC. The electrokinetic introduction of sample is in general more variable than hydrodynamic introduction but incorporation of an internal standard, which is required to accommodate variable extraction efficiencies, results in within day and day to day precision better than that achieved by normal-phase LC. The limits of quantification quoted are indicative only, since time of sampling could be considerably increased over the 10 s at 5 kV used. The significance is that there is more than adequate sensitivity to allow detection of these particular opiates at the required levels for elimination of common opiates that have resulted in so-called false positives during immunoassay screening. The elec-

trokinetic method of sample introduction, however, does require that the solution for injection into the capillary be of low ionic strength. Consideration must be given to the overall sample matrix pretreatment to ensure this. In addition, at very low concentrations, there is considerable sample depletion from the small sample volumes used and care has to be taken to limit the number of successive injections from a given sample vial.

## Conclusions

There are many opiates for which a multitude of assays have been determined in a variety of matrices for often very widely different purposes. In screening for drugs of abuse, chromatographic methods are extensively used to increase the information gained by rapid immunoassay. The method of choice for unequivocal identification of particular opiates is likely to continue to be GC-MS on the basis of the mass spectral information obtainable. The present article indicates that, at least for the specified relevant group of opiates chosen, separations in the liquid phase are capable of achieving adequate separations and that these can be applied with variable success to the determination of the specified compounds in urine. The results quoted for the reversed-phase LC method show an inherent limitation of this approach when determining a number of compounds differing widely in hydrophobicity. Such methods, however, can usually be manipulated to allow good resolution and quantification for more restricted mixtures of compounds. The widely

**Table 3** Analytical characteristics of opiate determination in urine by CE

| | Pholcodine | Levorphanol | Dihydrocodeine | Codeine | Morphine | 6-MAM | Heroin |
|---|---|---|---|---|---|---|---|
| Migration time (min) | 9.18 | 10.18 | 10.68 | 10.99 | 11.21 | 11.55 | 11.87 |
| (%RSD) | (1.1) | (0.80) | (0.71) | (0.77) | (0.79) | (0.79) | (0.74) |
| Resolution | | 13.4 | 4.7 | 3.7 | 2.3 | 3.9 | 3.4 |
| Peak efficiency $\times 10^{-5}$ | 2.5 | 2.7 | 2.8 | 2.6 | 2.2 | 2.2 | 2.8 |
| Recovery (%) | 94.8 | | 90.7 | 92.4 | 88.5 | 91.8 | 96.4 |
| Calibration correlation coefficient | 0.9917 | | 0.9924 | 0.9944 | 0.9966 | 0.9967 | 0.9980 |
| Slope of calibration line $\times 10^3$ | 1.3 | | 1.6 | 1.5 | 1.4 | 1.5 | 1.2 |
| (%RSD) | (5.1) | | (5.2) | (4.2) | (3.4) | (3.2) | (2.6) |
| Limit of quantification at $S/N = 6$ (ng cm$^{-3}$) | 6.0 | | 16 | 14 | 16 | 18 | 16 |
| Within day precision as %RSD at 300 ng cm$^{-3}$ | 3.8 | | 1.8 | 1.4 | 2.1 | 0.6 | 3.4 |
| Day to day precision as %RSD at 300 ng cm$^{-3}$ | 2.9n | | 2.2 | 2.1 | 2.5 | 1.4 | 2.8 |

Reprinted in part from Taylor *et al.* (1996) with permission from Elsevier Scientific.

differing retention times for codeine and morphine tends to make reliable quantification of the ratio of these compounds difficult. The normal-phase method offers a viable alternative to GC as a separation procedure and is capable of quantifying the codeine-to-morphine ratio and of detecting 6-monoacetylmorphine at useful concentration levels. More importantly, the normal-phase method is capable of detecting and quantifying the commonly ingested legal opiates that can obscure or delay the final results of an immunoassay opiate screen. The separation achieved by CE arguably offers the most realistic alternative to GC. Method development is rapid and the plate numbers are comparable with those achievable by capillary GC. The underlying difficulty of lack of concentration sensitivity can be overcome by electrokinetic injection techniques if suitable sample pretreatment methods are developed with appropriate sample solvent composition. The eventual linking of CE with MS on a commercial basis should result in a combined technique capable of challenging GC-MS in this area of drug analysis.

*See Colour Plate 86.*

*See also:* **III/Alkaloids:** Gas Chromatography; Liquid Chromatography. **Drugs of Abuse: Solid-Phase Extraction.**

## Further Reading

Adamovics JA (ed.) (1994) *Analysis of Addictive and Misused Drugs.* New York: Marcel Dekker.

Braithwaite RA, Jarvie DR, Minty PSB, Simpson D and Widdop B (1995) Screening for drugs of abuse. I: Opiates, amphetamines and cocaine. *Annals of Clinical Biochemistry* 32: 123–153.

Gough TA (ed.) (1991) *The Analysis of Drugs of Abuse.* Chichester: John Wiley & Sons.

Low AS and Taylor RB (1995) Analysis of common opiates and heroin metabolites in urine by HPLC. *Journal of Chromatography B* 663: 225–233.

Tagliaro F, Turrina S and Smith FP (1996) Capillary electrophoresis: principles and applications in illicit drug analysis. *Forensic Science International* 77: 211–229.

Taylor RB, Low AS and Reid RG (1996) Determination of opiates in urine by capillary electrophoresis. *Journal of Chromatography B* 675: 213–223.

# HOT-PRESSURIZED WATER: EXTRACTION

*See* **III / SUPERCRITICAL FLUID EXTRACTION-SUPERCRITICAL FLUID CHROMATOGRAPHY**

# HUMIC SUBSTANCES

## Capillary Zone Electrophoresis

**J. Havel**, Masaryk University, Brno, Czech Republic
**D. Fetsch**, Surface Measurement Systems,
London, UK

Humic substances (HS) which are widespread in the soil, present structural complexity and polyelectrolyte properties. Many techniques and methods have been used to obtain more information about them, as reviewed by Davies and Ghabbour in 1999. The characterization of HS has been the focus of intense research for many years because the organic matter in soil contributes to the quality of the soil more than the other constituents. Nowadays, one of the main goals is the separation of HS into fractions which can then be studied independently. The aim is to determine HS structure, which remains unknown, despite the enormous efforts and remarkable progress in this field. HS exhibit properties similar to weak acid polyelectrolytes, presenting a wide range in molecular weights, solubilities and acid strength. They are sparingly soluble in acid and slightly acidic solutions, but solubility increases with pH and they are only fully soluble alkaline solutions. Many general hypothetical models of HS structure have been proposed, and most of these have been discussed by Shevchenko and Bailey.

As HS play an important role in environmental chemistry, it is important to know their properties. HS are very important as regards the quality and productivity of a soil and in retention of metalions and pollutants by the environment. Dissolved humic material has a tendency to interact with these compounds, complexing metal ions or pollutants, which may alter their fate and transfer in the soil. The mechanisms involved in the interaction of these compounds with HS are not clear and may vary depending on the physicochemical properties of the compounds, soil pH, redox status and the heterogeneous structure of HS. The effect of pH, for example, is related to the degree of humic R-COOH and R-OH dissociation; deprotonation increases the polarity of the humic material and altering its structure.

Electrophoretic methods have been applied for the purpose of HS studies since the 1960s. The first work on the separation of HS by a capillary technique was performed in 1991 using capillary isotachophoresis. However, few really suitable methods for the separation of HS exist.

Nevertheless, to understand HS behaviour, it is useful to perform separation into fractions and determine the structure of each fraction. Capillary zone electrophoresis (CZE) has been applied by Rigol *et al.* in 1994 for this purpose.

## HS Properties Studied by CZE

A short overview will be given here because the understanding of the properties of HS is the key to their successful separation.

### Adsorption of HS

In the literature, most of authors have used high HS concentrations in order to reach or observe CZE separation patterns; however, no explanation was given (**Table 1**). In fact, it has been observed that when using a background electrolyte that did not interact with HS (rimantadine, for example), high adsorption of HS on bare fused silica capillary wall occurs. This explains why previous authors used rather high concentrations of HS solutions for analysis, because a considerable part of the HS was lost from the solution during the separation process. This observation is in agreement with the fact that HS are highly adsorbed on to silica rocks and silicates or induce interactions with metal ions on clay surface. Such adsorption processes of HS have already been studied on different silica-based materials with other methods. Adsorption on the surface of the silica capillary complicates CZE separation. It was demonstrated recently that the adsorption can be reduced using additives such as magnesium(II) ($>14$ mmol L$^{-1}$) in the background electrolyte (BGE). In such magnesium-doped electrolytes, more fractions of HS can be observed, but also a secondary hump pattern. Via study of HS adsorption, a new method of CZE separation has been developed, which allows the concentration of HS in solution needed for CZE separation to be lowered to around 35–50 mg L$^{-1}$.

### Oligomerization of HS

It has long been known that HS aggregate in solution. Various authors describe this property differently. Von Wandruszka *et al.* described the

**Table 1**  Review of background electrolytes used for HS separation by CZE, reported by Fetsch *et al.* in 1997 and references therein

| Background electrolyte | Number of peaks | Hump | pH | [HS] (g L$^{-1}$) | Reference |
|---|---|---|---|---|---|
| 10–200 mmol L$^{-1}$ tetraborate | 1–4 | + | 8.3–10.0 | 0.01–1 | All authors |
| 6 mmol L$^{-1}$ tetraborate–3 mmol L$^{-1}$ dihydrogenphosphate | 3–6 | + | 8.9 | 0.05–0.5 | Pompe *et al.* Fetsch *et al.* |
| 10 mmol L$^{-1}$ tetraborate–5 mol L$^{-1}$ urea | 3–5 | + | 7.4 | High | Dunkelog *et al.* |
| 10 mmol L$^{-1}$ tetraborate–10% v/v acetonitrile | 1–2 | + | 9.0 | 1 | Norden *et al.* |
| 10 mmol L$^{-1}$ tetraborate–10% v/v acetone | 1 | + | 9.0 | 1 | Norden *et al.* |
| 10 mmol L$^{-1}$ tetraborate–10% v/v isopropanol | 2–3 | + | 9.0 | 0.01–0.1 | Norden *et al.* |
| 10 mmol L$^{-1}$ tetraborate–10% v/v propanol | 2–3 | + | 9.0 | 1 | Norden *et al.* |
| 10 mmol L$^{-1}$ tetraborate–10% v/v 2-propanol–5 mmol L$^{-1}$ urea | 2–6 | + | 9.0 | 1 | Norden *et al.* |
| 10 mmol L$^{-1}$ tetraborate–10% v/v tetrahydrofuran | 2–3 | + | 9.0 | 1 | Norden *et al.* |
| 20 mmol L$^{-1}$ tetraborate–5 mmol L$^{-1}$ CDTA | 2–4 | + | 8.6 | 0.01–0.1 | Norden *et al.* |
| 20 mmol L$^{-1}$ tetraborate–100 mmol L$^{-1}$ boric acid | 2–5 | + | 8.45 | 0.1–0.25 | Fetsch *et al.* |
| 8 mmol L$^{-1}$ L-Alanine | 2–4 | + | 3.17 | High | Dunkelog *et al.* |
| HCl–59.8 mmol L$^{-1}$ L-Alanine | 3 | + | 3.17 | 0.1 | Rigol *et al.* |
| HCl–60 mmol L$^{-1}$ DL-Alanine | 3–4 | + | 3.2 | 0.1–0.25 | Fetsch *et al.* |
| HCl–60 mmol L$^{-1}$ DL-Alanine–10% v/v methanol | 2–5 | + | 3.2 | 0.1–0.25 | Fetsch *et al.* |
| HCl–60 mmol L$^{-1}$ L-Alanine | 3 | + | 3.2 | 0.1–0.5 | Fetsch *et al.* |
| HCl–60 mmol L$^{-1}$ D-Alanine | 4 | + | 3.2 | 0.1–0.5 | Fetsch *et al.* |
| HCl–$\beta$-Alanine | 1–2 | + | $\approx 3$ | 0.1 | Fetsch *et al.* |
| HCl–60 mmol L$^{-1}$ $\beta$-Alanine | 1–2 | + | 3.2 | 0.1–0.25 | Fetsch *et al.* |
| HCl–30 mmol L$^{-1}$ $\beta$-Phenylalanine | 1 | + | 3.2 | 0.1–0.25 | Fetsch *et al.* |
| HCl–30 mmol L$^{-1}$ L-Cystine | 1–2 | + | 3.2 | 0.1–0.25 | Fetsch *et al.* |
| HCl–L-Leucine | 1–2 | + | 3.17 | 0.1 | Rigol *et al.* |
| HCl–L-Lysine | 1–2 | + | $\approx 3$ | 0.1 | Rigol *et al.* |
| HCl–L-Serine | 1–2 | + | 3.17 | 0.1 | Rigol *et al.* |
| HCl–10 mmol L$^{-1}$ DL-Serine | 1–2 | + | 3.2 | 0.1–0.25 | Fetsch *et al.* |
| HCl–15 mmol L$^{-1}$ DL-Proline | 1–2 | + | 3.2 | 0.1–0.25 | Fetsch *et al.* |
| HCl–L-Aspartic acid | 1–2 | + | $\approx 3$ | 0.1 | Rigol *et al.* |
| HCl–60 mmol L$^{-1}$ glycoclic acid | 1 | + | 3.2 | 0.1–0.25 | Fetsch *et al.* |
| HCl–100 mmol L$^{-1}$ boric acid | 5–15 | − | 3.15 | 0.5 | Fetsch *et al.* |
| HCl–350 mmol L$^{-1}$ boric acid | 10–30 | − | 3.15 | 0.25–0.5 | Fetsch *et al.* |
| HCl–500 mmol L$^{-1}$ boric acid | 3–8 | + | 3.38 | 0.15 | Fetsch *et al.* |
| 25–50 mmol L$^{-1}$ dihydrogenphosphate | 5–15 | − | 9.2 | 0.01–0.5 | Fetsch *et al.* |
| 50–100 mmol L$^{-1}$ phosphate | 5–15 | − | 6.3–9.2 | 0.1–1 | Garrison *et al.* Fetsch *et al.* |
| 100 mmol L$^{-1}$ dihydrogenphosphate–5 mmol L$^{-1}$ phosphate–250 mmol L$^{-1}$ boric acid | 3–8 | + | 3.3 | 0.15 | Fetsch *et al.* |
| 67 mmol L$^{-1}$ dihydrogenphosphate–3.3 mmol L$^{-1}$ phosphate–167 mmol L$^{-1}$ boric acid–3.3 mmol L$^{-1}$ wolframate | 3–8 | + | 5.0 | 0.15 | Fetsch *et al.* |
| 20 mmol L$^{-1}$ rimantadine hydrochloride | 1–3 | + | 3.40 | 0.01–0.5 | Fetsch *et al.* |
| 20 mmol L$^{-1}$ rimantadine hydrochloride–2 to 50 mmol L$^{-1}$ MgCl$_2$ | 2–9 | + | 3.40 | 0.01–0.5 | Fetsch *et al.* |
| 50 mmol L$^{-1}$ carbonate | 1–4 | + | 9.0–11.4 | $\geq 0.05$ | Schmitt *et al.* |
| HCl–glycylglycine | 1–2 | + | $\approx 3$ | 0.1 | Rigol *et al.* |
| HCl–glycine | 1–2 | + | 3.17 | 0.1 | Rigol *et al.* |
| Citric acid–citrate | 1–2 | + | $\approx 3$ | 0.1 | Rigol *et al.* |
| HCl–imidazole | 1–2 | + | $\approx 3$ | 0.1 | Rigol *et al.* |
| 5 mmol L$^{-1}$ imidazole–acetic acid–20 mmol L$^{-1}$ boric acid | 2–5 | + | 4.5 | 0.01–0.1 | Norden *et al.* |
| 50 mmol L$^{-1}$ acetate | 5–10 | + | 4.6–5.15 | 0.01–1 | All authors |
| 103 mmol L$^{-1}$ urea | 5–12 | + | 3.65–6.6 | 0.15 | Fetsch *et al.* |
| 20 mmol L$^{-1}$ 2-(*N*-morpholino)ethanesulfonic acid (MES) | 1–3 | + | 6.15 | High | Dunkelog *et al.* |
| 2-(*N*-morpholino)ethanesulfonic acid (MES)–NaOH | 1–3 | + | $\approx 3$ | 0.1 | Rigol *et al.* |
| 20 mmol L$^{-1}$ *tris*(hydroxymethyl)aminomethane (TRIS) | 1–3 | + | 8.30 | High | Dunkelog *et al.* |
| HCl–*tris*(hydroxymethyl)aminomethane (TRIS) | 1–3 | + | $\approx 3$ | 0.1 | Rigol *et al.* |
| 20 mmol L$^{-1}$ 3-(cyclohexylamino)-1-propanesulfonic acid (CAPS) | 1–2 | + | 10.4 | High | Dunkelog *et al.* |
| 20 mmol L$^{-1}$ 2-(*N*-cyclohexylamino)ethanesulfonic acid (CHES) | 1–2 | + | 9.5 | High | Dunkelog *et al.* |

[HS], concentration of humic substances; + hump present; − no hump.

aggregates as pseudomicelles, whereas Wershaw and Chien *et al.* used humic membrane-micelle, Dachs *et al.* introduced the term fractal aggregates, Shevchenko *et al.* used polymers and finally, the formation of oligomers was proposed by Havel *et al.* in 1997.

The most interesting theory from the last decade is that of Wershaw, who introduced a model for humic materials which provides a mean of understanding the interactions of hydrophobic compounds and HS. This model considers humic material to be constituted of a number of different oligomers and simple compounds. The resulting structures are similar to micelles or membranes in which the interiors are hydrophobic and exterior surfaces hydrophilic. Thus, the enhancement in the solubility of DDT was observed by Wershaw in concentrated HS solutions which implies the formation of HS micelles. According to Karckhoff *et al.* and Wang *et al.*, the adsorption of hydrophobic organic compounds by organic matter involves weak mechanisms of adsorption such as hydrogen bonding. Few papers concerning aggregation of HS have been published as yet. Schmitt *et al.*, using micellar electrokinetic chromatography (MECK), observed that some HS behave like ionic micelles and defined a humic initial micellar concentration (CMC) around $30 \text{ mg L}^{-1}$. The micellar properties of HS are due to both hydrophilic and hydrophobic sites which are responsible for the enhancement of the solubility of organic compounds in aqueous media or the lowering of the surface tension of water.

In 1998, the aggregation properties of HS were studied by CZE. Oligomerization of HS was observed during CZE when the effect of HS concentration on CZE separation patterns was studied. The phenomenon was described for several BGE: DL-alanine, phosphate and rimantadine systems. When either concentration or sample injection time were increased, significant changes in HS separation patterns were observed. The CMC of humic acid was estimated to be around $35 \text{ mg L}^{-1}$.

# Interaction of HS with Metal Ions and Organic Compounds

## Metal Ions

The interaction of metal ions with natural soil particles is complex, involving various mechanisms Binding between trace metals and dissolved HS is important in controlling the chemical speciation, toxicity and bioavailability of trace metals. For example, it has been observed that, in the presence of HS dissolved in seawater, the accumulation rates of cadmium in organisms were faster in comparison to accumulation rates of cadmium from seawater without HS.

Different models of metal–HS interactions have been described. Due to the broad spectrum of binding sites reported by Ephraim, HS can interact

with metal ions by adsorption, ion exchange, precipitation and/or surface complexation. Specific adsorption involves several heavy metal ions such as $Cd^{2+}$, $Ni^{2+}$, $Co^{2+}$, $Zn^{2+}$, $Cu^{2+}$, $Pb^{2+}$ and $Hg^{2+}$. Co-precipitation of trace metals with carbonates is very important for semi-arid soils and in soils formed from limestone. Techniques like synchronous and time-resolved fluorescence, luminescence, anodic stripping voltammetry and modified carbon paste electrodes have been used in the study of the fate and distribution of inorganic pollutants in the environment. Many papers concerning the binding of metal ions to HS have been published.

In 1997, Nordén and Dabek-Zlotorzynska applied CZE to study $Sr^{2+}$, $Pb^{2+}$, $Cu^{2+}$, $Hg^{2+}$ and $Al^{3+}$. Their interactions with HS were confirmed. Study of the influence of pH on the complexation showed that complexation can be better followed by direct rather than indirect detection. The following order of complexation was obtained by CZE: $Al^{3+} > Hg^{2+} > Cu^{2+} > Pb^{2+} > Sr^{2+}$. This result is in agreement with other work on metal ion–HS interaction strengths. Furthermore, when pH is increased, complexes are more stable and, ageing of HS solutions showed an increase of more than 25% in the amount $Hg^{2+}$ bound to HS after 5 months. This increase is supposed to be connected to the possible time dependence of the reduction process, and/or to be related to the macromolecule nature of HS. Changes in HS separation pattern were also observed when adding iron(III) to HS and forming $Fe^{3+}$–HS complexes, when the study was performed at pH 12.

## Radionuclides

The other important interactions which have been studied and are highly important with respect to environmental protection policies are those between HS and radionuclides like $^{60}Co$, $^{85}Sr$, $^{137}Cs$, $^{237}Np$, $^{241}Am$, etc. For example, it has been shown that samarium(III) ($10^{-10}$ to $10^{-5} \text{ mol L}^{-1}$) and americium(III) ($10^{-11} \text{ mol L}^{-1}$) form 1:1 complexes with HS; their pH-independent complexation constants, $\log \beta$, are respectively $7.1 \pm 0.2$ and $6.6 \pm 0.2$. Spiking actinide elements (Th, U, Np, Pu, Am) in HS fractions, it has been shown that these fractions vary greatly in their effectiveness and selectivity as ligands for early actinides.

Rigol *et al.* used CZE to study humic fractions in organic soil and their relationship with radiocaesium mobility. The quantification of the radiocaesium confirmed that there may be some organic matter–radionuclide interactions other than those originated by HS and which may govern radionuclides retention in soils with a high content of organic matter.

### Organic Compounds

It has been known for many years that organic compounds interact with HS. There have been studies on the interaction of pesticides (chlorodimeform and lindane) and herbicides (paraquat, 2,4-dichlorophenoxyacetic acid and atrazine) with HS. Later, the effect of HS was studied on solute transport in clays and it was suggested that humic and fulvic acids facilitate the transport of small organic molecules by encapsulation. Over the last 5 years, the effects of dissolved HS on the bioconcentration of xenobiotics have been studied by several authors. It was observed that dissolved HS can change the physicochemical properties of these pollutants in aqueous environments by modifying the hydrolysis kinetics, enhancing their water solubility or decreasing the toxicity of the organic chemicals. The binding of hydrophobic organic compounds to humic polymers has been studied, using a predictive thermodynamic HS–organic solute interaction model. A distributed reactivity model was studied for phenanthrene sorption and desorption equilibria by soils and sediments. Relationships were observed between the chemical and structural characteristics of associated organic matter and it was noted that hysteresis is important to the fate and transport of organic contaminants in environmental systems.

It was proposed to use the potential of dissolved HS to enhance the desorption of hydrophilic pollutants in remediation processes for soils and waters. Following this idea, some authors used a complexation–flocculation method to determine binding coefficients of phenanthrene, anthracene, pyrene and fluoranthene to dissolved HS. In some cases, pH dependence of the sorption was observed. While naphthalene binding to dissolved HS was not decreased at lower pH, dichlorodiphenyl trichloroethane (DDT) and polyaromatic hydrocarbons (PAHs) showed the opposite effect.

As reviewed in Table 1, the effect of organic solvents on the migration behaviour of HS in CZE was also studied. Migration time values are usually shifted to longer values compared to aqueous BGE. This effect is attributed to either a decrease in electroosmotic flow (EOF) or an increase in the electrophoretic mobility of HS compounds, or even to both effects together. Furthermore, some additional fractions were distinguished when applying mixed solvents. They are attributed to the formation of supramolecules between the organic solvent and HS.

Paramagnetic studies have been carried out in the study of atrazine solubilized by humic micellar solutions. This work suggests that atrazine is preferably present in the hydrophobic interior of HS micelles. Other techniques, such as light and X-ray scattering also give evidence of molecular aggregation in humic solutions. Using ultraviolet-visible (UV/Vis), Fourier transform infrared and electron spin resonance spectroscopy, a group of authors studied the mechanism of atrazine sorption on HS and confirmed that hydrogen bonding is responsible for the interaction. Proton transfer and possibly also hydrophobic bonding are involved in the interactions between atrazines, and HS and CZE studies of the binding between S-triazines and HS confirmed that interactions between ionizable pesticides and HS occur: differences in the electropherograms were observed.

## Capillary Electrophoretic Separation of HS

The major advances in the separation of HS are listed in Table 2.

CZE separations of HS have been performed using UV/Vis, diode array or laser-induced fluorescence detection. In Table 1, the BGE used for this purpose are listed; mainly borates, amines and amino acids have been applied. Nevertheless, even if a large variety of BGE were used, four main types of electropherogram separation patterns can be observed and the three most important types of HS separation are shown in Figure 1. The first type is obtained in a tetraborate BGE and represents separation patterns with one or more broad peaks, called usually the humic hump.

**Table 2**  Important advances in HS separation

| | |
|---|---|
| 1960–70 | First applications of electrophoretic methods to HS |
| 1980–90 | Application of size exclusion chromatography to HS |
| 1991 | First HS separation by capillary technique (Kopáček *et al.*) |
| 1994 | First application of capillary zone electrophoresis (CZE) to HS separation (Rigol *et al.*) |
| 1996–1997 | CZE study of HS interactions with some metal ions and organic compounds |
| 1998 | First separation of HS on to 15–30 peaks or fractions (Fetsch and Havel) |
| | Study of HS properties by CZE as adsorption on capillary wall and oligomerization process (Fetsch *et al.*) |
| | Combination of ultrafiltration and CZE (Rigol *et al.*) |

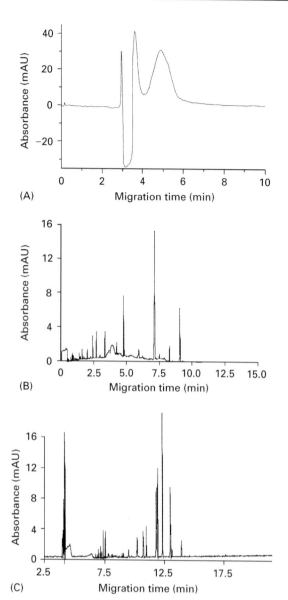

**Figure 1** Three BGEs for coal-derived humic substances. Humic acid: Fluka no. 53680 (analysis number 38537/1 293) supplied by Fluka Chemika (Switzerland). Fused silica capillary: L = 43.5 cm (l = 35.5 cm) × 75 μm i.d. (A) BGE 1 and conditions: 60 mmol L$^{-1}$ DL-alanine adjusted by HCl at pH 3.20; 15 kV, 20 s hydrodynamic injection, 40°C, 220 nm and C$_{HA}$ = 100 mg L$^{-1}$. (B) BGE 2 and conditions: 50 mmol L$^{-1}$ phosphate adjusted by NaOH at pH 9.20; 20 kV, 15 s hydrodynamic injection, 40°C, 220 nm and C$_{HA}$ = 100 mg L$^{-1}$. (C) BGE 3 and conditions: 350 mmol L$^{-1}$ boric acid adjusted by HCl at pH 3.20; 20 kV, 15 s hydrodynamic injection, 40°C, 220 nm and C$_{HA}$ = 530 mg L$^{-1}$.

Sometimes, these humps do show shoulders and some resolution. It is usually suggested that the hump corresponds to the average electrophoretic mobility of the HS polymeric mixture.

The second type of electropherograms in L- or DL-alanine BGE consists of three fractions. These electropherograms can distinguish between HS of different origin via changes in migration times of the fractions and in the number of peaks. It was shown in these studies that temperature plays an important role in the patterns obtained. The best separation is obtained for temperatures in the range 30–40°C and the highest peak area and peak height are obtained at ≈ 40°C. An example of separation under optimized conditions is shown in Figure 1A. Three main fractions were found and, for the wavelength lower than 220 nm, a negative peak was observed. Separation also depends on pH and buffer concentration. As the EOF changes with pH, the migration times of the separated components are directly affected and generally decrease as pH increases. In the pH range 5.0–9.5, HS constituents are moving as anions. In the case of alanine-based BGE, it appears that one fraction is moving as a neutral molecule, while the others are moving as anions. Under the presence of HS, it is difficult to determine the EOF simultaneously with HS separation because common neutral markers used (mesityl oxide, methanol, acetone and water) interact with HS. Independent determination of the EOF is necessary.

In the case of borate BGE, when increasing BGE concentration, migration times increase but no important changes in separation patterns are observed.

The next type of HS separation gives a humic hump with small peaks on it. This effect is generally observed with dihydrogen phosphate BGE (Figure 1C) or dihydrogenphosphate combined with borate BGE.

On the other hand, quite important changes in the separation patterns and number of peaks, were observed using slightly acid BGE, which consists of a high concentration of boric acid. This gives the best HS separation resolving from 10 to 30 peaks depending on the type of humic acid sample. In this case, highly concentrated (100–350 mmol L$^{-1}$) boric acid BGE is used as a complexing BGE (Figure 1C). The possible explanation for this excellent separation pattern is as follows: functional groups of hydroxycarboxylic acids, oxalic acid and oligoalcohols are present in the HS structure. Boric acid interacts with phenolic and carboxylic groups and therefore also reacts in a similar way with HS. This reaction was also proved by spectrophotometric and potentiometric studies. It was suggested that the large number of peaks observed in highly concentrated boric acid BGE is due to the breaking of HS oligomers into real fractions of humic substances (**Figure 2**). Humic substances contain a limited number of chemical entities (even as low as perhaps only 3–7). However, when the concentration of HS is increased, these species do oligomerize. Complexation of boric acid with phenolic or carboxylic groups prevents the formation of oligomers and thus the individual chemical compounds can be separated by CZE. The species formed must be kinetically robust, which is the condition to observe separated peaks. If just three fractions, $HA_1$, $HA_2$ and $HA_3$, are considered, oligomerization might take place according to eqns [1–3]:

$$m(HA_1) \rightleftharpoons (HA_1)_m \qquad [1]$$

$$n(HA_2) \rightleftharpoons (HA_2)_n \qquad [2]$$

$$o(HA_3) \rightleftharpoons (HA_3)_o \qquad [3]$$

In addition, mixed oligomers can be formed according to the general reaction:

$$p(HA_1) + q(HA_2) + r(HA_3) \rightleftharpoons (HA_1)_p(HA_2)_q(HA_3)_r$$
$$\text{(oligomer)} \qquad [4]$$

If reaction [4] is quantitative, instead of the peaks of $HA_1$, $HA_2$, $HA_3$ species, just one peak (one hump) of the oligomer will be observed.

This oligomer will be broken down by reaction with boric acid

$$(HA_1)_p(HA_2)_q(HA_3)_r + sB(OH)_3$$
$$\rightleftharpoons pHA_1(B(OH)_3)_t + qHA_2(B(OH)_3)_u$$
$$+ xHA_3(B(OH)_3)_v + (s - pt - qu - xv)B(OH)_3$$
$$[5]$$

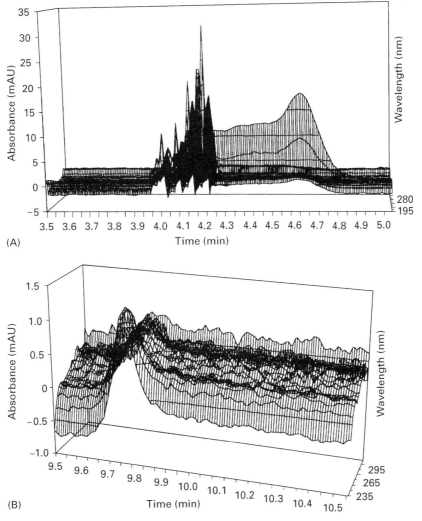

**Figure 2** Fluka humic acid three-dimensional electropherogram at higher HS concentration in boric acid BGE. Humic acid: Fluka no. 53680 (analysis number 38537/1 594) supplied by Fluka Chemika (Switzerland). Fused silica capillary: L = 43.5 cm (l = 35.5 cm) × 75 μm i.d. BGE: 350 mmol L$^{-1}$ H$_3$BO$_3$ adjusted by HCl at pH 3.15. Conditions. 20 kV, 40°C, 15 s hydrodynamic injection and C$_{HA}$ = 533 mg L$^{-1}$. (A) HA Cathedrale; (B) HA Volcano; (C) HA Alpes; (D) HA Mt Everest (Reproduced with permission from Fetsch and Havel, 1998.)

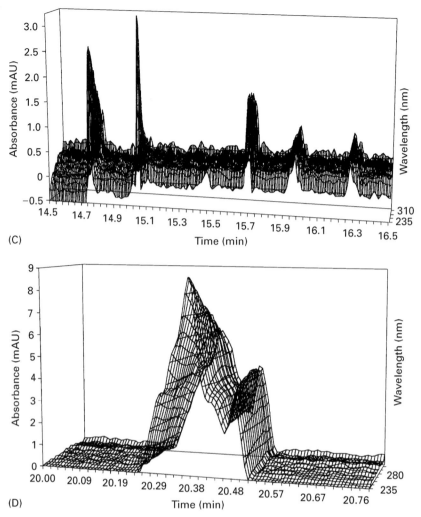

**Figure 2** *Continued*

Oligomers are broken down and three monomers will be observed in this theoretical case. Thus, it seems that for the first time separation of HS into the real fractions, real chemical entities forming the humic substances mixture, has been achieved in highly concentrated boric acid BGE. Some other examples of fingerprints are presented in **Figure 3**, showing the separation of humic acids of peat, soil, oxyhumolite and tschernozem origin.

In conclusion, different CZE separations have been presented with very different results; the most interesting and efficient were the results of Fetsch and Havel in highly concentrated boric acid solution.

## Future Developments

CZE has been shown to be the most powerful tool for the separation and characterization of HS due to their ionic and/or polyelectrolyte properties. The CZE separation patterns of HS obtained may, in the future, find a direct application in forensic science.

Nevertheless, even if several different models of HS are proposed and various properties of HS intensively studied, the real structure of humic, fulvic, humin and hymatomelanic acids is still unknown. The latest results obtained by CZE showing separation into 10–30 fractions present an optimistic insight into the humic substances puzzle. On the basis of recent results, fraction collection in order to perform studies on individual fractions by gas chromatography–mass spectrometry, matrix assisted laser desorption–time of flight (MALDI-TOF) mass spectrometry and nuclear magnetic resonance is beginning to appear feasible. It is possible that the problem of HS structure is at last on the way to being resolved.

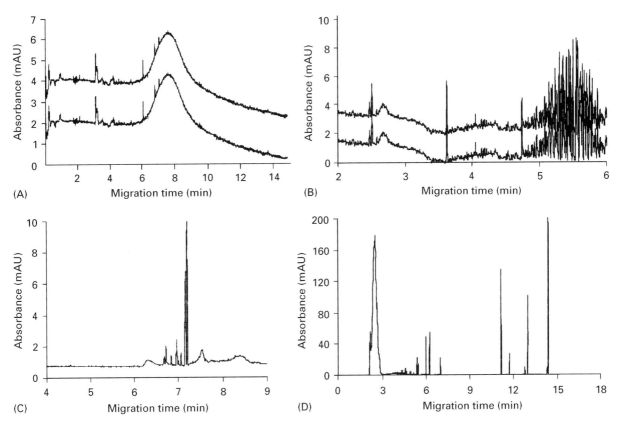

**Figure 3** Electropherogram fingerprints of humic substances of different origin in boric acid BGE. (A) International humic substances society peat reference humic acid; (B) Desmonte soil humic acid, Argentina; (C) oxyhumolite-derived humic acid, Bilina, Czech Republic; (D) Tschemozem humic acid, Chotesov, Czech Republic. Fused silica capillary: L = 43.4 cm (l = 35.4 cm) × 75 μm i.d. BGE: 350 mmol L$^{-1}$ H$_3$BO$_3$ adjusted by HCl at pH 3.20. Conditions: 20 kV, 40°C, 215 nm and 1 s hydrodynamic injection. Humic acid concentration: C$_{HA}$ ≅ 600 mg L$^{-1}$.

## Further Reading

Davies G and Ghabbour EA (1999) *Humic Substances: Structures, Properties and Uses*. Special publication no. 228. Letchworth, UK: Royal Society of Chemistry.

Dunkelog R, Rüttinger H-H and Peisker K (1997) Comparitive study for the separation of aquatic humic substances by electrophoresis. *Journal of Chromatography A* 777: 355.

Fetsch D and Havel J (1998) Capillary zone electrophoresis for the separation and the characterization of humic acids. *Journal of Chromatography A* 802: 189.

Fetsch D, Hradilová M. Peña Méndez EM and Havel J (1998) Capillary zone electrophoresis study of aggregation of humic substances. *Journal of Chromatography A* 817: 313.

Kopáček P, Kaniansky D and Hejzlar J (1991) Characterization of humic substances by capillary isotachophoresis. *Journal of Chromatography A* 545: 461.

Nordén M and Dabek-Zlotorzynska E (1997) Characterization of humic substances using capillary electrophoresis with photodiode array and laser-induced fluorescence detection. *Electrophoresis* 18: 292.

Rigol A, Vidal M and Rauret G (1998) Ultrafiltration-capillary zone electrophoresis for the determination of humic acid fractions. *Journal of Chromatography A* 807: 275.

Schmitt-Koplin Ph, Garisson AW, Perdue EM, Freitag D and Kettrup A (1998) Capillary electrophoresis in the analysis of humic substances facts and artifacts. *Journal of Chromatography A* 807: 101.

Shevchenko SM and Bailey GW (1996) Life after death: lignin–humic relationships reexamined. *Critical Review of Environmental Science and Technology* 26: 95.

Stevenson FJ (1982) *Humus Chemistry: Genesis, Composition, Reactions*. New York: Wiley-Interscience.

# Gas Chromatography

**J. Pöerschmann**, Centre for Environmental Research, Leipzig, Germany

The main mass of organic carbon in the aquatic environment and in soils/sediments is located in humic organic matter (HOM). A typical agricultural soil contains about 2–7% HOM in the upper level. Depending on the type of water (freshwater lake, marine origin, bog water, etc.), the content of aquatic HOM may range from 50 to almost 100% of the dissolved organic carbon. Terrestrial HOM acts mainly as a pollutant sink; aquatic HOM has a distinctive vehicle function. Thus, HOM influences fate, bioavailability and transport behaviour of organic and inorganic pollutants.

The detailed structural characterization of the ubiquitous HOM is an ambitious task because these polymers do not possess the uniform structure observed with other natural polymers. The extreme heterogeneity of both soil-derived and aquatic HOM in terms of chemical composition, monomer unit sequence, molar mass and functionality is associated with their genesis: for example, soil organic matter is composed of a variety of plant and animal residues in different stages of decomposition, of (radical) metabolites possessing mesomeric forms, and of products formed from these breakdown products. During the genesis, the chemical identity of the precursors is lost by abiotic and/or biotic condensation reactions.

The combined application of powerful analytical techniques may reveal some basic principles in HOM structure, but will probably never give a complete structure. Therefore, HOM analysis should not be targeted at the elucidation of discrete HOM structure but to recognizing substructures and functional groups (including their linkages), on the basis of which the fundamental interactions of HOM with environmental contaminants of organic and inorganic origin may be better understood. Nondestructive nuclear magnetic resonance (NMR) and Fourier transform infrared spectroscopy (FTIR) have proved to be very useful in providing information on functional groups and on the neighbourhood of substructures. Small scale analytical pyrolysis and controlled wet chemical degradation, in which the polymeric network is broken down to smaller subunits, constitute efficient supplements to nondestructive spectroscopic methods, in which an averaged structure is studied. Thermal and wet chemical degradation can give volatile products amenable to gas chromatography (GC) analysis, including hyphenated techniques, such as gas chromatography–mass spectrometry (GC-MS) and gas chromatography–atomic emission detection (GC–AED). The combination of data obtained from the structure-related mass spectrum and the element-related atomic spectrum is especially useful for the detection of heterocyclic compounds as well as chlorinated structures in humic-like chlorolignins. The basic assumption in this strategy is to correlate the destructive products with original moieties present in the starting HOM network.

In addition to structure-related analytical investigations, destructive methods can also be applied for elucidation of the carbon cycle in soils and water, for solving geochemical problems (such as the diagenesis of HOM), in recognizing pathways of emissions from biomass combustion, in revealing pesticide contamination of soils, in studying humification processes taking place in landfills, in studying chlorinated aqueous HOM, in screening residues in bioremediated soils, and for other applications.

## Conventional Analytical Pyrolysis

The traditional analytical pyrolysis, which can easily be performed using a flash pyrolyser or ferromagnetic wires with defined Curie temperatures, is useful in detecting certain building blocks, including polysaccharide-like substances, aromatic units, cyclopentenone units, N-containing units and aliphatic units in the HOM network, thus giving some evidence of the structure, origin, genesis, degree of decomposition and humification (condensation) of the polymer under study. As an example, pyrograms of marine samples are rich in protein products derived from phytoplankton, but are free from the lignin-derived guaiacyl (4-hydroxy-3-methoxyphenol) and syringyl (4-hydroxy-3,5-dimethoxyphenyl) compounds which are common in terristrial and freshwater samples. Although there are many sources of typical fragments in the pyrolysate, it is generally accepted that furfural, levoglucosan (and other anhydrosugars), pyranones and acetic acid originate from carbohydrates, methoxyphenols from wood and lignin, ammonia, acetonitrile and pyrrole from proteins, while alkanes, alkylbenzenes (-naphthalenes, -phenanthrenes, etc.) and fatty acids originate from fossil fuels and biomass. Pyrograms of soil- and sediment-derived HOM are mainly characterized by the classical $n$-alkane/$n$-alk-l-ene/$n$-$\alpha,\omega$-alkadiene triplet series.

They may originate from esters, the saturated alkanes originating from decarboxylation of the fatty acid moiety, and the olefins from the alcohol moiety as a result of $\beta$-scission. Another source may be alkylbenzenes with long side chains.

Generally, pyrograms are very complex, composed of several hundreds of compounds. **Figure 1** shows two extracted ion chromatograms used to trace thiophenes and benzenediols in the complex pyrogram. The (anthropogenic) fulvic acid under study was isolated from a coal wastewater pond, where it was formed spontaneously from coal wastewater components. High concentrations of phenols and benzenediols as well as of heterocyclic compounds are striking in comparison with pyrograms of HOM isolated from natural sources. The comparison between these patterns in the fulvic acid isolated from the coal wastewater, in the HOM of the associated sediment on the bottom of the pond, in the organic solvent extract of the sediment (obtained by means of accelerated solvent extraction or ASE) and the pollutant pattern in the wastewater allows conclusions to be drawn on the humification process and on the

pathways of contaminants present in the coal wastewater.

Reproducibility of the pyrograms is mainly determined by the steps in sample preparation, pyrolysis and pyrolysate transfer to the GC column rather than by the pyrolysate analysis by GC-MS (AED). Prior to pyrolysis, the HOM sample is usually subjected to an intense degreasing procedure, e.g. by Folch's chloroform–methanol (2 : 1) mixture to avoid ambiguous results with regard to the source of the pyrolysate. The extract from exhaustive solvent extraction contains the same compounds which are released during thermal desorption at subpyrolysis temperatures (e.g. 300°C). Both methods reveal contaminants attached to the HOM rather than covalently bound. To obtain a more homogeneous material and to simplify the pyrogram, a combination of pyrolysis and chemical modifications may be used: for example, acid hydrolysis removes carbohydrates and proteins/peptides, whereas the more resistant hydrocarbon fraction remains unchanged.

In general, pyrograms of HOM reveal overwhelmingly nonpolar hydrocarbons and less polar

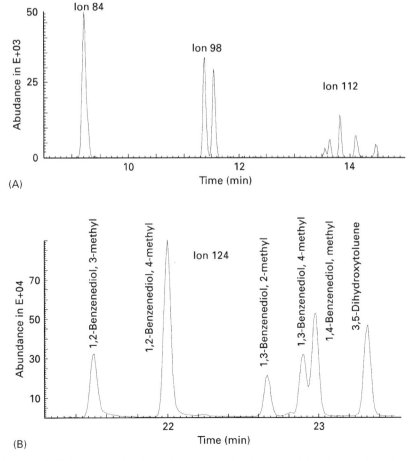

**Figure 1**  (A) Thiophene and (B) benzenediol pattern of an aquatic humic acid isolated from coal wastewater on an inert HP-5 capillary column.

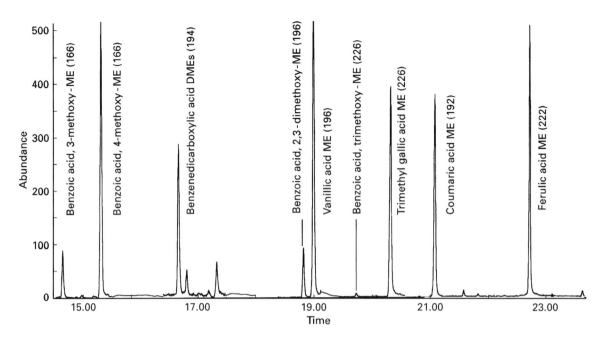

**Figure 2**    Aromatic methyl esters (ME) obtained in TMAH-assisted *in situ* methylation of a humic acid isolated from peat. Selected ion traces are given in parentheses.

heterocyclic compounds (e.g. benzofuranes, chinolines, benzothiophenes, etc.). Evidently, there are large discrepancies between NMR results and common titration results on the one hand, which indicate high carboxyl and hydroxyl group contents, and the basically nonpolar pyrolysates on the other hand. This finding is due to pitfalls occurring during pyrolysis, like the decarboxylation of fatty acids to give alkanes and the formation of polar, less volatile products inaccessible to GC (see below). Some efforts have been made to overcome these drawbacks. The coupling of pyrolysis with the 'soft' field ionization MS to give overwhelmingly molecular ions of thermal degradation products, including polar and higher molecular ones, turned out to be very useful in rapid profiling of HOM. The significance of this strategy can be further enhanced by using high resolution MS. The hyphenated technique thermogravimetry-MS, which allows running experiments in a time-resolved manner, can give further information on the binding state of the compounds released from the matrix. Another approach is to produce appropriate derivatives to preserve the analytical integrity of HOM in a better way.

## Thermochemolysis with TMAH

*Challinor* was the first to introduce *in situ* methylation using a methanolic solution of tetramethylammonium hydroxide (TMAH) with biopolymers to convert polar carboxylic and hydroxyl groups to less polar and GC-accessible methyl derivatives so as

to avoid decarboxylation and dehydration. Methodological studies revealed that simultaneous pyrolysis/methylation (SPM) is not a pyrolysis but a thermally assisted chemolysis. This saponification/esterification reaction can be performed at subpyrolysis temperatures in both an online and offline (batch) approach, the latter including flushing of the thermochemolysis products into an organic solvent spiked with internal standards.

In contrast to conventional pyrolysis, some compound classes of diagnostic value can be detected by means of the SPM approach:

1. Aromatic acids, including mono-, di- and trimethoxybenzenecarboxylic acid methyl esters, benzenedicarboxylic acid dimethyl esters, mono- and dimethoxybenzenepropenoic acid methyl esters (**Figure 2**). The presence of trimethylgallic acid (3,4,5-trimethoxybenzoic acid) is of special diagnostic value, because it indicates the final step in the oxidation of side chains in lignin units[1]. To verify the ambiguous origin of methoxy compounds (real methoxy moieties or resulting from originally free hydroxyls), deuterated TMAH can be used[2]. Another approach is the application of

---

[1] The 3,4,5-trimethoxybenzoic acid methyl ester may also be derived from syringic acid (4-hydroxy-3,5-dimethoxybenzoic acid).

[2] Analogously, to investigate the role of methanolysis, the reagent TMAH/CD₃OH may be used.

tetrabutylammonium hydroxide, giving rise to O-butyl ethers only from phenolic groups, whereas aliphatic hydroxyls do not undergo a reaction with the weaker acidic reagent.

2. Long chain fatty acid methyl esters (FAMEs). A pronounced even-over-odd discrimination constitutes a strong evidence of biogenic input, while uniform FAME patterns indicate anthropogenic origins. Iso- and anteiso FAMEs (frequently $C_{15}$ and $C_{17}$) are related to Gram-positive bacterial activities, whereas cis-vaccenic acid ($C_{18:1}$, double bond in $\Delta 11$ position) is related to Gram-negative bacteria.

3. The homologous series of $\alpha,\omega$-dicarboxylic acid methyl esters (DME). The $\alpha,\omega$-alkanedioic acids are of distinctive diagnostic value because they are thought to act as bridges in the HOM network and thus are considered to be markers for the cross-linking ratio. However, other pathways cannot be ruled out: the $C_9$-DME is widely distributed in both aquatic and terrestrial HOM of natural origin. It may originate from fatty acids with double bonds in $\Delta 9$ position, e.g. oleic acid, which have undergone oxidation. Analogously, the $C_{11}$-DME is common in pyrolysates of bacteria, the lipids of which contain distinctive cis-vaccenic acid moieties.

4. The pristenes refer to chemically bound phytol and tocopherols.

5. 1-Methoxyalkanes formed by thermochemolysis represent mainly alkanols bound by ester groups.

SPM at 'mild' temperatures followed immediately by thermal treatment under more severe conditions can be used to estimate the share of ester, ether and C–C bonds in the HOM polymeric network. Pyrograms of HOM of both aquatic and terrestrial origin reveal a pronounced FAME pattern under mild conditions at 500°C, but do not reveal any alkanes. SPM at 750°C after the treatment at 500°C gives typical n-alkane patterns originating either from ether linkages or from alkylbenzenes. These findings may be attributed to the fact that ester bonds are cleaved at 500°C, whereas the more stable C–O–C and C–C bonds are resistant to decomposition at this 'mild' temperature.

## Limitations of Pyrolysis in Structure Elucidation of HOM

Generally, pyrolysis work is aimed at obtaining a maximum yield of pyrolysis products to be analysed by means of GC and at ensuring that these volatile products reflect the structure of the polymer

from which they are generated as truly as possible[3].

It was not until the mid 1990s that Kopinke carried out the first quantification experiments with HOM pyrolysis using a two-detector technique (MS to identify the pyrolysate, flame ionization detector (FID) to quantify organic carbon). Before that time, quantification was done in a semiquantitative way at best (+ + + most abundant, + + less abundant, etc.). Calibration can be done offline using on-column injection of calibration mixtures, or can be performed by vaporizing internal standards in the hot pyrolysis interface during equilibration just before the beginning of the pyrolysis process. Three fractions can be distinguished in the pyrolysate, the sum of which amounts to almost 100% of the original HOM organic carbon:

1. A solid residue, which consists almost exclusively of carbon, due to charring reactions resulting from the elimination of functional groups and additional cross-linking. This residue is about 30% of the original organic carbon content, largely independent of the kind of HOM studied. Thus, considering the fact that HOM consists of about 40–50% organic carbon, over 50% of the HOM's carbon, which is expected to carry substantial structural information, is lost on pyrolysis. Unfortunately, this residue cannot be significantly reduced by increasing the pyrolysis temperature, and it is also in the same order of magnitude when pyrolysing under in-source vacuum conditions. The latter finding indicates that the 'coke' formation is a solid-phase reaction rather than a transport-limited step. When pyrolysing soils, the tendency to form solid carbonaceous residue is correlated to organic carbon content.

2. Compounds in the pyrolysate with little structural significance. The remaining carbon is detected mainly as nonspecific light gases, including carbon dioxide, carbon monoxide and methane eluting at the beginning of the chromatogram. Although carbon dioxide is assumed to be a rough measure of decarboxylation and methane a rough measure of methyl substitution, etc., all these light gases are of minor diagnostic relevance to the HOM structure.

3. Compounds with supposed structural significance ($C > 4$, assumed arbitrarily), comprising about

---

[3] Losses due to the limitations of the analytical system are outside this consideration. They may be circumvented by pyrolysis inside a GC precolumn; all volatile pyrolysates are completely transferred to the column.

6–10% of the organic carbon. A slight increase in the yield of structure-related compounds can be observed when turning from the conventional pyrolysis to SPM (15% at best).

However, the situation gets even worse when considering the real significance of the compounds in group 3 above. Whereas long aliphatic chains definitely originate from the HOM skeleton, aromatic and cyclic compounds may undergo unwanted thermal modifications, resulting in their misinterpretation, as already discussed above with alkanes/fatty acids. Further, the occurrence of furanes in the pyrolysate does not necessarily mean that furane moieties are present in the HOM backbone. It can also be related to thermal rearrangements and reactions originating from polysaccharide-related units (e.g. xylans, cellulose from terrestrial plants) or OH-substituted carbon chains with at least four carbons. Hence, if pyrolysis products are considered to be building blocks, cellulose should be made up of furans, pyranones, anhydrosugars, etc. Similar pitfalls occur with alkylbenzenes. They can be formed in the cracking of triglycerides and fatty acids (catalytically stimulated, e.g. by elemental sulfur). Most recent findings also indicate the formation of alkylbenzenes via a metal-catalysed alkylation of aromatic hydrocarbons, the cations being complexed in the HOM network. Therefore, structural models based on conventional pyrolysis studies are largely biased.

SPM is considered to be less error-prone. However, this approach has not been scrutinized as closely as conventional approach up until now. Saiz-Jimenez, who pioneered the SPM approach, found that model phenolic acids underwent decarboxylation reactions. Most recent results indicate that the occurrence of benzene carboxylic acids in the pyrogram does not necessarily indicate their presence in the HOM backbone: lignin models free of carboxylic groups may yield MEs of benzenecarboxylic acids during SPM. On the other hand, decarboxylation reactions and isomerization of unsaturated fatty acids have not been under intense study yet, but presumably they cannot be completely excluded. Hydroxyl- and carbonyl-containing resin acid MEs have been shown to undergo side reactions with the reagent, resulting in the formation of nitrogen-containing derivatives. Likewise, aldehydes have been recently shown to undergo a Cannizarro-like reaction, in which the products are methylated to esters and ethers. In addition to that, the quantity of FAMEs released in 500°C thermochemolysis surpasses by far the quantity of hydrocarbons (including alkanes, alkenes and alkylbenzenes) released in conventional pyrolysis, pointing to further aliphatic precursors. Moreover, the FAME pattern does not necessarily coincide with the alkane pattern.

Apart from the ambiguous origin of markers in the pyrograms of HOM, these markers cannot be assigned to defined biological precursors by themselves. As an example, *n*-alkanes can be derived from anthropogenic sources (fuels), from direct input of biosynthesis (higher plants), from reduction of fatty alcohols (bacteria, higher plants), from decarboxylation of almost ubiquitous fatty acids and from depolymerization of aliphatic biopolymers. To assign these sources, isotope ratio GC-MS is very useful.

## Controlled Wet Chemical Degradation

The ultimate aim of this approach (mostly conducted in an offline mode), elucidation of the HOM structure on the basis of degradation products, is similar to that of pyrolysis. As with HOM pyrolysis, large quantities of nondiscriminated building blocks are expected from an ideal chemical degradation method in a mechanistically predictable manner[4]. Common methods to obtain defined degradation products include mainly traditional approaches, such as:

1. Reductive degradation, e.g. using catalytic hydrogenation or alkali metals such as sodium or sodium amalgam in liquid ammonia, or lithium in liquid ethylamine. (The free or solvated electrons are provided in metallic reductions by the conversion of metal atoms to their cations.) In all cases, the carbon skeleton of aliphatic substructures is claimed to remain intact, whereas bonds between carbon and heteroatoms (in particular oxygen) are cleaved. A reagent with reactivity slightly higher than that in catalytic hydrogenolysis is iodotrimethylsilane, capable of cleaving ether and ester bonds. **Figure 3** shows the fatty acid pattern of a terrestrial HOM isolated from the sediment at the bottom of a coal wastewater pond after this treatment followed by methylation. The fatty acid pattern reflects both natural sources, expressed by the even-over-odd discrimination in the $C_{14}$–$C_{18}$ range, and the presence of bacterial iso- and anteiso-acids and anthropogenic sources, expressed by very long chain acids and a less pronounced even-over-odd discrimination at higher carbon numbers.

2. Oxidative and hydrolytic degradation. Common oxidative reagents include cupric oxide, ruthenium tetroxide and potassium permanganate. The

---

[4] The derivatization of HOM has also been used for subsequent analysis by NMR or infrared spectroscopy to obtain well-resolved spectra, e.g. by silylation of groups with acidic protons.

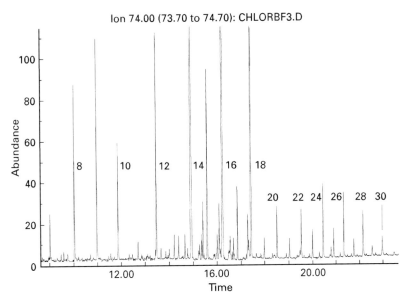

**Figure 3** Fatty acid pattern (as methyl esters) of an HOM isolated from a sediment associated with coal wastewater, obtained after iodotrimethylsilane cleavage (using chlorotrimethylsilane–sodium iodide) and boron trifluoride catalysed methylation. Numbers indicate the carbon chain length of the homomorphic fatty acid.

hydrolysis is mainly catalysed either by sodium hydroxide or sulfuric/hydrochloric acid[5]. Another direction includes mild and selective enzymatic hydrolysis. As with thermochemolysis with TMAH, these approaches revealed products, including benzene mono-, di-, tri- and tetra-carboxylic acids, furane di- and tricarboxylic acids, aliphatic mono-, di- and tribasic acids and (di-, tri- and tetracarboxyphenyl) glyoxylic acids, all of them preferably subjected to GC as MEs or trimethylsilyl ethers. CuO is known to oxidize the lignin macromolecule (cleavage of the $\beta$ O-4 ether bond, as with TMAH thermochemolysis) and produces a series of p-hydroxyphenyl, guaiacyl and syringyl compounds.

3. Transesterification, e.g. using boron trifluoride–methanol. The yields from this milder approach are quite low, but the formation of by-products and the destruction of labile structures are minimized.

As with pyrolysis, chemical degradation may be full of pitfalls. Hexose sugars are known to dehydrate to hydroxymethylfurfural; pentoses give furfural in strong acids at elevated temperatures. Moreover, melanoidins may be formed from sugars in the presence of amino acids leading to enhanced (apparent)

aromaticity. Diols can be oxidized, including ring rupture, to form low molecular weight aliphatic acids. Gallic acid can be decarboxylated on heating in water, and so forth. Although the loss in weight of the HOM on chemical degradation gives yields between 3 and 30%[6] of the mass of the starting polymer, the yields of products accessible to GC are significantly lower, in the range of 0.5–3%. A large percentage (about 25%) of the total organic carbon is lost as $CO_2$ by $KMnO_4$ oxidation. FTIR spectra give strong evidence that permethylation cannot remove all acidic protons. Thus, extrapolation to the macromolecular skeleton is quite biased.

Selective chemical degradation has also been used in a more restricted way to elucidate the fate of organic pollutants and metabolites in soils/sediments, which cannot be released by means of organic solvent extraction. This results in a better understanding of their incorporation into the macromolecular HOM and their remobilization potential. Pioneering work has been done by *Michaelis*, who cleaved bound polycyclic aromatic hydrocarbon (PAH) metabolites such as hydroxynaphthoic acids from the macromolecular HOM network using boron trichloride (capable of degrading ester and ether bonds) as well as by alkaline hydrolysis, and degraded phenolic entities from the HOM matrix by rhodium on charcoal

---

[5] Hydrolysis proceeds faster and more efficiently if the HOM is dissolved in the hydrolytic reagent; thus, alkaline hydrolysis should be used with humic acids, which by definition are insoluble in acids.

[6] Empirical findings suggest that a methylation procedure prior to degradation is beneficial towards yields of oxidative products of HOM.

hydrogenolysis. The type of bonding and linkage of pollutants/metabolites can be better elucidated using isotope-labelled Na$^{18}$OH, because only the carboxylic entity of the ester carries the heavier oxygen isotope, thus referring to products attached to the HOM matrix by ester bonds.

Although wet chemical degradation techniques as well as pyrolysis techniques involve some pitfalls and limitations, they can contribute to elucidation of the fundamentals of the diagenesis of organic compounds in soil and water. $^{13}$C NMR results yield strong evidence that a large percentage (in general, about 60% in case of the HOM under study) of methyl groups is attached to hydrogen-free paraffinic carbon. These findings cannot be explained by building blocks consisting of alkanes or alkylbenzenes. Steranes and hopanes, considered to be the end products of a complex web of diagenetic reactions starting from functionalized bacteriohopanepolyols and serving as biomarkers in ancient sediments and petroleum, are assumed to account for that. Indeed, pentacyclic terpanes with 32–35 carbon atoms, confirmed by tracing $m/z = 191$ in the GC-MS mode, can be detected using both chemical and thermal degradation.

## Future Developments

Conventional and TMAH pyrolysis, as well as controlled chemical degradation methods, applied together in combination with highly efficient GC-MS can give insight to the building blocks and the linkages between them in the polymeric HOM network. However, analytical pyrolysis has a significantly higher potential in revealing the chemical nature of simpler polymers, e.g. polyethylene and polystyrene, which give no or minor solid residue. Traditionally performed pyrolysis work with HOM conceals much significant information on the chemical nature of HOM by thermal degradation of functional groups and thermal rearrangements. Thermochemolysis with TMAH gives rise to complementary and more independent reactions in comparison

with conventional pyrolysis. Not much attention has been paid to improved specificity of bond cleavage in the structures. It will be important to understand the mechanisms of the cleavages, and to be able to relate the products identified to possible structures in the parent macromolecule.

*See also:* **II/Chromtography: Gas:** Derivatization; Detectors: Mass Spectrometry; Pyrolysis Gas Chromatography.

## Further Reading

Abbt-Braun G, Frimmel FH and Schulten HR (1989) Structural investigations of aquatic humic substances by pyrolysis-field ionisation mass spectrometry and pyrolysis-gas chromatography/mass spectrometry. *Water Research* 23: 1579.

Hautala J, Peuravuori J and Pihlaja K (1998) Organic compounds formed by chemical degradation of lake aquatic humic matter. *Environment International* 24: 527.

Hayes MHB, MacCarthy P, Malcolm RL and Swift RS (1989) The search of structure: setting the scene. In: Hayes MB *et al.* (eds) *Humic Substances II*, chap. 1, pp. 1–31. John Wiley.

Kopinke FD and Remmler M (1995) Reactions of hydrocarbons during thermodesorption from sediments. *Thermochimica Acta* 263: 123.

Lighthouse E (1998) Isotope and biosynthetic evidence for the origin of long-chain aliphatic lipids in soil. *Naturwissenschaften* 85: 76.

Poerschmann J, Kopinke FD, Balcke G and Mothes S (1998) Pyrolysis pattern of anthropogenic and natural humic organic matter. *Journal of Microcolumn Separations* 10: 401.

Richnow HH, Seifert R, Hefter J *et al.* (1997) Organic pollutants associated with macromolecular soil organic matter: mode of binding. *Organic Geochemistry* 26: 745.

Saiz-Jimenez (1996) The chemical structure of humic substances: recent advances. In: Picollo A (ed.) *Humic Substances in Terrestrial Ecosystems*, ch. 1, pp. 1–44. Amsterdam: Elsevier.

Schulten HR and Leinweber P (1996) Characterisation of humic and soil particles by analytical pyrolysis and computer modelling. *Journal of Analytical and Applied Pyrolysis* 38: 1.

# Liquid Chromatography

**D. K. Ryan**, University of Lowell, Lowell, MA, USA

Humic substances are complex mixtures of compounds and in order adequately to elucidate and characterize the properties and reactions of humic substances, separation techniques are an absolute requirement. Initial work in this area employed low pressure liquid chromatography (LC) with extensive use of size exclusion chromatography (SEC). The advent and acceptance of high performance liquid chromatography (HPLC) as well as its miniaturiz-ation and the development of column technologies, throughout the 1970s, has resulted in HPLC becoming the most widely used chromatographic technique for the separation of humic substances. The focus here will be on the HPLC of humic materials with only a passing reference to low pressure chromatographic methods.

The physical and chemical properties of humic substances are of obvious importance in their interaction with column materials and subsequent separation in chromatography. A brief description of germane humic characteristics will be presented first, followed by a discussion of the various chromatographic modes of separation, the important detectors employed and new horizons in the LC of humic substances.

## Humic Properties

Humic substances are known by many names and can be found in measurable concentrations in almost every soil, water or sediment system on earth. Carbon-containing compounds are the common denominator from both plant and animal sources which break down under the normal sequence of death and microbial decomposition in the environment. All forms of organic carbon (OC) or organic matter (OM) that are dissolved and of natural origin are of interest here and form the broadest category, of which humic substances (HS) or the synonymous humic materials (HM) are the largest part. Nonhumic compounds under the OC classification include identifiable species such as amino acids, sugars, fatty acids and the like. Humic acids (HA) and fulvic acids (FA) are the two subunits of humic substances making up all water-soluble material in this class. The relationship between all of these categories is depicted in Figure 1.

Humic substances are macromolecular species ranging in molecular weight from about 1 to 100 kDa. They have historically not been considered polymeric in nature, lacking any confirmed monomeric unit, although this view has been challenged in recent years with the proposal of an approximately 700 Da monomer. Figure 2 shows a molecular model of the lowest energy building block conformation computed for a humic acid heximer. Humic substances exhibit significant polydispersity and are polyelectrolytic, containing numerous carboxylic acid and ionizable phenolic groups. Very small amounts of amine functionality are sometimes present; however, essentially no sulfur groups are found associated with humic samples. Properties of humic substances that are relevant to chromatographic sep-

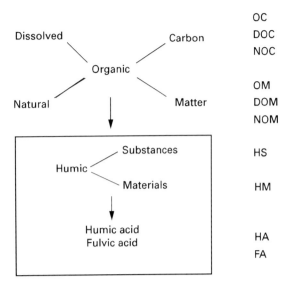

**Figure 1** Terminology, acronyms and relationship for soluble organic carbon (OC) of natural origin. Humic and fulvic acid are the two categories of soluble humic substances or humic materials which are a subset of all organic materials in the environment.

arations are summarized in **Table 1**. Significant among these are the hydrophobicity or surface activity of humic substances causing them to adsorb and partition with appropriate materials, their acidity and polyelectrolytic character imparting water solubility and charge to the molecules allowing for ion exchange, and their range of molecular sizes and possibly shapes that makes feasible separations based on size.

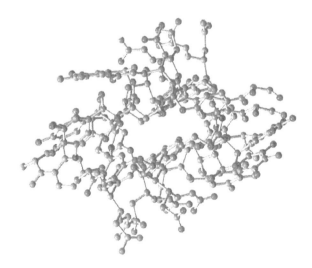

**Figure 2** (See Colour Plate 87). Molecular model of the lowest energy conformations of humic acid building blocks linked to form a hexamer. Carbon atoms are green, oxygen atoms are red, nitrogen atoms are blue and hydrogen atoms are not shown. Reproduced with permission from Davies and Ghabbour (1999).

**Table 1** Summary of characteristics for typical humic substances

| | |
|---|---|
| Elemental composition | Approximately 50% C, 5% H, 45% O; very little N, no S |
| Ash content | Typically less than 0.5% |
| Molecular weight range | 1–100 kDa |
| Purity | Complex mixture |
| Structure | Difficult to specify exactly |
| Functional groups | Aromatic and aliphatic carboxylic acids |
| | Phenolic OH and aliphatic OH |
| | Carbonyl groups |
| | Numerous aromatic rings |
| | Aliphatic chains |
| | Quinone/hydroquinone present |
| | Traces of bound metals, particularly Fe and Al |
| Acidity | Approximately 6 mmoL g$^{-1}$ from carboxyl groups |
| Solubility | Very soluble in water; generally, solubility increases with pH |
| | Poor solubility in most organic solvents |
| Other characteristics | Polydisperse |
| | Polyelectrolytic |
| | Surface active: hydrophobic portion with hydrophilic groups |
| | Metal complexation: binding of numerous metal ions |
| | Binding of organic molecules |

## Chromatographic Modes

### Size Exclusion Chromatography

Some of the earliest LC separations of humic substances were based on the size exclusion mode of separation using distilled water, salt solutions or aqueous buffers as mobile phases. SEC studies showed that size separations were possible but, unfortunately, most attempts at SEC of humic substances gave poor resolution and marginal results. Chromatograms typically exhibited only one or two broad peaks with an occasional shoulder. Experiments resulting in several peaks by SEC have often been found by subsequent analysis to be the result of a mixed-mode separation. This comes about in low ionic strength mobile phases because of the repulsive forces between negatively charged humic substance molecules and negative charges on the surface of the column-packing materials. The actual separation is based partially on size and partially on charge, resulting in poor reproducibility and dramatic changes in the chromatography with slight changes in ionic strength, pH or solution composition. This complication completely eliminates the possibility of obtaining any molecular weight information from size exclusion – something that is normally a strong point of the method.

Once the problems with SEC of humic substances were better understood, improved packing materials such as the TSK gels (Toyo Soda, Tokyo, Japan) provided somewhat better resolution and reduced surface charge. However, the SEC of humic substances did not improve dramatically with these materials, primarily because of the nature of the humic substances themselves. Although humic substances have a seemingly broad size range, most samples have been preselected for size because of environmental conditions or by a particular isolation procedure, i.e. the method by which the sample was extracted from soil, sediment or water. In addition, each humic sample may contain a fairly uniform distribution of molecules over that range. These factors result in simple one- or two-peak chromatograms consisting of broad or poorly resolved peaks, such as those shown in **Figure 3**. The recent literature on humic substances has revealed little application of the SEC technique.

### Ion Exchange

Ion exchange chromatography has found fairly limited application in the separation of humic

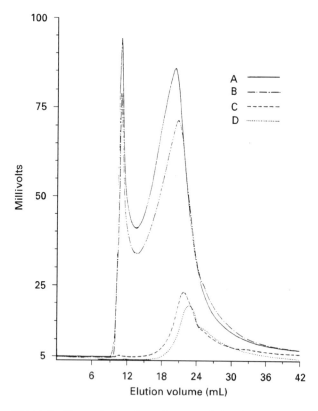

**Figure 3** Size exclusion chromatograms of a humic acid sample on a TSK G3000SW (600 × 7.5 mm i.d.) with UV detection at 280 nm. Mobile-phase conditions are as follows: A, 0.05 moL L$^{-1}$ NaNO$_3$, pH 7; B, same as A with 4.6 × 10$^{-7}$ moL L$^{-1}$, pH 6.97; C, same as A with pH adjusted to 5.54 with HCl; D, same as A with 4.6 × 10$^{-7}$ mol L$^{-1}$ acetic acid, pH 5.69. Reproduced with permission from Conte and Piccolo (1999).

substances, possibly because of the very strong retention of humic molecules for most ion exchange packings. Early work with diethylaminoethyl (DEAE) functionalized supports demonstrated that ion exchange of humic substances is feasible, but rather harsh conditions must be used to approach quantitative elution of the retained material from the column. Some researchers advocated the use of 0.5 mol L$^{-1}$ NaOH to obtain the best recoveries of humic substances from DEAE columns. Concern about the possible alteration of the humic substance molecules (hydrolysis, etc.) under these conditions has restricted its use. Ion exchange separations have been used extensively in the past for isolation and purification of humic substances from the environment. However, strongly basic conditions are also required for elution in this application and there has been a concerted effort in the area of humic substance isolation and purification to move towards less severe methods.

### Reversed-phase HPLC

The general popularity of the reversed-phase mode for HPLC (RP-HPLC) separations in analytical chemistry is paralleled by its prominence as the chromatographic mode most applied to humic separations. One important reason for the development of this situation is the availability of several variables that can be adjusted to influence the separation. The percentage of organic modifier such as methanol, 2-propanol or acetonitrile can be regulated and held constant in isocratic separations or varied (either stepwise or continuously) in gradient elution. Ionic strength and buffer pH are commonly adjusted to improve resolution and even the length (C$_1$ to C$_{18}$) or type of stationary phase can be changed from alkyl to phenyl, diol or other types of functionality.

With all the flexibility available in RP-HPLC separations, it is often unclear upon initial examination why essentially all of the published chromatograms of humic substances exhibit broad, poorly resolved peaks that by some chromatographic standards would be unacceptable. The reason for this dilemma once again is linked to the nature of the mixture of molecules in humic substances. One view suggests that, although varied in properties, the molecules form a near continuum of species with characteristics so similar to one another that they are difficult to separate. The appearance of isolated peaks in the chromatogram is a function of greater numbers of molecules of certain types (in the continuum) over other molecules which elute in the valleys between peaks.

Since RP-HPLC separates on the basis of polarity, most chromatograms show two or three distinct regions of peaks. Early in the chromatogram the most polar compounds elute, often as a jumble of sharp, but unresolved bands. Late in the chromatogram, the nonpolar species appear usually as very broad peaks with unresolved shoulders or side bands. Sometimes, a band of intermediate polarity is present midway through the chromatogram.

An important consideration in selecting a column for separating humic substances by RP-HPLC is the pore diameter of the packing material. Reversed-phase supports are available with a variety of pore sizes; however, larger pore diameters are more suitable for larger molecules such as humic substances. A common choice is a 30 nm pore diameter sold as a reversed-phase column for protein separations.

### Ion Pair Reversed-phase Chromatography

Improvements in the RP-HPLC separation of many charged species can be realized by adding an ion-pairing reagent to the mobile phase. The ion-pairing reagent is commonly a tetrabutylammonium salt (in the cationic case) which can form an ion pair with the species of interest. Humic substances contain many ionizable carboxylic acid and phenolic groups, making them very suitable for this mode of chromatographic separation. **Figure 4** shows a representative ion pair reversed-phase (IP-RP-HPLC) chromatogram for the fulvic acid fraction of humic substances derived from soil. Although IP-RP-HPLC represents some improvement over RP-HPLC, the nature of the humic substances still gives rise to broad peaks that are less than completely resolved.

With regard to the exact mechanism of the separation in IP-RP-HPLC, at least two models exist. One perspective is that the ion pairing alters the polarity of the humic substance molecules, dramatically changing their retention characteristics on the reversed-phase column. A second theory supposes that the ion-pairing reagent partitions into the column leaving charged sites that allow for an ion exchange process with the humic substances as a means of separation. Reality is most probably somewhere between these two viewpoints. It should be noted that, once a RP-HPLC column has been subjected to an ion-pairing reagent, it is always an ion-pairing column.

**Table 2** gives an overview of the HPLC modes discussed above. Representative publications are included for further reading in this area.

## Detectors

Humic substances absorb UV and visible radiation at essentially all wavelengths and can therefore be readily monitored by absorbance detectors in HPLC. The

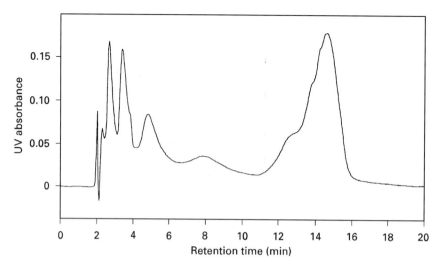

**Figure 4**  Ion pair reversed-phase chromatogram of Swannee Stream reference fulvic acid (SSRFA) separated on an SEG C$_{18}$ column using gradient elution beginning wih 37% CH$_3$CN for 8 min followed by a linear gradient to 68.5% CH$_3$CN over 8 min at a flow rate of 1 mL min$^{-1}$. Tetrabutylammonium perchlorate was used as the ion-pairing reagent at a concentration of 50 mmoL L$^{-1}$. UV absorbance at 254 nm was monitored for an 8 µL injection of 0.048% (w/v) SSRFA.

absorbance spectrum of a typical humic substance sample is essentially featureless, looking somewhat like an exponential curve. Absorbance is high in the short wavelength region of the UV spectrum and drops quickly with increasing wavelength, tailing off into the red region of the visible spectrum. Therefore, almost any wavelength can be used to detect humic substances, but shorter wavelengths are more sensitive. Fixed wavelength measurements at 254 nm are the most common; however, variable wavelength detectors have been used at a variety of wavelengths with good success.

Photodiode array (PDA) monitoring of the complete UV-visible spectrum of the column effluent adds an additional dimension to HPLC, providing several advantages over conventional single wavelength detectors. PDA systems repeatedly collect spectra at a specified time interval (often once every second) throughout the life of the chromatogram. This spectral information is invaluable in method development, particularly when many unknown compounds are present. Once the data are collected, several chromatograms can be generated, each with a different monitoring wavelength revealing features not visible in a single wavelength chromatogram. For example, a single well resolved peak at one wavelength may show a poorly resolved doublet at another wavelength. Similarly, peak purity can be determined by examining the spectra at several time intervals across a peak.

Finally, PDA detection can aid in the qualitative identification of the separated components by comparison of a peak's spectrum with the spectrum of known compounds. This approach has been used in some applications to elucidate certain structural features in separated fractions of humic substances. The general drawback with this application is that UV-Vis absorbance spectra are not conclusive evidence of a compound's identity and can only provide possible structural clues. When applied to the complex mixture of humic substances, only vague inferences can be made about possible structural components in

**Table 2**  Modes of liquid chromatography employed for separations of humic substances (HS)

| Mode | Mechanism of separation | Reference |
| --- | --- | --- |
| Size exclusion | Exclusion of larger HS molecules from certain pores based on size | Chin *et al.* (1994) Huber and Frimmel (1994) |
| Reversed-phase | Partitioning of hydrophobic portion of HS molecules into nonpolar stationary phase | Saleh *et al.* (1989) |
| Ion exchange | Exchange of anionic HS on cationic stationary phase | Andres *et al.* (1987) |
| Ion pair reversed-phase | Retention of ion-paired HS on nonpolar phase or ion exchange of HS on ion-pairing reagent-loaded phase | Butler and Ryan (1996) |

the macromolecules that are contained within a peak. This type of data gives further evidence that moieties such as vanillic acid, catechol or other substituted aromatics are part of the structure of humic substances, but cannot provide unequivocal results.

The second and only other major detection method used for the HPLC of humic substances is fluorescence. Humic substances emit a broad fluorescence peak centred around 450 nm when excited by radiation in the 330–350 nm range. This fluorescence is not considered to be a very strong signal (high quantum yield) in comparison to other fluorescent species; however, sufficient sensitivity can be obtained with standard commercial fluorescence detectors to measure easily humic substances down to the low part-per-million level sufficient for most environmental measurements. Fluorescence is a selective means of detection because not all molecules fluoresce – a situation that is clearly true of the mixture of molecules in humic substances. HPLC experiments with both absorbance and fluorescence detection have shown that some components of humic substances absorb but do not fluoresce. This lack of fluorescence may result from one of at least two possibilities. The absence of the appropriate structural features (usually extended $\pi$ electron system) in certain fractions of humic substances will render them non-fluorescent or the presence of fluorescence-quenching agents in a normally fluorescent fraction will likewise remove fluorescence. The most common fluorescence quenchers for humic substances are the paramagnetic metal ions, such as $Cu^{2+}$, $Fe^{3+}$, $Ni^{2+}$, $Co^{2+}$, $Mn^{2+}$, which form complexes at appropriate binding sites on the molecule. Recent studies have also shown that certain organic molecules can bind to humic substances either electrostatically or through hydrophobic interactions and also quench fluorescence.

## Other Considerations

Humic substances are a fairly difficult class of compounds to work with because they are poorly characterized with respect to their chemical structure and many of their properties are not yet fully understood. Intra- and intermolecular hydrogen bonding, hydrophobic interactions and electrostatic effects surely play a role in what can be called the tertiary structure or shape of humic substance molecules. The variables of pH and ionic strength clearly play a role in determining molecular shape, which is very important with respect to HPLC separations; however, an understanding of the factors influencing shape is only now yielding to analysis through the use of molecular modelling.

Irreversible adsorption, concentration effects and solubility problems also plague the analysis of humic substances by HPLC. Although the fulvic acid fraction of humic substances is soluble at any pH, humic acid is not soluble at low pH and may have varying solubility depending on the exact solution conditions. Samples of isolated humic substances must be carefully dissolved and filtered before HPLC analysis and even then may not remain soluble throughout the separation because of changing gradient conditions.

Concentration effects are a related problem since high concentrations tend to favour aggregation and coagulation of humic substances, resulting in precipitation of the most nonpolar components in the sample. An important consideration when examining the published results in chromatographic studies of humic substances is the concentration of the solution injected into the chromatographic system. Many studies are conducted using as much as $500–1000$ mg L$^{-1}$ of humic substances – a concentration level that is orders of magnitude higher than typical environmental levels in most cases. The results obtained in these instances may be influenced by concentration artifacts and are not likely to be representative of environmental conditions.

Irreversible adsorption is another difficulty that frequently occurs at the top of guard or analytical columns because of a high affinity of the column packing material for the humic substances. This mechanism may result in significant losses of sample material to the column and may skew results by selectively removing some compounds while not affecting others. The presence of bare silica for silica-based packings or other active sites can influence adsorption. In addition, the presence of sample components or impurities such as metal ions can facilitate interactions between humic substances and active sites on a column.

## New Horizons

Although HPLC may be considered a mature technique, many advances have been made in recent years, primarily in the area of detectors, that have an impact on the chromatography of humic substances. One example is the use of carbon analysers as detectors for HPLC. Instruments that measure organic carbon in solution have long been used for the measurement of humic substances and other natural organics in nonchromatographic applications. It is only in the last decade that these instruments have been modified for use in HPLC. The advantages of dissolved organic carbon (DOC) detectors include their universal detection of all organic compounds and their specificity for only carbon-containing species.

Two new approaches in the area of fluorescence detection of humic substances are the coupling of a fluorescence lifetime instrument to HPLC and post-column fluorescence quenching titrations as a means of measuring metal complex equilibria. Fluorescence lifetime measurements add yet another means of increasing the specificity of analysis, allowing improved qualitative determinations and generally increasing the amount of information obtained from fluorescence analysis of a sample. The coupling of phase modulation spectrofluorometry to HPLC results in a powerful method for measuring and resolving overlapping peaks as well as adding a dimension for identifying unknowns in a chromatogram. Although this type of detector is inherently expensive and will not see routine use, its application to humic substances, which have been shown to have three distinct fluorescence lifetimes, seems logical.

Post-column fluorescence quenching titration experiments have recently been employed in metal complexation studies to elucidate binding of metals to fractions of humic substances separated by HPLC. Quenching of the natural fluorescence by certain metal ions has become an accepted method for measuring equilibrium constants and other binding parameters for complexation of the metal by sites on humic substance molecules. Studying these equilibrium processes by HPLC has heretofore been very difficult because the chromatographic separation of the humic–metal complexes is adversely influenced by the presence of the metal ion. By separating the humic substances first without metal present, then adding metal ion via a post-column inline mixing tee, the complexes are formed just prior to their measurement by a conventional fluorescence detector. This approach alleviates any chromatographic problems associated with the presence of metal ions and allows valuable binding studies of metal ions with separated fractions of humic substances in a time-saving online mode.

Two of the most powerful instruments for measuring chemical structures of organic molecules are Fourier transform infrared spectroscopy (FTIR) and mass spectrometry (MS). A relatively new form of MS that has a major advantage in its application to macromolecules such as proteins is matrix-assisted laser desorption and ionization (MALDI) MS. The immense benefit of coupling a powerful separation technique such as HPLC with the unparalleled structural capabilities of FTIR and MS is obvious, particularly when applied to complex mixtures. However, there are significant difficulties encountered in the interfacing of these techniques which have hindered advances in this area. A novel approach has been developed to meet this challenge which centres on the inherent ability of FTIR and MS to analyse solid samples. Column effluent which exits from the chromatograph is sprayed on to a slowly moving solid substrate under controlled conditions. Solvent is evaporated and the sample dried as the separated components are deposited sequentially. Once the chromatographic run is complete, the solid substrate with the deposited sample components is placed in a modified FTIR or MALDI MS unit for analysis. Sequential analysis along the track of deposited sample components yields a series of spectra similar in concept to PDA detection in HPLC, but this approach is far more useful in elucidating structural features and providing absolute qualitative confirmation of component's identity. The application of this technology to the analysis of humic substances with RP-HPLC has been pioneered.

## Conclusion

The use of HPLC for separation of humic substances has been, and will continue to be, a challenging area in separation science. Additional information about the nature of humic substances will aid in the fine-tuning of this application of HPLC, as will improvements in column technology. Probably the most significant advances in the HPLC of humic substances will come with the utilization and development of more powerful detectors, particularly spectroscopic detectors. No doubt the complex, heterogeneous nature of humic substance samples will require a host of separation and detection strategies to be applied in the years to come before their analysis is considered in any way routine.

**See Colour Plate 87.**

*See also:* **II/Chromatography: Liquid:** Detectors: Fluorescence Detection; Detectors: Ultraviolet and Visible Detection; Ion Pair Liquid Chromatography; Mechanisms: Reversed Phases; Mechanisms; Size Exclusion Chromatography.

## Further Reading

Andres JM, Romero C and Gavilan JM (1987) Ion exchange chromatography of fulvic acids from lignite. *Fuel* 66: 827.

Butler GC and Ryan DK (1996) Investigation of fulvic acid-Cu$^{2+}$ complexation by ion-pair reversed-phase high-performance liquid chromatography with post-column fluorescence quenching titration. In: Gaffney JS, Marley NA and Clark SB (eds) *Humic and Fulvic Acids: Isolation, Structure and Environmental Role*, p. 140. Washington, DC: ACS Symposium Series 651.

Chin YP, Aiken G and O'Loughlin E (1994) Molecular weight, polydispersity, and spectroscopic properties of aquatic humic substances. *Environmental Science and Technology* 28: 1853.

Conte P and Piccolo A (1999) Conformational arrangement of dissolved humic substances. Influence of solution composition on association of humic molecules. *Environmental Science and Technology* 33: 1682.

Davies G and Ghabbour EA (1999) Understanding life after death. *Chemistry and Industry* 7 June: 426.

Huber SA and Frimmel FH (1994) Direct gel chromatographic characterization and quantification of marine dissolved organic carbon using high-sensitivity DOC detection. *Environmental Science and Technology* 28: 1194.

Liu X and Ryan DK (1997) Analysis of fulvic acids using HPLC/UV coupled to FTIR spectroscopy. *Environmental Technology* 18: 417.

Ryan DK, Thompson CP and Weber JH (1983) Comparison of $Mn^{2+}$, $Co^{2+}$, and $Cu^{2+}$ binding to fulvic acid as measured by fluorescence quenching. *Canadian Journal of Chemistry* 61: 1505.

Saleh FY, Ong WA and Chang DY (1989) Structural features of aquatic fulvic acids. Analytical and preparative reversed-phase high-performance liquid chromatography separation with photodiode array detection. *Analytical Chemistry* 61: 2792.

Sein LT, Varnum JM and Jansen SA (1999) Conformational modeling of a new building block of humic acid: approaches to the lowest energy conformer. *Environmental Science and Technology* 33: 546.

Smelley MB, Shaver JM and McGown LB (1993) On-the-fly fluorescence lifetime detection in HPLC using a multiharmonic Fourier transform phase-modulation spectrofluorometer. *Analytical Chemistry* 65: 3466.

# HYDRODYNAMIC CHROMATOGRAPHY: PRACTICAL APPLICATIONS

**A. Revillon**, Centre National de la Recherche Scientifique, Vernaison, France

## Introduction

Either in research, the control or the production of chemicals, or during general observations in biochemistry, characterization is a primary necessity. Depending on the product, the nature, structure, size, shape, and molecular weight are some of the important parameters to be measured. For instance, the molecular weight distribution of polymers and particle size of latex/colloids (MWD and PSD, respectively) have to be known so that they can be correlated with properties. The data may be obtained using many techniques. These techniques are governed according to the various properties of a material and depend on the size range of the investigated material. Hydrodynamic chromatography (HDC) is one of these techniques and has found applications for sizing soluble or dispersed solid components. This article will discuss the size distribution of organic latex particles from the theoretical and practical points of view. Progress in packing columns with fine materials or in the use of fine capillary tubes has allowed rapid separation of species with high resolution. A second field of interest is polymers in solution. The combined effects of hydrodynamic and exclusion chromatography have extended possibilities for the separation of high-molecular-weight materials.

Hydrodynamic chromatography is used for diameter determination in the micron range (some nm to some μm). It will work for both solid and soluble samples, which are eluted according to their decreasing size. This leads to a visual picture of the size distribution. The main interest of HDC lies in the rapid separation (fractionation, which is an alternative name for this method: HDF) of the liquid or solid components present in the sample. Often low peak capacity and poor resolution are its limitations and involve the necessity for peak dispersion correction. Moreover, a quantitative study requires a double calibration. The first relates to elution volume and diameter of the analysed particles; the second gives a correspondence between the signal intensity and size of particles. This intensity depends on the nature of the detector and the operating conditions, for instance the choice of wavelength in UV. As an example, **Figure 1** shows the different absorbance curves of polystyrene (PS) latexes of different sizes. **Figure 2** shows the change in absorbance and scattering with wavelength and illustrates different peak contributions of PS particles to the chromatogram.

HDC operates similarly to size exclusion chromatography (SEC) and field flow fractionation (FFF), but needs one (inert) mobile phase and one (hydrodynamic) field only. The interesting principle of particles separation is the difference in their transport rates in a capillary, related to their location in the eluent. Large particles are preferentially in the centre of the capillary, where the flow rate is max-

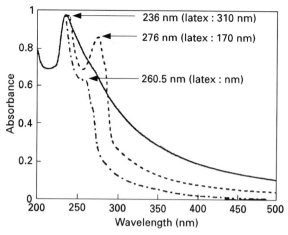

**Figure 1** UV absorbance curves for polystyrene latex particles. (Reprinted from Nagasaki S, Tanaka S and Suzuki A (1993) Fast transport of colloidal particles through quartz-packed columns. *Journal of Nuclear Science and Technology* 30(11): 1136–1144.)

imum, so they are eluted faster than small ones, which are closer to the wall of the capillary, where the flow rate is zero. The ratio of the highest elution volume for a small molecule – called a marker – to that of a given particle is the separation factor ($R_f$). Besides the simplicity of the process, secondary advantages are the rapidity of measurement performed directly on the untreated medium and ease of operation of the equipment. Variation of operating parameters (flow rate, nature of eluent and additives, size of the capillary, etc.) allows a considerable range of possible applications. Chromatography is performed

on two types of columns, open capillary or packed, where the interstitial space defines the channels. The use of very small particles (diameter 2 μm) or fine capillaries (diameter 4 μm) allows a high resolution. Relatively short capillary columns (3 m) lead to a rapid and efficient separation (5 min), as shown in **Figure 3** for monodisperse polystyrene latexes.

This article focuses on the main applications of HDC and which are under constant evaluation. There is no effect of density and no limitation on the nature of the sample, but most studies are related to the field of polymers. Since there is no solvent limitation (soluble samples) or no solubility condition (solid samples), all polymer families have been examined. Most of the applications are relative to synthetic organic colloid separation in latex production and diameter measurements for quality control. As mentioned above, interesting perspectives are apparent by combining HDC and SEC with porous packings.

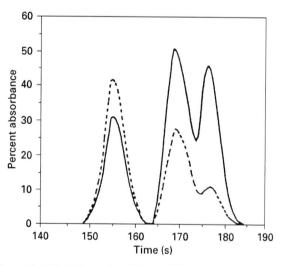

**Figure 2** Effect of wavelength (——, 220 nm; - - - -, 254 nm) on the chromatogram of a trimodal mixture of polystyrene latexes (357, 176, 109 nm) on a capillary column 2 m × 4 μm, ionic strength 1 mM. (Reprinted from Dos Ramos JG and Silebi CA (1990) The determination of particle size distribution of submicrometer particles by capillary hydrodynamic fractionation. *Journal of Colloid and Interface Science* 135(1): 165–177.)

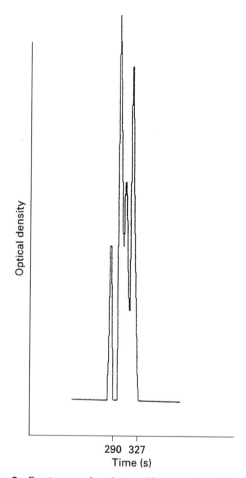

**Figure 3** Fractogram of a mixture of four samples of Dow polystyrene latex of diameters 357, 234, 176 and 109 nm. 0.1 mM SLS, capillary 3 m × 4 μm. (Reprinted from Dos Ramos JG and Silebi CA (1989) An analysis of the separation of submicron particles by capillary hydrodynamic fractionation. *Journal of Colloid and Interface Science* 133(2): 302–320.)

# Applications in Polymer Chemistry

## Colloids/Latex

Some methods of polymer synthesis offer the possibility of preparing polymers with a defined average molecular weight with a narrow distribution (anionic initiation or modern controlled free-radical polymerization). This is important since some polymer properties are dependent on molecular weight and its distribution.

For certain applications it is the size which is the key parameter. For instance, it is the particle size which is important for rheological properties, film formation or protecting ability in relatively low-cost industrial coatings such as aqueous-phase paints and inks. Latexes of small particles have a lower viscosity than those of larger particles with the same percentage solids. They also have a better storage stability against sedimentation and further aggregation. In other applications, these 'water-borne' particles may also be high-value colloids for model compounds and reference materials. For instance, they are used as standards for calibration (membranes) or for packing chromatography columns. Surface modification allows many applications in biochemistry for diagnostic aids and purification. It is thus evident that accurate monitoring of diameter is important.

These particles are obtained in the free-radical emulsion polymerization heterogeneous process. Their diameter is around 100 nm, which can be dealt with in HDC. This process tends to be substituted by bulk and organic solvents processes, in order to reduce the use of solvents in the production of polymers. General interest in these particles is 'triple' fold: their regular and spherical shape, and the possibility of obtaining a predetermined diameter of a given value. This diameter does not depend on mechanical treatment, such as milling and sieving, but depends on the chemical and thermodynamic values of the process: size measurement is a way to determine the polymerization mechanism in relation to these parameters.

**Polymerization kinetics**  As has been mentioned, a major use of HDC is related to colloids obtained by emulsion polymerization, a process which allows simultaneous high-polymerization rates and high-molecular weights. The rapidity of the HDC measurement is compatible with kinetic studies and monitoring during the polymerization. Eluted fractions may be characterized further by transmission electron microscopy (TEM) and analytical centrifugation to verify the effective existence of detected particles and to obtain a calibration curve. Photon correlation spectroscopy (PCS) may be used for a simultaneous rapid analysis.

A normal emulsion system can be depicted as follows: a large fraction of the monomer is in large droplets (diameter around 1 μm); a small part of the monomer is dispersed in the aqueous phase either as an individual soluble molecule or as small aggregates stabilized by the surfactant, and called micelles (diameter around 5 nm). Various ionic and nonionic surfactants are used, besides steric ones, which are not as sensitive to electrolyte and pH effects. The initiator is in the aqueous phase: it is decomposed to active radicals by thermal activation or redox reaction. The primary radicals may add to soluble monomers molecules to form oligoradicals and/or enter into the micelle to give active particles. This is the first step of the polymerization up to a conversion of 5–10%: initiation or nucleation. The second step is the growth of these particles by consumption of the monomer contained in the large droplets. The number of particles remains constant up to 90% conversion, but their size increases. The rate of polymerization is constant during this second step. This rate ($R_p$) and number of particles ($N$) are governed by the concentration of monomer (M), initiator (I) and surfactant (S):

$$R_p = k_p \, [M] \, N/2 \quad \text{and} \quad N[I]^{0.4}[S]^{0.6}$$

$k_p$ is the propagation rate constant.

Measurement of conversion and of particle size allows determination of the number of particles. A constant number of particles and an increase in their size are observed. The values may be correlated to concentration. Exponents may differ to those expected in the classical scheme, where particles are spherical and uniform in size. Some deviations may occur, such as additional nucleation by coagulation of precursor particles generating new particles and a second size distribution. They are of lower stability and have a slower rate of polymerization. Large aggregates may be formed by association of particles, as a result of insufficient stabilization. This leads to a broad or multimodal size distribution.

When industrial processes are discontinuous, the particles size of successive batches may differ. In order to correct this and obtain a constant quality, several batches are blended to deliver a uniform size distribution, corresponding to specific properties. Instead of performing physico-chemical measurements (rheology, surface tension) and end use tests, HDC is a rapid means of diagnosis.

Most of the literature on HDC is related to the separation of latexes. Conditions of elution – particularly the ionic strength – affect the separation, but the surface charge density of the particles does not affect the separation factor. In contrast with polymers, the shear rate has little or no effect on particle size characterization. No deformation occurs when the particle is relatively rigid.

**Modification of latexes**   To obtain large size latex, colloids obtained in a first step are polymerized further: successful results have been obtained with the method of Ugelstad.

Latexes are reactive to an extent depending on their stability. For instance, changes in size by the swelling of carboxylated styrene-butadiene latexes has been measured, according to changes in pH.

Flocculation of colloids in the presence of water-soluble ionic polymers or inorganic oxides has been observed by HDC in relation to other methods. Association of particulates under the effect of a thickener has been clearly demonstrated, though this association can be broken by intensive shear, and the same applies to aggregates.

### Mini-emulsions

Mini-emulsions are oil droplets (50–500 nm) in water, stabilized by a surfactant (such as sodium lauryl sulfate) and a co-surfactant, which may be a long-chain alkane or a fatty alcohol with low water solubility. Their action is to reduce the oil diffusion from smaller to larger droplets. These mini-emulsions may also constitute an interesting polymerization system. Their stability depends on the relative concentration of components. An exponential decrease of droplet size has been observed when the concentration of cetyl alcohol is increased. The effect of the nature of the co-surfactant and ageing time has been studied. **Figure 4** shows chromatograms of a mini-emulsion at different times, corresponding to size differences: increase of diameter and decrease of dispersity occurs when cetyl alcohol is used. This is not observed when hexadecane is the co-surfactant.

### Polymers

Dimensions of soluble polymers depend on molecular weight, nature of polymer, solvent type and temperature. A linear polymer is represented by an ideal random coil of a flexible chain, the radius of gyration being $r_g$. For actual polymers, requirements mean taking account of bond angles, hindered rotations and short-range intrachain interactions. Since polymer segments occupy space, this generates an 'exclusion volume', corresponding to an increase of dimensions; however 'unperturbed' dimensions, $\langle r_g^2 \rangle_0$, do not take solvent into account. Both interactions between

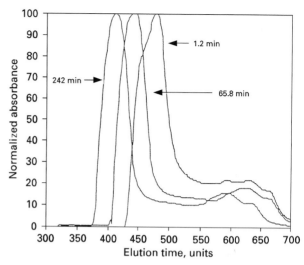

**Figure 4**   Fractograms at three different times for a toluene mini-emulsion prepared from a gel phase consisting of 10 mM SLS and 30 mM cetyl alcohol. (Reprinted from Miller CM, Venkatesan J, Silebi CA, Sudol ED and El-Aasser MS (1994) Characterisation of miniemulsion droplet size and stability using capillary hydrodynamic fractionation. *Journal of Colloid and Interface Science* 162: 11–18.)

the chains and between chain and solvent modify $r_g$. In a 'poor' solvent, these unperturbed dimensions mean that square of the radius of gyration $\langle r_g^2 \rangle_0$ is valid, since the polymer sphere is poor in the solvent. But in a 'good' solvent, it becomes $\langle r_g^2 \rangle = \alpha^2 \langle r_g^2 \rangle_0$, due to the expansion of the coil by the internal solvent molecules. For a given polymer, the unperturbed dimensions depend on solvent and temperature, corresponding to the $\theta$ conditions of Flory, with $\alpha = 1$. In these conditions, the $a$ exponent of the Mark–Houwink viscosity law $[\eta] = KM^a$ is 0.5. Finally, applying classical theories of polymers in solution allows a relationship between size and molecular weight to be obtained. But the question is: what radius is to be considered in a chromatography experiment? In other words, is the apparent radius obtained in HDC a useful answer? Viscometric measurements show that polymer chains behave rather as non-free-draining (solvent immobilized in the interior of the chain) rather than a permeable coil. This means that the hard-sphere model is a reasonable approach. It may be represented by the hydrodynamic radius $r_h$, the value of which is about $0.7\ r_g$.

Polymer molecules which are in a laminar flow near the wall in a tube are in a high stress domain and can be elongated and oriented. Those in the centre of the tube are in a low stress region, so that they differ in entropy. The entropy gradient in the tube leads to a migration of the molecules away from the wall, called 'stress-induced diffusion'. This is favoured by a high flow rate, small tube diameter and high molecular weight. A second feature is a result of elongation: the cross-section of the chain is decreased and the molecule is eluted later. A third factor may affect the elution time: the fluid inertia induces a radial migration of the molecules towards an annular region of equilibrium, which is about at $0.6R$ ($R$ is the capillary radius) of the centre of the tube (also called the 'tubular pinch effect'). This is more effective in capillary tubes than in packed columns.

Excellent results have been obtained for the separation of numerous polymers in solution, on columns packed with very small non-porous silica particles. As an example, a set of four polystyrenes (molecular weight differing by a ratio of 2) in THF was fully separated in less than 2 min (**Figure 5**). Two sets of polybutadienes (PB) and polyisoprenes (PIP) in THF at a lower flow rate were completely separated in 11 min. The maximum separation factor was about 1.3. An experimental calibration curve (molecular weight as a function of $R_f$) may be obtained for each polymer and fits theoretical predictions. A universal calibration curve is obtained when plotting relative size (chain radius $r_p/R$) as a function of $1/R_f$ (**Figure 6**). Alternatively, this means that different polymers of similar molecular weight may be separated, as shown in **Figure 7**. Other solvents of different thermodynamic properties are dioxane (PS, PB, PIP), toluene, methanol, ethyl methyl ketone and acetonitrile (PMMA), used in $\theta$ conditions (minimum interactions between chain and solvent, for a min-

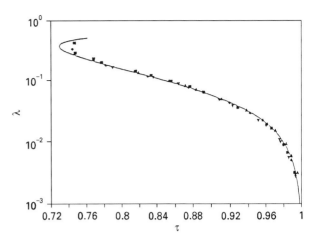

**Figure 6** Universal calibration curve: $\lambda = r_p/R$ and $\tau = 1/R_f$. (Reprinted from Stegeman G, Kraak JC, Poppe, H and Tijssen R (1993) Hydrodynamic chromatography of polymers in packed columns. *Journal of Chromatography A* 657: 283–303.)

imum coil size). The polymer size is effectively decreased, and the elution time is very sensitive to small changes in temperature around $\theta$.

Combined HDC and SEC fractionation has been shown to expand the elution scale and allow complete separation of a wide range of polystyrenes, using porous silica (**Figure 8**) and polymeric packings. This offers interesting possibilities for elution of very high-molecular-weight materials. The

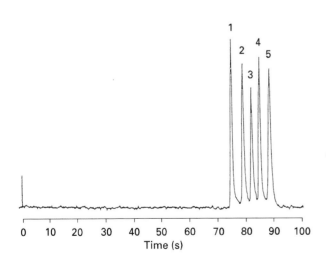

**Figure 5** High-speed HDC of polystyrenes (1: 775 000; 2: 336 000; 3: 127 000; 4: 43 900; 5: toluene as marker) dissolved in THF on a column 150 × 4.6 mm, packed with 1.50-μm nonporous silica particles. UV detection. (Reprinted from Stegeman G, Kraak JC, Poppe H and Tijssen R (1993) Hydrodynamic chromatography of polymers in packed columns. *Journal of Chromatography A* 657: 283–303.)

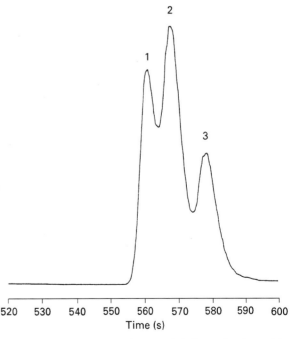

**Figure 7** Three different polymers of similar molecular weight in THF. 1: PB 330 000; 2: PIP 295 000; 3: PS 336 000. (Reprinted from Stegeman G, Kraak JC, Poppe H and Tijssen RJ (1993) Hydrodynamic chromatography of polymers in packed columns. *Journal of Chromatography A*, 657: 283–303.)

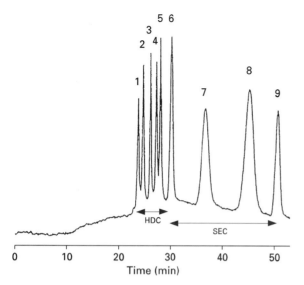

**Figure 8** HDC–SEC separation of polystyrenes ($10^{-3}$ molecular weight 1: 4000; 2: 2200; 3: 775; 4: 336; 5: 127; 6: 43.9; 7: 12.5; 8: 2.2; 9: toluene as marker) on three columns of 15 cm each, packed with Hypersil 3-μm porous particles. (Reprinted from Stegeman G, Kraak JC, and Poppe H (1991) Hydrodynamic and size-exclusion chromatography of polymers of porous particles. *Journal of Chromatography* 550: 721–739.)

calibration curve on silica (**Figure 9**) clearly shows HDC and SEC domains (higher molecular weight and steeper slope for HDC). With narrow-pore cross-linked polystyrene, only one classical sigmoidal SEC curve is obtained.

Some authors have attempted to determine molecular weight or size for very large polymers, for instance water-soluble ones, which are used as viscosifiers in

oil-recovery wells. The porous structure induces converging and diverging flow channels, where polymer solutions have non-Newtonian behaviour. At high flow rate, under pressure and passing through small pores, polymer coils may undergo chain extension and orientation which modify their size and viscosity.

Flow and dynamic behaviour of flexible and rigid macromolecules in a packed bed has been studied comparatively. Partially hydrolysed (10%) polyacrylamide (molecular weight higher than 12 million) was a flexible model of a random coil, xanthan polysaccharide (molecular weight over 2 million) a model for a rigid backbone and tobacco mosaic virus as a rodlike structure (length 0.7 μm, diameter 15 nm). They have been eluted on a column packed with an ion exchange resin (diameter 11 μm). Deformation of the initial flexible linear polymers occurs, since their apparent size changes with flow rate, at a value corresponding to unity for De, the Deborah number, defined as the ratio of hydrodynamic forces to Brownian forces. The polymers which have been strongly sheared or sonicated before the chromatographic analysis do not show this change in size, since the chain has been extensively broken. The effective size of rodike polymers decreases when the flow rate increases: this is accounted for by orientation. For xanthans, two domains are observed when the flow rate increases: constant size, then a decrease after a flow value corresponding to the chain elongation (**Figure 10**). The slope (−0.5) of the high flow rate region fits that of the dumbbell model. Xanthan

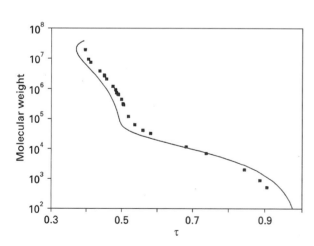

**Figure 9** HDC–SEC calibration graph (molecular weight as a function of the ratio of the solvent to polymer migration velocity) with polystyrene on three columns of 15-cm each, packed with Hypersil 3-μm porous particles. (Reprinted from Stegeman G, Kraak JC and Poppe H (1991) Hydrodynamic and size-exclusion chromatography of polymers on porous particles. *Journal of Chromatography* 550: 721–739.)

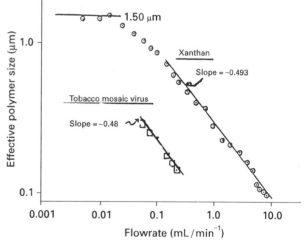

**Figure 10** Effective size of xanthan and tobacco mosaic virus as the flow rate is varied in a column 248 × 10 mm packed with 11-μm ion exchange particles; eluent with 2 g L$^{-1}$ nonionic surfactant and 2 mM NaN$_3$. (Reprinted from Hoagland DA and Prud'homme RK (1989) Hydrodynamic chromatography as a probe of polymer dynamics during flow through porous media. *Macromolecules* 22: 775–781.)

polysaccharide behaviour has also been studied on nonporous SiC particles.

## Biomaterials

Biomembranes are efficient separating materials. They are found in cells and are an excellent model for synthetic chemistry. They have many applications such as the separation of enantiomers, isotope enrichment, photosynthesis, and catalysis. Liposomes are used as drug-delivery systems and membranes, so their size and the amount of the transported material are important factors. Elution of liposomes from egg yolk lecithin has been performed on a column packed with porous inert glass, at different ionic strengths. An equivalent capillary radius $R$ is given by the formula $R = d/[2 - 2(2 - R_f)^{1/2}]$, where $d$ is the diameter of a colloidal particle, known from calibration with polystyrene standards. Recycling of the sample improves the separation (**Figure 11**).

Separation of natural products such as milk or globular proteins is also of interest. Preliminary results have been obtained for very large proteins on small nonporous glass spheres (**Figure 12**). Decreasing the ionic strength improves the separation. For common proteins, which require smaller particle sizes, the difficulty lies in the affinity of silica to proteins. The alternatives are to modify chemically the OH groups by grafting an inert moiety, or using an aqueous buffer, to repel proteins from the surface.

## Inorganic Compounds

Silica or carbon black are widely used in various industries as fillers, the effect of which depends on the size. A variety of other compounds has

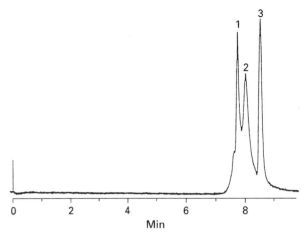

**Figure 12** Separation of proteins on nonporous silica gel 2.1 µm particles, column 250 × 4.6 mm, sodium borate 5 mM, pH = 9. 1: thyroglobulin; 2: γ-globulin; 3: glycine. (Reprinted from Kraak JC, Oostervink R, Poppe H, Esser U and Unger KK (1989) Hydrodynamic chromatography of macromolecules on 2 µm nonporous spherical silica gel packings. *Chromatographia* 27(11/12): 585–590.)

been examined: paper fibres, cement, clay, metals and metal oxides of Fe, Ti, Al, silver halides. It has been found that a difference in density does not alter the elution order. The calibration based on polystyrene latexes remains valid for carbon black particles of density 1.86, and the results agree with those obtained by photon correlation spectroscopy.

Applications are also found in geology.

## Conclusion

Despite its ability to provide useful results in size separations and determination, HDC is still only infrequently used. However, efforts are being made to improve its resolution and ease of use. Progress in synthesizing small size particles for packings and in narrow capillaries effectively allows rapid and excellent separation. Combined HDC and SEC offers new possibilities for determination of size and molecular weight of polymers in solution. HDC is also of interest in fundamental studies of flow behaviour in tubing or pores which are encountered in 'transport technology of materials': various fluids, solid particles, waste in rocks and soils, as well as in factories or pipes. In such work silica has the interesting advantage of being a hard spherical model for HDC mechanism studies.

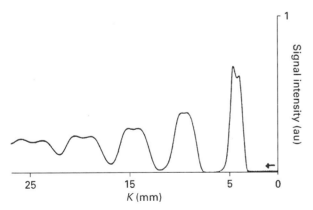

**Figure 11** HDC of liposomes from egg yolk lecithin on inert glass particles (125–180 µm) in a 300 × 15-mm column, eluent with 0.02 mM NaCl; filtration on 0.3-µm pore size, four recycles. (Reprinted from Molina FJ, Vila AO, Martos MJ and Figueruelo JE (1991) Estimation of the size of liposomes by modified HDC. *Journal of High Resolution Chromatography* 14: 590–592.)

*See also:* **II/Chromatography: Liquid:** Mechanisms: Size Exclusion Chromatography. **Particle Size Separation:** Theory and Instrumentation of Field Flow Fractionation.

## Further Reading

Barth HG (ed.) (1984) *Modern Methods of Particle Size Analysis*. New York: John Wiley.

Dos Ramos JG and Silebi CA (1990) Size analysis of simple and complex colloids in the submicron range using capillary hydrodynamic fractionation (CHDF). *Polymer Material Science Engineering* 62: 73–76.

Guillaume JL, Pichot C and Revillon A (1985) Approaches cinétiques du mécanisme de la copolymérisation styrène-acrylate de butyle. *Die Makromoleculare Chemie* Suppl. 10/11 69–86.

Revillon A and Boucher P (1989) Capillary hydrodynamic chromatography: optimization study. *Journal of Applied Polymer Science Symposium Edition* 43: 115.

Revillon A, Boucher P and Guilland JF (1991) Comparison of packed and capillary columns in hydrodynamic chromatography. *Journal of Applied Polymer Science*: *Symposium Edition* 48: 243–257.

Ugelstad J, Kaggerud KH, Hansen FK and Berge A (1979) Absorption of low-molecular weight compounds in aqueous dispersions of polymer-oligoner particles. 2. A two step swelling process of polymer particles giving an enormous increase in absorption capacity. *Die Makromolekulare Chemie* 180: 737–744.

# HYDROGEN RECOVERY USING INORGANIC MEMBRANES

**R. Hughes**, University of Salford, Salford, UK

## Introduction

Global annual production and utilization of hydrogen now total some $6 \times 10^{11}$ Nm$^3$. Of this amount, approximately 50% is produced by steam reforming or partial oxidation of natural gas. Other sources include electrolysis of water and recovery from refinery off-gases. The main use of bulk hydrogen includes synthesis of ammonia, hydrotreating of heavy petroleum feedstocks, hydrogenation of vegetable oils and manufacture of transformer steels and other metallurgical heat treatment operations. More recent applications where high purity hydrogen is required include semiconductor processing and fuel cells. To obtain high purity several methods may be adopted, such as pressure swing adsorption and cryogenic technology. However, both these methods have disadvantages, including cost and degree of purity obtained; the cryogenic method gives purities ranging from 90 to 98% only.

Membranes can provide an efficient low cost means of separation and purification for hydrogen. Although polymeric membranes are well proven for gas separation, they are limited to temperatures not much greater than 250°C whereas, in many applications, high temperature processing is required. Inorganic membranes have shown considerable develop-ment in recent years and, apart from their high temperature stability, they have in general much higher fluxes than polymeric membranes.

Inorganic membranes suitable for the recovery and purification of hydrogen may be divided into two classes, namely, porous and dense membranes. The former includes materials such as alumina, silica, zirconia and porous metals, for example, stainless steel. The dense membranes include palladium and its alloys, in which the unique permeation properties of palladium to hydrogen are utilized.

### Porous Membranes

To achieve appropriate separations with this type of membrane, the pore size needs to be small. However, to achieve a suitable gas permeation rate through the membrane, the membrane should be as thin as possible. To meet this requirement, a composite structure is usually adopted in which a thin finely microporous separation layer is supported on a thicker more open microporous material. Such structures have now been developed by a number of procedures.

The various possible gas permeation mechanisms applicable to porous membranes are illustrated in **Figure 1**.

In this figure, the progression from Knudsen diffusion to molecular sieving is in parallel with increasing permselectivities (it should be noted that viscous flow gives no separation). The separation factor for all the processes depends strongly on the pore size and its distribution, the temperature, pressure and the

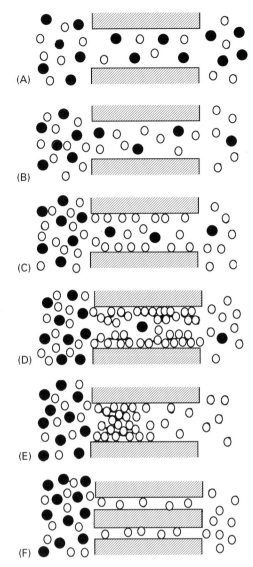

**Figure 1** Transport mechansim in porous membranes. (A) Viscous flow. Pore diameter is greater than gas mean free path – no separation. (B) Knudsen flow. Pore diameter is smaller than gas mean free path – separation is proportional to the square root of Ma/mB. (C) Surface diffusion. Absorbed gas on the pore water contributes to total gas flow. Transport of condensable vapour is enhanced. (D) Multilayer diffusion. Transport of absorbed gases is dominant and so condensable vapours have high transport rates. (E) Capillary condensation. Absorbed gas completely blocks pores. There is no transport of nonabsorbed gas. (F) Molecular sieving. Pores are so small that they begin to filter small molecules from large ones. Transport of small molecules is preferred. Modified with permission from Saracco G and Specchia V (1994) Catalytic inorganic membrane reactors: present experience and future opportunities. *Catalysis Review, Science and Engineering* 36: 305.

separation of gases other than just hydrogen and oxygen only. Conversely, porous membranes show significantly lower selectivities compared with dense membranes.

Viscous flow occurs when the mean free path of the gaseous molecules is much less than the pore diameter. Under these conditions, molecule–molecule collisions are much more frequent than collisions between the molecules and the walls of the pores. The mean free path, $\lambda$, of a gas molecule is given by:

$$\lambda = kT/(2\sigma p) \qquad [1]$$

where $k$ is the Boltzmann constant, $T$ the absolute temperature, $p$ the absolute pressure and $\sigma$ the collision diameter of the molecule. Thus, $\lambda$ will increase with increase in temperature and with decrease in pressure. Calculated values of $\lambda$ are shown in **Table 1** for a number of gases. Since the pore diameter of many separation membranes is of the order of 1–4 nm it can be seen that mean free paths under conditions often encountered in many catalytic processes are greater than the pore diameter and therefore Knudsen diffusion can be the operating mechanism for many current membranes.

In Knudsen flow, as shown in Figure 1B, the molecules collide on average more frequently with the pore walls than with one another when the mean free path becomes much greater than the pore diameter and absolute pressure does not affect the flux if this flow is fully developed. With a binary mixture, the highest separation factor achievable is given by the ratio of the inverse of the square roots of the molecular weights. Hence, small molecules are preferentially transported across the membrane and the greatest potential is for separation of hydrogen from other gases. Even in this case, however, separation is limited: that for a hydrogen–nitrogen mixture is only 3.74 at best.

nature of the membrane and the permeating molecules. Because of these factors, porous membranes are more versatile in their applications compared with dense metal membranes, because they can be used for

**Table 1** Mean free paths for representative gases (nm)

| Gas | Gas diameter (nm) $\sigma$ | Temperature 500 K | | Temperature 800 K | |
|---|---|---|---|---|---|
| | | 0.1 MPa | 1.0 MPa | 0.1 MPa | 1.0 MPa |
| $H_2$ | 0.29 | 183 | 18.3 | 293 | 2.9 |
| CO | 0.37 | 113 | 11.3 | 181 | 18.1 |
| $N_2$ | 0.37 | 111 | 11.1 | 177 | 17.7 |
| $CO_2$ | 0.39 | 102 | 10.2 | 164 | 16.4 |
| $C_4H_{10}$ | 0.50 | 62 | 6.2 | 100 | 10.0 |
| $C_6H_{12}$ | 0.61 | 42 | 4.2 | 67 | 6.7 |

Surface flow occurs (Figure 1C) when one of the permeating molecules can be preferentially physisorbed or chemisorbed on the pore walls and migrates along the surface. Surface diffusion increases the flux of the more strongly adsorbed components while reducing the contribution of the gas phase diffusion to total gas transport by decreasing pore diameter. Surface flow becomes more important as the pore size is reduced. Also, the effective pore diameter is further decreased by adsorption of the relevant species on the walls of the pores and thus obstructing the transfer of the other species through the free volume of the pores. As the temperature increases, most species will desorb from the surface and surface diffusion becomes less important.

Multilayer diffusion has been postulated to occur when molecule–surface interactions are very strong. The process is shown diagrammatically in Figure 1D. It may be regarded as an intermediate flow regime between surface flow and capillary condensation.

When the pores are small enough and one of the components of the gaseous mixture to be separated is a condensable vapour, absorbed vapour on the surface of the pore walls may be sufficient for the condensate to block gas-phase diffusion through the pore (Figure 1E). The condensate fills the pores and then evaporates at the permeate side, which has to be maintained at low pressure. In this case, only the condensed vapour permeates the membrane and nonabsorbed gases are almost totally retained by the membrane.

Molecular sieve transport occurs when pore diameters are small enough to permit only smaller molecules to permeate, while larger ones are excluded from entering these molecular-sized pores (Figure 1F). This type of process is frequently referred to as shape-selective diffusion. A necessary condition for effective separation by this means is that the pore size distribution is monodisperse and that the pores are very small – of the order of 0.5–1.0 nm.

From the above considerations it can be seen that a small pore size is necessary to obtain adequate separations. Since very finely porous membranes must necessarily have low permeabilities, useful fluxes are only obtained if the separating layer is made very thin. This has led to the evolution of asymmetric membranes in which the thin separation layer is supported on a wide pore matrix of similar material. Composite membranes have been developed in which different materials are used for the separation and support layers of the membrane. Molecular sieve membranes have been produced in this manner. An interesting development is the use of nanoporous carbon membranes for gas separation.

These membranes, which are produced by carbonization of polymers, have pore sizes in the molecular sieve range 0.5–0.6 nm) and operate by a combination of selective adsorption and surface flow. Rao and Sircar have shown that these nanoporous membranes can separate hydrogen from carbon dioxide and hydrocarbons with high selectivity. Experiments with silicalite zeolite for the separation of hydrogen and butane mixtures by Moulijn's group at Delft have shown that butane excludes hydrogen from the pores of the silicalite at room temperature by butane adsorption. However, at high temperature where adsorption effects are reduced, hydrogen permeation now predominates. These examples show that there is considerable scope for hydrogen separation from gas mixtures by using membranes with very small pore sizes and utilizing the adsorption and surface diffusion behaviour of the species involved.

### Dense Membranes

A second type of hydrogen-separating membrane is the dense membranes made from various metals, metal alloys and metal oxides. Hydrogen permeates a number of metals at high temperatures, including tantalum, niobium, vanadium nickel, iron, copper, cobalt and platinum. Alloys of these and other metals also possess high permeation rates for hydrogen together with certain metal oxides such as perovskites, which permeate hydrogen ions at temperatures in excess of 700°C. However, the main materials developed for hydrogen permeation are palladium and its alloys. This is because of the very high level of hydrogen absorption possible with these materials. Pure palladium absorbs 600 times its volume of hydrogen at room temperature, with the various palladium alloys absorbing comparable quantities. Although metals such as niobium and tantalum, also possess the ability to absorb large quantities of hydrogen, the formation of surface oxide films inhibits the ingress of hydrogen into the bulk metal and consequently these metals cannot be used to permeate hydrogen as such. Attempts have been made to overcome this problem by deposition of thin layers of platinum or palladium but the results have not been entirely successful, due in part of the formation of intermetallic compounds.

Consequently, palladium and/or palladium alloys represent the main dense metals currently favoured for membranes for selective permeation of hydrogen because of the high hydrogen permeation rates achievable at temperatures less than 500°C. However, below a critical temperature of 300°C and a critical pressure of 20 bars, the hydrogen–palladium system exhibits two-phase behaviour. Early experiments with palladium membranes resulted in mechanical failure due to the expansion and contraction of the metal lattice as hydrogen was absorbed and released. Alloying palladium with other elements such as silver, yttrium or cerium enables this difficulty to be overcome and an alloy of palladium with, for example, about 23% silver has been shown to be stable and have the optimum permeation rate for hydrogen.

### Hydrogen Transport in Metal Membranes

Hydrogen transport through metal membranes is a multi-step process, as illustrated in **Figure 2**. Hydrogen molecules in the feed gas are adsorbed on to the metal surface where they dissociate into atoms. These atoms then diffuse through the metal to the downstream (permeate) side of the membrane where they recombine to form hydrogen molecules on the surface. These molecules are finally desorbed into the permeate gas stream. The process is selective, in that only hydrogen is transported through the membrane; other gases are excluded, so that in theory the membranes should have infinite selectivity. However, due to a number of factors, this does not occur, although very high selectivities can be attained.

Depending on which of the above steps is controlling, the exponent of the pressure driving force for the overall permeation will change. For diffusion control by hydrogen atoms, which process is normally attained at high temperatures, the overall hydrogen flux ($J$) will be given by:

$$J = K(p_h^{0.5} - p_l^{0.5}) \qquad [2]$$

where $p_h$ and $p_l$ are the feed and permeate side pressures of hydrogen respectively.

However, if adsorption and/or dissociation of hydrogen into atoms is the controlling step then the pressure exponents will tend to a value close to unity.

### Metal Composite Membranes

Palladium–silver alloys have been used for a number of years for preparation of ultra-pure hydrogen for laboratory use. In recent years larger units have been made for the electronics industry. These membrane modules typically contain a number of small diameter alloy tubes with a metal wall thickness of about 50 μm. Sheet palladium alloys have also been used successfully for separation of hydrogen in reaction systems by Gryaznov and co-workers in Russia. However, the use of metal membranes alone has two serious disadvantages. If the permeation rate of hydrogen through the metal is diffusion-controlled, then the hydrogen flux is inversely proportional to metal thickness; most pure metal–alloy systems require a finite thickness to withstand the necessary pressure differential inherent in the separation process. Associated with this increased metal thickness is the material cost of palladium and its alloys.

Because of these factors, a number of methods have been proposed to overcome this problem, but essentially the common feature is to deposit the palladium on to a porous support material. The aim has been to obtain thin films of thickness ranging from 1 to 10 μm, but much thinner films have been produced by some investigators. The danger is that, with very thin films, the likelihood of pinholes being present increases considerably. Thin films in the micron range will give greatly increased fluxes, but the main difficulty is to produce films without cracks and to have good adherence to the support material. Substrates which have been used include porous glass, aluminas, silica, zirconia and stainless steel, although other porous materials have also been tried. The main

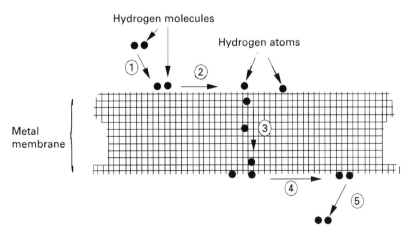

**Figure 2**   Permeation of hydrogen through metal membranes. 1, Sorption; 2, dissociation; 3, diffusion; 4, reassociation; 5, desorption.

techniques which have been developed for the deposition of thin films of palladium and palladium alloys on to porous substrates include:

1. magnetron sputtering
2. electroless plating
3. chemical vapour deposition
4. physical vapour deposition (thermal evaporation)
5. flame pyrolysis

There is some distinction in the way which palladium is deposited by these methods. The magnetron sputtering and vapour deposition methods tend to produce surface films whereas the chemical process of electroless plating can also act as a pore-plugging process. It has been found by a number of investigators, including the present author, that in comparing the suitability of α- and γ-alumina substrates for electroless plating, the wider pore α-alumina gives a more pinhole-free deposition. This has been attributed to the tendency of the electroless plating process to plug the pores; the more accessible wider pore mouths of the α-alumina facilitate entry of the depositing palladium.

In recent years the use of porous stainless steel as a support medium has been investigated since, although the pore sizes of stainless steel are currently larger than those of ceramic materials, this is compensated for by the robustness of the stainless steel and its potentially easier fabrication into an appropriate membrane module.

Use of composite membranes based on either ceramic or stainless steel supports is able to provide hydrogen permeation fluxes of $10^{-6}$ mol m$^{-2}$ s$^{-1}$ Pa$^{-1}$ or better with hydrogen–nitrogen selectivities in the range of 1000–5000. These results are suitable for a number of processes for hydrogen recovery, which are currently under evaluation.

### Technical Challenges to the Use of Membranes for Hydrogen Recovery

Membranes used for separation of hydrogen from other gases at high temperature should possess the following features:

1. thermal and chemical resistance
2. crack-free
3. small pore diameters (porous membranes)
4. large surface areas for membrane modules
5. require high temperature seals
6. composite membranes must retain adherence of the separation layer under thermal cycling
7. fouling of the membranes must be avoided

Some of these problems are being overcome currently. A major problem is that of sealing the individual membrane units into a module. If cylindrical tubes are to be used as membranes, then a shell-and-tube configuration as in conventional heat exchangers would seem to be the most favourable configuration, with the individual membrane tubes sealed into the end-plates of the shell. A stainless steel supported membrane tube would clearly have advantages in this case and would be preferable in terms of robustness compared to ceramic tubes.

Cost is a further factor. Although palladium is frequently cited as a major cost, it should be pointed out that, at present, the costs of porous ceramic or stainless steel support tubes are significant. No doubt these costs will decrease with increased production, but these represent a current constraint on applications.

## Present Status and Potential Developments

Because of the increased demand for hydrogen as a clean fuel and for newer applications where very high purity is required (current semiconductor processing requires hydrogen with impurities of 10 p.p.b. or less), there will be a continued demand for processes to fill this need. Membrane technology can provide a significant input into this demand. Both porous and dense membranes can be utilized to provide recovery of hydrogen from process streams, but dense membranes are necessary at present to provide hydrogen of high purity. Although palladium and palladium alloys have been used almost entirely for production of pure hydrogen, other metals such as niobium, tantalum and vanadium possess, in theory, good permeating properties and are less expensive than palladium. The problem of the surface oxide layer inhibiting permeation of hydrogen is being tackled with some success and further efforts will undoubtedly lead to better performance from these materials.

With porous membranes, perhaps the most important developments are likely to occur with zeolite membranes. The development of these, either as individual membranes or in composite form on another support material, will undoubtedly open up a new area of application for inorganic membranes for hydrogen recovery and purification.

## Further Reading

Buxbaumb RE and Kinney AB (1996) Hydrogen transport through tubular membranes of palladium-coated tantalum and niobium. *Industrial and Engineering Chemistry Research* 35: 530.

Dixon AG (1999) *Innovations in Catalytic Inorganic Membrane Reactors, Catalysis*, vol. 14. Cambridge: Royal Society of Chemistry.

Edlund DJ and McCarthy J (1995) The relationship between intermetallic diffusion and flux decline in compartmental membranes: implications for achieving long membrane lifetime. *Journal of Membrane Science* 107: 147.

Gryaznov VM (1986) Hydrogen permeable membrane catalysts. An aid to the efficient production of ultrapure chemicals and pharmaceuticals. *Platinum Metals Review* 30: 68.

Hughes R (1996) Applications in gas and vapour phase separations. In: Scott K and Hughes R *Industrial Membrane Separation Technology*, pp. 114–150. London: Blackie Academic and Professional.

Kapteijn F, Bakker WJW, Van der Graaf J *et al.* (1995) Permeation behaviour of a Silicalite-1 membrane. *Catalysis Today* 25: 213.

Keizer K, Ulhorn RJR and Burggraaf TJ (1995) Gas separation using inorganic membranes. In: Noble RA and Stern SA (eds) *Membrane Separations Technology, Principles and Applications*, pp. 553–588. Amsterdam: Elsevier Science.

Knapton AG (1977) Palladium alloys for hydrogen diffusion membranes – a review of high permeability materials. *Platinum Metals Review* 21: 44.

Lewis FA (1967) *The Palladium–Hydrogen System*. London: Academic Press.

Li A, Liang W and Hughes R (1998) Characterisation and permeation of palladium/stainless steel composite membranes. *Journal of Membrane Science* 149: 259.

Rao MB and Sircar S (1993) Nanoporous carbon membrane for gas separation. *Gas Separation and Purification* 7: 279.

# IMMOBILIZED BORONIC ACIDS: EXTRACTION

**P. Martin and I. D. Wilson,**
AstraZeneca Pharmaceuticals,
Macclesfield, Cheshire, UK

## Introduction

With the exceptions of antibody and molecular imprint-based methods, most solid-phase extractions rely on relatively nonspecific nonpolar van der Waals or ionic interactions. Another exception to the use of nonspecific interactions involves the use of reversible covalent bond formation with vicinal diols, or similar structures, in the target analyte with immobilized boronic acids. Clearly, the potential to exploit this type of interaction is limited but, where it can be exploited, highly specific solid-phase extraction (SPE) methods can result. Such methods, based on the use of boronic acids immobilized on materials, such as sepharose gels or phenylboronic acid (PBA) covalently linked to silica gel, have provided the basis for a number of SPE methods, as described below.

## Mechanism of Interaction of Analytes with Immobilized PBA

The extraction mechanism that results in the formation of cyclic boronates is illustrated in **Figure 1**. In order for the reaction to proceed, the boronate must be in the reactive $-B(OH)_3^-$ form, which is readily obtained by equilibrating the phase with an alkaline buffer. When the sample is applied, the covalent bond forms only with analytes possessing suitable functional groups, e.g. vicinal diols, found in sugars or catechols. Other functional groups which can form covalent bonds with boronic acids include α-hydroxy acids, aromatic O-hydroxyacids and amides, 1,3-dihydroxy-, diketo-, triketo- and aminoalcohol-containing compounds. With formation of the covalent bond, the analyte is strongly bound to the phase, which may then be washed with strongly eluotropic solvents to remove nonspecifically retained contaminants (an alkaline pH must be maintained). Analytes can then be recovered using an acidic buffer/solvent, which hydrolyses the covalent bonds to liberate retained compounds and return the PBA to the $-B(OH)_2$ form.

As well as these specific interactions with PBA, a number of nonspecific interactions can also occur with residual silanols, the aminopropyl group via which the PBA is attached to the silica, and the phenyl ring itself, which offers the opportunity for π–π interactions. In addition, the boronic acid can act as a hydrogen bond donor, cations can also bind to the boronic acid, and there is the potential for the formation of charge-transfer complexes with unprotonated amines. All of these interactions may happen when performing an extraction, and care must be taken to ensure that the extraction scheme is optimized to the required boronate retention mechanism if the maximum specificity is to be obtained.

## Buffer Selection

The first criterion to ensure a good extraction efficiency is to select a buffer for extraction on to

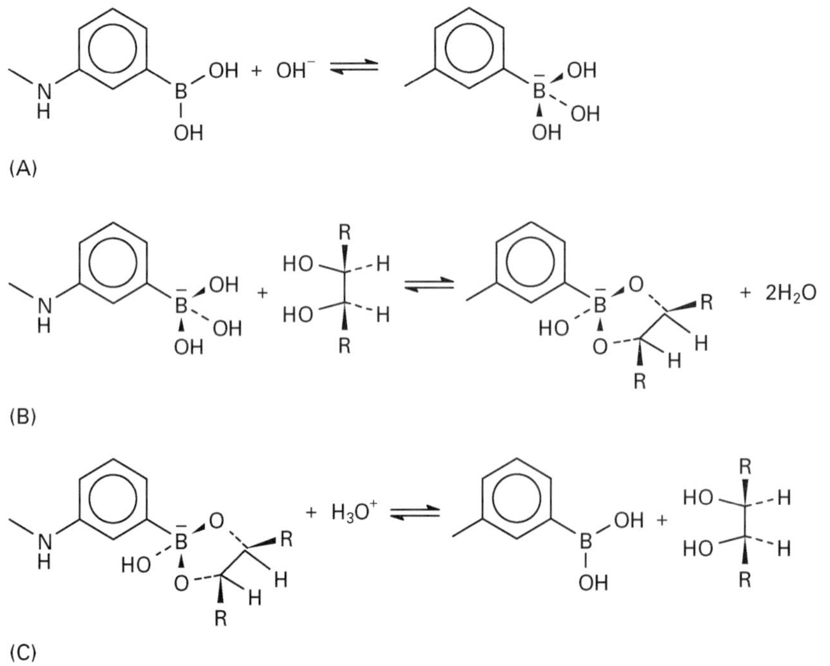

**Figure 1**  Mechanism of boronic acid extraction. (A) Activation of boronic acid in the presence of alkali; (B) formation of a covalent bond with a vicinal diol; (C) the use of acid to break the covalent bond and regenerate the boronic acid and the free diol.

PBA with an alkaline pH for the conditioning and sample application steps. Equilibration can be performed with, e.g. 0.1–1.5 mol L$^{-1}$ buffer at pH 10–12 and then 0.01–0.05 mol L$^{-1}$ buffer at pH 8–8.5. Zwiterionic buffers such as HEPES, glycine, diglycine and morpholine offer advantages and all have been used in this type of application. Clearly, buffers which can form covalent adducts with PBA are to be avoided, for the obvious reason that they eliminate covalent bond formation with the analyte. Such buffers include bicine, tricine, tris and 1',2',3'-ethanolamine.

Having obtained retention, analytes can be recovered by reducing the pH of the eluent to 5 or below. Typically acetic, trifluoroacetic and phosphoric acids may be used, with the addition of an organic modifier to help overcome nonpolar or silanophilic secondary interactions. Occasionally, the covalent adduct is sufficiently stable to require the addition of lactic or salicylic acid to the eluting buffer. Alternatively, elution with borate-containing buffers can be employed when acid-labile analytes are present.

## Applications

### Endogenous Biochemicals

**Adenosine, catecholamines, dopamine, DOPA and related substances**    A major field of application in the use of SPE with immobilized boronates (both gels and silica-based materials) is the extraction of various endogenous substances, especially catecholamines, l-dihydroxyphenylalanine (DOPA) and related materials, from biofluids. However, here we have concentrated on descriptions of the more recent procedures, generally involving the use of silica-based materials.

PBA gels have been employed in the extraction of nucleosides from biological samples for several studies, including the isolation of inosine and adenosine from human plasma. Methods have also been described for the simultaneous extraction of the adenosine and dopamine from human urine using a silica-based PBA phase. In this case extraction was performed using 100 mg PBA, activated by washing first with 5 mL of 0.1 mol L$^{-1}$ formic acid followed by 5 mL pH 8.8 ammonium acetate buffer (0.25 mol L$^{-1}$). Urine (0.5 mL, pH 8.8) containing ($\pm$)-isoproterenol and 2-chloroadenosine, was allowed to flow through the sorbent bed until the liquid meniscus just reached the top of the layer, at which point 1 mL of pH 8.8 ammonium acetate (0.25 mol L$^{-1}$) was also applied to the cartridge. Elution was achieved with 1 mL of 0.1 mol L$^{-1}$ HCl–methanol (4 : 1 v/v). This methodology allowed recoveries of 88–104% to be attained with good coefficients of variation (less than 5%).

An interesting two-stage SPE method has been devised for the isolation of DOPA from plasma and urine using $[^{14}C]$-DOPA as an internal standard. Interfering urinary pigments (urochromes) were eliminated via an initial extraction on to a dual-layer cartridge consisting of an upper layer of strong cation exchanger (SCX) and a lower layer of Cl silica. After this cartridge had been conditioned with 5 mL methanol and HCl ($0.2 \, mol \, L^{-1}$) the urine sample (2.8 mL) was applied followed by a further two bed volumes of $0.2 \, mol \, L^{-1}$ HCl. The eluate was taken to pH 7.5–7.7 (1.5 mL of $2 \, mol \, L^{-1}$ Tris buffer), after which an aliquot was passed through 200 mg of PBA SPE column (conditioned with 1 mL of methanol and 1 mL of $0.2 \, mol \, L^{-1}$ Tris buffer). After removing of interferences with methanol (2 mL) and $0.1 \, mol \, L^{-1}$ Tris (1 mL), the analyte was eluted in 0.3 mL of $0.1 \, mol \, L^{-1}$ HCl. For plasma samples, it was necessary to remove plasma proteins prior to extraction. This was done via precipitation using ice-cold perchloric acid. The supernatant obtained after centrifugation was then passed through the PBA cartridge (at pH 7.5–7.7 with $2 \, mol \, L^{-1}$ Tris) and the analyte subsequently eluted, as described above. Overall recoveries of 80% for urine and 84% for plasma (SD 2–3%) were obtained (allowing 10–15 pg to be detected with high performance liquid chromatography (HPLC) electrochemical detection).

As well as DOPA itself, several methods have used PBA for extracting 5-S-L-cysteinyl-L-dopa (5-SCD) from urine. These methods have included a dual extraction, first on to a cation exchanger and then a PBA gel with 5-S-D-diastereoisomer as an internal standard. A more recent method using PBA on silica was also based on dual extraction with SCX and PBA. The urine samples, to which 5-S-D-cysteinyl-L-dopa (D–CD) had been added as an internal standard, were applied to an SCX column that had been washed sequentially with methanol (1.0 mL) and HCl ($0.1 \, mol \, L^{-1}$, 1 mL). After washing with HCl ($0.1 \, mol \, L^{-1}$) the SCX cartridges were placed in series with a PBA column (pretreated with methanol and then $1 \, mol \, L^{-1}$ dipotassium phosphate buffer). The target compounds were eluted from the SCX using the dipotassium hydrogen phosphate buffer. Compounds retained on the PBA cartridge were then eluted with 0.5 mL of $0.1 \, mol \, L^{-1}$ HCl (containing $10 \, mg \, L^{-1}$ of ascorbic acid), following a water wash step. Aliquots of the eluate were then analysed by HPLC.

More recently a new, fully automated method for 5-S-cysteinyldopa in plasma and urine has been proposed. In this method, the PBA cartridges were treated sequentially with heptane and acetone (1 mL of each) to remove impurities and were then conditioned with methanol (1 mL), HCl ($0.3 \, mol \, L^{-1}$,

1 mL) and Tris buffer ($10 \, mmol \, L^{-1}$, 2 mL) followed by application of the appropriately buffered sample (1 mL of plasma or 1 : 100 diluted urine). The cartridges were then washed with 4 mL of $10 \, mmol \, L^{-1}$ Tris buffer prior to elution of the analyte with HCl (1 mL of $0.3 \, mol \, L^{-1}$). Typical chromatograms from this work are illustrated in **Figure 2**.

Boric acid gels and silica–PBA have both been used to extract catecholamines from biological samples. The silica-based material has also been employed for the determination of noradrenaline and adrenaline in urine and found to be superior to conventional methods. In this method two SPE cartridges, a pentanesulfonic acid phase (PSA) and PBA were connected in series (the PBA cartridge had previously been washed with methanol (1 mL) and then HCl ($0.1 \, mol \, L^{-1}$, 1 mL)). Both cartridges were then washed sequentially with methanol (2 mL) and then aqueous ammonia (4 mL) and finally phosphate buffer (4 mL, $5 \, mmol \, L^{-1}$ pH 8.5). The urine sample (1 mL, pH 5 using ammonia) plus internal standard (dihydroxybenzylamine) were then applied followed by a wash with phosphate buffer (4 mL of pH 8.5 phosphate). After washing with a further 2 mL of the alkaline phosphate buffer, the PSA column was removed and the PBA cartridge was washed first with methanol (1 mL) and then acetonitrile–phosphate buffer (1 mL, 1 : 1 v/v). Recovery of the analytes was achieved using HCl (1 mL, $0.1 \, mol \, L^{-1}$) and these were then analysed by HPLC with electrochemical detection.

In a similar method, urine (1 mL, diluted to 5 mL with water, pH 6.5–7.0) being assayed for dopamine, adrenaline and noradrenaline with 3,4-dihydroxybenzylamine added as internal standard was extracted using a combination of SCX and PBA. The SCX and PBA cartridges were initially treated with $1 \, mol \, L^{-1}$ HCl followed by methanol and $0.01 \, mol \, L^{-1}$ ammonium acetate buffer (pH 7.3), following which the sample was applied to the SCX column. This was washed with methanol and then ammonium acetate ($0.01 \, mol \, L^{-1}$): recovery of the analytes from this phase was achieved with $3 \times 500 \, \mu L$ of perchloric acid. This eluate was neutralized using a saturated solution of sodium carbonate and was then loaded on to the PBA cartridge which was first washed with methanol and then water. Finally the analyte was recovered for analysis, by HPLC with electrochemical detection, by elution with $2 \times 500 \, \mu L$ of $0.1 \, mol \, L^{-1}$ perchloric acid. Limits of detection of $1 \, \mu g \, L^{-1}$ for noradrenaline and $2 \, \mu g \, L^{-1}$ for dopamine were claimed.

**Glycosylated amino acids** In diabetes the glycosylated amino acid glucitollysine is formed when the

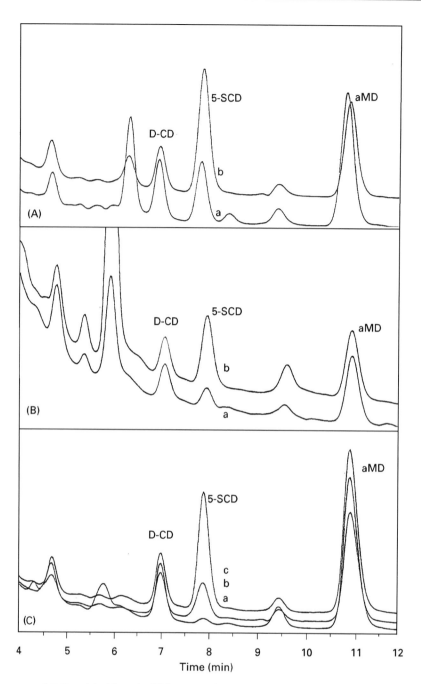

**Figure 2**  Chromatograms of 5-*S*-cysteinyldopa in PBA-extracted aqueous calibration standards plasma and urine. (A) Urine samples: (a) normal 5-SCD concentration (320 $\mu$g L$^{-1}$) and (b) pathological 5-SCD concentration (1310 $\mu$g L$^{-1}$). (B) Plasma samples: (a) nomal 5-SCD concentration (1.4 $\mu$g L$^{-1}$) and (b) pathological 5-SCD concentration (5.3 $\mu$g L$^{-1}$). (C) Extracted aqueous calibration standards: 5-SCD concentration: (a) 0.4 $\mu$g L$^{-1}$; (b) (3.2 $\mu$g L$^{-1}$; (c) 8.0 $\mu$g L$^{-1}$. aMD, 2-methyl-3-C3,4-dihydroxyphenyl)-L-alanine Reproduced with permission from Hartleb *et al.* (1999).

amino acid lysine in proteins reacts with glucose. An online extraction method for glucitollysine in protein hydrolysates with 'on-column' reaction with *o*-phthaldialdehyde (OPA) to allow HPLC with fluorescence detection has been described.

Glucitollysine extraction was achieved by washing the PBA phase with 0.1 mol L$^{-1}$ HCl to remove contaminants, followed by 0.1 mol L$^{-1}$ NaOH, equilibration with pH 8.5 phosphate buffer (0.1 mol L$^{-1}$) and then application of the sample, also at pH 8.5. With the protein hydrolysates, interfering co-extracted amino acids were removed by washing with water or methanol. Recovery of glucitollysine was achieved by lowering the pH of the mobile phase.

**Reduced oligosaccharides** PBA has also been applied to the separation of oligosaccharides from their alditols and interfering amino acids and glycopeptides. The purification of an oligosaccharide–lipid conjugate (neoglycolipid) formed by the reductive amination of the sugar lactose with phosphatidylethanolamine dipalmitoyl (PPEADP) has also been demonstrated.

The columns were activated by treatment with methanol, HCl (0.1 mol L$^{-1}$), water and then NaOH (0.2 mol L$^{-1}$). Following washing with water (2 × 1 mL) samples were applied as aqueous solutions. Elution was performed by washing with water, acetic acid (0.1 mol L$^{-1}$) and finally HCl (0.1 mol L$^{-1}$) with the fractions eluted from the PBA analysed by thin-layer chromatography on silica. Under these conditions, oligosaccharides with glucose at the reducing end were not retained by the PBA columns, but the corresponding alditols were retained. Similar results were obtained for glycoprotein-derived octasaccharides and their corresponding alditols. The application of samples under alkaline conditions enabled the separation of the analytes from amino acids, peptides and glycopeptides, although there was some retention of nonreducing oligosaccharides.

In addition, methods were also provided that enabled the purification of oligosaccharide derivatives formed by reductive amination (resulting in the ring-opened sugars giving acylic vicinal hydroxyl groups). Column activation in this example was performed using water, methanol and a 1 : 1 (v/v) mixture of methanol–chloroform to wash the column.

Reaction mixtures obtained after reductive amination were applied in methanol–chloroform (1 : 1 v/v), with subsequent elution (after various washes) in chloroform–methanol–0.1 mol L$^{-1}$ acetic acid (30 : 70 : 30 v/v).

### Natural Products

**Polyhydroxyflavones** HPLC with sample preparation via extraction on to PBA cartridges has recently been applied to the analysis of a variety of dietary polyhydroxyflavones (quercetin, kaempferol, fisetin, rutin, myricetin and morin; see **Figure 3** for structures) present in vegetables, red wine and human blood plasma. The extraction involved conditioning the cartridges with aqueous acetonitrile (1 mL, 28% v/v) containing 1% trifluoroacetic acid followed by 1 mL water and 1 mL of phosphate buffer (0.5 mol L$^{-1}$, pH 8.5). The samples were then loaded on to the cartridge in phosphate buffer (pH 8.5). Following a wash step (1 mL of 10 mmol L$^{-1}$ phosphate buffer, pH 8.5) the analytes were recovered in 2 mL of the aqueous acetonitrile solution used in the first step of the cartridge conditioning process (applied in four 0.5 mL aliquots). In general, good recoveries of the target compounds were obtained from matrices such as red wine and onions; recovery of, for example, quercitin was always greater than 90%. From human plasma the recovery of this compound was reduced to c. 80%, but with quite acceptable inter- and intraassay coefficients of variation. In **Figure 4** chromatograms for a variety of sample types are illustrated following sample preparation on PBA.

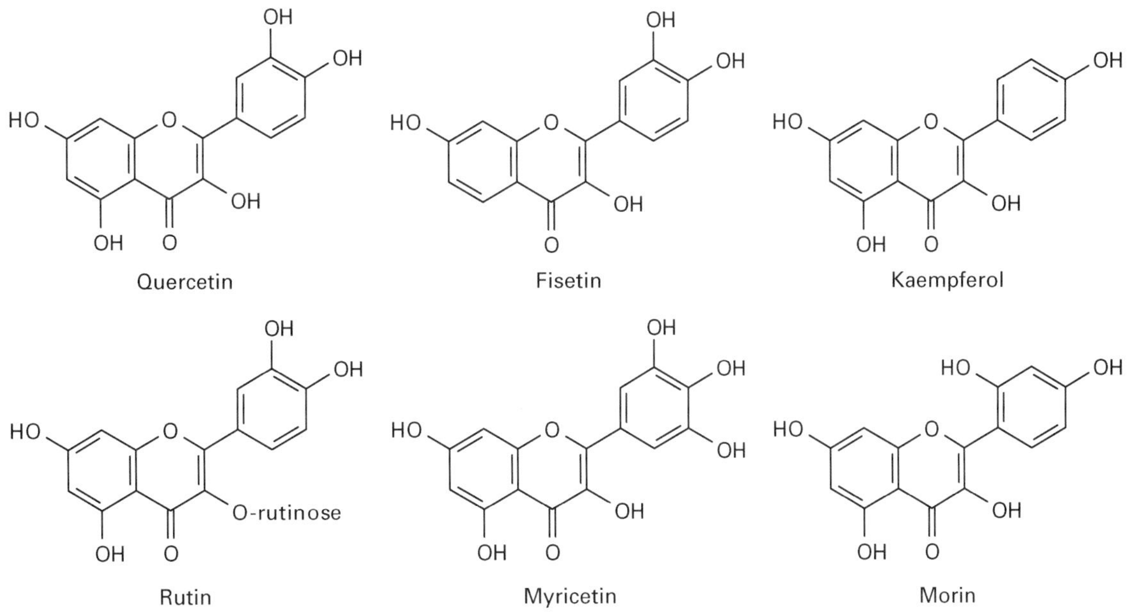

**Figure 3** Structures of the dietary polyhydroxyflavones.

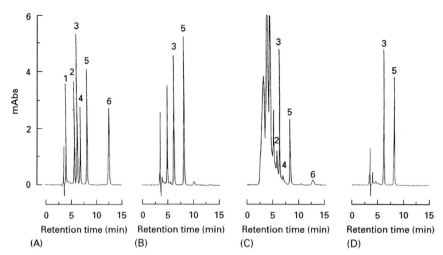

**Figure 4**   Chromatograms for a variety of sample types containing dietary polyhydroxyflavones following sample preparation on PBA. (A) Standard polyhydroxyflavones; (B) onion skin extract; (C) wine; (D) plasma spiked with quercetin. Peaks: 1, rutin; 2, myricetin; 3, fisetin; 4, morin; 5, quercetin; 6, kaempferol. Structures given in Figure 3. Reproduced with permission from Tsuchiya (1998).

It was noted that the absolute recoveries of quercetin, fisetin and rutin were better than those for the other compounds, and it was suggested that this could be explained by differences in the boronate complexes formed by 1,2 as opposed to 1,3 diols.

**Ecdysteroids**   The ecdysteroids are the moulting hormones of insects and crustaceans. They are relatively polar polyhydroxy steroids (the structure of 20-hydroxyecdysone is given in **Figure 5**) which are widely distributed in nature. Indeed, over 250 ecdysteroids have been isolated from various sources, particularly plants, where they probably function as chemical defences against predatory insects. Many of the ecdysteroids contain one or more vicinal diols, most often encountered at C-2 and C-3 of the A ring of the steroid nucleus, and in the side chain at C-20 and C-22. PBA has been found to provide the basis for the selective extraction of those compounds in possession of a C-20,22 diol group, but not ecdysteroids containing only a C-2,3 structure.

In this instance the extraction of the C-20,22 diol-containing compounds involved activation of the PBA with ethanol (5 mL) and then an alkaline buffer (5 mL, c. pH 8). Borate (100 mmol L$^{-1}$ pH 8.0 or pH 8.2) or phosphate (100 mmol L$^{-1}$, pH 8.0) buffers gave essentially the same result. Under these conditions compounds such as ecdysone and 2-deoxyecdysone, which lack the C-20,22-diol, were extracted from the matrix but were readily recovered using alkaline methanol (70% methanol). The C-20,22-containing compounds were, in contrast, surprisingly well retained and even eluents composed of 90% methanol–1% trifluoroacetyl failed to recover more than 20% of these substances. Quantitative

recoveries of these strongly adsorbed ecdysteroids required buffers that contained either salicylic acid (25 mmol L$^{-1}$) or lactic acid (3% w/v) in 50–70% methanol. A typical chromatogram for 20-hydroxyecdysone from a plant extract is shown in Figure 5.

These differences in the extraction properties of the C-20,22-diol and the C-2,3-diol-containing ecdysteroids is interesting and may well result from the difference in the O–O atomic distances in the two structures. Thus, with the C-2,3 compounds, this distance is 28 pm but with the C-20,22 structures this narrows to 25.2 pm. It is thus possible therefore that a rigid 2,3-diol would be unable to form a cyclic boronate whereas with the less rigid C-20,22 a cyclic boronate is possible.

## Drugs and Metabolites

### β-Blockers

PBA has been used to extract β-blockers (a class of aminoalcohol-containing drugs) from aqueous solution, rat, and human plasma. The analytes included propranolol, epanolol, ICI 118551 and practolol (see structures in inset to **Figure 6**). The cartridges (100 mg PBA) were first conditioned with 1 mL methanol followed by 5 mL of glycine buffer, following which SPE was performed using 0.1 mol L$^{-1}$ glycine buffer at pH 8.2. Following sample application, nonspecifically retained substances were removed by washing with 1 mL of deionized water followed by 3 mL of methanol–water (40 : 60 v/v). The analytes were then recovered in 3 mL methanol–water trifluoroacetyl (50 : 50 : 1 v/v). The extraction was pH-dependent, with the greate

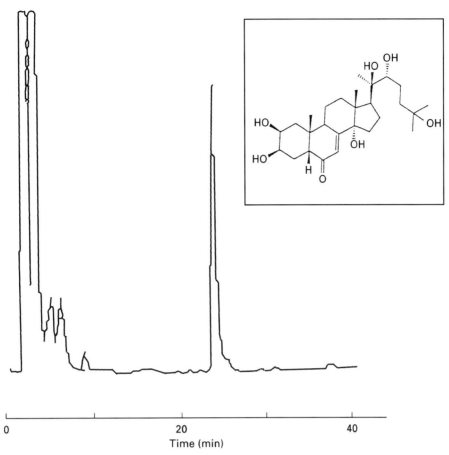

**Figure 5**  A typical chromatogram obtained for a PBA extract of a plant sample containing 20-hydroxyecdysone (see inset for structure).

extraction efficiency observed at pH 8 (Figure 6) but, in addition, structural features were also important. The extraction was most efficient for propranolol and ICI 118,551 (greater than 90%) with only small losses at the application and wash steps. With practolol, losses at the application step were high ( > 7%), and both practolol and epanolol showed losses at the wash step (8.8 and 16.5% re-

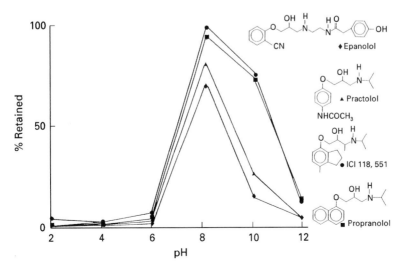

**Figure 6**  The effect of pH on the extraction of the β-blockers, propranolol, epanolol, ICI 118,551 and practolol on to PBA.

spectively). The latter could be reduced by decreasing the proportion of methanol in the solvent used for elution.

Some matrix effects were also noted for epanolol (but not propranolol, practolol and ICI118,551) when extraction was performed from rat plasma where losses at the application and wash steps were greater than from buffer. This effect, which probably resulted from protein binding, was reduced by diluting the sample with glycine buffer prior to extraction.

## Glucuronides

Glucuronides are an important class of metabolites for xenobiotics such as drugs. In many analytical methods these conjugates are hydrolysed back to the aglycone followed by extraction. However, the glucuronides have the potential for SPE on PBA, enabling glucuronide-specific assays to be developed. A limited number of studies into the potential of this type of SPE using a range of model phenolic glucuronides, spiked into urine, have been performed.

The test analytes in these studies were phenolphthalien glucuronide, $p$-nitrophenylglucuronide, $\alpha$-naphthylglucuronide and 6-bromo-2-naphthylglucuronide present in human urine at a concentration of 5 mmol L$^{-1}$. The extraction protocol developed for these compounds involved mixing 500 μL urine with 1.5 mL of 100 mmol L$^{-1}$ 8.5 glycine buffer (pH 8.5) which was then applied to a PBA column that had been conditioned first with pH 10 glycine buffer (5 mL, 100 mmol L$^{-1}$) and then equilibrated with a further 5 mL glycine buffer at pH 5. Glucuronides that were retained on the PBA were then eluted with 5 mL of methanol–1% HCl (90 : 10 v/v). With this protocol the selective retention of some of the test compounds was demonstrated depending upon the structure of the analyte and the amount of PBA employed. With phenolphthalien glucuronide, good extraction was obtained with 300 mg PBA and complete extraction was demonstrated with cartridges containing 400 mg. Extraction on 100 and 200 mg cartridges was, however, incomplete. Good recoveries were seen in the methanol–HCl elution step. Phenolphthalien glucuronide and phenolphthalien sulfate were readily separated from each other using PBA. Good results for the extraction of 6-bromo-2-naphthyl-$\beta$-D-glucuronide were also obtained on 600 mg cartridges.

However, with both $p$-nitrophenol glucuronide and $\alpha$-naphthylglucuronide extraction efficiency was not as good (20 and 50% respectively with 500 mg PBA cartridges).

Thus, whilst it was possible for certain structures to extract phenolic glucuronides on to PBA, and selectively to fractionate sulfates and glucuronides, the presence of glucuronic acid is not of itself sufficient to ensure extraction. Indeed, glucuronic acid itself was not retained under the extraction conditions used for the conjugates. Clearly the structure of the aglycone on to which the glucuronic acid is attached is also important. The exact structural features that would ensure good extraction of glucuronides have still to be elucidated, but $\pi$–$\pi$ interactions may be important. It should also be noted that, whilst glucuronides possess vicinal diol groups with which to form boronate esters, it is also possible that the carboxylic acid and its adjacent hydroxyl group are responsible for the observed extraction.

### Miscellaneous

**Alizarin**  The tricyclic anthraquinone dye alizarin contains a *cis*-diol and has been used as a model compound for studying SPE with PBA. The cartridges were conditioned with methanol and then pH 8.6 HEPES buffer (0.1 mol L$^{-1}$). An aqueous solution of the dye (0.1%) was quantitatively retained, as a sharp band on the cartridge with elution subsequently achieved with methanol–HCl (0.1 mol L$^{-1}$) (3 : 1). Efficient extraction was only achieved at pH 7–10; however it was also shown that good extractions were also possible from solutions containing up to 70% of an organic solvent as long as they were alkaline. Indeed, as long as wash solvents were alkaline, it was possible to use solvents such as methanol, ethanol or acetonitrile without loss of the dye from the cartridge. The extraction of the analyte from plasma was only efficient if the sample was first extracted on to a C$_{18}$ phase. Elution from C$_{18}$ with an alkaline methanol buffer on to the PBA then resulted in good recoveries. Presumably protein binding was responsible for the poor result (see the $\beta$-blockers above).

## Conclusions

For those compounds with the structural features that permit it, the possibility of selective extraction using immobilized phenylboronic acid may have potential benefits. Such a sorbent clearly has the potential to result in a relatively specific clean-up and it is perhaps surprising that there are relatively few published applications. However, it is evident from some of the examples provided above that the suitability of a particular analyte for extraction on to PBA can only be determined by experiment as the apparent possession of a suitable structure for cyclic boronate formation (e.g. a *cis*-diol) is no guarantee of success.

See also: **II/Affinity Separation:** Immobilized Boronates and Lectins. **Chromatography: Liquid:** Derivatization. **Extraction:** Solid-Phase Extraction. **III/Ecdysteroids: Chromatography. Appendix 1/Essential Guides for Isolation/Purification of Drug Metabolites.**

## Further Reading

Benedict CR and Risk M (1984) Determination of urinary and plasma dihydroxyphenylalanine by coupled-column high-performance liquid chromatography with C8 and C18 stationary phases. *Journal of Chromatography* 317: 27–34.

Echizen H, Itoh R and Ishizaki T (1989) Adenosine and dopamine simultaneously determined in urine by reversed-phase HPLC, with on-line measurement of ultraviolet absorbance and electrochemical detection. *Clinical Chemistry* 35: 64–68.

Hartleb J, Damm Y, Arndt R, Christophers E and Stockfleth E (1999) Determination of 5-s-cysteinyldopa in plasma and urine using a fully automatal solid-phase extraction-high-performance liquid chromatographic method for an improvement of specificity and sensitivity of this prognostic marker of malignent melanoma. *Journal of Chromatography* B 727: 31–42.

Huang T, Wall J and Kabra P (1988) Improved solid-phase extraction and liquid chromatography with electrochemical detection of urinary catecholamines and 5-S-L-cysteinyl-L-dopa. *Journal of Chromatography* 452: 409–418.

Imai Y, Ito S, Maruta K and Fujita K (1988). Simultaneous determination of catecholamines and serotonin by liquid chromatography after treatment with boric acid gel. *Clinical Chemistry* 34: 528–530.

Kagedal B and Pettersson A (1983) Liquid-chromatographic determination of 5-S-L-Cyseinyl-L-dopa with electrochemical detection in urine prepurified with a phenylboronate affinity gel. *Clinical Chemistry* 29: 2031–2034.

Kupferschmidt R and Schmid R (1986) Organic dye compounds as an evaluation tool for sample extraction using bonded silicas. *Proceedings of the 3rd Annual International Symposium*, Tschiya H, High-performance liquid chromatographic analysis of polyhydroxyflvones using solid-phase borate-complex extraction. Journal of Chromatography B, 720 (1998) 225–230.

Martin P, Leadbetter B and Wilson ID (1993) Immobilized phenylboronic acids for the selective extraction of β-blocking drugs from aqueous solution and plasma. *Journal of Pharmaceutical and Biomedical Analysis* 11: 307–312.

Maruta K, Fujita K, Ito S and Nagatsu T (1984). Liquid chromatography of plasma catecholamines, with electrochemical detection after treatment with boric acid gel. *Clinical Chemistry* 30: 1271–1273.

Murphy SJ, Morgan ED and Wilson ID (1990) Selective separation of 20,22-dihydroxyecdysteroids from insect and plant material with immobilized phenylboronic acid. In: McCaffery A and Wilson ID (eds) *Chromatography and Isolation of Insect Hormones and Pheromones*, pp. 131–136. New York: Plenum.

Oka K, Sekiya M Osada H, Fujita K, Kato T and Nagatsu T (1982) Simultaneous fluorimetry of urinary dopamine, norepinephrine, and epinephrine compared with liquid chromatography with electrochemical detection. *Clinical Chemistry* 28: 646–649.

Pfadenhauer EH and Tong S-D (1979) Determination of inosine and adenosine in human plasma using high-performance liquid chromatography and a boronate affinity gel. *Journal of Chromatography* 162: 585–590.

Schmid R and Pollak A (1985) Specific extraction of a glycosylated amino acid from protein hydrolysates using boronic acid derivatised silica gel. Sample preparation and isolation using bonded silicas. In: *Proceedings of the 2nd International Symposium*, pp. 15–20, Analytichem International.

Stolowitz ML (1985) Covalent chromatography: immobilized phenylboronic acid for sample preparation. Sample preparation and isolation using bonded silicas. *Proceedings of the 2nd International Symposium*. Analytichem International 41–44.

Tschiya H (1998) High-performance liquid chromatographic analysis of polyhydroxyflavones using solid-phase borate complex extraction. *Journal of Chromatography* B 720: 225–230.

Tugnait M, Ghauri FYK, Wilson ID and Nicholson JK (1992) NMR monitored solid-phase extraction of phenolphthalein glucuronide on phenylboronic acid and C18 bonded phases. *Journal of Pharmaceutical and Biomedical Analysis* 9: 895–899.

Tugnait M, Wilson FYK and Nicholson JK (1994) High resolution [1]H NMR spectroscopic monitoring of extraction of model glucuronides on phenylboronic acid and C18 bonded phases. In: Stevenson D and Wilson ID (eds) *Sample Preparation for Biomedical and Environmental Analysis*, pp. 127–138. New York: Plenum.

Wu A and Gornet TG (1985) Preparation of urine samples for liquid-chromatographic determination of catecholamines bonded-phase phenylboronic acid, cation-exchange resin, and alumina adsorbents compared. *Clinical Chemistry* 31: 298–302.

# IMMUNOAFFINITY EXTRACTION

**D. Stevenson**, University of Surrey, Guildford, UK

## Introduction

The mechanisms of separation in liquid chromatography are often classified as adsorption, partition, ion exchange and size exclusion. A further category could be included – affinity separations. Affinity chromatography uses very specific interactions between the compound of interest and a ligand bound to a chromatographic support to obtain separations. An early example of affinity separations was the use of an enzyme and its substrate. One particular type of affinity separation is immunoaffinity chromatography. In this case antibody–antigen interactions are used to obtain the separation. Either the antibody or the antigen can be bound to a support (immobilized). The current use of immunoaffinity extraction usually has the antibody immobilized (**Figure 1**). Immunoaffinity chromatography has often been used in the preparative mode where molecules of biological interest, which are difficult to recover by other methods, have been purified. Exam-

ples include enzymes, hormones, vaccines, interferons and antibodies.

Many modern analytical methods involve at least two distinct stages: preparation of a sample in a relatively clean form followed by instrumental analysis. This is particularly the case for the measurement of low concentrations of organic compounds in complex biological matrices such as blood, plasma, serum, urine, tissues and environmental matrices such as water, air, soil, foods, etc. One reason for the current interest in immunoaffinity extraction is its potential use as a highly specific variant of traditional solid-phase extraction in such analyses. In an ideal immunoextraction the sample is added to the column and only the target analyte is retained on the column. A wash step is then incorporated and potentially interfering material in the sample is washed from the column and discarded. The solvent is then changed and the elution solvent removes the target analyte from the column. The clean eluent is then analysed, usually by a modern instrumental method such as high performance liquid chromatography (HPLC) or gas chromatography (GC). This principle is shown in **Figure 2**. Immunoaffinity extraction is thus an attempt to combine the specificity of antibody-based methods with the separation and selective detection that can be obtained from instrumental chromatographic methods.

## Solid-Phase Extraction

Solid-phase extraction is one of the most common forms of sample preparation in current use. In its usual format, it involves introducing a liquid sample to an extraction cartridge in a small syringe-shaped container. The cartridge contains a solid phase capable of extracting the analytes of interest and retaining them on the solid phase. The analyte is thus removed from a 'dirty' matrix. It is then eluted from the solid phase and injected into a GC or an HPLC. Such a procedure produces a cleaner sample and therefore less likelihood of peaks co-eluting with the analyte. The liquids are normally drawn through the cartridge under vacuum using a purpose-designed vacuum box, or using positive pressure at the head of the column. Solid-phase extraction is a simple form of liquid chromatography. A range of phases is commercially available, such as silica, $C_{18}$, $C_8$, $C_5$, $C_2$, phenyl, diol, amino bonded silica, ion exchange phases and polymer phases. Conventional solid-phase extraction

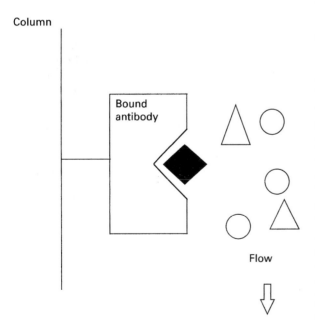

**Figure 1** Principle of affinity extraction. Only analyte binds to the antibody. ◆, analyte; △, ○, unrelated compounds.

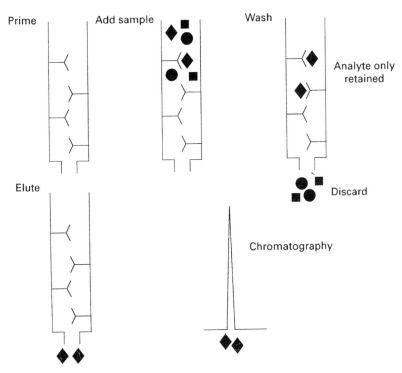

**Figure 2** Idealized immunoextraction. Only analyte (♦) is retained through the wash step. This is then eluted and subsequently injected onto HPLC, GC, etc.

is easy to automate both online and offline. Commercially available phases have been used to analyse thousands of different compounds, but generally they are nonselective about which analytes they extract. A range of tailor-made phases have been developed, designed to extract only one or a few closely related analytes. Immunoaffinity extraction is an example of an attempt to develop highly specific solid-phase extraction procedures.

## Antibodies

The key reagent for immunoextraction is the antibody which is immobilized on to a support. Antibodies are large biological molecules present in the serum of animals. They are produced by the immune system in response to foreign compounds, the so-called antibody–antigen response. Antibodies belong to a group of proteins called the immunoglobulins and have a relative molecular mass of about 150 000–900 000. Antibodies are normally only produced in response to compounds with a molecular mass of 1000 or above. As many of the compounds of interest in analytical chemistry are much smaller than this, they are chemically bonded to a carrier protein in order to elicit the immune response. For the antibody to be useful it must respond to the analyte, not

to the analyte–protein complex alone. In cases where the analyte does not contain a functional group suitable for bonding to a carrier protein, a structural analogue to the analyte is sometimes evaluated. As serum containing the antibodies is collected, it is referred to as antiserum. It will contain a number of different antibodies and is known as a polyclonal antibody.

In practice these antibodies will bind compounds bearing a close structural relationship to the compound of interest. This is known as cross-reactivity, and can be useful in immunoextraction as a group of compounds, such as phenylurea pesticides, can be extracted and then subsequently separated by HPLC. The forces involved in the antibody–antigen interaction are a mixture of ionic attraction, hydrogen bonding, hydrophobic attractions and van der Waals forces. Although individually they are relatively weak forces, in combination a relatively strong attraction is achieved. As a chemical reagent, antibodies are not very stable. They are easily denatured by extremes of pH and by organic solvents. They are much more stable under physiological conditions (i.e. close to pH 7 and in saline at about 1%). The main attraction of biological antibodies in analytical chemistry is their specificity, which arises due to biological recognition at the molecular level.

**Table 1**   Support materials used for antibody immobilization

Dextran ($\alpha$-1,6-linked glucose)
Agarose (poly galactose and anhydro-galactose)
Cellulose (1,6-linked glucose chains)
Polyacrylimide
Alumina
Silica
Controlled pore glass

## Immobilization of Antibody

Immunoextraction columns require the bonding of an antibody on to a suitable support while retaining the maximum amount of antibody activity. Some of the support materials used are shown in **Table 1** (for their particular characteristics, see Godfrey in Further Reading). Ideally, supports should show good flow characteristics, good chemical and mechanical stability, low nonspecific adsorption, low cost and suitable functional groups for bonding the antibody. As with conventional HPLC and solid-phase extraction, silica-based sorbents are the most popular for immunoextraction.

The methods used to couple antibodies to sorbents usually involve reaction with the carboxyl or amino groups on the antibodies. A range of different reagents is used to activate the sorbent on to which the antibody is bound. These include cyanogen bromide, carbonyl diimidazole, 1,4-butanediol diglycidoxy ether, divinyl sulfone, tresyl chloride and glutaraldehyde. Sorbents need to be thoroughly washed after bonding to remove residual reagents.

## Optimization of Immunoextraction

Once an immunosorbent has been prepared, a known amount (by weight or volume) is added to a plastic or glass column with a retaining frit. As immunosorbents are relatively expensive to prepare (compared with commercially available solid-phase extraction columns), the minimum amount needed for a particular assay is used. Conditions are also chosen to allow the columns to be used many times. This means that gentle extraction conditions are favoured, otherwise the antibody will be denatured. However, when using immunoaffinity extraction as a clean-up method for chromatography, it is also desirable to elute the analyte in as small a volume as possible so that further preconcentration is unnecessary. A further consideration is the nature of the desorbing eluent with respect to the possibility of direct injection into an HPLC or GC. In principle there are many different variables requiring optimization for a successful immunoextraction (**Table 2**).

As the actual column preparation follows established protocols, much of the effort of developing a successful extraction protocol concentrates on the solvents used for conditioning and washing the column and on the solvent used for elution of anlyte. Columns typically contain between 50 and 500 $\mu$L of antiserum. As antibodies orginate from animal serum, physiological conditions are favourable for column washing.

A typical protocol would prime the column with phosphate-buffered saline at neutral pH. The sample would be loaded at a pH adjusted to fall in the range pH 5–9. Large sample volumes can be added and this allows concentration of the analyte on the column. The capacity of immunocolumns is usually dictated by the mass of analyte rather than the volume of sample. Biological samples such as plasma or serum often cause column blockage unless proteins are precipitated before the sample is added to the column. Flow rates up to about 5 mL min$^{-1}$ are acceptable; otherwise, with higher flow rates the antibody–antigen interaction does not have sufficient time to ensure binding.

Once the analyte is bound on to the column it can be washed with phosphate-buffered saline at neutral pH. In order to elute the analyte in as small a volume as possible, without damaging the antibody, elution solvents are typically composed of phosphate-buffered saline at a low pH (down to pH 2) with the addition of a water-miscible solvent such as methanol or ethanol at a concentration of up to about 50%. At higher pH or lower concentration of organic modifier, the analyte elutes in a larger volume of elution solvent, which necessitates further concentration. This type of desorption is known as nonselective desorption. An alternative approach is to try selective desorption by adding a compound very similar in structure to the analyte in the elution solvent. This

**Table 2**   Parameters for optimization of immunoaffinity extraction

Type of support
Activation of support
Particle size of support
Pore size of support
Amount of antibody to immobilize
Immobilization chemistry
Quality and purity of antibody
Column dimensions
Column priming
Volume of sample to load
pH of sample
Wash solvent composition including pH
Elution solvent composition including pH
Flow rate
Regeneration conditions

approach has generally necessitated larger desorption volumes to obtain quantitative recovery than non-selective desorption and hence is not common.

One of the advantages of immunoaffinity extraction is that these procedures are usually carried out under aqueous conditions. As reversed-phase HPLC is often the favoured technique for subsequent analysis, the introduction of analyte dissolved in an aqueous medium is compatible with the mobile phase. Most applications of the technique involve extraction of drugs, endogenous compounds and pesticides. Examples of compounds for which immunoaffinity extraction has been reported are shown in **Table 3**.

One of the most successful applications of immunoaffinity has been the extraction of pesticides from water. Immunoextraction columns have been shown to be capable of preconcentrating up to 1 L of water containing the herbicides chlortoluron and isoproturon, yet still capable of desorption into low volumes (even as low as 1 mL) of elution solvent. Although lower sample volumes (such as 50–100 mL) are more likely to be used in practice, this feature of immunocolumns offers the possibility of large concentration factors and low overall detection limits. The cross-reactivity of antibodies to triazines and phenylureas has been used to immunoextract several compounds which were then subsequently separated and measured by HPLC.

The capacity of the immunocolumns is governed by the mass of analyte that can be retained before the column is overloaded rather than the volume of water passed through. The mass capacity of the column can be assessed by loading 1 mL aliquots of a standard solution of analyte and analysing fractions eluting from the column until the presence of analyte is detected. A simple calculation of the number of addi-

tions to the column times the amount added each time gives the mass breakthrough of analyte. An alternative approach involves overloading the column but then washing out excess analyte in solution, leaving bound analyte on the column. This is then eluted with the desorbing solution and the concentration and volume measured. A simple calculation gives the amount of analyte required to saturate the column.

The major advantage of immunoaffinity columns is the specificity that can be obtained. This is utilized to give cleaner chromatographic traces than using nonselective extraction such as liquid–liquid extraction or solid-phase extraction on silica or non-polar bonded silica.

## Other Formats of Immunoaffinity Extraction

Immunoaffinity extraction has been carried out in formats other than solid-phase extraction. This has included high performance immunoaffinity chromatography and online HPLC column switching. In the former, immunosorbents are used as HPLC columns whereas in the latter, samples are extracted on an HPLC immunosorbent, preconcentrated and the flow then switched to a conventional HPLC column for analysis. Both methods attempt to base the separation on antibody–antigen interactions. In the case of column switching, complete automation can be achieved.

## Molecular Imprinted Polymers

The major disadvantage with immunoaffinity extraction is the difficulty and expense in obtaining biological antibodies. An alternative approach is the use of molecular imprinted polymers as antibody mimics. These are synthesized in the chemistry laboratory and are consequently easier and less expensive to obtain. The target analyte (template) is mixed with a monomer such as methyl acrylic acid and a cross-linking agent such as ethylene glycol dimethacrylate. They are dissolved in a suitable solvent such as acetonitrile along with an initiator such as 2,2′-azobis-(2-methylpropionitrile) and heated or subjected to UV radiation. The polymer forms around the template within about 16 h. The polymer is ground into fine particles and then washed to remove the analyte template, thereby leaving cavities where the analyte can subsequently be bound. This polymer can then behave as an affinity column, mimicking the biologically derived immunoaffinity solid-phase extraction columns.

**Table 3**  Examples of immunoaffinity extraction

Aflatoxins
Albuterol
Atrazine
Carbofuran
Chloramphenicol
Ciprostene
Clenbuterol
Cytokinins
Isoproturon
Morphine
Ochratoxin A
Propranolol
Prostaglandins
Steroids
Thromboxane metabolites
Tenbolone
Zeranol

Columns derived from molecular imprinted polymers often show secondary interactions arising from the monomer, e.g. with methyl acrylic acid, cation exchange can occur. They show best specificity in the solvent in which they were originally dissolved, hence they are used with organic rather than aqueous solvents. Although easy to obtain and more stable to extremes of pH and organic solvents, they are not as specific as columns utilizing biologically derived antibodies. One problem with molecular imprinted polymers is the difficulty in washing out all traces of the analyte template. Remaining template leaches out when the columns are used for analysis, giving falsely high results. This problem is partially overcome by using a structural analogue to the analyte as the template. Provided the template can be separated from the analyte by HPLC, GC, etc. it will not interfere with the analysis. This approach does require cross-reactivity of the polymer, i.e. it must retain the analyte as well as the template.

The use of molecular imprinted polymers is an emerging field and new synthetic methods may improve the performance of these columns as well as other uses of the polymers. Examples of solid-phase extraction using molecular imprinted polymers include atrazine, pentamidine, propranolol, sameridine and tamoxifen.

## Future Developments

Immunoaffinity extraction has been demonstrated as being capable of selectively capturing analytes from complex matrices using antibody–antigen interactions. Techniques for preparing the columns and procedures for optimizing the retention and desorption of analyte are now well established. The availability of antisera to more compounds will expand the use of immunoextraction. As better procedures to produce antibodies or antibody fragments become available, the cost of antisera should come down. Although much of the work to date uses low molecular weight compounds as target analytes, immunoextraction might be even more valuable for the new products emerging from biotechnology which may present different problems with extraction using conventional liquid–liquid or solid-phase extraction methods. Better specificity from synthetic polymer antibody mimics should also see a growth in their utility in immunoaffinity-type extractions. Polymers that show specificity for analyte under aqueous conditions would be an advantage. Selective extraction at present comes at an extra cost in the production of the columns and is not yet available 'off the shelf'. It is likely to prove most useful

where simpler procedures cannot be used due to analyte instability or where particularly low detection levels are required. It should also be remembered that immunoaffinity extraction need not only be used with HPLC, GC or capillary electrophoresis.

## Further Reading

Burrin D (1995) Immunochemical techniques. In: Wilson K and Walker J (eds) *Principles and Techniques of Practical Biochemistry*, 4th edn, pp. 65–109. Cambridge: Cambridge University Press.

de Frutos M (1995) Chromatography–immunology coupling, a powerful tool for environmental analysis. *Trends in Analytical Chemistry* 14: 133–140.

Farjam A, Brugman EA, Henk L and Brinkman UAT (1991) On-line immunoaffinity sample pre-treatment for column liquid chromatography: evaluation of desorption techniques and operating conditions using an anti-estrogen immuno-pre-column as a model system. *Analyst* 116: 891–896.

Godfrey MAJ (1997) Immunoaffinity and IgG receptor technologies. In: Matejtschuk P (ed.) *Affinity Separations: A Practical Approach*, pp. 141–195. Oxford: IRL Press.

Holme DJ and Peck H (eds) (1993) *Analytical Biochemistry*, 2nd edn, pp. 233–260. Harlow: Longman.

Janis LJ and Regnier FE (1988) Immunological-chromatographic analysis. *Journal of Chromatography* 444: 1–11.

Martin-Esteban A, Kwasowski P and Stevenson D (1997) Immunoaffinity-based extraction of phenylurea herbicides using mixed antibodies against isoproturon and chlortoluron. *Chromatographia* 45: 364–368.

Martin P, Wilson ID, Morgan DE *et al.* (1997) Evaluation of molecular-imprinted polymer for use in solid phase-extraction of propranolol from biological fluids. *Analytical Communications* 34: 45–47.

Phillips TM (1989) High-performance immunnoaffinity chromatography. *Advances in Chromatography* 29: 134–173.

Pichon V, Chen L and Hennion M-C (1995) On-line pre-concentration and liquid chromatographic analysis of phenylurea pesticides in environmental water using a silica-based immunosorbent. *Analytical Chimica Acta* 311: 429–436.

Rashid BA, Aherne GW, Katmeh MF *et al.* (1998) Determination of morphine in urine by solid-phase immunoextraction and HPLC with electrochemical detection. *Journal of Chromatography A* 797: 245–250.

van Ginkel LA, Stephany RW, van Rossum HJ and Zoontjes PW (1992) Perspectives in residue analysis; the use of immobilised antibodies in (multi) residue analysis. *Trends in Analytical Chemistry* 11: 294–298.

Walt DR and Agayn VI (1994) The chemistry of enzyme and protein immobilisation with glutaraldehyde. *Trends in Analytical Chemistry* 13: 425–430.

# IMPREGNATION TECHNIQUES: THIN-LAYER (PLANAR) CHROMATOGRAPHY

**I. D. Wilson**, AstraZeneca Pharmaceuticals, Macclesfield, Cheshire, UK

## Introduction

The use of impregnated phases in thin-layer chromatography (TLC) has a long history, beginning in the very earliest days of the technique. As a consequence, much of the key literature in this field dates back to the 1960s and 1970s. The literature also contains a considerable number of examples of impregnation reagents where subsequently the methods have received little attention except in reviews of the subject.

The impregnation of the stationary phase in thin-layer, or planar, chromatography can be used to achieve a number of ends. These include changing the mode or selectivity of the chromatographic system, improving chromatographic performance and enhancing detectability. An example of the use of impregnation to change the type of chromatography would be the use of a nonpolar, water-immiscible material such as paraffin oil which can be used to enable reversed-phase separations to be performed on silica-gel TLC plates. Alternatively, plates impregnated with silver nitrate can be used to improve the separation of compounds containing double bonds (argentation chromatography). Similarly, boric acids or boronates are useful for compounds such as carbohydrates and sugars containing vicinal diols, and ion pair reagents provide a means for successfully separating polar ionic species. Although the use of impregnation techniques for enhancing detectability will not be considered in depth here, examples would include the incorporation of ammonium acetate followed by heating, which can result in the formation of intensely fluorescent derivatives, whilst paraffin oil impregnation can stabilize or enhance fluorescence and the radiolabelled substances can be detected using impregnation with a scintillant.

There are a variety of methods for obtaining impregnated layers, including the preparation of the TLC plate with slurries containing both the stationary phase and the impregnating reagent, pre- or post-chromatographic dipping, pre- or post-chromatographic development of the plate in a solvent containing the impregnating agent or spraying, and these are described below.

## Impregnation Techniques

All of the techniques for impregnating TLC plates described below have been used at one time or another and are illustrated with examples in the subsequent sections on applications.

### Spraying

Spraying the TLC plate is a simple and convenient method for applying the reagents used for detection in TLC and indeed is widely used for this type of application. It is, however, much less well suited to the impregnation of TLC plates in order to modify chromatography because of the difficulty of ensuring uniform coverage. It also requires much skill and expertise in order to ensure reproducible results from day to day. Both of these technical difficulties are clearly disadvantages, and spraying is therefore not recommended for routine applications in this area. Spraying is therefore probably best used, if at all, for one-off applications or for screening phases and impregnating reagents for a particular property during method development.

### Dipping

Dipping plates in solutions of the impregnating reagent in a suitable (usually volatile) solvent is probably the simplest and most efficient method of ensuring even coverage of the stationary phase with the reagent of choice. Suitable dipping chambers, of the type used for immersing plates in solutions of chromogenic visualization reagents, are available from a number of manufacturers at modest cost. These chambers have relatively small volumes and so are not expensive in reagent. Some automated devices are also available that can be used to control the impregnation time accurately. Dipping, in general, because of the ease of control over parameters such as reagent concentration and impregnation time, should be considered to be the preferred method for impregnating TLC and HPTLC plates.

### Pre-development

It is also a relatively easy matter to prepare TLC plates impregnated with the desired reagent by pre-development in a solution of reagent in a suitable solvent. This technique is also referred to as irrigation of the plates. The solvent used for this purpose needs to be sufficiently eluotropic to ensure that any affinity of the reagent for the stationary phase is overcome or else, at worst, the plate will not be impregnated with the reagent which will stay at or near the origin (or else be distributed as a gradient of decreasing concentration up the plate). Nonmigration of the impregnating reagent is seen with, for example, the ion pair reagent cetrimide if dissolved in water and then used to coat reversed-phase TLC plates in the pre-development method. Because the reagent is relatively nonpolar it has a strong affinity for the stationary phase and remains at the origin. In such cases dipping is to be preferred.

### Mixed Phases

Although with the advent of good quality commercial TLC and high performance TLC (HPTLC) products the preparation of TLC plates in the laboratory is no longer widely practised, one method of producing impregnated plates was to produce mixed phases. Thus, the bulk stationary phase (generally silica gel) was mixed with an appropriate amount of the impregnating reagent, a binding agent, slurried in a suitable solvent and spread as a layer on glass plates. After drying and activating (if necessary), the plates were then used as required. Some manufacturers provide a range of mixed phases (e.g. silver nitrate-impregnated plates, buffered layers, ammonium acetate-impregnated layers, etc.).

## Specific Examples of Impregnation in TLC

### Impregnation to Form Lipophilic Stationary Phases for Reversed-phase Separations

Before the introduction of good-quality bonded phases, the preparation of suitably hydrophobic stationary phases used to be a very popular method for obtaining layers with which to perform reversed-phase separations, and still has many advantages. This is readily achieved using liquid paraffin oil, undecane, *n*-decane, nitromethane, propylene glycol, silicone oil, decalin, Carbowax 400, formamide, 2-phenoxy- and 2-methoxyethanol, various mineral oils, and substances such as ethyl oleate and similar materials. In addition to modifying chromatography, impregnation with these materials can also be used to enhance or stabilize the fluorescence of suitable analytes, thus improving detection and quantification. However, it is the use of these materials to provide reversed-phase separations that is of interest here.

A typical example of the use of paraffin oil is provided by work on the reversed-phase separation of ecdysteroids (polar, polyhydroxylated steroids found in plants and arthropods). Here silica-gel TLC and

HPTLC plates were impregnated with a solution of 7.5% paraffin oil in dicholoromethane (v/v) by dipping. The plates were air-dried and chromatography was performed using methanol–water mixtures. In an interesting variation on this general technique of impregnation, a variety of normal-phase separations, for nonsteroidal anti-inflammatory drugs, ecdysteroids, antipyrine and aminophenols, on silica gel were performed with the paraffin added to the mobile phase. Addition of 7.5% (v/v) of paraffin to the normal-phase solvent system did not affect chromatography but did enable the plate to be impregnated so that a subsequent reversed-phase separation in a second dimension could be undertaken immediately that the solvent from the normal-phase separation had evaporated. An example of this for the nonsteroidal anti-inflammatory compounds is shown in **Figure 1**. It should be noted that, with this type of impregnated plate it is also possible to perform the reversed-phase separation first, remove the nonpolar impregnating reagent with a suitable solvent (i.e. one that does not affect the analytes but does remove the impregnating reagent) and then perform a normal-phase separation in the second dimension. Alternatively, a similar outcome can be achieved by impregnation of that portion of the plate not used for

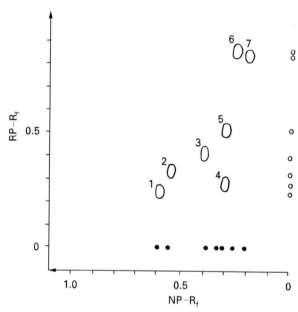

**Figure 1**  Two-dimensional separation of nonsteroidal drugs with paraffin impregnation during normal-phase (NP) separation in the first dimension to enable reversed-phase (RP)-TLC to be carned out in the second. 1, Ibuprofen; 2, ibufenac; 3, methyl analogue of isoxepac; 4, indomethacin; 5, isoxepac; 6, salicylic acid; 7, 5-methoxysalicylic acid. Details in Wilson ID (1984) Normal-phase thin-layer chromatography on silica gel with simultaneous paraffin impregnation for subsequent reversed-phase thin-layer chromatography in a second dimension. *Journal of Chromatography* 287: 183–188.

the separation in the first dimension, and examples of a normal-phase separation followed by impregnation with 2-phenoxyethanol or undecane for reversed-phase chromatography in the second dimension have been described.

### Impregnation with Silver Nitrate

Argentation TLC also represents an important methodology and is considered in detail elsewhere in this work and so will only be briefly described here. The impregnation of TLC plates with silver nitrate has been used for many years as a means of improving the separation of unsaturated compounds, particularly certain lipids, based on the ability of silver ions to form charge transfer complexes with the $\pi$ electrons of the carbon–carbon double bonds. In general, silica gel is used: the amount of silver nitrate used varies from as little as 2% up to 20 or 30% w/w depending upon the author and application. With silica gel, impregnation by both spraying (with a 10–20% solution in either water or methanol) and the preparation of mixed phases have been described. Following preparation it seems to be good practice either to use the plates the same day or else to store them in a sealed container in the dark until required. In general, non-polar solvent systems are employed (e.g. pentane, hexane-diethyl ether, chloroform-methanol, diethyl ether, light petroleum, etc.).

Whilst the use of silver nitrate-impregnated plates has generally been with normal-phase separations on silica gel, there has been recent work using reversed-phase ($C_8$-bonded) layers dipped in solutions containing between 0.5 and 4% silver nitrate for 10 s. These plates were investigated for their ability to separate the cis/trans isomers of capsaicin, and comparison was made simply using the silver nitrate in the mobile phase (60 : 40 methanol–water v/v). With impregnation no effects were observed until the 2% (w/v) impregnating solvent was used, and even with 4% this was insufficient to provide the required resolution. In contrast, when present as a mobile-phase additive, even as low a concentration as 0.5% w/v was sufficient to give baseline resolution. It appears likely from this result that the high solubility of the silver nitrate in the mobile phase led to its rapid elution from the impregnated plate, with consequent loss of effect. Thus, interestingly, it seems from this work that the use of the silver nitrate in reversed-phase systems is only practicable when it is present as a mobile-phase additive.

### Impregnation with Polyol and Sugar Complexing Reagents

The ability of borate ions to complex with suitable polyhydroxylated compounds such as carbohydrates is well known and has provided the basis of a number of methods for their separation. TLC with layers impregnated with sodium arsenite, phosphotungstic, tungstoarsenate and molybdic acids have also been shown to have useful properties for the resolution of mixtures of oligo and monosaccharides.

Various methods have been used to prepare such layers, including both spraying and the preparation of mixed layers. Thus, in an early example of the use of boric acid, sodium borate and sodium arsenite plates were either sprayed with methanolic or aqueous solutions containing 10–20% of the impregnating reagent or plates were prepared by mixing 2.8 g of the reagent in 50 mL of water with 25 g of silica gel G to give a 10% (w/w) mixed layer. These layers were then used to separate the erythro and threo isomers of a variety of di- and trihydroxy long chain fatty acid esters.

Similar work on phosphotungstic and molybdic acid impregnated silica gel TLC plates showed them to be particularly useful for the separation of oligosaccharides giving complexes with higher $R_f$ values than the corresponding boric acid complexes under similar conditions. In this case the plates were made by mixing 35 g of the chromatographic stationary phase (e.g. silica gel–alumina 1 : 1 or 3 : 1 or alumina) with 70 mL of an aqueous solution containing an appropriate amount of the impregnating reagent and spreading the plates as a 0.4 mm layer on to glass. The resulting plates were then dried at room temperature for 24 h, and heated for 1 h at 110°C before use. In general, the best results were obtained with TLC plates treated with phosphoric acid–sodium tungstate or saturated molybdic acid as impregnating reagents.

### Impregnation with Liquid Ion Exchangers and Neutral Organophosphorous Compounds for Metal Ion Separations

The separation of inorganic ions has been an important application of TLC. Another area that has proved to be of some interest for the use of the impregnation technique as a means of improving analyte resolution in TLC has been the employment of the so-called liquid ion exchangers. This developed from the widespread use of this type of reagent in paper and column chromatography. A range of these liquid ion exchangers were used in early examples of this type of application. However, extensive experimentation suggested that adogen 464, alamine 336, amberlite LA1 and primene JM-T were suitable for this type of application. In these early studies a 0.1 mol $L^{-1}$ solution in chloroform was used for impregnation of the silica gel, with plates prepared from a suspension of the silica gel in this solution. Metal ions on TLC have also been resolved on layers impregnated with neutral organophosphorous compounds such as tri-*n*-butyl phosphate and tri-*n*-octylphosphine oxide. The bulk of the early literature in this area was collated by Brinkman *et al.* (see Further Reading).

More recent examples of the use of impregnation for metal ion separations have included further examples of the use of primene JM-T, amberlite LA-1 and LA-2, alamine 336 and aliquat 336, tri-*n*-octylamine, tri-*n*-butyl phosphate and tri-*n*-butyl amine-impregnated silica gel TLC plates.

## Impregnation with Ion Pair Reagents

The use of ion pair reagents is also discussed in detail elsewhere in this work and will therefore only be briefly described here. A number of workers have shown that ion pair reagents can be used as mobile-phase additives or following impregnation in the stationary phase for the subsequent chromatography of polar organic compounds. The methodology used depends to a large extent on the nature of the reagent, but also to some degree on the stationary phase. So, whilst both silica gel and alkyl-bonded layers can be treated with these reagents, the results are generally better with the bonded phases. In addition, it should be noted that low molecular mass ion pair reagents such as tetramethylammonium salts are so soluble that simply impregnating the stationary phase is generally ineffective as, with aqueous solvents, the reagent rapidly dissolves in the mobile phase when chromatography is initiated. This rapidly depletes the amount of reagent available for ion-pairing and results in a generally unsatisfactory chromatographic result. Other reagents, however, typically long chain sulfonic acids or quaternary ammonium compounds (e.g. sodium dodecyl sulfate or cetrimide) are only effective when the plate has been impregnated with the reagent (as a solution in a volatile organic solvent such as chloroform or ethanol) prior to chromatography. We have found that dipping is the most effective means of ensuring an even coating of the layer with these substances.

## Impregnation with Chiral Selectors

Chiral separations in TLC have been accomplished using chiral stationary phases, chiral mobile-phase additives and by impregnating suitable TLC phases with chiral selectors. Probably the best characterized separation of this type is based on chiral ligand exchange where the chiral selector (2S, 4R, 2'RS)-4-hydroxy-1-(2'-hydroxydodecyl)proline/copper [II] acetate impregnated into $C_{18}$-bonded TLC or HPTLC plates. These plates are excellent for the separation of chiral amino acids and related compounds, and sub-

stances such as α-hydroxycarboxylic acids. Plates of this type are commercially available from several manufacturers (as the CHIR and Chiral Plate from Merck and Macherey Nagel, respectively). A typical series of separations on the CHIR HPTLC plate is shown in **Figure 2**. Alternative systems based on a similar mechanism involve the use of N,N-di-*n*-propyl-L-alanine or poly-1-phenylalaninamide copper complexes.

Another example of the impregnation approach to the chiral TLC separations involves the impregnation of diol-bonded HPTLC plates with the chiral ion pair reagent N-benzoxycarbonyl-glycyl-L-proline (ZGP). These plates are prepared by first washing with dichloromethane, then dipping in a 4 mmol $L^{-1}$ solution of ZGP and 0.4 mmol $L^{-1}$ ethanolamine. Following drying these plates can be used to separate the enantiomers of β-blockers such as popranolol using chloroform–ethanol solvent systems.

It is also possible to prepare a Pirkle-type stationary phase for chiral separations by dipping aminopropyl-bonded silicia gel HPTLC plates into solutions of substances such as (R)-N-(3,5-dinitrobenzoyl)-phenylglycine or (L)-N-(3,5-dinitrobenzoyl)leucine

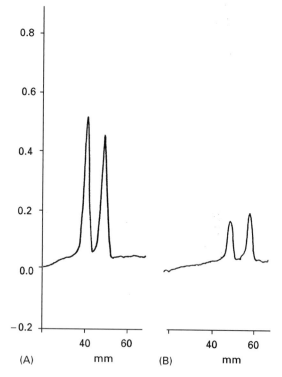

**Figure 2** Separation of the enantiomers of (A) isoleucine and (B) leucine on the CHIR HPTLC plate (detection using ninhydrin) with a water–methanol–acetonitrile mobile phase (50 : 50 : 200 v/v). Details in Wilson ID, Spurway TD, Witherow L, Ruane RJ and Longden K (1990) Chiral separations by thin-layer chromatography. *Recent Advances in Chiral Separations*, pp. 159–168. New York: Plenum.

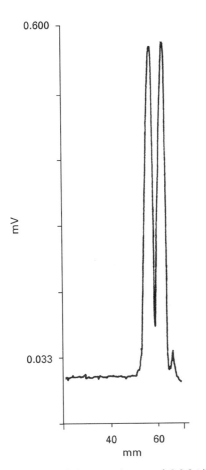

**Figure 3** Separation of the enantiomers of 2,2,2-trifluoro-(9-anthryl)ethanol on aminopropyl bonded HPTLC plates modified by impregnation with *N*-(3,5-dinitrobenzoyl)-L-leucine to form an ionically bonded Pirkle-phase. Details in Witherow L, Spurway TD, Ruane RJ and Wilson ID (1991) Problems and solutions in chiral thin-layer chromatography: a two-phase 'Pirkle' modified amino-bonded plate. *Journal of Chromatography* 553: 479–501.

(0.5 mol L$^{-1}$). Since what is being formed is essentially an ionically bonded stationary phase, it is arguable that the plate is not being impregnated. However, the process that is performed, and the overall effects, are indistinguishable from conventional impregnation and the methodology is included for completeness. An example of a typical separation on such a modified plate is shown in **Figure 3**.

## Impregnation with Oxalic Acid

The modification of TLC plates with oxalic acid has been performed to good effect for a range of solutes, including fatty acids, insecticides, nonionic detergents, azulenes and amines. In the example of the azulenes, the TLC plates were prepared by spreading the silica gel in a aqueous solution of 0.19 mol L$^{-1}$ oxalic acid, as immersion was found to give less satisfactory results. The plates were then activated by allowing them to dry for several hours over desiccant silica gel (activation using heating in an oven gave similar but irreproducible results). This activation was quite critical for the achievement of a good separation, as plates that were too damp gave no resolution. For the TLC of certain acidic mycotoxins, the use of silica gel TLC plates that had been immersed in a 10% solution of oxalic acid in methanol for 2 min and then activated by heating at 110°C for 2 min enabled tailing to be eliminated and resulted in well-defined spots.

## Impregnation with Inorganic Buffer Salts

The use of phases impregnated with sodium or potassium salts has been described for a number of solutes. Thus, conjugated dihydroxy bile acids were separated on silica gel plates that had been impregnated by dipping in a solution of potassium dihydrogen phosphate. However, a major application of this type of impregnation has been for carbohydrates. As examples, glucose, fructose and sucrose in molasses were resolved on silica gel HPTLC plates dipped in 0.2 mol L$^{-1}$ aqueous solution of monobasic potassium phosphate. In contrast, when this separation was attempted on nonimpregnated plates it failed. Similar results have been observed by other workers with this type of sample and in addition to potassium-based buffers, plates buffered with 0.15 mol L$^{-1}$ sodium dihydrogen phosphate (prepared by mixing Kieselghur with the buffer prior to spreading the plates) have also been used to separate sugars (fucose, xylose, ribose, etc). Other workers have also examined impregnation of silica gel, prepared by spreading the adsorbent in a solution of the appropriate salt, with a range of sodium salts (phosphates, acetate, sulfate, phenyl phosphate) for a wide range of carbohydrates, including mono- and oligosaccharides and uronic acids. From these studies the best separations of monosaccharides and uronic acids were obtained with plates impregnated with 0.2–0.3 mol L$^{-1}$ salt concentrations, whilst oligosaccharides gave the best results with 0.05–0.1 mol L$^{-1}$ salt solutions. Overall, from the published examples, it seems clear that the TLC separation of carbohydrates does benefit from the use of this type of impregnating reagent.

## Charge Transfer Complexes for the Resolution of Polynuclear Aromatic Hydrocarbons

A range of compounds have been used to impregnate silica gel TLC plates in order to improve the resolution of compounds such as the polynuclear aromatic compounds (fluoranthene, benzopyrine, etc.) based on the formation of charge transfer complexes with different electron acceptors. These reagents have included caffeine, tetracyanoethylene, 1,3,5-trinitrobenzene, picric acid, chloranil, bromanil, benzoquinone and similar compounds, 2,4,7-trinitrofluorenone, teramethyluric acid, urea, pyromellitic-dianhydride, 9-dicyanomethylene-2,4,7-trinitrofluorenone, sodium desoxycholate, dimethylformamide, styphnic acid and various amino acids and nucleic acid bases. However, although a wide range of reagents have been impregnated into TLC plates in an attempt to enhance the separation of polynuclear aromatic hydrocarbons, not all have been equally effective: in one study picric acid was described as values, whilst styphnic acid showed some effect and trinitrofluorenone (on alumina) proved to be excellent. Some of the reagents listed above also proved to be heat- or light-sensitive further restricting their utility.

A recent example of the use of caffeine, shown by several groups to have a profound effect on the TLC of polynuclear aromatic hydrocarbons, as

a means of improving the HPTLC resolution of the polynuclear aromatic hydrocarbons, by charge transfer complex formation, involved preparing a solution of 4 g of the reagent in 96 mL of dichloromethane. Silica gel plates (with a preconcentration zone) were then dipped in this solution for 4 s and dried at 110°C for 30 min. Then, before sample application, the plates were pre-washed by running a blank chromatogram with dichloromethane as the mobile phase with subsequent reactivation at 110°C and then preconditioning the impregnated plate for 30 min at 100% relative humidity prior to sample application and chromatography. Following sample application, development was performed with diisopropyl ether–n-hexane (4 : 1) as the solvent in an unsaturated TLC tank. Although chromatography could be performed at room temperature, the best results (especially where quantification was to be performed) were obtained when the plates were developed at 22°C. The results of this type of separation, at room temperature and − 22°C, are illustrated in **Figure 4**.

An added benefit of this type of system is that some of the reagents used for charge transfer chromatography also result in enhanced detection of the compounds of interest because of the highly coloured or even fluorescent complexes that are formed with reagents such as chloranil or pyromellitic dianhydride.

## Miscellaneous Impregnation Reagents

As well as the areas described above, there are a large number of other applications of impregnation to modify the properties of the stationary phase in TLC. Some of these are briefly outlined below in order to give an indication of the extent of this type of work, but the list is by no means exhaustive.

Plates impregnated with ethylenediaminetetraacetic acid have been employed for the chromatography of certain antibiotics (e.g anthracyclines and tetracyline), mycotoxins, citrinin and 8-hydroxyquinolone derivatives, and may confer benefits when the TLC of metal chelating compounds is performed.

TLC plates have also been impregnated with a variety of metal salts for particular compounds or classes of compounds. These include ferric chloride on Kieselguhr (oxine derivatives), zinc salts on silica gel and silanized silica gel (chlorinated anilines, carbamates), cadmium salts on silica gel (aromatic amines), manganese salts on silica gel (aromatic amines), copper sulfate on silica gel (hexosamines. glycosamines, barbiturates), thallium nitrate on silica gel (monoterpene hydrocarbons) and lithium

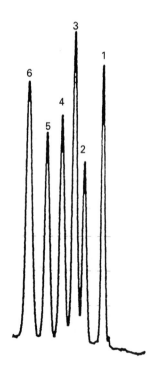

**Figure 4** Fluoresence scan of a separation of polynuclear aromatic hydrocarbons (2 ng per spot, except 6 which was 10 ng per spot) on a caffeine-impregnated silica gel layer with chromatography performed at − 20°C. 1, Benzo (ghi) perylene; 2, benzo (a) pyrene; 3, benzo (b) fluoranthene; 4, benzo(k)fluoranthene; 6, fluoranthene. Reproduced with permission from Funk W, Gluck V, Schuch B and Donnevert G (1989) Polynuclear aromatic hydrocarbons (PAHs): charge transfer chromatography and fluorimetric determination. *Journal of Planar Chromatography* 2: 28–32.

chloride-impregnated silica gel (pyrrolizidine alkaloids). Magnesium acetate has been used for phospholipids. In addition, lead salts have been employed as impregnating reagents to modify the separations of sugars and polyols. Recently, ammonium cerium (IV) nitrate was used for the separation of aromatic amines.

Other types of impregnation include the use of silica gel with phenol for aliphatic and aromatic amines following impregnation with 2% aqueous solution of the reagent. For the aromatic amines, phenol itself was found to be the most useful reagent, providing a significant improvement in spot shape was noted for a wide range of anilines compared to chromatography on the untreated stationary phase. In the case of aliphatic amines o-chlorophenol gave the best result. For both classes of compound this improvement in chromatography was assumed to be due to hydrogen bond formation between the reagent and the solutes. The chromatography of phenols on cellulose has also been modified by impregnation with 10% polyamide, and on silica gel with 20% polyamide. In addition, aniline-impregnated layers have been used for phenol derivatives whilst sodium nitrite-impregnated silica gel has also been shown to provide separations that could not be achieved on native silica gel. Carbonyl compounds were modified by derivatization to 2,4-dinitrophenylhydrazones on alumina TLC plates impregnated with silver nitrate (44%, w/w). Urea impregnation, described above for polynuclear aromatic hydrocarbons and related compounds, has also been used for the separation of lipid classes on silica gel plates.

## Concluding Comments

As indicated in the introduction, impregnation techniques have been employed since the earliest days of TLC and their use greatly extends the versatility of TLC by enabling more selective separations or detection. To some extent the increasing availability of bonded layers has reduced the need for the preparation of silica gel layers impregnated with nonpolar materials such as paraffin oil in order to perform reversed-phase separations. However the usefulness of silver nitrate-impregnated phases for the separation of compounds containing double bonds remains undiminished, and similar observations could be made for many of the impregnation reagents described above. A continued role for impregnated stationary phases in TLC therefore seems likely.

## Further Reading

Brinkman UA Th, De Vries G and Kuroda R (1973) Thin-layer chromatographic data for inorganic substances. *Journal of Chromatography* 85: 187–526.

Funk W, Gluck V, Schuch B and Donnevert G (1989) Polynuclear aromatic hydrocarbons (PAHs): Charge transfer chromatography and fluorimetric determination. *Journal of Planar Chromatography* 2: 28–32.

Gocan S (1990) Stationary phases in thin-layer chromatography. In: Grinberg N (ed.) *Modern Thin-layer Chromatography*. Chromatographic Science Series, vol. 52, pp. 5–138. New York: Marcel Dekker.

Kirchner J (1967) *Thin-layer Chromatography*: *Techniques of Organic Chemistry*, vol. XII. New York: John Wiley.

Morris LJ (1964) Specific separations by chromatography on impregnated adsorbents. In: James AT and Morris LJ (eds) *New Biochemical Separations*. London: Van Nostrand.

Stahl E (1966) *Thin-layer Chromatography, a Laboratory Handbook*, 2nd edn. Berlin: Springer-Verlag.

Wall P (1987) Argentation HPTLC as an effective separation technique for the *cis/trans* isomers of capsaicin. *Journal of Planar Chromatography* 10: 4–9.

Wilson ID, Spurway TD, Witherow L *et al.* (1990) Chiral separations by thin-layer chromatography. In: Stevenson D and Wilson ID (eds) *Recent Advances in Chiral Separations*, pp. 159–168. New York: Plenum.

# IMPRINTED POLYMERS: AFFINITY SEPARATION

*See*  **III / SELECTIVITY OF IMPRINTED POLYMERS: AFFINITY SEPARATION**

# IN-BORN METABOLIC DISORDERS: THIN-LAYER (PLANAR) CHROMATOGRAPHY

**E. Marklová**, Charles University, Hradec Králové, Czech Republic

The term inborn errors of metabolism or, more precisely, inherited metabolic diseases (IMD), is usually applied to a large group of relatively rare genetic disorders in the process of intermediary metabolism, transport defects or impaired receptors. The diagnostic procedure is complicated; the difficulty is the lack of sharp criteria for differential diagnosis, since the attendant symptoms are usually nonspecific. Comprehensive and specialized biochemical investigations therefore are the basis for the diagnosis of IMD. Using this approach, selective laboratory screening programmes can be prepared. The analytical programme includes a three-stage systematic procedure. Qualitative or semiquantitative procedures for urine metabolites are used when starting the investigation, the second step includes quantitative methods, while enzyme analysis or DNA testing belongs to the third stage, and completes the process of examination.

Multicomponent analysis of body fluids using gas and liquid chromatography combined with mass spectroscopy is the most important procedure used in the screening of IMD. However, at first one cannot do without simple and inexpensive methods like colour tests and thin-layer chromatography (TLC), which may give rapid qualitative information on metabolic conditions.

Selective laboratory screening concerns the analysis of individual groups of metabolites, usually those of amino acids and small peptides, catabolites of tryptophan, sugars, oligosaccharides, glycosaminoglycans, organic acids, purines and pyrimidines.

Separations are performed on pre-coated cellulose or silica gel high performance TLC (HPTLC) plastic or glass plates with or without fluorescent indicator $F_{254}$. The plates lay on a temperature-controlled surface (110–115°C), rapidly evaporating the elution solvent in the process of sample spotting. A volume, equivalent to a chosen amount of creatinine is used, when urine is applied.

Plates are developed with a mobile phase in either a horizontal DS chamber (Chromdes, Poland) or in a vertical pre-saturated glass tank. Chromatograms are dried and visualized under ultraviolet (UV) light of $\lambda = 254$ and 366 nm and/or by spraying with a detection reagent. Reference standards are used for metabolite identification. Quantification is carried out by linear scanning with a TLC scanner, operated with a PC software package. Interpretation of results is made according to age, diet, therapy, clinical symptoms and elementary biochemical results.

## Amino Acids

Two-dimensional ascending TLC on cellulose with ninhydrin detection is used as the first approach in the screening for amino acid metabolic disorders. As a rule, 10 μL of plasma, serum, cerebrospinal or amniotic fluids (previously deproteinized with solid sulfosalicylic acid, 50 mg mL$^{-1}$) and a volume of urine, equivalent to 20 nmol of creatinine (previously desalted on Dowex-50WX8 in H$^+$ form, eluted with 2 mol L$^{-1}$ ammonia) are applied to the $50 \times 50$ mm HPTLC cellulose layer. The plate is then developed in the solvent systems 2-propanol–formic acid–water, 80 : 4 : 20, in the first dimension and *tert*-butanol–acetone–25% ammonia–water, 50 : 30 : 10 : 10, in the second dimension. Chromatograms are sprayed with ninhydrin reagent, observed within 1 h and the next day, then heated for 3 min at 80°C (**Figure 1**). Aspartylglycosamine (in patients with lysosomal storage disease) can be detected as a blue-green spot when lightly overstained with a mixture of concentrated acetic and hydrochloride acids, 4 : 1 and heated at 60°C for 5 min. For detection of proline and hydroxyproline isatin (0.2 g in 100 mL acetone + 5 mL acetic acid) is recommended. When abnormalities are suspected, multiple spraying, in addition to ninhydrin, can be used: Ehrlich reagent (for tryptophan, hydroxyproline, citrulline and homocitrulline), Pauly reagent (for histidine and tyrosine metabolites), Sakaguchi reagent (for arginine) or platinic iodide (for sulfur amino acids).

## Mono- and Disaccharides

The best results of screening for sugar defects have been achieved using two-dimensional vertical TLC on silica gel glass plates with orcinol detection.

Filtered urine is diluted with 2-propanol, 1 : 1, and an aliquot, equivalent to 5 nmol of creatinine (or alternatively 5 μL equivalent of plasma, serum or cerebrospinal fluid, deproteinized with solid sulfo-

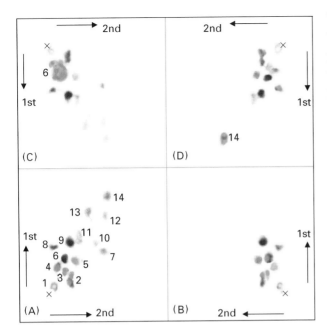

**Figure 1** TLC of amino acids. The original 100 × 100 mm glass plate was divided into four by scraping off material. (A) Normal plasma; (B) normal urine (infant); (C) nonketotic hyperglycinaemia (urine, neonate); (D) leucinosis (urine, neonate). 1, Cys; 2, His + MeHis; 3, Lys; 4, Gln; 5, Ser; 6, Gly; 7, Tau; 8, Glu; 9, Ala; 10, Thr; 11, Tyr; 12, Phe; 13, Val; 14, Leu.

salicylic acid, 50 mg mL$^{-1}$) applied on the 50 × 50 mm layer. The plate is consecutively developed, with 0.5% boric acid in water–n-butanol–2-propanol, 20 : 30 : 50 (w/v) in the first direction and ethyl-acetate–acetic acid–water, 60 : 20 : 20 in the second direction. A freshly prepared mixture of orcinol (0.4% in ethanol, w/v) and concentrated sulfuric acid, 19 : 1, is used for detection. The plate is examined by transmitted light after 10 min warming at 100°C. Identification is made by comparison with standards, separated on the same plate (**Figure 2**).

## Oligosaccharides

One-dimensional horizontal TLC on silica gel glass plates with fluorescent indicator F$_{254}$ combined with orcinol or resorcinol detection is used for screening of certain lysosomal storage diseases and adenylosuccinate lyase deficiency. A supernatant of centrifuged native urine is applied in an amount, derived from urinary creatinine and the age (μL of sample = 40 × F per concentration of creatinine in mmol L$^{-1}$; F = 0.75, 1, 1.5 and 2 for the ages < 1, 1–2, 2–8 and > 8 years, respectively). The plate is developed twice under saturated conditions with a freshly prepared mixture of n-butanol–acetic acid–water, 4.5 : 2 : 2, with drying in between. The chromatogram is observed under UV 254 nm light to look for two dark bands of succinyl purines (absorbing at 254 nm)

below raffinose. The oligosaccharides are then visualized by spraying with orcinol in the same way as described for mono- and disaccharides (**Figure 3**). Positive finding in UV light leads to rechromatography and detection with Pauly reagent and naphthoresorcinol, warmed for 10 min at 100°C. The same chromatographic procedure is convenient for screening of sialurias if using resorcinol reagent. For this a mixture of 1% resorcinol in 95% ethanol, w/v, and 2 mol L$^{-1}$ HCl, 1 : 9, with the addition of 0.1 mol L$^{-1}$ CuSO$_4$.5H$_2$O in water (0.025 mL per 10 mL of the mixture) is used. After spraying, the chromatograms are covered with glass and warmed at 120°C for 30 min. Pathological glycopeptides in urine (on α-N-acetylgalactosaminidase deficiency, aspartylglycosaminuria and fucosidosis) should first be detected by ninhydrin (see section on amino acids, above), then overstained with orcinol.

## Glycosaminoglycans

One-dimensional multisolvent TLC on cellulose plastic sheet is one of the methods used for qualitative analysis of urinary glycosaminoglycans (GAGs, acid mucopolysaccharides), whenever the photometric screening test with azure A + B is repeatedly positive. GAGs are isolated from the sediment of centrifuged urine (10 mL aliquots of 24 h urine, adjusted to

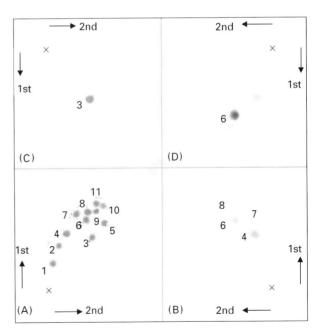

**Figure 2** Separation of sugars (and oligosaccharides). The original 100 × 100 mm glass plate was divided into four by scraping off material. (A) Standard mixture; (B–D) urine samples: (B) normal neonate; (C) fructose intolerance (infant); (D) galactosaemia (neonate). 1, stachyose; 2, raffinose; 3, fructose; 4, lactose; 5, ribose; 6, galactose; 7, saccharose; 8, glucose; 9, xylose; 10, mannose; 11, arabinose.

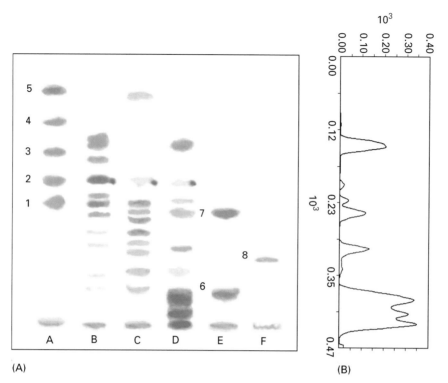

**Figure 3** (A) TLC of oligosaccharides (and sugars), overlapping lactose as a standard. A, E, F, standards; B–D urine samples; A, normal infant; B, normal neonate; D, juvenile $G_{M1}$-gangliosidosis; F, overstained with resorcinol. 1, raffinose; 2, lactose; 3, glucose; 4, xylose; 5, ribose; 6, $G_{M1}$-octosaccharide; 7, glucotetrasaccharide; 8, sialic acid. (B) Densitogram of line D.

pH 5.5) by precipitation with 0.2 mL 5% aqueous cetylpyridinium chloride for 4 h in an iced-water bath. The dry precipitate is washed using 10 mL of 95% ethanol (saturated with sodium chloride) then diethyl ether, with centrifugation, decantation and drying in between. The precipitate is dissolved in $0.6 \, mol \, L^{-1}$ sodium chloride (100 µL) and 10 µL aliquots are spotted on the layer. GAGs are successively separated (incremental distances of 2 cm for each run without intermediate drying) according to the solubility of their calcium salts in six solvents of decreasing ethanol concentration ($1 \, mol \, L^{-1}$ acetic acid–calcium acetate (g)–95% ethanol, v/w/v; I, 30:1:70; II, 40:2.5:60; III, 50:2.5:50; IV, 60:2.5:40; V, 70:2.5:30; VI, 100:5:0). After the sixth run, the plate is dried and immediately stained by immersing in toluidine blue in ethanol–acetic acid for 3 min. Excess stain is removed by rinsing in 10% acetic acid and the air-dried plate is evaluated by comparing with standards and using a densitometer (**Figure 4**).

## Organic Acids

One-dimensional sequential TLC on cellulose glass plates in a horizontal arrangement with aniline–xylose detection is used in the screening for pathological organic acidurias.

Plasma, cerebrospinal fluid or vitreous humour is deproteinized with 95% ethanol, the supernatant is evaporated at 25°C under nitrogen, dissolved in water and the equivalent of 1 mL of the native material is further processed. A volume of urine, equivalent to 2.2 µmol of creatinine, is made up to 1 mL with deionized water. All samples are then spiked with phenylbutyric acid as an internal standard, acidified to pH 1 with concentrated HCl and saturated with NaCl. Organic acids are extracted with 6 mL of diethyl ether–ethyl acetate mixture, 1:1 (vortexed, three times for 30 s). The supernatant is mixed with 100 µL of $1 \, mol \, L^{-1}$ ammonia solution in ethanol (to protect the volatile organic acids), concentrated under nitrogen at 25°C to 1 mL, and 20 µL aliquots applied on the plate. Development is performed under saturated conditions in four consecutive steps (development distances increased in 1 cm steps in each run with intermediate drying), using the mobile-phase n-propanol–$2 \, mol \, L^{-1}$ ammonia, 7:3. The plate is sprayed with aniline–xylose reagent (xylose and aniline in methanol) and evaluated with a densitometer (**Figure 5**).

## Purines and Pyrimidines

Two-dimensional TLC on cellulose glass plates in the horizontal arrangement with UV detection at

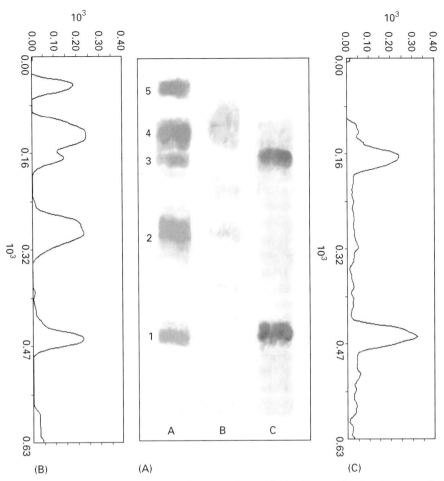

**Figure 4**  (A) TLC of glycosaminoglycans, multiple development. A, standards; B, control urine; C, mucopolysaccharidosis I–H. 1, dermatan-; 2, chondroitin-4-; 3, heparan-; 4, chondroitin-6-; and 5, keratan sulfates. (B) Densitogram of lane A; (C) densitogram of lane C.

254 nm is used to screen purine and pyrimidine defects. After 3 days on a low purine diet, 24 h urine is collected, warmed for 30 min at 50°C to dissolve precipitates and a filtered sample, equivalent to 50 μmol creatinine, spotted on the layer. The mobile phase is *n*-butanol–methanol–water–25% ammonia, 40 : 20 : 20 : 1, developed twice with 15 min drying between each run for the first direction and 2 mol L$^{-1}$ ammonium sulfate in water for the second direction. Chromatograms are evaluated under UV light and by comparison with an age-control urine and the nucleoside and base standards, separated in parallel (**Figure 6**). For further identification, chromatograms are sprayed with mercuric acetate in 95% ethanol, then immediately with diphenylcarbazone in 95% ethanol and heated at 120°C for 10 min.

## Tryptophan and its Metabolites

Screening method for both indolic and kynurenine metabolites of tryptophan (Trp) in urine is based on Sep-Pak C$_{18}$ pretreatment, two-dimensional TLC on cellulose and detection at 254 and 366 nm, followed by staining with Ehrlich reagent. A volume of urine, equivalent to 2 μmol of creatinine, is acidified to pH 3.5 and the clear supernatant applied on the Sep-Pak cartridge. Impurities from the sample are washed out, successively with sodium dodecyl sulfate (SDS) and SDS–methanol. The Trp metabolites are eluted with a mixture of 1 mol L$^{-1}$ ammonia and methanol (8 : 2). After evaporating the solvent under nitrogen the residue is dissolved in 250 μL methanol. An aliquot of 20 μL is applied on the dry cellulose layer, previously washed with deionized water. The plate is subjected to ascending development under saturated conditions at 4°C with two 0.2 mol L$^{-1}$ sodium acetate buffers: pH 6 for the first and pH 3.3 for the second direction. The air-dried plate is examined under UV light (**Figure 7**) and then sprayed with Ehrlich reagent. Comparing the urine sample with a mixture of standards, identification of 13 metabolites of tryptophan is possible.

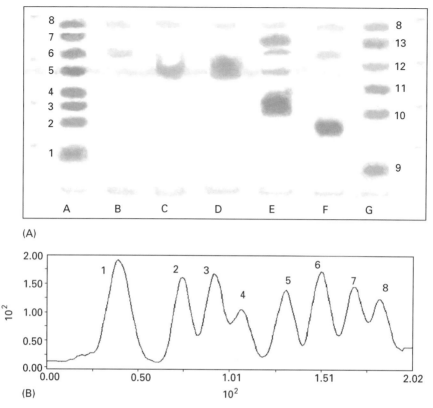

(A)

(B)

**Figure 5**    (A) TLC of organic acids, multiple development. A, standard mixture I; B, normal urine; C, normal plasma; D, lactic aciduria; E, 3-hydroxy-3-methylglutaric aciduria; F, methylmalonic aciduria; G, standard mixture II. 1, citric; 2, methylmalonic; 3, 3-hydroxy-3-methylglutaric; 4, ascorbic; 5, lactic; 6, hippuric; 7, isovaleric; 8, phenylbutyric (internal standard); 9, phosphoric; 10, adipic; 11, suberic; 12, sebacic; 13, 3-hydroxyisovaleric acids. (B) Densitogram of lane A.

## Discussion

The protein-free filtrate from plasma and other material must be prepared rapidly to avoid binding of sulfur-containing amino acids to proteins. Some changes occur rapidly at room temperature or even when stored at $-20°C$ for a week (glutamine and asparagine, in particular, disappear).

There are several amino acid metabolic disorders which can easily be detected solely by TLC screening

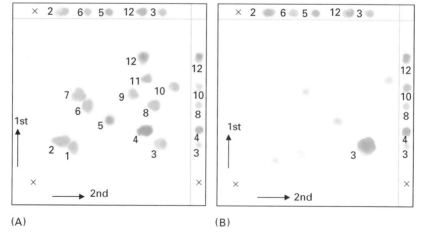

(A)                                    (B)

**Figure 6**    TLC of purines and pyrimidines. (A) Standards; (B) uric aciduria (adult). 1, xanthine; 2, guanine; 3, uric acid; 4, xantosine; 5, hypoxanthine; 6, adenosine; 7, adenine; 8, orotic acid; 9, uracil; 10, cytosine; 11, inosine; 12, thymine.

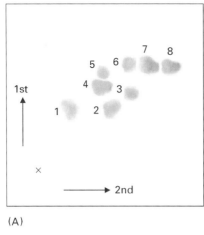
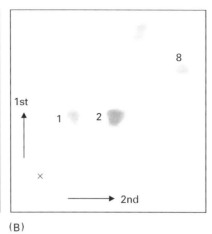

(A) (B)

**Figure 7** Separation of some indolic and kynurenine metabolites of tryptophan, UV 254 nm detection. (A) Standards, (B) xanthurenic aciduria (urine, child). 1, indolylacryloylglycine; 2, xanthurenic acid; 3, kynurenine; 4, kynurenic acid; 5, 5-hydroxyindolic acid; 6, 3-hydroxyanthranilic acid; 7, anthranilic acid; 8, indoxyl sulfate.

of urine, plasma or cerebrospinal fluid, such as leucinosis, phenylketonuria (where infants have not had newborn mass screening), hypermethioninaemic or -uric conditions, iminoglycinuria or cystinuria. An increase of glycine indicates primary nonketotic hyperglycinaemia or can be a secondary indication of an organic acid disease. Plasma glutamine is an important indicator of disorders of ammonia removal. Amniotic fluid is useful for prenatal detection of citrullinaemia, argininosuccinic aciduria or homocystinuria. High glycine in urine during therapy with valproate or ninhydrin-positive metabolites of antibiotics may confuse the interpretation of results.

On other occasions TLC only leads to a suspicion of IMD and a more precise method (amino acid analysis or HPLC) must be used for final diagnosis.

TLC is a simple screening technique for sugars, provided that their precursors are present in food. Under normal conditions no melituria is detectable. In neonates, traces of lactose and galactose appear in urine. Malabsorption, sugar intolerance, type I tyrosinosis and liver diseases, besides the genetic defects in metabolism of sugars, are the conditions related to pathological melituria. Two ribose-containing succinylpurines appear as blue-grey spots close to saccharose and galactose on chromatograms from urine and cerebrospinal fluid of patients with adenylosuccinase deficiency (IMD of purines).

Further investigation of sugar defects can be performed by specific enzyme assays, GC and HPLC.

Only slight banding with the most prominent glucose below raffinose-tetrasaccharide is detectable in normal urine when analysing oligosaccharides. Diagnosis may be difficult in neonates, where physiological lilac bands should be differentiated from the brownish ones excreted in mannosidosis, or from the pink-brown stripes seen under fucosidosis.

It is possible to combine a successive detection in UV light and spraying by ninhydrin, orcinol and resorcinol without losing sensitivity substantially. Some authors emphasize the importance of sample desalting; we have not found this necessary, desalting may lose some metabolites.

The final diagnosis can be achieved by enzyme activity assessment or structural analysis of the excreted oligosaccharides by gas chromatography–mass spectrometry (GC-MS) or nuclear magnetic resonance spectroscopy.

Most of the mucopolysaccharidoses (MPS) known so far present patterns of urinary GAGs which are clearly different from normal, even if the total urinary GAGs are not always present in excess (some cases of type III MPS), or the typical component does not react (keratan sulfate in MPS type IV). On the other hand, some pathological chromatograms may not reliably differentiate between MPS types I and VI, for example, without enzyme analysis. The main advantage of TLC of GAGs is to facilitate selection of the probable defective enzyme to be analysed for the definite diagnosis.

Urinary organic acid analysis is a vital diagnostic tool in the investigation of patients with suspected IMD, although not many laboratories perform TLC of organic acids. We consider this approach to be an important and convenient way of getting prompt orientation, allowing exclusion of negative samples in the first place. In healthy controls, organic acids are present at trace levels, the most prominent being hippuric acid. Pathological specimens are conspicuous by either unusual or very dark bands. Unclear or suspicious results are further checked by GC-MS, employing the remaining sample which is immediately lyophilized after an aliquot has been applied on the TLC layer.

Organic acid analysis may help in the diagnosis of saccharides, amino acids, fatty acids and respiratory chain defects.

Using two-dimensional TLC, some purine and pyrimidine defects can be revealed. Unfortunately, specific and sensitive visualization agents of these strongly UV-absorbing substances are not available. Large unusual spots in urine are easily detectable, but some false positives, due to other metabolites, drugs or food additives, should be carefully interpreted and verified by HPLC. HPLC-MS combined with an ion exchange pretreatment of urine is a promising method for the final diagnosis of these defects.

Metabolites of Trp are difficult to analyse because of their instability and low concentration in urine. To minimize losses, bright light should be avoided throughout the procedure. The method used is simple and reliable enough to detect increased excretion of Trp metabolites. In the case of any abnormality observed, the same pretreated urine sample is further processed by HPLC.

## Conclusion

TLC might seem to be slightly overshadowed since the introduction of more sophisticated methods in the screening of IMD. However, these methods are not available everywhere and their cost may discourage their use for screening, so that they are only applied in rather advanced cases. Using TLC, more samples can be separated simultaneously on one plate under identical conditions, which decreases the price per analysis. Horizontal arrangement of plates can be economically employed for routine use because mobile-phase consumption is very low. Availability of pre-scored layers makes analyses more flexible, more rational and economical. Advantages of HPTLC precoated layers, compared to the standard ones which were previously used, consist in sharper separation, shorter migration distance, minimal diffusion and increased detection sensitivity.

Various TLC techniques in screening for IMD have been described. In general, we prefer such procedures, when the sample, analysed by TLC, may be subject to more detailed analysis by HPLC or GC-MS, if necessary. Using creatinine concentration as a basis for the volume option of the urine processed, misinterpretation of results can be minimized.

In children with unexplained neurological disease it is highly recommended that amino acid and sugar analysis of cerebrospinal fluid and urine before and after acid hydrolysis is performed systematically. In this way defects of purines or (N-acetylated) amino acids and peptides can be detected by virtue

of a large increase in aspartate, glycine or ribose levels.

TLC represents only one part of the whole IMD screening procedure and it is necessary to emphasize that negative TLC results do not prevent more detailed investigation if the patient shows continuous signs of an acute disease or progression of neurological symptoms.

It should be emphasized that negative findings are also important, because metabolic disease can be excluded.

## Acknowledgement

This work has been supported in part by the Ministry of Health, Czech Republic, Project No. 4097-3.

*See also:* **II/Chromatography: Thin-Layer (Planar):** Densitometry and Image Analysis; Instrumentation; Modes of Development: Conventional. **III/Acids:** Thin-Layer (Planar) Chromatography. **Amino Acids:** Thin-Layer (Planar) Chromatography. **Carbohydrates:** Thin-Layer (Planar) Chromatography. **Clinical Chemistry: Thin-Layer (Planar) Chromatography. Nucleic Acids:** Thin-Layer (Planar) Chromatography.

## Further Reading

Bender DA, Joseph MH, Kochen W and Steinhart H (eds) (1986) *Progress in Tryptophan and Serotonin Research.* Berlin: Walter de Gruyter.

Chalmers RA and Lawson AM (1982) *Organic Acids in Man: Analytical Chemistry, Biochemistry and Diagnosis of Organic Acidurias.* London: Chapman & Hall.

Duran M, Dorland L, Wadman SK and Berger R (1994) Group tests for selective screening of inborn errors of metabolism. *European Journal of Pediatrics* 153 (suppl 1): S27–S32.

Hommes FA (ed.) (1991) *Techniques in Diagnostic Human Biochemical Genetics. A Laboratory Manual.* New York: Wiley-Liss.

Kelly S (1977) *Biochemical Methods in Medical Genetics.* American Lecture Series. Springfield, IL: Charles C Thomas.

Scriver CR, Beaudet AL, Sly WS and Valle D (eds) (1995) *The Metabolic and Molecular Basis of Inherited Disease.* New York: McGraw-Hill.

Sherma J and Fried B (eds) (1990) *Handbook of Thin-layer Chromatography.* New York: Marcel Dekker.

Shih V (1973) *Laboratory Techniques for the Detection of Hereditary Metabolic Disorders.* Cleveland, OH: CRC Press Uniscience.

Smith I (ed.) (1969) *Chromatographic and Electrophoretic Techniques, Vol. I, Chromatography,* 3rd edn. Bath, UK: Heinemann-Medical.

Touchstone JC (1992) *Practice of Thin Layer Chromatography.* New York: Wiley-Interscience.

# INCLUSION COMPLEXATION: LIQUID CHROMATOGRAPHY

**S. R. Gratz, B. M. Gamble and A. M. Stalcup**,
University of Cincinnati, Cincinnati, OH, USA

## Introduction

Liquid chromatography (LC) may be simply described as an analytical technique used to separate the individual components of a solution based on their relative affinities for a liquid mobile phase and a stationary phase. There are numerous modes of LC (i.e. reversed-phase, normal-phase, ion exchange) that are commonly classified according to the mechanism by which separation occurs. One such mechanism involves the formation of inclusion complexes. An inclusion complex can be defined as an entity, consisting of two or more molecules, in which one molecule, the 'host', noncovalently encapsulates or includes a 'guest' molecule. In chromatography, differences in the strength of binding between guest molecules or analytes and a specific host allow for separation. The purpose of this article is to review recent applications of inclusion complexation in liquid chromatography. Although a variety of small molecule interactions with macromolecules or polymers have been ascribed to inclusion complexation, the focus of this article is on well-defined host molecules.

## Host Molecules

Currently, there is a variety of compounds that can serve as a host molecule in an inclusion complex, most of which may be grouped into one of four classes. These classes include cyclodextrins, crown ethers, calix[n]arenes and macrocyclic antibiotics. A representative structure from each class of hosts is shown in **Figure 1**. One characteristic shared by each class of host molecules is the ability to include selectively compounds on the basis of size and shape.

It is important to point out that many inclusion host molecules possess at least one chiral centre (e.g. macrocyclic antibiotics and cyclodextrins) and that the vast majority of applications lie in the area of chiral separations. Chiral molecules are those that can exist as nonsuperimposable mirror images, or enantiomers. It has been demonstrated that the enantiomers of many chiral compounds exhibit different bioactivities and/or biotoxicities. It has also been demonstrated that enantiomers may exhibit different complex stabilities when included in a given chiral host. This fact, combined with an increased awareness of the implications of chirality, has led to the exploitation of inclusion complexation for chromatographic chiral separations. For this reason the emphasis of this article is placed on chiral applications, but some achiral applications are also presented.

Briefly, cyclodextrins (CDs), which are natural products, are cyclic oligosaccharides consisting of D-glucose rings linked through α-(1, 4)-bonds. The glucose units are arranged in a toroidal structure best described as a truncated cone. The most common cyclodextrins are α-, β- and γ-CD which consist of six, seven and eight glucose rings, respectively. The interior of the cyclodextrin molecule is hydrophobic while the exterior is hydrophillic. Because a cyclodextrin consists of chiral D-glucose units, it constitutes a chiral host.

Crown ethers are perhaps best described as synthetic, heterocyclic macrocycles with repeating units of $(-X-C_2H_4-)$ where X is the heteroatom. In contrast to cyclodextrins, crown ethers are characterized by a hydrophobic exterior and a hydrophilic cavity. The cavity typically has a strong affinity for cationic species owing to electrostatic interactions between the cation and the heteroatoms of the crown ether. These characteristics of crown ethers have led to their use as phase transfer catalysts in organic synthesis.

Calix[n]arenes are synthetic, cyclic oligomers composed of phenolic units linked by methylene bridges. They may contain four to eight aryl moieties forming a macrocyclic structure with a central cavity. The inclusion interaction between calixarenes and various guest molecules is determined both by the cavity size and by the nature of the functional groups that act as binding sites. Although still relatively unexplored, the ability to functionalize calixarenes offers yet another method for the tailoring of chromatographic selectivity.

Finally, macrocyclic antibiotics typically possess numerous chiral centres that surround several pockets or cavities bridged by a series of aromatic rings. Like cyclodextrins, the macrocyclic antibiotics are also natural products. In contrast to other host molecules, native macrocyclic antibiotics incorporate ionizable moieties.

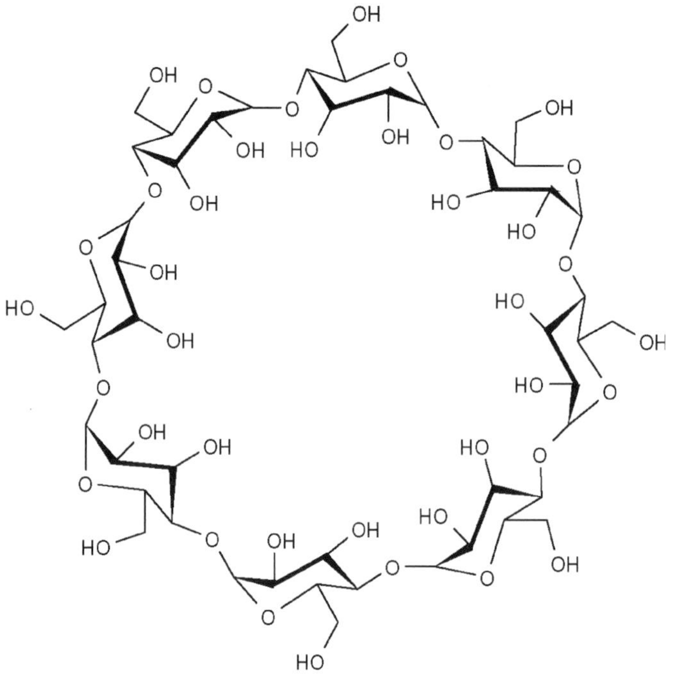

**Figure 1** Representative structures of inclusion complex host molecules: [I], $\gamma$-CD; [II], a chiral crown ether; [III], a calix[4]arene; and [IV], vancomycin.

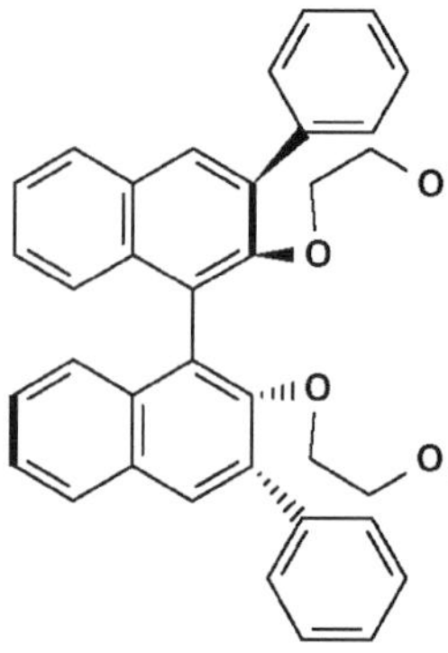

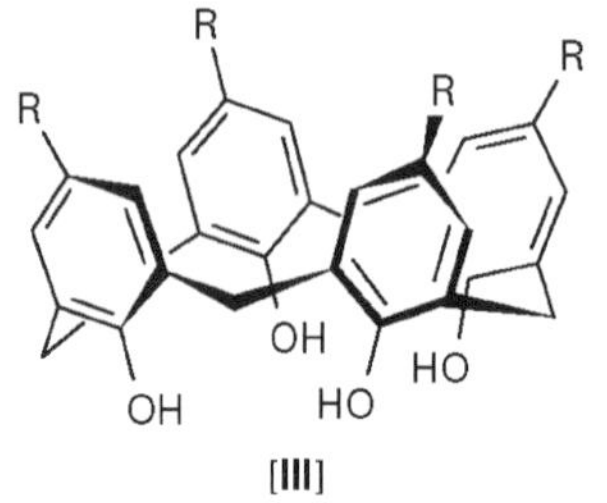

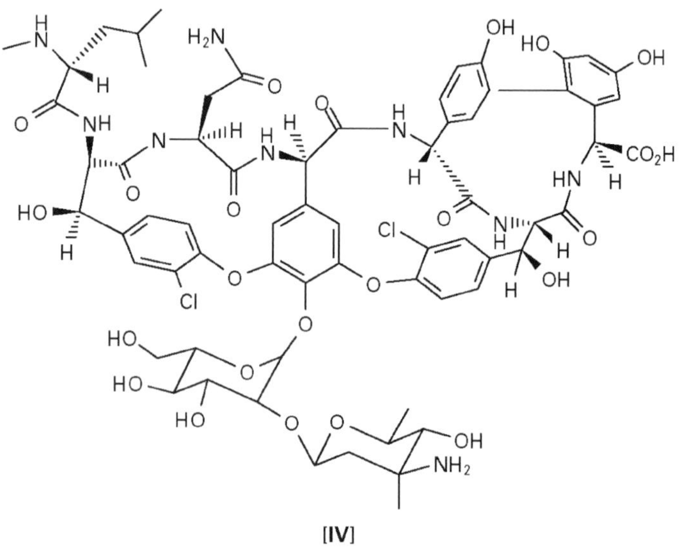

Currently there are two primary methods for exploiting inclusion complexation in LC. One method is the use of an inclusion host molecule as an immobilized ligand on a chromatographic sorbent. Stationary phases modified by cyclodextrin, crown ether, calixarene and macrocyclic antibiotic have all been reported. Alternatively, inclusion complex host molecules may be employed as mobile phase additives.

Historically, many of the hosts used to modify stationary phases originated as mobile phase additives, and thus will be considered first. Some advantages and disadvantages of each method are outlined in **Table 1**.

## Inclusion Complexation in the Mobile Phase

One approach to exploiting inclusion complexation in chromatography involves dissolving the host in the mobile phase. As indicated in Table 1, use of the host as a mobile-phase additive (MPA) instead of the cor-

**Table 1** Advantages and disadvantages of using inclusion complex host molecules as mobile phase additives

| Advantages | Disadvantages/limitations |
|---|---|
| Less expensive, conventional packed columns can be used | Possible interference with detection of analyte |
| Type and concentration of host are easily changed | Potential solubility problems |
| Wider variety of additives compared with stationary phases | Effect on mobile phase viscosity |
| Different selectivities relative to corresponding stationary phase | Impact on analyte recovery |
| | Large amounts of additive needed for column pre-equilibration |

responding bonded stationary phase has a number of advantages and disadvantages, depending on the application. One of the primary reasons for utilizing the host as a MPA is that it offers more flexibility for the application, both in the type of host used and the concentration of the host. Unfortunately, the amount of additive needed to equilibrate chromatographic columns and to separate the samples can be very large, making this approach very costly. This is not a problem in thin-layer chromatography, since TLC plates do not require equilibration with the mobile phase. In either case, the presence of the additive in the mobile phase can complicate detection and analyte recovery.

Both cyclodextrins and macrocyclic antibiotics have been used extensively as MPAs, with cyclodextrins being the most widely used. Calixarenes and crown ethers have found more limited use as MPAs. The large absorption of the phenyl groups in the calixarene molecule makes it very difficult to detect the analyte by UV spectrophotometry owing to the high background. Crown ethers have been used to aid in the dissolution of buffers in strong organic systems and to assist in the partitioning of analytes into the mobile phase. In one study, 18-crown-6 was used in the separation of carboxylated porphyrins. It was suggested that the crown ether could facilitate the partition of the porphyrins into the mobile phase through hydrophobic and polar interactions.

**Cyclodextrins**

Cyclodextrins are commonly used for the separation of enantiomers (chiral applications), but have also found use in the separation of geometric and structural isomers (e.g. $\alpha$-pinene and $\beta$-pinene). Cyclodextrins can be used in their native form or as chemically modified additives. By derivatizing the outer rim hydroxyls of the cyclodextrin with various functional groups, the properties of the native cyclodextrins can be changed. Several representative examples of native and derivatized cyclodextrins, together with the types of molecules separated, are outlined in **Table 2**.

One of the main advantages to derivatization is the change in the solubility properties of cyclodextrins. Native $\beta$-CD, which is one of the most commonly used inclusion-type additives, has a very low solubility in aqueous–organic systems (solubility = 1.85 g

**Table 2** Typical applications of native and derivatized cyclodextrins in liquid chromatography

| Cyclodextrin | Abbreviation | Applications |
|---|---|---|
| Hydroxypropyl $\beta$-CD | HPBCD | Barbiturates, chlorophenols, amino acids, laser dyes, chlorophenols, hydantoins |
| Carboxymethyl $\beta$/$\gamma$-CD | CMCD | Steroidal drugs, amino acids |
| Water-soluble $\beta$-CD polymer | SCDP | Pesticides, barbiturates, chlorophenols, nitrostyrenes, tensides, cyclic hydrocarbons |
| Dimethyl carbamate $\beta$-CD | DMP | Radioligands |
| 2,6-di-O-methyl $\beta$-CD | DIMEB | Barbiturates |
| Heptakis(2,3,6-tri-O-methyl) $\beta$-CD | TRIMEB | Tensides |
| Naphthylethyl carbamate $\beta$-CD | NEC-CD | Benzodiazepines, anesthetics, amines, amino acid esters, diuretics |
| Cationic $\beta$-CD | | Amino acids, hydantoins |
| Acetyl $\beta$-CD | | Anticancer drugs, aminoalkyl phosphonic acids, bronchodilators |
| Permethylated $\beta$-CD | | Amino acids |
| Maltosyl $\beta$-CD | | Amino acids, alkaloids |
| Sulfated $\beta$-CD | | Anesthetics, antiarrhythmics, antimalarials, catecholamines, antidepressants, anticonvulsants, antihistamines |
| Native $\alpha$-CD | | Barbiturates, ArOH, tensides, terpenes |
| Native $\beta$-CD | | Barbiturates, amino acids, PAHs, ArOH, aflatoxins, hydantoins, antiepileptic drugs, tensides, terpenes, alkaloids |
| Native $\gamma$-CD | | Barbiturates, tensides, porphyrins, $\beta$-blockers |

ArOH, aromatic alcohol.

per 100 mL of water, $\approx 16$ mmol L$^{-1}$ solution). Solutions of derivatized cyclodextrins, however, can be made at concentrations as high as 400 mmol L$^{-1}$. It is important to note that these functionalized cyclodextrins are complex mixtures of reaction products. The increased solubility is a direct consequence of this product impurity. Unfortunately, the more concentrated cyclodextrin solutions have increased viscosity. Derivatization of cyclodextrins also affects its chromatographic selectivity, as a result of additional interactions made possible by the presence of various functional groups.

**Liquid chromatography**   Thus far, of all the potential host molecules, cyclodextrins are the most extensively used MPAs in liquid chromatography. Most commonly, cyclodextrins are used as MPAs in the reversed-phase mode. The mobile phase typically consists of an aqueous buffer and an organic modifier. The cyclodextrin additive is included in the buffer component at concentrations appropriate to the application. In general, as the concentration of the cyclodextrin is increased, the enantioselectivity increases until an upper limit is reached. This is illustrated in **Figure 2** for the separation of $(\pm)$-$\alpha$-pinene. The organic component of the mobile phase should not form strong inclusion complexes with cyclodextrin as this will limit the interactions with the analyte(s) of interest. Large, bulky organic solvents should be avoided for this reason. Solvents typically used are methanol and acetonitrile, both of which form relatively weak inclusion complexes with cyclodextrins. Percentages of organic solvent used can be as low as 3% and as high as 45% in an isocratic mixture where gradients are not used. Typically, octadecylsilane (ODS) columns are used but other

bonded phases can be employed, depending on the application.

As mentioned, the use of cyclodextrins as MPAs allows the separation of enantiomers (e.g. $(\pm)$-$\alpha$-pinene) and structural isomers. This allows the application of this method to many pharmaceuticals, pesticides and natural products. By varying the experimental parameters (type and concentration of cyclodextrin, percent organic solvent, etc.), information about the strength and stability of the complex can be obtained.

Mechanistic information about complex formation can also be obtained from the data. Frequently, a 1 : 1 host–guest complex is formed, but 2 : 1 complexes have been reported in the separation of terpenes and polycyclic aromatic hydrocarbons (PAHs).

**Thin-layer chromatography**   Cyclodextrins can also be used as complex hosts in TLC. They are almost exclusively employed as MPAs rather than as bonded phases because there are no commercially available cyclodextrin bonded-phase TLC plates. The success of cyclodextrins may be partially attributed to their UV transparency.

Commercially available, reversed-phase plates are typically used. However, a 5% paraffin-in-hexane solution can be used to impregnate standard TLC plates for use as reversed-phase plates. The mobile phase is usually a water/organic solvent mixture with occasional use of buffers in the aqueous portion. As in column chromatography, the organic component is typically methanol or acetonitrile, and in some cases ethanol. The concentration of cyclodextrin in the mobile phase is usually less than 30 mmol L$^{-1}$ and the choice of cyclodextrin concentration is governed by the same factors as discussed above for liquid chromatography. In general, as the concentration of the cyclodextrin is increased, the development time will increase due to the increased viscosity of the resulting solution.

The use of MPAs in TLC is often used to study the binding characteristics of various compounds to inclusion complex host molecules. This is typically accomplished using the following general equation (eqn [1]):

$$R_M = R_{M0} + bC \tag{1}$$

where $R_M$ is the actual $R_M$ value of an analyte determined at $C$ (mmol L$^{-1}$) additive concentration; $R_{M0}$ is the $R_M$ value of an analyte extrapolated to zero additive concentration; $b$ is the decrease in the $R_M$ value caused by a 1 mmol L$^{-1}$ increase of the additive concentration in the eluent (indicator of the complexing capacity); and $C$ is the additive concentration in the eluent (mmol L$^{-1}$).

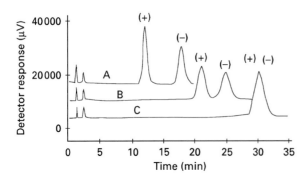

**Figure 2**   Separation of $\alpha$-pinene enantiomers using $\alpha$-CD as a mobile phase additive: (A) 16 mmol $\alpha$-CD; (B) 8 mmol $\alpha$-CD; (C) No $\alpha$-CD. (Adapted from Moeder C, O'Brien T, Thompson R and Bicker G (1996) Determination of stoichiometric coefficients and apparent formation constants for $\alpha$- and $\beta$-CD complexes of terpenes using reversed-phase liquid chromatography. *Journal of Chromatography* A 736: 1–9).

The slope value, $b$, is linearly related to the stability of the host–guest complex. For example, a wide variety of pharmaceuticals, nonionic surfactants (tensides) and substituted aromatics have been investigated by this method. Frequently, information is needed about the steric and hydrophobic character of various analytes. Host–guest binding constants can provide information about inclusion within the cyclodextrin cavity. Cyclodextrins have also been used to study the effects of inclusion on the stability and transport of drugs. It is known that cyclodextrins can influence many of the biological properties of drugs when a host–guest complex is formed. The uptake, adsorption properties and degradation of the drug can be modified by complex formation.

Cserháti and Forgács recently studied complex formation between carboxymethyl-$\gamma$-cyclodextrin and steroidal drugs. They found that the cyclodextrin mediated the hydrophobicity of the drugs suggesting that the complexed drug may have a modified efficacy.

## Macrocyclic Antibiotics

In contrast to cyclodextrins, which have been used for chiral as well as achiral applications, macrocyclic antibiotic MPAs have been applied exclusively to enantiomeric separations. Macrocyclic antibiotics such as vancomycin and erythromycin have been used to resolve enantiomers by TLC. The first macrocyclic antibiotic to be used as a MPA was vancomycin by Armstrong and Zhou in 1994. It was used as a chiral selector in the TLC resolution of dansyl-amino acids, 6-aminoquinolyl-$N$-hydroxysuccinimidyl carbamate (AQC)-derivatized amino acids, and several pharmaceutical compounds. The composition of the mobile phase as well as the nature of the stationary phase were found to affect chiral resolution for these compounds. The mobile phase consisted of $0.6 \text{ mol L}^{-1}$ NaCl with acetonitrile concentration between 17 and 40 vol%. The purpose of NaCl was to stabilize the binder on the TLC plates. For some of the derivatives, a 1% triethylammonium acetate buffer (pH 4.1) was also needed. It was found that diphenyl-type bonded phases provided the best resolution of the compounds studied. For all the amino acid derivatives, the D-enantiomer had a larger $R_F$ value than the L-enantiomer.

# Inclusion Complexation in the Stationary Phase

As mentioned previously, no commercially available TLC plates incorporate inclusion complexation ligands. However, conventional TLC plates have been coated with various inclusion host molecules. For example, erythromycin was used to resolve a variety of dansyl-DL-amino acids. By impregnating the plate with 0.05% erythromycin, resolution of the enantiomers of the amino acids was achieved in a relatively short time (20–25 min). A mobile phase consisting of $0.5 \text{ mol L}^{-1}$ NaCl, acetonitrile and methanol was used for the separation with one amino acid requiring the presence of acetic acid for separation. The amount of NaCl solution used was generally 7–25 times the amount of acetonitrile used. Methanol was required in very small amounts in some cases. The resolution of the enantiomers was affected by small changes in the mobile phase composition, as is the case with most chiral separations.

## Cyclodextrin Phases

Among the inclusion-type stationary phases, the cyclodextrin phases have been the most widely used and the most successful, specifically in the area of chiral separations. Most chiral separations reported on native CD phases ($\alpha,\beta,\gamma$) have been accomplished in the reversed-phase mode using aqueous buffers with small amounts of organic modifier. $\beta$-CD is the most commonly used of the native cyclodextrins. The applicability of CD phases has increased with the development of strategies for derivatizing these compounds. Functionalized cyclodextrins have also been used in the reversed-phase, normal-phase, and nonaqueous reversed-phase modes.

The functional groups on derivatized CDs can play a variety of roles in enhancing enantioselectivity. In some cases the substituent provides additional sites for interaction or may enlarge the CD cavity, allowing inclusion of larger analytes. Reported applications of the cyclodextrin phases are seemingly endless. However, several general conclusions can be drawn based on applications along with the theoretical work done in this area. First of all, inclusion complexation is believed to be a significant interaction only when CD-bonded phases, native or derivatized, are used in the reversed-phase mode. Next, inclusion complexation requires that at least some portion of the analyte fits structurally into the CD cavity. Typically, the presence of an aromatic ring in the analyte molecule satisfies this condition. Results indicate that chiral recognition is enhanced if the stereogenic centre lies between two $\pi$-systems or is incorporated in a ring. The presence of secondary hydroxyl groups at the rim of the CD cavity allows for hydrogen bonding with polar molecules. Amine and carboxyl functional groups have been shown to interact strongly with those hydroxyl groups on native CD phases, while nitrate, sulfate, phosphate and hydroxyl functional groups seem to prefer inclusion. Finally, both flow rate and temperature are known to affect CD-based chiral recognition in the reversed-phase mode. The enantioselective inclusion properties of native CDs have rendered them quite useful as chromatographic substrates, and the ability to derivatize these macrocyclic compounds has made them extremely versatile as inclusion complex host molecules.

## Crown Ether Phases

The use of crown ethers as inclusion complexation stationary phases for LC originated with the work of Cram and co-workers during the mid-1970s. They observed that these compounds, specifically tetra-naphthylated 18-crown-6-ethers, could form inclusion complexes with alkylammonium compounds and used this idea to develop a chiral stationary phase for the separation of enantiomeric amines. Initially, crown ethers were covalently attached to a silica gel or polystyrene matrix until it was realized that they could be dynamically coated onto reversed-phase packing materials. Eventually, this chiral crown ether stationary phase was made commercially available. Numerous applications have demonstrated the ability of this phase to resolve most primary amino acids, amines, amino esters and amino alcohols. **Figure 3** shows a chromatogram in which a series of three racemic amino acids are separated using an ODS column coated with the chiral crown ether, compound [II]. With respect to enantiomeric separations, the crown ether phase is unique in that it does not require that analytes possess an aromatic ring for the achievement of chiral separation. Optimum performance of the crown ether is observed in acidic mobile phases that ensure protonation of the primary amine

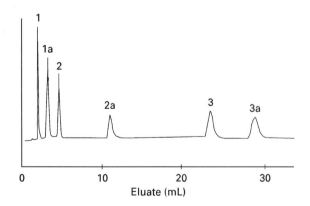

**Figure 3** Enantiomeric separation of racemic amino acids: 1, D-arginine; 1a, L-arginine; 2, D-methionine; 2a, L-methionine; 3, D-tyrosine; and 3a, L-tyrosine. (Adapted from Shinbo et al., 1987.)

functionality, which is included in the crown ether cavity. Typically, perchloric acid ($\cong 0.01 \, \text{mol L}^{-1}$) with small amounts of acetonitrile or methanol is recommended as it provides better resolution and exhibits low UV absorption. A notable advantage of the crown ether phases is the commercial availability of both chiral configurations, so that the enantiomeric elution order can be readily changed.

## Calix[*n*]arene Phases

Much like cyclodextrins, calixarenes are macrocyclic compounds that may be easily functionalized, which gives them great potential as stationary phase material. Unlike CDs, however, calixarenes have not yet been extensively employed in liquid chromatography, and therefore only a limited number of applications have been reported. The most successful applications have come with the calixarene chemically immobilized onto a silica gel matrix using a short hydrophillic spacer.

Successful separations of several classes of compounds including amino acids, polyaromatics, nucleosides and nucleobases using a *p*-tertbutyl calixarene-bonded phase indicated that this phase behaves primarily as a reverse-phase material. A typical chromatogram illustrating the separation of nucleosides on this phase is shown in **Figure 4**. In the same study, calixarene phases were reported to form strong inclusion complexes with sodium ions. The complexation with sodium presumably induces a conformational reorientation of the calixarene that increases its hydrophobic surface area, allowing for stronger interaction with hydrophobic analytes.

In another study using a calix[4]arene tetra-diethylamide phase, selective retention of Na$^+$ over other alkali metals and Ca$^{2+}$ over Mg$^{2+}$ was observed using only water as the mobile phase. Furthermore, addition of methanol to the mobile phase resulted in increased retention of NaCl while the retention of LiCl, KCl and CsCl was unaffected by methanol.

## Macrocyclic Antibiotic Phases

Stationary phases based on macrocyclic antibiotics (MAs) such as vancomycin and teicoplanin were introduced by Armstrong in 1994. Now commercially available, and used almost exclusively for enantiomeric separations, these phases are typically referred to as the Chirobiotic™ phases. **Figure 5** illustrates a chiral separation of bromocil, devrinol and coumachlor on a vancomycin stationary phase. Chiral selectivity with these phases is demonstrated in

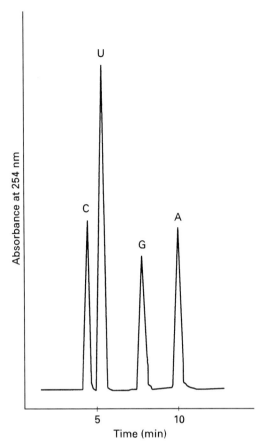

**Figure 4** Separation of four nucleosides using a calixarene-bonded phase: cytidine (C), uridine (U), guanosine (G) and adenosine (A). (Adapted from Friebe *et al.*, 1995.)

the normal, polar organic and reversed-phase modes, allowing for the separation of a large variety of chiral analytes. The complexity of the antibiotic structures provides a number of potential interactions that may assist in separation, only one of which is inclusion complexation. Furthermore, an inclusion mechanism is likely to play a role only in the reversed-phase

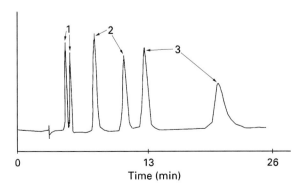

**Figure 5** Enantiomeric separation of (1) bromocil, (2) devrinol and (3) coumachlor using a vancomycin stationary phase. (Adapted from Armstrong *et al.*, 1994.)

mode, hence this discussion will be limited to the reversed-phase characteristics of these phases.

A great deal of success has been attained with MA phases for the chiral separations of neutral molecules as well as amides and esters. **Table 3** lists the conditions for several representative examples of chiral separations accomplished on a vancomycin-bonded phase.

In terms of optimization and method development, retention can be controlled by adjusting the amount of organic modifier in the mobile phase. Selectivity is controlled by the type of buffer, the type of organic modifier and the pH. Both efficiency and selectivity are affected by the buffer type, ionic strength, flow rate and temperature. Tetrahydrofuran is reported to be the best overall starting organic modifier for efficiency and selectivity, but acetonitrile, methanol and ethanol have also been used with success.

It is important to note that these MAs can also be derivatized to alter enantioselectivity, which will undoubtedly increase the number of potential applications. Furthermore, the relatively small size of these molecules together with the fact that their structures are known increases the feasibility of performing mechanistic studies on chiral recognition.

## Conclusions and the Future of Inclusion Complexation in Liquid Chromatography

The inclusion properties of several macrocyclic compounds with guest molecules have been extensively utilized in liquid chromatography. This has been accomplished by employing potential inclusion hosts as mobile-phase additives or as immobilized ligands on the chromatographic stationary phase. The presence of these macrocyclic host molecules in the mobile phase can interfere with analyte detection. Hence, most applications of inclusion complexation in liquid chromatography employ and will continue to employ immobilized ligands. While the commercial availability of the cyclodextrin, macrocyclic antibiotics and crown ether phases has no doubt contributed to their widespread exploitation, the calixarene phases may become more useful in the future as their properties become better understood. Certainly, the calixarene structure is more amenable to intelligent design than the naturally derived macrocyclic antibiotics or cyclodextrins. Inclusion complexation technology has been most extensively used in the area of chiral separations. Derivatization of macrocyclic compounds, such as cyclodextrins, macrocyclic antibiotics, crown ethers and calix[*n*]arenes has further enhanced their chromatographic utility.

**Table 3** Applications of macrocyclic antibiotic-bonded phase

| Compound | Structure | k (1st enantiomer) | α | Mobile phase | pH |
|---|---|---|---|---|---|
| Warfarin | | 1.98<br>2.27 | 1.70<br>1.44 | 10% ACN/<br>90% buffer | 7.0<br>4.1 |
| Temazepam | | 1.17 | 1.06 | 10% ACN/90% buffer | 7.0 |
| Aminoglutethimide | | 0.79 | 1.15 | 10% ACN/90% buffer | 7.0 |

The buffer is 1% triethylammonium acetate. The column used was a 25 cm × 4.6 mm i.d. vancomycin stationary phase.

*See also:* **III/Chiral Separations:** Capillary Electrophoresis; Cyclodextrins and Other Inclusion Complexation Approaches; Liquid Chromatography; Thin-Layer (Planar) Chromatography. **Essential Oils:** Gas Chromatography.

## Further Reading

Armstrong D and Zhou Y (1994) Use of a macrocyclic antibiotic as the chiral selector for enantiomeric separations by TLC. *Journal of Liquid Chromatography* 17: 1695–1707.

Armstrong D, Tang Y, Chen S, Zhou Y, Bagwill C and Chen J (1994) Macrocyclic antibiotics as a new class of chiral selectors for liquid chromatography. *Analytical Chemistry* 66: 1473–1484.

Bhushan R and Parshad V (1996) Thin-layer chromatographic separation of enantiomeric dansylamino acids using a macrocyclic antibiotic as a chiral selector. *Journal of Chromatography A* 736: 235–238.

*Chirobiotic*<sup>TM</sup> *Handbook* (1996) Advanced Separation Technologies, Inc.

Cserháti T and Valkó K (1994) *Chromatographic Determination of Molecular Interactions: Applications in Biochemistry, Chemistry and Biophysics.* Boca Raton, FL: CRC Press.

Friebe S, Gebauer S, Krauss G, Goermar G and Krueger J (1995) HPLC on calixarene bonded silica gels. I. Characterization and applications of the *p-tert*-butyl-calix[4]arene bonded material. *Journal of Chromatographic Science* 33: 281–284.

Glennon J, Horne E, Hall K et al. (1996) Silica-bonded calixarenes in chromatography. II. Chromatographic retention of metal ions and amino acid ester hydrochlorides. *Journal of Chromatography* 731: 47–55.

Shinbo T, Yamaguchi T, Nishimura K and Sugiura M (1987) Chromatographic separation of racemic amino acids by use of chiral crown ether-coated reversed-phase packings. *Journal of Chromatography* 405: 145–153.

Snopek J, Smolkaová-Keulemansová E, Cserháti T, Gahm K and Stalcup A (1996) In: Atwood J, Davies J, Macnicol D and Vogtle F (eds) *Comprehensive Supramolecular Chemistry*, pp. 515–573. Oxford: Pergamon.

Stalcup A and Gahm K (1996) A sulfated cyclodextrin chiral stationary phase for high-performance liquid chromatography. *Analytical Chemistry* 68: 1369–1374.

# IN-DEPTH DISTRIBUTION IN QUANTITATIVE TLC

**I. Vovk and M. Prošek**, National Institute of Chemistry, Ljubljana, Slovenia

Among many techniques used for quantification of thin-layer chromatograms, slit-scanning densitometry is the most common. It gives the opportunity to choose between either reflectance or transmission mode according to the nature of the supporting material. For instance, glass-backed plates can only be used for transmission measurements above 320 nm due to absorption of the UV radiation, while aluminium-backed plates are opaque at all wavelengths.

The matrix of sample and stationary phase, which consists of small particles, is optically opaque and strongly light-scattering. Therefore, densitometric measurements of separated substances are much more difficult than equivalent photometric measurements in solution. A further problem with densitometry is the fact that it is unable to detect the vertical and radial concentration profiles of the separated substance within the spot as a result of diffusion and the chromatographic process.

Reflectance and transmission are particularly sensitive to changes in sorbent quality, thickness, spot shape and size, eluent and development conditions. To reduce the errors due to plate-to-plate variation, standards should always be applied to the same plate as the unknown samples. The errors due to migration differences as a result of edge effects, deviations in layer thickness and nonlinear solvent fronts can be further minimized using the data-pair technique introduced by Bethke and co-workers in 1974. This technique is based on an internal compensation, by pairing up the measurements of two spots on the same plate.

Most thin-layer chromatography (TLC) analysis is performed in the UV and visible spectral region by applying reflectance scanning densitometry. However, the results of reflectance measurements are restricted to the surface of the sorbent on the TLC plate and are therefore strongly dependent upon the in-depth distribution of the analysed compound inside the sorbent. Nonuniformity of the in-depth distribution of a compound inside the sorbent as a result of secondary chromatography can occur during the evaporation–drying stage. Differences in the in-depth concentration profiles of the samples and stan-

dards can cause significant errors and cannot be completely eliminated by the data-pair technique.

## Theoretical Problems of Optical Scanning Densitometry

All optical methods for the quantitative evaluation of planar chromatograms are based on measuring the difference in optical response between blank regions of the stationary phase and regions with a separated substance. When monochromatic light falls on an opaque medium, some light may be reflected from the surface, some may be absorbed by the medium and converted to heat, and the remainder is diffusely reflected or transmitted by the medium. Regularly (specularly) reflected light does not give any useful information of the sample distributed within the sorbent, however, it can contribute to the noise signal in scanning densitometry as it cannot be distinguished from the diffusely reflected light. Quantification in TLC is based on measuring the diffusely reflected or transmitted light and assuming that the specularly reflected component from the sorbent is very small.

The propagation of light within an opaque medium is a very complex process that can be mathematically solved only by assuming certain simplifications. The transmission and reflectance of light in highly scattering media has been discussed by Chandrasekhar (1950) in his book on radiative transfer, where he gives the basic integro-differential radiation transport equation. As this equation has no analytical solution, all useful equations and theories have been developed by simplifying the actual case.

Continuum theories of absorption in opaque media, such as Kubelka-Munk's theory, are not appropriate for quantitative TLC. Although they describe the absorption and scattering properties of the medium, they do not take into account the interaction of light with individual particles in the layer and the very important problem of nonuniform in-depth distribution of a compound within the sorbent. Bodo, Johnson and Melamed have developed well-known discontinuum theories for the determination of absolute optical constants from the properties of individual sample particles. In 1968 Goldman and Godal published their paper on the theoretical basis of measurements by optical transmission and reflection in silica gel layers using a modified

Kubelka-Munk equation. The theoretical background of *in situ* evaluation of thin-layer chromatograms was described by Kortüm in 1969.

Theoretical studies of the problems in light scattering are important in order to reach an understanding of the measuring principle and to help the analyst to find more precise methods of evaluating the data. However, one of the main difficulties is the fact that the adsorbed substance is never homogeneously distributed over the whole sorbent which is required in theory for a relationship between reflectance and concentration of the adsorbed substance.

## Multilayer Models

The effect of the concentration gradient of the substance on the intensity of the signal in reflectance and transmission measurements has been studied by a number of authors. Most have studied the theoretical aspect of this problem. Nevertheless, using different approaches, they have all come to the same conclusion. They showed that densitometric transmittance measurements and fluorescence measurements from the far (nonilluminated) side of a plate yield results which are almost independent of the in-depth distribution of the analysed material. In contrast, the densitometric reflectance and fluorescence measurements from the near (illuminated) side are strongly dependent on that distribution. They suggested using transmittance or fluorescence transmittance mode whenever there are reasons to suspect a non-homogeneous distribution of concentration or a changing coefficient of fluorescence of the separated material in the depth of the chromatogram.

However, as already pointed out, most densitometric measurements are preformed in reflectance mode, and are therefore restricted only to the surface of the sorbent. It is for this reason that differences in the in-depth distribution of compounds (samples and standards) can cause erroneous results. Prošek and his co-workers have investigated the effect of the concentration gradient on the intensity of the diffusely transmitted and reflected light. They solved multilayer models using the mathematical theory of Bodo and Markov chains. In such a model a chromatographic band was placed in different sublayers (**Figure 1**). Although their equations do not offer any improvement from a practical point of view, they show the effect of the concentration gradient in the sorbent layer that the Kubelka-Munk theory cannot. The practical application of their theory was confirmed using the real models prepared from different kinds of layers (paper and TLC sorbents). The results obtained from these real models were in good agreement, confirming theoretical

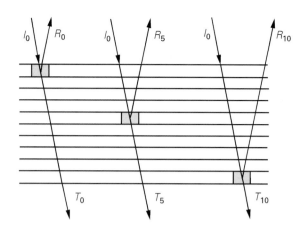

**Figure 1** Schematic presentation of a multilayer model. $I$, incident beam; $R$, reflected beam; $T$, transmitted beam.

results and showing that reflectance densitometric measurements do not take into account the vertical subsurface concentration distribution.

Further investigations have been made with image analysing systems, an emerging detection technique in TLC. In the case of image analysing systems, higher radiation fluxes must be taken into account as the result of the illumination of a whole plate. Compared to slit-scanning densitometry, the intensity of diffuse light inside the illuminated layer is much higher, as higher numbers of reflected beams come from all parts of the layer. The results obtained by measuring multilayer models with a charge coupled device (CCD) camera (**Figure 2**) showed the effect of the position of the spot inside the layer on the signal. As can be seen from **Figure 3**, the intensity of the signal in the reflectance measurements is highly dependent on the position of the spot inside the layer. This has confirmed that differences in the in-depth distributions of standards and samples can cause significant errors in reflectance quantitative TLC. On the other hand, the position of the spot inside the layer has almost no effect on the signal in transmittance measurements.

## Secondary Chromatography

While drying a TLC plate, molecules of compounds are moving with the evaporating mobile phase in the direction of the sorbent's surface. This process is called secondary chromatography and results in inhomogeneous vertical distribution of compounds inside the sorbent. This effect is bigger for compounds having high $R_F$ values in the solvent being evaporated.

Variations of in-depth distribution have a big impact on densitometric reflectance measurements. This

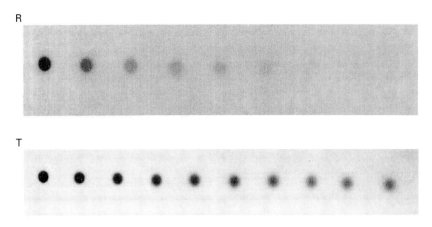

**Figure 2**   CCD images of a 10-layer model while measuring reflectance (R) and transmission (T) with a Camag video documentation system.

is one of the reasons for the nonlinearity of calibration curves and may lead to the erroneous interpretation of TLC chromatograms by reflectance densitometry. It is therefore very important to find drying conditions which will give TLC plates with the most homogeneous and consistent depth distribution of compounds.

## Depth Profiling of TLC Plates by Photoacoustic Spectroscopy

Preliminary investigations of nondestructive depth profiling of TLC plates have been performed. The results obtained using different photothermal techniques, showed that photoacoustic spectroscopy (PAS) is most suitable for characterization of TLC

plates. Although PAS has been used previously for the qualitative and quantitative spectroscopic analysis of TLC plates, the first results of photoacoustic depth profiling of TLC plates have only recently been published.

PAS is a photothermal technique, which relies on the detection of pressure waves (sound) generated by the absorption of radiation in a periodically irradiated sample. It has the capability for *in situ* and nondestructive depth profiling of solid samples. This unique feature is due to the fact that the magnitude of the induced photothermal effect depends on the concentration and on the thermal diffusivity of a compound. The plot of the dependence of the photoacoustic (PA) signal on the modulation frequency provides information about the depth profile of the analysed compound.

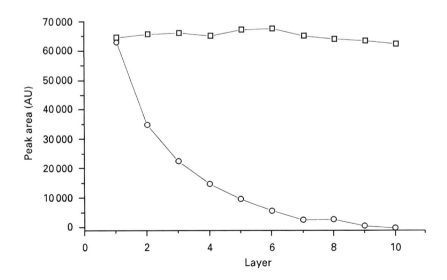

**Figure 3**   The effect of the spot position in the 10-layer model on the reflectance (circles) and transmission (squares) measurements with a Camag video documentation system.

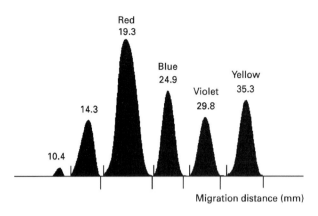

**Figure 4** Densitogram obtained from Camag test dye mixture III. (Reproduced with permission from Vovk I, Franko M, Gibkes J, Prošek M and Bicanic D (1997) Depth profiling of TLC plates by photoacoustic spectroscopy. *Journal of Planar Chromatography* 10: 258. Copyright Research Institute for Medicinal Plants in cooperation with Springer Hungarica.)

The PA signals were analysed using the theory for a two-layer model consisting of the sorbent and the glass as the supporting material. Thermal diffusivity values ($\alpha$) of the spots on TLC plates were obtained by curve fitting of normalized phase lags of PA signals. Different thicknesses of sample, corresponding to the thermal diffusion length $\mu$, given as $\mu = \sqrt{(\alpha/\pi f)}$, were probed by varying the frequency ($f$) of the laser beam modulation.

The PA signals originating from different layers of a TLC plate were calculated by subtracting the values obtained at higher modulation frequencies

from those obtained at lower frequencies. From these two modulation frequencies and the previously obtained thermal diffusivities of each spot, the depth and thickness of each layer can be determined. Depending on the available range of modulation frequencies used in the PA measurements, the thickness of the probed layers varied from 23 to 37 μm. All the results were corrected for differences in layer thickness.

## In-Depth Distribution of Compounds Inside the Sorbent and Quantitative TLC

The effect of nonhomogeneous in-depth distribution of compounds on quantitative TLC has been studied on TLC and high performance TLC (HPTLC) plates using the separation of a test dye mixture (**Figure 4**). The results of PA investigations showed that all compounds exhibited nonhomogeneous concentration profiles in a vertical direction. Although all the investigated compounds tended to concentrate in the upper 25% of a 250 μm thick layer, secondary chromatography can lead to deviations from this behaviour. Additionally, differences in the in-depth concentration distribution of different compounds on the same TLC plate were seen (**Figures 5** and **6**). While the yellow dye spots concentrated in the top 31 μm layer, the highest concentration in the violet spots was observed in the region between 39 and 61 μm. This indicates that the effects of secondary chromatography depend strongly on the properties of the compounds in each spot. The situation is different in the case of HPTLC plates, where

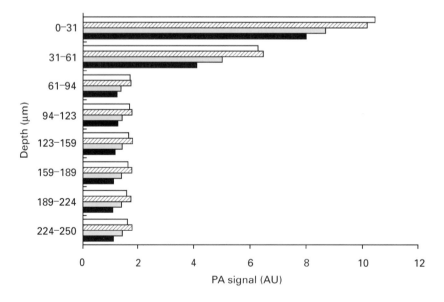

**Figure 5** Depth distribution of the compound in yellow spots of equal concentration. Open column, track no. 4; cross-hatched column, track no. 3; dotted column, track no. 2; filled column, track no. 1. (Reproduced with permission from Vovk I, Franko M, Gibkes J, Prošek M and Bicanic D (1997) Photoacoustic investigations of secondary chromatographic effects on TLC plates. *Analytical Science* 13 (Suppl.): 191. Copyright The Japan Society for Analytical Chemistry.)

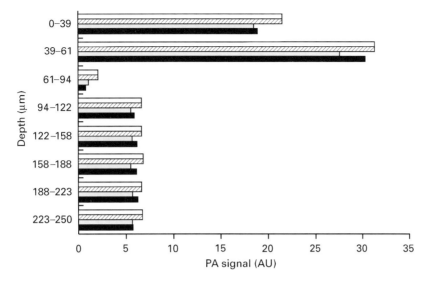

**Figure 6** Depth distribution of the compound in violet spots of equal concentration. Open column, track no. 4; cross-hatched column, track no. 3; dotted column, track no. 2; filled column, track no. 1. (Reproduced with permission from Vovk I, Franko M, Gibkes J, Prošek M and Bicanic D (1997) Photoacoustic investigations of secondary chromatographic effects on TLC plates. *Analytical Science* 13 (Suppl.): 191. Copyright The Japan Society for Analytical Chemistry.)

even the violet spots tends to concentrate in the upper, 0–37 µm layer (**Figure 7**). Different in-depth distribution of compound inside the sorbent of TLC and HPTLC plates can be explained by 50 µm differences in the layer thickness. In the case of the thinner layer of the HPTLC plate, the evaporation of the mobile phase is faster and causes faster movement of the substance to the surface of the plate. Differences in PA signals from the same depths were observed for spots from different tracks. This

indicated that nonuniformity within one TLC plate could be the source of erroneous quantification in scanning densitometry (Figures 5–7). Additionally, when monitoring only the surface of the TLC plate, by reflectance densitometry, such irregular vertical concentration distribution can result in the nonlinearity of calibration curves (**Figure 8**). For the same reasons a nonlinear calibration curve is obtained by PAS, when the probed sorbent layer is too thin (top curve in **Figure 9**). Taking into account all the con-

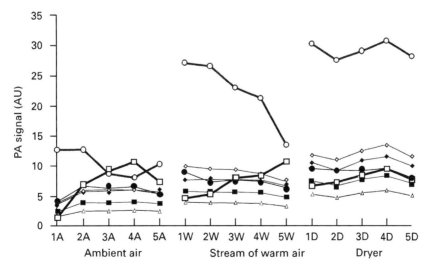

**Figure 7** Depth distribution of the compound in violet spots on HPTLC plates dried in ambient air (A), in a stream of warm air (W) or in a dryer (D). Open circles, 0–37 µm; open squares, 37–62 µm; filled circles, 62–88 µm; open diamonds, 88–125 µm; open triangles, 125–148 µm; filled squares, 148–176 µm; filled diamonds, 176–200 µm. (Reproduced with permission from Vovk I, Franko M, Gibkes J, Prošek M and Bicanic D (1998) The effect of drying conditions on the in-depth distribution of compounds on TLC plates investigated by photoacoustic spectroscopy. *Journal of Chromatography* 11: 379. Copyright Research Institute for Medicinal Plants in cooperation with Springer Hungarica.)

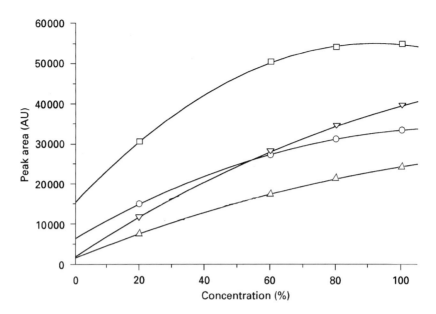

**Figure 8** Calibration curves for the Camag test dye mixture III (squares, red spots; circles, blue spots; triangles, violet spots; inverted triangles, yellow spots) obtained by reflectance densitometry. (Reproduced with permission from Vovk I, Franko M, Gibkes J, Prošek M and Bicanic D (1997) Photoacoustic investigations of secondary chromatographic effects on TLC plates. *Analytical Science* 13 (Suppl.): 191. Copyright The Japan Society for Analytical Chemistry.)

siderable irregularities in vertical concentration distribution (61 μm) by PA probing of thicker sorbent layers (lower curves in Figure 9) leads to improved linearity of calibration curves compared to those obtained by reflectance densitometry.

Our investigations confirmed that in TLC not only accurate application, development and quantitative evaluation, but also accurate drying are essential to obtain good results. Unfortunately, up to now the importance of this procedure has not been fully appreciated. The effects of drying process on the in-depth distribution of compounds inside the sorbent on HPTLC plates has therefore been investigated for drying in a dryer, in a stream of warm air

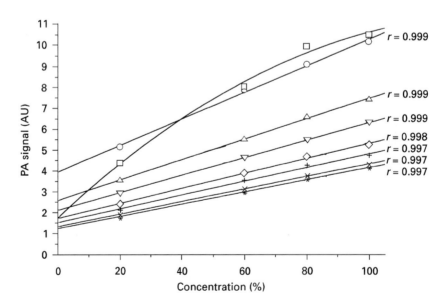

**Figure 9** Calibration curves for the yellow spots obtained by PAS when probing different thicknesses of the sorbent. Squares, 0–36 μm; circles, 0–61 μm; triangles, 0–94 μm; inverted triangles, 0–123 μm; diamonds, 0–159 μm; crosses, 0–185 μm; multiplication signs, 0–225 μm; asterisks, 0–250 μm. (Reproduced with permission from Vovk I, Franko M, Gibkes J, Prošek M and Bicanic D (1997) Photoacoustic investigations of secondary chromatographic effects on TLC plates. *Analytical Science* 13 (Suppl.): 191. Copyright The Japan Society for Analytical Chemistry.)

**Table 1**  Peak areas obtained by densitometric measurements of violet spots on three different HPTLC plates, dried in the ambient air (A) or in a stream of warm air (W) or in a dryer (D)

| | Peak areas (a.u.) for violet spots | | | | | | |
|---|---|---|---|---|---|---|---|
| | Track 1 | Track 2 | Track 3 | Track 4 | Track 5 | Mean | RSD (%) |
| Ambient air | 39 037 | 38 992 | 38 876 | 36 908 | 36 250 | 38 012.6 | 3.50 |
| Warm air | 38 242 | 38 886 | 38 788 | 37 889 | 35 311 | 37 823.2 | 3.86 |
| Dryer | 36 707 | 36 089 | 36 857 | 36 445 | 36 755 | 36 570.6 | 0.84 |

and in ambient air. The results obtained by PAS studies from different HPTLC plates were compared to those obtained by reflectance densitometry.

A comparison of the response areas was obtained for five violet dye spots from HPTLC plates dried in ambient air (A), in a stream of warm air (W) and in a dryer (D). The highest relative standard deviation (RSD) values were obtained for the plate dried in a stream of warm air (**Table 1**). The RSD for drying in a dryer was significantly lower than the RSD values for drying in the ambient air or in the stream of warm air. This can be explained by the nonuniform drying conditions across the HPTLC plate while drying in a stream of warm air.

The results of PA measurements are presented in Figure 7. For most of the tracks on all three HPTLC plates, the highest PA signals were detected in the top 37 μm thick layer. Significant differences in PA signals were observed under different drying conditions in the top 62 μm layer of the sorbent. Differences in the in-depth distribution of compounds due to variations of drying conditions across the same HPTLC plate are especially remarkable when the results for all five tracks from one plate (A, W or D) are compared (Figure 7). The largest variation was observed between tracks on the HPTLC plate dried in the stream of warm air. The most reasonable explanation for the large differences in PA signals observed between the plates dried in ambient air and those dried in a dryer or in the stream of warm air is the difference in radial distribution of the compound within the spots. When a TLC plate is dried in ambient air, the compound is radially distributed over a larger area as a result of diffusion. In the case of drying in the dryer and drying in a stream of warm air, the spots are shrinking due to migration of molecules in a radial direction towards the centre of the spot. This results in an increased concentration, and therefore higher PA signals (Figure 7), when the radius of the laser beam is smaller than the spot radius, as was the case. In the case of densitometric

measurements the probed area is larger than the diameter of a spot so that radial changes in concentration cannot be observed.

The magnitude of secondary chromatography, and consequently the vertical as well as the radial concentration distribution of compounds in the sorbent, depends on drying conditions. It also depends on the properties of the analytes and the type of TLC plate (TLC or HPTLC). Drying in a dryer gives the most reproducible results, while drying in a stream of warm air gives the least reproducible results.

## Future Trends

We anticipate that in-depth distribution will have less effect on quantitative TLC in the future, because developments in the field of TLC plates is focused on producing thinner layers by using stationary phases with a smaller and more defined particle size. However, until this is achieved, further studies of secondary chromatography are necessary.

In addition, the wider use of image-analysing systems, mostly working in the visible part of the spectrum, will increase the number of methods using postchromatographic derivatization and quenching techniques. These procedures should also be tested by PAS to elucidate their effect on the in-depth and radial distribution of compounds on TLC plates.

*See also:* **II/Chromatography: Thin-Layer (Planar):** Densitometry and Image Analysis; Instrumentation.

## Further Reading

Bein K and Pelzl J (1989) In: Analysis of surface exposed to plasmas by nondestructive photoacoustic and photothermal techniques. Auciello O and Flamm DL (eds) *Plasma Diagnostics*: *Surface Analysis and Interactions*, p. 211. New York: Academic Press.

Gibkes J, Vovk I, Bolte J *et al.* (1997) Photothermal characterisation of TLC plates. *Journal of Chromatography A* 786: 163.

Prošek M and Pukl M (1996) Basic principles of optical quantitation in TLC. In: Sherma J and Fried B (eds) *Handbook of Thin-layer Chromatography*, p. 273. New York: Marcel Dekker.

Prošek M, Medja A, Kučan E *et al.* (1979) Quantitative evaluation of thin-layer chromatograms: The calculation of remission and transmission using multilayer model. *Journal of High Resolution Chromatography & Chromatographic Communications* 2: 517.

Prošek M, Medja A, Kučan E *et al.* (1980) Quantitative evaluation of thin-layer chromatograms: The calcu-

lation of fluorescence using multilayer models. *Journal of High Resolution Chromatography & Chromatographic Communications* 3: 183.

Vovk I, Franko M, Gibkes J *et al.* (1997) Depth profiling of TLC plates by photoacoustic spectroscopy. *Journal of Planar Chromatography* 10: 258.

Vovk I, Franko M, Gibkes J *et al.* (1997) Photoacoustic investigations of secondary chromatographic effects on TLC plates. *Analytical Science* 13(suppl): 191.

Vovk I, Franko M, Gibkes J *et al.* (1998) The effect of drying conditions on the in-depth distribution of compounds on TLC plates investigated by photoacoustic spectroscopy. *Journal of Planar Chromatography* 11: 379.

# INDUSTRIAL ANALYTICAL APPLICATIONS: SUPERCRITICAL FLUID EXTRACTION

**M. E. P. McNally**, DuPont de Nemours, Wilmington, DE, USA

By the mid-1980s, routine analysis was accomplished in most industrial laboratories with the help of automated sampling and data-handling systems. The next major bottleneck caused by the burden of large sample loads that existed in many industrial environments was sample preparation and the method development time associated with it. Robots and automated systems were introduced to address these bottlenecks. Supercritical fluid extraction (SFE) was one of the automated systems introduced.

The initial introduction of SFE was unique, in that supercritical fluid chromatography (SFC) was developing at a parallel pace. Both introductions into the analytical community produced scientists who needed to learn the nuances of a totally new fluid, its capabilities, limitations, likeness to and difference from both gases and liquids.

SFE was embraced more readily in the industrial community than SFC. The reasons for this, in its utility as well as its developments, are considered in more detail in this overview.

Actual industrial applications of any analytical technology are sometimes well-guarded secrets. References can come from a variety of sources, not necessarily the published literature. Trends in the applications of SFE are no different. The most use has been in the food industry and the environmental area. Additional ap-

plications to a lesser extent are in polymer analysis, consumer products and pharmaceuticals.

Beyond the ability to conduct an analysis previously unachievable, the driving force for the adaptation of a technique into the industrial laboratory depends on a financial benefit. SFE has solved some problems that could not previously be solved. But it has mostly gained an industrial foothold by offering worthwhile cost savings over the already acceptable methodology, such as liquid extraction.

## Instrumental Developments

In the development of SFE, industrial applications were the driving force to the advancement of commercial instrumentation. Initial offerings of commercial equipment involved cumbersome sample vessels with a myriad of problems. Not uncommonly, these were placed in single-vessel extraction units. At most, the instrumentation was capable of conducting SFC following the supercritical extraction. Speeds of analysis were typically measured by actual extraction time, and not overall sample handling, operator time or productivity. In this early equipment, productivity was not truly improved and industrial acceptance was slow. This equipment was only viable for those extraction methods that could not be conducted by another extraction solvent, i.e. a liquid or a gas. This led SFC, where both liquid and gas chromatography had a strong hold in the separation field, to be labelled as a niche technique. In the sample preparation field, where alternative automated extraction methods with liquids and gases were not as firmly established, the competition was not as fierce for SFE.

Two paths were chosen. Work ensued to optimize the capabilities of supercritical fluids to encompass those compounds that might be more readily extracted by a liquid or a gas as opposed to a supercritical fluid. In addition, to compete with existing technology, albeit even classical liquid extraction, SFE needed to be able to routinely handle a larger volume of samples than was possible with these early instruments.

The first path necessitated a study of the theory and principles of supercritical fluids in terms of their solubilizing powers, matrix interactions, change with modifier or additive additions and their handling. These are generally considered academic endeavours, but much of this occurred in the industrial environment strictly out of need. Simply, it was cost-beneficial to use SFE over the classical technology. SFE was demonstrated to be a factor of 2–5 times cheaper per analysis but, more importantly, it was demonstrated to reduce classical liquid extraction method development time by 3–5 months. Reducing man-months in the process of reaching a product to market is a critical success factor in business. Not only is labour costly, but timing influences the introduction of product to market, the return of capital investiture and the ability to outpace the competition. This tremendous cost benefit resulted in industrial work on understanding some of the theoretical principles which would have otherwise been more slowly developed in academia.

Beyond the theoretical principles, the biggest single improvement, achieved through the second path, was the incorporation of the sample carousel. Easily adapted from the principle of commercial autosamplers, the carousel not only increased the total number of samples that could be analysed, but enabled unattended method development in the most labour-intensive portion of chemical analysis, the extraction step. Replicates of this have already spilled into other technologies, such as pressurized liquid extraction, but SFE was the first to utilize this feature in sample preparation.

A technique such as SFE is unlikely to survive in the industrial environment without the development of commercially available instrumentation. In the late 1980s, the market place had several key manufacturers developing and selling suitable equipment. More than a decade later, some of these have merged, some have reduced their efforts, some have retreated and concentrated on more commercially profitable areas. What remains are some well-designed pieces of instrumentation, directed at specific application areas, as in the fat analysers currently offered by Leco and Isco/Suprex. In addition, there are small suppliers whose instrumentation tends to be less expensive but less amenable to adapting to large routine applications.

The development of SFE is by no means complete, nor can the technology be considered mature. Much of the theoretical questions have returned to academia, progress will be slower, but breakthroughs can still be expected. Much of the development in industrial supercritical fluids is currently in process and larger scale and analytical developments will follow these trends.

## Industrial Applications

Successful applications have been readily demonstrated in some fields, specifically with environmental and food samples. Other applications have been more difficult to achieve and reproduce. The following applications outline some of the well-known uses of SFE.

### Food Industry

**Fat analysis**    Fats, lipids and oils have been extracted from a wide variety of foods from soybeans to oats, corn chips to brownies and meat to meat by-products and fish. Literature reports optimize the pressure and temperature of the extraction, the length of the extraction, the sample weight, the flow rate and the collection mode. Dynamic and static modes of extraction have also been investigated. Percentages of fat have been removed from sample matrices up to the 35% level; typically, gravimetric analysis follows the extraction procedure at these levels. Commercial manufacturers of foodstuffs are required by law to report percentage of fat on the package label. SFE has revolutionized this methodology. An American Society for Testing and Materials (ASTM) method for determining fat in certain selected foodstuffs has been accepted.

The nonpolar nature of fat, lipids and oils makes them extremely amenable to SFE. Liquid extraction of fat from foodstuffs is conducted using hexane solvent. Supercritical carbon dioxide has equivalent solubility properties, i.e. Hildebrand solubility parameter, to liquid hexane at low densities. Total fat analyses as well as the relationship of different lipid classes in foodstuffs have been determined. **Figure 1** shows the dynamic extraction profiles for the removal of triglycerides from 0.5 g of pork loin. As expected, higher densities give higher extraction efficiencies in shorter time periods. Although a time of 100 min may seem excessively long, compared to the classical techniques the time was orders of magnitude less. **Figure 2** compares SFE technology

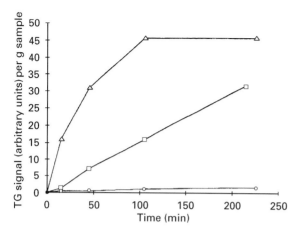

**Figure 1**  Extraction profiles for different densities. Circles, 105 bar (0.45 g mL$^{-1}$); squares, 134 bar (0.65 g mL$^{-1}$); triangles, 370 bar (0.91 g mL$^{-1}$). Pork loin 0.5 g. Dynamic extraction with pure $CO_2$. Flow rate: 2 mL min$^{-1}$, temperature: 50°C, trap: octadecyl-silica, trap temperature: 50°C, nozzle temperature: 55°C, rinse solvent: cyclohexane at a flow rate of 1.0 mL min$^{-1}$, eluted volume: 1.4 mL.

versus the two classical methods for fat analysis: SBR (Schmid, Bondzynski and Rutzlaff) and the popular Bligh & Dyer (B&D) for six different samples. A paired sample $t$-test demonstrated no significant difference at the 95% confidence level for SFE versus the classical methods.

**Decaffeination**  Commercial decaffeination of coffee, tea and cocoa using supercritical carbon dioxide has been the object of huge industrial effort. Patents have been issued since 1974 in this area, to assignees such as General Foods and Société d'Assistance Technique pour Products Nestlé. In Germany, there is a plant able to decaffeinate approximately 27.3 million kg of product per year.

There is a substantial effort to make sure the process works and is cost-effective. It is also the first example of a commercial scale supercritical fluid process. Without a doubt, the analytical departments of the commercial operations are examining the effectiveness of supercritical fluids in both analysis and process development, but the work is rarely talked about outside individual commercial enterprises.

Not surprisingly, then, the first example of routine SFE in analysis was in the removal of caffeine from food products. Caffeine, a moderately polar molecule, would not structurally be chosen as likely to have a high solubility in supercritical carbon dioxide. The commercial and analytical scale extraction of caffeine from coffee requires the addition of water as does the extraction of caffeine using methylene chloride. Reports have suggested that there is a chemical bonding of caffeine to the coffee bean or that the moisture instigates a swelling to free the entrapped compound. In either case, this landmark extraction process had opened up the understanding of SFE; overcoming chemical interactions and access to the matrix in total are important features in the investigation of SFE.

### Environmental Analysis

**Soil analysis**  Environmental applications of SFE have attracted the largest effort. Soil has received the most attention; it is one of the most difficult matrices and one of the most frequently analysed. Soil types vary widely in both their chemical and physical properties, depending on their geographical origin and their unique history. All extraction methods for soil show wide variability of recoveries and standard deviations of $\pm 30\%$ are not uncommon. In general, the advent of SFE, which was the first analytical

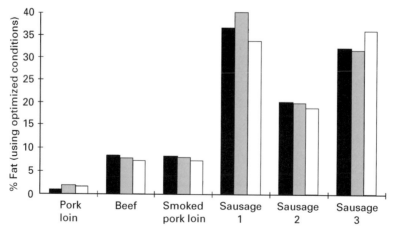

**Figure 2**  Comparison of total fat determination between SBR (filled columns), SFE (cross-hatched columns) and B&D (open columns). The samples were mixed with Hydromatrix (1 : 2 w/w) and 1 mL cyclohexane was added. Flow rate was 4.0 mL min$^{-1}$. 8% ethanol was used as modifier; the analytes were collected directly in vials; the rinse solvent pump was operated at 2 mL min$^{-1}$; other conditions as outlined for Figure 1.

technology for automated soil extraction, reduced this variability to 10–20% within individual soil types. This reduced variability makes differences in extraction from different soils more apparent and gives rise to an understanding of the individual soil parameters controlling extraction.

Analytes of different polarities have been extracted with a wide range of supercritical fluid polarities. Unmodified carbon dioxide showed the best extraction reproducibility, while modified carbon dioxide demonstrated the ability to overcome active sites in the matrix or contribute to desirable matrix swelling. Additives, which show drastic effects in chromatographic retention, also show effects in SFE.

As with other extraction technologies, analytes in spiked samples are easier to recover than are real residues. Bound residues still exist with SFE, but with tighter precision a greater understanding of bound versus available residues has been obtained.

In terms of applications, wide variety of analytes from very nonpolar polychlorinated biphenyls (PCBs) to very polar herbicides have been extracted successfully. Again, optimized conditions show the wide range of polarizabilty that can be obtained with modified carbon dioxide mobile phases. **Table 1** gives a small example of the wide range of compounds that have been extracted from soils and the pertinent conditions. Multi-analyte methods are the norm, and coupled with liquid, gas and SFC, the methods are rugged.

**Water analysis** Aqueous samples have been extracted using supercritical fluids in several distinct methods. The first method is to place the liquid in the extraction vessel. The supercritical fluid either percolates through the matrix from a bottom entry port to a top exit or, in an opposite fashion, is introduced through the top of the vessel, contacting the sample for a specified period of time before it exits. Sulfonylureas have been extracted from aqueous matrices using a system with a bottom entry and top exit. Flavour components of orange and lemon juice have used the top entry bottom exit format. Alternatively, some workers have passed the liquid either through a filter, pre-coated disc, or a separation cartridge. These trapping materials are then placed separately into extraction cartridges before introduction of the supercritical fluid. Success has been reported with all these techniques for a wide variety of environmental samples.

**Agricultural products** In this application area, matrices are complex and varied, detection levels are frequently at the parts per billion (p.p.b.) and parts per trillion (p.p.t.) level, and analyte similarity to the matrix is likely. All of these features lead to complex sample preparation and time-consuming method development. Food safety, environmental fate and product stewardship are major concerns for a competitive, consolidating industry. Successful extraction technologies need to demonstrate ease of use on

**Table 1** Range of analytes successfully extracted by SFE from soils

| Analyte | Soil | Supercritical fluid | Temperature | Pressure | Multi-residue method? |
|---------|------|---------------------|-------------|----------|-----------------------|
| Total petroleum hydrocarbons (TPHs) | Diesel and gas contaminated | $CO_2$ | 80°C | 340 atm | Yes |
| Sulfonylurea herbicides | All types | Methanol and water-modified $CO_2$ | 60°C | 300 atm | Yes |
| Ureas | All types | Methanol-modified $CO_2$ | 100°C | 300 atm | No |
| Organotin | Marine sediment, clay, topsoil | Methanol-modified $CO_2$ | 60°C | 450 atm | Yes |
| Polychlorinated biphenyls (PCBs) and polycyclic aromatic hydrocarbons (PAHs) | Railroad bed, storage dumpster soils and sediment | $CO_2$ | 150°C | 400 atm | Yes |
| Nitroaromatic and polycyclic aromatic hydrocarbons (PAHs) | Polluted industrial site and former ammunition plant soils | $CO_2$ modified with triethylamine and trifluoroacetic acid in toulene | 90°C | 395 atm | Yes |
| Nonylphenol polyethoxylates (nonionic surfactants) and carboxylic acid metabolites | Sediment and sewage treatment plant sludge | Water-modified $CO_2$ | 80°C | 340 atm | Yes |

a routine basis, ruggedness, reproducibility, accuracy and commercial availability of equipment. In addition, the current regulatory environment is insistent on advances in multi-residue analysis. These requirements and directives suggest that SFE is an obvious choice.

Herbicide, insecticide and fungicide extraction from soils by SFE has been demonstrated widely; conditions for some of these analytes are listed in Table 1. In addition to the work conducted on soils, plant and raw agricultural commodities have been routinely analysed. Most of the agrochemical classes have been extracted: organophosphates, phthalimides, organochlorines, phenols, carbamates, nitroanilines, oxazolidine, benzimidazoles, triazoles, sulfite esters, pyrethroids, imidazolines and sulfonylureas.

Frequently, sample pretreatment or processing must occur before sample extraction. Although this is time-consuming, this typically also occurs in most classical extraction methodologies. Grains and seeds are ground or milled into powders to expose the internal surface area to the extractant fluid; fruits and vegetables are chopped or diced; straws and hays may be cut or shred; oils may be treated as liquids or deposited on adsorbent surfaces. For some high water content species, drying agents such as Celite 545, Hydromatrix®, magnesium sulfate and molecular sieves have been used to lower the overall water content. They have been added to the extraction vessel before extraction or mixed with the sample during preparation. For very low moisture samples, the opposite sample preparation scheme has been pursued. As in the analysis of coffee beans described above, water has been added to dry samples such as wheat straw to swell the matrix and provide access to interstitial regions.

TCP, 3,5,6-trichloro-2-pyridinal, the major product of chlorpyrifos and chlorpyrifos methyl insecticides and trichlopyr herbicide, has been extracted from soil using SFE followed by detection via immunoassay. Effects of modifiers and additives with this extraction were dramatic, as seen in **Table 2**.

## Pharmaceutical Applications

Some of the first reports of the use of analytical scale SFE were in the pharmaceutical industry. Specifically, benzoquinones, known as ubiquinones or co-enzyme Q (with a number which represents the number of side chain isoprenoid groups that are present, e.g. Q-6 has six side chains), were extracted from bacterial cell extracts. They have physical properties analogous to synthetic nonionic detergents, in that there is a polar aromatic end-group and a nonpolar oligomeric side chain. Because of the nature of these

**Table 2** Effect of a co-solvent on the recovery of TCP from soil with SC-$CO_2$[a]

| Additive | Recovery (%) |
| --- | --- |
| None | 15.4 |
| Methanol | 57.4 |
| Methanol and ion-pairing reagent[b] | 80.0 |

[a] 1 mL min$^{-1}$ $CO_2$ flow rate; 383 bar; 40°C; 30 min extraction; 1 mL methanol, 0.5 mL ion-pairing reagent.
[b] Ion-pairing reagent: 0.1 mol L$^{-1}$ methanolic solution of R-10 camphorsulfonic acid ammonium salt (Aldrich).

molecules, their solubility range is large; they are soluble in aqueous and hydrocarbon solvents. This property makes them easily amenable to extraction with either pure carbon dioxide or carbon dioxide systems.

As an example, an experimental human immunodeficiency virus (HIV) protease inhibitor drug (**Figure 3**), has been extracted from animal feed. **Figure 4** shows the effects of extraction temperature and percent ethanol modifier on recovery, while **Figure 5** illustrates the high performance liquid chromatography (HPLC) chromatograms of the animal feed extracts. These chromatograms illustrate the relatively clean background that can be obtained with SF extractions.

Workers from Agriculture and Agri-Food Canada gave an excellent example of a multi-residue methodology in the extraction of 22 organochlorine pesticides from eggs with a programme to screen routinely for these compounds. **Figure 6** shows recoveries of the pesticides using $CO_2$ with and without methanol. Both solvents yielded adequate extraction results as well as drastic improvements in extraction time and solvent consumption over the existing Soxhlet extraction method. Background was significantly reduced with pure carbon dioxide.

Literature references show that fat- and water-soluble vitamins have been extracted from pharmaceutical formulations as well as food matrices. Traditionally, fat-soluble vitamin analyses require saponification followed by extraction with an organic

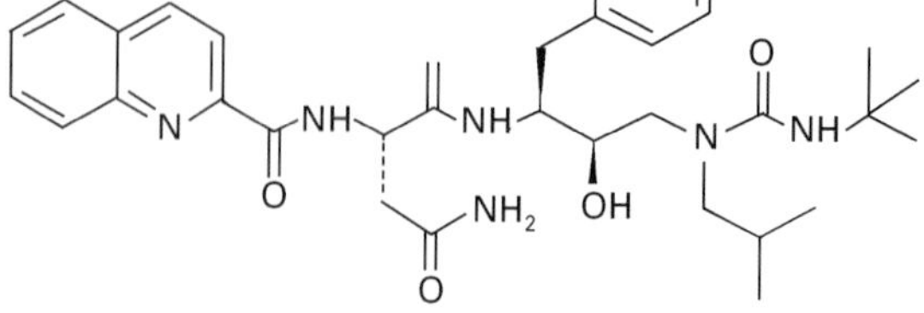

**Figure 3** Structure of experimental drug SC-52151 extracted from animal feed using SFE.

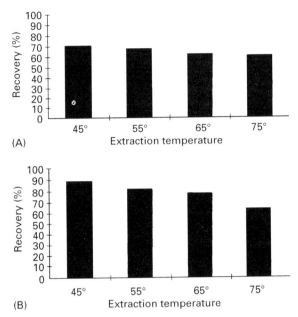

(A)

(B)

**Figure 4** Extraction recoveries for SC-52151 from animal feed spiked at the 0.05% level. (A) 5%; (B) 10% ethanol modifier.

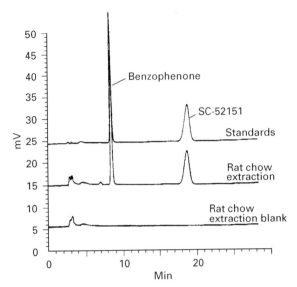

**Figure 5** Reversed-phase HPLC traces of SFE animal feed extracts. Chromatographic conditions: column, Supelco C-8 DB (250 × 4.6 mm i.d.; 5 μm particle size); mobile-phase flow rate, 1.0 mL min⁻¹; injection volume, 10 μL, detection wavelength, 205 nm; mobile phase, 0.04 mol L⁻¹ sodium pentanesulfonate.

solvent. Solvent reduction and clean-up are generally also needed. SFE eliminates these steps. The water-soluble vitamins are more readily extracted with modified carbon dioxide and minimal clean-up is required. Vitamins $A_1$, $B_6$, C, $D_2$, $D_3$ and K have all been successfully extracted using SFE.

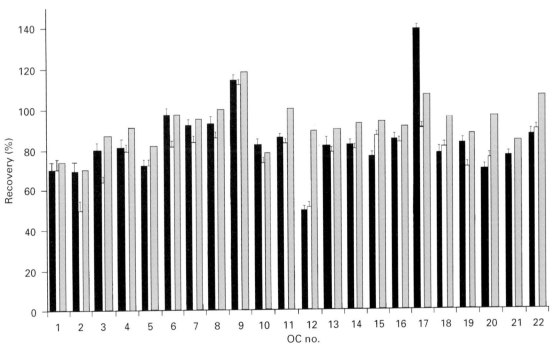

**Figure 6** Recoveries of organochlorine pesticides from eggs using SFE with (filled columns) and without methanol (open columns) as modifier. Hatched columns represent spiked blanks. Error bars represent standard deviation of five replicates for methanol-modified extractions at three levels 3, 5 and 10% where concentration showed no significant difference. Organochlorine no. 1, hexachlorobenzene; 2, α-hexachlorobenzene; 3, lindane; 4, heptachlor; 5, aldrin; 6, ronnel; 7, β-hexachlorocyclohexane; 8, chlorpyrifos; 9, dicofol; 10, oxychlordane; 11, heptachlor epoxide; 12, α-endosulfan, 13, *trans*-chlordane; 14, *cis*-chlordane; 15, *p,p'*-DDE, 16, dieldrin; 17, endrin; 18, *o,p'*-DDT; 19, *p,p'*-tetrachlorodiphenylethane; 20, *p,p'* DDT; 21, mirex; 22, methoxychlor.

Herbal medicines have been characterized for industrial marketing in Japan using online SFE/SFC. A typical example is the case for atractylon, a characteristic component of atractylodes rhizome. Rapid oxidation and degradation in UV light make manual methods difficult to execute with acceptable precision. The speed of the analysis by SFE coupled online with SFC make unstable, thermally labile and light-sensitive compounds easier to analyse.

### Polymer Applications

The uses of supercritical fluids in industrial polymerization processes are many and varied. They include polymer fibre spinning, polymer-organic solvent-phase separation, fractionation, extraction of low molecular weight oligomers from polymers, high pressure polyethylene polymerization and fractionation and monomer purification. Backbone structure and slight differences in molecular weight can cause changes in polymer solubility properties. These differences are present in commercially prepared polymers and enable fractionation or extraction to be conducted on a large scale.

On the analytical scale, process monitoring of purity, oligomer content, oligomer and antioxidant extraction, additive concentration and removal of impurities have been conducted by SFE. Under supercritical conditions, the polymer matrix has been found to swell, in the same manner as has been described for soils. This has enabled SFE to accomplish analyses that were not previously possible by liquid extraction methods.

The most frequently reported area of SFE applications for polymers is the analysis of additives. Additives influence the physical nature of the polymer and come from a wide variety of classes. They can be low molecular weight and volatile or greater than 1 kDa with solubility in only some liquid solvents. Their analysis generally requires long, tedious multi-step procedures, which may include sample preprocessing, Soxhlet extraction, concentration and clean-up. The low viscosity and high solute diffusivity of supercritical fluids aids in the reduction and, in some cases, the elimination of these steps. The ability to determine the distribution of the additives within the polymeric grid has been demonstrated by several reports along with routine quantitative analysis. Commercial competitive product analysis is standard in industry and SFE has aided in the speed, ease and accuracy of this type of analysis.

The use of supercritical fluids in the polymer industry has opened up an area of knowledge that was previously closed; work continues to be actively conducted.

### Additional Industrial Applications

In the fuel industry, stripping organics from minerals and shale rock has been conducted by SFE. Frequently, these methods used classical extraction methods that required several days. Equivalent results have been obtained in rapid extraction times of 15–30 min. Aromatic and aliphatic hydrocarbons ($C_{10}$–$C_{35}$), as well as naphthalene and asphaltene, are typical analytes that have been removed.

In the fibre industry, treatments with UV-stabilizers, removal and quantitation of residual dye components, and characterization of coatings, binders and adhesives have been conducted with the aid of SFE. Subsequent quantitative analysis either with SFC, LC, GC or mass spectrometry have all been successfully reported. The suggested procedures appear to be rugged enough to use for routine sample analysis.

## Future Developments

The amount of research on SFE in the industrial environment is cyclical. Currently more work is being conducted on a larger scale in the areas of high temperature reactions, heterogeneous catalysis, reaction/separation schemes, antisolvent recrystallization and microencapsulation. However, research at the analytical scale continues. Without a doubt success with larger scale industrial processing leads to process monitoring and analysis. Analysis schemes, of necessity, frequently mimic the larger scale. This is one of the areas where SFE, and most likely supercritical chromatography, will continue to evolve with the pace controlled by need and economics. SFE, though novel in some locations, is no longer a new technology.

Low waste generation in SFE was originally touted as a huge payoff for exploiting the technique. Reduction in waste has been achieved with all automated sample preparation technologies, although the analytical laboratory has not been pushed to reduce solvent waste to the point that large scale processes have. The ability of a supercritical fluid to conduct the extraction of an analyte and leave no organic waste is a feature that can be exploited. The potential that it can be cleaned and recycled with minimum energy input makes it still more desirable. This is an area which could force the acceptance of supercritical technologies, including extraction, where it has been slow to take off.

*See also:* **II/Extraction:** Supercritical Fluid Extraction.

## Further Reading

Bright FV and McNally MEP (eds) (1992) *Supercritical Fluid Technology: Theoretical and Applied Approaches to Analytical Chemistry*. Washington: American Chemical Society.

Charpentier BA and Sevenants MR (eds) (1988) *Supercritical Fluid Extraction and Chromatography: Techniques and Applications*. Washington: American Chemical Society.

Johnston KP and Penninger JML (eds) (1989) *Supercritical Fluid Science and Technology*. Washington: American Chemical Society.

Lee ML and Markides KE (eds) (1990) *Analytical Supercritical Fluid Chromatography and Extraction*. Provo: Chromatography Conferences.

McHugh MA and Krukonis VJ (1994) *Supercritical Fluid Extraction: Principles and Practice*, 2nd edn. Stoneham: Butterworth-Heinemann.

Smith RM and Howthorne SB (eds) (1997) Supercritical Fluids in Chromatography and Extraction. Reprinted from *Journal of Chromatography A 785*. Amsterdam: Elsevier.

Westwood SA (ed.) (1993) *Supercritical Fluid Extraction and its Use in Chromatographic Sample Preparation*. Boca Raton: CRC Press.

# INKS: FORENSIC ANALYSIS BY THIN-LAYER (PLANAR) CHROMATOGRAPHY

**L. W. Pagano, M. J. Surrency and A. A. Cantu**, US Secret Service, Washington DC, USA

## Introduction

Because of the immense number of documents written, printed and copied each year, suspect documents generated by these methods have become common subjects for forensic examinations.

The forensic analysis of inks is performed to determine if inks are similar or different; to establish the authenticity of a document; to establish whether a document could have been produced on the purported date; or to determine the origin of a document. Although the chromatographic analysis of writing and stamp pad inks have been extensively documented, little has been written on the analysis of ink jet inks (jet inks) or photocopier/laser printer/LED printer toners, despite the volumes of such evidential documents under investigation.

All inks are composed of colorants in a vehicle. Today's inks include: writing, typewriting, stamp, printing, printer, jet and toner. The consistency of the vehicle ranges from liquids, such as those used in writing and jet inks; to paste, characteristic of some printing inks; to solids, such as those used in type-writing ribbons and toner.

This article only deals with inks whose colorants are separable by thin-layer chromatography (TLC). Such colorants include dyes and pigments that have some solubility in the extraction solvent. Since most printing inks use insoluble pigments as colorants, and therefore are not analysable by TLC, such inks are not considered here. Additionally, the jet inks and

toner under consideration are those used in full colour systems (YMCK).

The TLC separation of soluble inks involves six steps: sampling, extraction, spotting, developing, visualization/detection and interpretation.

A TLC method successfully used for the comparison of writing inks has translated well for the analysis of today's modern imaging media – jet ink and toner. Although colour toners are very different from classically defined 'ink', they are made up of separable dyes and TLC has proved useful in their analysis.

TLC is not an identification method unless used for comparison with a complete collection of standards. TLC is used in conjunction with a reference specimen library to match the manufacturer of an ink. These reference collections must contain samples of all inks manufactured throughout the world. Deficiencies in the library weaken the interpretation of a match and increase the number of nonmatches.

The most important criterion in the application of TLC to matching inks with library standard inks is that the inks under investigation and library inks be chromatographed under identical conditions using identical methods. A match becomes an identity only if the match is known to be unique and/or the library is complete.

Described in this article are procedures for analysing inks. The first step in an ink analysis is the identification of the ink type. This is best determined microscopically (**Figures 1–5**). This determination is critical because different chromatographic methods are used for different inks.

Once an ink is identified, the manufacturer may provide crucial information regarding the ink's formulation, earliest possible production date, and distribution.

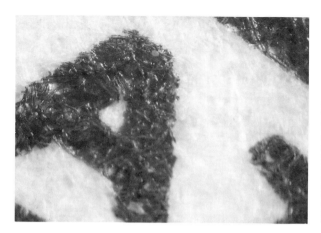

**Figure 1** (See Colour Plate 88). Offset lithography photographed at 25 ×.

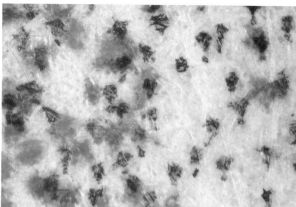

**Figure 3** (See Colour Plate 90). Full colour ink jet photographed at 25 ×.

## Jet Ink

### Ink Jet Technology

Surprisingly, ink jet technology has been around for over 35 years, but only recently has it gained widespread use, most notably in desktop computer printers. This digital printing technology directs small droplets of ink to the substrate surface. Although there are several methods for accomplishing this, the two most common methods are continuous jets and drop-on-demand. Continuous jets create a stream of uniformly sized and spaced drops, which are deflected to produce an image. Drop-on-demand generates only the drops needed for image creation.

### Fluid Jet Ink Chemistry

Most copiers/printers, which use ink jet technology, utilize a fluid, water-based ink. During the developmental stages of ink jet technology, researchers initially used fountain pen inks. These fountain pen inks caused corrosion on the print head surfaces, clogged

jet orifices, and had long drying times and excessive ink bleeding. Although these inks provided a starting point, it became apparent that this new technology would require specially formulated inks.

The early dyes were borrowed from the textile industry, then reworked and purified. Most of the organic dye counterions were replaced with larger ions, decreasing the activity and pH to a more suitable level and improving solubility, thus minimizing corrosion and crusting of the jet orifices. Along with the direct dyes, acid dyes and basic dyes used in textiles and food dyes were also adapted for ink jet systems. Organic dyes generally consist of three-dimensional compounds with ring structures having electron configurations that interact with incident radiation in the visible light range. Although these dyes must be soluble in water, they must be water-fast enough on the printed substrate to be used in an office environment. Amines and carboxylic functional groups were added to improve the binding of the dyes to the paper, and thus increase water-fastness.

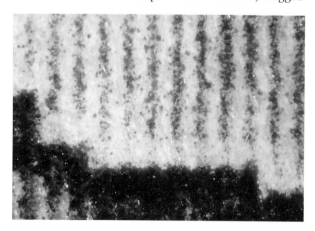

**Figure 2** (See Colour Plate 89). Full colour toner photographed at 25 ×.

**Figure 4** (See Colour Plate 91). Letterpress ink photographed at 25 ×.

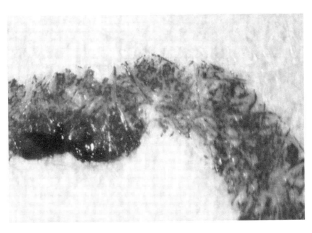

**Figure 5** (See Colour Plate 92). Writing ink photographed at 25×.

Today, the majority of commercially available jet inks are specifically formulated by the manufacturers of the print-head and delivery systems, to several performance and aesthetic requirements. In addition, the ink must be chemically compatible with the components used to construct the ink reservoir and delivery and head assemblies; it must have sufficient relative density and surface tension to remain in the nozzle without leaking out of the orifice; and when ejected it must produce the desired drop shape and radius. The ink also needs to have good wetting properties, while maintaining rapid drying times, and must have chroma, hue and optical density suitable for producing colours pleasing to the end user. In addition, the ink must be nontoxic and environmentally safe.

Dyes are the most useful colorants for jet inks; however, pigments also have properties valuable to these systems. Pigments, which are insoluble and must be suspended in the ink, have many advantages over dyes, including optical density, light- and water-fastness and stability. The main disadvantage is the lack of commercially available pigments with particle sizes less than 1 μm; most pigments are simply too large for use in ink jet systems. Advances in micropigments are producing more suitable pigments for use in ink jet systems.

To enhance ink performance, additives are incorporated. Often a low vapour pressure solvent, such as polyethylene glycol, is added to inhibit crusting. Other additives aid in buffering the solution, stabilizing the dyes (i.e. pyrollidone), and wetting the paper (i.e. glycols, butyl ethers). To inhibit organic growth, biocides and fungicides are carefully selected and added to the formula. Like other water-based ink systems, bacteria and fungus thrive in jet inks.

The ink solution is a dynamic environment. Upon entering the delivery system or cartridge, the ink reacts with the polymers, foams and metals of the container. Therefore, the ink in the printer or cartridge is no longer in its original state. The ink also changes composition, depending on its location in the cartridge and what materials are used. In systems that use low vapour pressure solvent(s) for crusting prevention, the actual chemical composition of the ink changes near the orifice as the solvent(s) evaporate.

### Ink–Paper Interaction

In addition to the critical compatibility of the ink and print system, ink and substrate compatibility also plays a crucial role in print quality. The print quality is partially determined by the surface tension, viscosity, delivery angle and velocity of the ink and the sizing, surface energy and topography of the substrate. When paper is used as the substrate, print problems such as spread (dot-gain), absorption (bleed through the paper) and paper curl are quite apparent. Initially these problems were addressed by developing special paper coatings to control these factors, as well as to enhance contrast and to improve the colour gamut, optical density and water-fastness. Many of these coatings contain silica and starch and have a large surface pore volume to absorb large quantities of ink quickly. Other ink jet papers include multilayer configurations and polymerized coatings, designed to interact with the dyes, to adjust and render them insoluble in water and adjusting for the pH of the ink.

### Analysing Jet Ink

**Extraction solvent**   The solubility of the jet inks in their sampled state determines the solvent used for their extraction. Since the ink to be forensically examined is almost always on a document, the primary focus is the solubility of the ink in its dried state on a particular substrate. The most useful method is classical solubility testing by sampling and determining solubility in pure solvents, then binary combinations, in dimple plates. Once the best extraction solution has been determined, the solvents used should be chromatographic grade to avoid contamination. For jet inks an extraction solution of ethanol/water (1 : 1) has been found to be the most successful.

**Solvent (mobile phase)/stationary phase**   Because of the complexity of the jet inks, the colorants generally cannot be separated using a single solvent. Therefore, solvent mixtures must be used. Chromatographically these solvents are chosen for their selectivity and strength. Additionally, the solvent should be made just before chromatographing. If the mobile phase is not made fresh, the concentrations will change as the higher vapour pressure solvents escape into the gas-

**Table 1** Ink solvent system I

| Solvent | Eluotropic value ($\varepsilon°$) solvent strength on silica gel | Snyder group (selectivity) | Parts |
|---|---|---|---|
| Ethyl acetate | 4.4 | VI | 70 |
| Absolute ethanol | 4.3 | II | 35 |
| Water | 10.2 | VIII | 30 |

eous portion of the container. One solvent system, widely used for examining writing inks, has been found useful in the analysis of jet inks; this is solvent system I (**Table 1**). Additionally, solvent system V (**Table 2**) has been found to be another successful system.

The stationary phase, which is as important as the mobile phase, dictates the interactions between itself and the sample and solvent. Although there are several different stationary phases available (e.g. diol, reversed-phase, etc.) silica gel has proved to be most useful. Since some dyes have fluorescent characteristics, fluorescent indicators should be absent from the plate. It is highly advisable to clean the plates to remove any contaminants. This is performed by running the plate in the chosen solvent for the length of the plate and drying prior to spotting.

**Sample preparation** The substrate from which the inks are sampled is critical. Inks sampled from coated papers, which chemically and/or physically bind the ink, will chromatograph differently from the same ink printed on noncoated paper. Further, samples of ink taken directly from the cartridge versus drying the ink on a neutral substrate may also affect the chromatograms. Therefore, it may not be possible to analyse reliably documents produced on certain substrates by TLC.

Using a scalpel, ink should be removed from the uppermost layer of paper containing the ink. This method not only increases the ink-to-paper ratio, but also reduces potential contamination from ink on the reverse of the document. Samples must contain all of the imaging colours on the document, preferably in

**Table 2** Ink solvent system V

| Solvent | Eluotropic value ($\varepsilon°$) solvent strength on silica gel | Snyder group (selectivity) | Parts |
|---|---|---|---|
| Water | 10.2 | VIII | 32 |
| Acetic acid | 6.0 | IV | 17 |
| N-Butanol | 3.9 | II | 41 |
| Butyl acetate | – | – | 10 |

equal amounts. If the document does not contain all four of the process colours, changes to the library comparison and confirmation test methods are performed with only those colours found in the sample in question. Additionally, samples of plain areas of the paper must also be taken to identify any components contributed by the paper. Depending upon the subject of the document, little if no plain areas may exist for a proper substrate blank. This is often the case when the background of the original is not white. The scalpel method should help to minimize paper influence.

**Extraction/spotting/development** The amount of ethanol/water is dependent on the size of the sample. The concentration of the samples should be within the relative range of the library specimens. The extract is spotted using a Camag Nanomat with 1.0 µL micropipettes at 1 cm from the bottom of a Whatman polyester silica gel plate. Once spotted, the plate is placed in a 95°C oven for 3 min to remove the extraction solvent. The plate is cooled to room temperature, then placed in a saturated vertical chamber containing solvent system I. The plate is allowed to develop to 4 cm from the origin, removed and dried.

**Comparison/interpretation** The developed plate is then compared with the chromatograms within the jet ink library and a list of possible matches is recorded. The possible library specimens are then sampled using the above mentioned method. The questioned ink and library specimens are spotted on a precleaned Whatman HPKF silica gel 60 plate. The plate is developed in a Camag saturated horizontal chamber containing solvent system I. After developing, the plate is dried. Once dry, the plate is interpreted under both UV and visible light (**Figure 6**).

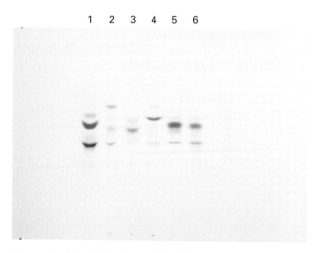

**Figure 6** (See Colour Plate 96). Thin-layer chromatogram of full colour jet inks, solvent system I.

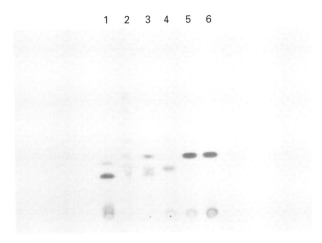

**Figure 7**  (See Colour Plate 97). Thin-layer chromatogram of full colour jet inks, solvent system V.

It should be noted that the interpretation of jet ink chromatograms is very different from interpretation of standard writing inks such as ball-point ink. Writing ink is eliminated if the relative dye concentrations are different between the questioned ink and a library specimen. This elimination is based on the fixed dye ratios within a writing ink, which is as characteristic as the dye composition itself. In ink jet systems, the concentrations of each of the four process colours varies according to the colour being reproduced; the relative concentrations of the dyes within the chromatograms will therefore also vary. As a result, relative concentration differences are not grounds for eliminating a jet ink specimen for consideration, as in the case of writing inks.

To confirm any matches to the library, a second plate with both the questioned sample and the matching library specimen is chromatographed using a Whatman HPKF silica gel 60 plate in solvent system V (Table 2). The match is then confirmed by interpreting the plate under both UV and visible light (**Figure 7**; see also **Table 3**).

If no matches are found, there may be deficiencies within the library or this may be caused by the use of post-market inks, i.e. inks made to refill original cartridges. In the case of such inks, a new range of reactions take place. Even though post-market inks may be well represented in the library, when they are added to existing cartridges the interaction with the original ink and container components may alter their chromatograms to the extent that they do not resemble either the original ink or the post-market ink. For example, the post-market ink will initially force the original ink to the print-head. Farther back in the cartridge the old and new inks mix, forming gradients. The changes that occur are often greater than the sum of the components.

## Writing Ink

### Writing Ink Chemistry

All writing inks consists of colorants and a vehicle containing solvents and resin binders (ballpoint). Writing inks can generally be placed into two categories: ballpoint (**Figure 8**) and non-ballpoint inks (**Figures 9** and **10**). Non-ballpoint inks can be further subdivided into inks that are water-based (i.e. fountain pens, felt-tip markers, roller-ball) and those that are solvent-based ('permanent' markers). For ballpoint inks, the vehicle is commonly a mixture of glycol(s) (i.e. 2,3-butanediol, 1,2-propanediol), alcohols with low relative molecular mass (i.e. 2-phenoxyethanol, benzyl alcohol) and resin binders, which result in a paste-like consistency.

Most writing inks contain complex mixtures of mostly soluble dyes and occasionally suspended pigments (carbon or inorganic pigment), which result in unique formulations. Some of the more popular dyes and pigments are pthalocyanine blue, rhodamine, nigrosine and methyl violet.

### Analysing Writing Ink

**Extraction solvent**   The method for choosing an extraction solvent for writing ink is the same as that used for jet inks. For writing inks, extraction solvents

**Table 3**  Thin-layer chromatography of full colour jet inks as shown in Figures 6 and 7

| Spot | Post-market Co. | Manufacturer | Model | Colours |
|------|-----------------|--------------|-------|---------|
| 1 | | Canon | BJC 620 | YMCK |
| 2 | | Epson | Stylus Colour 400 | YMCK |
| 3 | | Hewlett Packard | DeskJet 600 | YMCK |
| 4 | | Lexmark | 2050 | YMCK |
| 5 | American Ink Jet | Used in Hewlett Packard | DeskJet Series | YMCK |
| 6 | High Resolution | Used in Hewlett Packard | DeskJet Series | YMCK |

Extraction solvent: ethanol/water.
Whatman HPKF Silica Gel 60.

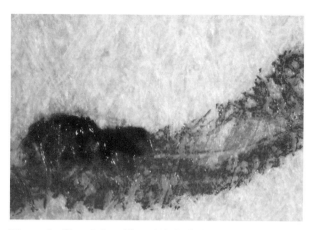

**Figure 8** (See Colour Plate 93). Ballpoint writing ink photographed at 25 ×.

**Figure 10** (See Colour Plate 95). Non-ballpoint writing ink (fountain pen) photographed at 25 ×.

include ethanol/water for water-based and some solvent-based inks and pyridine for ballpoint and most solvent-based inks.

**Solvent (mobile phase)/stationary phase**   Writing inks are complex mixtures and generally cannot be separated using a single solvent. Therefore, solvent mixtures are used to enhance separation of the colorants. Several different systems have been developed and tried. Solvent systems I and II (Table 1 and **Table 4**) are the most successful. Other solvent systems such as III and IV (**Tables 5** and **6**) can be more effective when inks are highly polar or contain Nigrosine.

Silica gel is the most useful stationary phase. The dyes and pigments used in writing inks may have fluorescent characteristics that are very helpful in their examination; fluorescent indicators should therefore be absent from the plate. Once again, the plates should be cleaned to remove any contaminants.

**Sample preparation**   Sampling is often accomplished by removal of the ink by use of a scalpel, a blunted 16–20 gauge hypodermic needle, or a commercially available forensic document sampling punch. If a scalpel is used, approximately 1 cm of an ink line is removed and placed into a vial. If the hypodermic needle or sampling device is used approximately ten plugs of ink ($\sim 0.1 \mu g$ of ink) are removed from the document and transferred to a vial. As with ink jet analysis, samples of paper/substrate should be taken and examined to identify its contribution to the ink sample extract.

**Extraction/spotting/development**   The ink is extracted by adding a minimum of appropriate extraction solvent and spotted onto the plate with 1.0 µL micropipettes at 1 cm from the bottom of a Whatman polyester silica gel plate. Once spotted, the plate is placed in a 95°C oven for 3 min to remove the extraction solvent. The plate is cooled to room temperature, then placed in a saturated vertical chamber containing solvent system I. The plate is allowed to develop to 4 cm from the origin. The plate is removed and allowed to dry by evaporation.

**Comparison/interpretation**   The developed plate is then compared with the chromatograms within the writing ink library and a list of possible matches is

**Figure 9** (See Colour Plate 94). Non-ballpoint writing ink (solvent-based) photographed at 25 ×.

**Table 4**   Ink solvent system II

| Solvent | Eluotropic value ($\varepsilon°$) solvent strength on silica gel | Snyder group (selectivity) | Parts |
|---------|---------|---------|-------|
| N-Butanol | 3.9 | II | 41 |
| Absolute ethanol | 4.3 | II | 35 |
| Water | 10.2 | VIII | 32 |

**Table 5**   Ink solvent system III

| Solvent | Eluotropic value ($\varepsilon°$) solvent strength on silica gel | Snyder group (selectivity) | Parts |
|---|---|---|---|
| Cyclohexane | 0.04 | VII | 10 |
| Chlorobenzene | 0.3 | – | 2 |
| Absolute ethanol | 4.3 | II | 1 |

recorded. The possible matches are sampled using the above mentioned method. The questioned ink and possible matching standards are then spotted on a precleaned Whatman HPKF silica gel 60 plate. The plate is developed in a horizontal chamber containing solvent system I. After developing, the plate is dried by evaporation and interpreted under both UV and visible light (**Figure 11**).

Unlike ink jet ink interpretation, the concentrations of the dye components from nonrefillable cartridge pens will not vary between the questioned document and the library samples. Therefore, relative concentration differences are grounds for eliminating an ink specimen.

To confirm any matches to the library, a second plate with both the questioned sample and the matching library specimen is chromatographed on a precleaned HPKF plate in solvent system II (Table 4). The match is then confirmed by interpreting the plate under both UV and visible light (**Figure 12**; see also Table 7).

## Toner

### Toner Technology

Electrophotography dates back to 1938 and has become the most widely used plateless printing technology. Electrophotography is a process that consists of a photoreceptive drum that is charged in darkness with a corona discharge and exposed to an original document through an optics system. The light reflected from the original image discharges the photoconductor where the light strikes it, creating a latent image on the drum corresponding to the dark areas on the original. This latent charged image is de-

**Table 6**   Ink solvent system IV (modified)

| Solvent | Eluotropic value ($\varepsilon°$) solvent strength on silica gel | Snyder group (selectivity) | Parts |
|---|---|---|---|
| Ethyl acetate | 4.4 | VI | 2 |
| Absolute ethanol | 4.3 | II | 2 |
| Chlorobenzene | 0.3 | – | 10 |

Two-phase liquid; the top phase is used as the solvent system.

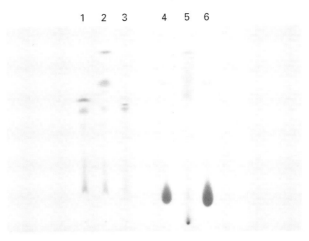

**Figure 11**   (See Colour Plate 98). Thin-layer chromatogram of ballpoint inks, solvent system I.

veloped with oppositely charged toner. The toner is transferred to a substrate and fixed. Today's digital electrophotographic machines use lasers or light emitting diodes (LED) driven by digital data to create the latent image on the photoreceptive drum.

### Dry Toner Chemistry

Toners are electrostatically transferred to the paper or print substrate and are fused by heat, pressure or a combination of both. As with other inks, toners must meet certain print parameters, such as water- and light-fastness, but because of their unique way of imaging they must also meet a unique set of chemical and physical parameters, such as triboelectric properties.

Dry toners are very fine powders consisting primarily of a polymer resin binder and colorants. These colorants, pigments and dyes, are chosen primarily for their chroma, hue and colour purity. The most common pigments and dyes are carbon black,

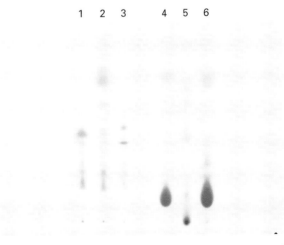

**Figure 12**   (See Colour Plate 99). Thin-layer chromatogram of ballpoint inks, solvent system II.

**Table 7** Thin-layer chromatography of ballpoint inks as shown in Figures 11 and 12

| Spot | Manufacturer | Model | Colours |
|------|--------------|-------|---------|
| 1 | Papermate | Flexgrip | Blue |
| 2 | Skilcraft | Stick | Blue |
| 3 | Bic | Round Stic | Blue |
| 4 | Papermate | Flex Grip | Black |
| 5 | Skilcraft | Stick | Black |
| 6 | Bic | Round Stic | Black |

Extraction solvent: pyridine.
Whatman HPKF Silica Gel 60.

nigrosines, copper phthalocyanines, azo-pigments and quinacridone.

The polymer or copolymer resin binder is chosen for its thermal characteristics such as glass transition temperature and flow viscosity. Polystyrene acrylates and epoxy polymers are used for hot roll and flash fusing. Small chain homologue polymers, e.g. polyethylene and polypropylene, and vinyl acetates are used for cold pressure and roll fusing. Polyester is used primarily for radiant fusing. Often additives are incorporated into the toners to alter physical properties: pigments and dyes are added for colour; magnetite salts are added to enhance control of the toner; ammonium salts are used for positive charge control; acidified carbon black and metal complexes are added for negative charge control; silica and zinc sterate are added as a lubricant and flow enhancement; and silicone oils and low weight polyethyl and polypropyl waxes are used as release agents to prevent the toner from adhering to the fusing roller.

### Analysing Toner

**Extraction solvent**  The method for choosing an extraction solvent for toner is the same as that used for jet and writing inks. It must first be determined in which solvent(s) the toner is soluble, then whether mixtures of solvents are needed to enhance solubility. Pure chloroform was found to be the most successful.

**Solvent (mobile phase)/stationary phase**  Like jet and writing ink, toners generally cannot be separated using a single solvent. Therefore, solvent mixtures must be used to accomplish separation of the colorants. The most successful solvent system for toners are solvent systems I (Table 1) and modified IV (Table 6).

Silica gel is the most useful stationary phase. The dyes and pigments used in toners often have fluorescent characteristics that are very helpful in their examination, hence fluorescent indicators should be absent from the plate. Once again, the plates should be cleaned to remove any contaminants.

**Sample preparation**  Sampling toner documents has advantages over jet ink documents. Although, as with ink jet documents, toner documents may have little or no clean paper areas, this is of little concern with toner sampling because the toner is removed from the document thermally. To remove the sample, a clean scanning electron microscope (SEM) aluminium stub placed over the sampling area is heated with a soldering iron. The heat causes some of the toner to be transferred to the stub. Since this method removes only the toner, paper interactions are avoided. Again, as with jet inks, all of the process colours should be sampled from the document in equal proportions. If the document does not contain all four of the process colours, changes to the library comparison and conformation test methods are performed with only those colours found in the sample.

**Extraction/spotting/development**  The toner is washed from the stub with chloroform into a sampling vial. The toner wash is spotted onto the plate using 2.0 μL micropipettes at 1 cm from the bottom of a Whatman polyester silica gel plate. Once spotted, the plate is placed in a 95°C oven for 3 min to remove the extraction solvent. The plate is cooled to room temperature, then placed in a saturated vertical chamber containing solvent system I. The plate is allowed to develop to 4 cm from the origin, removed and allowed to dry by evaporation (**Figure 13**).

**Comparison/interpretation**  The developed plate is then compared with the chromatograms within the toner library and a list of possible matches is recorded. The possible library specimens are sampled using the abovementioned method. The questioned toner and library specimens are then spotted on a precleaned Merck silica gel 60 plate. The plate is

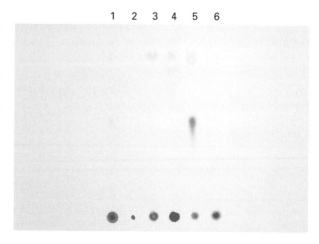

**Figure 13**  (See Colour Plate 100). Thin-layer chromatogram of full colour toner, solvent system I.

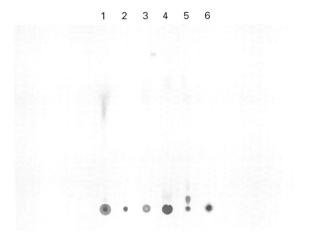

1    2    3    4    5    6

**Figure 14** (See Colour Plate 101). Thin-layer chromatogram of full colour toner, solvent system IV.

developed in a Camag saturated horizontal chamber containing solvent system I. After developing, the plate is dried by evaporation. Once dry, the plate is interpreted under both UV and visible light.

As with jet ink interpretation, the concentrations of the dye components will vary owing to the varying concentrations of the four process colours between the questioned document and the library samples. Therefore, relative concentration differences are not grounds for eliminating a toner specimen, as in the case of writing inks.

To confirm any matches to the library, a second plate with both the sample and the matching library specimen is chromatographed using a Merck silica gel 60 plate, without fluorescent indicator, in modified solvent system IV (Table 6). The match is then confirmed by interpreting the plate under both UV and visible light (**Figure 14**; see also **Table 8**).

## Conclusions

Through the use of colorant separation by TLC in conjunction with comparison libraries, the various original manufactures' inks can generally be distinguished from each other. At the time of this article,

**Table 8** Thin-layer chromatography of full colour toner as shown in Figures 13 and 14

| Spot | Manufacturer | Model | Colours |
|------|-------------|--------|---------|
| 1 | Canon | CLC 1000 | YMCK |
| 2 | Konica | 7728 | YMCK |
| 3 | Minolta | CF 900 | YMCK |
| 4 | Ricoh | 8015 | YMCK |
| 5 | Sharp | CX 7500 | YMCK |
| 6 | Xerox | 5775 | YMCK |

Extraction solvent: chloroform.
Merck Silica Gel 60.

total distinguishability is obtained for full colour jet inks and toners. The comparison of the colour components has proved to be more discriminating than the comparison of other components, such as resins within ballpoint inks and toner. The ability to discriminate inks, coupled with the low cost, ease and multiple samples per run, make TLC a powerful forensic tool for the analysis of modern documents produced by traditional and new inks.

**See Colour Plates 88, 89, 90, 91, 92, 93, 94, 95, 96, 97, 98, 99, 100, 101.**

*See also:* **II/Chromatography: Thin-Layer (Planar): Instrumentation; Layers.**

## Further Reading

American Society for Testing and Materials (ASTM) (1991) Standard Guide for Test Methods for Forensic Writing Ink Comparison. *ASTM STANDARD E1422-91.*

Aginsky VN (1993) Comparative examination of inks by using instrumental thin-layer chromatography and microspectrophotometry. *Journal of Forensic Sciences* 38(5): 1111–1130.

Andrasko J (1994) A simple method for sampling photocopy toners for examination by microreflectance Fourier transform infrared spectroscopy. *Journal of Forensic Sciences* 39(1): 226–230.

Brunelle RL and Reed WR (1991) *Forensic Examination of Ink and Paper*, pp. 165–170. Springfield, IL: Charles C Thomas.

Cantu AA (1995) A sketch of analytical methods for document dating. Part I. The static approach: determining age independent analytical profiles; and Part II. The dynamic approach: determining age dependent profiles. *International Journal of Forensic Document Examiners* 1(1): 40–51; and 2(3): 192–208.

Diamond AS (1991) *Handbook of Imaging Materials*, pp. 548–562. New York: Marcel Dekker.

Fishman DH (1997) Ink jet technology. *American Ink Maker* June: 36–39.

Hackleman D (1985) Where the ink hits the paper .... *Hewlett-Packard Journal* May: 32.

Maze C, Loren EJ, Kearl DA and Shields JP (1992) Ink and print cartridge development for the HP DeskJet 500 C/DeskWriter C Printer Family. *Hewlett-Packard Journal* August: 69–76.

Pagano B (1985) *Identification of Photocopies by their Physical and Chemical Properties.* Canadian Society of Forensic Science.

Pagano L (1991) *Colour Photocopy Analysis by Thin Layer Chromatography.* Mid-Atlantic Association of Forensic Science, Bethesda, Maryland, May 1991.

Snyder LR (1968) *Principles of Adsorption Chromatography*, pp. 193–221. New York: Marcel Dekker.

Solodar W (1998) Designing dyes for ink jets. *International Journal of Forensic Document Examiners* 4(1): 22–24.

Tsujita J (1998) Copier/printer market research. J & F Associates, Inc., Great Neck, New York.

# INORGANIC EXTRACTION: MOLECULAR RECOGNITION TECHNOLOGY

*See* **III/MOLECULAR RECOGNITION TECHNOLOGY IN INORGANIC EXTRACTION**

# INSECTICIDES

## Gas Chromatography

**P. Brown,** Central Science Laboratory, York, UK

Developments in insecticide analysis have been closely linked to developments in gas chromatography (GC). Indeed, the invention of the electron-capture detector led to the discovery of residues of the persistent organochlorine insecticides in dead birds and this in turn raised concerns regarding pesticide residues in food and the environment. The range of applications of GC to insecticide analysis is very wide and the choice of technique must be appropriate for the intended purpose. The first requirement is to detect the compounds of interest; most GC detectors have some use in insecticide analysis. Different types of column are needed for different applications and the choice of injection technique is related to the analyte and the type of column. The ability to obtain a chromatogram from a solution of insecticides is just a starting point. The insecticides have to be extracted from a variety of samples and separated from many other co-extracted compounds which can interfere with the separation or degrade the column. There are four major classes of insecticides, each of which has its particular analytical requirements.

## Applications of GC in Insecticide Analysis

Insecticides are biologically active compounds used principally in agriculture but also in public health applications. There is public concern over their production and use, their presence in food and their persistence in the environment. The requirement for insecticide analysis is wide-ranging. GC is better suited to residue analysis than measuring the high concentrations encountered in formulations. For GC analysis, the insecticides must be sufficiently volatile and thermally stable.

A manufacturer may be interested in a particular compound together with its metabolites and degradation products in biological systems and the environment. In cases of animal poisoning by an insecticide, the identification of the poison is the main task. Foods are monitored for many different insecticides and some degradation products; small residues around statutory limits must be identified and measured accurately in a wide variety of foods, each of which poses particular analytical challenges. Residues in water and air are even smaller and must be trapped and concentrated prior to measurement.

With such a variety of applications, it is essential that the analytical method is fit for its purpose.

## Detectors for GC of Insecticides

A list of detectors and their application to insecticide analysis is shown in **Table 1**.

## Columns and Injection Ports for GC of Insecticides

### Packed Columns

GC of insecticides began with packed columns that gave low resolution and were limited by the temperature at which the nonbonded stationary phase began to bleed. To achieve sufficient resolution for identification and to obtain good peak shapes, a wide range of stationary phases was needed. Packed columns are not well suited to temperature programming and the sample is usually introduced by vaporization of a solution in a heated injection port. This is unsuitable for the thermally labile carbamates which have to be converted to more stable derivatives.

### Capillary Columns

The introduction of fused silica capillary columns and bonded stationary phases brought higher resolution GC to routine analysis of insecticides. Higher

**Table 1** Detectors used in the gas chromatography of insecticides

| Detector | Species detected | Insecticide classes | Comments |
|---|---|---|---|
| FID (flame ionization) PID (photoionization) | Carbon compounds | All | With these detectors, difficult to identify the insecticides amongst the peaks from co-extracted compounds in residue analysis |
| ECD (electron capture) | Halogens and other electron-capturing species | Organochlorines, most pyrethroids, some OPs | Sensitive but not highly selective. Responds to some nonhalogenated compounds. Response affected by co-extractives |
| Electrolytic conductivity | Chlorine or sulfur or others | Organochlorines | More selective than ECD but less sensitive. Less robust. Requires more maintenance |
| NPD (nitrogen phosphorus) | Nitrogen and phosphorus | Organophosphates, carbamates | Responds to both elements. Slightly 'tunable'. Response changes slowly as salt bead ages. Response affected by co-extractives |
| Chemiluminescence | Nitrogen | Carbamates | Rarely used to date |
| FPD (flame photometric) | Phosphorus (with 526 nm filter) | Organophosphates some have phosphorus and sulfur | Very selective for phosphorus but large amounts of sulfur compounds may interfere |
| | Sulfur (394 nm) | Others with sulfur | Nonlinear response. Affected by bleed |
| AED (atomic emission) | Carbon, chlorine, nitrogen, sulfur, phosphorus, oxygen (not all at once) | All (but selective for chosen elements) | Versatile selective multielement detector but not as sensitive as some and costs more to buy and run |
| FTIR (Fourier transform infrared) | Functional groups that absorb infrared | Various, but only at high levels | Not sensitive enough for residue analysis |
| MS (mass spectrometry) Total ion chromatogram | All that ionize in MS | All | Same disadvantages as FID (above), but other modes are selective (below) |
| Selected ion(s) | Selected ions | All chosen compounds | Selective for compounds producing the chosen ions |
| Full spectrum | Mass spectrum | Identification of peak | Good identification if sufficiently sensitive |
| Negative-ion CI | Halogens (as ECD) | OCs, pyrethroids | Limited application |
| MS-MS | Fragmentation of ion | Co-eluting compounds or sparse spectrum | Aids identification in these difficult situations |

OPs, organophosphates; OCs, organochlorines; CI, chemical ionization.

resolution, less active sites, temperature programming, cool on-column injection and the development of sensitive mass spectrometric detectors enabled many insecticides, including some that are thermally labile, to be analysed on a low polarity stationary phase (dimethyl polysiloxane, DB-1 or methylphenyl polysiloxane, DB-5, DB-17, etc.)

For fast analyses or for thermally labile compounds, a short wide-bore column (10 m × 0.53 mm i.d. with a 1 μm stationary phase thickness) works well in combination with semi-selective detectors (electron-capture, nitrogen-phosphorus and flame photometric detectors). If the analyte is sufficiently stable, a packed column injector can be converted for direct injection into a 0.53 mm capillary column by the addition of a liner with a volume of about 1 mL and an appropriate reducing adapter. For cool on-column injection, a purpose-built port is necessary. A carrier gas (nitrogen or preferably helium or hydrogen) flow of about 5 mL min$^{-1}$ and a temperature gradient from 50 or 100°C at 5°C min$^{-1}$ to 250°C will be suitable for the analysis of most insecticides.

For higher resolution, a smaller bore (e.g. 0.25 mm internal diameter) and/or longer column is needed. Also, if the detector is a mass spectrometer, a low carrier flow around 1 mL min$^{-1}$ is preferred. Concentrated sample solutions are best analysed with a split injection technique where only a small proportion of the injected extract goes on to the column. For most insecticide analysis, splitless injection is used together with an initial column temperature low enough to cold-trap the analytes and to condense the solvent if the solvent effect is to be employed.

Cool on-column injection is ideal for compounds that start to decompose in a hot injection port, but nonvolatile co-extracted solutes remaining in sample extracts will rapidly contaminate the column. A wide-bore retention gap will trap nonvolatiles and allow injection with a conventional syringe needle (instead of needing a fragile thin needle to inject directly into a small-bore column). Some carbamate insecticides can be analysed without derivatization on a nonpolar column with on-column injection.

Large volume injection is becoming popular for the analysis of low concentrations of insecticides in water samples. Volumes in excess of 100 µL of extract are injected into packing in an injection port liner from which the solvent is evaporated. The solutes are thermally desorbed on to the column where they are cold-trapped prior to the start of temperature-programmed GC analysis.

A thermal desorption and cold-trapping technique is also used for analysis of air samples from which contaminants have been trapped by passage through a tube of adsorbent.

Volatile insecticides used as fumigants in grain and other stored commodities may be analysed by headspace analysis; the volatile compounds are sampled from the air above the samples and passed in a gas stream to be trapped on the top of the GC column.

## Analysis for Insecticides in Biological and Environmental Samples

Before insecticide residues can be analysed by GC, they have to be extracted from the samples containing them. These extracts contain many other co-extractives in addition to the insecticides and require a clean-up.

### Extraction

The simple – and usual – way to check whether an analytical procedure will extract compounds of interest is to add a solution of the compounds to the exposed surfaces of sample (whole, chopped up or finely ground), allow the solvent to evaporate fully, then apply the analytical procedure. Residues that have been acquired by a plant or animal by systemic processes (transport via a plant's vascular system or the processes through gut and internal organs following ingestion by an animal) may be more difficult to extract than surface residues. Studies using radiolabelled compounds may be needed to determine whether such residues are extractable or bound.

Extracts are usually obtained by cutting the sample very finely (blending, homogenizing, milling, grinding, etc.) and extracting with an organic solvent. Extraction may be a one- or two-stage process aided by agitation (shaking, ultrasonicating, homogenizing) followed by filtration or centrifugation (to separate solid and liquid). It may be continuous as in liquid–liquid extraction (for liquid samples) or in Soxhlet extraction, accelerated solvent extraction or supercritical fluid extraction (for solid samples).

### Clean-up

The effects of co-extractives on GC include:

1. visible additional peaks on the chromatogram, perhaps unresolved from the analyte peak;
2. volatile substances that, although not themselves showing a detector response, elute with the analyte and affect the chromatography or the response of the detector to the analyte;
3. nonvolatile solutes deposited on injection, coating the injection liner and top of the column, changing the characteristics of the stationary phase in that region.

A clean-up from which a wide range of insecticides can be recovered will also recover many other co-extracted compounds. Some analysts use extracts with little or no clean-up, and selective detection such as GC-mass spectrometry (GC-MS). They compensate for chromatographic effects from co-extractives by the use of carefully chosen internal standards or prepare matrix-matched external standards in extracts of analyte-free sample material. An instrument used in this way will require frequent maintenance, but the overall result may be cost-effective.

Clean-up is a source of potential loss of analyte and adds extra time to analyses (but in recent years equipment for partial automation of this process has been introduced).

Gel permeation chromatography (GPC) is a technique that delays small molecules such as insecticides as they flow in and out of pores in the gel while larger molecules flow past in the solvent stream. The large molecules mainly form the nonvolatile material deposited at the GC injection port, so GPC is potentially a useful clean-up for insecticides in a wide range of sample matrices. The disadvantage of traditional GPC is that it uses large quantities of solvents (hundreds of millilitres) which have to be removed by evaporation. High performance GPC works on a smaller scale and can be automated, but the columns are expensive compared to analytical high performance liquid chromatography columns.

Column adsorption chromatography using quantities of between 1 g and 50 g of materials such as alumina, silica gel and Florisil was the traditional method for clean-up of extracts containing insecticides. These traditional columns have been largely superseded by small particle size materials in quantities from 100 mg to 1 g in cartridges, usually requiring a small pressure or vacuum to give a suitable flow rate. Cartridges of this type, including those with reversed-phase ($C_{18}$) and ion exchange packings, are now known as solid-phase extraction (SPE) cartridges. Equipment for automated operation of these cartridges is widely available.

Partition between two immiscible solvents, such as water and hexane, may be used to separate polar co-extractives from insecticides that are generally of low polarity. Originally separating funnels were used, but there are now cartridges of inert material that will absorb an aqueous solution but still allow efficient partition of insecticides into a water-immiscible solvent poured through the cartridge. A partition mechanism also operates with SPE cartridges containing bonded $C_{18}$ packing material. Insecticides in aqueous solution are extracted by a partition process into the $C_{18}$ and then eluted selectively from the cartridge by organic solvent mixtures of appropriate polarity.

There are other clean-up techniques developed for specific purposes. Sweep co-distillation is a process analogous to steam distillation, in which insecticides and other volatile materials are distilled out of heated liquid fat samples. Some of the organochlorine insecticides are resistant to chemical attack, so reaction of interfering compounds with strong acids or oxidizing agents has been used as a clean-up for these compounds.

The extraction and clean-up methods to be employed depend on the insecticide(s) to be analysed and the nature of the sample material. The analyst should consult the literature for a particular application, but if no suitable technique is described, some experimentation will be necessary.

## Determination of Insecticides by GC

The requirements for GC determination depend on the purpose of the analysis. Generally, GC determination with a semi-selective detector or a single ion on a mass spectrometer requires additional confirmation, by an alternative technique or by measurement of other ions.

### Organochlorines

Most of the insecticides of this type have been withdrawn from use in many countries, but residues persist in the environment. Organochlorine insecticides are of several different chemical types but all contain chlorine atoms and are fat-soluble (nonpolar). Lindane (hexachlorocyclohexane (gamma isomer), $\gamma$-HCH), dichlorodiphenyltrichloroethane (DDT) and dieldrin are examples (**Figure 1**). They are very slightly soluble in water, but are readily extractable into hexane from water or homogenized water-containing samples. Diethyl ether or ethyl acetate will also extract them from dry materials on which they are adsorbed. An alternative is to use a water-miscible solvent such as acetone to extract from water-containing samples; this is followed by partition between water and hexane. Supercritical fluid extraction with carbon dioxide as solvent and a $C_{18}$ trap has also been used to extract organochlorine insecticides from relatively dry commodities. For clean-up of hexane extracts of fatty samples, chromatography on deactivated alumina or Florisil with further hexane elution recovers most of the organochlorines while leaving most co-extractives behind. GPC and sweep co-distillation are sometimes used to clean up extracts containing organochlorines. Although these insecticides are solids, care must be taken when evaporating extracts as, for example, $\gamma$-HCH is easily lost if a solution is evaporated to dryness.

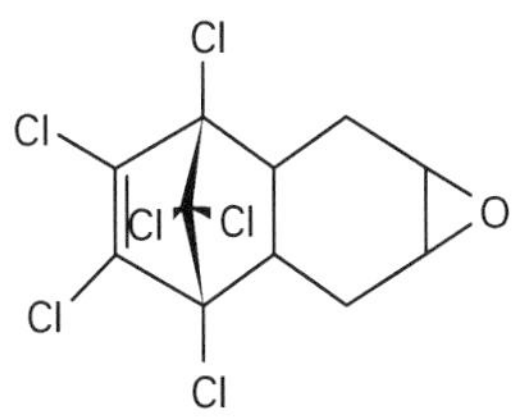

All the organochlorine insecticides chromatograph well on dimethyl silicone stationary phases (DB-1, OV-1, etc.). Endrin and 1,1,1-trichloro-2,2-bis(4-chlorophenyl)ethane ($p,p'$-DDT) may break down or

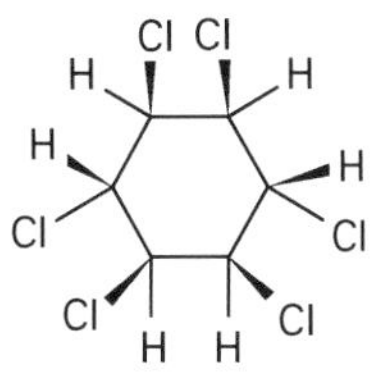

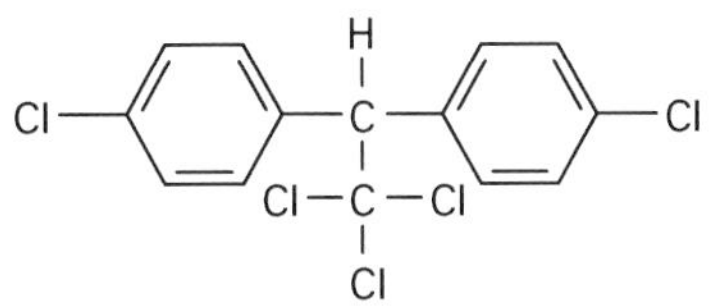

**Figure 1** Organochlorine insecticides.

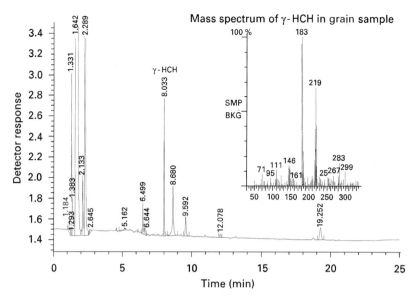

**Figure 2** Chromatogram of extract of grain sample containing $\gamma$-HCH. Chromatogram of a cleaned-up extract of grain showing a residue of $\gamma$-HCH (0.7 mg kg$^{-1}$). 1 µL direct injection at 150°C on to 30 m × 0.53 mm i.d., DB-1 column (1.5 µm film thickness). Gradient from 100°C (1.0 min) at 25°C min$^{-1}$ to 225°C then at 2.5°C min$^{-1}$ to 265°C (3 min). Helium carrier gas. Hewlett Packard 5890 GC with ECD. GC-MS confirmation on 30 m × 0.25 mm i.d. BPX5 column (0.25 µm film thickness) in Finnigan GCQ instrument.

adsorb strongly if the injection liner and top of column become dirty or active. Splitless injection is usually used for residue analysis. A 20 m long, 0.53 mm i.d. column gives satisfactory resolution with a temperature programme from 100°C to 225°C. A chlorine-selective detector such as the very sensitive electron-capture detector or the less sensitive but more selective atomic emission detector may be used. For certainty of identification, a bench-top mass spectrometer is used to obtain a full mass spectrum; the chromatogram is monitored for selected ions at particular retention times to locate any organochlorine peaks (**Figure 2**). A capillary column has to be used to suit the maximum gas flow requirements of the mass spectrometer. For some of the organochlorines, negative-ion chemical ionization mass spectrometry is a successful detection technique.

## Synthetic Pyrethroids

Synthetic pyrethroid insecticides have certain structural features in common. Permethrin, bifenthrin and fenvalerate are typical examples (**Figure 3**). They are only slightly more polar than the organochlorines and may be extracted in a similar way. Mixtures of, for example, hexane and ether will elute pyrethroids from Florisil clean-up columns or cartridges. Those synthetic pyrethroids that do not contain chlorine atoms contain other halogens or chemical groups that are electron-capturing. Retention times are longer than for the organochlorine compounds, so the temperature programme has to rise to 275°C (**Figure 4**). However, resolution of the isomers of some pyreth-

roids (e.g. cypermethrin) requires a higher resolution column (e.g. 30 m × 0.25 mm i.d.).

## Carbamates

The carbamate insecticides are of two types, esters of N-methyl (or N,N-dimethyl) carbamic acid with

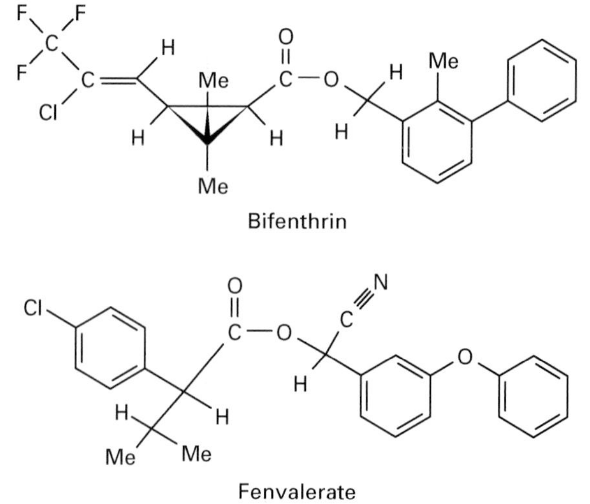

Permethrin

Bifenthrin

Fenvalerate

**Figure 3** Synthetic pyrethroid insecticides.

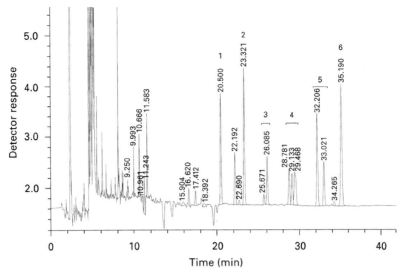

**Figure 4** Chromatogram of six synthetic pyrethroids in honey bee extract. Chromatogram of cleaned-up supercritical fluid extract of honey bees spiked before extraction with six pyrethroids, each at 0.1 mg kg$^{-1}$. 1, Bifenthrin; 2, $\lambda$-cyhalothrin; 3, permethrin; 4, cypermethrin; 5, fenvalerate; 6, deltamethrin. 1 μL direct injection at 175°C on to 30 m × 0.53 mm i.d., DB-1 column (1.5 μm film thickness). Gradient from 50°C (1.0 min) at 25°C min$^{-1}$ to 225°C then at 2°C min$^{-1}$ to 275°C (9 min). Helium carrier gas. Hewlett Packard 5890 GC with ECD.

either a phenol or an oxime. Carbaryl and pirimicarb are examples of the phenolic type. Aldicarb and methomyl are examples of the oxime type (**Figure 5**). They are extractable with diethyl ether, dichloro-

Carbamate insecticide: general structure

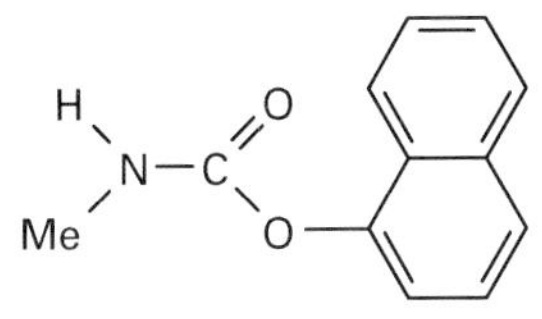

R is PhR′ or N=CR′

Carbaryl

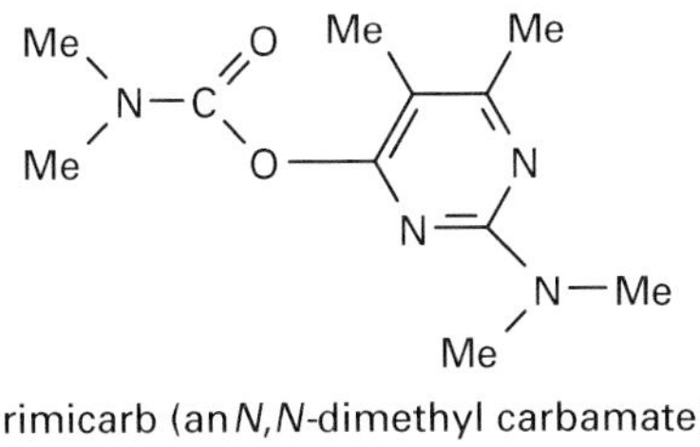

Pirimicarb (an *N,N*-dimethyl carbamate)

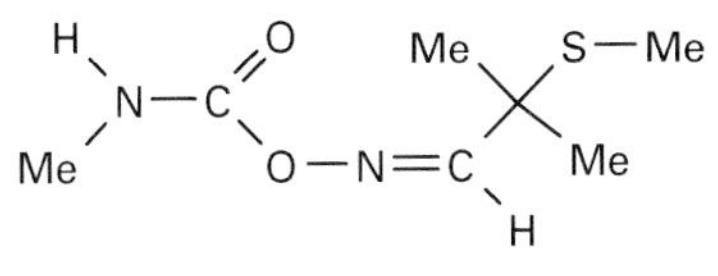

Aldicarb (an *N*-methyl oxime carbamate)

**Figure 5** Carbamate insecticides.

methane, ethyl acetate or acetone. Supercritical fluid extraction with carbon dioxide as extractant and a $C_{18}$ trap is effective for dry or easily dried samples. Extracts containing carbamates may be cleaned up on columns or cartridges of silica gel, Florisil or very deactivated alumina eluted with hexane–diethyl ether mixtures or hexane with small additions of acetone. Gel permeation is an alternative clean-up technique that is useful for multiclass analyses.

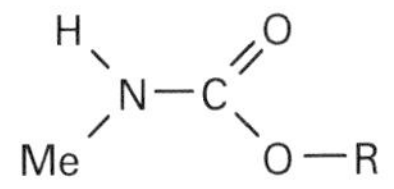

The carbamate insecticides in general, and the oxime carbamates in particular, break down when heated. Before fused silica capillary GC columns and cool on-column injection became available, the carbamates were usually chromatographed as derivatives. A widely used derivatization procedure involved hydrolysis with hot aqueous sodium hydroxide solution, reaction of the phenol with 1-fluro-2, 4-dinitrobenzene and analysis on a dimethyl silicone packed column with electron-capture detection. Another procedure formed derivatives from intact carbamates and trifluroacetic anhydride prior to GC analysis. Aldicarb, an oxime carbamate, and its toxic breakdown products may all be oxidized to aldicarb sulfone using peracetic acid or potassium permanganate. Aldicarb sulfone is chromatographed on a nonpolar column with a nitrogen-phosphorus detector (NPD), but with an injection port packed with glass wool at a temperature of 300°C where the carbamate breaks down and forms a nitrile from the oxime part of the molecule.

Carbamates of the phenolic type may be analysed successfully by using a nonpolar column of moderate length (10–15 m × 0.53 mm i.d.) together with cool

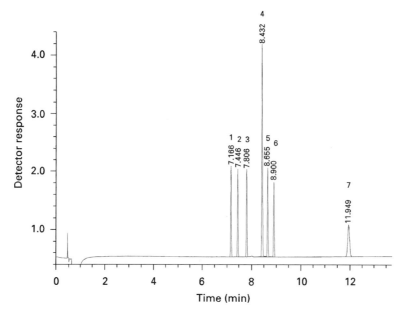

**Figure 6** Chromatogram of seven *N*-methyl carbamates. 1, Propoxur; 2, bendiocarb; 3, carbofuran; 4, pirimicarb; 5, carbaryl; 6, methiocarb; 7, carbosulfan, 2 μg mL$^{-1}$ solution of each, 1 μL cool on-column injection at 50°C on to 15 m × 0.53 mm i.d., DB-1 column (1.5 μm film thickness). Gradient from 50°C (0.5 min) at 20°C min$^{-1}$ to 235°C (4 min). Helium carrier gas. Hewlett Packard 5890 GC with NPD. Note: Pirimicarb has four nitrogen atoms, so gives a larger response than the others.

on-column injection, a gentle temperature gradient and NPD (**Figure 6**). For analysis by GC-MS, small bore capillaries are used, coupled with a 0.53 mm i.d. retention gap. GC of most oxime carbamates is unsuccessful unless they are converted to derivatives. For oxime carbamates, and perhaps for many phenol carbamates, the preferred method of analysis is by liquid chromatography-MS (LC-MS) or by liquid chromatography with post-column conversion to a fluorescent derivative.

**Organophosphates**

The organophosphorus insecticides are esters of phosphoric acid (or its sulfur analogues) and are therefore often referred to as organophosphates to distinguish them from compounds with a C–P bond. There are several groups of chemically related organophosphates. Dichlorvos, malathion and parathion are examples (**Figure 7**). Organophosphates cover a wide polarity range from persistent and fat-soluble compounds such as carbophenothion to water-soluble and easily hydrolysed compounds such as mevinphos. Most are extractable from dried samples with dichloromethane, diethyl ether or ethyl acetate. As with carbamates, supercritical fluid extraction with carbon dioxide and a C$_{18}$ trap will extract most of them from dry samples (or from samples that can be dried with sufficient drying agent). Clean-up depends on the compounds being analysed and the nature of the sample material. Silica

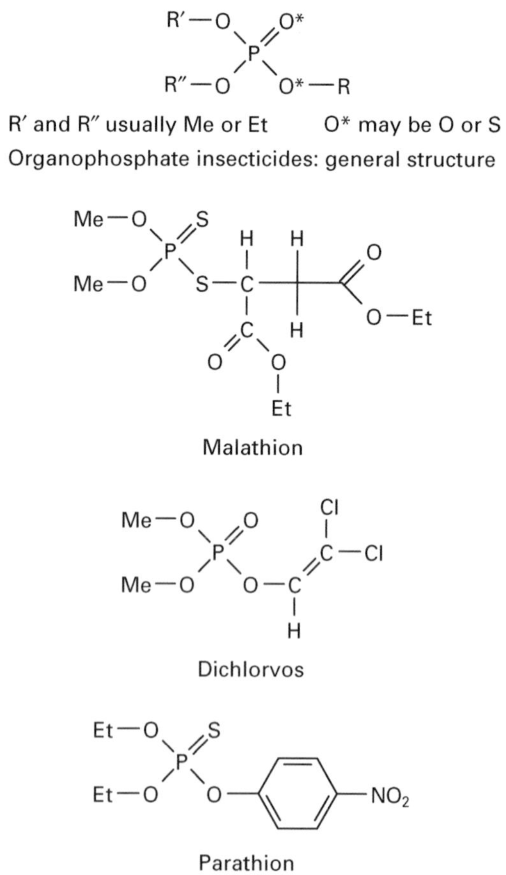

R′—O, O*
      P
R″—O, O*—R

R′ and R″ usually Me or Et     O* may be O or S
Organophosphate insecticides: general structure

Malathion

Dichlorvos

Parathion

**Figure 7** Organphosphate insecticides.

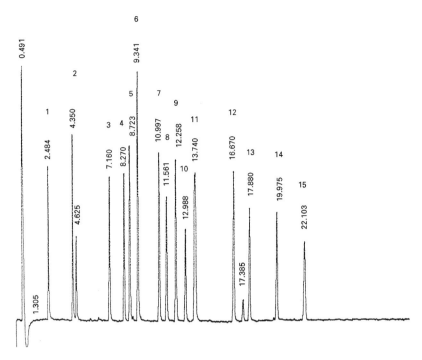

**Figure 8** Chromatogram of 15 organophosphates in pheasant gizzard. Chromatogram of diethyl ether Soxhlet extract of pheasant gizzard contents, spiked before extraction with a mixture of 15 organophosphates, each at approximately 4 mg kg$^{-1}$. 2 μL direct injection at 225°C on to 15 m × 0.53 mm i.d., DB-17 column (1.0 μm film thickness). 1, Dichlorvos; 2, mevinphos (E + Z); 3, phorate; 4, diazinon; 5, fonofos; 6, dimethoate; 7, pirimiphos-methyl; 8, malathion; 9, fenthion; 10, chlorfenvinphos; 11, fosthiazate; 12, carbophenothion; 13, triazophos, 14, phosalone; 15, azinphos-methyl. Gradient from 140°C (0.5 min) at 10°C min$^{-1}$ to 180°C then at 6°C min$^{-1}$ to 270°C (10.5 min). Helium carrier gas. AI Cambridge Ai93 GC with Tracor FPD.

gel does not give a very effective multiresidue clean-up from fatty materials as many organophosphates are eluted with co-extractives. Active alumina will break down some organophosphates. Gel permeation is probably the best choice when analysing organophosphates covering a wide range of polarities.

Residue-level GC may be carried out using splitless or cool on-column injection with detection by mass spectrometry, flame photometric or atomic emission detector or NPD (of these, NPD is the least selective). Medium polarity columns (such as 50% phenyl) give good separation and peak shapes, but a more polar phase such as cyanopropyl is sometimes useful (avoid using it with an NPD). As for column length and diameter, although a 15 m × 0.53 mm i.d. column will separate many organophosphates over a long slow temperature gradient, there are so many compounds of this type that some are bound to co-elute whatever the resolving power of the column (**Figure 8**).

## Future Trends in the Analysis of Insecticides by GC

Capillary columns have largely replaced packed columns and precise electronic control of temperatures and gas pressures are standard features of modern gas chromatographs. The bench-top mass spectrometer is now firmly established as the detector of choice for most applications. As computers increase in power, software for interpreting chromatograms and spectra and selecting results of interest will continue to develop and will become integrated with laboratory information management systems. The atomic emission detector may gradually displace the well-established element-selective detectors in a role complementary to GC-MS-MS in the larger analytical laboratories. The move away from GC for the thermally less stable compounds will continue as LC-MS is now sufficiently sensitive and robust for routine residue analysis. Large volume injection is becoming established as a standard technique in water analysis. The slow trend towards the analysis of smaller and smaller residues in food and environmental samples may take a step forward when large volume injection is used together with improved cartridge clean-up methods. For some sample types, solid-phase microextraction will routinely provide rapid transfer of analyte to the GC.

Automation of extraction and clean-up operations has been predicted as a major area of advance for the past 15 years. The laboratory robot has not taken over all laboratory work and the trend is towards laboratory instruments with added versatility and capacity for automation.

Supercritical fluid extraction is making slow progress, probably because of its high cost compared to traditional extraction methods, Perhaps eventually the gas chromatograph will control the extraction and clean-up of its own samples, recording all steps with a thorough audit trail. Trends in insecticide analysis will also depend on what new classes of compounds are approaching commercial use. Element-selective detectors may not be suitable, but GC-MS-MS and LC-MS-MS should be able to cope with just about anything.

*See also:* **III/Herbicides:** Gas Chromatography; Solid-Phase Extraction; Thin-Layer (Planar) Chromatography. **Insecticides:** Solid-Phase Extraction.

## Further Reading

Bottomley P and Baker PG (1984) Multi-residue determination of organochlorine, organophosphorus and synthetic pyrethroid pesticides in grain by gas-liquid and high-performance liquid chromatography. *Analyst* 109: 85–90.

Brown P, Charlton A, Cuthbert M, Barnett L, Green M, Gillies L, Shaw K and Fletcher M (1996) Identification of pesticide poisoning in wildlife. *Journal of Chromatography A* 754: 463–478.

Chamberlain SJ (1990) Determination of multi-pesticide residues in cereals, cereal products and animal feed using gel-permeation chromatography. *Analyst* 115: 1161–1165.

Cunniff P (ed.) (1995) *Official Methods of Analysis of AOAC International* (16th edn). Arlington: AOAC International.

McMahon BM and Hardin NF (eds) (1994) *Pesticide Analytical Manual* (3rd edn). USA: US Food and Drug Administration.

Meloan CE (ed.) (1996) *Pesticides Laboratory Training Manual*. Gaithersburg: AOAC International.

Tomlin CDS (ed.) (1999) *The Pesticide Manual* (11th edn). Farnham: British Crop Protection Council.

# Solid-Phase Extraction

**A. Przyjazny**, Kettering University, Flint, MI, USA

## Introduction

Synthetic insecticides represent an almost universal environmental pollutant. Chemical structures of insecticides are very diverse, but major groups include organochlorine and organophosphorus compounds, methylcarbamates, and synthetic pyrethroids.

Organochlorine insecticides have been largely phased out of general use because of their toxicities and especially their persistence and accumulation in food chains. The structures of several common organochlorine insecticides are shown in **Figure 1**.

Organophosphorus insecticides have largely replaced organochlorine insecticides, because organophosphorus compounds readily undergo biodegradation and do not bioaccumulate. The structural formulas of some common organophosphorus insecticides are shown in **Figure 2**.

Carbamate insecticides are widely used for crop protection. Methylcarbamates are of environmental concern because of their high acute toxicity. **Figure 3** depicts the structures of several common carbamate insecticides.

Synthetic pyrethroids have been widely produced as insecticides during recent years. They have several advantages over organochlorine and organophosphorus insecticides, including greater photostability, enhanced insecticidal activity, and relatively low toxicity. The structures of common pyrethroids are shown in **Figure 4**.

As a result of their toxicity and carcinogenicity, insecticides are hazardous to human health and life. The major route of human exposure to insecticides is the gastrointestinal system. Besides regular use of polluted water, humans eat large amounts of food in which these pollutants have been accumulated, e.g., milk and dairy products, fish, poultry and meat, and fruits and vegetables.

Prior to 1960, an individual analytical procedure was used for almost every insecticide. As the number of insecticides in use increased, it became impractical to apply a large number of individual methods for all the insecticides that may be present. This has led to the development of multiresidue methods for the analysis of environmental samples. Ideally, multiresidue methods should provide rapid identification and quantification of as many different insecticides as possible at the required detection level.

The concentrations of insecticides in the environment are very low, typically in the parts-per-trillion to the parts-per-billion range. Furthermore, the sample matrices in which insecticides are usually determined are very complex in most cases. Consequently, extensive sample extraction, clean-up, and preconcentration are often required prior to the analysis.

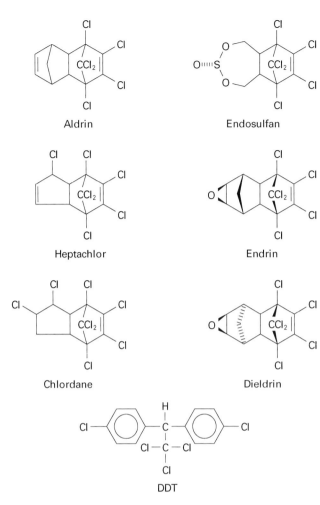

**Figure 1**  Structures of common organochlorine insecticides.

# Comparison of Methods of Extraction for Insecticides

The most difficult and time-consuming step in the determination of insecticides in environmental samples is the extraction of the analytes from the matrix. Several methods are used to accomplish this task, including liquid–liquid extraction (LLE), solid-phase extraction (SPE), solid-phase microextraction (SPME), and supercritical fluid extraction (SFE).

Semivolatile compounds, such as insecticides, have been traditionally extracted by LLE using an organic solvent, such as methylene chloride. The main advantages of LLE are its simplicity and inexpensive equipment used. The procedure suffers from a number of disadvantages, including the use of large volumes of organic solvent which must be very pure (pesticide-grade), tediousness, difficulty with automation, and the formation of emulsions which are difficult to break. Also, the LLE is a multistep procedure and is therefore prone to loss of analytes and/or contamination.

There is no doubt that solid-phase extraction has now become the method of choice for the extraction, clean-up, and preconcentration of insecticides from environmental, food, or biological samples. SPE drastically reduces amounts of organic solvents used and is not as time-consuming. It enables field sampling and does not suffer from emulsion problems. Other reasons for the growing number of procedures using SPE are the large choice of sorbents, including new polar sorbents capable of retaining more polar insecticides, the introduction of apparatus for automated SPE, which reduces time of analysis and increases sample throughput, and the possibility of automation of analytical procedures. However, SPE is not completely free from problems such as column overloading by passing samples with high content of contaminants, or early breakthrough caused by clogging of the pores by solids present in a sample. The cost of disposable SPE cartridges or discs may be significant in the case of a large number of analyses. Also, the interaction between sample matrix and analytes may result in low recoveries. SPE can carry contaminants into the final sample producing a high background.

Solid-phase microextraction is a new and simple extraction method which uses fused silica fibres coated with a polymeric liquid phase, such as polydimethylsiloxane, or a solid adsorbent, to extract analytes from gaseous, liquid or solid samples. The SPME process has only two steps: (1) partitioning of analytes between the fibre coating and the sample matrix, followed by (2) desorption of extracts into an analytical instrument: gas chromatograph or high-performance liquid chromatograph. SPME is different from SPE in that SPE isolates the majority of the analyte ( > 90%) from a sample but only a small fraction of the sample (1–2%) is analysed, while SPME isolates only 2–20% of the analyte, but all of that sample is used in the analysis. SPME is solvent-free and it requires small amounts of samples. The SPME devices enable simple and direct introduction of concentrated samples into the analytical instruments. These devices are commercially available and the technique has been automated. The SPME technique has a wide linear dynamic range and low detection limits. Since SPME is usually used in the equilibrium rather than exhaustive extraction mode, recoveries are not 100%, but the precision and accuracy of the analytical methods using SPME are similar to those using other extraction procedures. The current limitation of SPME is a limited choice of available fibres, especially for the extraction of the more polar insecticides.

In the last six years, supercritical fluid extraction has proved to be an appropriate replacement for the

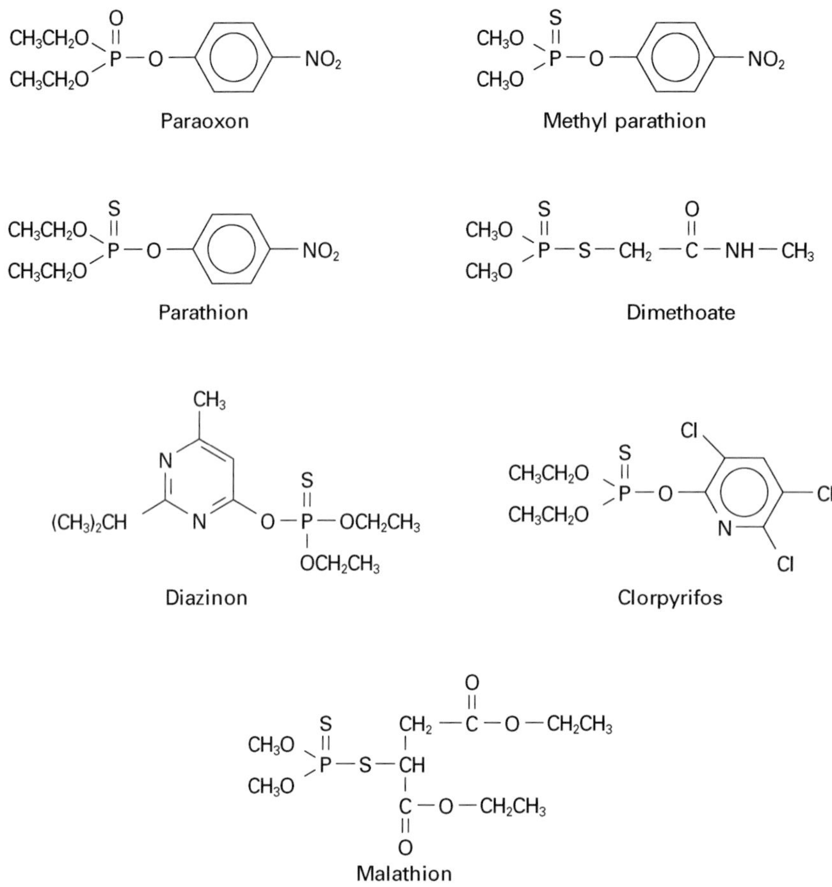

**Figure 2**  Structures of typical organophosphorus insecticides.

solvent extraction of insecticides from solid samples. SFE has gained acceptance in environmental analyses because it is a rapid, nonorganic solvent extraction technique that gives recoveries equal to, or even better than, traditional extraction procedures. SFE can be used directly or it can be coupled with solid-phase extraction (SPE-SFE). Direct SFE is mostly used for solid matrices (soil, sediment, food), while the tandem SPE-SFE is usually used for the extraction of insecticides from aqueous samples. The disadvantages of SFE include high costs associated with high-pressure fluid delivery system and high-purity gas source, both of which are heavy equipment that make field analysis difficult. However, the combination of solid-phase extraction and supercritical fluid extraction enables the performance of SPE directly in the field, using SPE sorbents, and carrying out the SFE step in the laboratory, thus eliminating the need for transporting the SFE equipment to the field.

In the near future, further improvements are to be expected with SPME, SPE, SFE and their combination (SPE-SFE). This should lead to the development of more standard methods for the determination of insecticides using these extraction techniques.

# Application of SPE and SPME to Extraction and Preconcentration of Insecticides

## Air Analysis

Increasing concern over the presence of insecticides in ambient and indoor air has led to the development of specific extraction methods for these pollutants. As a result of very low concentrations of insecticides in air ($ng\,m^{-3}$ to $\mu g\,m^{-3}$), air sampling methods call for large sample volumes, ranging from about $1\,m^3$ to close to $1000\,m^3$. In order to achieve those sample volumes in reasonable time, high sampling rates are required. Consequently, the sorbents used for air sampling should have low pneumatic resistance. The most typical extraction procedure makes use of polyurethane foam (PUF) as a lightweight, easy-to-use sampling material for these semivolatile compounds. Cylindrical PUF plugs effectively trap insecticides without creating excessive back pressure. Thus, high sampling rates are possible, which allows shorter sampling times and ensures more representative samples. These high sampling rates may also be required

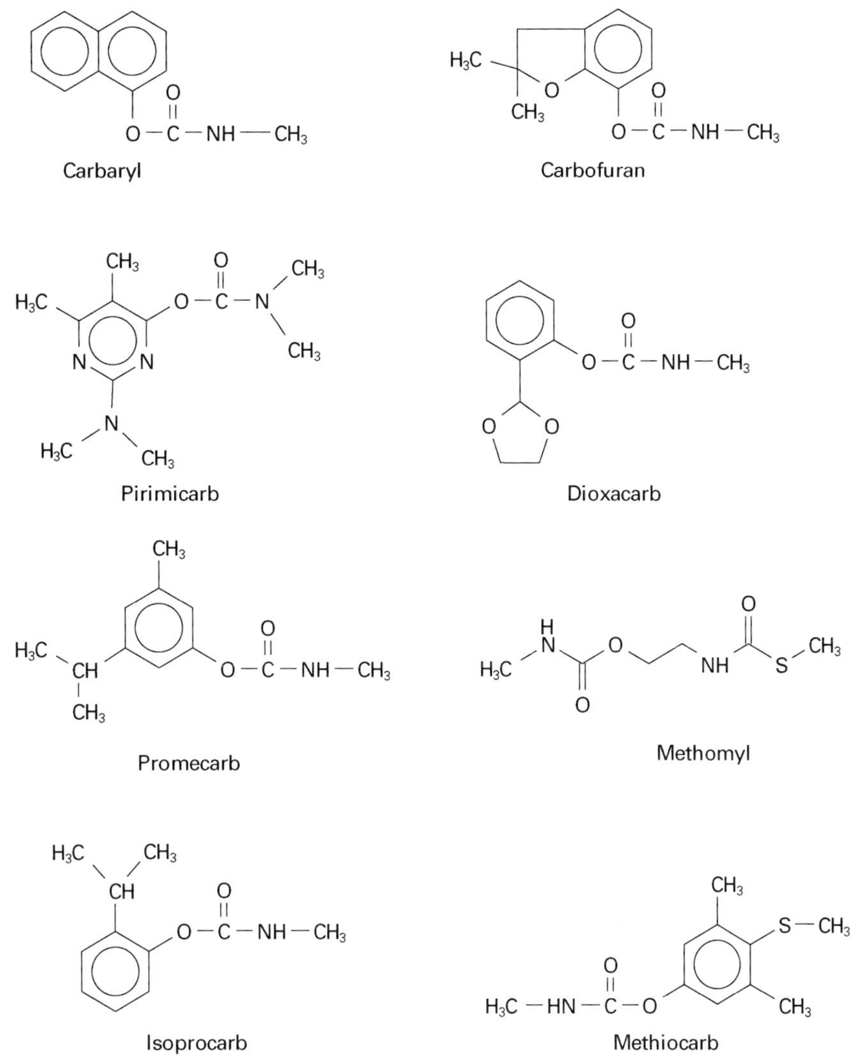

**Figure 3** Structures of commonly used carbamate insecticides.

to attain needed detection limits. PUF plugs are sometimes used with an additional sorbent, such as Tenax TA or XAD-2 resin, to collect more volatile analytes. The air sample is usually first filtered through a microfibre filter, which traps aerosols and particulates. Then the air is drawn through the PUF cartridge. Following sample collection, the PUF sorbent is extracted by Soxhlet extraction with 5% diethyl ether in hexane and the insecticides are determined by gas chromatography with selective detection. For some insecticides, high performance liquid chromatography (HPLC) with an ultraviolet (UV) detector or electrochemical detector may also be the method of choice. If necessary, the extract may be cleaned up by SPE using Florisil or alumina and concentrated using a Kuderna-Danish apparatus. The above procedure can be applied to the multiresidue analysis of insecticides.

Applications using PUF to extract insecticides from air samples include a number of standard methods, including EPA Method IP-8, Method TO-4A and TO-10A. **Table 1** summarizes SPE conditions used for the extraction of insecticides from air samples.

XAD and Tenax resins have also been used for the isolation and analysis of insecticides from air samples. A method based on a high volume sampler with an XAD-2 resin trap was used to monitor 39 pesticides in ambient air. The volume of air sampled was $700 \, m^3$. After the extraction, pesticides were separately extracted with methylene chloride from the filter and XAD-2 resin. Extracts were concentrated and cleaned up by silica gel column chromatography and analysed by capillary gas chromatography/mass spectrometry with selected ion monitoring (cGC-MS-SIM). Organophosphorus and carbamate insecticides were detected at the $10 \, ng \, m^{-3}$ level.

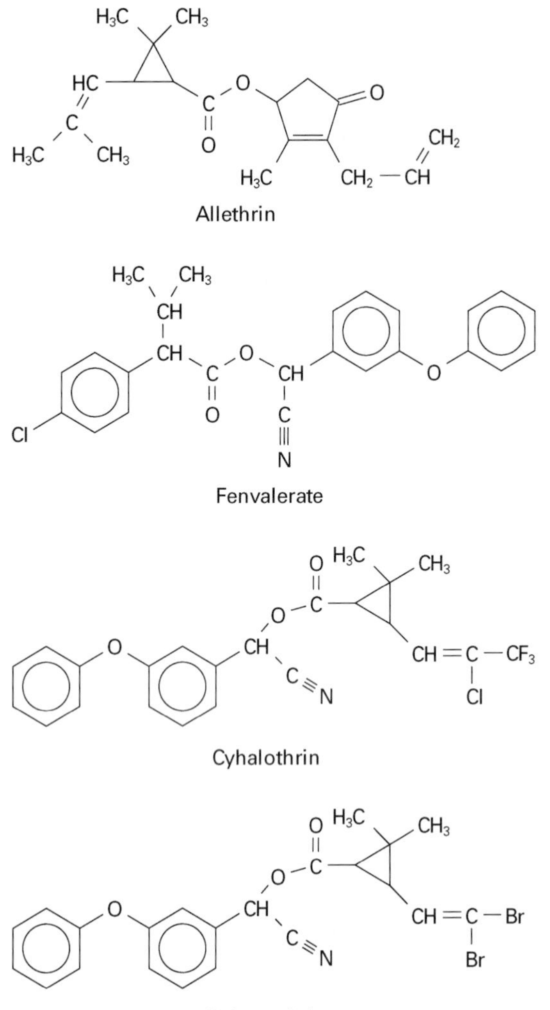

**Figure 4**  Structures of common pyrethroid insecticides.

An improved sampling method using SPE was developed for the multiresidue determination of pesticides in indoor air. The method involves adsorption of the pesticides in $1\ m^3$ of air onto Tenax TA via an air-sampling pump, desorption with acetone, and determination by GC-MS. Limits of detection for the 23 pesticides studied (including organophosphorus insecticides) were on the order of $ng\,m^{-3}$.

Recoveries of insecticides from XAD and Tenax resins can be improved by using supercritical fluid extraction to elute the analytes.

## Water Analysis

The determination of insecticides in water samples is carried out by gas chromatography, liquid chromatography, or thin-layer chromatography. These chromatographic techniques require efficient isolation and concentration procedures, such as solid-phase extraction and solid-phase microextraction. Generally, the major role of SPE in water analysis is for trace enrichment of insecticides. The usual sample volume for trace enrichment varies from 100 mL to 1 L, although large-volume analysis by SPE is also possible. In the latter case, 10 to 100 L of water is passed through a column containing 1 to 10 L of XAD resin. This approach has been used to isolate and determine DDT from natural waters and the detection limit was in the order of $pg\,L^{-1}$.

The mechanism of sorption in trace enrichment is typically reversed phase, and the commonly used sorbents are C-18, C-8 or styrene-divinylbenzene (SDB) porous polymer. Trace enrichment using SPE can be accomplished in two modes: cartridge or disc. Both modes can be automated and be performed

**Table 1**  SPE conditions used in extraction of insecticides from air

| | |
|---|---|
| Sample | Air ($0.9\ m^3$) at $1-5\ L\,min^{-1}$ for 4-24 h (low volume sampling (LVS)). Air ($>300\ m^3$) at $0.225\ m^3\,min^{-1}$ for 24 h (high volume sampling (HVS)). |
| Analytes | Organochlorine (OC), organophosphorus (OP), methylcarbamates (MC), pyrethroids (PY). Concentration: $0.001-50\ \mu g\,m^{-3}$. |
| Sorbent | 7.6 cm × 22 mm ID. PUF plug (density $0.0225\ g\,m\,L^{-1}$) (LVS). 2.5 cm × 65 mm ID. PUF plug (HVS). |
| Sorbent preparation | Using Soxhlet extractor, wash plug with acetone for 16 h, followed by ether-hexane (5 : 95) for 16 h. Vacuum dry 2-4 h at room temperature. Place in glass sampling cartridge and seal until sample collection. |
| Elution of analytes | Using Soxhlet apparatus, extract plug with 300 mL of ether-hexane (5 : 95) for 16 h. Concentrate the extract to 5.0 mL using a Kuderna-Danish apparatus. |
| Extract clean-up | For OC analysis, remove OP and MC with alumina. Use Florisil to achieve class separation. |
| Determination | OC       cGC/ECD<br>OP       cGC/FPD or NPD<br>MC       cGC/NPD or reversed-phase HPLC<br>Multiresidue       GC-MS |

**Table 2** SPE conditions used in extraction of organochlorine insecticides from water

| | |
|---|---|
| Sample | River water (100 mL) with 1% methanol added |
| Analytes | 28 organochlorine insecticides and their metabolites |
| Sorbent | 500 mg C-18 cartridges |
| Sorbent preparation | Precondition with 5 mL each of acetone, methanol, and distilled water |
| Elution of analytes | Rinse the cartridge with 3 mL water, apply vacuum to remove water, and elute the analytes with 3 mL acetone |
| Determination | GC/ECD |

either online or offline. The major advantage of discs over cartridges is that the former can use higher sample flow rates, which reduces the time necessary for extraction. On the other hand, the advantage of cartridges over discs is that the cartridges require a smaller volume of solvent to elute the analytes, which simplifies subsequent steps in the analytical procedure.

A typical SPE sequence using a C-18 cartridge consists of four steps. First, the SPE column is prepared to receive a water sample, by wetting with an organic solvent and by conditioning with water. Then, the aqueous sample is applied, and often the insecticides of interest are retained together with interferences from the sample matrix. Some of these interferences can then be removed by application of a washing solution. In the last step, the concentrated insecticides are desorbed with a small volume of organic solvent, which can then be partially evaporated to increase the enrichment factor. An example of such a procedure used for the extraction of organochlorine insecticides from water is shown in **Table 2**.

The US EPA has a number of standard methods for the analysis of organic pollutants in water, including insecticides, which make use of SPE. In Method 525.1, for example, analytes are extracted from a 1 L water sample using a C-18 SPE cartridge or disc. Next, the analytes are eluted from the cartridge or disc with a small quantity of methylene chloride, and concentrated further by evaporation of some of the solvent. The analytes are then quantified using GC-MS.

The US EPA has also approved various methods based on SPE discs containing either C-18 silica or styrene-divinylbenzene sorbent for the determination of organonitrogen and organophosphorus pesticides (Method 507) and of organochlorine insecticides (Method 508.1) in drinking and source waters. The use of SPE discs is particularly easy. The disc is placed in a filtration apparatus attached to a water-aspirator

vacuum source. Next, it is conditioned with 10 mL of methanol and 10 mL of distilled water, and the water sample is filtered through it. Then the extraction funnel and frit assembly is transferred to a second vacuum filtration flask containing a test tube. Three 5 mL aliquots of the eluting organic solvent are then drawn through the disc. The combined eluates are concentrated by evaporation and analysed by GC or HPLC.

There are many choices for eluting solvents in trace enrichment by C-18 solid-phase extraction. One of the most common solvents, removing the majority of hydrophobic insecticides sorbed on the resin, is ethyl acetate. If this eluent is followed by methanol, then this combination is compatible with both GC-MS and HPLC analysis. For very hydrophobic analytes, such as dichlorodiphenyltrichloroethane (DDT), a mixture of ethyl acetate–methylene chloride (1 : 1) is more effective than ethyl acetate alone.

Reversed-phase SPE is best suited for the extraction of nonpolar insecticides, such as organochlorine compounds. For polar pesticides, such as methylcarbamates or some organophosphorus compounds, different sorbents are preferred. Of those, two types are most common: graphitized carbon blacks (GCB) and styrene-divinylbenzene resins. Graphitized carbon black was successfully used to isolate carbamate and organophosphorus insecticides from water. Multi-residue methods have been developed for both polar and nonpolar insecticides in water for subsequent analysis by HPLC. Detection limits of 0.003 to 0.007 $\mu g\,L^{-1}$ were reported for a number of the analytes. A typical procedure for insecticide extraction from drinking water using a GCB sorbent is shown in **Table 3**. Usually, the preferred eluting solvent is a mixture of methylene chloride–methanol (80 : 20).

**Table 3** SPE conditions used in extraction of carbamates and organophosphorus insecticides from water

| | |
|---|---|
| Sample | Drinking water (100 mL to 1 L) |
| Analytes | Carbamates and organophosphorus insecticides |
| Sorbent | 0.25 g ENVI-Carb |
| Sorbent preparation | Precondition with 5 mL methylene chloride–methanol (80 : 20), 1 mL methanol, and 10 mL 2% acetic acid in water |
| Elution of analytes | Elute the analytes with 0.8–1 mL methanol, followed by 2 × 3.5 mL methylene chloride–methanol (80 : 20). Dry eluate to 400–500 μL. Reconstitute samples to 1 mL with methanol |
| Determination | HPLC with UV, DAD, or MS detection |

DAD, diode array detector.

New polymeric sorbents, based on styrene-divinyl-benzene copolymer, have proved their usefulness in the isolation of polar insecticides from water. These polymers (SDB from International Sorbent Technology or the Oasis HLB from Waters) have some hydrophilic nature to improve their wetting characteristics for good mass transfer, but they still have high capacities for polar analytes. The Oasis sorbent was used to isolate organophosphorus insecticides from tap water. Recoveries of the analytes ranged from 87 to 112%.

SPE clean-up and trace enrichment methods can be automated. In semiautomated methods, some operator intervention is required. In fully automated procedures, the entire SPE operation is carried out without user intervention, including online analysis by gas chromatography (GC) and HPLC. Automated online methods using SPE for the determination of carbamates and organophosphorus insecticides in water coupled with HPLC or GC have been developed. The methods offer all the inherent advantages of automatic methods, i.e. low sample and reagent consumption, minimal manipulation and contact with the reagents, accurate and reproducible results, and high throughput. The detection limits for the automated procedures were in the ng L$^{-1}$ range for organophosphorus insecticides and between 0.01 and 1 μg L$^{-1}$ for carbamates.

The majority of environmental applications of solid-phase microextraction have dealt with the isolation and determination of organic analytes in water samples. SPME has been successfully used for the determination of carbamates, as well as organochlorine and organophosphorus insecticides, in aqueous matrices. Usually, poly(dimethylsiloxane) (PDMS) fibres are used in the analysis of organochlorine insecticides; for organophosphorus insecticides, polyacrylate (PA) fibres are preferred, whereas for carbamates, either PDMS or PDMS/divinylbenzene (DVB) fibres are applied. A multiresidue procedure for the simultaneous determination of 60 pesticides in water has been developed. Either a PDMS-coated or a PA-coated fibre may be used to achieve detection limits in the low μg L$^{-1}$ range. The method had adequate precision and good linearity over the range 0.1–100 μg L$^{-1}$.

SPME is characterized by its simplicity, low cost, rapidity and sensitivity. An example of SPME procedure for the determination of organochlorine insecticides in water is shown in **Table 4**.

The compact nature of the SPME device and the simplicity of the procedure allow this method to be readily automated. The automated SPME instrument is available from Varian. Two online methods of determination using SPME have been described: one

**Table 4** SPME conditions used in extraction of insecticides from water

| | |
|---|---|
| Sample | 4 mL water |
| Analytes | Organochlorine insecticides |
| Sorbent | 100 μm polydimethylsiloxane fibre |
| Sorbent conditioning | Expose the fibre to the hot GC injection port (250°C) for at least 3 h |
| Extraction | Immerse the fibre into the sample for 15 minutes, stir rapidly |
| Elution of analytes | After extraction, insert the fibre into the GC injector port held at 260°C for five minutes |
| Determination | GC with ECD detection |

for organochlorine insecticides and the other for organophosphorus insecticides. The former method used GC with electron-capture detector (ECD), while the latter method employed GC with nitrogen–phosphorus detector (NPD).

**Soil and Sediment Analysis**

Soils and sediments have much more complex matrices than air or groundwater. Consequently, the analysis of soil making use of SPE must be preceded by the extraction of the insecticides from a soil sample by an organic solvent or a mixture of solvents. A methanol/water mixture is often used for this purpose. Alternatively, accelerated solvent extraction can be used for the extraction of soils and sediments. Recently, however, supercritical fluid extraction has also been tried and found to be superior to liquid extraction in terms of reproducibility. The use of microwave-assisted solvent extraction (MASE) for the extraction of organochlorine insecticides from soil samples has been investigated. Compared to a conventional liquid extraction method, MASE yielded better or equal recoveries and superior repeatability. Extraction of insecticides from soil and sludge samples can also be accomplished by using subcritical water. Subcritical water is an excellent solvent to quantitatively extract both polar and nonpolar analytes from soils. Furthermore, subcritical water extractions can be highly selective and can be coupled with both SPE and SPME.

Following the extraction of insecticides from soils or sediments, SPE is used for both clean-up of the extract and for trace enrichment of the analytes. The procedures employed will be different for clean-up compared to those of trace enrichment. Two strategies are possible. The first is to remove interfering compounds by trapping them on the SPE cartridge and allowing the insecticides to pass through the cartridge for direct analysis or further enrichment (SPE clean-up). The second approach is to isolate the insecticides directly by SPE and elute them free from the interferences or to wash the interferences off the SPE sorbent prior to elution (trace enrichment).

Polar sorbents are usually used to accomplish SPE clean-up of soil extracts. The most common sorbent is Florisil, which has been used widely for clean-up of soil extracts for the determination of various insecticides. Silica gel has also been used for this purpose, mostly in the determination of nonpolar analytes, such as organochlorine insecticides or pyrethroids. Alumina has been used less often, and only for the determination of organochlorine analytes. The clean-up step can also use SPE cartridges packed with silica gel modified with polar aminopropyl groups. This procedure has been applied to the determination of N-methylcarbamates in soils.

Clean-up of soil extracts can also be based on size-exclusion chromatography (SEC), which separates species by size rather than polarity. Clean-up by sorbents such as Florisil does not remove interferences of high molecular mass and polarity similar to that of the insecticides. In contrast, SEC removes materials of high molecular mass, such as humic substances from soil extracts. Currently, polystyrene columns are the most often used SEC sorbents. They are eluted with a number of different solvents, such as cyclohexane or cyclohexane–ethyl acetate.

When nonpolar sorbents, such as C-18 or C-8, are used, the clean-up of soil extracts can be incorporated in the SPE sequence just before the insecticide desorption. In this strategy, the soil extract is passed through the SPE cartridge or disc. Next, the sorbent is rinsed with a small volume of water containing an organic modifier, usually methanol. The interferences are removed, but the insecticides are retained. This step can only remove interferences that are more polar than the analytes, so the method can only be applied for the determination of nonpolar insecticides. Following the clean-up step, the insecticides are eluted from the SPE cartridge or disc using a small volume of an organic solvent, typically ethyl acetate, and this extract is further preconcentrated or directly analysed.

Table 5 summarizes SPE conditions used for the extraction of insecticides from soils and sediments.

When using polar sorbents, such as silica gel, alumina, Florisil, or silica gel with chemically bonded aminopropyl groups, SPE procedures can combine trace enrichment of insecticides and removal of interferences in soil extracts. In this strategy, the soil extract in a nonpolar solvent is passed through a column or cartridge containing a polar sorbent. Next, the analytes are eluted from the sorbent in a series of fractions using solvents of increasing polarity. For example, the effectiveness of Florisil, silica gel and alumina for the clean-up and trace enrichment of soil extracts containing organochlorine and organophosphorus insecticides has been compared. The results showed that silica gel was the best adsorbent. Nonpolar interferences were removed by a rinse with cyclohexane. An elution with ethyl acetate–hexane (5 : 95) followed by ethyl acetate resulted in the separation of the two classes of insecticides. The organochlorine insecticides were present in the first fraction, and the organophosphorus insecticides in the second fraction.

SPME has been used for the extraction of organochlorine insecticides from soil samples and soil solutions. The fibre was coated with polydimethylsiloxane. The method linearity and detection limits were tested in the 0.1–20.0 ng g$^{-1}$ range. SPME was found to be useful for screening of insecticides in contaminated soil samples, offering a simple alternative to established methods of analysis of insecticides in soil.

### Analysis of Biological Materials

Biological matrices consist of body fluids, such as urine or plasma, and tissues. Biological fluids are viscous and may require pre-treatment, e.g. centrifugation, dilution or buffer addition, before SPE can be applied. Prior to extraction, tissue samples generally need to be blended or ground in order to disrupt the general architecture of the sample. Because of the complexity of matrices, SPE may have to

**Table 5** SPE conditions used in extraction of insecticides from soil and sediment

| | |
|---|---|
| Sample | Soil or sediment (20 g) |
| Analytes | Organochlorine (OC), organophosphorus (OP), methylcarbamates (MC), pyrethroids (PY) |
| Extractant | Water–methanol (10 : 90), water–acetone (50 : 50), hexane–acetone (90 : 10). Extract twice using Soxhlet, heated vial or accelerated solvent extractor. Dilute with distilled water and then process by SPE |
| Sorbent | 360 mg of C-18 in a cartridge or 47 mm C-8 discs |
| Sorbent preparation | Precondition with 2 mL each of methanol, ethyl acetate, methanol, and distilled water |
| Elution of analytes | After sample addition, remove water from the sorbent by air, and elute with 2 mL of ethyl acetate |
| Determination | OC | cGC/ECD or GC/MS |
| | OP | cGC/FPD or NPD or GC/MS |
| | MC | cGC/NPD or reversed phase HPLC |
| | Multiresidue | GC-MS |

be used for clean-up of sample extracts, as well as for trace enrichment.

Reversed-phase sorbents, such as C-18, are most commonly used for SPE of biological samples. The determination of insecticides in fluids does not require such extensive sample preparation as the analysis of insecticides in tissues. For example, in SPE methods for isolation and clean-up of organochlorine insecticides or synthetic pyrethroids from human urine or plasma, the samples only needed to be diluted prior to extraction, and the eluent did not require any further clean-up prior to gas chromatographic determination (see **Table 6** for details). High recoveries, ranging from 90–102% for urine and 81–93% for plasma, were obtained.

The extraction of insecticides from tissue samples requires large volumes of organic solvents (100–500 mL), and extensive extract clean-up. The procedure is also very time-consuming. A novel procedure, so-called matrix solid-phase dispersion (MSPD), can reduce solvent use by 98% and increase sample throughput by 90%. This process involves the grinding of biological samples with bulk C-18 sorbent. The MSPD method consists of adding 0.5 g of sample to 2.0 g of C-18 packing and grinding the sample until a nearly homogeneous blend of sample components adsorbed onto the SPE material is obtained. The packing material is then transferred into a syringe barrel plugged with a filter paper disc. The column head is covered with a second disc and the contents are compressed with a plunger to a volume of 4.5 mL. The column may then be eluted with a solvent or a series of solvents for the analytes of interest. This procedure was used for the isolation of organochlorine and organophosphorus insecticides from a number of animal tissues. The analytes were eluted from the MSPD column with 8.0 mL of acetonitrile or acetonitrile–methanol (9 : 1) through a Florisil co-column. The resulting eluate was analysed directly by GC. The recoveries of the analytes ranged from 60–114% for the concentrations examined.

## Food and Natural Product Analysis

The determination of insecticides in food and natural products requires extracting the analytes from a complex matrix, either liquid or solid. Solids, such as food, must be homogenized and extracted with organic solvents or aqueous buffers before SPE. Interferences from the food products must be removed either during extraction or during SPE isolation. Homogenization can be carried out by grinding followed by solvent extraction and filtration, or by grinding and Soxhlet extraction. A recent improvement is accelerated solvent extraction (ASE), in which high temperature and pressure are used to push an organic solvent through a solid sample and to collect the eluate in a vial. Automated instrumentation capable of running 30 samples at once is available commercially. One sample is processed in 15 minutes with extraction efficiency equal to that produced by Soxhlet extraction in 12 h. Supercritical fluids can also be used for extraction of food. In this procedure, supercritical carbon dioxide is used to remove analytes from food without dissolving the matrix. Methanol is often added as a modifier to the supercritical $CO_2$ to enhance the solubility of analytes. Microwave-assisted solvent extraction and matrix solid-phase dispersion, fast and safe alternatives to traditional solvent extraction, can also be applied to food analysis.

The solvent used to extract insecticides should be compatible with the SPE method being used. In addition, it should not co-extract interferences. For samples with high content of nonpolar components (such as fats or oils), a nonpolar solvent such as hexane should be chosen. This, in turn, determines normal-phase SPE with Florisil, silica, alumina or a cyano (CN) sorbent. Samples with high water content are best extracted with polar solvents such as methanol, acetone, or acetonitrile. In this case, reversed-phase SPE (C-18 or C-8) is preferable.

A multiresidue method for the determination of 43 organophosphorus, 17 organochlorine, and 11 N-methyl carbamate insecticides in 10 g of plant or animal tissues has been developed. The insecticides are extracted with 5% ethanol in ethyl acetate. Samples with high lipid content are cleaned up by automated gel permeation chromatography with a 30% ethyl acetate in hexane eluent and inline silica gel minicolumns. Highly pigmented samples are cleaned up with class-specific SPE columns: Florisil for organochlorine insecticides, and aminopropyl for N-methyl carbamates. No further clean-up of

**Table 6** SPE conditions used in extraction of insecticides from biological fluids

| | |
|---|---|
| Sample | Human urine or plasma diluted with water (OP) or with 70% methanol (pyrethroids) |
| Analytes | Pyrethroids, organophosphorus (OP) |
| Sorbent | C-18 cartridges |
| Sorbent preparation | Precondition with 2 mL each of methanol, ethyl acetate, methanol, and distilled water |
| Elution of analytes | 2 mL chloroform (pyrethroids) or chloroform–isopropanol (9 : 1) (OP). |
| Determination | Pyrethroids    cGC/FID |
| | OP    cGC/FID |

organophosphorus insecticide was necessary. Recovery of 71 insecticides ranged from 77 to 113%. The concentrated extracts were analysed by GC or liquid chromatography (LC) with specific detection.

Many multiresidue methods use acetonitrile extraction of the homogenized sample, reversed-phase SPE extraction of the analytes, and clean-up with an aminopropyl cartridge. In other cases, SPE is used for clean-up only. In this case, preparation of samples involves extraction with acetone and partitioning into methylene chloride–petroleum ether. This extract is cleaned up with an SPE aminopropyl column. The latter procedure has been automated by Gilson (Automated SPE clean-up (ASPEC)).

MSPD with C-18 silica has been used in multiresidue insecticide analysis in fruits and vegetables. The procedure is simple, inexpensive and rapid. Low detection limits (in the ppb range) and high recoveries (67–105%) were found for the investigated analytes.

In addition to using reversed-phase and normalphase SPE for the clean-up and isolation of insecticides from foods and natural products, graphitized carbon black has also proved an excellent sorbent for this purpose. Table 7 shows SPE conditions for a rapid multiresidue clean-up and analysis of over 200 organochlorine, organophosphorus and

**Table 7** SPE conditions used in extraction of insecticides from fruits and vegetables

1.  Homogenize 50 g of chopped sample with 100 mL of acetonitrile
2.  Add 10 g of NaCl. Homogenize for 5 minutes. Discard the lower aqueous layer
3.  Condition C-18 tube with 5 mL of acetonitrile
4.  Add 2 mL of acetonitrile extract from the sample; discard
5.  Pass 13 mL of acetonitrile sample extract through the tube; collect
6.  Add enough $Na_2SO_4$ to C-18 extracted sample to reach 15-mL mark. Cap the tube, shake well, and centrifuge for 5 minutes
7.  Evaporate 10 mL of centrifuged sample to 0.5 mL
8.  Add 1 cm $Na_2SO_4$ to the top of an ENVI-Carb (graphitized carbon black) tube
9.  Condition the ENVI-Carb and LC-$NH_2$ tubes separately with 5 mL acetonitrile–toluene (3 : 1)
10. Connect the LC-$NH_2$ tube to the outlet of the ENVI-Carb tube
11. Condition with 5 mL of acetonitrile–toluene (3 : 1)
12. Add 0.5 mL of C-18 cleaned sample; allow to drain through both tubes by gravity; collect
13. Rinse the tubes with 1 mL acetonitrile–toluene (3 : 1). Continue eluting with 5 mL of solvent mixture
14. Rotoevaporate to approximately 2 mL. Add 10 mL of acetone. Evaporate again. Reconstitute to the desired volume
15. Organochlorine and organophosphorus insecticides are determined by GC/MS. Carbamates are determined by HPLC/postcolumn derivatization/fluorescence detection.

methylcarbamate insecticides in fruits and vegetables. Recoveries of the analytes ranged from 65 to 99.3%.

## Future Developments

It is reasonable to expect continued development of the methods of isolation and preconcentration of insecticides based on SPE and SPME. The future for the latter technique looks particularly interesting, because simplification and increasing automation of preliminary analytical operations, particularly the extraction steps, is one of the modern trends in analytical chemistry. The analytical procedures using SPME have only two steps, can be easily automated, are rapid, simple, inexpensive, sensitive, and suitable for field analysis. New fibre coatings, more suitable for polar analytes, including those containing in-fibre derivatization reagents, should further extend the applicability of this technique.

In the extraction of insecticides from solid samples, such as food or biological materials, more procedures using MSPD are expected.

Future developments in solid-phase extraction will involve miniaturization of SPE procedures and considerably more online use of GC and HPLC. Sample handling will be minimized with automated systems. The use of automation will result in fast, easy, and reliable methods for SPE. To this end, syringe barrel designs and 96-well SPE plates will be used more extensively. Also, disposable pipette tips holding an extraction disc in the tip end, which are robotics compatible, will become more popular.

New SPE phases taking advantage of specific interactions will be introduced more widely. These include affinity SPE with antibodies bound to a solid substrate (immunosorbents) and the molecularimprinted polymers.

Other uses of SPE, such as derivatization on SPE sorbents, will become more widely used. Also, the use of supercritical fluid extraction and subcritical water extraction for the elution of insecticides from sorbents will be given more attention. Compared to conventional solvents, SPE will enable more selective removal of interferences and more selective extractions of insecticides from matrix, and will minimize preparation of some complex samples (e.g. biological tissues) with the potential for developing completely automated SFE-SPE-SFE methods.

*See also:* **II/Extraction:** Analytical Extractions; Solid-Phase Extraction. **III/Airborne Samples: Solid Phase Extraction. Environmental Applications:** Pressurized Fluid Extraction; Solid-Phase Microextraction; Supercritical Fluid Extraction. **Pesticides:** Extraction from Water. **Solid-Phase Micro-Extraction:** Environmental Applications.

## Further Reading

Barcelo D (1993) *Environmental Analysis; Techniques and Instrumentation in Analytical Chemistry 13.* Amsterdam: Elsevier.

Barcelo D and Hennion M-C *Trace Determination of Pesticides and Their Degradation Products in Water.* Oxford: Elsevier.

Chau ASY and Afgan BK (1982) *Analysis of Pesticides in Water,* vol. 2, *Chlorine- and Phosphorus-containing Pesticides.* Boca Raton: CRC Press.

Chau ASY and Afgan BK (1982) *Analysis of Pesticides in Water,* vol. 3, *Nitrogen-containing Pesticides.* Boca Raton: CRC Press.

Das KG (1981) *Pesticide Analysis.* New York: Marcel Dekker.

Font G, Mañes J, Moltó JC and Picó Y (1993) Solid-phase extraction in multi-residue pesticide analysis of water. *Journal of Chromatography* 642: 135.

McDonald PD and Bouvier ESP (1995) *Solid phase Extraction: Applications Guide and Bibliography. A Resource for Sample Preparation Method Development,* 6th edn. Milford: Waters.

McMahon BM and Sawyer LD (eds) (1985) *FDA Pesticide Analytical Manual,* vol. 1, Washington: FDA.

Marcotte AL and Bradley M (eds) (1985) *FDA Pesticide Analytical Manual,* vol. 2, Washington: FDA.

Pawliszyn J (1997) *Solid-Phase Microextraction: Theory and Practice.* New York: Wiley-VCH.

Preston ST Jr and Pankratz R (1981) *A Guide to the Analysis of Pesticides by Gas Chromatography,* 3rd edn. Niles: Preston Publishers.

Sherma J (1988) *Analytical Methods for Pesticides and Plant Growth Regulators,* vol. 16, *Specific Applications.* San Diego: Academic Press.

Sherma J (ed.) (1989) *Analytical Methods for Pesticides and Plant Growth Regulators,* vol. 17, *Advanced Analytical Techniques.* San Diego: Academic Press.

Simpson N (1997) *Solid Phase Extraction: Principles, Strategies, and Applications.* New York: Marcel Dekker.

Thurman EM and Mills MS (1998) *Solid-Phase Extraction: Principles and Practice.* New York: John Wiley & Sons.

# INSECTICIDES IN FOODSTUFFS

*See* **III/CARBAMATE INSECTICIDES IN FOODSTUFFS: CHROMATOGRAPHY AND IMMUNOASSAY**

# ION ANALYSIS

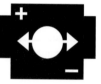

## Capillary Electrophoresis

**M. Macka and P. R. Haddad**, University of Tasmania, Hobart, Tasmania, Australia

## Introduction and Scope

Before the advent of capillary electrophoresis (CE), some impressive separations of inorganic species had been achieved by other electrophoretic methods, such as the separation of lanthanoids by paper electrophoresis using complexation with 2-hydroxyisobutyric acid (HIBA). With the introduction of capillary electrophoresis in 1981 by Jorgenson and Lukacs, separations of inorganic anions and cations started to appear quite early, but only sporadically until the late-1980s. There were several reasons for this. First, CE was a very new analytical technique and some time was needed for the theoretical background and instrumentation to mature. The second factor was competition with alternative analytical techniques, mainly spectroscopic methods (atomic absorption spectrometry, inductively coupled plasma spectrometry), in the area of determinations of metals, and ion chromatography in the area of separation of inorganic anions. Third, the main potential of CE was seen to be in the separation of biopolymers and biologically active compounds including drugs, with inorganic analysis being regarded as a relatively minor application area of CE. Although the last two factors remain valid, CE has matured to the stage where applications are sought in all areas of analysis, including inorganic analysis. This development has been characterized by more research oriented towards solving the practical requirements of inorganic analysis. The important developments in inorganic analysis by CE are summarized in **Table 1**.

Anions and cations migrate in opposite directions when placed in an electric field. Typical commercial

**Table 1**  Some important developments in inorganic analysis by CE

| Time | Development |
| --- | --- |
| 1967 | First separation of cations ($Br^{3+}$ $Cu^{2+}$) by free solution electrophoresis (0.1 mol $L^{-1}$ lactate) in a 3 mm axially rotating tube |
| 1974 | Separation of alkali metal cations in 200–500 mm Pyrex capillaries with potentiometric detection |
| 1979 | Separation of inorganic anions using 200 mm PTFE capillary and conductometric detection |
| 1983 | Separation of $Cu^{2+}$, $Fe^{3+}$ using a simple acetic acid electrolyte and direct photometric detection at 254 nm |
| 1987 | Indirect detection of anions |
| 1989 | Use of pre-capillary formed PAR complexes for separation of $Co^{II}$, $Cr^{III}$, $Ni^{II}$, $Fe^{III}$ MEC (electrolyte containing SDS, pH 8, 0.1 mmol $L^{-1}$ PAR) and using direct detection in visible |
| 1990 | Indirect photometric detection of anions using chromate electrolyte at pH 8 and reversed EOF |
| 1990 | Separation of rare earths metals and Li, Na, K, Mg using HIBA and indirect UV detection using an electrolyte of 0.03 mol $L^{-1}$ creatinine-HAc pH 4.8, 4 mmol $L^{-1}$ HIBA |
| 1995 | Hyphenation with ICP-MS: separation and detection of Sr, Cu, $Fe^{III}$, $Fe^{II}$, Cr, As, $Sn^{II}$, $Sn^{IV}$ |
| 1997 | ITP-CE online: separation of $Fe^{III}$ as pre-capillary formed EDTA complex; BGE: 25 mmol $L^{-1}$ MES + 10 mmol $L^{-1}$ bis-tris-propane, pH 6.6; leading: 10 mmol $L^{-1}$ HCl + 20 mmol $L^{-1}$ L-histidine + 0.1% HPMC, pH 6.0; terminating: 5 mmol $L^{-1}$ MES |

MEC, micellar electrochromatography

CE instruments have one point of sample introduction and one point of detection, so the choice must be made as to the polarity of the electrodes placed at the injection and detection ends of the capillary in order to establish the direction of movement of ions. Further, the electroosmotic flow (EOF) caused by the application of the separation potential will sweep the ions either towards or away from the detector, according to the polarity of the electrodes and the surface charge on the capillary. While it is theoretically possible to design a CE system in which a high EOF flowing towards the detector is established, causing both anions and cations to flow towards the detector, the electrophoretic mobilities of most inorganic anions and cations are too high to allow this approach to be applied. It is therefore more common for inorganic anions and cations to be analysed separately and for their electrophoretic movement to be in the same direction as the EOF (i.e. towards the detector). This is termed 'co-electroosmotic' separation.

## Separation of Inorganic Anions

### Separation Strategy

The electrophoretic mobilities of inorganic anions are quite large in comparison to those of most organic anions (**Figure 1**). Consequently, the normal instrumental arrangement is to assign the negative side of the separation voltage to the electrode at the injection end (so that the electrophoretic migration of the analyte anions is towards the positive electrode at the detection end). Moreover, since the normal flow of EOF in a fused silica capillary is towards the cathode (i.e. in the direction opposite to the electrophoretic migration of anions in the above case), it is normally

necessary to suppress or reverse the EOF to give a rapid co-electroosmotic separation. Such EOF reversal can be achieved by dynamic modification of the fused silica capillary inner wall by adsorption of cationic compounds or by permanent (covalent bonding) of cationic groups onto the fused silica capillary inner wall to provide an overall positive charge on the wall. The dynamic modification method is used most commonly and can be achieved by adsorption of suitable cationic surfactants (such as $C_{12}$ to $C_{16}$ alkyltrimethylammonium salts) or of large molecules such as cationic polymers (e.g. polybrene, poly(N,N,N′,N′-tetramethyl-N-trimethylenehexamethylenediammonium dibromide) by adding these to the background electrolyte (BGE) or by flushing the capillary between the runs. This is illustrated schematically in **Figure 2**.

After introduction of the sample into the capillary and application of the separation voltage, the anions migrate towards the cathode at the detection end in the order of their effective electrophoretic mobilities, which means that the anions migrating fastest will be detected first. The flow of the bulk BGE driven by the EOF towards the detector (co-EOF separation) allows anions of low electrophoretic mobility to be carried to the detector point more rapidly than would occur based only on their electrophoretic mobility in case of a suppressed-EOF separation. Therefore typical co-EOF separations of anions are characterized by rapid analysis times, as illustrated in **Figure 3**.

### Separation Selectivity

The electrophoretic mobilities of inorganic anions cover a wide range from about $30 \times 10^{-9}$ $m^2$ $V^{-1}$ $s^{-1}$ to over $100 \times 10^{-9}$ $m^2$ $V^{-1}$ $s^{-1}$ (Figure 1). Inspection

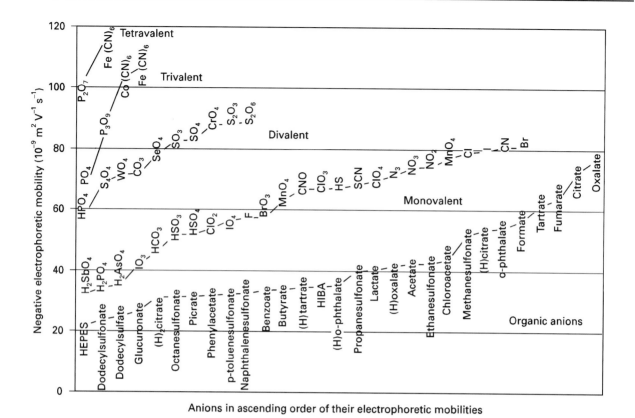

**Figure 1** Electrophoretic mobilities of inorganic and some organic anions in ascending order of their mobilities. Key: (H), protonated form; (H)$_2$, diprotonated form, charges left out for simplicity. Electrophoretic mobilities were calculated from tabulated ionic conductances or taken from published values.

of Figure 1 shows that most inorganic anions have a unique value of electrophoretic mobility and should be straightforward to separate without the need to optimize the separation selectivity. However, in some cases electrophoretic mobilities of ions are similar

and steps need to be taken to manipulate selectivity in order to achieve a separation. The approaches available to achieve this goal are summarized in **Table 2**.

Probably the most universal tool for manipulating selectivity is to alter the solvation of the analyte anion

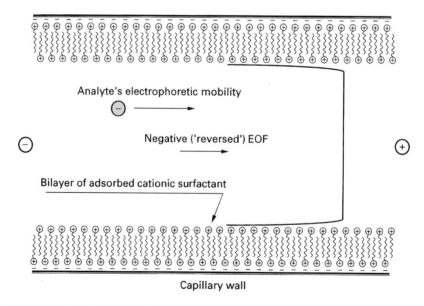

**Figure 2** Schematic representation of co-electroosmotic separation of anions using EOF reversal.

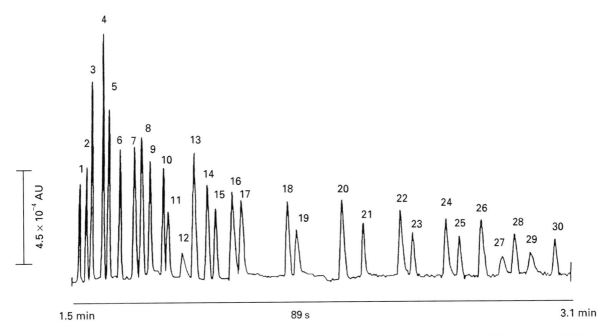

**Figure 3** Separation of 30 anions using chromate BGE. Conditions: capillary, fused silica 75 μm i.d., 0.600 m length, 0.527 m to detector; BGE, 5 mmol L$^{-1}$ chromate, 0.5 mmol L$^{-1}$ tetradecyltrimethylammonium bromide (TTAB), pH 8.0; separation voltage, – 30 kV; detection, indirect at 254 nm; injection, electrokinetic at 1 kV for 15 s; sample, 0.3–1.7 ppm of each anion. Peak identification: 1, thiosulfate; 2, bromide; 3, chloride; 4, sulfate; 5, nitrite; 6, nitrate; 7, molybdate; 8, azide; 9, tungstate; 10, monofluorophosphate; 11, chlorate; 12, citrate; 13, fluoride; 14, formate; 15, phosphate; 16, phosphite; 17, chlorite; 18, galactarate; 19, carbonate; 20, acetate; 21, ethanesulfonate; 22, propionate; 23, propanesulfonate; 24, butyrate; 25, butanesulfonate; 26, valerate; 27, benzoate; 28, L-glutamate; 29, pentanesulfonate; 30, D-glucanate. (Reproduced with permission from Jones WR and Jandik P (1991) Controlled changes of selectivity in the separation of ions by capillary electrophoresis. *Journal of Chromatography* 546: 445–458.)

by adding organic solvents. For those anions that exhibit protonation equilibria in the pH range used in CE, substantial selectivity changes can be achieved by pH variations. The effective mobility of the analyte is defined as the weighted average of all the mobilities of each of the forms, which for an acid H$_n$A undergoing protonation equilibria is

$$\bar{\mu}_A = \sum_{i=0}^{n} \mu_i \alpha_i \qquad [1]$$

**Table 2** Equilibria utilized for governing the separation selectivity

| Anions/cations | Source of separation selectivity | Examples of BGE additives |
|---|---|---|
| Anions | Changes of effective mobility by utilizing protonation equilibria | pH optimization |
| | Changes of effective mobility by changes in solvation of the anions | Organic solvents |
| | Ion association of anions capable of hydrophobic interactions (e.g. I$^-$, SCN$^-$), with amphiphilic cations | Hexadecyltrimethylammonium |
| | Ion exchange interactions with polycationic molecules | Polybrene, polyethyleneimine |
| Cations | Complexation with an auxiliary ligand in partial complexation mode | HIBA, lactate, etc. |
| | Influence of the metal on the p$K_a$ of a ligand group (total complexation mode) | PAR |
| | Dissociation of bound water molecules to mixed hydroxo–ligand –metal complexes (total complexation mode) | EDTA, CDTA and analogues |
| | Ion association of anionic complexes with amphiphilic cations, e.g. TBA$^+$ or hexamethonium (total complexation mode) | DHABS, CN$^-$ |
| | IEEC (ion exchange electrochromatography), e.g. with poly(diallyldimethylammonium chloride), (total complexation mode) | EDTA, Quin2 |
| | Changes of effective mobility by changes in solvation of the cations | Organic solvents |

HIBA, 2-hydroxyisobutyric acid; PAR, pyridylazoresorcinol; EDTA, ethylenediaminetetraacetic acid; CDTA, trans-1,2-diamino-cyclohexane-N,N',N'-tetraacetic acid; DHABS, 2,2'-dihydroxyazobenzene-5,5'-disulfonate; Quin, 2,8-amino-2-[(2-aminomethyl-phenoxy)methyl]-6-methoxyquinoline-N,N,N',N'-tetraacetic acid.

where $\mu_i$ is mobility and $\alpha_i$ is the fraction of the anion ion existing in form $i$.

For this approach to be successful, the BGE must be well buffered to avoid pH inhomogeneity within the migrating sample zone. Finally, ion exchange-type interactions between the analyte anions and some cationic water-soluble polymers may also be used to manipulate separation selectivity.

### Detection of Anions

**Direct and indirect photometric detection** Most commercial CE instruments are equipped with a photometric detector and therefore direct and indirect photometric detection are the most commonly used detection methods in CE. Probably the most simple and robust detection technique in terms of baseline stability and lack of system peaks is direct photometric detection. Unfortunately this can be applied only to a few inorganic anions, such as iodate, bromide, thiocyanate, nitrate, nitrite, and some others.

Indirect detection has the prime advantage of being universal in its applicability and is the most frequently applied detection mode for CE separations of both inorganic anions and cations. An absorbing co-ion (commonly referred to as the probe ion) is added to the BGE and the detector monitors a suitable absorbing wavelength of the probe. Migrating bands of analytes displace the probe from the BGE and indirect detection is possible due to the resulting decrease in absorbance. The limit of detection (LOD) of a nonabsorbing analyte ion when detected by this process is given by:

$$c_{\mathrm{LOD}} = \frac{c_{\mathrm{P}}}{R \times D} = \frac{c_{\mathrm{P}} \times N_{\mathrm{BL}}}{R \times A} = \frac{N_{\mathrm{BL}}}{R \times \varepsilon \times l} \quad [2]$$

where $c_{\mathrm{LOD}}$ is the concentration LOD, $c_{\mathrm{P}}$ is concentration of the probe co-ion in the BGE, $R$ is the transfer (or displacement) ratio (that is, the average number of

probe co-ions displaced by one analyte ion), $D$ is the dynamic reserve (the ratio of absorbance caused by the probe in the BGE to baseline noise), $N_{\mathrm{BL}}$ is the baseline noise, $A$ is the absorbance caused by the probe in the BGE, $\varepsilon$ is the molar absorptivity of the probe, and $l$ is the effective pathlength.

The factors influencing detection sensitivity in indirect photometric detection are summarized in **Table 3**. As a general rule, the BGE should be well buffered to ensure optimal reproducibility, but at the same time the buffer must not add any competing co-ions to the BGE so that the probe ion is the only species displaced by the analyte (maximizing $R$) and detection sensitivity is therefore maximized. For indirect detection of anions it has been shown that this can be achieved either by using a buffering counter ion (i.e. one having the opposite charge sign to the probe) or by using buffering species of extremely low mobility, such as buffering ampholytes employed at a pH close to their p$I$. A further consideration in maximizing detection sensitivity is to ensure that the probe and the analyte have similar mobilities. **Figure 4** shows the electrophoretic mobilities of some inorganic anions and some anionic probes. Typical concentration detection limits for indirect detection are in the low $\mu$mol L$^{-1}$ region, although sub-$\mu$mol L$^{-1}$ detection limits have been achieved by using highly absorbing dyes as probes.

**Other detection methods** Electrochemical detection methods, such as conductometric, amperometric and potentiometric detection methods, can be used in CE of inorganic anions with the first being the most successful. End-capillary detection, in which the detection electrodes are placed at the capillary outlet, has proved to be the optimal configuration. Commercial instrumentation for end-capillary conductometric detection has been introduced and is a useful approach for highly conductive analyte ions separated in a low conductivity BGE, such as 2-(N-

**Table 3** Factors influencing sensitivity of indirect photometric detection

| Factor | Methods of achievement |
|---|---|
| Minimal analyte peak width | Matching the mobilities of the analyte and the probe co-ion(s); minimizing solute–wall interactions |
| Maximal transfer ratio ($R$) | Matching the mobilities of analyte and probe co-ion(s); choice of BGEs without further co-ions which can compete with the probe (especially of similar mobilities) |
| Minimal baseline noise ($N_{\mathrm{BL}}$) | Minimal detector noise (usually by ensuring that the background absorbance is not too high); minimal 'chemical noise' (i.e. minimal solute–wall interactions of the analytes and/or probe) |
| Maximal absorptivity ($\varepsilon$) of the probe | Correct choice of the probe |
| Maximal effective pathlength ($l$) | Use of larger rather than smaller i.d. capillaries, use of extended pathlength capillary or z-cell |

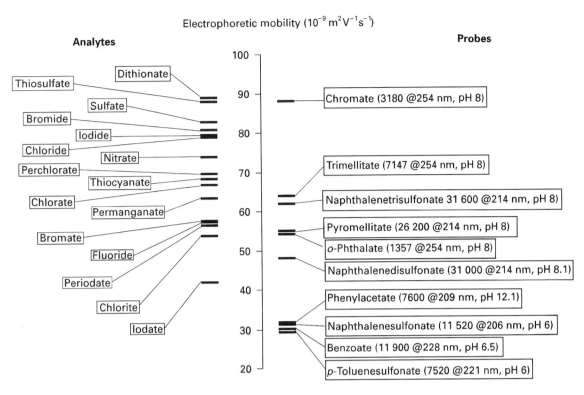

**Figure 4** Matching of electrophoretic mobilities of some inorganic anionic analytes and probes. Given in brackets are molar absorptivity (L mol$^{-1}$ cm$^{-1}$) at a wavelength (nm), and optionally a pH. (Data taken from Doble P and Haddad PR (1998) Indirect photometric detection of anions in capillary electrophoresis. *Journal of Chromatography A* 834: 189–212.)

cyclohexylamino)ethanesulfonic acid (CHES) buffer. The detection sensitivity is best for the most mobile anions and typical LODs are in the low µmol L$^{-1}$ range. End-capillary amperometric detection based on oxidation of anions such as nitrite, iodide, thiocyanate, azide or sulfite on gold, platinum or carbon fibre electrodes has been applied, with LODs down to the low nmol L$^{-1}$ region. Finally, potentiometric detection using liquid membrane, solid-state coated wire or metallic copper electrodes has been applied to a range of anions. Concentration detection limits in the µmol L$^{-1}$ range have been achieved.

The use of information-rich detection techniques, such as mass spectrometry (MS), provides additional information that can be of advantage for analyte identification or to enhance the separation selectivity. The intriguing task of coupling CE to ICP-MS has been solved with the design of special nebulizers, such as the direct injection nebulizer, which introduces 100% of the sample to the plasma and does not cause any detectable peak broadening. Despite the disadvantages of high running costs and equipment complexity, CE interfaced to inductively coupled plasma mass spectrometry (CE-ICP-MS) has been successfully applied to speciation studies, such as of inorganic and organic species of selenium or arsenic,

for which LODs in the low ppt range have been reported.

**Method Optimization**

An important factor in the attainment of a robust CE method, delivering reproducible migration times and possessing some matrix tolerance, is correct buffering of BGEs. **Figure 5** shows the difference in tolerance towards an alkaline sample matrix for a buffered BGE and an unbuffered BGE, from which it can be seen that buffering is essential in the analysis of samples of this type.

With appropriate knowledge of the underlying principles of CE, such as methods for governing the separation selectivity or maximizing the sensitivity of indirect detection, methods for the separation of a limited number of analytes can often be developed without use of computer-based optimization procedures. In the case of inorganic anions the most frequently used BGE is sodium chromate (pH 8.0) containing a low concentration of a cationic surfactant such as tetradecyltrimethylammonium bromide for reversal of the EOF. The chromate ion acts as the probe for indirect photometric detection and has an electrophoretic mobility that is similar to those of many inorganic anions. Sensitive indirect

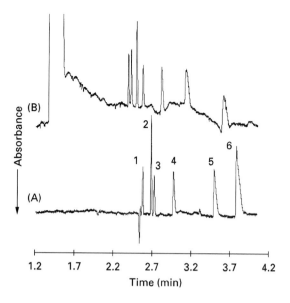

**Figure 5** Tolerance to alkaline matrix illustrated using buffered (A), and unbuffered (B) chromate electrolytes. Conditions: capillary, fused silica 75 µm i.d., 0.600 m length, 0.500 m to detector; BGE, 5 mmol L$^{-1}$ chromic trioxide, 20 mmol L$^{-1}$ Tris, 0.5 mmol L$^{-1}$ TTAB, pH 8.5 (a) or 5 mmol L$^{-1}$ sodium chromate, 0.5 mmol L$^{-1}$ TTAB, pH 8.5 (b); separation voltage, − 20 kV; detection, indirect at 254 nm; injection, hydrostatic at 100 mm for 10 s; temperature, 25°C; sample, 0.1 mmol L$^{-1}$ of each anion in 50 mmol L$^{-1}$ sodium hydroxide. Peak identification: 1, chloride; 2, sulfate; 3, nitrate; 4, chlorate; 5, phosphate; 6, carbonate. (Reproduced with permission from Doble P, Macka M, Andersson P and Haddad PR (1997) Buffered chromate electrolytes for separation and indirect absorbance detection of inorganic anions in capillary electrophoresis. *Analytical Communications* 34: 351–353.)

photometric detection at 254 nm can be achieved. Recently, buffered chromate electrolytes have been introduced using counterionic (cationic) buffers such as Tris or diethanolamine (Figure 5).

## Sample Introduction and Sample Pretreatment

Sample introduction by electromigration methods is known to be matrix-dependent, therefore hydrostatic/hydrodynamic sample introduction is typically used. If electromigration injection is to be employed, matrix effects should be examined and standard addition rather than an external standard method should be used for calibration. Field-amplified stacking effects can be used for samples of lower ionic strength than the BGE, leading to lower concentration detection limits.

Real samples often require the application of simple procedures such as extraction, filtration to remove particular matter, or dilution, prior to the CE step and an online or at-line combination of such procedure(s) with CE is desirable. A dialysis/flow-

injection analysis (FIA) sample clean-up system coupled online to a CE allowed analysis of a number of anions in a variety of samples with complex matrixes such as milk, juice, slurries or liquors from the pulp and paper industry. In some cases, matrix removal (sample clean-up) combined with preconcentration of the analyte(s) may be necessary. Online isotachophoresis (ITP)-CE systems are capable of analysing anionic analytes in complex matrixes, but commercial instrumentation is not available widely. A general rule for the applicability of CITP-CE for sample clean-up is that the mobility of the analyte(s) should differ from the matrix ion(s) to be removed.

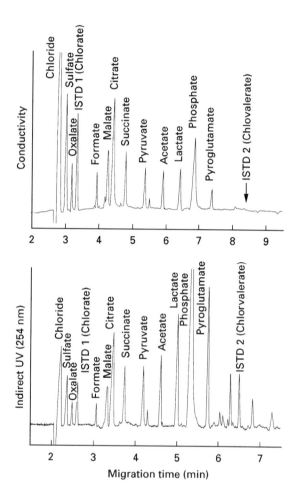

**Figure 6** Separation of inorganic and anions and carboxylic acids in a beer sample using simultaneous nonsuppressed conductivity and indirect conductivity detection. Conditions: capillary, fused silica 50 µm i.d., 0.600 m length to end-capillary conductivity detector, 0.480 m to photometric detector; BGE, 7.5 mmol L$^{-1}$ 4-aminobenzoic acid, 0.12 mmol L$^{-1}$ TTAB, pH 5.75; separation voltage, − 30 kV; detection, indirect at 254 nm; injection, hydrodynamic at 25 mbar for 12 s; sample, 10 × diluted stout. (Reproduced with permission from Klampfl CW and Katzmayr MU (1998) Determination of low-molecular-mass anionic compounds in beverage samples using capillary zone electrophoresis with simultaneous indirect ultra-violet and conductivity detection. *Journal of Chromatography A* 822: 117–123.)

## Applications

A high proportion of the real samples analysed for inorganic anions by CE have been water samples (drinking water, mineral water, river water, ground water, well water, etc.), with fruit juices and beverages being another common type of sample. The fact that CE is not tolerant to high ionic strength samples is reflected in the infrequent application of CE to samples such as seawater, unless considerable dilution of the sample is undertaken. **Figure 6** shows a successful application of CE to the determination of inorganic anions using simultaneous indirect photometric and conductometric detection. This example also illustrates the fact that real samples usually require separation of inorganic anions from organic anions.

## Cations

### Separation Strategy

As for inorganic anions, most inorganic cations have electrophoretic mobilities that are too high to permit counter-EOF separation, so these species are usually separated co-electroosmotically. In this case the electrode with positive polarity is placed at the injection

side and when a bare fused silica capillary is used, the EOF is towards the detection side (cathode). Under these conditions the migration order is such that the analyte with the highest positive electrophoretic mobility migrates first and the analyte having the lowest positive mobility migrates last.

In contrast to the case for inorganic anions, many inorganic cations exhibit very similar mobilities, such as the whole group of rare earth metals or numerous transition metal ions (**Figure 7**). Therefore additional sources of separation selectivity to those available for anions are needed in order to separate these species. The main approach used is the addition of an auxiliary ligand to the BGE in order partially to complex the analyte cations (**Figure 8**). Provided the degree of complexation is different for each analyte cation, the effective charge and hence the effective mobility of each analyte will be unique and separation should be possible. These auxiliary ligands can be divided qualitatively into two groups according to the thermodynamic stability of the complexes formed with the analyte cations. **Table 4** summarizes the important characteristics of CE separation of metal ions using either weakly or strongly complexing auxiliary ligands.

Weakly complexing auxiliary ligands (such as HIBA or 19-crown-6) usually serve the sole purpose

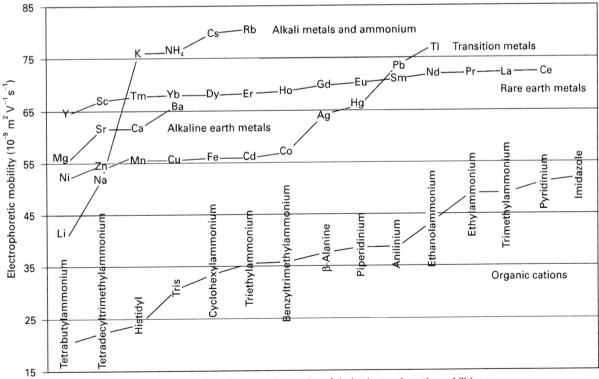

**Figure 7** Electrophoretic mobilities of metal ions and some organic cations in ascending order to their mobilities. Ammonium was included with the alkaline metals because of its frequent importance in analysis of metal ions. Charges of metal cations have been left out for simplicity; transition metal ions divalent apart from monovalent Ag and Tl; organic cations all bear a charge of $+1$.

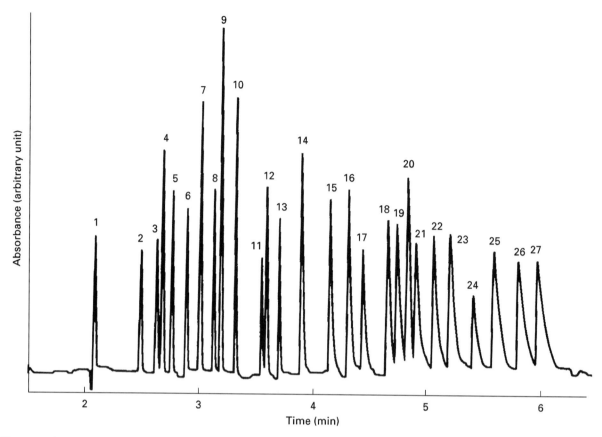

**Figure 8**  Separation of 27 alkali, alkaline earth, transition and rare earth metal ions in a single run using lactate as auxiliary ligand. Conditions: capillary, fused silica 75 μm i.d., 0.600 m length, 0.527 m to detector; BGE, 15 mmol L$^{-1}$ lactic acid, 8 mmol L$^{-1}$ 4-methylbenzylamine, 5% methanol, pH 4.25; separation voltage, 30 kV; detection, indirect at 214 nm; injection, hydrostatic at 100 mm for 30 s; sample, 1–5 ppm or each metal. Peak identification: 1, K$^+$; 2, Ba$^{2+}$; 3, Sr$^{2+}$; 4, Na$^+$; 5, Ca$^{2+}$; 6, Mg$^{2+}$; 7, Mn$^{2+}$; 8, Cd$^{2+}$; 9, Li$^+$; 10, Co$^{2+}$; 11, Pb$^{2+}$; 12, Ni$^{2+}$; 13, Zn$^{2+}$; 14, La$^{3+}$; 15, Ce$^{3+}$; 16, Pr$^{3+}$; 17, Nd$^{3+}$; 18, Sm$^{3+}$; 19, Gd$^{3+}$; 20, Cu$^{2+}$; 21, Tb$^{3+}$; 22, Dy$^{3+}$; 23, Ho$^{3+}$; 24, Er$^{3+}$; 25, Tm$^{3+}$; 26, Yb$^{3+}$; 27, Lu$^{3+}$. (Reproduced with permission from Shi YC and Fritz JS (1993) Separation of metal ions by capillary electrophoresis with a complexing electrolyte. *Journal of Chromatography* 640: 473–479.)

of manipulating the separation selectivity and normally form part of the BGE into which samples containing free metal ions are injected. That is, *on-capillary complexation* is utilized (Figure 8). Since a substantial part of the metal exists as an uncomplexed cation, indirect absorption detection using a cationic absorbing probe contained within the BGE

is normally applied. Rapid complexation equilibria are necessary to enable manipulation of the selectivity of separation by varying parameters of the BGE such as the concentration of the auxiliary ligand and pH (Figure 8).

A completely different strategy is to use ligands forming stable, typically anionic complexes,

**Table 4**  Typical method properties according to the stability of the complex formed betweeen the auxiliary ligand and the metal ion analyte

| Thermodynamic stability of the metal complex with the auxiliary ligand | Low | High |
|---|---|---|
| Typical examples of the auxiliary ligand | HIBA, citric acid, lactic acid | CN$^-$, polydentate ligands such as EDTA and its analogues, metallochromic ligands |
| Degree of metal ion complexation | Partial | Total |
| Method of complex formation | On-capillary | Pre-capillary, often with on-capillary also used |
| Method of photometric detection | Indirect | Direct |
| Ease of selectivity manipulation | Straightforward | More difficult |
| Polarity of separation voltage on injection side | + | +, − |

which are then separated as anions using the same approaches discussed earlier. Auxiliary ligands such as ethylenediaminetetraacetic acid (EDTA) or 2-pyridylazoresorcinol (PAR) may be used for this purpose and are usually added to the sample (that is, *precapillary complexation* is used). Often the BGE also contains a low concentration of the ligand in order to prevent the decomposition of less stable complexes. Most (or all) of the metal is complexed, so that detection using direct photometry is usually applicable. The complexation equilibria can be slow, but this then makes it difficult to alter the selectivity of separation and to retain good peak shapes. These separations are carried out in bare fused silica capillaries, fused silica capillaries with reversed EOF, or using other capillary wall chemistries. Depending on the charge of the metal complexes, the separation can be performed in either the co- or counter-electroosmotic modes, with either positive or negative separation potential applied on the injection side.

Metal ions are known to adsorb onto the surface of silica particles through interaction with silanol groups. In the case of polyvalent metal cations, this adsorption can be considered to be irreversible. Although the surface density of silanols on fused silica is lower by approximately an order of magnitude than for porous silica, it is well documented in CE using bare fused silica capillaries that metal ions present in the BGE (even as an impurity) can adsorb onto the capillary wall. The extent of adsorption is often reflected in changes to the EOF, with both suppression and reversal of EOF having been demonstrated. Despite these facts, coated capillaries have found relatively little use in separations of metal ions. In bare fused silica capillaries, the risk of adsorption of metal ions is counteracted by typically weakly acidic and complexing electrolytes.

## Separation Selectivity

The most powerful and straightforward tool governing separation selectivity is control of the degree of complexation with the auxiliary ligand. For a metal ion present in several forms that are in equilibrium with rapid kinetics of interchange between the forms, the effective mobility of the analyte is given by the weighted average of the mobilities of each of the forms. For a metal ion M migrating in a BGE containing a ligand L forming complexes ML, ML$_2$, ..., ML$_i$, the effective mobility of the metal can then be expressed as:

$$\bar{\mu}_M = \sum_{i=0}^{n} \mu_i \alpha_i = \frac{\sum_{i=0}^{n} \mu_i [ML_i]}{\sum_{i=0}^{n} [ML_i]} = \frac{\sum_{i=0}^{n} \mu_i \beta_i [L]^i}{\sum_{i=0}^{n} \beta_i [L]^i} \quad [3]$$

where $\mu_i$ is mobility and $\alpha_i$ is the fraction of metal ion existing in form $i$, [ML$_i$] is the concentration of the complex ML$_i$, [L] is the concentration of the form of ligand forming the complex and $\beta$ is the overall stability constant (ML$_0$ = M, $\beta_0$ = 1).

A different situation occurs when utilizing auxiliary complexing ligands that form strong complexes with the metal ions. Since most of the metal ion is complexed under all BGE conditions, the complex formation/dissociation cannot be used to govern the separation selectivity as in the case of weakly complexing ligands. Apart from minor factors influencing the selectivity, such as solvation changes in various media, there are very few means to bring about a change in the charge/mass value of the analytes (and consequently a substantial change in selectivity). These include: (1) dissociation of functional groups on the ligand exhibiting protonation equilibria; (2) exchange of the remaining water molecules on the metal ion coordination sites not saturated by the ligand; and (3) ion association or ion exchange equilibria in the BGE solution (Table 4). The separation selectivity therefore varies considerably with the nature of the ligand used.

## Detection

**Direct and indirect photometric detection**  As with inorganic anions, direct and indirect photometric detection are the most commonly used detection methods in the separation of metal ions, being employed in about 85% of publications. Since most hydrated metal ions do not absorb at all, or have only weak absorption bands in the UV region above 185 nm, direct detection is normally possible only when the metal ions are complexed with an auxiliary ligand. Thus the auxiliary ligand, which in earlier discussion has been shown to play a crucial role in the separation of metal ions, also enhances the detectability of these species. Metallochromic ligands form highly absorbing ($\varepsilon \sim 10^5$ L mol$^{-1}$ cm$^{-1}$) coloured complexes with metal ions. When these complexes are stable and the separation results in well-shaped peaks, very good detection sensitivity can be obtained. For example, determination of transition metals in the form of complexes with PAR achieved concentration LODs in the order of $10^{-7}$ mol L$^{-1}$ or absolute LODs at fmol levels. Even lower LODs (80 nmol L$^{-1}$ or 0.5 fmol for Zn) have been obtained for porphinate complexes ($\varepsilon \sim 10^5$ L mol$^{-1}$ cm$^{-1}$), but the detection method is less universal because different metal ions exhibit a range of absorption maxima and absorptivities.

The underlying principles for indirect photometric detection of inorganic cations are the same as for inorganic anions (see above). **Figure 9** illustrates matching of electrophoretic mobilities of some metal cations as analytes with some cationic probes.

**Other detection methods** Conductometric, amperometric and potentiometric detection methods can be utilized for inorganic cations and the same principles discussed apply. End-capillary conductometric detection has been used for a range of cations using low-conductive histidine-MES BGEs with similar success to anion analysis (LODs are in low $\mu mol\,L^{-1}$ range). End-capillary amperometric detection utilizing reduction of metal ions such as thallium, lead, cadmium, etc., and inorganic and organic mercury compounds on gold, platinum or mercury films, has been applied with low $nmol\,L^{-1}$ range LODs. Potentiometric detection can also be used for metal ions and, as for anions, the solid-state electrodes show greatest promise.

CE coupled with electrospray mass spectrometry has achieved detection of about 30 metal ions by positive ion MS. Although chemical noise does not allow detection below ppb levels, LODs are generally below those obtained for indirect photometric detection. Impressive LODs have been obtained by CE-ICP-MS, e.g. of 0.06 ppb for $Sr^{II}$ (8 fg or 90 amol).

**Method Optimization**

The separation of up to 30 metal ions in one run using weakly complexing auxiliary ligands is a challenging analytical task and optimization of the BGE composition often requires the use of computer methods based on thermodynamic complexation models or the use of artificial neural networks. The BGEs used most frequently for metal ion analysis would normally utilize indirect photometric detection and a typical BGE contains a weakly complexing auxiliary ligand (such as HIBA) for governing separation selectivity for most transition metal ions and rare earth metals. It often also contains a crown-ether as a second auxiliary ligand to separate ammonium from potassium (see Figure 7) and an indirect detection cationic probe (such as imidazole, benzylamine, or creatinine) at pH $\sim 4.5–5$.

While most authors have demonstrated the application of their developed separation to a real

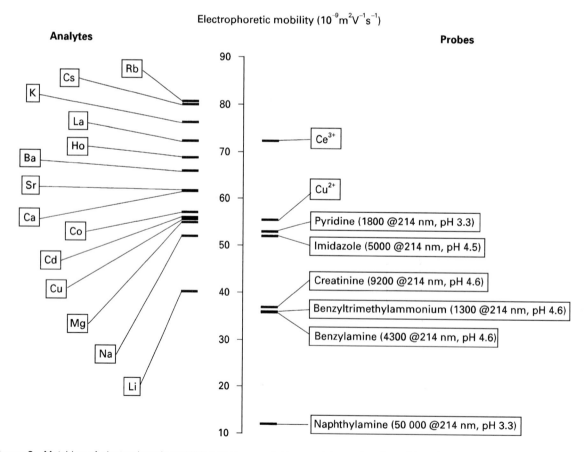

**Figure 9** Matching of electrophoretic mobilities of some metal ion analytes and probes. Given in brackets are molar absorptivity ($L\,mol^{-1}\,cm^{-1}$) at a wavelength (nm), and optionally a pH. (Source of data: Beck W and Engelhardt H (1992) *Chromatographia* 313; Chen M and Cassidy RM (1993) Separation of metal ions by capillary electrophoresis. *Journal of Chromatography* 640: 425–431.)

sample of some kind, there have been relatively few studies on aspects of method validation, such as precision and accuracy. Typical precision for migration times, peak areas and peak heights show a relative standard deviation ranging from 3 to 10%.

### Sample Introduction and Sample Pretreatment

In addition to the principles discussed earlier for anions, stacking effects can be achieved in the separation of metal ions using on-capillary complex formation in which oppositely migrating metal cations and ligand anions converge and react at the boundary between the injected sample plug and the BGE. A necessary condition to utilize the stacking by on-capillary complexation is fast complex formation. This approach was first demonstrated for four divalent metal complexes with PAR in which a plug of 1 mmol L$^{-1}$ PAR was first injected into the capillary are then electromigration injection from a sample without added ligand was performed. This method with an optimized stacking procedure gave detection limits in the range of 10 nmol L$^{-1}$.

Ion exchange and chelating resins may be used to preconcentrate metal ions prior to CE determination in an offline configuration. Ion exchange materials are less successful for the selective preconcentration of heavy metals than chelating (iminodiacetate or dithiocarbamate) resins, or by the formation of metal chelates with an excess of the reagent, followed by adsorption of the chelate on a hydrophobic column.

### Applications

As with anions, many of the real samples analysed have been various water samples, fruit juices and beverages and food. An example of an unusual application of CE to the analysis of high salinity samples is shown in **Figure 10**.

## Simultaneous Separations of Anions and Cations

Despite the potential advantages of reduced analysis time and costs that arise from analysing both anions and cations in one run, separate analysis of both groups is still much easier and more robust. The simple fact that anions and cations migrate in opposite directions in an electric field means that a simultaneous analysis cannot be realized with one point of sample introduction and one point of detection.

However, there are two approaches that have proved successful. The first uses a sample injection simultaneously in two capillaries operated with the same BGE, with both capillaries equipped with independent detectors located near their ends. A relatively

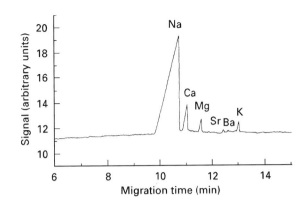

**Figure 10** Separation of metal ions in a mixture of salt water and formation water using indirect photometric detection. Conditions: capillary, fused silica 75 μm i.d., 0.600 m length, 0.520 m to detector; BGE, 6.5 mmol L$^{-1}$ HIBA, 5 mmol L$^{-1}$ UVCAT-1 (Waters), 6.2 mmol L$^{-1}$ 18-crown-6, 25% (v/v) methanol, apparent pH 4.8; separation voltage, 20 kV; detection, indirect at 185 nm; injection, hydrodynamic at 98 mm for 20 s; sample, diluted by a factor of 125. (Reproduced with permission from Tangen A, Lund W and Frederiksen RB (1997) Determination of sodium, potassium, magnesium and calcium ions in mixtures of sea water and formation water by capillary electrophoresis. *Journal of Chromatography A* 767: 311–317.)

easier approach uses just one capillary, but sample is injected at both ends and the detector is located approximately in the middle of the capillary. This approach has been applied to the analysis of a range of real samples; an example is given in **Figure 11**.

It should be noted that when applying indirect detection using both an anionic and a cationic probe, the probes should be mixed in their free acid and free base forms to avoid the presence of competing ions in the BGE.

## Future Developments

Control of the separation selectivity will be an area of development in CE of inorganic species. Utilization of ion exchange-type interactions with pseudostationary phases, and particles of sub-μm size, can introduce new selectivity into a CE separation. It is instructive to examine the separation selectivities achieved by ion chromatography and CE under standard conditions, as revealed by the elution or migration order of common analytes. In the case of IC, anions are eluted in the following order of retention times: $F^- < Cl^- < NO_2^- < Br^- < NO_3^- < PO_4^{3-} < SO_4^{2-} < I^-$. However, the migration times of these species in co-EOF CE are $Br^- < Cl^- < SO_4^{2-} < NO_2^- < I^- < NO_3^- < F^- < PO_4^{3-}$. A similar pattern emerges when the same comparison is made for inorganic cations, for which the IC retention times follow the order: $Li^+ < Na^+ < NH_4^+ < K^+ < Mg^{2+} < Ca^{2+} < Sr^{2+} < Ba^{2+}$, while migration times for co-EOF CE are

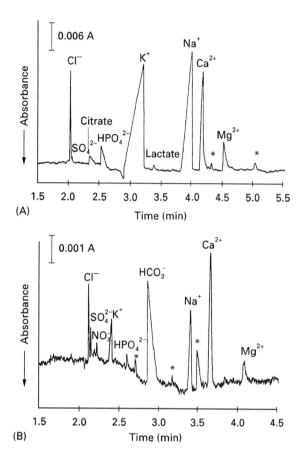

**Figure 11** Simultaneous determination of inorganic anions and cations in milk (A) and mud (B) after offline dialysis. Conditions: capillary, fused silica 50 μm i.d., 0.500 m total length, 0.200 m to detection window from the anodic end; BGE, 6 mmol L$^{-1}$ 4-aminopyridine, 2.7 mmol L$^{-1}$ H$_2$CrO$_4$, 30 μmol L$^{-1}$ CTAB, 2 mmol L$^{-1}$ 18-crown-6, pH 8; separation voltage, 20 kV; detection, indirect at 262 nm; injection, hydrostatic at 50 mm for 10 s (cathode end) and 50 mm for 10 s (anode end) (A) or at 50 mm for 20 s (cathode end) and 100 mm for 40 s (anode end) (B); time between the injections, 60 s (A) or 30 s (B). (Reproduced with permission from Kuban P and Karlberg B (1998) Simultaneous determination of small cations and anions by capillary electrophoresis. *Analytical Chemistry* 70: 360–365.)

NH$_4^+$ < K$^+$ < Ba$^{2+}$ < Sr$^{2+}$ < Ca$^{2+}$ < Na$^+$ < Mg$^{2+}$ < Li$^+$. Complementary selectivities are again apparent. These selectivities suggest that a mixed-mode separation system in which the movement of analytes is influenced both by electromigration effects and ion exchange interactions might provide a means to manipulate selectivity in order to solve existing separation problems.

New developments can also be expected in the area of detection techniques. For instance, the use of highly absorbing cationic probes in carefully formulated BGEs that avoid competitive displacement has the potential to increase the sensitivity using indirect photometric detection. Further development

of other detection techniques can be anticipated, especially those compatible with the capillary dimensions and suitable for on- or end-capillary use and which do not necessitate elaborate changes to the capillary.

Finally, more advances in the area of online sample treatment and/or preconcentration techniques are likely to occur as CE becomes more of a routine tool for the determination of inorganic species in complex samples.

## Further Reading

Chiari M (1998) Selectivity in capillary electrophoretic separations of metals and ligands through complex formation. *Journal of Chromatography A* 805: 1–15.

Dabek-Zlotorzynska E, Lai EPC and Timerbaev AR (1998) Capillary electrophoresis – the state-of-the-art in metal speciation studies. *Analytica Chimica Acta* 359: 1–26.

Doble P and Haddad PR (1998) Indirect photometric detection of anion in capillary electrophoresis. *Journal of Chromatography A* 834: 189–212.

Foret F, Křivánková L and Boček P (1993) *Capillary Zone Electrophoresis*. Weinheim: VCH.

Fritz JS (1998) Determination of inorganic anions and metal cations. In: Camilleri P (ed.). *Capillary Electrophoresis, Theory and Practice*. Boca Raton, FL: CRC Press.

Haddad PR (1997) Ion chromatography and capillary electrophoresis: a comparison of two technologies for the determination of inorganic ions. *Journal of Chromatography A* 770: 281–290.

Haddad PR, Doble P and Macka M (1999) Developments in sample preparation and separation techniques for the determination of inorganic ions by ion chromatography and capillary electrophoresis. *Journal of Chromatography A* 856: 145–177.

Jandik P and Bonn G (1993) *Capillary Electrophoresis of Small Molecules and Ions*. New York: VCH.

Jones P (ed.) (1999) Electrophoresis of inorganic species, *Journal of Chromatography A* v. 834, Parts 1 + 2.

Li SFY (1992) *Capillary Electrophoresis, Principles, Practice and Applications*. Amsterdam: Elsevier.

Macka M, Haddad PR (1997) Determination of metal ions by capillary electrophoresis. *Electrophoresis* 18: 2482–2501.

Mazzeo JR (1998) Capillary electrophoresis of inorganic anions. In: Khaledi MG (ed.). *High Performance Capillary Electrophoresis, Theory, Techniques and Applications*. New York: Wiley-Interscience.

Pacáková K and Štulík V (1997) Capillary electrophoresis of inorganic anions and its comparison with ion chromatography. *Journal of Chromatography A* 789: 169–180.

Timerbaev AR (1997) Strategies for selectivity control in capillary electrophoresis of metal species. *Journal of Chromatography A* 792: 495–518.

Timerbaev AR (1997) Analysis of inorganic pollutants by capillary electrophoresis. *Electrophoresis* 18: 185–195.

# Electrophoresis

*See*  **III / ION ANALYSIS / Capillary Electrophoresis**

# High-Speed Countercurrent Chromatography

**E. Kitazume**, Iwate University, Morioka, Iwate, Japan

High-speed countercurrent chromatography (HSCCC), developed by Ito, is a useful method of separating many organic materials, such as biologically active substances, natural and synthetic peptides and various plant hormones. Like other countercurrent chromatography (CCC) methods, it is also free from problems based on solid supports, such as adsorption or irreversible binding and contamination of the sample.

HSCCC has been applied to preconcentration and separation of inorganic elements since the late 1980s and since then, inorganic elements, including rare earth elements, have been separated by HSCCC. Also the preconcentration and separation of inorganics from geological samples have been studied. The liquid systems for inorganics are more complicated than those for separation of organics, because they usually contain significant amounts of an extracting agent, which influences kinetic properties and viscosities of the whole two-phase system.

In order to achieve high sensitivity for analysing trace inorganic elements in a solution using atomic absorption spectrometry (AAS) or inductively coupled plasma atomic emission spectrometry (ICP-AES), conventional preconcentration methods such as evaporation, ion exchange and solvent extraction have been used. However, there are several problems in these methods for the determination of ultra trace elements, for example, peak broadening for ion exchange and a small enrichment factor for solvent extraction. It is difficult to achieve under 0.5 mL concentrated sample solutions by conventional methods. If there were effective methods to concentrate traces into 0.1 mL volume or less, absolute detection limits for trace analysis such as AAS, ICP-AES and ICP–mass spectrometry (ICP-MS) would be greatly decreased, and matrix effects would be eliminated.

Recently, pH-zone refining CCC, which is a unique technique based on the neutralization reaction between the mobile and stationary phase, has been developed for the separation and enrichment of organic acids, basic derivatives of amino acids and acidic peptide derivatives. It can initiate chemical reactions in quite a limited thin area, the interface between the organic and aqueous phase. Therefore, if there is a reversible pH area between the mobile and stationary phase, the pH in the column can be continuously controlled. This means that another flow rate, concerned with pH and different from real flow rate, can be realized in the column. Impurities in sample solutions can be concentrated in the pH boundary in the column. As it can be successfully applied to the enrichment of inorganic trace elements in solution, it has great potential for on-line enrichment and subsequent analysis, when HSCCC is combined with instruments such as AAS, ICP-AES and ICP-MS.

## Mechanism of Two-phase Separation in HSCCC

In HSCCC, a stationary sun gear is mounted around the central stationary axis of the centrifuge to prevent the flow tubes twisting. This gear arrangement gives a planetary motion of the column holder – one rotation about its own axis for one revolution around the central stationary axis of the centrifuge in the same direction.

**Figure 1** shows a schematic diagram of two-phase separation in an HSCCC column. The heavier mobile phase (black) is introduced into the column from the right side (column head). The upper stationary phase (grey) is retained in the column by a rotational force field and Archimedean screw effect (ASE), in spite of being pushed to the column end by the mobile phase. For separation and enrichment of inorganic elements, the stationary phase commonly contains one or more extracting reagents, such as di(2-ethylhexyl)phosphoric acid (DEHPA), dissolved in the appropriate stationary organic phase. The mobile phase is commonly composed of inorganic acids and their salts. Water-soluble complexing reagents forming

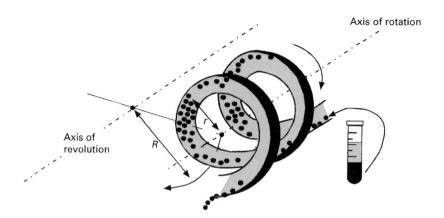

**Figure 1**   Schematic diagram of two-phase separation in HSCCC column. r, Rotation radius; R, revolution radius.

stable compounds with specified elements in the sample can also be used as an aqueous mobile phase. The column is revolved in the radius $R$ and rotated with the radius $r$ at the same time. When the coil column is facing the axis of revolution (top left in Figure 1), the two phases are vigorously mixed because of the weak force field. In contrast, when facing the outer side of the column, the mobile phase is clearly separated by a stronger force field than that at the inner side. The inner quarter turn of the coil is the most vigorous mixing area. As one phase usually occupies the column head by an Archimedean screw effect, another phase can be introduced as an eluent. After being introduced from the column head, the heavier mobile phase emerges from the end of the column (bottom left in Figure 1). When the column makes one revolution around the axis, it rotates once on the orbital of revolution. Therefore, as a specific point in the column, one turn in the coil proceeds towards the column head per revolution. At the same time, one cycle of each mixing and separation process at the specific point occurs in the column. As the HSCCC column rotates at over several hundred rpm, this is a rapid process. For example, there are 13 processes per second taking place at 800 rpm, giving efficient separation and mixing in the column.

## Extracting Reagent for Separation of Inorganic Elements

As mentioned above, the existence of extracting reagent in the mobile phase is an essential factor in the separation of inorganic elements. It complicates the determination of several important factors, such as distribution coefficients, peak resolution and separation efficiency. Research reveals that the kinetic properties of a specific system used in HSCCC affect the separation efficiency. The mass transfer rate into the organic stationary phase is responsible for using either stepwise or isocratic elution. In addition, the value of the distribution coefficients, determined by batch extraction measurements on systems, are sometimes considerably different from the dynamic distribution coefficients calculated from the elution curve. Further theoretical and basic investigations are necessarily concerned with extraction kinetics, as well as the hydrodynamic behaviour of the two phases in the HSCCC column.

In **Table 1**, typical extracting reagents used for separation and enrichment of inorganic elements are summarized. Organophosphorus extractants are often used because of their solubility properties.

**Table 1**   Typical extracting reagents for separation and enrichment of inorganic elements using high-speed countercurrent chromatography

| Extracting reagent | Two-phase system | Inorganic element |
|---|---|---|
| Di(2-ethylhexyl)phosphoric acid (DEHPA) | HCl, organic acid–heptane | Rare earth, heavy metals |
| 2-Ethylhexylphosphonic acid mono-2-ethylhexyl ester (EHPA) | Carboxylic acid–toluene | Rare earth |
| Dinonyltin dichloride | HCl, $HNO_3^-$ Methylisobutylketone (MIBK) | Ortho- and pyrophosphate |
| Cobalt dicarbolide | $HNO_3$-nitrobenzene | Cs and Sr |
| Tetraoctylethylenediamine (TOEDA) | HCl, $HNO_3$, organic acid–chloroform | Alkali, alkaline earth, rare earth, heavy metals, Hf, Zr, Nb, Ta |

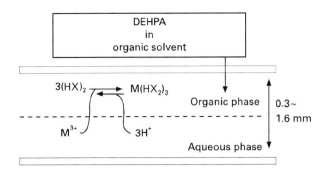

**Figure 2** Extraction equilibrium in HSCCC column using di (2-ethylhexyl) phosphoric acid (DEHPA) as an extracing reagent for trivalent metal ion ($M^{3+}$). $(HX)_2$, dimer of DEHPA.

Figure 2 shows the equilibrium for extraction of trivalent metal ions, such as lanthanide, using the extracting reagent (DEHPA). DEHPA is commonly applied in industrial separation due to its high extractability and high separation factors between many inorganic elements, especially for rare earth elements. Trivalent metal ions ($M^{3+}$) are extracted as shown in Figure 2 in PTFE tubing of 0.3–1.6 mm in diameter. The long thin rectangles in Figure 2 show the wall of the PTFE tubing. $(HX)_2$ is a dimer of DEHPA. Metal ions in the aqueous phase are extracted into organic phase as $M(HX_2)_3$.

## Mutual Separation of Inorganic Elements

Figure 3 shows a typical flow diagram of the instrumentation assembly for the separation of rare earth elements and other inorganic elements by HSCCC. In this case, the spectrophotometer is used as a detector for each element. If an alternative measurement system such as AAS, ICP-AES or ICP-MS is used, all the equipment depicted after the splitter is not required. Each separation is usually initiated by filling the en-

tire column with the stationary nonaqueous phase, followed by injection of an appropriate volume of the sample solution through the sampling port. Then, the mobile phase is eluted through the column at a flow rate of 0.1–5 mL min$^{-1}$ while the apparatus is rotated at several hundred rpm. In Figure 3, continuous detection of the inorganic elements is effected by means of a post-column reaction with an appropriate compound such as arsenazo III, and the elution curve is obtained by monitoring the effluent with a spectrophotometer. The effluent is divided into two streams with a tee adapter and a low-dead-volume pump (pump II). Pump II delivers a portion of the effluent at the required flow rate to the spectrophotometer and pump III adds reagent for post-column reaction to the effluent stream. The resulting stream first passed through a narrow mixing coil and then leads through an analytical flow cell in a spectrophotometer. The other effluent stream through the tee adapter is either collected or discarded.

Figure 4 shows a one-step separation of all 14 lanthanides (except for promethium) performed by applying an exponential gradient of hydrochloric acid in the mobile phase. The main problem in gradient elution is that the optimum range of the ligand concentration in the stationary phase is substantially different between the lighter and heavier groups of the rare earth elements. Because the separation of the heavy lanthanide elements, including thulium, ytterbium and lutetium, is more difficult, a ligand concentration of 0.003 mol L$^{-1}$ was selected for best resolution. With an isocratic separation mode, using a constant eluent concentration, even heavy lanthanide elements such as Tm, Yb and La are well separated.

Figure 5 shows a typical chromatogram of Ni(II), Co(II), Mg(II) and Cu(II) obtained by eluting with

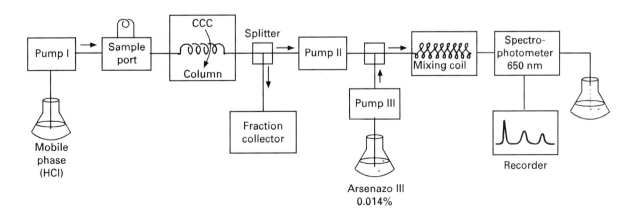

**Figure 3** Flow diagram of instrumentation assembly for separation of rare earth elements by HSCCC. CCC, countercurrent chromatography.

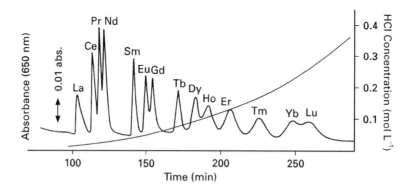

**Figure 4**   Gradient separation of 14 lanthanides obtained by HSCCC. Apparatus: HSCCC centrifuge with 7.6 cm revolution radius; column: three multilayer coils connected in series, 300 m × 1.07 mm i.d., 270 mL capacity; stationary phase: 0.003 mol L$^{-1}$ DEHPA in *n*-heptane; mobile phase: exponential gradient of hydrochloric acid concentration from 0 to 0.3 mol L$^{-1}$, as indicated in the chromatogram; sample: 14 lanthanide chlorides, each 0.001 mol L$^{-1}$ in 100 μL water; speed: 900 rpm; flow rate: 5 mL min$^{-1}$; pressure: 300 p.s.i.

7 mmol L$^{-1}$ citric acid at a flow rate of 5 mL min$^{-1}$. The efficiencies range from 1600 theoretical plates (Ni) to 200 (Cu).

## Group Separation of Inorganic Elements

There are numerous methods, such as X-ray fluorescence (XRF), spectroscopy, AAS, AES and ICP-MS, for the determination of trace inorganics including rare earth elements. Of these methods, ICP-AES is one of the most popular for the determination of metals such as Ta, Zr, Hf, etc., including rare earth elements. However, there may be problems with spectral interference because of the many spectral lines. To ensure sufficient precision and accuracy,

especially for the determination of trace elements, separation of matrix elements, which interfere with the determination, must be undertaken. HSCCC has great potential as a separation technique which preconcentrates trace elements before determination.

**Figure 6** shows an example of the quantitative group separation of the rare earth elements from the constituents of rocks, which interfere with the determination of trace elements by ICP-AES. A 0.5 mol L$^{-1}$ solution of DEHPA in decane was used as the stationary phase. Alkaline, alkaline earth elements and iron(II) were separated from the total amount of rare earth elements at the first stage of the eluent concentration with 0.1 mol L$^{-1}$ hydrochloric acid. Then the total amount of rare earth elements, including yttrium, was eluted from the stationary

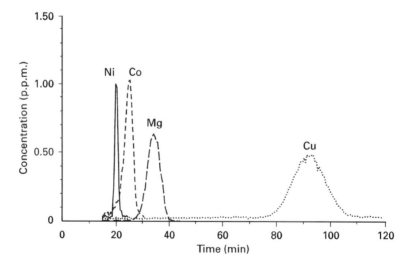

**Figure 5**   Isocratic separation of nickel, cobalt, magnesium and copper by HSCCC. Apparatus: HSCCC centrifuge with 10.0 cm revolution radius; column: one multilayer coil, 150 m × 1.6 mm i.d., 300 mL capacity; stationary phase: 0.2 mol L$^{-1}$ DEHPA in heptane; mobile phase: 7 mmol L$^{-1}$ citric acid; sample 10 μg Ni, each 20 μg Co and Mg, 40 μg Cu; speed: 800 rpm; flow rate: 5.0 mL min$^{-1}$; pressure: 140 p.s.i.; observed wavelength: 341.4 nm (Ni), 345.3 nm (Co), 280.2 nm (Mg), 324.7 nm (Cu).

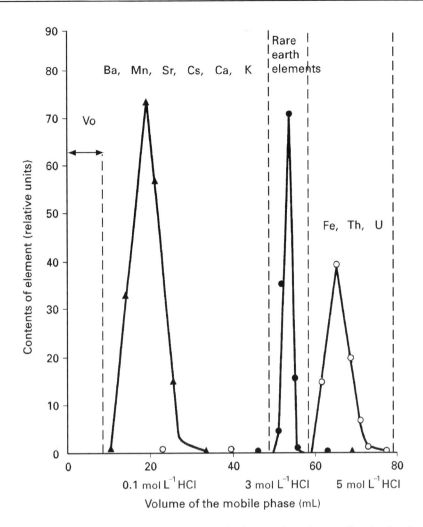

**Figure 6** Example of quantitative group separation of the rare earth elements from the constituents of rocks. Apparatus: HSCCC centrifuge with 14.0 cm revolution radius; column: one monolayer coil, 1.13 m × 1.5 mm i.d., 20 mL capacity; stationary phase: 0.5 mol L$^{-1}$ DEHPA in decane Kr: 30%; mobile phase: 1–0.1 mol L$^{-1}$ HCl, 2–3 mol L$^{-1}$ HCl, 3–5 mol L$^{-1}$ HCl; sample: Mn 0.3, Fe 6.4, Ca 8.2, Na 4.2, K 4.1 (%) and Ba 440, La 1350, Ce 2290, Nd 760, Pr 239, Sm 97, Eu 18.3, Gd 120, Tb 15, Dy 139, Ho 29, Er 88, Tm 12.5, Yb 71, Lu 8.6, Y 716, Cs 2.6, Sr 307, U 682, Th 940 (p.p.m.); speed: 350 rpm; flow rate: 2.0 mL min$^{-1}$. Kr, the retained stationary phase volume relative to the total column volume.

phase with 3 mol L$^{-1}$ hydrochloric acid. To elute other elements, including iron(III), from the stationary phase, 5 mol L$^{-1}$ hydrochloric acid was introduced into the column. Separation of the mixture was performed within 40 min at a pumping rate of the mobile phase of 2 mL min$^{-1}$.

## Enrichment of Inorganic Elements

### Large-scale Enrichment of Inorganics

HSCCC has great potential as an enrichment technique for trace elements before determination. Enrichment prior to determination can overcome problems such as interference, toxic or radioactive samples.

The p.p.b. level of metal ions in a 500 mL of the mobile phase was continuously concentrated into a small volume of the stationary phase retained in the column. Concentrated metal ions were simultaneously eluted with nitric acid and determined by a plasma atomic emission spectrometer. The recoveries of Ca, Cd, Mg, Mn, Pb and Zn were over 88% at the 10 p.p.b. level in 500 mL of the sample solution. The versatility of this method has been further demonstrated by the determination of trace metals in tap water and deionized water.

Rare earth elements have been enriched in a stationary phase of toluene: 2-ethylhexylphosphonic acid mono-2-ethylhexyl ester (EHPA) from a litre of aqueous solution, and eluted with a stepwise pH gradient.

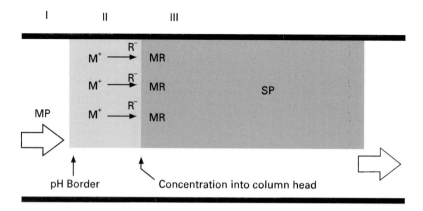

**Figure 7**  Concentration procedure just after revolving the HSCCC column. I, Mobile phase (MP) with HCl; II, sample with NH$_3$; III, stationary phase (SP) with NH$_3$.

Large-scale enrichment is very useful to determine extremely low levels of metals in solution, when a large amount of sample is available.

### Enrichment Using the pH-zone Refining Technique

Even if a large volume of sample is not available, enrichment techniques that concentrate trace metals in microlitre samples are sometimes quite useful because modern instrumental methods do not need a large sample size. Moreover, if trace metals separated from their major matrices can be concentrated in an extremely small area of the polytetrafluoroethylene (PTFE) tube in HSCCC, this would be ideal for flow-injection analysis. From this point of view, the recently developed pH-zone refining technique has great potential for enrichment of trace inorganic elements.

In pH-zone refining, a basic organic solution containing a complex-forming reagent such as DEHPA as

a stationary phase is used. After sample introduction into the column, metal ions stay close to the sharp pH border region in the small-bore PTFE tube. Then the trace inorganic ions in the sample are moved by the acid effluent (dilute HCl or HNO$_3$) to the tail of the column while concentrating in the sharp-moving .pH interface, and finally eluted as small fractions containing concentrated inorganic ions.

This concentration procedure is shown in **Figures 7** and **8**. In Figure 7, sample ion is concentrated into the column head soon after the concentration procedure is started. After the stationary phase and the sample are introduced into the system, HSCCC is started at an appropriate revolution rate, followed by the mobile-phase pump. Metal or inorganic ions (M$^+$) are concentrated into the column head as MR. Zone III shows the stationary phase, which includes DEHPA and ammonia in an organic solvent such as ether or heptane. Zone II shows a sample phase with the pH adjusted by ammonia. When the column revolution is started, as the organic stationary phase is lighter than

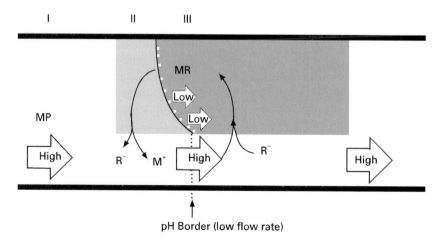

**Figure 8**  Concentration procedure at the pH interface in the HSCCC column. I, Mobile phase (MP) with HCl; II, stationary phase after being neutralized with HCl; III, stationary phase with NH$_3$.

the mobile phase (diluted acid solution), it moves to the head of the column by ASE. The driving force based on ASE increases with the revolution speed of the column. Therefore, the stationary phase can be retained in the column by selecting an appropriate speed of revolution and pump rate, even if the mobile phase is introduced from the head into the column. The retention ratio of the stationary phase to the whole column can be varied from 20 to 70%, but is stable when all conditions including pump or revolution rate are constant. So, the position of the mobile phase is stable in the column when in operation. Inorganic ions $(M^+)$ form complexes with ligand ions $(R^-)$ and are mainly concentrated on the column head, shown as MR in Figure 7. Hydrogen and counter ions of the ligand are not shown in the figure.

Figure 8 shows a concentration procedure in the HSCCC column after all the ions in the sample are extracted on top of the zone III in Figure 7. After Zone II in Figure 7 (sample) has completely passed through the top of the stationary phase, ammonia in the stationary phase begins to be neutralized with hydrochloric acid in the mobile phase. The neutralization area, where reaction between the acid and the base has just finished, is shown as the pH border in Figure 8. The pH border proceeds from left to right. Also the concentrated ions proceed circulating (extracting and back-extracting process in Figure 8) around the pH interface. The rectangular zone (II + III) shows the total stationary phase.

The mobile phase that includes acid solution (hydrochloric acid in Figures 7 and 8) moves at a constant flow rate from the head of the column (left) to the tail of the column (right). The three arrows labelled High show a constant stream of the mobile phase.

As the mobile phase is an acidic solution, it reacts with ammonia in the stationary phase as it proceeds to the tail of the column in Figure 8. The left area (II) in the rectangular zone shows the stationary phase that has been saturated with hydrochloric acid. In the right area (III) in the rectangular zone, ammonia is not yet neutralized with hydrochloric acid. The dark gradation zone between areas II and III shows a pH interface, where the neutralization between acid and base has just finished. As the stationary phase and the mobile phase are mixed well in the HSCCC column, there must be a pH interface in the mobile phase.

The pH interface moves to the right (the same direction as the mobile phase); however, its flow rate is lower than that of the mobile phase because of the delay based on the neutralization. So, concentrated ions shown as MR on the top left of the zone III in

Figure 7 move slowly as shown by the flow rate arrow Low in Figure 8, and circulate around the pH interface by repeating extraction and back-extraction into the stationary phase and the mobile phase, respectively. The flow rate of the pH interface in the mobile phase is shown as the Low arrow in the stationary phase in Figure 8. The flow rate of the pH interface in the mobile phase may be made the same as that in the stationary phase by vigorous mixing in HSCCC column. The pH interface in the mobile phase is shown by the dotted line at the pH border in Figure 8. When the inorganic ions combined with ligand (MR) pass through the pH interface from zone III to zone II in the stationary phase, they are back-extracted into the mobile phase. Then, as the flow rate of the mobile phase is faster than that of the pH interface, the ions $(M^+)$ pass through the pH border (shown by the black dotted line in Figure 8) from left to right in the mobile phase. Just after passing the pH border, the pH is rapidly rising, so, the ions react with the ligand again and are extracted back into the stationary phase. By repeating this cycle, the ions are moved to the tail (right) of the column without diffusion.

In the concentration mechanisms described above, there is no diffusion process observed as in elution procedures with ion exchange or other chromatographic separation methods, such as conventional HSCCC and HPLC. If there is no basic compound such as ammonia in the stationary phase, ions move to the tail with a different flow rate as a function of the distribution ratio between the stationary phase and the mobile phase. Many ions can be separated from each other in this way but because there is no sharp pH interface in the column, concentration is not effected, but only separation with diffusion.

Figure 9 shows the typical concentration results for a 10 p.p.m. solution of cadmium, magnesium and zinc. The injected sample solution contained 50 µg of each in 5 mL of 0.1 mol $L^{-1}$ tartaric acid solution adjusted to pH 8.8. The mobile phase was pumped at a flow rate of 0.05 mL min$^{-1}$. Revolutional speed was 950 rpm. The eluent was collected every 2 min (0.1 mL fractions). The fractions were diluted 1 : 10 with water and the emission intensity for each element was measured by direct current plasma atomic emission spectrometer. The emission intensities for each element were increased 20-fold compared to the original sample solution. The results of this study demonstrate the high performance capabilities of the pH-zone refining technique. Trace elements in the sample solution can be successfully concentrated into a small volume with enormous enrichment.

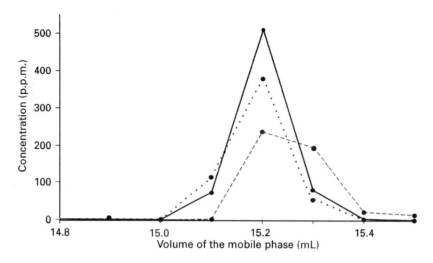

**Figure 9**  Typical concentration results for 10 p.p.m. solution of cadmium (continuous line), zinc (dashed line), magnesium (dotted line). Apparatus: HSCCC centrifuge with 10.0 cm revolution radius; column: one multilayer coil; sample: 5 mL of each 10 p.p.m. solution (pH 9.25) in 0.1 mol L$^{-1}$ tartaric acid; mobile phase: 0.1 mol L$^{-1}$ HCl saturated with ether; stationary phase: 6 mL of 0.2 mol L$^{-1}$ DEHPA and 0.18 mol L$^{-1}$ ammonia in ether; column: 0.5 mm i.d. × 32 m; flow rate: 0.05 mL min$^{-1}$; speed: 950 rpm.

## Conclusions

In contrast to HPLC, the unique feature of HSCCC is that there is no solid support in the column. As the distribution abilities including the capacity of the stationary phase are easy to control, HSCCC can be applied to the separation, enrichment and purification of inorganics over a wide range of concentration. In particular, enrichment of trace elements using pH-zone refining technique will be an ideal preconcentration method for subsequent determination by modern instrumental methods. HSCCC can be combined directly with the flow injection technique, and shows great potential for preconcentration of selected desired trace element prior to detection. On-line enrichment and subsequent analysis may thus take the place of conventional sample preparation using a beaker and separating funnel, in future investigations in this field.

*See also:* **III/Ion Analysis:** Capillary Electrophoresis; Liquid Chromatography; Thin-Layer (Planar) Chromatography.

## Further Reading

Conway WD (1990) *Countercurrent Chromatography: Apparatus, Theory and Applications*. New York: VCH.

Fedotov PS, Maryutina TA, Grebneva ON *et al*. (1997) Use of countercurrent partition chromatography for the preconcentration and separation of inorganic compounds: group extraction of Zr, Hf, Nb, and Ta for their subsequent determination by inductively coupled plasma atomic emission spectrometry. *Journal of Analytical Chemistry* 52: 1034–1038.

Ito Y and Conway WD (1995) *High-Speed Countercurrent Chromatography*. New York: John Wiley.

Ito Y and Ma Y (1996) pH-Zone-refining countercurrent chromatography. *Journal of Chromatography* A 753: 1–36.

Kitazume E, Sato N, Saito Y and Ito Y (1993) Separation of heavy metals by high speed counter current chromatography. *Analytical Chemistry* 65: 2225–2228.

Kitazume E, Sato N and Ito Y (1998) Concentration of heavy metals by high-speed countercurrent chromatography. *Journal of Liquid Chromatography* 21: 251–261.

Mandava NB and Ito Y (1988) *Countercurrent Chromatography: Theory and Practice*. New York: Marcel Dekker.

Nakamura S, Hashimoto H and Akiba K (1997) Enrichment separation of rare earth elements by high-speed countercurrent chromatography in a multilayer column. *Journal of Liquid Chromatography* A 789: 381–387.

# Liquid Chromatography

C. A. Lucy, The University of Alberta, Edmonton, Alberta, Canada

## Introduction

In general, inorganic ions cannot be detected using absorbance spectroscopy or electrochemistry. As a result, determinations of inorganic ions by liquid chromatography (LC) possess many characteristics distinct from determinations of organic compounds by LC. These differences are most noticeable in the detection methodologies used, but also feed back into the column design and operation. This article focuses on the distinct aspects of LC determinations of inorganic ions, and provides an introduction to approved methodologies for inorganic ion analysis. For further reading, the monograph by Haddad and Jackson and the recent special issue of *Journal of Chromatography A* (volume 789) devoted to chromatography of inorganic ions are strongly recommended.

## Instrumentation for Inorganic Ion Analysis

### General

Figure 1 shows a schematic diagram of a high performance liquid chromatograph (HPLC) for inorganic ion analysis. This instrument is commonly referred to as an ion chromatograph. The most distinguishing difference from a typical HPLC is the post-column reactor. Nonetheless, there are a number of other subtle differences. The pump, injector and tubing are essentially standard HPLC components. The only significant difference is that for inorganic ion analysis these components are generally constructed of metal-free materials, such as poly-etheretherketone (PEEK). PEEK is extremely chemically inert, flexible, inexpensive and capable of withstanding high pressures.

### Column

Separations of inorganic ions are performed using ion exchange. However, classical gel type (polystyrene-divinylbenzene) ion exchangers are not used. Such resins possess too high a capacity and the mass transfer is too slow to allow the low eluent strengths and high separation efficiencies required. Rather, columns for inorganic ion analysis are generally of two types. Firstly, and most commonly, the columns are packed with pellicular particles. Pellicular pickings have a solid core with ion exchange sites only on the outer surface of the particle. Alternatively, dynamic ion exchange columns can be prepared from reversed-phase packings ($C_{18}$) by addition of an ion pair reagent to the mobile phase. In either case the ion exchange capacity of the columns is low, typically 20–200 µeq per column.

### Post-column Reactor

As stated above, the most distinguishing feature of an HPLC for inorganic ion analysis is the post-column reactor. Its purpose is to facilitate detection of the inorganic ions. The simplest post-column reaction is the formation of a coloured product which can be measured with a spectrophotometer. For example, transition and lanthanide metal ions (M) can be monitored using post-column reaction with 4-(2-pyridylazo)resorcinol (PAR):

$$M^{2+} + 2\,PAR^- \rightleftharpoons M(PAR)_2 \qquad [1]$$

A reagent stream of excess PAR is added to the effluent from the ion exchange column. The

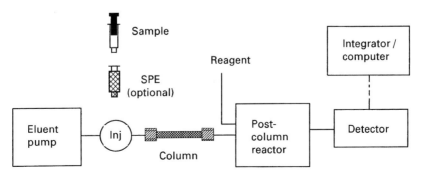

**Figure 1**  Schematic diagram of a high performance liquid chromatograph for analysis of inorganic ions. SPE, solid-phase extraction.

post-column reaction should ideally be instantaneous, as with PAR, allowing use of a simple T-connector as the post-column reactor. Once the colourless metal ions present are complexed by the PAR, the PAR absorbance undergoes a bathochromic shift. That is, in the absence of metal ions, the PAR reagent absorbs little light at ~510 nm, whereas metal–PAR complexes absorb intensely at this wavelength. This absorbance can be directly related to the metal ion concentration.

A more complex post-column reaction is suppression, which is used in conjunction with conductivity detection. All inorganic ions conduct electricity. Thus, conductivity is a universal detector for these species. However, the eluents used in ion exchange chromatography are also conducting. The function of the post-column suppressor is to eliminate (suppress) the background conductivity of the eluent, or at least reduce it to an insignificant level. Typical suppressors are membrane-based, although fibre- and column-based suppressors were used in older systems. The suppression process can be illustrated using the example of the determination of inorganic anions in a sodium carbonate eluent, as is used in Environmental Protection Agency (EPA) Method 300.0 and American Society for Testing and Materials (ASTM) Method D 4327. The membrane in a suppressor for anion chromatography is a cation exchange membrane. Eluent from the column flows on one side of the membrane, while a regenerant flows on the other side. The regenerant is typically $H_2SO_4$. $H^+$ from the regenerant and $Na^+$ from the eluent freely exchange through the cation exchange membrane. Anions cannot pass through the membrane due to charge repulsion. As a result of exchange of $H^+$ for $Na^+$ in the eluent stream, the carbonate eluent is protonated, minimizing the background conductivity.

$$Na^+ + CO_3^{2-} \xrightarrow{+H^+/-Na^+} H_2CO_3 \quad [2]$$

Carbonic acid is however a weak acid and does undergo some dissociation back to $H^+$ and $HCO_3^-$. However, this dissociation only results in a small residual ( ~15 μS) background conductivity. Samples do not possess this background conductivity. Thus, a 'water dip' is observed at the dead volume of the column (0.7 min in **Figure 2**) when the water from the sample elutes from the column. The weak acid character of carbonic acid can also cause some nonlinearity in calibration curves for inorganic anions when suppressed conductivity detection is used.

**Figure 2** Separation of standard inorganic anions using preconcentration as per ASTM method D 5542. Peaks: 1, 5 μg L⁻¹ chloride; 2, 20 μg L⁻¹ nitrate; 3, 20 μg L⁻¹ phosphate; 4, 20 μg L⁻¹ sulfate. Experimental conditions: column, Dionex AS4A; eluent, 0.75 mmol L⁻¹ NaHCO₃/2.2 mmol L⁻¹ Na₂CO₃; flow rate, 2.0 mL min⁻¹; detection, suppressed conductivity; injection volume, 10 mL on to a TAC-1 concentrator column. (Courtesy of ASTM.)

## Special Considerations for Inorganic Ion Analysis

### Water

Water used in the preparation of standards and eluents for routine determinations should be freshly distilled and deionized, and meet the specifications for ASTM Type II water (1 MΩ·cm). However, 1 MΩ·cm water may contain up to 200 p.p.b. of any anion. Therefore, for trace anion analyses, ASTM Type I water (18 MΩ·cm) is recommended. Modern deionization systems using nuclear-grade ion exchange resins to polish out inorganic impurities provide water of this grade which contains down to 20 p.p.t. chloride, 100 p.p.t. sodium and 50–60 p.p.t. ammonium.

## Eluents

All eluents should be filtered through 0.2 μm membranes before use. Many ion chromatographic eluents are excellent growth media for algae. Therefore eluents should be stored at 4°C and not kept for longer than 1 month.

## Sample Handling

Many samples possess acidic, alkaline or saline matrices which can disrupt liquid chromatographic determinations of inorganic ions. In such cases it is essential that the matrix components be removed prior to analysis. Solid-phase extraction (SPE) is a convenient means of eliminating such interferents. A SPE device is generally a small column or filter containing a chromatographic medium which fits on

to a sample syringe. Typical SPE cartridges for inorganic ion analysis contain cation exchange resin in the $H^+$ form to reduce sample pH, in the $Ag^+$ form to selectively precipitate halides and in the $Ba^{2+}$ form to precipitate sulfate; anion exchange resin in the $OH^-$ form to neutralize acidic samples; and hydrophobic materials such as polystyrene to remove hydrophobic matrix components. Passage of the sample through the SPE and directly into the injector provides a quick and convenient means of eliminating deleterious matrix components.

## Methods for Inorganic Ion Analysis

A reliable source of methods for inorganic ion analysis is the approved regulatory methodologies. **Table 1** provides a brief summary of approved methods from the US EPA and the ASTM. Other agencies, such as the National Institute for Occupational Safety and Health (US) and the Occupational Safety and Health Administration (US), also have approved methods for determination of air-borne species such as $SO_2$, $Br_2$, $Cl_2$, $NO_x$ and organoarsenics by conversion into aqueous inorganic ions and determination by ion chromatography.

Another rich source of information are vendor application notes. On-line listings of these application

**Table 1**  Regulatory methods for inorganic ion analysis by liquid chromatography

| Method | Analytes[a] | Specified matrices |
|---|---|---|
| *EPA[b]* | | |
| 218.6 | Cr(VI) | Drinking water, groundwater, industrial wastewater effluents |
| 300.0 A | $F^-$, $Cl^-$, $NO_2^-$, $Br^-$, $NO_3^-$, $HPO_4^{2-}$, $SO_4^{2-}$ | Drinking water, surface water, wastewater, groundwater, reagent water, solids (after extraction), leachate |
| 300.0 B | $ClO_2^-$, $ClO_3^-$, $BrO_3^-$ | Drinking water and reagent water |
| 300.6 | $Cl^-$, $NO_3^-$, $HPO_4^{2-}$, $SO_4^{2-}$ | Wet deposition (rain, snow, sleet, hail) |
| 300.7 | $Na^+$, $NH_4^+$, $K^+$, $Mg^{2+}$, $Ca^{2+}$ | Wet deposition (rain, snow, dew, sleet, hail) |
| A-1000 | $Cl^-$, $SO_4^{2-}$ | Drinking water |
| B-1012 | $NO_2^-$, $NO_3^-$ | Wastewater |
| *ASTM[c]* | | |
| D4327 | $F^-$, $Cl^-$, $NO_2^-$, $Br^-$, $NO_3^-$, $HPO_4^{2-}$, $SO_4^{2-}$ | Drinking water and wastewater |
| D4856 | $H_2SO_4$ | Workplace air samples |
| D5085 | $Cl^-$, $NO_3^-$, $SO_4^{2-}$ | Wet deposition (rain, snow, dew, sleet, hail) |
| D5257 | Cr(VI) | Waste, drinking and surface waters |
| D5542 A | $Cl^-$, $HPO_4^{2-}$, $SO_4^{2-}$ | High purity water |
| D5542 B | $F^-$, acetate, formate | High purity water |
| E1787 | $F^-$, $Cl^-$, $Br^-$, $ClO_3^-$, $NO_3^-$, $HPO_4^{2-}$, $SO_4^{2-}$ | Caustic soda (NaOH) and caustic potash (KOH) |

[a] Analytes listed in order of elution. Only analytes for which the method has been approved are listed. Other ions may be separated and detected using these methods.
[b] EPA refers to the United States Environmental Protection Agency, Environmental Monitoring and Systems Laboratory, Cincinnati, OH 45268 USA. Copies of methods can be obtained from Superintendent of Documents, US Government Printing Office, Washington, DC 20402, USA.
[c] STM refers to the American Society for Testing and Materials, 1916 Race Street, Philadelphia, PA 19103-1187 USA (telephone: (215) 299-5400; facsimile: (215) 977-9679). Methods are published in the *Annual Book of ASTM Standards*.

notes are available from a number of companies, such as Alltech Associates, Dionex Corporation, Metrohm and Waters Corporation.

## Anions by Suppressed Conductivity

Many of the regulatory procedures listed in Table 1 for standard anions use suppressed conductivity ion chromatography and low capacity (20–35 µeq per column) ion exchange columns. Bicarbonate/carbonate eluents are used for the inorganic acid anions (EPA 300.0 and 300.6; ASTM D4327, D5085, D5542 A and E1787). Figure 2 shows the separation of standard anions using ASTM Method D 5542 with a bicarbonate/carbonate eluent on a Dionex AS4A column. Similar elution orders are observed with other ion chromatographic columns. However, separation selectivities do vary with the column capacity, hydrophobicity and structure. Thus, the precise eluent concentrations and selectivities depend upon the specific column used.

Method B of the ASTM D 5542 protocol recommends the use of 5 mmol $L^{-1}$ sodium tetraborate, for separation of fluoride, acetate and formate. These anions are very weakly retained, and so a weak eluent is required to separate them from each other and from the water dip.

Detection limits for these regulatory methods are listed in **Table 2**. Using a 50 µL injection loop, low parts per billion (p.p.b., µg $L^{-1}$) are achieved using suppressed conductivity. This equates to less than

**Table 2**  Performance of regulatory methods for inorganic ion analysis by liquid chromatography with conductivity detection

| Ion | Suppressed conductivity | | Direct conductivity detection limit (p.p.m.) |
| --- | --- | --- | --- |
| | Approved range (p.p.m.) | Detection limit (p.p.m.) | |
| *Direct injection* | | | |
| Fluoride | 0.3–8[a] | 0.01[a] | 0.04[b] |
| Chloride | 0.8–26[a] | 0.02[a] | 0.02[b] |
| Nitrite | 0.4–12[a] | 0.004[a] | 0.04[b] |
| Bromide | 0.6–21[a] | 0.01[a] | 0.06[b] |
| Nitrate | 0.4–14[a] | 0.002[a] | 0.04[b] |
| *o*-phosphate | 0.7–23[a] | 0.003[a] | 0.15[b] |
| Sulfate | 2.9–95[a] | 0.02[a] | 0.10[b] |
| Chlorite[c] | Not stated | 0.01 | |
| Bromate[c] | Not stated | 0.02 | |
| Chlorate[c] | Not stated | 0.003 | |
| Sodium[d] | 0.03–1.00 | 0.03 | |
| Ammonium (as $NH_4^+$)[d] | 0.03–2.00 | 0.03 | |
| Potassium[d] | 0.01–1.00 | 0.01 | |
| Magnesium[d] | 0.02–1.00 | 0.02 | |
| Calcium[d] | 0.02–3.00 | 0.02 | |
| *Preconcentration*[e] | | | |
| Fluoride | 0–0.014 | 0.0007 | |
| Acetate | 0–0.4 | 0.007 | |
| Formate | 0–0.35 | 0.006 | |
| Chloride | 0–0.024 | 0.0008 | |
| Phosphate | 0–0.04 | Not determined | |
| Sulfate | 0–0.6 | 0.002 | |

[a] From ASTM D5085 and EPA 300.0 A Column, Dionex AS4A; eluent, 1.7 mmol $L^{-1}$ $NaHCO_3$/1.8 mmol $L^{-1}$ $Na_2CO_3$; flow rate, 2.0 mL $min^{-1}$; detection, suppressed conductivity; injection volume, 50 µL.
[b] From EPA A-1000. Column, Waters IC-Pak Anion; eluent, 1.5 mmol $L^{-1}$ sodium gluconate/5.8 mmol $L^{-1}$ boric acid/1.3 mmol $L^{-1}$ sodium tetraborate/0.5% glycerine/2% butanol/6% acetonitrile (pH 8.5); flow rate, 1.2 mL $min^{-1}$; detection, direct conductivity; injection volume, 100 µL.
[c] From US EPA method 300.0 B. Column, Dionex AS9A; eluent, 1.7 mmol $L^{-1}$ $NaHCO_3$/1.8 mmol $L^{-1}$ $Na_2CO_3$; flow rate, 1.0 mL $min^{-1}$; detection, suppressed conductivity; injection volume, 50 µL.
[d] From US EPA method 300.7. Column, Dionex CS1; eluent, 0.005 mol $L^{-1}$ HCl for monovalent ions and 0.005 mol $L^{-1}$ HCl/0.004 mol $L^{-1}$ meta phenylenediamine dihydrochloride (MPDA-2HCl); flow rate, 2.3 mL $min^{-1}$; detection, suppressed conductivity; injection volume, 100 µL.
[e] From ASTM D5542. Column, Dionex AS4A; flow rate, 2.0 mL $min^{-1}$; detection, suppressed conductivity; injection volume, 20 mL on to an AG-4A concentrator column. Eluent, 5 mmol $L^{-1}$ sodium tetraborate for $F^-$, $CH_3COO^-$ and $HCOO^-$, 0.75 mmol $L^{-1}$ $NaHCO_3$/2.2 mmol $L^{-1}$ $Na_2CO_3$ for all other ions.

$1~\mu mol~L^{-1}$ concentration in solution, or to less than 1 ng of anion injected. Much lower concentration detection limits can be achieved by preconcentrating the sample onto a small low capacity pre-column. In ASTM Method D5542 (Figure 2), 10 mL of high purity water is concentrated onto a pre-column. Such preconcentration can yield sub-p.p.b. detection limits. At such levels, proper handling of ultra pure water samples is essential to avoid contamination. Appropriate handling procedures are covered in ASTM Method D4453.

## Anions by Direct Conductivity

Suppressed conductivity is the most common means of monitoring inorganic anions. However, approved methods also exist which do not use suppression. Rather these procedures measure the conductivity of the column effluent directly. EPA Method A-1000 has been approved for the determination of chloride and sulfate in drinking water. **Figure 3A** shows the separation achieved using a Waters IC-Pak anion column with a pH 8.5 borate/gluconate eluent.

Detection limits for direct conductivity detection are given in Table 2. In the analysis of drinking waters or other environmental samples with this direct conductivity method, high levels of carbonate, interfering metals (e.g. magnesium and calcium) and excess base may cause baseline disturbances. These interferences can be eliminated using an $H^+$ form SPE system.

## Anions by Absorbance Detection

**Anions by direct absorbance detection** A number of inorganic anions absorb light in the low UV (205–215 nm) range. In such cases it is possible to analyse these anions using a standard high performance liquid chromatograph possessing a UV detector. EPA Method B-1011 describes the analysis of nitrite and nitrate in water using absorbance detection at 214 nm. The molar absorptivity of these ions is quite strong (e.g. $9000~M^{-1}~cm^{-1}$ for nitrate at 210 nm), yielding detection limits of 0.003 and $0.0003~mg~L^{-1}$ for nitrite and nitrate respectively. UV detection provides comparable precision and

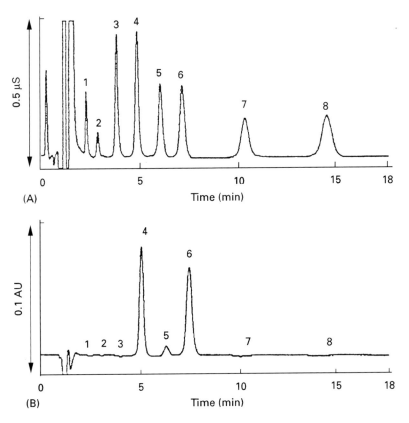

**Figure 3** Separation and detection of standard inorganic anions using: (A) direct conductivity detection; (B) direct absorbance detection at 214 nm. Peaks: 1, $1~\mu g~L^{-1}$ fluoride; 2, unknown concentration of bicarbonate (due to contamination by $CO_2$ in air); 3, $2~\mu g~L^{-1}$ chloride; 4, $4~\mu g~L^{-1}$ nitrite; 5, $4~\mu g~L^{-1}$ bromide; 6, $4~\mu g~L^{-1}$ nitrate; 7, $6~\mu g~L^{-1}$ phosphate; 8, $4~\mu g~L^{-1}$ sulfate. Experimental conditions: column, Waters IC-Pak A HR (75 × 4.6 mm; 6 μm); eluent, borate–gluconate (pH 8.5); flow rate, $1.0~mL~min^{-1}$; detection, direct absorbance at 214 nm followed in series by direct conductivity at 35°C; injection, 100 μL. (Courtesy of Waters.)

accuracy to suppressed conductivity for nitrate. Other anions such as $F^-$, $Cl^-$, $ClO_4^-$, $HPO_4^{2-}$ and $SO_4^{2-}$ do not absorb significantly above 195 nm, and so do not interfere, as can be seen in Figure 3B.

Other ions which absorb in the UV include $Br^-$ (214 nm; 50 p.p.b. detection limit); $BrO_3^-$ (210 nm; 50 p.p.b.); $CrO_4^{2-}$ (1600 $M^{-1}$ $cm^{-1}$ at 365 nm; < 1 p.p.b.); $I^-$ (226 nm; 50 p.p.b.); and $IO_3^-$ (210 nm; 50 p.p.b.). However, there are no approved methods using UV absorbance detection for these ions.

**Anions by indirect absorbance detection** Indirect UV absorbance may also be used to detect inorganic anions. In indirect detection, the eluent is a strongly absorbing anion such as phthalate. Within the eluting peak, the analyte anion displaces an equivalent amount of the absorbing eluent anion. Thus, the absorbance decreases as each analyte elutes from the column. Often the leads on the detector are reversed when performing indirect detection, such that the peaks appear positive. Standard anions can be determined in 15 min with low p.p.b. detection limits using 0.8 mmol $L^{-1}$ phthalate at pH 6.8 as eluent and indirect detection at 265 nm.

**Post-column reaction detection** In EPA Method 218.6 and ASTM Method D5257, hexavalent chro-mium Cr(VI) is separated from Cr(III) and other reactive metal ions by anion exchange chromatogra-phy. Detection is by absorbance at 520 nm after post-column reaction with diphenylcarbohydrazide. This procedure is suitable for monitoring Cr(VI) from 1 to 1000 p.p.b.

Dasgupta has provided an exhaustive review of other methods for post-column reaction detection of anions.

**Cations by suppressed conductivity** EPA Method 300.7 is currently the only approved procedure for determination of inorganic cations using ion chromatography. This procedure was developed to determine monovalent and divalent cations in pre-cipitation using ion chromatography with suppressed conductivity detection. As indicated in Table 2, de-tection limits for this procedure are in the low p.p.b. ($\mu g$ $L^{-1}$) range. Cation exchange with suppressed conductivity is not limited to metal ions. **Figure 4** shows a separation of alkali metals, alkaline earth metals and some volatile amines.

### Cations by Direct Conductivity

Alkali metals and alkaline earth metals can be detected using nonsuppressed conductivity detection. Using a

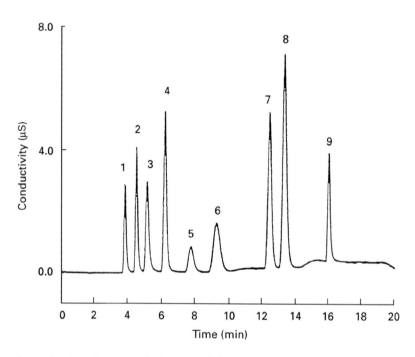

**Figure 4** Separation of group I and II cations and volatile amines. Column: 25 cm × 4 mm, 8–9 μm IonPac CS12A; mobile phase, step gradient of 16 mmol $L^{-1}$ sulfuric acid–4% acetonitrile to 28 mmol $L^{-1}$ sulfuric acid–4% acetonitrile to 50 mmol $L^{-1}$ sulfuric acid–5% acetonitrile; flow rate, 1.0 mL $min^{-1}$; detection, conductivity with suppression from a CSRS-II autosuppressor unit; temperature, 40°C. Sample: 1, lithium (0.5 p.p.m.); 2, sodium (2.0 p.p.m.); 3, ammonium (2.5 p.p.m.); 4, potassium (5.0 p.p.m.); 5, morpholine (10.0 p.p.m.); 6, 2-diethylaminoethanol (10.0 p.p.m.); 7, magnesium (2.5 p.p.m.); 8, calcium (5.0 p.p.m.); 9, cyclohexylamine (15.0 p.p.m.). (Courtesy of Dionex.)

3 mmol $L^{-1}$ $HNO_3$/0.1 mmol $L^{-1}$ ethylenediaminetetra-acetic acid (EDTA) eluent on a Waters IC-Pak cation M/D column, linear response is observed down to 0.05 p.p.m. for $Li^+$, $Na^+$, $NH_4^+$, $K^+$, $Mg^{2+}$ and $Ca^{2+}$ in a 15 min separation. Other alkali metals, alkaline earth metals and amines can also be detected in this fashion.

### Inorganic Cations by Absorbance Detection

**Cations by direct and indirect absorbance detection** Simple inorganic cations generally cannot be effectively monitored by direct UV absorbance due to their low molar absorptivities. However, metal complexes with complexing ligands such as dithiocarbamates provide high molar absorptivites which yield low p.p.b. detection limits. Furthermore, these hydrophobic complexes can be separated using conventional reversed-phase columns. For instance, pre-derivatization of $Ru^{3+}$, $Pt^{2+}$, $Pd^{2+}$ and $Rh^{3+}$ with diethyldithiocarbamate enables their separation on a standard $C_{18}$ reversed-phase column with a methanol–water eluent and sensitive (p.p.b.) detection at 260 nm. This approach is most attractive when a post-column reaction detection system cannot be used and for the analysis of precious metals.

Indirect absorbance detection of alkali metals can be performed using eluents such as 4-methylbenzylamine (262 nm) and $Cu^{2+}$ (252 nm). However, the detection limits are inferior to those obtained with suppressed conductivity.

### Inorganic cations by post-column reaction detection

The most common means of separating the first row transition metals by LC is on reversed-phase ($C_{18}$) columns dynamically coated with $C_8SO_3^-$ or low capacity ion exchangers. Complexing eluents such as tartrate and pyridine-2,6-dicarboxylate (PDCA) are used. Post-column reaction detection with PAR can be used for a wide range of metal ions including $Pb^{2+}$, $Fe^{3+}$, $Cu^{2+}$, $Ni^{2+}$, $Zn^{2+}$, $Co^{2+}$, $Cd^{2+}$, $Mn^{2+}$, $Fe^{2+}$ and $Hg^{2+}$ with detection limits in the low p.p.b. range for typical injection volumes. This is more than 100 times the sensitivity achievable with conductivity. **Figure 5** shows a typical separation of transition metals by HPLC. A number of species ($Fe^{3+}$, $Cr^{3+}$, $Sn^{4+}$) elute at the void volume (1.6 min in Figure 5) and so cannot be quantified. Degradation of the peak shape for copper is observed if trace organic acids are present in the water used to prepare the eluent.

High efficiency lanthanide separations are achieved using reversed-phase $C_{18}$ columns dynamically coated with $C_8SO_3^-$ with gradient elution by $\alpha$-hydroxyisobutyric acid. Elution is in the order of decreasing atomic mass (Lu → La). Post-column reaction detection with Arsenazo III with detection at 658 nm yields 1–5 p.p.b. detection limits using a 20 $\mu$L injection.

Simultaneous separation of transition and lanthanide metals has been achieved based on the charge of the metal–PDCA complexes. Transition metals form monovalent or divalent anionic complexes with PDCA and so can be separated with a PDCA eluent on a Dionex CS5 column. Once all of the transition metals have eluted, the lanthanides, which form strongly retained trivalent anionic complexes with PDCA, are eluted with an oxalate–diglycolate eluent. Detection limits of 20–40 p.p.b. are observed for post-column reaction detection with PAR.

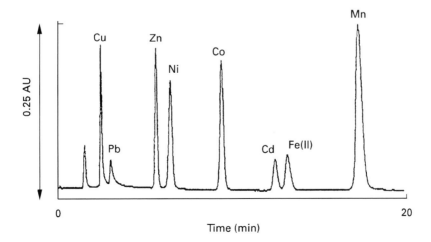

**Figure 5** Separation of transition metals with post-column reaction detection. Experimental conditions: column, Waters Delta Pak $C_{18}$ (10 nm, 5 $\mu$m); eluent, 2 mmol $L^{-1}$ sodium octanesulfonate, 35 mmol $L^{-1}$ sodium tartrate, 5% acetonitrile adjusted to pH 3.65 with NaOH; flow rate, 0.8 mL $min^{-1}$; detection, 500 nm after post-column addition of 0.5 mL $min^{-1}$ of 0.2 mmol $L^{-1}$ 4-(2-pyridylazo) resorcinol (PAR), 1 mol $L^{-1}$ acetic acid, 3 mol $L^{-1}$ ammonium hydroxide. (Courtesy of Waters.)

See also: II/Chromatography: Liquid: Derivatization; Mechanisms: Ion Chromatography.

## Further Reading

Chauret N and Hubert J (1989) Characterization of indirect photometry for the determination of inorganic anions in natural water by ion chromatography. *Journal of Chromatography* 469: 329–338.

Dasgupta PK (1989) Postcolumn techniques: a critical perspective for ion chromatography. *Journal of Chromatographic Science* 27: 422–448.

Haddad PR and Jackson PE (1990) *Ion Chromatography: Principles and Applications.* Amsterdam: Elsevier.

Henderson IK, Saari-Nordhaus R and Anderson JM (1991) Sample preparation for ion chromatography by solid-phase extraction. *Journal of Chromatography* 546: 61–71.

Krol J, Benvvenuti M and Romano J (1997) *Ion Analysis Methods for IC and CIA® and Practical Aspects of Capillary Ion Analysis Theory.* Milford, MA: Waters.

Lucy CA (1996) Practical aspects of ion chromatographic determinations. *LC-GC* 14: 406–415.

Rey MA and Pohl CA (1996) Novel cation-exchange stationary phase for the separation of amines and of six common inorganic cations. *Journal of Chromatography A* 739: 87–97.

Robards K, Starr P and Patsalides E (1991) Metal determination and metal speciation by liquid chromatography: a review. *Analyst (London)* 116: 1247–1273.

Romano JP and Krol J (1992) Regulated methods for ion analysis. *Journal of Chromatography A* 602: 205–211.

# Thin-Layer (Planar) Chromatography

**A. Mohammad**, Aligarh Muslim University, Aligarh, India

## Introduction

Thin-layer chromatography (TLC) is a subdivision of liquid chromatography in which the mobile phase (a liquid) migrates through the stationary phase (thin layer of porous sorbent on a flat inert surface) by capillary action. In 1938, two Russian workers, Izmailov and Schraiber, separated certain medicinal compounds on binder-free horizontal thin layers of alumina spread on a glass plate. Since the development was carried out by placing solvent drops on the glass plate containing sample and sorbent, their method was called drop chromatography. This method remained unnoticed for 10 years until two American chemists, Meinhard and Hall, used a mixture of aluminium oxide (adsorbent) and celite (binder) in a layer on a microscope slide to separate $Fe^{2+}$ from $Zn^{2+}$. They called this technique surface chromatography and this was the first application of planar chromatography to the separation of inorganic ions. The real impetus for advancement of TLC started in 1951 with the work of Kirchner and his associates. Stahl introduced the term thin-layer chromatography in 1958 and standardized procedures, materials and nomenclature. The rapid growth of TLC slowed down during the 1970s with the rise in popularity of high performance liquid chromatography (HPLC) and ion chromatography (IC). However, recent improvements in TLC have removed many of its limitations.

High performance TLC (HPTLC) layers, being thinner and made of more uniform particle size sorbents, provide faster separations, reduced zone diffusion, lower detection limits, less solvent consumption and better separation efficiency. The distinct advantages of TLC over HPLC have been identified as low solvent consumption, low operational cost, easier sample preparation, more rapid throughput, greater detection possibilities and the use of disposable plates. TLC permits the simultaneous analysis of many samples in the same time required for one HPLC analysis and samples and standards are analysed by TLC under exactly the same conditions rather than serially, as in HPLC. Typically, 18–36 samples can be run on a single HPTLC plate with a development time of 3–20 min over a migration distance of 2–7 cm. However, the influence of environmental conditions on the reproducibility of $R_F$ values and poor separation efficiency have been major disadvantages of TLC compared with HPLC and gas chromatography (GC).

TLC can be used for *qualitative* analysis, to identify the presence or absence of a particular substance in a mixture; *quantitative* analysis, to determine precisely and accurately the amount of a particular substance in a sample mixture; and *preparative* analysis, to purify and isolate a particular substance for subsequent use. All three applications require the common procedures of sample application, chromatographic separation and component visualization. However, analytical TLC differs from preparative TLC in that volumes and/or weights of samples are applied to thicker layers in the latter case.

## Inorganic TLC

Though the first published reports on TLC described the separation of inorganic species, the importance of inorganic TLC did not receive recognition until the beginning of the 1960s when Seiler separated inorganic substances. After the work of Seiler, TLC of metal ions received a great impetus. Some of the major fields in which inorganic TLC has found applications include the analysis of biological, food, geological, industrial, pharmaceutical, soil, water and industrial waste water samples. The purpose of this article is to present briefly the current state-of-the-art procedures of TLC/HPTLC as applied to the analysis of inorganic ions.

## Procedure

TLC is an offline process in which various steps are carried out independently (**Figure 1**). The basic TLC procedure involves the spotting of sample mixture (5–10 μL for conventional TLC and 1–2 μL for HPTLC) at about 1.5–2 cm above the lower edge of the layer, drying the spot completely at room temperature or at an elevated temperature, development of the plate, usually by one-dimensional ascending technique in a closed chamber (cylindrical or rectangular) to a distance of 8–10 cm, removing the plate from the developing chamber, removal of mobile phase from the layer by drying, detection of spots on the plate using a suitable detection reagent/procedure, measurement of $R_F$ values of resolved spots and determination of the separated analyte. The differential migration of components in a mixture is due to varying degrees of affinity of the components for the stationary and mobile phases.

## Sample Preparation

The sample solution to be analysed must be sufficiently concentrated to provide clear detection and/or be pure enough so that it can be separated as a discrete and compact zone. For low concentrations of analyte in a complex sample, preconcentration and clean-up procedures must frequently precede TLC.

Metal solutions are generally prepared by dissolving their corresponding salts in distilled water or 0.1 mol L$^{-1}$ HCl (or HNO$_3$) to a final metal concentration of 0.1–0.2 mol L$^{-1}$. Rare earth oxide solutions are prepared by fusion followed by dissolution in 0.5–6 mol L$^{-1}$ HNO$_3$ and anion solutions in distilled water, dilute acid or alkali are prepared from sodium, potassium and ammonium salts of the corresponding acids. Metal complexes are used as freshly

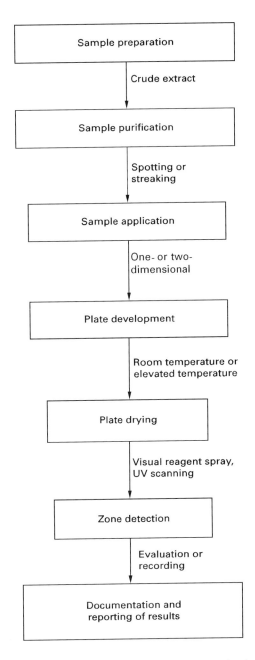

**Figure 1** Schematic diagram showing the steps involved in a TLC process.

prepared solutions in ethanol, acetone, chloroform or distilled water. Specific standard methods are followed for sample preparation to identify and determine metal ions in biological, environmental, alloys, plants, foods, textile and geological samples.

## TLC Plate Preparation

The current trend is to use commercially available pre-coated plates. The manual preparation of layers involves the coating of a slurry of the adsorbent (silica gel, alumina, etc.) on to glass, aluminium or plastic sheets ($20 \times 20$ or $20 \times 10$ cm) with the help of an applicator. The thickness of the dried layer of analytical purposes is kept to 0.1–0.3 mm. A binder (starch, gypsum, dextrin, polyvinyl alcohol) is usually added to the adsorbent to provide better adhesion, mechanical stability and durability. The addition of a fluorescent indicator compound is optional. Compared to conventional TLC, HPTLC layers are produced from sorbents of smaller and more uniform particle size (5–10 μm instead of 10–20 μm).

## Sample Application

Definite volumes of samples are applied as spots or streaks using a micropipette, microsyringe or glass capillaries. A number of automatic spotters of various designs are available for sample application. The nano-applicator (Nanomat) is an example of a micrometer-controlled syringe which has a dynamic volume range of 50–230 nL. Another applicator (Linomat) allows sample application in narrow bands by a spray-on technique. The application of a sample as a streak or band provides more efficient separations because the efficiency of the separation on a TLC plate depends on the diameter of the spot along the direction of development. Thus, the best efficiency is achieved with the smallest diameter spot. Alternatively, the use of TLC plates with concentration zones converts the sample from the original spot into a band or streak. These plates consist of a bottom layer (2–2.5 cm) of a chromatographically inactive adsorbent (e.g. Kieselguhr) followed by the layer of an active adsorbent (silica or bonded silica). The advantages of sample application as a streak and the use of TLC plates with concentration zones are illustrated in **Figure 2**. The general aspects

of a sample application have been reviewed by Jaenchen.

## Development Techniques

The migration of the mobile phases through the stationary phase (or sorbent layer) to effect separation of samples is called development. One-dimensional ascending development is the most commonly used mode in inorganic TLC. Other development techniques, such as multiple, stepwise, circular and two-dimensional development, have also been used to a limited extent. The migration distance for the mobile phase is kept to 10–12 cm for conventional TLC and 2–6 cm for HPTLC. Dzido has described the variants of the development techniques.

## Chromatographic Systems

A combination of stationary and mobile phases constitutes a chromatographic system. The proper selection of stationary and mobile-phase conditions determines the effectiveness of a separation.

### Stationary Phase (Layer Sorbent)

Many materials have been used as stationary phase in inorganic TLC but silica gel, an amorphous and porous adsorbent, has been the most favoured material, followed by alumina and cellulose. Thin layers of silica gel G (gypsum binder) and S (starch binder) with or without fluorescent indicator are frequently used. Silica gel is slightly acidic in nature and the silanol groups (Si–OH) can interact with solute molecules. On the other hand, alumina (aluminium oxide) is basic and more reactive than silica gel. Adsorption is the separation mechanism with both alumina

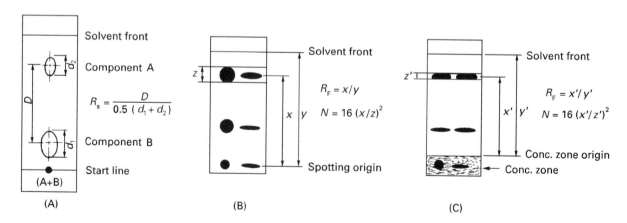

**Figure 2** (A) Resolution of a two-component mixture on a TLC plate; (B) improved efficiency due to streaking; and (C) utility of concentration zone as a promoter of efficiency.

and silica gel. Cellulose, an organic material, is used as a sorbent to perform separations with increased sensitivity of detection and decreased development time compared to paper chromatography. The various layer materials may be broadly classified as:

1. nonsurface-modified or untreated sorbents;
2. impregnated or treated sorbents (organic and inorganic impregnants);
3. bonded or chemically modified sorbents (hydrophobic modified or reversed-phase and hydrophilic modified);
4. inorganic ion exchangers;
5. mixed sorbents.

More detailed information on pre-coated layers and sorbents that are commonly used in TLC is available elsewhere (see Further Reading).

The selective aplication of these different types of layer materials in inorganic TLC is summarized below.

**Anions** Silica gel, silica gel silica impregnated with fluorescein or inorganic salts, silica gel + antimonic acid/alumina/hydrous antimony(V) oxide/Zr(IV) molybdate, Sephadex, microcrystalline cellulose, surface-modified cellulose, cellulose + Kieselguhr/alumina, Kieselguhr, alumina, kaolin, hydrated stannic oxide and polyamide have all been used for the separation of anions.

**Metal ions and rare earth elements** The various layer materials used for the separation of metal ions, metal complexes and rare earth elements (REE) may, broadly, be classified as follows.

*Nonsurface modified or untreated sorbents* These include silica gel, alumina, cellulose, polyamide, polyacrylonitrile, Sephadex and Kieselguhr. Layers of chitin and its deacetylated derivative chitosan have been used to separate metal ions.

*Impregnated or treated sorbents* In general, silica gel impregnated with aqueous salt solutions, high molecular weight amines, organophosphorous compounds and organic chelating agents has been widely used for the separation of metal ions and REE. Metal complexes have been separated on silica gel impregnated with chlorobenzene, *p*-toluidine or surfactants and on layers of egg shell powder impregnated with Triton X-100.

*Bonded or chemically modified sorbents* Lipophilic $C_{18}$-bonded silica gel for REE and metal complexes, aminopropyl silica gel ($NH_2$) and octadecyl silica gel

($C_{18}$) for lanthanide complexes of tetraphenyl porphine and surface-modified cellulose as well as cellulose derivatives for several metal ions have been used.

*Mixed sorbents* Combinations of silica gel and microcrystalline cellulose for noble metal ions and transition metal chlorosulfates, silica gel and microcrystalline cellulose containing ammonium nitrate for REE and silica gel and inorganic ion exchange gel mixtures for transition metal ions have been used.

*Other sorbents* Synthetic inorganic ion exchangers, porous glass sheets, soil and soil–flyash mixture, polychrome A, carbamide–formaldehyde copolymer and immobilized analogue of dibenzo-18-crown-6 on silica support for metal ions and diatomite for REE have also been used, but to a lesser extent.

The adsorbents in different forms (untreated, impregnated, chemically modified or mixed) preferred in the analysis of inorganic ions since 1973 are shown in **Table 1**.

**Anions** Silica gel > cellulose > alumina > inorganic ion exchangers > Kieselguhr > polyamide.

**Cations/metal ions** Silica gel > cellulose > inorganic ion exchangers > alumina > chitin and chitosan > diatomite > soil.

**Metal complexes** Silica gel > alumina > cellulose > polyacrylonitrile > Sephadex = polyamide.

The salting-out efficiency of mixed aminocarboxylato Co(III) complexes obtained on different adsorbents shows the sorbent precipitation in the order (with small discrepancies) polyacrylonitrile > cellulose > silica gel. Polyacrylonitrile is the most suitable sorbent for the separation of complexes. The HPTLC results of rare earth–tetraphenylporphine complexes show an increase in $R_F$ values in the order of the atomic number on $NH_2$-bonded silica plates and a reverse trend, i.e. decrease in $R_F$ values, with an

**Table 1** Number of publications (%) on some sorbent layers appearing from 1973 to 1996

| Layer material | Per cent publications | | |
|---|---|---|---|
| | Metal ions | Metal complexes | Anions |
| Silica gel | 49.6 | 75.6 | 50.5 |
| Alumina | 6.4 | 10.7 | 9.3 |
| Cellulose | 29.3 | 4.8 | 29.3 |
| Inorganic ion exchanger | 10.0 | | 4.0 |
| Chitin and chitosan | 2.0 | | |
| Polyacrylonitrile | | 4.8 | |
| Polyamide | | ~1.0 | 1.3 |

increase in atomic number is obtained on $C_{18}$-bonded silica layers.

### Mobile Phase (Solvent System)

With a particular sorbent layer, the separation possibility of a complex mixture is greatly improved by the proper selection of the mobile phase. Mixtures of organic solvents containing an aqueous acid, base or a buffer are, in general, well suited for the separation of ionic species whereas anhydrous organic solvents are more useful for separating nonionic species. The following mobile phases have been used for inorganic TLC.

**Inorganic solvents**　This includes solutions of mineral acids, bases, salts and mixtures of acids, bases and/or their salts.

**Organic solvents**　This includes acids, bases, hydrocarbons, alcohols, amines, ketones, aldehydes, esters, phosphates and their mixtures in different proportions.

**Mixed aqueous–organic solvents**　This includes organic solvents mixed with water, mineral acids, inorganic bases or dimethyl sulfoxide and buffered salt solutions.

**Complexing solvents**　Solutions of surfactants (sodium dodecyl sulphate (SDS), cetyltrimethylammonium bromide (CTAB), Triton X-100) and ethylenediaminetetraacetic acid.

TLC with mobile phases of lower volatility gives better reproducibility compared to volatile mobile phases, although the latter have the advantage of quick evaporation from the sorbent layer after development.

## Visualization

For the visualization of separated zones, physical, chemical or biological detection methods are commonly used. The physical detection methods are based on substance-specific properties and the most commonly employed methods of this group include the absorption or emission spectrophotometry, autoradiography and X-ray fluorescence. The chemical methods of detection involve spraying reagents capable of forming coloured compounds with the separated species on the plate or exposing the plate to vapour. Alternatively, the reagent can be incorporated in the mobile phase or in the adsorbent. In some cases, detection is completed by inspecting the TLC plate under ultraviolet light, after spraying with

a suitable reagent, or by exposing the plate to ammonia vapour. Bioautographic analysis, reprint methods and enzymatic tests can also be applied for detection purposes. Immunostaining and flame ionization detection methods have also been reported.

The detection methods and reagents used in inorganic TLC are summarized below.

### Metal Ions

In addition to using conventional detection reagents such as dithizone, dimethylglyoxime, potassium ferrocyanide and 8-hydroxyquinoline, for cations, new reagents such as sulfochlorophenol azorhodamine, phenolazotriaminorhodamine and benzolazobenzolazorhodamine have been proposed for selective detection of toxic heavy metals at nanogram levels. Radiometry is used to detect Pr(III), Pr(IV) and Tb(III). An elegant fluorescent method for the detection of Mg, Al, Ca, V, Cu, Zn, Ge, Y, Zr, Mo, Ag, Cd, In, La, Ce, Eu, Tb, Tl, Pb, and Bi at $3–3 \times 10^{-6}$ μmol levels has been reported whereby these cations are detected as coloured fluorescent zones generated simply on heating the chromatograms on porous glass sheets at 100–700°C for 15 min. Typical representative fluorescence spectra are shown in **Figure 3**.

### Anions

For the detection of anions, saturated silver nitrate solution in methanol, 0.2–0.5% diphenylamine solution in $4$ mol $L^{-1}$ $H_2SO_4$, 1% aqueous solution of potassium ferrocyanide, 0.5% alcoholic solution of pyrogallol, 10% $FeCl_3$ solution in $2$ mol $L^{-1}$ HCl, 1% KI in 1.0 mol $L^{-1}$ HCl and mixture of aqueous KSCN and $SnCl_2$ in 1.0 mol $L^{-1}$ HCl, ammoniacal $AgNO_3$, aqueous bromocresol green, $FeSO_4 + FeCl_3$, alizarin, alizarin-zirconium lake, benzidine solution, $(NH_4)_2 MoO_4 + SnCl_2$ and 0.1% bromocresol purple containing dil. $NH_4OH$ have been used. Autoradiography, scintillation counting and radiometric detection methods have also been applied. Several anions are detected on the basis of quenching effects: dark spots of the anions appeared on the bright greenish fluorescent background when the chromatograms are sprayed with aluminium(III)–morin fluorescent complex (prepared by dissolving 5 mg $AlCl_3$ and 5 mg of 2′,3,4′,5,7-pentahydroxyflavone (morin) in a mixture of 10 mL 30% $CH_3COOH$, 20 mL 98% ethanol and 20 mL water). The detected anions are $IO_3^-$, $IO_4^-$, $CrO_4^{2-}$, $PO_4^{3-}$, $Cr_2O_7^{2-}$, $NO_2^-$, $NO_3^-$, $SO_4^{2-}$, $SO_3^{2-}$, $Fe(CN)_6^{3-}$, $Fe(CN)_6^{4-}$, and $BrO_3^-$ (strong violet spots), $F^-$, $Cl^-$, $Br^-$, $I^-$, $S^{2-}$, $S_2O_3^{2-}$, $VO_3^-$, $VO_4^{3-}$, and $MoO_4^{2-}$ (medium blue spots) and $ClO_2^-$, $ClO_3^-$, $ClO_4^-$, $SCN^-$ and $WO_4^{2-}$ (weak yellow spots). Several anions producing intense blue colour with

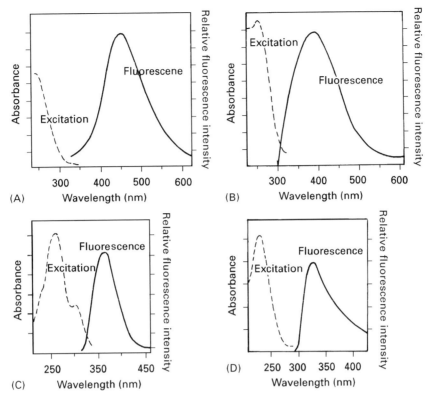

**Figure 3**    Fluorescence spectra of (A) copper, (B) lead, (C) cerium and (D) thallium ions heated on porous glass sheet. (Reproduced with permission from Yoshioka M *et al.* (1992) Fluorescence reactions of inorganic cations heated on a porous glass sheet for thin-layer chromatography. *Journal of Chromatography* 603: 223–229.)

0.2% $Ph_2NH$ in $H_2SO_4$ are detected down to 0.1 µg on silufol layers. Halides have been detected with alizarin-zirconium lake and $AgNO_3$.

### Rare Earth Elements

The REEs have been detected by first, spraying the plate with 0.1% arsenazo(III) solution and then with aqueous ammonia followed by gentle heating; second, heating at 70°C for 10 min after spraying with 0.02% chlorophosphonazo solution; and third, exposure of the plate to $NH_3$ after spraying with tribromochlorophosphonazo or xylenol orange solution. Saturated ethanolic solution of alizarin and dilute solutions (0.2–1%) of tribromoarsenazo have been used to detect REE.

### Metal Complexes

Most of the complexes being coloured are visible without further treatment, e.g. $Fe(phen)_3^{2+}$; $Fe(bpy)_3^{2+}$; metal glyoxaldithiosemicarbazone; thiocarbonate complexes of Cu, Ni and Co; anil complexes of Cu, Mn, Fe and Zn; metal chelates of V, Co, Cr, Ni and Mn with 2,2'-dihydroxy-5,5'-dimethylazobenzene; trifluoroacetylacetonates of rate

earths; metal xanthates; metal chelates of 1-hydroxyphenazine; metal diethyldithiocarbamate complexes; metal dithizonates; transition metal monothio-$\beta$-diketonates; dithioacetylacetone; metal acetothioacetanilide. Metal oxinates, some geometrical isomeric complexes of Rh, Pt and Co, methylbenzyldithiocarbonate metal chelates, Cu(II) carboxylates and Co (gly) are self-coloured but located under ultraviolet light. Trisethylenediamine Co(III) complexes are detected with sodium sulfide solution. $\beta$-Diketonates of Fe, Cr and Co, organotin compounds, alkali metal xanthates and piperidine dithiocarbamate complexes can be detected with iodine vapour. A fluorometric method has been used for the detection of heavy metal complexes with pyrene-substituted N-acylthiourea. Sometimes spots are detected by spraying coloured reagents such as pyrocatechol violet and copper sulfate as well as by immersing the TLC plates in a dilute solution of phenylfluorone reagent. N,N-diethyl-N'-benzoylthiourea-metal chelates have been detected by graphite furnace atomic absorption spectrometry and by UV detection. These techniques permit the sensitive detection of enriched platinum metals with detection limits in the nano- and picogram range (**Table 2**).

**Table 2**   Absolute detection limits for N,N-di ethyl-N'-benzoyl-thiourea chelates in chromatography

| Element | HPTLC (ng) | HPLC (ng) | $\lambda$ (nm) |
|---|---|---|---|
| Ru | 0.22 | 10.018 | 275 |
| Rh | 0.21 | 10.006 | 261 |
| Pd | 0.08 | 10.008 | 274 |
| Os | 0.46 | 10.021 | 244 |
| Ir | 0.22 | 10.042 | 252 |
| Pt | 0.25 | 10.010 | 249 |

Data from Sehuster M (1992) Selective complexing agents for the trace enrichment of platinum metals. *Fresenius Journal of Analytical Chemistry* 342: 791–794.

## Qualitative Analysis

### Identification

In TLC the identification of separated compounds is primarily based on their mobility in a suitable solvent which is described by the $R_F$ value of each compound, where:

$$R_F = \frac{\text{distance of spot migration from the origin}}{\text{distance of solvent front from origin}}$$

The factors which influence the magnitude of the $R_F$ include the nature of the sorbent, layer thickness, activation temperature, chamber saturation, nature of the mobile phase, pH of the medium, room temperature, sample volume, relative humidity and mode of development. Another term, $R_M$, which is the logarithmic function of the $R_F$ value, i.e. $R_M = \log(1/R_F - 1)$, is more useful as it bears a linear relationship to some TLC parameters or structural elements of the analyte. However, in cases of continuous and multiple development, where the solvent front is not measured, the term $R_X$:

$$R_X = \frac{\text{distance travelled by solute}}{\text{distance travelled by standard}}$$

is used.

If the retention data ($R_F$, $R_M$, or $R_X$ values) of the compound to be identified are identical with those of the reference substance in three different solvent systems but on the same stationary phase or with the same solvent but on three different types of stationary phases, the two compounds can be regarded as identical with a good probability. However, for correct identification the chromatographic retention data are not enough and at least one spectroscopic method is necessary.

**Table 3**   $hR_F$ ($R_F \times 100$) values, standard deviation (SD) of $R_F$ values of metal ions present in industrial wastewater and dilution limits of metal ions in standard spiked water

| Metal ion | $hR_F$ value | SD of $R_F$ value | Detection limit ($\mu$g) |
|---|---|---|---|
| $Pb^{2+}$ | 00 | $34 \times 10^{-4}$ | 7.78 |
| $Cd^{2+}$ | 62 | $44 \times 10^{-4}$ | 6.00 |
| $Zn^{2+}$ | 70 | $12.2 \times 10^{-3}$ | 0.46 |
| $Cu^{2+}$ | 97 | $26 \times 10^{-4}$ | 5.23 |
| $Co^{2+}$ | 97 | $12.4 \times 10^{-3}$ | 7.40 |
| $Ni^{2+}$ | 97 | $58 \times 10^{-4}$ | 3.22 |

Stationary phase: silica gel G; mobile phase: 1.0 mol L$^{-1}$ sodium formate + 1.0 mol L$^{-1}$ KI (1 + 9). (Reproduced with permission from Mohammad A (1995) Identification, quantitative separation and recovery of copper from spiked water and industrial wastewater by TLC-atomic absorption and TLC-titrimetry. *Journal of Planar Chromatography – Modern TLC* 8: 463–466.)

### Separation

The separated components of a mixture are detected and their $R_F$ values recorded from the values of $R_L$ ($R_F$ of leading front) and $R_T$ ($R_F$ of trailing front). Some of the basic requirements for a good separation are (a) each spot should be compact ($R_L - R_T \leq 0.3$), (b) the difference in $R_F$ values of two adjacent spots should be at least 0.1, (c) no complexation should occur between/among separable species and (d) chromatography of reference compounds and the

**Table 4**   Separation of thorium from uranium in presence of common anions

| Anions | $\Delta R_F$ | $K_{Th}$ | $\alpha$ | $R_S$ |
|---|---|---|---|---|
| $I^-$ | 0.50 | 1.22 | 24.40 | 2.56 |
| $IO_3^-$ | 0.41 | 0.96 | 10.60 | 2.34 |
| $Br^-$ | 0.51 | 1.27 | 25.40 | 2.62 |
| $BrO_3^-$ | 0.40 | 1.00 | 9.09 | 2.22 |
| $NO_3^-$ | 0.40 | 1.00 | 9.09 | 2.00 |
| $Cl^-$ | 0.48 | 1.38 | 27.60 | 2.74 |
| $SCN^-$ | 0.25 | 0.54 | 4.91 | 1.25 |
| $S^{2-}$ | 0.30 | 0.66 | 6.00 | 1.20 |

$\Delta R_F$ = Difference in the $R_F$ values of $UO_2^{2+}$ and $Th^{4+}$. $K_{Th}$ = Capacity factor of $Th^{4+}$ [$K_{Th} = (1 - R_F)/R_F$ for thorium]. $\alpha$ = Separation factor ($\alpha = K_{Th}/K_{UO_2}$). $R_S$ = Resolution for the separation of $Th^{4+}$ from $UO_2^{2+}$. [$R_S = D/0.5\,(d_1 + d_2)$]: $D$ = distance between the centres of separated spots of $Th^{4+}$ and $UO_2^{2+}$ whereas $d_1$ and $d_2$ are their respective diameters. Stationary phase: silica gel; mobile phase: dimethylamine–acetone–formic acid (2 + 6 + 2, v/v). Reproduced with permission from Mohammad A and Fatima N (1988) A new solvent system for the separation of $Th^{4+}$, $UO_2^{2+}$ and $Zr^{4+}$ in the presence of common anions by thin layer chromatography. *Chromatographia* 25: 536–538.

**Table 5** Separation of Zn–Cd–Hg mixture in spiked environmental samples

| Metal ion | $R_F$ values of separated ions | | | |
|---|---|---|---|---|
| | Seawater | Industrial wastewater | River water | Soil |
| $Cd^{2+}$ | 0.79 | 0.81 | 0.81 | 0.80 |
| $Hg^{2+}$ | 0.96 | 0.90 | 0.96 | 0.98 |
| $Zn^{2+}$ | 0.10 | 0.11 | 0.11 | 0.12 |

Stationary phase: silica gel G impregnated with 0.1% thorium nitrate; mobile phase: 1.0 mol L$^{-1}$ aqueous solution of sodium formate (pH 7.65). The presence of pesticides (malathion, carbaryl, carbofuran, bavistin and 2,4-dichlorophenoxy acetic acid) and anions ($Cl^-$, $Br^-$, $I^-$, $SCN^-$, $MoO_4^{2-}$ and $CrO_4^{2-}$) did not hamper the separation of $Zn^{2+}$ from $Hg^{2+}$ and $Cd^{2+}$. Reproduced with permission from Mohammad A and Majid Khan MA (1992) *Proceedings of National Conference on Clean Environment Strategies, Planning and Management*. Lucknow, India, pp. 191–193. Lucknow: Legend India Environment Protection Pvt. Ltd.

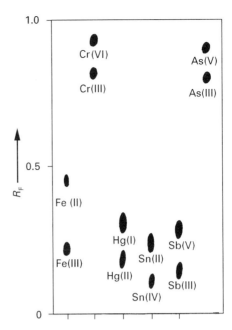

**Figure 5** Chromatogram of the separation of some ions of different valency states on aminoplast layers developed with ethanol-2-propanol-5 mol L$^{-1}$ HCl (2 : 1 : 2). (Reproduced with permission from Perisic-Janjic NU, Petrovic SM and Podunavac S (1991) Thin-layer chromatography of metal ions on a new carbamide–formaldehyde polymer. *Chromatographia* 31: 281–284.)

mixture should be performed under identical experimental conditions. Selected examples of TLC/HPTLC separation of inorganic ions are given in **Tables 3–5** and **Figures 4–10**. Bonded silica C$_{18}$ reversed-phase layers in combination with a methanol–lactate medium as mobile phase have been found to be most suitable for the separation of adjacent rare earths of middle atomic weight group (**Figure 4**). Carbamide–formaldehyde polymer (aminoplast) with acidic eluents is useful for the separation of metal ions of different valency states (Figure 5). Figures 6 and 7 show the separation of

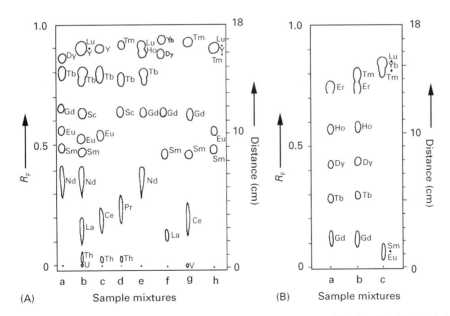

**Figure 4** TLC separation of rare earth element on C$_{18}$-bonded silica with eluent systems (A) 1.0 mol L$^{-1}$ lactate in 50% methanol (pH adjusted to 6.35 before mixing with methanol) and (B) 0.5 mol L$^{-1}$ lactate in 50% methanol (pH adjusted to 6.35 before mixing with methanol). (Reproduced with permission from Kuroda R, Adachi M and Oguma K (1998) Reversed-phase thin-layer chromatography of rare earth elements. *Chromatographia* 25: 989–992.)

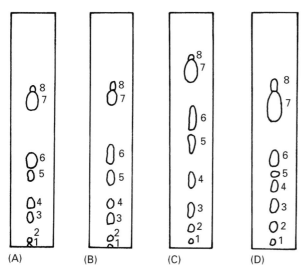

**Figure 6** Separation of a mixture of diantipyrilmethane (DAM) salts. Mobile phase: acetone–chloroform (3 : 1 v/v). Reagents: (A) DAM; (B) MDAM (C) HDAM; (D) PDAM. Anions: 1, $SO_4^{2-}$; 2, $Cl^-$; 3, $Br^-$; 4, $NO_3^-$; 5, $SCN^-$; 6, $I^-$; 7, $ClO_4^-$; 8, reagent. (Reproduced with permission from Shadrin O, Zhivopistsev V and Timerbaev A (1993) Thin-layer chromatographic determination of inorganic anions as counter-ions of metal diantipyrilmethane cationic complexes and diantipyrilmethane cations. *Chromatographia* 35: 667–670.)

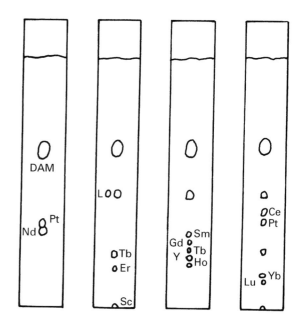

**Figure 8** Separation of the mixture of diantipyril methanates of rare earth elements. TLC plate: silufol; mobile phase: *n*-propanol–0.7 mol $L^{-1}$ HCl (9 : 2). (Reproduced with permission from Timerbaev A, Shadrin O and Zhivopistsev V (1990) Diantipyrilmethane as complex-forming reagents in the thin-layer chromatographic determination of metals. *Chromatographia* 30: 436–441.)

common anions in the form of metal diantipyrilmethane (DAM) complexes and salts of protonated DAM. The separation takes place on silica gel plates with elution by organic solvent–mineral acid or bi-

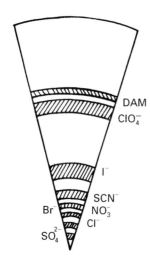

**Figure 7** Sector of the radial thin-layer chromatogram of the mixture of diantipyrilmethane (DAM) salts. (Reproduced with permission from Shadrin O, Zhivopistsev V and Timerbaev A (1993) Thin-layer chromatographic determination of inorganic anions as counter-ions of metal diantipyrilmethane cationic complexes and diantipyrilmethane cations. *Chromatographia* 35: 667–670.)

nary organic solvent mixtures using ascending or radial development. Being highly coloured (iron complexes) or fluorescent under ultraviolet light (terbium complexes), chromatographic zones can easily be detected. Mixtures of REEs as diantipyrilmethanates are well resolved on silufol plates (Figure 8). A four-component mixture consisting of 4-methyl-2-pentanone, tetrahydrofuran, nitric acid and mono-2-ethylhexyl ester of 2-ethylhexylphosphoric acid (P 507) has been used for HPTLC resolution of 10 rare earths (Figure 9). HPTLC allows fast and effective separation of platinum group metals with N,N-diethyl-N'-4-(1-pyrene)butyrylthiourea (DE Py BuT) on silica gel layers using toluene as the mobile phase (Figure 10).

## Quantitative Analysis

Methods for the quantitative evaluation of thin-layer chromatograms may be divided into two main categories: quantitation after elution from the layer and *in situ* quantitation on the layer. In the first, quantitation is performed after scraping off the separated analyte zone, collecting the sorbent and recovery of the substance by elution from the sorbent. Thereafter, the eluates are analysed by applying any suitable method of analysis, such as GC,

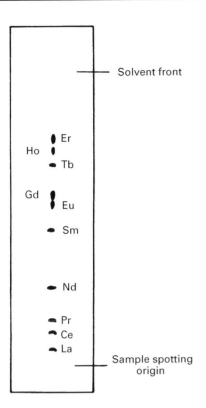

**Figure 9** Chromatogram of the 10 rare earths using optimum mobile-phase conditions: 4-methyl-2-pentanone-THF-HNO$_3$-P 507 (3 : 1.5 : 0.46 : 0.46). (Reproduced with permission from Wang QS and Fan DP (1991) Optimization of separation of rare earths in high-performance thin-layer chromatography. *Journal of Chromatography* 587: 359–363.)

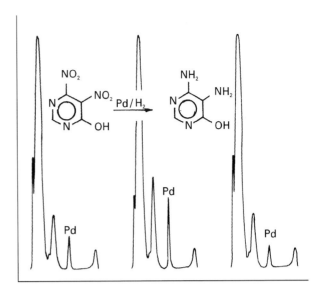

**Figure 11** HPTLC determination of Pd in a synthesis solution. Ligand: *N,N*-di-*n*-hexyl-*N'*-benzoylthiourea. Stationary phase: silica gel 60, mobile phase: chloroform, relative humidity: 20%, separation distance: 3 cm detection: reflectance at 280 nm. (Reproduced with permission from Schuster M (1992) Selective complexing agents for the trace enrichment of platinum metals. *Fresenius Journal of Analytical Chemistry* 342: 791–794.)

spectrophotometry or titrimetry. In the second, solutes are assayed directly on the layer with the help of visual, manual or instrumental measurement methods.

*In situ* densitometry, a preferred technique for quantitative TLC, involves the measurement of visible or ultraviolet absorbance, fluorescence of fluorescence quenching directly on the layer. The measurements are made either through the plates (transmission), by reflection from the plate, or by reflection and transmission simultaneously, using either single-beam, double-beam, or single-beam–dual-wavelength

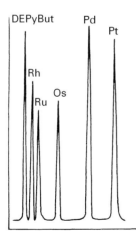

**Figure 10** HPTLC separation of platinum group metals with DEP$_Y$ BuT. Eluent: toluene; relative humidity 5.0%, separation distance, 6 cm (Reproduced with permission from Schuster M and Unterreitmaier E (1993) Fluorometric detection of heavy metals with pyrene substituted *N*-acylthioureas. *Fresenius Journal of Analytical Chemistry* 346: 630–633.)

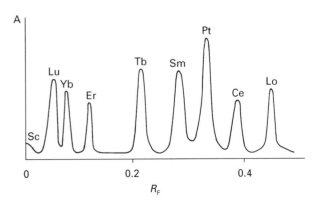

**Figure 12** Densitogram of a mixture of rare earth element complexes. Stationary phase: silufol, mobile phase: *n*-propanol–0.7 mol L$^{-1}$ HCl (9 : 2): spraying reagent: 0.1% arsenazo III; wavelength: 590 nm. (Reproduced with permission from Timerbaev A, Shadrin O and Zhivopistsev V (1990) Diantipyrilmethane as complex-forming reagents in the thin-layer chromatographic determination of metals. *Chromatographia* 30: 436–441.)

**Table 6**  Quantitative and semiquantitative determination methods for inorganic ions

| Method | Species determined |
|---|---|
| TLC-densitometry | B, Pb, Fe, Cu, Mn, Co, Ni, Mg, Zr, Mo, Ba, La, Ce, Sr, Se, Tl, Hg, phosphates, selenocyanate, U, Cd, Y, $NO_2^-$, $NO_3^-$, $Fe(CN)_6^{3-}$, $SCN^-$, rare earth elements, Co(III)-1-(2-pyridylazo)-2-naphthol complex and bis carboxy ethyl germanium sesquioxide |
| TLC-spectrophotometry | Pb, Zn, Ni, Cu, Co, Zr, Pd, Mo, U, Hg(II), metal complexes, $SCN^-$, La, Ce, Pr, Nd, Cr(III), Fe(II), $Cr_2O_7^{2-}$ and $Fe(CN)_6^{3-}$ |
| TLC-UV spectroscopy | Hg(II), Cu, Cd |
| TLC-titrimetry | Cu, $SCN^-$ |
| TLC-atomic absorption spectroscopy (AAS) | Cu, Zn, Fe, Ni, Pb, Mn |
| ICP-TLC-atomic emission spectrometry (AES) | Rare earth elements, heavy metals |
| Neutron activation-circular TLC | Rare earth elements |
| Square wave-stripping voltammetry-TLC | Pb, Cd, Zn, Cu |
| TLC-photoacoustic measurement | Cobalt complex of 1-(2-pyridylazo)-2-naphthol |
| TLC-emission spectrometry | Alkaline earth metals |
| Visual colorimetry and peak spot-area measurement | Ti, Ni, Cu, Cr, Pb, Ag, Th, Tl(I), Al, Co, Zr, Fe(II), Fe(III), V, U, Cd, Zn, Ce, Hg, $IO_4^-$, $Br^-$, $I^-$, $NO_2^-$, $Cl^-$, $NO_3^-$, $SCN^-$, $SO_4^{2-}$, $BrO_3^-$ |

Source: *Chemical Abstracts* (1973–1996), USA.

scanning instruments. Modern optical densitometric scanners are linked to a computer and are capable of automated peak location, multiple wavelength scanning and spectral comparison of fractions in several operating modes (reflectance, absorption, transmission and fluorescence). Representative examples of HPTLC determination and densitometry are shown in **Figures 11** and **12**.

The combination of TLC with other analytical techniques has proved useful for the analysis of complex samples. Spectrophotometry, HPLC, inductively coupled plasma–mass spectrometry and voltammetry in conjugation with TLC are the most commonly used techniques. However, infrared, thermal anlaysis and mass spectrometry have also been used.

For semiquantitative analysis, visual comparison and spot-size measurement methods are used. A definite volume of sample is chromatographed alongside standards containing the analyte. After detection, the amount of analyte in the sample is estimated by visual comparison of the size and intensity of the sample zone with the standards. This method works

**Table 7**  TLC results of the determination of Ti, Zr, Hf, La and Tb in various samples

| Technique | Sample | Analyte | Determination results | |
|---|---|---|---|---|
| | | | Amount found (mean and SD) | Known amount (concentration) |
| Spot-area measurement method | Sulfite-cellulose liquor | Ti | $31 \pm 4$ p.p.m. | 35 p.p.m. |
| | High speed steel | Ti | $25 \pm 3$ mg g$^{-1}$ | 29.3 mg g$^{-1}$ |
| TLC-spectrophotometry | Synthetic mixture (Zr + Hf) | Zr | $10.4 \pm 0.3$ µg | 10 µg |
| | | Hf | $9.7 \pm 0.2$ µg | 10 µg |
| | Mg–Al–Zr–Hf | Zr | $1.1 \pm 0.2$ mg g$^{-1}$ | 1.1 mg g$^{-1}$ |
| TLC-densitometry | Synthetic mixture | La | $1.1 \pm 0.1$ µg | 1.2 µg |
| | | Tb | $1.4 \pm 0.1$ µg | 1.6 µg |
| | Lanthanum glass | La | $179 \pm 5$ mg g$^{-1}$ | 174 mg |
| | | Tb | | |
| | Monazite | La | $4.0 \pm 0.2$ mg g$^{-1}$ | 4.3 mg g$^{-1}$ |
| | | Tb | $1.1 \pm 0.2$ mg g$^{-1}$ | 1.2 mg g$^{-1}$ |

Reproduced with permission from Timerbaev A, Shadrin O and Zhivopistsev V (1990) Diantipyrilmethane as complex-forming reagent in the thin layer chromatographic determination of metals. *Chromatographia* 30: 436–441.

**Table 8** Comparison of results of anions in water samples by TLC and ion chromatography (IC)

| Sample | Technique | Determination results (p.p.m.) | | |
| --- | --- | --- | --- | --- |
| | | $SO_4^{2-}$ | $Cl^-$ | $NO_3^-$ |
| Ground water | TLC | $34 \pm 5$ | ND | $10 \pm 2$ |
| | IC | 38.3 | ND | 12.8 |
| Lake water | TLC | $25 \pm 4$ | $29 \pm 5$ | ND |
| | IC | 27.2 | 28.0 | ND |
| River water | TLC | $9.8 \pm 1.6$ | $6.3 \pm 1.2$ | ND |
| | IC | 9.3 | 7.0 | 2.9 |

ND, not detected. Reproduced with permission from Shadrin O, Zhivopistsev V and Timerbaev A (1993) Thin-layer chromatographic determination of inorganic anions as counter-ions of metal-diantipyrilmethane cationic complexes and diantipyrilmethane cations. *Chromatographia* 35: 667–670.

well if the applied amounts of sample are kept close to the detection limit and the sample is accurately bracketed with standards. The shape and size of the spot produced are significantly influenced by the amount of analyte. A linear relationship between the size of the spot and the amount of analyte has been observed. This method has been used for semiquantitative estimation of titanium in steel and inorganic anions in water.

TLC techniques used and the species determined are listed in **Table 6**. TLC results of determination of certain species in real samples are shown in **Tables 7** and **8**, which indicate the versatility and accuracy of TLC.

## Application

Some applications of inorganic TLC have been covered above. **Tables 9–11** list representative applications of TLC as used for the analysis of anionic, cationic and metal complex mixtures. In **Table 12** selective TLC applications related to the analysis of real samples (biological, food, geological, industrial, pharmaceutical, soil, water, wastewater and irradiated products) have been given.

## Further Developments

Looking to the future, it is reasonable to expect increasing use of TLC and HPTLC for more application-oriented research in several fields (pharmaceutical, environmental, geological, forensic, agrochemical, textile, cosmetic and food sciences) because of the continued development of computer and microprocessor-based instrumental TLC. Layers with immobilized phases of wider range of selectivity, automation of sample application, hybrid mobile phases with improved chromatographic performance, online

**Table 9** Stationary and mobile phases used in the analysis of certain anionic species

| Anions | Stationary phase | Mobile phase |
| --- | --- | --- |
| Hexacyanoferrate (II) and (III) | Silica gel G, alumina G | Polyhydric alcohols, formamide, DMF, methylamine, pyridine, water, ketones, esters and their mixtures in various ratios |
| Sulfur oxyanions | Silufol UV 254 | Ethanol–dioxane–water–NH$_4$OH  $(30 + 60 + 50 + 25)$ |
| Oxyanions | Cellulose, microcrystalline cellulose containing fluorescent indicator | 28% Aqueous ammonia–acetone–*n*-butanol $(6 + 13 + 3)$, 28% aqueous ammonia–acetone $(2 + 3)$, dioxane–water $(3 + 2)$, acetone–acetic acid–water $(20 + 1 + 20)$ |
| Halogen anions | Silica gel G, zirconium(IV) molybdate, cellulose | Basic and polar solvent systems, various alcohols mixed with aqueous ammonia |
| Phosphorous anions | Cellulose | Water–ethanol–2-methylpropanol–2-propanol–aqueous ammonia–trichloroacetic acid $(150 + 175 + 75 + 107 + 2 + 25)$ |
| Arsenate, arsenite | Aluminium oxide | Aqueous solutions of KNO$_3$, Na$_2$CO$_3$, NaOAc, HOAc, K$_2$SO$_4$, K$_3$PO$_4$, NaF, acids and bases and buffer solutions |
| Oxyanions, chromate, dichromate, oxalate | Silica gel G impregnated with 0.1% aq. solutions of CuSO$_4$, ZnSO$_4$, NiCl$_2$, CoCl$_2$, Co(NH$_3$)$_6$Cl | Acetone–DMSO or formic acid, acetone–DMSO–formic acid and acetone–mineral acid mixtures |

DMF, dimethylformamide; DMSO, dimethylsulfoxide.

**Table 10** Stationary and mobile phases used in the analysis of some cationic species

| Metal ions | Stationary phase | Mobile phase |
|---|---|---|
| Alkali and alkaline metals | Microcrystalline cellulose, cellulose (MN-300) microcrystalline cellulose Lk, zinc ferrocyanide, tin(IV) arsenosilicate and arsenophosphate | Methanol–conc.HCl–water $(8 + 1 + 1, 7 + 1 + 2)$, dioxane-2-propanol-conc. HCl–H$_2$O $(7 + 7 + 5 + 8)$. HNO$_3$–methanol mixtures, buffered EDTA solutions, aqueous ammonium nitrate |
| Transition metals | Silica gel G, DEAE cellulose in chloride form, chitosan, chitin, silica gel impregnated with high molecular weight amines, cellulose | Aqueous solutions of HCl $(1–5$ mol L$^{-1})$, LiCl $(1–9$ mol L$^{-1})$, MgCl$_2$ $(2.5$ mol L$^{-1})$, CaCl$_2$ $(2.5$ mol L$^{-1})$ and saturated NaCl, Me$_2$CO–EtOAc–C$_6$H$_6$ $(7 + 1 + 3)$, $0.03$ mol L$^{-1}$ citric acid, HNO$_3$ at different concentrations |
| Light rare earth metals | Mixture of silica gel, starch and NH$_4$NO$_3$; silica gel–NH$_4$NO$_3$–CM cellulose $(5:0.64:0.16,$ w/w) | P$_{204}$-dioxane–EtOAc–HNO$_3$ $(1.1 + 2 + 2.4 + 4)$ tributylphosphate–THF–diethyl ether–HNO$_3$ $(1 + 9 + 9 + 1.5)$, trialkylmethyl ammonium chloride–$n$-octyl alcohol–petroleum ether–conc. HNO$_3$ $(60 + 7 + 25 + 1)$ |
| Heavy rare earth metals | Silica gel–NH$_4$NO$_3$–CM-cellulose $(5.0 + 0.64 + 0.16,$ w/w); silica gel–starch–ammonium rhodanate $(2.8 + 0.15 + 0.5,$ w/w) | Et$_2$O–THF–bis (2-ethylhexyl)phosphate–HNO$_3$, trimethylammonium chloride–$n$-octyl alcohol–petroleum ether–HCl $(2 + 10 + 30 + 1, 2 + 6 + 30 + 1.2$ and $2 + 7 + 30 + 1)$, bis (2 + ethylhexyl)phosphate–diisopropyl ether–diethyl ether–nitric acid $(1 + 10 + 6 + 1.1)$, mono(2-ethylhexyl) phosphate–isopropyl ether–diethyl ether–nitric acid $(1 + 8 + 8 + 1.1)$ |
| Lanthanides | Silanized silica gel (Merck) and polygram plate, silica gel impregnated with different concentrations of mono (2-ethyl hexyl) phosphate | 3 mol L$^{-1}$ Tributyl phosphate in HNO$_3$–isooctane $(2 + 8)$; THF–paraldehyde–HNO$_3$ $(2 + 7 + 1)$; $0.3$ mol L$^{-1}$ triphenylphosphine oxide in paraldehyde–HNO$_3$ $(9 + 1)$; di-isopropyl ether–THF–HNO$_3$ $(10 + 6 + 1)$; HNO$_3$ $(0.05–3.0$ mol L$^{-1})$ |
| Radionuclides $^{90}$Sr, $^{90}$Y, $^{140}$Ba, $^{140}$La | Silica gel (with and without gypsum binder) | Aqueous solutions of NaCl, KCl, NH$_4$Cl, CaCl$_2$, SrCl$_2$, and BaCl$_2$ |
| High valence metal ions | Silica gel impregnated with crystalline antimonic(V) acid-$p$-sulfochlorophosphanazo | Potassium pyrophosphate solution |
| Mn(II), Mn(III) | Silica gel G | Eleven organic mobile phases consisting of hydrocarbons and their derivatives in different ratios |
| Transition metals | 1,10-Phenanthroline (1%), DMG (1%), EDTA (2%) or $\beta$-naphthol (0.1%) impregnated silica gel | Pyridine–benzene–HOAc–H$_2$O $(6 + 5 + 8 + 4)$, BuOH–benzene–formic acid $(5 + 10 + 9)$, pyridine–benzene–HOAc–H$_2$O $(5 + 5 + 4 + 1)$ |

DEAE, diethylaminoethyl; THF, tetrahydrofuran; DMG, dimethylglyoxime; EDTA, ethylenediaminetetraacetic acid.

coupling of TLC with other sensitive techniques, better approaches in method development and application of detection reagents and greater use of instrumental densitometry can be expected. It is hoped that the forced-flow layer methods with increased automation will be developed in the near future for faster separation of inorganic ions in real samples.

**Table 11** Stationary and mobile phases used in the analysis of certain metal complexes

| Metal complexes | Stationary phase | Mobile phase |
|---|---|---|
| Diethylthiocarbamates of Bi, Cu, Co, Ni | Silica gel G | Dichloromethane–petroleum ether (5 + 3) |
| Sulfate complexes of Pt, Pd, Rh, Ir | Commercial silufol plates (Kavalier, CSSR), silica gel KSK, silica gel | 0.1, 1.0 and 6.0 $NH_4SO_4$, 0.2 or 0.5 mol $L^{-1}$ tetraoctylamine in benzene |
| $\beta$-Diketonates of Fe, Cr and Co | Silica gel (Merck) | $CCl_4$, toluene, benzene, dichloromethane, di-ethyl ether |
| Xanthates of Cu, Ni, Co, Mo, Bi, Pb, Zn | Silica gel G | $CCl_4$–$CHCl_3$ (10 + 1), toluene–benzene (10 + 1) |
| Anil complexes of Mn, Fe, Cu, Zn | Silica gel–starch (19 : 1) | Acetonitrile, methanol, ethanol, butanol, acetic acid or butanol–acetic acid (4 + 1, 3 + 2, 2 + 3, 1 + 4) |
| Metal oxinates (OX), (HOX), Cu $(OX)_2$, Zn $(OX)_2$, Al$(OX)_3$, Ga$(OX)_3$, In $(OX)_3$ | Styragel 60A (polystyrene–divinylbenzene copolymers), Merckogel OR-PVA 2000 | $CHCl_3$, $p$-dioxane, benzene, $10^{-3}$–$10^{-1}$ mol $L^{-1}$ HOX in $CHCl_3$, $10^{-2}$ mol $L^{-1}$ HOX in dioxane, $10^{-1}$ mol $L^{-1}$ pyridine in $CHCl_3$ |
| Mn, Co, Ni, Zn, Rh, Pd, Pt, Chelates of dithioacetyl-acetone | Silica gel | $CCl_4$, 1,1,1-trichloroethane |
| Mixed amino carboxylato cobalt(III) complexes | Polyacrylonitrile | Aqueous ammonium sulfate solutions (1.1–3.48 mol $dm^{-3}$) |
| Chlorosulfate complexes of transition metals and rare earths | Silica gel G–cellulose (2 : 1) | Sodium formate (1 mol $L^{-1}$) and ammonium sulfate (1 mol $L^{-1}$) |
| Mercapto-4-methyl-5-phenylazopyrimidine complexes of Co, Ni, Pb, Cd, Cu | Surfactant-impregnated silica gel plate | Acetonitrile–xylene (70 : 30) |

**Table 12** Application of TLC to the analysis of real samples

| Species | Sample | Remark |
|---|---|---|
| Se | Biological (tissues, blood, serum) environmental (drinking and surface water) and food stuffs | Isolation of Se from matrix, derivatization, extraction of Se-contaning complex, TLC separation and fluorimetric determination. Detection limit 250 fg Se per spot |
| Cu, Fe | Serum | TLC separation and densitometric determination of $Fe^{2+}$, $Fe^{3+}$ and $Cu^{2+}$ complexed with 2-[(5-bromo-2-pyridinyl)azo]-5-(diethylamino) phenol in serum |
| Cd, Hg, Pb | Urine, blood | Separation on silica gel layer prior to spectrophotometric determination of $Cd^{2+}$, $Pb^{2+}$ and $Hg^{2+}$ in blood and urine samples |
| Heavy metals | Urine, blood, excrement | A sensitive detection of $Zn^{2+}$, $Hg^{2+}$, $Cu^{2+}$, $Co^{2+}$, $Pb^{2+}$, $Ni^{2+}$, $Ag^+$ and $Bi^{3+}$ on silica gel layers (sensitivity $10^{-3}$ g $L^{-1}$) as dithizonates in biological samples |
| Hg | Animal, fish and plant tissues | Quantitative determination of mercury species (inorganic and methyl mercury) after extraction from tissues |
| Cu, Mg, Zn, Fe, Mn | Human faeces | Detection of metal–EDTA complexes on silica gel and cellulose layers |
| $BrO_3^-$ | Food stuff | Extraction and purification using alumina column, separation on silica gel layer and densitometric determination of $BrO_3^-$ in bread and flour dough. Detection limit of $BrO_3^-$ was 0.1 $\mu$g $g^{-1}$ of bread |
| Fe, Mn, Co | Human milk | Extraction of metals from human milk with isobutyl methyl ketone–amyl acetate (2 : 1) and identification on cellulose layer |

**Table 12**  *Continued*

| Species | Sample | Remark |
|---------|--------|--------|
| Polyphosphoric acids | Cheese, milk | Extraction with 25% trichloroacetic acid, separation and identification on polyamide layer developed with *n*-butanol–formic acid (1 : 1) |
| Polyphosphates | Soft drinks | Separation and determination of ortho- and polyphosphates in soft drinks using two-dimensional TLC and ion-exchange column chromatography |
| Si | Edible oils | Detection with rhodanine on silica gel layers developed with light petroleum–diethyl ether (98 : 2) |
| Rare earth metals | Rocks, monazite sand, ores, irradiated nuclear fuels | Determination by neutron activation analysis (determination limits $0.05$–$10\ \mu g\ g^{-1}$ for 10–30 mg sample) after preconcentration by circular TLC on Fixion $50 \times 8$. Use of ICP-AES for determination of rare earths in concentrates obtained by means of TLC. Densitometric determination in monazite sand (linear range $0.015$–$0.60\ \mu g$ of individual rare earth) after TLC separation. The limits of detection for rare earths were 9–12 ng |
| Fe, Mn, Cu | Cotton materials | Detection of metal ions on microcrystalic cellulose plates developed with acetone–HCl–$H_2O$ (8 : 1 : 2). TLC-spectrophotometry for determination of Mn traces in textile materials |
| $NO_3^-$, Fe $(CN)_6^{3-}$ | Molasses | TLC separation on silufol 254 plates developed with propanol–ammonia solution (2 : 1), detection with acidified diphenylamine and densitometric determination |
| La, Y | Alloys | Circular TLC with trioctylamine-treated cellulose layers as stationary phase and aqueous HCl as mobile phase for separation and spectrophotometric determination of La and Y (0.01–1.0%) in Mo-based alloys |
| Co | White wine | Fixation of $Co^{2+}$ as 1-(2-pyridylazo)-2-naphthol complex on membrane filter followed by densitometric determination (concentration range in white wine $2.5$–$4.5\ \mu g\ L^{-1}$) by reflection absorbance of the complex |
| Fe(II) | Pharmaceuticals | Separation on microcrystalline cellulose layers, detection by 1,10-phenanthroline and determination by spectrophotometry |
| Mercury salts | Homeopathic drugs | Identification of chloride, nitrate, cyanide, sulfate and sulfide of Hg on silica gel G plates developed with acetone–chloroform–conc. $HNO_3$ (4 : 5 : 1) |
| Mn | Pharmaceuticals | TLC separation and photodensitometric determination of Mn in vitamins and pharmaceuticals as PAN complex |
| Cr | $Na_2^{51}\ CrO_4$ injection | Determination of radiochemical purity of $Na_2^{51}\ CrO_4$ injections and $\gamma$-scintillation for quantitation of Cr-containing zones |
| $NO_3^-$ | Feeds | Separation on alumina layer and spectrophotometric determination at 430 nm after extracting the coloured product formed with 3,4-xylenol |
| Rare earths and fission products | Irradiated nuclear fuels | Two-dimensional TLC separation and enrichment followed by quantification by $\gamma$-spectroscopy |

**Table 12** *Continued*

| Species | Sample | Remark |
|---|---|---|
| Co, Fe(III) | Cosmic dust | TLC separation and semiquantitative determination on the basis of spot size and colour intensity or by reflectance densitometry |
| $^{89}Sr$, $^{90}Sr$ | Soil | Extraction of strontium from soil with dicyclohexano-18-crown-6 in chloroform and TLC separation using circular procedure with silufol layer |
| Ta | Molybdenum-based alloys | TLC-AES for determination of Ta ( $\geq 0.5\%$ ) in molybdenum alloys |
| Zn, Pb, Cu, Cd, Hg (organic and inorganic) | Plants | Detection (10–500 ng) using azo rhodanines as detector and nanogram determination by visual sorption-photometric method |
| Hg (organic and inorganic) | Water | HPTLC separation on silica gel layers and densitometric determination as dithizonates |
| Al, Be, Cr, Bi, Cu, Hg | Water | Extraction, TLC separation, detection (as Bi, Cu, Hg dithizonates, detection limit, 0.5 p.p.m.) and determination fluorimetrically (Al and Be as oxinates) and photometrically (Cr as diphenylcarbamide complex) |
| $CuSO_4$, $CdSO_4$, $HgCl_2$, $AgNO_3$ | Fresh and seawater | Evaporation, precipitation, micro-TLC separation and enzymatic detection |
| Fe, Cu, Hg, Cd, Co, Ni, Bi, Mn, Pb, Zn, Sn, Al, Cr, Be | Water | Extraction, TLC separation on silica gel layer developed with EtOH–1.0 mol $L^{-1}$ $HNO_3$ (99 : 1) and photometry in transmission mode. Detection limits 0.1–1 p.p.m. except Cr (8 p.p.m.) |
| Cu, Zn, Cr, Fe, Ni, Co, V | Electroplating wastewater | Simultaneous separation of metals and semiquantitative determination of Ni, Cu, Cr in wastewaters using cellulose with azopyrocatechol groups as layer material. Detection limit of metals in coloured zone was 0.05–2.0 µg |
| Co, Ni, Cu, Fe | Industrial and wastewaters | Visual semiquantitative determination of concentration of total heavy metals in wastewaters according to the intensity of coloured metal diethyldithiocarbamates formed directly on silufol plates. Determination of Cu and Ni at 5 mg $L^{-1}$ in electroplating wastewater |
| Cu, Ni, Fe, Co | Electroplating wastewater | Quantitative trace analysis using 2-(-5-bromo-2-pyridylazo)-5-diethyl aminophenol as complexing agent |

EDTA, ethylenediaminetetraacetic acid; ICP-AES, inductively coupled plasma–atomic emission spectroscopy; PAN, 1-(2-pyridylazo)-2-naphthol.

*See also:* **II/Chromatography: Thin-Layer (Planar):** Densitometry and Image Analysis; Layers; Spray Reagents. **III/Impregnation Techniques: Thin-Layer (Planar) Chromatography. Ion Analysis:** Capillary Electrophoresis; High-Speed Counter Current Chromatography; Liquid Chromatography.

## Further Reading

Fried B and Sherma J (eds) (1994) *Thin Layer Chromatography: Techniques and Applications.* New York: Marcel Dekker.

Gocan S (1990) Stationary phases in thin-layer chromatography. In: Grinberg N (ed.) *Modern Thin Layer Chromatography*, pp. 5–137. New York: Marcel Dekker.

Jork H, Funk W, Fisher W and Wimmer H (1990) *Thin Layer Chromatography, Reagents and Detection Methods.* Weinheim, Germany: VCH Verlagsgesellschaft.

Kuroda R and Volynets MP (1987) Thin-layer chromatography. In: Qureshi M (ed.) *CRC Handbook of Chromatography: Inorganics.* Boca Raton, FL: CRC Press.

Lederer M (1994) *Chromatography for Inorganic Chemistry*. New York: John Wiley.

MacDonald JC (ed.) (1994) *Inorganic Chromatographic Analysis*. New York: John Wiley.

Mohammad A (1996) Inorganics and organometallics. In: Sherma J and Fried B (eds) *Handbook of Thin Layer Chromatography*, 2nd edn. New York: Marcel Dekker.

Mohammad A and Tiwari S (1995) Thirty-five years thin-layer chromatography in the analysis of inorganic anions. *Separation Science Technology* 30: 3591–3628.

Mohammad A, Fatima N, Ahmad J and Khan MAM (1993) Planar layer chromatography in the analysis of

inorganic pollutants. *Journal of Chromatography* 652: 445–453.

Mohammad A, Ajmal M, Anwar S and Iraqi E (1996) Twenty-two years report on the thin layer chromatography of inorganic mixtures: Observations and future prospects. *Journal of Planar Chromatography – Modern TLC* 9: 318–360.

Poole CF and Poole SK (1991) *Chromatography Today*. Amsterdam: Elsevier.

Touchstone JC (1992) *Practice of Thin Layer Chromatography*, 3rd edn. New York: Wiley-Interscience.

# ION EXCHANGE RESINS: CHARACTERIZATION OF

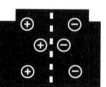

**L. S. Golden**, Purolite International Ltd, Pontyclun, Mid-Glamorgan, Wales, UK

## Introduction

An ion exchange resin is an insoluble polymer matrix containing labile ions which are capable of exchanging with ions in the surrounding medium without any major physical change taking place in its structure. They are of two basic types, cation and anion exchangers. So-called cation exchange resins are in fact polymeric anions to which the labile cation is bound, and it is this cation which exchanges with other cations in solution. Likewise, anion exchange resins are polymeric cations with a labile, exchangeable anion.

The first synthetic ion exchange resins were developed by Adams and Holmes in 1935, based on a phenol-formaldehyde structure. The next most important development was the introduction of commercial ion exchange resins based on a cross-linked polystyrene matrix, and these resin types today still represent about 90% of the commercial resin market. More recently, polyacrylic resins have been introduced which have widened the scope and versatility of the synthetic ion exchange resins, and these represent most of the remaining 10% of the commercial market.

The development of synthetic polystyrene and polyacrylic resins has enabled these ion exchangers to be produced in spherical bead form, unlike the irregular-shaped particles of the phenol-formaldehyde types. In the majority of applications, the most efficient treatment of a solution is obtained by passing this solution through a bed of ion exchange resin. A spherical bead shape offers optimum contact with the percolating solution without undue pressure drop across the bed, requiring only minimal inlet pressures.

The first polystyrene resins, introduced in 1947, were of what is now known as a *gel* type. Exchange takes place by diffusion of the ions through the resin structure to and from the ion exchange sites. The polymer chains are only separated by molecular distances, and the ease of penetration of the ions is very much influenced by the amount by which the resin structure can be swelled by the contacting solution. In 1956, Mikes and co-workers discovered a means of introducing pores (or holes) into the resin structure, which led to the introduction of *macroporous* resins, which further increased the scope of ion exchange techniques. These resins comprise a continuous polymer matrix interspersed with a continuous pore matrix.

## Synthesis

### Polymerization

Generally speaking, the synthesis of a modern ion exchange resin is a two-step process. In the first step, the spherical bead is produced by polymerization of styrene (or an acrylic monomer, usually methyl acrylate) plus a cross-linking agent (usually divinylbenzene, DVB) in an aqueous suspension. This technique of suspension polymerization is used extensively in the production of pearl (or bead) polymers, not specifically to make ion exchange resins, and is dependent on the fact that the monomer is essentially insoluble in water. Thus, when stirred with water, the monomer will disperse into spherical droplets. Small quantities of various stabilizing ingredients are added to the water to make this dispersion more permanent.

An initiator is added to the monomer mixture which, on the application of heat, produces free radicals. These start a chain reaction with the monomer in which monomer units are progressively added to the growing polymer chains. Styrene contains just one double bond capable of reacting into the polymer chain. As a consequence, a polymer produced from styrene alone will comprise a large number of unconnected chains, and this hydrophobic polymer will be soluble in organic solvents such as aromatic hydrocarbons. Cross-linking agents, such as DVB, contain two or more reactive double bonds, each of which is capable of reacting in a separate polymer chain, ultimately leading to an 'infinite' single polymer chain network. Solvents will still be absorbed by the polymer chain and swell the polymer, but will not be able to dissolve it. The extent of the swelling is dependent on the proportion of DVB in the monomer: the more DVB, the less the polymer will swell.

The invention of macroporous resins has been briefly mentioned. The continuous pore structure within the resin matrix is produced by polymerizing the monomers in the presence of an inert diluent which is miscible with the monomers but essentially immiscible with the growing polymer chains. As the polymerization progresses, the mixture separates into two phases, one phase being the growing polymer, the other being the diluent plus a continuously decreasing amount of monomer. At the end of the polymerization, the diluent, which has not actually polymerized into the structure, is removed by distillation or washing, leaving the interconnecting network of pores.

### Activation

The second stage of synthesis is to add a functional group to the polymer, which will contain the labile ion capable of exchange with ions in solution. The effect of this functional group is to make the polymer structure hydrophilic. The resin will now be swollen by water, but again cannot dissolve due to the polymer chain structure. The extent of this swelling by water is an important characteristic of an ion exchange resin, and controls its behaviour in many applications.

The sulfonic acid group of the strong acid cation resins is added by reaction of the polystyrene polymer with sulfuric acid at temperatures between 90 and 140°C. The amino group of the strong and weak base resins is added in two stages, firstly a chloromethylation to add a chloromethyl group to the polystyrene chain followed by addition of an amine to give the final quaternary or tertiary amino functionality. The choice of amine affects the resin properties, which will be discussed shortly.

Polyacrylic resins are produced as either weak acid cation resins with a carboxylic acid functional group (to which there is no polystyrene resin equivalent), or as strong or weak base anion resin whose properties are similar, but not identical, to those produced from polystyrene.

## Classification of Ion Exchange Resins

Both cation and anion resins are available with strong and weak functional groups. The strong functional groups are ionized species within which the ionized labile ion can be replaced with an ion in solution. These resins will therefore easily exchange ions with dissolved salts. On the other hand, the weak functional groups are themselves nonionized and will therefore have very little ability to exchange with salts. Weakly functional anion resins will however readily remove anions from acidic solutions, and weakly acidic resins will readily remove calcium and magnesium from solutions of carbonates and bicarbonates.

Strongly basic resins are further subdivided into three categories. The most commonly encountered of these is the *strong base anion Type I* resin (often abbreviated to SBA I). The second most common category is the *strong base anion Type II* (SBA II), which is a weaker base than the Type I but still not weakly basic enough to be categorized as a weakly basic resin. The third category, *strong base anion Type III* (SBA III), is actually intermediate between these two types, and, although not as yet widely used, is being shown to combine useful characteristics of both.

In more recent years, a number of resins with special functional groups have been produced commercially. These are designed for removal of specific ions for special applications. One particular application is the use of a polystyrene resin with an aminophosphonic chelating group – this resin, in the sodium form, is used to remove calcium and other divalent ions from brine prior to the production of caustic soda by the membrane cell process. A conventional cation resin would be completely unable to remove these alkaline earths from concentrated salt solutions.

Resins in each category can be produced in both gel and macroporous forms.

# Fundamental Ion Exchange Resin Characteristics

Irrespective of the resin type, there are certain fundamental parameters which characterize ion exchange resins, and determine how that resin will perform in its applications.

## Moisture Retention

As mentioned earlier, an ion exchange resin is swollen by water but will not dissolve in it. The extent of this swelling is one factor which controls how the resin will perform in specific applications. A resin will absorb a certain amount of water, depending on the functional group, the ionic form and the amount of cross-linking. For a given resin type in a given ionic form, the swelling is entirely dependent on the amount of cross-linking.

Moisture retention is defined as the amount of water which is in equilibrium with the dry resin matrix. This water can be removed by hot air drying, but will be readsorbed when the resin is put back in contact with excess water. The value of the moisture retention is easily determined by measuring the loss in weight of a fully swollen resin on drying, and expressing the moisture loss as a percentage of the total wet weight.

The importance of moisture retention in the performance of a resin should be fairly obvious. For the resin to operate, the ions in solution must have access to the active sites throughout the polymer structure, just as the ions which are displaced from these sites must be able to pass out of the resin back into solution. The more open the structure (i.e. the higher the moisture retention), the easier this will be.

The labile ion which is attached to the resin influences the hydrophilic nature of the structure, and so the moisture retention. For instance, at a given amount of cross-linking, a strong acid cation resin in the sodium form (i.e. with a $Na^+$ group attached to each sulfonic acid site on the structure) will have a significantly lower moisture retention than the same resin in the hydrogen form. Therefore, the ionic form must be stated along with the moisture retention.

## Resin Capacity

The ability of an ion exchange resin to exchange ions is a function of the number of active groups that have been placed on the resin during the activation process. The overall term for this is the *capacity* of the resin, but this can be expressed and determined in a number of ways.

**Dry weight capacity**   This measures the proportion of active sites in the dry resin matrix, and is expressed as equivalents per kilogram (or milliequivalents per gram, as some workers prefer). It is easily determined by weighing a dry sample of resin, rewetting it and displacing completely from the resin an ion which can be titrated in solution. This gives the total equivalents of sites on the dry resin, and is a direct measure of the extent of activation of the base polymer. For most applications, the figure obtained does not have any direct significance, since the resin is rarely used in its dry state except in certain specialized applications.

**Total volume capacity**   Since resins are normally supplied and used wet (i.e. fully swollen in water), the capacity based on a volume of wet resin is more useful. It is, however, more difficult to measure a wet resin volume as accurately as dry resin weight, but essentially the test method is the same apart from measuring a known volume of resin rather than weight. The data is expressed in terms of equivalents per litre (or milliequivalents per millilitre).

Although the test value can now be related to the volume of resin installed in the resin treatment plant, it will still not tell the plant designer or user the capacity achievable when the unit is in practical operation.

**Operating capacity**   This is the true practical parameter, but unfortunately it cannot be determined by a straightforward laboratory test. Since the operating capacity of a resin depends on many factors, such as the dimensions of the unit, the flow rate and quality of the liquid to be treated, the quality required of the effluent from the unit, the regenerant quantity and the operating temperature, etc., this capacity can only be calculated from data obtained by extensive laboratory testing of a resin type under simulated operating conditions. This data is usually provided in brochures supplied by the resin manufacturers, and more recently a number of these manufacturers have computer programs available to make these calculations quickly on data stored within the software.

## Particle Size Distribution

As already described, modern ion exchange resins are in the form of spherical beads. The suspension polymerization technique does not produce beads of a uniform particle size, rather a distribution of sizes to give a resin within the range 0.3–1.2 mm. Most ion exchange applications have been developed around resins of this particle size range, but some special applications do require the particles to be of a more specific size.

In the majority of applications, the ion exchange resin is contained in a vessel which is designed to allow the solution to be treated to pass through the resin bed and subsequently emerge from the unit. The resin has therefore to be retained within this unit by a suitable means, usually nozzles or slats of such an aperture that the solution can freely pass. Depending upon engineering design, these apertures will normally be about 0.2 mm wide, but in some designs can be as large as 0.4 mm. It is therefore essential that the ion exchange resin does not contain beads close to or less than this aperture size.

A number of resin applications require two resins to operate mixed together in the same unit. For regeneration, it is essential that these resins are separated. Strong acid cation resins have a density of about 1.25–1.30 g g$^{-1}$, whereas anion resins have a density of 1.07–1.10 g g$^{-1}$. Therefore, by applying backwash to lift and expand the resin bed, the cation resin beads will sink below the anion resin beads. A complete separation is readily achieved if the cation resin contains less coarse beads an the anion resin less fine beads.

Ions obviously have a greater path length to travel to the centre of a large bead compared to that of a small bead. This decreases the efficiency with which a larger bead can complete the ion exchange process. A number of applications have developed today in which beads of a more uniform size distribution, normally in the range 0.4–0.8 mm, are used, thereby eliminating the most coarse beads of the 'standard' size range.

For these reasons, measurement of particle size distribution of resins is important. The classical method is to use standard sieves, but due to limitations of sieve accuracy and availability, they are not best suited to measuring resins with narrower size distributions. Most resin manufacturers and larger users now use an instrumental technique based on a light extinction principle, such as the OMEGA (Fortress Dyamics, UK) or HIAC (Hiac-Rogco, USA).

### Resin Volume Change

It was observed earlier that a given ion exchange resin will be in equilibrium with a different volume of water depending on its ionic form. Consequently, when the resin is treating the feed solution, the bed of resin will slowly swell or shrink as the ions are being exchanged. It is important for the plant design engineer to know the extent of this volume change in order to allow sufficient space for any increase in vol-

ume to be accommodated. It has been known for poorly designed plants to smash a resin, or even for the plant itself to be damaged, if this swelling factor has not been properly accounted for.

### Other Resin Characteristics

There are a number of other parameters which have to be taken into account when designing an ion exchange plant, such as pH range and operating temperature. All resins of a given type will have virtually identical limitations on these parameters. The recommendations given by the resin manufactures should be followed. It is worthy of note that SBA Type II resins are less thermally stable than their Type I counterparts, and that polyacrylic anion resins are less thermally stable than polystyrene ones with the same functionality.

## Macroporous versus Gel Resins

In appearance, gel resin beads are usually transparent. Light entering the bead will pass through the homogeneous structure without being diffracted, therefore being visible from the other side of the bead. With a macroporous bead, there are numerous phase boundaries within the bead between the pores and resin matrix, at each of which the entering light will be refracted. Consequently, little or no light will emerge, giving the resin bead an opaque appearance.

Gel resins were in extensive use in ion exchange for many years before the invention of macroporous resins. These macroporous resins were found to have two main advantages over their gel counterparts – they were less susceptible to osmotic shock and less liable to organic fouling.

As just discussed, a resin will change in volume as it changes its ionic form. Moreover its volume will contract when it is in contact with strong electrolyte solutions which it will encounter during regeneration, and will rapidly swell again when this regenerant solution is washed off. These volume changes exert an osmotic stress across the boundary between resin and solution, which in the case of a gel resin is the resin bead boundary itself. These forces can weaken or even smash the resin bead. Although fragments of resin are still just as good as a whole bead in the exchange process, they will certainly impede the flow of liquid through the bed, increasing the pressure drop and decreasing the flow. Also, the small fragments can block the strainers of the unit, and even contaminate the treated solution.

A macroporous resin, by nature of its structure of interconnecting pores and matrix, comprises very many small resin–solute interfaces, so although the osmotic force across the bead as a whole will be the same, it will be dispersed across all these small boundaries. Therefore, the stress on the bead will be much less, and there will be very much less chance of the resin bead being weakened or broken.

Most natural waters, as well as many organic solutions which ion exchange resins are used to treat, contain large and complex organic molecules. These will slowly penetrate a resin bead, particularly if the organic molecule itself contains an ionic charge. Because of their size, these molecules become entangled with the resin structure, and are not easily removed when the resin is regenerated. Consequently, these molecules build up within the resin, blocking access of the ions in solution to the exchange sites, with a resulting drop in resin capacity. This is known as *organic fouling*. The pore structure of macroporous resins allows greater freedom of passage of these organic molecules, so they are more freely released during regeneration. Anion resins based on an acrylic matrix, due to the nature of this matrix, are also less susceptible to organic fouling.

Unfortunately, macroporous resins also have a major disadvantage. Part of the resin bead, the pores, is purely water with no ion exchange properties. In order that the overall number of ion exchange sites within a given volume of resin is similar to a gel resin, the resin matrix itself has to contain less water – it will be more highly cross-linked. With such a matrix having a lower amount of water associated with the structure, ions will be able to move less freely, so causing the resin to exchange ions more slowly. In practical terms, this results in either reducing the rate at which the solution to be treated can be fed to the resin, or reducing the efficiency of the ion exchange process.

Since optimum efficiency of ion exchange (known as ion exchange *kinetics*) is an important factor in the design and operation of a treatment plant, gel resins will normally be preferred except in circumstances where the advantages of macroporous resins outweigh this major disadvantage.

## Resin Regeneration

In most resin applications, the resin is used over many treatment cycles. Once the resin has become exhausted (in other words, when it is no longer removing the ions from solution at the threshold level required by the user), it is then regenerated with suitable chemicals to once again attach to the active group the mobile ion which will subsequently be released into the treated solution. In the case of cation resins in demineralization processes, this will be a mineral acid

and in the case of anion resins, an alkali, usually sodium hydroxide.

Not only do resins swell in water to a different extent when different ions are attached, they have a different affinity for the various ions, and in the case of treatment of solutions containing a mixture of ions, the resin sites will compete at different rates for the different ions. This is known as *selectivity*. Likewise, if an ion for which the resin has a higher selectivity is already attached to an active group, it will be much more difficult for an ion of a lower selectivity to displace it from the resin.

Consider a solution containing a cation B, and a cation resin in which all the sites are occupied by cation A. As a B ion penetrates through the bead, it will attach to an active site displacing an A ion in the process (**Figure 1**). Initially, since the resin contains only B ions, this released A ion can do nothing but emerge from the resin back into solution. However, as more and more sites within the bead take up B ions, there is a chance that the released A ion could displace another B ion which had already been taken up on a different site. The ease with which this would occur depends on the relative selectivity of ions A and B to the resin – if the resin is significantly more selective for the B ion, then this is less likely than if the resin were more selective for the A ion.

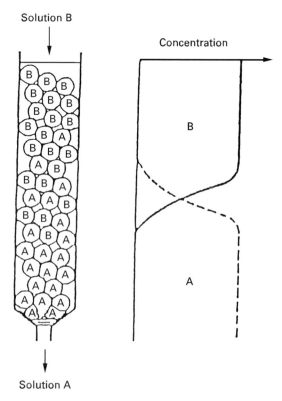

**Figure 1**  Ion exchange equilibrium.

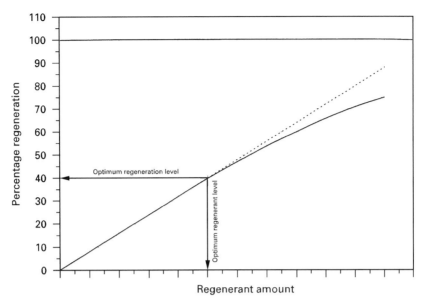

**Figure 2**   Typical regeneration curve. Dotted line, optimal; continuous line, practical.

When a resin is being regenerated, the more regenerant which is passed, the more completely will the resin be regenerated. However, because of this interchange of ions within the resin, the percentage of resin regenerated is not a linear function of the amount of regenerant (**Figure 2**). In other words, doubling the amount of regenerant used will not necessarily double the number of sites regenerated. Since the operational cost of an ion exchange resin plant is heavily influenced by the regenerant cost, it is not usually practicable to regenerate a resin fully, rather to balance the percentage regeneration to the most economic use of the regenerant. Consequently, a quantity of regenerant much more than the point at which the regeneration curve loses linearity is unlikely to be used.

The shape of this regeneration curve will vary with different resins, depending not only on the nature of the active group and the moisture retention, but also on the structure of the resin matrix itself.

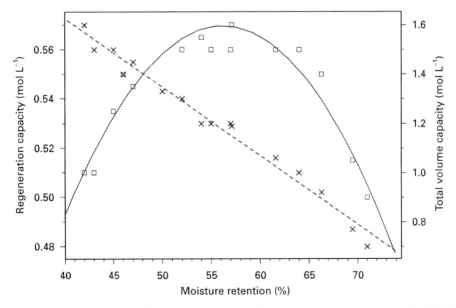

**Figure 3**   Comparison of regeneration (squares) and volume capacities (crosses) (on the same resin samples). SBA Type I gel resins regenerated with 65 g NaOH per litre resin.

**Table 1** Typical regeneration efficiencies of different resin types

| Resin | Regeneration efficiency (%) |
|---|---|
| SAC gel | 50–60 |
| SAC MP | 45–55 |
| WAC gel/MP | 80–95 |
| SBA I gel | 35–45 |
| SBA I MP | 30–35 |
| SBA II gel | 60–70 |
| SBA II MP | 55–65 |
| SBA III MP | 45–55 |
| WBA MP | 80–95 |

Even two resins of the same functionality and total exchange capacity might regenerate differently.

Strongly acidic or strongly basic resins readily exchange ions from neutral salts. The exchange process is therefore an equilibrium, and the ion displaced from a resin site will be capable of exchanging with the ion on another site. In the case of a weakly functional resin, the exchange is more of neutralization reaction, and there is little chance of the released ion exchanging with another site on the resin. For this reason, the regenerability of a resin increases as its acidity or basicity decreases.

### Regeneration capacity

In this test, a known volume of fully exhausted resin is put in a column, and a fixed amount of regenerant passed through. The type and quantity of regenerant would be appropriate to the application under consideration. The equivalents of ions displaced or adsorbed by the resin are determined, and the regeneration capacity can be expressed in terms of equivalents per litre of resin, or as a percentage of the total volume capacity of the resin (known as *regeneration efficiency*), which can conveniently be determined on the same measured volume of resin.

**Figure 3** clearly shows the significance of moisture retention on the kinetic performance of a resin. As discussed earlier, a gel resin of a given type with a lower moisture retention will have a higher total capacity, but this higher capacity is not reflected by the regeneration capacity, and under many conditions the operating capacity, of the resin.

For the reasons discussed earlier, macroporous resins will generally have a lower regeneration efficiency than their gel counterparts. **Table 1** gives an approximate indication of the regeneration efficiencies of the different categories of resin, using typical regeneration amounts of 60–80 g regenerant per litre of resin.

## Conclusions

The main characteristics of ion exchange resins are summarized in **Table 2**.

Most current developments in ion exchange resin manufacture have been based on modifying production techniques to give more uniform bead distributions at lower production costs. One exception has been the introduction of a range of highly porous Macronet adsorbent resins which are finding applications in the removal of trace levels of organic contaminants in aqueous, and even gaseous, feed stocks, as well as replacing activated carbon in the removal of coloured bodies from sugar syrups.

There are some new developments of resins as pharmaceutical products or in the extraction of precious metals from spent ores. Newer and more exotic applications are continually being found for ion exchange resins outside the conventional water treatment field, but these generally use existing resin types or only slight modifications thereof.

*See also:* **II/Ion Exchange:** Historical Development; Inorganic Ion Exchangers; Organic Ion Exchangers; Organic Membranes; Theory of Ion Exchange.

**Table 2** Typical characteristics of ion exchange resins

| | Strong acid | Weak acid | Strong base | Weak base |
|---|---|---|---|---|
| Functional group | $-SO_3^-H^+$ | $-COOH$ | $-CH_2N^+OH^-$ $\quad$ $(CH_3)_3$ | $-CH_2N(CH_3)_2$ $\quad$ $(CH_3)_2$ |
| Effect of pH on exchange capacity | Largely independent | Negligible in acid solutions | Largely independent | Negligible in alkaline solutions |
| Resin salts | Stable | Hydrolyse on washing | Stable | Hydrolyse on washing |
| Regeneration | Excess strong acid required | Readily regenerated | Excess strong base required | Readily regenerated |

## Further Reading

Adams, BA and Holmes EL (1935) *J. Soc. Chem. Ind.*, 54.

Dale JA and Irving J (1992) Comparison of strong base resin types. In: Slater MJ (ed.) *Ion Exchange Advances – Proceeding of IEX'92*, pp. 33–40. London/New York: Elsevier Applied Science.

Golden LS and Irving J (1972) Osmotic and mechanical strength in ion-exchange resins. *Chemistry and Industry*, 837–844.

Mikes J (1958) *J. Polym. Sci.*, 30, 615–23.

Williamson WS and Irving J (1996) A preliminary comparison of Type II and Type III strong base anion resins at the new Plymouth power station, New Zealand. In: Grieg JA (ed.) *Ion Exchange Developments and Applications – Proceedings of IEX'96*, pp. 43–50. UK: The Royal Society of Chemistry.

# ION EXCHANGE: ZEOLITES

*See* **III / ZEOLITES: ION EXCHANGERS**

# ION FLOTATION

**L. O. Filippov**, Laboratoire Environnement et Minéralurgie, INPL-ENSG, Nancy, France

## Introduction

Sebba published a paper in 1959 in which he discussed a new method (ion flotation) for recovering solute from dilute solutions by adding surfactant, with subsequent adsorption of the solute onto bubbles. The principles of the process and the characteristics of the solute–surfactant product formed in solution were discussed in his monograph on ion flotation published in 1962. The method rapidly became popular and researchers in several countries have studied various aspects of the separation of metallic ions, trace elements, molecules, inorganic anions and organic matter from aqueous solutions. Many laboratory-scale studies have been carried out, most of them aimed at development, analytical applications, water purification, resource recycling, removing radionuclides from solutions, and recovering metals from sea water.

A comprehensive development of all aspects of the subject was presented in a monograph on adsorptive bubble separation techniques, edited by Lemlich in 1972, in which details and applications of ion and precipitate flotation methods were reported by Pinfold.

The research group directed by Grieves made an important contribution to the theoretical and applied aspects of ion and precipitate flotation during the 1960s, particularly on wastewater treatment. They showed that flotation efficiency of long-chain surfactants was the result of physicochemical aspects of particle growth and dispersion. But the adsorption of surfactant onto the solid and gas phases was identified as a factor limiting bubble-particles attachment in some cases of ion flotation.

The adsorption of the surface-active solutes to the gas–liquid interface was studied by Rubin. An analysis based on the Gibbs and Langmuir isotherm and on an originally developed approach of long-chain ion adsorption in a solution containing several surface-active species was used to determine the effect of their concentrations on the ratio of distribution coefficients. This author also described the kinetic parameters for ion and precipitate flotation.

A detailed review of the precipitate and adsorbing colloid flotation technique with a comprehensive literature review appears in the monograph published in 1983 by Clarke and Wilson.

Golman has given a qualitative description of the chemical and kinetic aspects of ion flotation and some industrial applications, including the removal of molybdenum from solutions of hydrometallurgical flowsheets. He has also given methods for treating foam products and purifying process residual solutions.

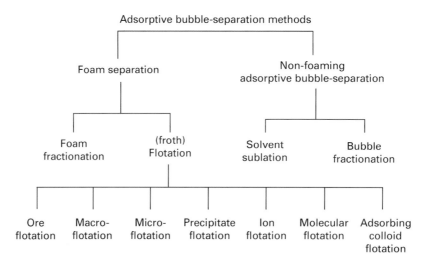

**Figure 1** Schematic classification of the adsorptive bubble-separation techniques. (Reproduced with permission from R Lemlich (1972) *Adsorptive Bubble Separation Techniques*. New York: Academic Press. Copyright.)

Relatively few studies on this subject have been published recently. They focus mainly on extending the applications of the method. The progress made in column flotation techniques, especially with bubble generation systems, offers hydrodynamic conditions favourable for ion flotation. Filippov has reported on the results of recent pilot-scale studies applying this technique to precipitate and ion flotation. It was demonstrated that aggregate formation and destruction under low dissipation energy are the parameters controlling precipitate flotation. The combined use of column and bubble spargers in ion flotation provided a removal rate for metals equivalent to that of laboratory-scale trials.

## Phenomenology and Classification

Separation by adsorption onto bubbles is based on the difference between the surface activities of the solute species. These species can be ionic, molecular, or colloidal, and their adsorption onto the bubble–liquid interface depends on their surface-active, adhesion or electric properties. A species that is not adsorbed onto bubbles can be made to do so by adding a surfactant to the solution.

The classification of methods based on these separation phenomena are shown in **Figure 1**. Lemlich called these processes 'adsbubble (adsorptive bubble) separation methods', while the term 'adbubble method' was recommended by Sasaki, since the method involves both adsorption and adhesion. The classification of these techniques in Figure 1 is based on the formation of the foam, which leads to two main groups:

- foam adsorptive separation and
- non-foaming bubble separation.

The nature of the entity is introduced later.

The Golman classification based on the phenomena at the various levels of process leads to another classification of ion recovery process (**Table 1**). This classification take into account:

- the type of phase that accumulates the floated species: foam, scum, organic or aqueous phase;
- the nature of the components to be adsorbed onto the bubble surface: ions or molecules, particles of the precipitate or carrier; and
- the collector use to modify the hydrophobic properties of the entities

## Ion Flotation

### Definition

Ion flotation, developed by Sebba, is a surface-inactive separation method that involves the removal of ions or molecules (*colligend*) from aqueous solution by adding surfactant, that is adsorbed onto the surface of rising bubbles. The surfactant–colligend product (*sublat*) may be formed in bulk solution or

**Table 1** Classification of ion-recovery method

| Flotation | Flotation | | | |
|---|---|---|---|---|
| | *Froth* | *Foam* | *Solvent sublation* | *Bubble fractionation* |
| Adsorptive | | Without collector With collector | | |
| Precipitate | | Hydrophobic Hydrophobized | | |
| Carrier | | Hydrophobic Hydrophobized | | |

only at the higher concentrations produced by preferential adsorption on the bubble surface. The process is called *adsorption* ion flotation if the sublat is a soluble complex or a pair of ions. The process is *adhesion* ion flotation if the sublat forms a new phase in aqueous solution. A hydrophobic product (*scum*) is formed at the surface of the solution by destruction of the rising bubbles.

The formation of foam is not necessary for ion flotation. The hydrophobic nature of the scum makes it stable on the solution surface. A foam thin-layer phase may be needed to isolate the scum from the liquid phase and to evacuate it later to avoid it redissolving. Stable foam is a factor limiting the application of ion flotation, because it requires less foam formed or a lower gas rate. Solution entrainment also decreases the colligend concentration in the foam.

If the formation of a foam during ion flotation is undesirable or impossible (i.e. recovery of organic ions or quantitative separation), the process of *solvent sublation* is used. This method involves spreading a thin immiscible organic solvent layer on the surface of the water causing dissolution of the floating sublat.

## Theory

There are several ways to describe ion flotation and to determine the quantity of surfactants required for optimal separation. One considers the bubble surface to be an ion exchanger owing to surfactant adsorption. The charge created is compensated for by the adsorption of inactive ions of opposite charge. Jorne and Rubin assumed that the radius of the hydrated ions determined the maximum approach of opposite ions to the bubble surface, based on the theory of a double electrical layer. Their theoretical calculations were confirmed experimentally. Another approach highlights the stability constants of soluble compounds (complexes and pairs of ions) formed by the surfactant with the colligend and an oppositely charged ion, according to Moore and Philipps. Some believe that a solid phase is formed (assumed to be two-dimensional) in the adsorption layer on the bubble surface. These assumptions mostly concern the adsorption mechanism of ion flotation.

Sebba and Golman used the product of the activities of the collector and colligend $(L_A)$ to explain adhesion. The equilibrium of the system is determined by the constant of stability of sublat $K_A$ and the solubility of its molecular form $S_{AM}$, when $L_A = K_A S_{AM}$. If we assume that a sublat flotation occurs as a colloid rather than molecules, the fraction $P$ of the stoichiometric ratio $\Phi$ of collector/colligend molar concentrations is given according to Golman by:

$$P = R + \frac{l}{m} \sqrt[m]{\frac{L_A}{f_{RX}^m f_A^l (1 - R)^l C_A^{m+l}}} \qquad [1]$$

where $R$ is the rate of colligend recovery; $f_{RX}$, $f_A$ are the surfactant RX and colligend A activity coefficients respectively; $C_A$ is the colligend concentration; $m$ and $l$ are the stoichiometric coefficients of the reaction of sublation of A by RX.

The parameter $(L_A)_{P/R}$ calculated from this formula allows the deduction of the value of $L_A$ that is needed to obtain a recovery $R$ for a given $\Phi$. This formula does not take into consideration changes in the ionic strength $I$ with collector concentration. However, for the characteristic colligend concentrations and for $\Phi$ being practically stoichiometric the parameter $I$ does not limit the application of this approach. This was confirmed by Golman for the ion flotation of various species.

The influence of the surface-active species concentration and solubility product $P_S$ of sublat $(P_S = \text{constant for } I = \text{constant})$ on the recovery of the colligend and the residual surfactant concentration is given in **Figure 2**. This confirms the experimental results. It is preferable to carry out ionic flotation in the concentration ranges of the collector so that $P = 1$ (values of $\Phi$ are stoichiometric). Colligend recovery when $P < 1$ is often reduced. The

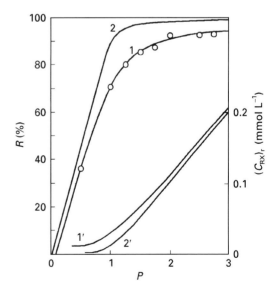

**Figure 2** Colligend recovery $R$ (1, 2) and surfactant residual concentration $C_{(RX)r}$ (1′, 2′) vs. surfactant consumption for a given sublat solubility product $P_S$: 1,1′ − $P_S = 10^{-9}$, 2,2′ − $P_S = 10^{-10}$. Solid lines, calculated results from eqn (1); ○, experimental results of ion flotation of ReO$_4^-$ with laurylammonium chloride, $L_A = 3.9 \times 10^{-7}$. (Adapted from AM Golman (1982) *Ionnaya Flotatsiya*, p. 42. Moscow: Nedra.)

considerable residual concentration of collector in the bulk solution when $P > 1$ decreases the process efficiency (economic). This makes it difficult to reuse a solution without purification and/or environmental problems. The excess collector also prevents flotation because of secondary adsorption onto bubbles and sublat surfaces if the sublat is solid, as demonstrated by Grieves and Golman.

### Colligend:Collector Ratio

Ion flotation operates with dilute solutions of the colligend ($10^{-5}$ to $10^{-3}$ mol $L^{-1}$). Higher colligend concentrations require significant collector consumption, increasing the operation costs. The ratio $\Phi$ of collector and colligend molar concentrations is one of the main parameters of ion flotation. The changes in the amount of colligend removed with $\Phi$ is shown in **Figure 3**, which are typical of ion-flotation systems.

As noted by Pinfold, the ratio $\Phi$ required for complete flotation must be at least stoichiometric ($\Phi_{st}$). This is true for the adhesion mechanism of ion flotation, but with adsorption, the amount of collector that can float without sublation depends on the bubble residence time in the liquid. The colligend cannot be completely removed in this case with a stoichiometric $\Phi$. The experimental results of Doyle, available in the literature, show that slightly more than stoichiometric amounts ($P = 1.1$–$1.2$) of sodium dodecyl sulfate were needed to reduce the heavy metals ion concentration to very low levels. Moreover, the curve behaviour around point $P = 1$ in Figure 3 could indicate the mechanism of ion flotation.

### Role of Electrolytes and Anions

The role of electrolytes is significant during the collector-colligend and sublat–bubbles interaction. Their role must be taken into account because of their significant quantities in real industrial solutions. Electrolytes modify the ionic strength and can react with the collector. For $I$ = constant, two cases are possible:

1. The opposite-charged ion forms a soluble product with the collector. The separation is more selective for higher colligend and for low opposite-charged ion concentrations. A higher collector concentration than that required by $\Phi$ also renders ion flotation more efficient.
2. The collector–colligend interaction product is insoluble. A critical concentration of the opposite ion can be defined, below which it does not react with the collector and consequently does not influence further colligend recovery.

Changing the ionic strength by adding NaCl allows the selective separation of metal oxyanions ($MeO^{4-}$ or $MeO^{2-}$) with hexadecyldimethylbenzylammonium chloride as collector (**Figure 4**). The recovery of oxyanion flotation as a function of p[NaCl] illustrates the phenomena described above.

The change in ionic strength of the medium caused by adding anions reduces the effectiveness of ion flotation. The flotation of dichromate with a quaternary amine (adhesion mechanism) is blocked in the sequence, according to Grieves: $PO_4^{3-} > SO_4^{2-} > Cl^-$. The recovery of $Fe(CN)_6^{4-}$ with a cationic collector is influenced by: $CN^- > NO_3^- > Cl^- > SO_4^{2-} > CO_3^{2-} > PO_4^{3-} > P_2O_7^{4-}$. This contradiction can be explained by the process taking place in each case.

The foam flotation of dichromate is controlled by the preliminary formation of a solid phase followed by adsorption onto bubbles (similar to precipitate flotation). The adsorption of differently charged ions influences surface hydration of the bubbles and the precipitate, the importance of which is directly related to the charge of the anion. Ferricyanide flotation is by solvent sublation. As was noted by Pinfold, the smaller ionic radius of the anion, the more effective the collector competes with the colligend. The use of 2-octanol to collect the sublat on the solution

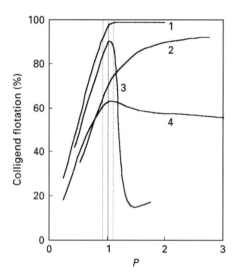

**Figure 3** Dependence of the flotation of various colligends on the fraction ($P$) of collector : colligend ratio ($\Phi$): (1) $Ge^{3+}$ + tetradecylammonium chloride ($\Phi_{st} = 2$); (2) $ReO^{4-}$ + laurylammonium ($\Phi_{st} = 1$); (3) $Cr^{6+}$ with hexadecylammonium bromide ($\Phi_{st} = 1.04$); (4) $Ga^{3+}$ + amide oxime ($\Phi_{st} = 3$). (Adapted respectively from: (1,2) AM Golman (1983) *Fiziko-khimitcheskie Aspekty Ionnoi Flotatsii*, p. 245, Moscow: Nauka; (3) *The Chemical Engineering Journal* 9: R Grieves Foam Separations: A Review, 93, Copyright (1975), with permission from Elsevier Science; (4) with permission from A Masuyama et al. *Industrial Engineering Chemical Research* 29: 290, Copyright (1990) American Chemical Society.)

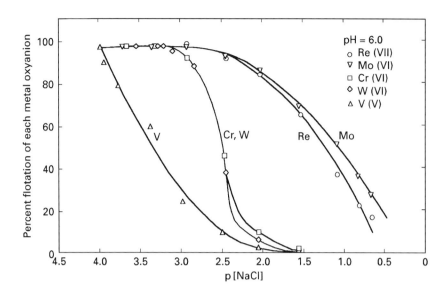

**Figure 4** Effect of the negative logarithm of the sodium chloride concentration on the percentage flotation of each of five metal oxyanions. (Reprinted from *The Chemical Engineering Journal* 9: R Grieves Foam Separations: A Review, 93, Copyright (1975), with permission from Elsevier Science.)

surface can cause solvent to dissolve and become adsorbed onto the bubble surface instead of the collector. Under these conditions the lower charged anions neutralize the adsorbed collector, so that it is no longer available to float the colligend.

### pH

Ion flotation is particularly sensitive to the pH because the pH determines the nature and the charge of the collector (the degree of ionization) and the colligend (hydrolysis), and causes variations in the ion-collecting mode. Pinfold also notes that the following phenomena that can take place when the pH changes:

- colligend hydroxides may form, and precipitate flotation may take place instead of ion flotation;
- extreme values of pH block flotation because the ionic strength is higher;
- the stability of scum could be affected because of sublat redispersion in the solution.

### Temperature

The temperature influences all aspects of ion flotation, i.e. the collector and sublat solubility, the sublat particle size and the flotation results. Its role were confirmed by Grieves for the flotation of $Cr^{6+}$ at pH = 4.1 with quaternary amine salts with hydrocarbon chains of $C_{14}$, $C_{16}$ and $C_{18}$. The collector used acted as a precipitant, dispersing and flocculating agent depending on the temperature, determining the orientation of the collector molecules adsorbed on the sublat surface. The surfactant also acted as foaming agent because of the free collector adsorbed on the bubbles.

Optimal metal removal with $C_{16}$- and $C_{18}$-amine salts occurred with $\Phi = 1.04$ and 33 or 40°C. The abrupt drop in removal rate at higher values of $\Phi$ was explained by excess surfactant adsorbing to the particles, stabilizing them and preventing further aggregation. One of the most significant conclusions deduced from these experiments is that the particles of the precipitate $> 25\ \mu m$ are completely removed from solution, while almost no particles $< 7\ \mu m$ are floated.

## Precipitate Flotation

According to Pinfold, precipitate flotation includes all processes in which an ionic species is precipitated in the liquid phase and is subsequently removed by attachment to the bubble surface. It is difficult to clearly distinguish between 'ion flotation' and 'precipitate flotation' when a collector is used as a precipitation agent.

If the colligend is first precipitated by a non-surface-active ion and made hydrophobic by adsorption of a surfactant, the process is termed *precipitate flotation of the first kind*. Many studies have been carried out on the removal of metal ions from aqueous solution (i.e. sea water) by this method. Heavy metals are generally precipitated with an alkali as the hydroxide and then removed by flotation with an ionic collector. The other insoluble salts (sulfide, carbonate, sulfate) can be precipitated.

The *precipitate flotation of the second kind* uses no surfactant for bubble-particle attachment because it suggests that the solid phase formed by interaction of two hydrophilic species (colligend and precipitation agent) is hydrophobic. The following precipitants have been used for various metal ions: benzoinoxime (Mo, Cu), benzoylacetone (U), $\alpha$-furyldioxime (Ni), hydroxyquinoline (Cu, Zn, U), $\alpha$-nitroso-$\beta$-naphthol (Ag, Co, Pd), dodecylpyridinium with a collector (Sr, V), etc.

## Kinetics

Ion flotation controlled by adhesion is a precipitate-like flotation because the sublat formed is a solid phase that can also be flocculated by adding active chemical agents or by the action of the collector. The two ionic flotation modes (adsorption and adhesion) may be distinguished at the level of particle–bubble interaction, depending on the nature of the sublat.

- The sublat formed is an ionic pair or a soluble complex (adsorption): the adsorption activity of these species on the bubbles determines the effectiveness of the process.
- The sublat is formed in the liquid phase as particles of $10^{-3}$ to $10^{-1}$ μm: the interaction of the sublat and the bubbles is controlled by the diffusion of the particles in the hydrodynamic fields of the bubbles.
- The sublat is a precipitate with micron- or millimetre-sized particle: the interaction results in sedimentation of the particles of negligible mass on the bubbles if the surface forces support the attachment of the entities.

The time necessary for 90–99% colligend recovery at colligend concentrations of $10^{-5}$ to $10^{-2}$ mol L$^{-1}$ indicates that sublat flotation as a precipitate is kinetically preferable than flotation of the molecules (**Figure 5**). The flotation mechanism was concluded from independent experiments on the solubility product, microscopy, separation of the precipitated phase by centrifugation and filtration.

Few studies on the kinetics of ion flotation are published, while there are many papers on mineral particle flotation. However the kinetic parameters of the process determine the practical applications of flotation because they determine the scale-up procedure adopted.

## Apparatus: Future Developments

Almost all ion flotation tests are carried out in cells (batch or continuous mode) equipped with a sintered-glass device to generate bubbles. The use of the column-type cells allows the user to: vary the introduction point of the collector and feed solution containing the colligend; vary the height of the foam;

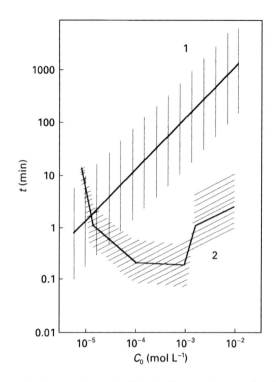

**Figure 5** Comparison of the kinetic of adsorption and adhesion mechanism of ion flotation. (1) Adsorption mechanism; (2) adhesion mechanism. (Adapted from AM Golman (1982) *Ionnaya Flotatsiya*, p. 29. Moscow: Nedra.)

| Colligend | Collector |
|---|---|
| Ba$^{2+}$ | Laurylsulfonate-Na |
| Ag(S$_2$O$_3$)$^-$ | Cetyldiethylbenzylammonium-Br |
| UO$_2$(CO$_3$)$_3^{4-}$ | Cetyltrimethylammonium-Br |
| H$_3$W$_6$O$_{21}^{3-}$ | Laurylammonium-Cl |
| H$_3$Mo$_7$O$_{24}^{3-}$ | Cetylpyridinium acetate |
| | Cetyltrimethylammonium-Br |
| | Amine-C14 |
| Ge$^{3+}$ | Laurylammonium-Cl |
| | Cetyltrimethylammonium-Br |

carry out sampling and/or *in situ* measurements if radioactive 'tracers' are used.

The most recent bubble-generating systems make column flotation a flexible tool for ion flotation.

A study of sublat formation and its structural organization showed that a column 75 mm in diameter and 3 m high could be used for ion flotation (Cr$^{6+}$) and for precipitate flotation (molybdenum). The bubble diameter determined the efficiency of separation by ion and precipitate flotation because of collision probability and aggregate stability in the microturbulence created by the rising bubbles. The flotation with small bubbles as in the dissolved-gas technique increases the collision probability. However, the low feed-flow rate of the process is a limiting parameter because of the low velocity of small rising

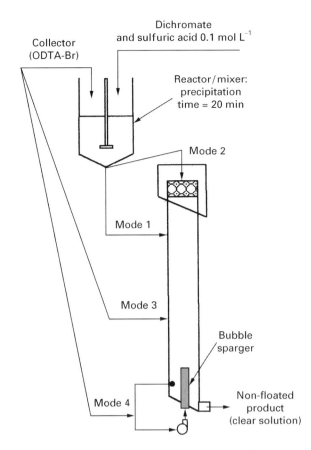

**Figure 6**  Ion flotation modes using column for $Cr^{6+}$ recovery with ODTA-Br. Mode 1: precipitation–flotation. Mode 2: foam fractionating. Mode 3: ion flotation. Mode 4: precipitation on bubbles. (Reprinted from Filippov LO *et al.* Physicochemical mechanisms and ion flotation possibilities using column for $Cr^{6+}$ recovery from sulphuric solutions, *International Journal Mineral Processing* 51: 229, Copyright (1997), with permission from Elsevier Science.)

bubbles. In addition, the flotation mechanism in this kind of technique, in which several small rising bubbles are trapped in a large precipitate floc structure, can cause transfer of the liquid present in the aggregate to the froth and reduce the separation efficiency.

## Ion Flotation in a Column

The ion flotation of $Cr^{6+}$ from sulfuric acid solutions (pH 1.5–2.0) with octadecyltrimethylammonium bromide (ODTA-Br) as collector is particularly difficult because a single chemical acts as precipitating, flocculating, dispersing and frothing agent. This is in addition to the problems of floating an element of negligible mass. It is thus nearly impossible for the precipitate to become adsorbed on bubbles because of an electrical charge of the same sign (owing to free collector) on the bubbles and precipitate particles. Several modes of ion flotation have been tested on a pilot scale (**Figure 6**).

In the classical mode of ion flotation (precipitation–flotation as described by Grieves), only the 200–240 μm diameter bubbles provide a low chromium recovery of 56.2–60.8%. An exceptionally stable, loaded froth with the liquid (7.0–19.0%) was carried over, which decreased the separation efficiency (**Table 2**). Some column/bubble generator assemblies can be used to solve the problems of bubble-precipitate electrostatic repulsion and collision. It has thus been possible to develop a new mode of ion flotation in which the chromium solution is introduced directly at the feeding point of column and the collector is introduced to the bubble generator. The strong surfactant properties of the collector caused to be adsorbed on the bubbles so that the chromium was precipitated on the gas phase. This new method gives about 81.6% $Cr^{6+}$ removal by column flotation for a $\Phi = 1.2$ and a liquid residence time of 15–20 min. Retreatment of the solution can increase the total Cr recovery to 91.6%, with a Cr residual concentration of 4.5 mg L$^{-1}$, which corresponds to separation results by filtration.

## Precipitate Column Flotation

The limiting conditions for molybdenum metal–organic precipitate flotation in columns are owing to aggregate stability under the turbulence created by upward movement of bubbles, which depends directly on the average bubble diameter. Pilot-scale flotation studies revealed the fundamental influence of the average bubble diameter and dissipation energy on molybdenum recovery in the form of precipitate obtained with the collector α-benzoin oxime for a molar concentration ratio of 2. It was therefore necessary to identify a bubble-size distribution and the gas hold-up in the column to ensure the flotation of low hydrophobic precipitate flocs, as they are extremely brittle with very low values of dissipated energy (0.01–0.05 W kg$^{-1}$). Destruction is conditioned by the aggregation mechanism (cluster–cluster type) and not only causes mean floc size to decrease from 150–350 μm to 30–50 μm, but also produces very fine particles that could elude collision with bubbles. Bubble spargers (Microcel, Flotaire, Imox) tested on a pilot scale provided the required values of these parameters for efficient column flotation (recovery up to 95%) of the precipitate for low superficial gas ($J_g = 0.22$–0.55 cm s$^{-1}$) and feed flow rates ($J_l = 0.19$–0.47 cm s$^{-1}$). An adjustment of the $J_g/J_l$ ratio to optimal hydrodynamic conditions in countercurrent pilot column corroborated laboratory-scale tests in the cell equipped with a fine porosity frit (No. 4).

**Table 2** Main results on the hexavalent chromium removal by column flotation according to Figure 6[a]

| Ion-flotation mode | Chromium concentration in initial solution | | Chromium residual concentration | Chromium removal (%) | Stoichiometric ratio ($\Phi$) | Average bubble diameter (mm) | Liquid entrainment (%) |
|---|---|---|---|---|---|---|---|
| | (mg L$^{-1}$) | (mol L$^{-1}$) | (mg L$^{-1}$) | | | | |
| Mode 1 | 52.3 | 10$^{-3}$ | 22.9 | 56.2 | 2.0 | 0.23 | 12.5 |
| Precipitation–flotation | 52.3 | 10$^{-3}$ | 20.5 | 60.8 | 2.0 | 0.21 | 7.5 |
| | 52.25 | 10$^{-3}$ | 25.7 | 50.4 | 2.0 | 0.21 | 19.0 |
| | 52.24 | 10$^{-3}$ | 21.0 | 59.8 | 1.2 | 0.22 | 6.6 |
| Mode 2 Foam fractionating | 52.52 | 10$^{-3}$ | 17.7 | 66.3 | 2.0 | — | 4.5 |
| Mode 3 | 32.1 | 6 × 10$^{-4}$ | 14.8 | 53.9 | 1.5 | 0.29 | 8.3 |
| Ion flotation | 52.38 | 10$^{-3}$ | 33.0 | 37.0 | 1.2 | 0.27 | 6.2 |
| Mode 4 | 23.6 | 4 × 10$^{-4}$ | 8.0 | 66.1 | 1.2 | 0.36 | 4.2 |
| Precipitation on bubbles | 34.9 | 6 × 10$^{-3}$ | 8.2 | 76.5 | 1.2 | 0.38 | 1.0 |
| | 52.33 | 10$^{-3}$ | 8.7 | 83.0 | 2.0 | 0.33 | 1.7 |
| | 52.33 | 10$^{-3}$ | 10.1 | 80.7 | 1.2 | 0.26 | 0.8 |
| | Retreatment of tail solution | | 4.4 | 91.6 | | 0.32 | — |

[a](Reprinted from Filippov LO *et al.* Physicochemical mechanisms and ion flotation possibilities using column for Cr$^{6+}$ recovery from sulphuric solutions, *International Journal Mineral Processing* 51: 229, Copyright (1997) with permission from Elsevier Science.)

## Other Ion Flotation Related Processes

### Bubble Fractionation

This is the partial separation of components within a solution by the selective adsorption of surfactants, colloid or ultrafine particle species onto the bubble. The effect of separation is demonstrated by a concentration gradient along the column-like cell that allows removal of a colligend-rich solution from the top and depleted solution from the bottom of the cell. Separation efficiency clearly decreases with increasing column diameter as a result of axial diffusion of rising bubbles, which breaks up the concentration gradient.

### Adsorbing Colloid Flotation or Carrier Flotation

This consists of the preliminary capture of colligend by the carrier particles (by adsorption, absorption, or co-precipitation), followed by charged-bubble flotation. Ion-exchange resin, activated charcoal, or the precipitate particles can be used as a carrier. The carrier particles can have flotation properties or be made hydrophobic by adding collector.

### Molecular Flotation (Koisumi)

This is the recovery of molecules using a surfactant. The name molecular flotation is use for all flotation processes that involve the recovery of the molecular colligend or those analogous to ion flotation.

## Further Reading

Clarke AN and Wilson DJ (1983) *Foam Flotation. Theory and Applications.* New York: Marcel Dekker.

Filippov LO, Joussemet R and Houot R (1997) Physicochemical mechanisms and ion flotation possibilities using column for Cr$^{6+}$ recovery from sulphuric solutions. *International Journal of Mineral Processing* 51: 229.

Filippov LO, Joussemet R and Houot R (2000) Bubble spargers in column flotation: adaptation to precipitate flotation. *Minerals Engineering* 13: 37.

Grieves RB (1975) Foam separations: A review. *The Chemical Engineering Journal* 9: 93.

Grieves RB, Bhattacharya D and Ghosal JK (1976) Surfactant–colligend particle size effects on ion flotation: Influences of mixing time, temperature, and surfactant chain length. *Colloid and Polymer Science*, 254: 507.

Karger BL, Grieves RB, Lemlich R, Rubin AJ and Sebba F (1967) Nomenclature recommendations for adsorptive bubble separation methods. *Separation Science* 2: 401.

Lemlich R (ed.) (1972) *Adsorptive Bubble Separation Techniques.* New York, London: Academic Press.

Nicol SK, Galvin KP and Engel MD (1992) Ion flotation – potential application to mineral processing. *Minerals Engineering* 5: 1259.

Sebba F (1959) Concentration by ion flotation. *Nature* 184: 1062.

Sebba F (1962) *Ion Flotation.* Elsevier: Amsterdam, New York.

# ION-CONDUCTING MEMBRANES: MEMBRANE SEPARATIONS

**J. A. Kilner**, Imperial College of Science, Technology and Medicine, London, UK

## Introduction

Ceramic membranes are commonly used in separation processes involving the filtration of particulate matter from a fluid stream. This involves the use of controlled porosity ceramic materials, which essentially act as inert filters. More recently, dense ceramic membranes fabricated from ionic conductors have been proposed as active membranes for the high-temperature separation of oxygen from air for a variety of purposes.

The principle of using a dense ceramic ionic conductor to separate the ion-conducting species is, in itself, not new. The idea has been around since the turn of the century when Nernst first investigated solid electrolyte compositions for use as incandescent filaments in his glower devices. Since then ion-conducting ceramics have been used to separate a number of elements, for example oxygen, hydrogen and gallium, but mainly as a scientific curiosity and only on a laboratory scale. The current industrial interest in membrane separators was instigated by a Japanese group, who, in 1985, investigated dense oxide membranes, which were able to permeate substantial fluxes of oxygen at temperatures of 800–1000°C, with 100% selectivity. Since then there has been a rapid growth of both scientific and industrial interest in the use of these membranes for the separation of oxygen. The industrial interest has been driven by the possibility of providing compact oxygen separation plants for a number of applications including aerospace and medical, petrochemical and manufacturing industries.

This article describes a novel oxygen separation process, based on dense ion-conducting ceramic membranes. It is important to note that two different but related devices can be made from such membranes, both of which can be used in the oxygen separation process. In the first device, the driving force is supplied electrically and the membrane is made of an ionic conductor. In the second device, the driving force is supplied by a gradient in oxygen activity and the membrane consists of a mixed ionic electronic conductor. The materials used in the construction of these devices and the devices themselves are described below.

## Materials for Conducting Membranes

Ionic conduction in solids can be achieved in two main ways. Both require crystal structures in which there are a number of equivalent, partially occupied atomic sites. This condition of partial occupancy can occur either by the disordering of a low-temperature structure over two approximately equivalent sub-lattices, or by promoting large deviations from stoichiometry. Such oxygen ion-conducting materials can be further sub-divided into two groups: *electrolytes* that exhibit predominantly ionic conduction; and *mixed conductors*, materials in which there are both significant electronic and ionic contributions to the total conductivity. A useful concept to introduce at this point is the ionic transference number $t_i$. The ionic transference number is defined as the fractional contribution of the ionic component to the total conductivity. For a mixed conductor with both ionic ($\sigma_i$) and electronic ($\sigma_e$) conductivity, $t_i$ is defined as:

$$t_i = \frac{\sigma_i}{\sigma_i + \sigma_e} \qquad [1]$$

Clearly for solid electrolytes we require $t_i$ to be very close to 1 and for mixed conductors $t_i < 1$.

The ionic conductivity of both types of material is thermally activated. The empirical relationship describing the ionic conductivity is given by:

$$\sigma = \frac{\sigma_0}{T} \exp\left\{\frac{-E_A}{kT}\right\} \qquad [2]$$

where $\sigma_0$ is a constant pre-exponential factor, $T$ is the absolute temperature, $k$ the Boltzmann constant and $E_A$ the observed activation energy for the process. Appreciable ionic conductivity for most oxygen conductors, greater than $10^{-1}\,S\,cm^{-1}$, is only achievable at high temperatures (600–900°C) as the activation energies are always substantial ($\sim 50$–$100\,kJ\,mol^{-1}$).

## Solid Electrolytes

Most practical oxide ion conductors are non-stoichiometric and are characterized by their ability

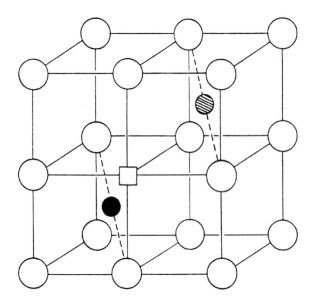

**Figure 1** A half-unit cell of the fluorite structure showing a dopant cation and oxygen vacancy. ○, O²-ion; ●, host cation (4⁺); □, vacancy; ◉, dopant cation (2⁺ or 3⁺).

to accommodate large departures from the ideal oxygen stoichiometry, without breakdown of the crystal structure. For example, oxides of the fluorite structure shown in **Figure 1** display high ionic conductivities when doped with lower valent cations. In one such material, gadolinia doped ceria, the trivalent gadolinium substitutes for the tetravalent cerium ion, causing a charge imbalance in the lattice. The deficit of positive charge is balanced by the formation of oxygen vacancies, which restores the charge balance (see Figure 1). This process can be formalized into a defect equation using Kröger–Vink notation:

$$Gd_2O_3 \rightarrow 2Gd'_{Ce} + V_O^{\bullet\bullet} + 3O_O^x \qquad [3]$$

Ion transport (and hence charge transport) can take place in such materials by a mechanism involving 'hopping' into neighbouring vacant oxygen sites, provided the ion has sufficient energy to overcome the activation barrier, $E_m$, associated with the ionic migration. High ionic conductivity can be achieved by ensuring an optimum level of doping. Because of the relatively stable nature of the cations used in these materials there is a negligible electronic component to the conductivity, especially at high oxygen pressures.

Of the fluorite oxides, considerable attention has been given to the cubic stabilized zirconia ($ZrO_2$) system for application in both the ceramic oxygen generator (COG) and a closely related device, the solid oxide fuel cell (SOFC). Zirconia with 8 mol% $Y_2O_3$ (8-YSZ) adopts the fluorite structure and is

a pure ionic conductor over a wide range of oxygen partial pressures; however, as mentioned above, substantial ionic conductivity is only obtained at temperatures above 900°C. Ceria-gadolinia (CGO) displays similar ionic conductivity to YSZ but at much lower temperatures ($\sim 700$°C), principally owing to a lower value of the activation energy observed for the conduction process. The main concern with CGO is that its electronic conductivity becomes substantial under reducing conditions due to the reduction of $Ce^{4+}$ to $Ce^{3+}$, which renders CGO unsuitable for high temperature applications where low partial oxygen pressures occur.

Of increasing interest is a new series of solid solutions which adopt the perovskite structure (see below) based on the parent compound $LaGaO_3$ (lanthanum gallates). Doped gallates of the type $La_{1-x}Sr_xGa_{1-y}Mg_yO_{3-\delta}$ (LSGM) have comparable ionic conductivity to CGO and do not appear to have any appreciable electronic conductivity. **Figure 2** shows a comparison of selected fluorite and perovskite ionic conductors as a function of temperature.

## Mixed Conductors

Most of the mixed conductors of technological interest adopt the perovskite ($ABO_3$) structure (**Figure 3**). The structure is able to accommodate the substitution of many different cations into its framework, assuming the necessary ion size constraints are met. Again, non-stoichiometry is the key to achieving high transport rates of oxygen; however, in these materials, by definition, the electronic component of the conductivity is not negligible. Taking the perovskite-structured material lanthanum cobaltate ($LaCoO_3$) as an example, when a divalent cation such as strontium is substituted for trivalent lanthanum on the A-site, charge compensation takes place by a dual mechanism. This involves the creation of oxygen vacancies and a change in valency of the cobalt from $Co^{3+}$ to $Co^{4+}$. As described earlier, this substitution can be expressed in a defect equation using Kröger–Vink notation of the type:

$$\tfrac{1}{2}O_2 + 4SrO \rightarrow 4Sr'_{La} + V_O^{\bullet\bullet} + 2h^{\bullet} + 5O_O^x \qquad [4]$$

where $h^{\bullet}$ represents an electronic hole ($Co^{4+}$). In this case both electronic compensation and vacancy compensation of the substitutional occurs, and $t_i$ is usually substantially less than one.

One problem associated with the mixed conducting perovskite and perovskite-related materials is that the dual compensation mechanism leads to changing non-stoichiometry with temperature and atmosphere,

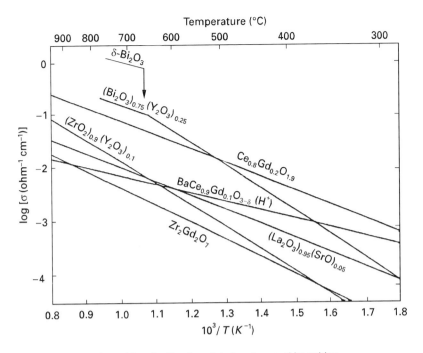

**Figure 2** Oxygen ion conductivity of selected fluorite, fluorite-related and perovskite oxides.

i.e. the balance of vacancy and electronic compensation of the dopant changes with conditions. This manifests itself in a loss or gain of oxygen from the lattice. Most notable is the loss of oxygen with tem-

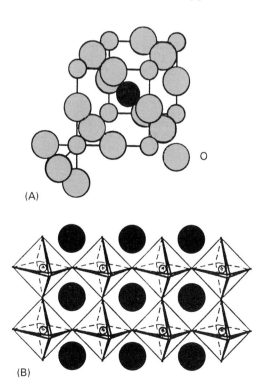

**Figure 3** The perovskite structure. (A) Unit cell and (B) extended structure showing the corner-sharing $BO_6$ octahedra. ●, cation A; ⬤ (grey), cation B.

perature in a constant environment, and/or the sensitivity of the materials to oxygen activity at constant temperature. This loss of oxygen causes a large effective expansion coefficient, which can cause cracking of the membranes upon heating and/or upon the imposition of an oxygen activity gradient.

### Dual Phase

Finding a single phase ceramic material that has high electronic and ionic conductivity while remaining mechanically and chemically stable is not a trivial matter – in fact, these requirements are often mutually exclusive. Materials that exhibit high oxygen ion fluxes also tend to possess high thermal expansion values, which can lead to catastrophic failure in a membrane subjected to a significant oxygen partial pressure gradient, as mentioned above. A way round this problem is to construct dual-phase membranes, effectively making a mixed conductor on a macroscopic scale (**Figure 4**). Such dual-phase materials consist of two separate phases, one an ionic conductor (e.g. YSZ) and the other an electronic conductor (Ag), which are mixed in suitable proportions to provide connectivity for both phases. The individual components are themselves stable at high temperature and in an oxygen pressure gradient. It is not as yet clear what is the best materials combination, or how to optimize the microstructures in order to maximize the oxygen flux that the membrane can transport.

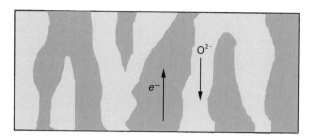

**Figure 4** Schematic of a dual-phase membrane incorporating an ionic and electronic conducting phase.

# Ion-Conducting Membrane Devices

## Pressure Driven Devices

Pressure driven devices are perhaps the simplest form of ceramic oxygen generator. The membrane consists of a dense, gas-tight mixed ionic–electronic conductor (MIEC), which allows the transport of both oxygen ions and electronic species. The driving force for oxygen transport is a differential oxygen chemical potential applied across the membrane. This is achieved by applying a higher partial pressure of oxygen on the membrane feed side than on the permeate side (see **Figure 5**). The whole device may be operated well in excess of atmospheric pressure to achieve a pressurized permeate stream. The most practical design of such a device would be a series of thin-walled tubes. Obviously, the mechanical integrity of such a system is a major concern when very thin membranes are used, and some type of porous support would have to be employed in this case.

The device operates in the following manner. The first step is a surface reaction at the high-pressure surface of the membrane. The gaseous oxygen molecules interact with electrons at 'active' sites on the surface, they dissociate, become ionized and finally are incorporated into the oxide. The rate of this reaction is governed by the gas–solid surface ex-

change coefficient, $k$. The second step is the diffusion of the oxide ions through the material to the lower pressure side where the reverse process occurs and oxygen is evolved, releasing electrons. The membrane in a pressure-driven device is electrically isolated, that is, there are no external electrical current paths. Thus, the membrane material must be a good electronic conductor to provide a return path for the electrons, providing a compensatory flux of electronic species to balance that of the oxygen ions. Normally, the level of electronic conductivity ($\approx 10^2$–$10^3$ S cm$^{-1}$) in these materials is much higher than the corresponding ionic conductivity ($\approx 1$–$10^{-1}$ S cm$^{-1}$), i.e. $t_i \ll 1$, consequently it is the oxygen transport parameters that determine the achievable oxygen fluxes.

In this case, a simplifying model can be applied which allows some insight into the operation of the membranes. The ionic current through the membrane can be described in terms of the simple equivalent circuit shown in **Figure 6**. The apparent potential $\eta$ is provided by the partial pressure drop across the membrane, and is defined using the Nernst equation:

$$\eta = \frac{RT}{4F} \ln \left\{ \frac{P'_{O_2}}{P''_{O_2}} \right\} \qquad [5]$$

where $P'_{O_2}$ and $P''_{O_2}$ refer to the partial pressure of oxygen on each side of the membrane, $T$ is the temperature, $F$ is Faraday's constant, and $R$ is the gas constant. The membrane resistance (expressed as area specific resistance R.A (ohm cm$^2$)) is then described in terms of two components, the resistance of the bulk of the membrane to the passage of the ionic current $R_0$ and the resistance of the surface of the materials caused by the oxygen exchange process. This latter term is expressed as an equivalent electrode resistance $R_E$, assigned equally to both high and low pressure surfaces. The chief utility of this model is that it allows an easy visualization of the processes involved and it can be applied to the electrically driven separator with a slight modification. The ionic current density, $J_{O^{2-}}$ (A cm$^{-2}$), through the membrane of thickness $L$ and of ionic conductivity $\sigma$ can be expressed as:

$$J_{O^{2-}} = \frac{\eta}{(2R_E + R_0)} \qquad [6]$$

$$= \frac{\eta}{\left( 2R_E + \dfrac{L}{\sigma} \right)} \qquad [7]$$

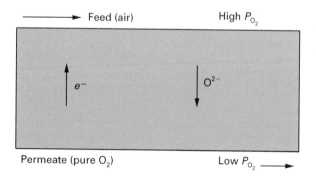

Feed (air)                              High $P_{O_2}$

$e^-$                    $O^{2-}$

Permeate (pure O$_2$)              Low $P_{O_2}$

**Figure 5** Schematic of a pressure-driven ceramic oxygen generator based on a mixed conducting oxide.

$R_E$ can be expanded in terms of the surface exchange coefficient for the oxygen exchange process, $k$,

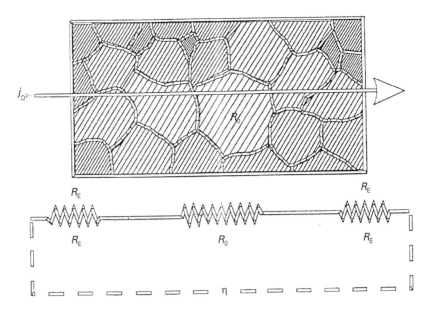

**Figure 6** Equivalent circuit for a ceramic membrane device showing the ionic current.

and the diffusion coefficient, $D$:

$$J_{O^{2-}} = \frac{\eta}{\left\{\dfrac{2D}{\sigma k} + \dfrac{L}{\sigma}\right\}} \qquad [8]$$

The flux of oxygen through a pressure-driven device can be increased by making the membrane thinner (reducing $R_0$), but this is true only up to a certain point. The surface reaction kinetics ($\equiv R_E$) limit the ultimate flux of oxygen through a thin membrane, and beyond a certain characteristic limiting thickness, $L_c$, the flux remains constant. The numerical value of $L_c$ is given by ratio $D/k$, which for most mixed conducting oxides of interest is approximately 100 µm, meaning that a supported thin film structure is probably required for an optimized permeation flux. A comparison of the oxygen fluxes achieved from a range of materials is discussed later.

## Electrically Driven Devices

The electrically driven device consists of a solid electrolyte membrane with electrodes applied to each side to form a tri-layer structure (**Figure 7**). When an electrical potential is applied to the tri-layer, oxygen is reduced at the cathode, passes through the electrolyte as an $O^{2-}$ ion and is evolved as oxygen at the anode. An external electrical connection allows the transfer of electrons from the anode to the cathode. The simple equivalent circuit used for the pressure-driven membrane is still applicable, however, $R_0$ now becomes the electrolyte ASR and $R_E$ the polarization resistance of the electrodes (assumed to be identical).

The flux of oxygen produced by an electrically driven device is directly proportional to the current passing through the membrane $(1\ \mathrm{A} \equiv 3.5\ \mathrm{mL}$ $O_2\ \mathrm{min}^{-1})$, provided the ionic transference number is close to unity. Thus, the flux of oxygen for a given applied potential is governed by the resistance of the membrane (the sum of the electrolyte and electrode polarization resistances) and may be increased by either increasing the potential across the membrane or by reducing the resistance of the membrane. The extent to which the voltage can be increased depends largely on the stability of the electrolyte material. High applied potentials lead to the partial reduction of the electrolyte and the consequent increase in electronic conductivity, leading to a loss of efficiency, for example, in a material such as CGO.

The resistance of the membrane may be reduced by decreasing its thickness. Self-supporting electrolyte

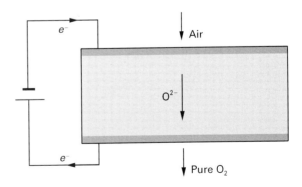

**Figure 7** Schematic of an electrically driven ceramic oxygen generator based on a solid electrolyte. □, Ionic conductor; ▨, mixed conductor.

membranes of the order of 100 μm are available. Thinner electrolytes are not sufficiently strong to support themselves, and therefore further reduction in electrolyte thickness can be achieved by preparing a dense film of the electrolyte on the surface of a porous electrode support. The gains that can be made from switching to thin layers are limited, analogous to the pressure-driven variant. This is because the largest contribution to the resistance of a membrane is usually the polarization resistance of the electrodes, $R_E$, particularly at the lower temperatures of operation. These are independent of the electrolyte thickness ( $\equiv R_0$) and thus again the flux of oxygen is limited by the value of $R_E$.

It is interesting to note that electrode compositions for the electrically driven separator are similar to compositions used as membranes in pressure-driven devices, because the requirements are identical, i.e. a high electronic conductivity and the fast transport of oxygen. Again, the limiting factor turns out to be the kinetics of the surface oxygen exchange reaction at the electrode.

The advantages of electrically driven oxygen separation devices are that large fluxes of oxygen per unit area are possible. At 800°C, an electrical potential of only 0.7 V is equivalent to an oxygen partial pressure gradient ratio of $7 \times 10^{30}$. For applications in which the volume of a device is a key constraint, such as

medical and aerospace applications, an electrically driven device would be the favoured option. An added advantage of an electrically driven COG is that the technology is closely related to that of the solid oxide fuel cell (SOFC) currently under development, and appreciable 'spin-off' is expected. The size of the device can be further reduced if a planar geometry is adopted, similar to that of the planar SOFC. Finally, an electrically driven device is also able to produce oxygen at a higher pressure than the air feed-stock, provided the applied potential across the membrane exceeds the back electromotive force due to the partial pressure gradient of oxygen.

The main disadvantage of electrically driven devices is that they are invariably complicated multicomponent devices. This has implications in terms of the thermal and chemical stability, and the compatibility of the various components at the elevated temperature of operation. The need to develop a high-temperature sealant, required for the planar geometry, is an added complication for these compact devices.

## Oxygen Fluxes

Having shown the principle of the devices and having discussed the materials involved, it is of interest to look at the 'state of the art' in terms of the fluxes of oxygen that can be achieved. **Figure 8** shows a range of fluxes

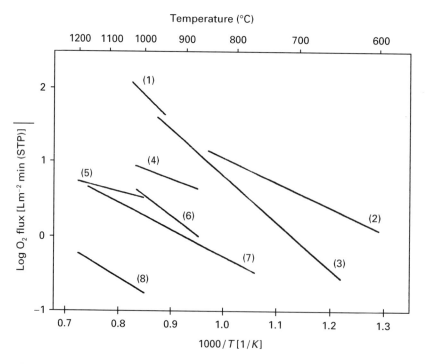

**Figure 8** Arrhenius plots of oxygen permeation for: (1) $SrFeCo_{0.5}O_{3.25-\delta}$; (2) $(Bi_2O_3)_{0.75}$-$(Y_2O_3)_{0.25}$-Ag (35 v/v), 90 μm; (3) $SrCo_{0.8}Fe_{0.2}O_{3-\delta}$; (4) $La_{0.2}Sr_{0.8}Co_{0.8}Fe_{0.2}O_{3-\delta}$; (5) $La_{0.3}Sr_{0.7}CoO_{3-\delta}$; (6) $La_{0.6}Sr_{0.4}Co_{0.2}Fe_{0.8}O_{3-\delta}$; (7) $La_{0.5}Sr_{0.5}CoO_{3-\delta}$; (8) YSZ-Pd (40 v/v), continuous Pd-phase.

measured from samples of single and dual-phase materials, operated in a pressure-driven mode, plotted as a function of inverse temperature. This figure is intended to give some appreciation of the fluxes that are attainable; however, they are not normalized to a given partial pressure gradient or thickness of membrane, and thus the fluxes are not directly comparable. Taking a value of between 10 and 100 L m$^{-2}$ min$^{-1}$ as the level of oxygen flux needed for practical applications, it can be seen that the cobalt-containing single-phase materials give appreciable oxygen fluxes above about 900°C. It is interesting to note on this figure that the dual-phase material, fabricated from $(Bi_2O_3)_{0.75}$ $(Y_2O_3)_{0.25}$–Ag (35 v/v), approaches the lower bound of the practical fluxes at temperatures of 800°C.

It is not sensible to put data for electrical driven COGs on the same figure, given the restrictions mentioned above, however, some comparable figures are interesting. An equivalent flux of 15.8 L m$^{-2}$ min$^{-1}$ is readily achievable with planar COG stack based on zirconia and operating at 1000°C. Similar performance has been reported for a system based on a CGO-operating temperature of 800°C.

## Further Reading

Bouwmeester HJM and Burggraaf AJ (1996) Dense ceramic membranes for oxygen separation. In: Burggraaf AJ and Cot L (eds) *Fundamentals of Inorganic Membrane Science and Technology*, pp. 435–528. New York: Elsevier.

Bouwmeester HJM and Burggraaf AJ (1997) Dense ceramic membranes for oxygen separation. In: Gellings PJ and Bouwmeester HJM (eds) *The CRC Handbook of Solid State Electrochemistry*, pp. 481–553. Boca Raton: CRC Press.

Steele BCH (1998) Ceramic ion conducting membranes and their technological applications. *C.R. Acad. Sci. Paris*, t.1, Serie II c, 533–543.

# ION EXCLUSION CHROMATOGRAPHY: LIQUID CHROMATOGRAPHY

**K. Tanaka**, The National Industrial Research Institute of Nagoya, Nagoya, Japan
**P. R. Haddad**, University of Tasmania, Hobart, Australia

## Introduction

Ion exclusion chromatography (IEC) is a relatively old separation technique, attributed to Wheaton and Bauman, which is now staging an impressive comeback for the simultaneous determination of ionic species. IEC provides a useful technique for the separation of ionic and nonionic substances using an ion exchange stationary phase in which ionic substances are rejected by the resin while nonionic or partially ionized substances are retained and separated by partition between the liquid inside the resin particles and the liquid outside the particles. The ionic substances therefore pass quickly through the column, but nonionic (molecular) or partially ionized substances are held up and are eluted more slowly.

IEC is also referred to by several other names, including ion exclusion partition chromatography, ion chromatography-exclusion mode, and Donnan exclusion chromatography. In this article we use the term ion exclusion chromatography.

Generally, anions (usually anions of weak acids) are separated on a strongly acidic cation exchange resin in the hydrogen form and are eluted as the corresponding fully or partially protonated acids, while cations (usually protonated bases) are separated on a strongly basic anion exchange resin in the hydroxide form and are eluted as the corresponding bases. The eluents used are usually water, water/organic solvent mixtures, dilute (high conductivity) acqueous solutions of a strong acid, or dilute (low conductivity) aqueous solutions of a weak acid. A conductivity detector is commonly used to monitor the column effluent and, when the eluent conductivity is extremely high, a suitable suppressor system is generally used. UV-visible detection is also used as a selective detector in the determination of some aliphatic and aromatic carboxylic acids and some inorganic anions, such as nitrite and hydrogen sulfide. Using IEC, it is posssible to separate weakly ionized anions such as fluoride, phosphate, nitrite, aliphatic carboxylic acids, aromatic carboxylic acids, bicarbonate, borate, aliphatic alcohols, sugars, amino acids, water, and others, as well as ammonium, amines, and others, based on a combination of the separation mechanisms of ion-exlusion, adsorption, and/or size-exclusion. Further discussion of these mechanisms may be found elsewhere in the encyclopedia.

More recently, a new concept in IEC has been developed in which a combination of a weakly acidic cation exchange resin and a weak-acid eluent is used for the separation of strong acid anions (such as sulfate, chloride and nitrate) and weak acid anions by an ion exlusion mechanism, together with the simultaneous separation of mono- and divalent cations by a cation exchange mechanism. The application of this method is described in this article.

Comprehensive reviews of IEC may be found in the texts of Haddad and Jackson, and of Gjerde and Fritz (see Further Reading). The goal of the present article is to explain the fundamental theory and some selected applications of IEC and to focus on some recent developments.

## Background

### Separation Mechanism

In conventional IEC of ionic and nonionic substances, a poly(styrene–divinylbenzene) (PS-DVB) based strongly acidic cation exchange resin in the hydrogen form is used exclusively as the separation column. The resin bed can be considered to consist of three distinct components:

1. a solid resin network with charged functional groups (the membrane);
2. occluded liquid within the resin beads (the stationary phase); and
3. the mobile liquid between the resin beads (the mobile phase or eluent).

The ion exchange resin acts as a hypothetical semipermeable membrane (a Donnan membrane) separating the two liquid phases (2) and (3). This membrane is permeable only for nonionic substances. When a mixture of analytes is injected onto the ion exchange column, anionic analytes are ion-excluded from the occluded liquid phase based on the Donnan membrane equilibrium established by the fixed negative charges on the cation exchange resin and therefore pass quickly through the column. On the other hand, nonionic substances may partition between the two liquid phases (2) and (3) and therefore pass more slowly through the column. Partially ionized analytes experience a lesser degree of repulsion by the membrane and are therefore eluted at retention times intermediate between fully ionized analytes and neutral analytes. This ion exclusion effect can be seen in **Figure 1A**, which shows the separation of aliphatic carboxylic acids.

In addition to this electrostatic ion exclusion effect, the separation process taking place on the surface of the resin particle may be influenced by hydro-phobic adsorption and size exclusion effects, depending on the nature of the solute. These effects can be seen in Figure 1B, which shows the separation of sugars such as mono- and disaccharides by size exclusion, and the separaton of alcohols such as methanol, ethanol, propanol and butanol by hydrophobic adsorption effects (Figure 1C).

Cation exchange resin columns with fairly large dimensions are often used for IEC because the retention volume ($V_r$) of the analyte is determined by the general equation:

$$V_r = V_0 + K_d V_i \qquad [1]$$

where $V_0$ is the interstitial volume, $V_i$ is the volume of eluent occluded within the pores of the resin beads, and $K_d$ is a distribution coefficient ranging from 0 to 1. A large $V_i$ value is needed to obtain good separations because of the narrow $K_d$ range, assuming that only the ion exclusion effect is predominant in the separation of ionic and nonionic substances. This equation is essentially the same as the general equation for size exclusion chromatography.

When the $V_r$ values of analytes measured on a strongly acidic cation exchange resin by elution with water are plotted against $pK_{a1}$ (first dissociation constant), the plot shown in **Figure 2** is obtained. $V_r$ values of strong acids, which are fully ionized, are independent of $pK_{a1}$, showing that the strong acid anions have been completely ion-excluded by the fixed sulfonate ions of the resin. $V_r$ values of the weak acids such as phosphoric, hydrofluoric, formic, and acetic acids increase proportionally with $pK_{a1}$, which shows that the weak acids have been partially ion-excluded by the fixed sulfonate ions of the resin and there has been some permeation of these analytes into the occluded liquid phase inside the resin. This permeation correlated with the $pK_{a1}$ values of the analytes between 1.3 and 6.4. The $V_r$ values of very weak acids such as carbonic and boric acids are independent of $pK_{a1}$. From Figure 2 and eqn [1], it is clear that the $V_r$ values of the strong acids correspond to $V_0$ and the difference between $V_r$ values of the strong acids and the very weak acids corresponds to $V_i$.

The $K_d$ values of strong acids, weak acids, and very weak acids calculated from eqn [1] are between 0 and 1, except for weak acids having a hydrophobic nature, such as propionic, butyric and hydrogen sulfide. For these species, an adsorption effect is evident. As an example, propionic acid is eluted at a larger retention volume than expected from consideration of its $pK_{a1}$ value alone, with the additional retention being attributable to hydrophobic adsorption of the analyte on the unfunctionalized regions of the stationary phase.

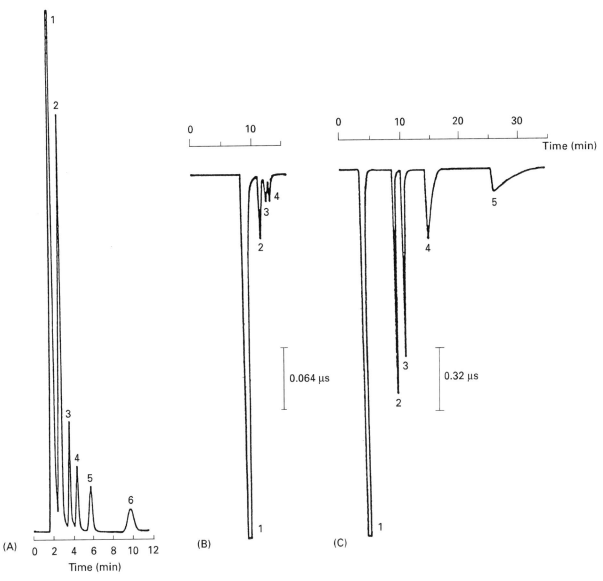

**Figure 1** IEC separation of (A) aliphatic carboxylic acids by elution with 0.5 mmol L$^{-1}$ benzoic acid/5% acetonitrile and (B) alcohols and (C) sugars by elution with 1 mmol L$^{-1}$ sulfuric acid/water on a PS-DVB-based strongly acidic cation exchange resin column (8 mm i.d. × 20 cm long).
(A) Peaks: 1, sulfuric acid ($V_0$); 2, formic acid; 3, acetic acid; 4, propionic acid; 5, butyric acid; 6, valeric acid. (B) Peaks: 1, dip; 2, sucrose; 3, glucose; 4, fructose. (C) Peaks: 1, dip; 2, methanol; 3, ethanol; 4, propanol; 5, butanol. (Figure 1A reproduced with permission from Fritz, 1988, and Figures 1B and 1C from Tanaka and Fritz 1986.)

In addition to the use of PS-DVB-based cation exchange resin, new IEC methods on polymethacrylate-based weakly acidic cation exchange resin and hydrophilic unfunctionalized silica gel have recently been developed to separate some of the more hydrophobic carboxylic acids. **Figure 3** shows the IEC separation of some hydrophobic aliphatic and aromatic carboxylic acids on PS-DVB (Figure 3A) and polymethacrylate-based cation exchange resins (Figure 3B), and on unfunctionalized silica gel (Figure 3C), using 5 mmol L$^{-1}$ sulfuric acid as eluent.

As can be seen from Figure 3, the $V_r$ values of hydrophobic carboxylic acids increase with increasing hydrophobicity of the stationary phase (PS-DVB > polymethacrylate > silica).

Turning now to IEC of bases (cations) performed on strong anion exchange resins in the hydroxide form, some similar retention trends to those described above can be noted. Ionic analytes such as sodium and potasium ions are completely ion-excluded from the fixed postive charged resin phase and are eluted at a retention volume of $V_0$, while nonionic

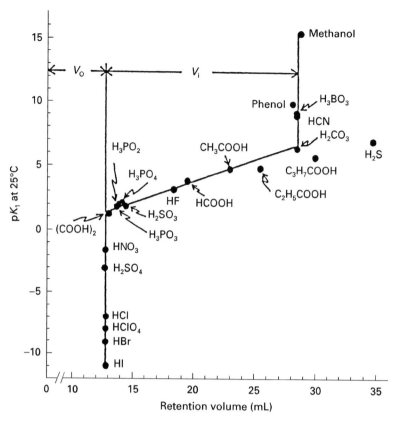

**Figure 2**  Relationship between retention volumes and their first dissociation constants for inorganic anions, carboxylic acids and nonionic substances on a PS-DVB-based strongly acidic cation exchange resin column (8 mm i.d. × 55 cm long) by elution with water. (Reproduced with permission from Tanaka *et al.*, 1979.)

substances will permeate into the resin phase and are eluted at a retention volume of $V_0 + V_i$. Weak bases, such as ammonia and amines, are eluted at intermediate retention volumes, depending on their $pK_b$ values and their hydrophobicity. As a result, weakly ionized cations can be separated from strongly ionized cations by the ion exclusion mechanism.

### Instrumentation, Stationary Phases and Eluents

Ion exclusion chromatographic systems consist of the same components as any high performance liquid chromatography instrument. A UV detector is often used for the sensitive and selective detection of UV-absorbing substances such as aliphatic and aromatic carboxylic acids. However, this detector is insensitive to some aliphatic carboxylic acids, sugars and alcohols. Although a refractive index detector can be used as a bulk detector, the detection response is not high and the detector is flow- and temperature-sensitive. The most popular and universal detection method for IEC is conductivity. In order to decrease or suppress the eluent background conductivity, a membrane

suppressor system can be used (normally when the eluent is highly conducting), or alternatively a weak-acid eluent (aliphatic or aromatic carboxylic acids) of low conductivity can be used.

The most common resins used in ion exclusion chromatography are high capacity PS-DVB-based strongly acidic cation exchange resins of 5 μm particle size. As discussed earlier, polymethacrylate-based weakly acidic cation exchange resin or un-functionalized silica gel can also be employed.

Although the IEC separation of ionic and nonionic substances may be carried out simply by using water as the eluent, dilute aqueous solutions of some mineral acids or weak carboxylic acids give greatly improved peak shape and are therefore preferred for high resolution separations. Decreasing the pH of the eluent increases the retention of weakly ionized analytes such as carboxylic acids owing to a decrease in the fraction of the ionized analyte present. Therefore, the eluent pH is a very important factor in regulating retention volumes in IEC. Organic modifiers such as methanol and acetonitrile are often used to reduce hydrophobic interactions of the analytes with the resin. A further approach that may be used

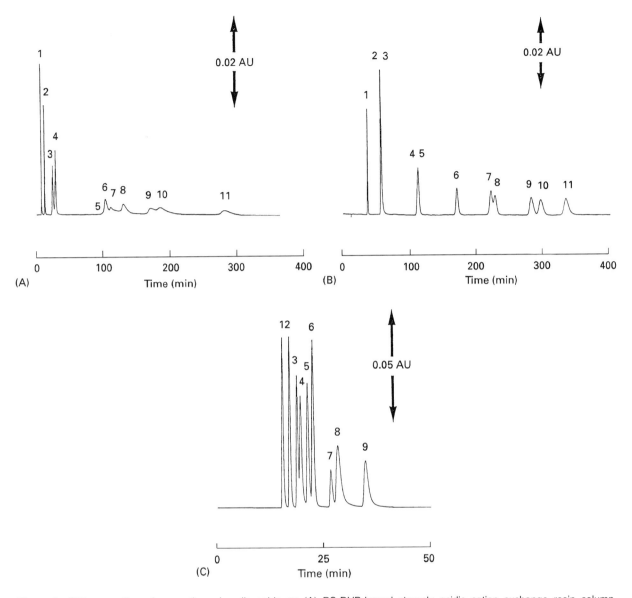

**Figure 3** IEC separation of aromatic carboxylic acids on (A) PS-DVB-based strongly acidic cation exchange resin column, (B) polymethacrylate-based weakly acidic cation exchange resin, and (C) unfunctionalized silica gel by elution with 5 mmol $L^{-1}$ sulfuric acid at 1 mL $min^{-1}$. Column size: 7.8 mm i.d. $\times$ 30 cm long for all.
(A) Peaks: 1, pyromellitic acid; 2, trimellitic acid; 3, hemimellitic acid; 4, phthalic acid; 5, trimesic acid; 6, *m*-hydroxybenzoic acid; 7, phenol; 8, *p*-hydroxybenzoic acid; 9, terephthalic acid; 10, isophthalic acid; 11, benzoic acid. (B) Peaks: 1, hemimellitic; 2, pyromellitic acid; 3, phthalic acid; 4, trimellitic acid; 5, phenol; 6, benzoic acid; 7, *m*-hydroxybenzoic acid; 8, *p*-hydroxybenzoic acid; 9, isophthalic acid; 10, salycylic acid; 12, trimesic acid. (C) Peaks: 1, pyromellitic acid; 2, trimellitic acid; 3, hemimellitic acid; 4, terephthalic acid; 5, isophthalic acid; 6, phthalic acid; 7, phenol; 8, salicylic acid; 9, benzoic acid. (Reproduced with permission from Ohta *et al.*, 1996.)

to decrease the hydrophobic adsorption of analytes onto the resin is the addition of hydrophilic species, such as sugars, polyols and polyvinyl alcohol, to the eluent.

## Optimization of Ion Exclusion Chromatographic Separation

To optimize an IEC separation, careful selection of the following experimental parameters must

be made:

1. the type of matrix used as the stationary phase (PS-DVB, polymethacrylate or silica);
2. the nature of the functional group (e.g. strong or weak acid);
3. the ion exchange capacity (low or high);
4. the nature of the eluent (e.g. strong or weak acids);
5. the pH of the eluent;

6. the amount of organic modifier present in the eluent; and

7. the type of detector used (universal or selective).

## Selected Applications

### Carboxylic Acids

Separation of carboxylic acids is probably the most common use of IEC. Carboxylic acids such as formic, acetic, propionic, butyric, valeric, citric, tartaric, oxalic, malonic, benzoic, salicylic, and others have been determined using UV and conductivity detection. **Table 1** lists some recent applications of IEC for these analytes. These methods have been applied to a wide variety of very complex sample martices, such as biological materials, foods, beverages, pharmaceuticals, environmental materials and others. The separation is almost always performed on a PS-DVB-based cation exchange resin in the hydrogen form; examples of such separations may be found elsewhere in the encyclopedia. As an example of the use of alternative stationary phases, the separation of aliphatic carboxylic acids on a silica gel column by elution with $5 \ \text{mmol L}^{-1}$ sulfuric acid is shown in **Figure 4**.

### Weak Inorganic Acids and Bases

IEC has found increasing use for the determination of weakly ionized inorganic anions such as fluoride, nitrite, phosphate, sulfite, arsenite, arsenate, bicarbonate, borate and cyanide. This approach is very effective for the determination of weakly ionized anions in samples containing a high concentration of

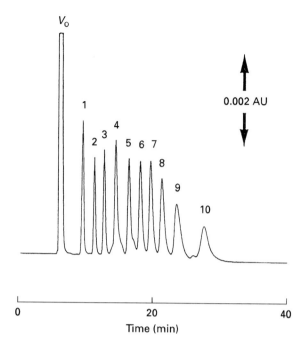

**Figure 4** IEC separation of aliphatic carboxylic acids on silica gel column (7.8 mm i.d × 30 cm long) by elution with 0.05% heptanol/$0.5 \ \text{mmol L}^{-1}$ sulfuric acid at $1 \ \text{mL min}^{-1}$. Peaks: $V_0$, nitric acid; 1, formic acid; 2, acetic acid, 3, propionic acid; 4, butyric acid; 5, valeric acid; 6, caproic acid; 7, heptanoic acid; 8, caprylic acid; 9, pelargonic acid; 10, capric acid. (Reproduced with permission from Ohta *et al.*, 1996.)

ionic species, e.g. seawater and wastewaters. **Figure 5** shows the separation of bicarbonate in tap waters by IEC with conductimetric detection by elution with water. The monitoring of bicarbonate ion is very important for the quality control of tap water and for

**Table 1** Some applications of the determination of carboxylic acids by IEC

| Sample | Column | Eluent | Detection[a] |
|---|---|---|---|
| Fermentation plants | Dionex HPICE-AS6 | 0.4/0.6 mmol L⁻¹ heptafluoro-butyric acid | CD/S |
| Wine | Waters Radial Pak 5 | 0.2 mmol L⁻¹ potassium dihydrogen phosphate | UV |
| Food | Shim-Pak SCR-102H | 2 mmol L⁻¹ toluenesulfonic acid | CD |
| Wine | TSKgel OA-Pak | 0.75 mmol L⁻¹ sulfuric acid | UV |
| Beverage | Shim-Pak IE | 5 mmol L⁻¹ sulfuric acid | UV |
| Silage liquor | Dionex IonPac-IEC AS-5 | 0.9/3.2 mmol L⁻¹ perfluorobutyric acid | CD/S |
| Air | Dionex HPICE-AS1 | 2 mmol L⁻¹ hydrochloric acid | CD/S |
| Ground water | Interaction Ion-300 | 0.2 mmol L⁻¹ octansulfonic acid | CD/UV |
| Wine | Bio-Rad Aminex HPX 87-H | 1 mmol L⁻¹ camphorsulfonic acid | P/UV |
| Rainwater | Dionex HPICE-AS1 | 0.05 mmol L⁻¹ sulfuric acid | CD |
| Rainwater | Hamilton PRPX-300 | 5 mmol L⁻¹ sulfuric acid | UV |
| Air | Aminex-HPX 87H | 0.25 mmol L⁻¹ sulfuric acid/benzoic acid | CD |
| Bread/cake | TSKgel SCX | 2 mmol L⁻¹ phophoric acid | UV |
| Antarctic ice | Bio-Rad HPX-87H | 5 mmol L⁻¹ methansulfonic acid | UV |
| Beverages | TSKgel SCX | 5/10 mmol L⁻¹ sulfuric acid | CD |
| Sewage | Yokogawa SCX-252 | 2 mmol L⁻¹ sulfuric acid | CD/S |
| Rat plasma | Hitachi Gelpak C-620-10 | 0.3% phosphoric acid | FL |

[a]CD, conductivity; S, suppressor; UV, UV spectrometry; P, potentiometry; FL, fluorimetry.

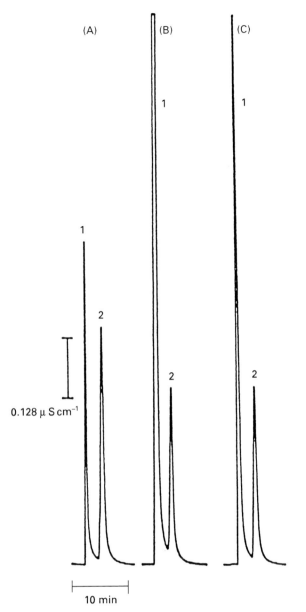

**Figure 5** IEC separation of bicarbonate in tap waters on a PS-DVB-based strongly acidic cation exchange resin column (7.5 mm i.d. × 10 cm long) by elution with water. (A) Raw tap water) (10-fold dilution); (B) tap water after softening treatment; (C) tap water (10-fold dilution). Peaks: 1, strong acid anions; 2, bicarbonate ion. (Reproduced with permission from Tanaka and Fritz, 1987.)

the evaluation of the buffering capacity of natural waters.

Ethanolamines and ammonium ion may be successfully determined by IEC with UV or conductimetric detection on a PS-DVB-based strongly basic anion exchange resin in the hydroxide form. Industrial and environmental applications of the determination of ammonium ion are also common and include samples such as biological treatment process waters and ur-

ban river waters, as shown in **Figure 6**. **Table 2** lists some of the recent applications of IEC in the analysis of weak inorganic acids and bases.

### Strong Inorganic Acids

More recently, a simple and highly sensitive method involving simultaneous ion exclusion/cation exchange chromatography with conductimetric detection on a polymethacrylate-based weakly acidic cation exchange resin in the hydrogen form has been developed for the determination of inogranic strong acid anions such as sulfate, nitrate and chloride ions, and strong base cations such as sodium, ammonium potassium, magnesium and calcium ions commonly found in acid rainwater. Use of a weak acid eluent (such as tartrate) permits both the anions and the cations to be determined, based on a simultaneous

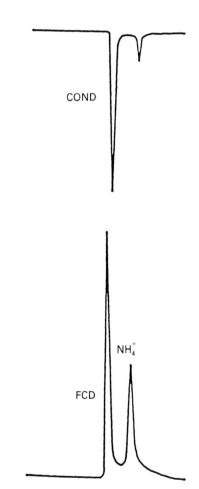

**Figure 6** IEC separation of ammonium ion in biological treatment process water on a PS-DVB-based strongly basic anion exchange resin column (8 mm i.d. × 550 mm long) by elution with water at 1 mL min$^{-1}$.
First peak is alkali- and alkaline earth metal cations. FCD, flow coulometric detector; COND, conductivity detector. (Reproduced with permission from Tanaka et al., 1979.)

**Table 2** Some applications of the determination of inorganic anions and cations by IEC

| Ion(s) | Sample | Column | Eluent | Detection[a] |
|---|---|---|---|---|
| $F^-$, $HCO_3^-$ | Beverages | Shim-pak SCR-102H | 0.5 mmol $L^{-1}$ toluene/sulfonic acid | CD |
| $F^-$ | Wastewater | Hitachi-2613 | 20% methanol/water | CL/CD |
| $HCO_3^-$ | Natural water | TSKgel SCX | Water | CD |
| $HCO_3^-$ | Natural water | Develosil 30-5 (silica gel) | Water/borate | CD |
| $NO_3^-$, $NO_2^-$ | Wastewater | Hitachi 2613 | 10% methanol/water | UV |
| $NO_2^-$ | Drinking water | Anion exclusion | 5 mmol $L^{-1}$ sulfuric acid | EC |
| As(V), As(III) | Sulfuric acid | Dionex HPICE AS-1 | 10 mmol $L^{-1}$ phosphoric acid | UV |
| As(III) | Mineral water | Aminex HPX-85H | 0.01 mol $L^{-1}$ phosphoric acid | EC |
| Silica | Seawater | Dionex Ionpak ICE-AS1 | Water | ICP-M |
| Silica | Natural water | Yokogawa SCX1-251 | 6.6 mmol $L^{-1}$ perchloric acid | CL |
| Borate | – | Excelpak ICS-R3G/ICS-R 35 | 1 mmol $L^{-1}$ sulfuric acid | VIS |
| Borate | Soil | Wescan Ion exclusion | 0.3 mol $L^{-1}$ D-sorbitol | CD |
| $CN^-$ | Tap water | Sulfonated PS/DVB gel | 1 mmol $L^{-1}$ sulfuric acid | VIS |
| $SO_3^{2-}$ | Beer | Dionex HPICE-AS1 | 10 mmol $L^{-1}$ sulfuric acid | EC |
| $PO_4^{3-}$ | Wastewater | Hitachi-2613 | 40% acetone/water | EC |
| $SO_4^{2-}$, $NO_3^-$, $Cl^-$, $Na^+$, $NH_4^+$, $K^+$, $Mg^{2+}$, $Ca^{2+b}$ | Rainwater | TSKgel OA-PAK | 5 mmol $L^{-1}$ tartaric acid/7.5% methanol | CD |
| $NH_4^+$ | Wastewater | Hitachi 2632 | Water | CD/E |

[a]CD, conductivity; UV, UV spectrometry; EC, electrochemical; ICP-MS, inductively coupled plasma–mass spectrometry; CL, chemiluminescence; VIS, visible spectrometry.
[b]Simultaneous ion-exclusion cation-exchange.

ion exclusion and cation exchange mechanism on the same stationary phase. The conductimetric detector responses are positive for the anions (which are separated by IEC and detected directly) and negative for the cations (which are separated by ion exchange and are detected indirectly). The effectiveness of this method has been demonstrated in its application to acid rain, as shown in **Figure 7**.

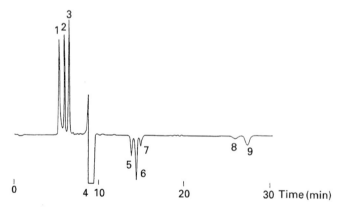

**Figure 7** Simultaneous ion exclusion/cation exchange chromatographic separation of strong acid anions and mono- and divalent cations in acid rain at pH 4.7 on a polymethacrylate-based weakly acidic cation exchange resin column (7.8 mm i.d. × 300 cm long) by elution with 6 mmol $L^{-1}$ tartaric acid–7.5% methanol/water at 1.2 mL $min^{-1}$. Peaks: 1, $SO_4^{2-}$; 2, $Cl^-$; 3, $NO_3^-$; 4, dip; 5, $Na^+$; 6, $NH_4^+$; 7, $K^+$; 8, $Mg^{2+}$; 9, $Ca^{2+}$. (Reproduced with permission from Tanaka *et al.*, 1994.)

**Table 3** Some applications of the determination of nonionic substances by IEC

| Analyte | Sample | Column | Eluent | Detection[a] |
|---------|--------|--------|--------|------------|
| Sugar | Wine, beer | TSKgel SCX | 5 mmol L$^{-1}$ sulfuric acid | CD |
| Sugar | – | Merck Polyspher OA-YH | 10 mmol L$^{-1}$ sulfuric acid | UV |
| Sugar | Juice, milk | Bio-Rad HPLX 87-H Aminex | 0.1 mol L$^{-1}$ sodium hydroxide | EC |
| Sugar | Corn syrup | PS-DVB sulfonate resin | Water | RI |
| Poloyol | – | Bio-Rad HPX-87H | Water | EC |
| Alcohol | Wine, beer | TSKgel SCX | 5 mmol L$^{-1}$ sulfuric acid | CD |
| Formaldehyde | Air | Rezex RFQ | 1 mmol L$^{-1}$ sulfuric acid | EC |
| Dimethyl sulfoxide | Seawater | Bio-Rad HPX-87H | 5 mmol L$^{-1}$ phosphoric acid | UV |
| Water | Organic solvents | Bio-Rad Aminex Q-150S | Acetonitrile/methanol | UV |
| p-Benzoquinone | Wastewater | TSKgel SCX | 20% methanol | UV |
| Ketone | – | TSKgel SCX | 5 mmol L$^{-1}$ sulfuric acid | CD |
| Trichloroethanol | Plasma, urine | Aminex A-15 | 10 mmol L$^{-1}$ K$_2$SO$_4$, 10 mmol L$^{-1}$ KOH | RI |

[a]CD, conductivity; EC, electrochemical; UV, UV absorption; RI, refractive index.

## Neutral Compounds

Neutral compounds such as sugars and alcohols can be separated by IEC, as shown earlier in Figure 1B and 1C. **Table 3** lists some of the recent applications of IEC in this field. One of the more significant applications of IEC is its use for the determination of water. Using a short column packed with PS-DVB-based cation exchange resin in the hydrogen form and eluting with methanol containing a small amount of strong acid, a peak for water can be obtained with a spectrophotometric detector at 310 nm. This method is applicable to the determination of water in some organic solvents.

## Conclusion

Despite the fact that IEC is a relatively old separation technique, new and diverse applications continue to emerge. IEC remains the method of choice for the separation of low molecular weight carboxylic acids. The separation mechanism of IEC is complicated by the fact that a wide range of processes are known to contribute to retention. At present there is no comprehensive mathematical retention model that accounts for all of the retention processes in IEC. For this reason, optimization of separations is generally performed on an empirical basis rather than through the use of computerized optimization routines such as those employed in many other forms of chromatography. However, there has been considerable recent activity in the study of retention processes in IEC and it can be expected that suitable computer optimization methods will soon appear.

*See also:* **II/Chromatography:Liquid:** Mechanisms: Ion Chromatography. **III/Acids:** Liquid Chromatography. **Porous Polymers: Liquid Chromatography.**

## Further Reading

Fortier NE and Fritz JS (1989) *Journal of Chromatography* 462: 323–332.

Fritz JS (1988) *Journal of Chromatography* 439: 3–11.

Gjerde D and Fritz JS (1987) *Ion Chromatography*, 2nd edn. New York: Huthig.

Haddad PR and Jackson PE (1990) *Ion Chromatography – Principles and Applications*. Amsterdam: Elsevier.

Haddad PR, Hao F and Glod BK (1994) *Journal of Chromatography A* 671: 3–9.

Ohta K, Tanaka K and Haddad PR (1996) *Journal of Chromatography A* 739: 359–365; 782: 33–40.

Tanaka K and Fritz JS (1986) *Journal of Chromatography* 361: 151–160.

Tanaka K and Fritz JS (1987) *Journal of Chromatography* 409: 271–279.

Tanaka K and Fritz JS (1987) *Analytical Chemistry* 59: 708–712.

Tanaka K and Haddad PR (1996) *Trends in Analytical Chemistry* 15: 266–273.

Tanaka K, Ishizuka T and Sunahara H (1979a) *Journal of Chromatography* 174: 153–157.

Tanaka K, Ishizuka T and Sunahara H (1979b) *Journal of Chromatography* 177: 21–27.

Tanaka K, Ohta K, Fritz JS, Miyanaga A and Matsushita S (1994) *Journal of Chromatography A* 671: 239–248.

Tanaka K, Ohta K, Fritz JS, Lee Y-S and Shim S-B (1995) *Journal of Chromatography A* 706: 385–393.

Tanaka K, Ohta K and Fritz JS (1996) *Journal of Chromatography A* 739: 317–325.

Tanaka K, Ohta K and Fritz JS (1997) *Journal of Chromatography A* 770: 211–218.

Weiss J (1995) *Ion Chromatography*. Weinheim: VCH Publishers.

# ISOTOPE SEPARATIONS

## Gas Centrifugation

**V. D. Borisevich**, Moscow State Engineering Physics Institute (Technical University), Moscow, Russia
**H. G. Wood**, University of Virginia, Charlottesville, VA, USA

### Introduction

The idea that a centrifugal field might be used to separate was first suggested by Lindemann and Aston in 1919. In 1934 Beams and others at the University of Virginia developed a convection-free centrifuge for isotope separations, shown in **Figure 1**. Two years later Beams and Haynes demonstrated practical separation of chlorine isotopes. In 1938 the Nobel prize winner Urey suggested multiplying the separating effect between axis and periphery of a rotor produced by centrifugal forces by introducing countercurrent convection (like a fractionating column) within the spinning tube. At the end of the 1930s Groth and co-workers in Germany started to construct a high speed centrifuge for uranium isotope separation with axial countercurrent flow induced

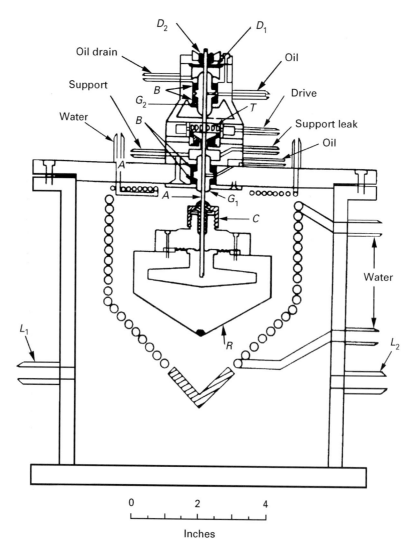

**Figure 1**   Schematic drawing of Beams' first successful evaporative centrifuge used to separate chlorine isotopes.

by heating the bottom of a rotor and cooling the top.

With the beginning of the Second World War in 1939, the separation of uranium isotopes became a subject of national importance. The experimental work on developing different types of gas centrifuges for uranium isotope separation was continued in Germany and was undertaken in the USA in the framework of the Manhattan Project. In that time, Dirac made fundamental contributions to the theory of the isotope separation process in a gas centrifuge, and Onsager's theory for calculation of separation efficiency in a thermal diffusion column was generalized for gas centrifuges by Cohen in 1951. They and their colleagues worked out a general mathematical model and demonstrated theoretically the ideal countercurrent flow profile to produce the maximum separation efficiency. Beginning in 1946, Steenbeck and Zippe contributed to the centrifuge development project in the USSR. Together with Russian co-workers, they developed a very elegant bearing system and adapted the scoop system of gas extraction not only to recover the gas from the rotor but also to generate the circulation flow to multiply the radial separating effect. After leaving the USSR in the mid 1950s, Zippe continued his activities in Germany and USA to reproduce the experiments that had been performed in USSR. In the USA, Zippe worked with Beams for about 2 years in the late 1950s. The Virginia short-bowl centrifuge, similar to the Russian design, was the result of their collaboration. A research and development programme for centrifugation was then pursued in the USA with the central effort located at the Department of Energy laboratories at Oak Ridge, Tennessee. In this programme, centrifuge rotors of an unprecedented length, of the order of 13 m were developed. The programme proved to be technically successful, but it was terminated in 1985 because of changes in US nuclear policies.

After the Second World War, centrifuge research programmes were initiated in several other countries: in the UK by Kronberger and Whitley; in France by Burgain and Le Manach; in Germany by Beyerle, Groth and Martin; in the Netherlands by Kistemaker and Los; in Japan by Kanagawa, Oyama and Takashima; and in Sweden by Landahl and associates.

The Soviet Union was the first country where this technology was developed on an industrial level. The first pilot plant, comprising 2500 gas centrifuges, was put into operation in 1957. The first industrial plant, which contained several tens of thousands of centrifuges, was commissioned in 1959. Soon after, USSR built an industrial plant with several hundred thousand centrifuges configured in three levels and commissioned it in 1962–1964. Besides Russia, this method was commercialized at the end of the 1970s by the UK, Germany and the Netherlands collaborating and in the 1980s by Japan. At present, gas centrifugation is the most efficient, economic and reliable technology for production of enriched uranium.

## Principles of Operation

The history of the development of a gas centrifuge for uranium isotope separation provides an excellent example of successfully overcoming numerous experimental and theoretical problems in the fields of physics of separation processes, gas dynamics, materials science, mechanical engineering and physical chemistry. The short-bowl centrifuge patented in 1957 by Zippe, Scheffel and Steenbeck is shown in **Figure 2**. The thin-walled vertical cylindrical rotor is suspended at the bottom by a low friction needle bearing and at the top by a frictionless magnetic bearing. It also uses damping bearings to resist vibrations at both ends of the rotor. In the case of the separation of uranium isotopes, uranium hexafluoride ($UF_6$) is introduced into the spinning rotor from a stationary central post and removed from stationary pipes called scoops located at either end of the

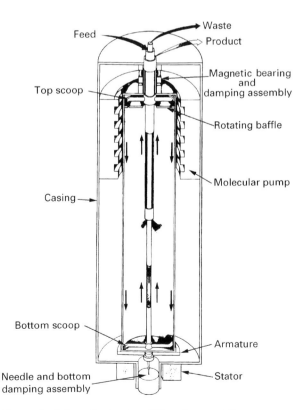

**Figure 2**  Schematic drawing of Zippe-type centrifuge.

rotor. In practice, the gas centrifuge is spun in a vacuum housing that is maintained by a Holweck-type spiral groove molecular pump. The very high peripheral speeds (for example, in the order of $600 \text{ m s}^{-1}$) generate centrifugal forces that compress the process gas into a thin stratified layer adjacent to the cylindrical wall of the rotor.

## Separation Factor

The action of centrifugal forces causes a partial separation of isotopes along the rotor radius. If we consider the process gas to be a binary mixture of the two isotopic species $^{235}UF_6$ and $^{238}UF_6$, then the heavier molecules containing $^{238}UF_6$ will tend to be concentrated near the cylinder wall and the lighter molecules containing $^{235}UF_6$ will tend to be concentrated near the axis. Considering $UF_6$ as an ideal gas, a pressure gradient is developed which is governed by eqn [1]:

$$\frac{dp}{dr} = \frac{Mp}{RT}\omega^2 r \qquad [1]$$

Here $p$ is the pressure, $M$ is the molecular weight of the gas, $R$ is the gas constant, $T$ is the absolute temperature, $\omega$ is the angular frequency of rotation and $r$ is the radial coordinate. In the case of an isothermal centrifuge, the equation above is readily integrated to yield the following relation:

$$p(r) = p(0)(M\omega^2 r^2/2RT) \qquad [2]$$

that gives the pressure $p(r)$ at any radial position $r$ in terms of the pressure at the axis $p(0)$. For a mixture of two ideal gases of molecular weights, $M_1$ and $M_2$, each gas would have a pressure governed by eqn [2] and the ratio of the two pressures gives the radial separation under equilibrium conditions (i.e. no axial gas circulation). An equilibrium separation factor between the two gases is therefore given by the expression:

$$\alpha_0 = \frac{x_1(0)}{x_2(0)} \bigg/ \frac{x_1(a)}{x_2(a)} = \exp[(M_2 - M_1)\omega^2 a^2/2RT] \quad [3]$$

in which $x_1$ and $x_2$ are the concentrations of species 1 and 2, respectively, and $a$ is the radius of the rotor.

The fundamental advantage of this technique over many other diffusion separation methods is that the primary isotope separation effect occurs at thermodynamic equilibrium. It should also be noted that the separation factor for the centrifuge process is a function of the absolute difference in the molecular weights of the components being separated. This is in contrast to various diffusion separation processes where it is a function of the ratio of the molecular weights.

## Separative Capacity

The stationary scoop at the top of the rotor induces a countercurrent flow by removing both mass and angular momentum, which induces pumping of the gas radially inward, forcing it to travel down near the axis and up along the cylinder wall. In order to prevent the influence of the scoop at the bottom, it is shielded by a baffle that rotates with the rotor and has holes which allow the gas to enter the scoop chamber and be removed from the centrifuge. This countercurrent flow produces a net transport of heavy isotopes to the top of the rotor and a net transport of light isotopes to the bottom, establishing a concentration gradient in the axial direction that is considerably greater than the primary (radial) isotope separation effect. Therefore, the gas removed by the bottom scoop is enriched in $^{235}UF_6$ (a product stream) and the gas removed by the top scoop is depleted in $^{235}UF_6$ (a waste stream).

As manufacturing technology has improved, higher rotational speeds have been achieved to obtain greater separation performance. At these higher speeds, shock waves develop in front of the stationary scoop, and the temperatures associated with these shock waves are high enough to cause decomposition of the process gas. It is important to note that the speed of sound in $UF_6$ at room temperature is approximately $90 \text{ m s}^{-1}$, and the peripheral speed of rotation of the process gas in modern centrifuges can be more than seven times greater than this. One design that avoids this decomposition problem is to shield the upper scoop with a rotating baffle with two concentric systems of holes. One system is located near the periphery to allow gas to enter the scoop, and the other system is located nearer the axis of rotation to induce internal circulation. Again, this internal circulation creates an axial separation factor many times that of the basic radial separation factor. The design of scoops and systems of holes in both baffles (their position and size) together with the temperature distribution on the side wall are used to control the internal circulation value and rate.

In 1941, Dirac demonstrated that a gas centrifuge, no matter how it is operated, has the maximum theoretical capacity:

$$\delta U_{max} = \rho D\left(\frac{\Delta M\omega^2 a^2}{2RT}\right)^2 \pi Z/2 \qquad [4]$$

where $Z$ is the length of the rotor, $\Delta M$ is the difference in molecular weights, $\rho$ is the density of the

**Table 1** Maximum pressure ratio, radial separation factor and separative work at various peripheral speeds

| $\omega a$ $(m\,s^{-1})$ | $p(a)/p(0)$ for $^{235}UF_6$ | $\alpha_0$ | $\delta U_{max}$ kg SW per year | $L/a^a$ |
|---|---|---|---|---|
| 300 | 573 | 1.056 | 2.15 | Does not exist |
| 400 | $8 \times 10^4$ | 1.101 | 6.78 | 0.571 |
| 500 | $4.6 \times 10^7$ | 1.162 | 16.55 | 0.309 |
| 600 | $1.1 \times 10^{11}$ | 1.242 | 34.32 | 0.202 |
| 700 | $1.0 \times 10^{15}$ | 1.343 | 63.59 | 0.144 |

$^aL/a$: Ratio of radial distance for pressure to fall by $10^4$ to centrifuge radius.

process gas, $D$ is the coefficient of self-diffusion, and $\delta U$ is the separative capacity in moles per unit time. Maximum pressure ratio, radial separation factor and separative work at various peripheral speeds of a gas centrifuge are presented in Table 1.

### Design Principles

As is evident from eqn [4], the most important parameters in centrifuge technology are peripheral rotor speed and rotor length. The peripheral speed is limited by the strength of the rotor material. Hence, gas centrifuges require materials which have a high strength to density ratio. Such materials are aluminium alloys, titanium alloys, alloy steels or fibre composites. The centrifuge rotor can be made of more than one layer; for example, aluminium alloy covered with fibre composites. The materials problems include long-term fatigue and creep at high speeds. Because uranium hexafluoride forms hydrofluoric acid even with only a little moisture, corrosion is also a challenging problem area that must be addressed in the design.

The spinning rotor has certain natural frequencies determined by the materials of construction, the rotor length to diameter ratio, and the damping characteristics of the suspension systems. Centrifuges that operate at rotational frequencies below the lowest natural flexural frequency of the rotor are called subcritical centrifuges, and those that operate at rotational frequencies above the first natural frequency are called supercritical ones. As a rule, the rotors of supercritical machines consist of sections connected by bellows that act to reduce vibrations caused by resonant frequencies at certain operating speeds.

The Dirac maximum separative capacity has been derived for the case of an ideal circulation profile in a gas centrifuge. However, this ideal profile cannot be achieved in practice. The flow patterns that actually occur within a centrifuge rotor are governed by the equations of fluid dynamics. The theoretical solution of these equations provides a basis for the optimization of the internal flow pattern that yields the maximum separation performance of a particular gas centrifuge design.

The flow pattern in a gas centrifuge can be divided into three regions in the radial direction, each with different flow features. Near the side wall, the flow is dominated by viscosity, and a strongly rarefied gas region (a vacuum core) is located near the axis of rotation. These two regions are connected by a transition region. The transition region as a rule occupies only a few per cent of the rotor radius, and its influence on separation is negligible. The vacuum core in modern centrifuges is spread over approximately three-quarters of the rotor radius. In this region the mean free path of the gas molecules is comparable with the rotor diameter. At these low pressures, the central core of the rotor cannot contribute to the isotope separation. As a result, the separation power of a gas centrifuge at high speeds increases only as the square of the peripheral speed instead of the fourth power, as given by the Dirac equation.

### Cascade of Centrifuges

The separative work output of a single gas centrifuge is generally small compared to the total desired separative work. Hence, it is necessary to combine many centrifuges into a cascade to achieve the desired separation. Each stage of a cascade may have many centrifuges connected in parallel. This arrangement of the large number of centrifuges allows for simple replacement of a faulty individual centrifuge in the cascade. The considerable advantage of the centrifuges is that they need no special compressors for pumping gas through the cascade. The pressure difference required for pumping the gas through the cascade is generated by the dragging action of the scoops on the rotating gas. The cascade contains only centrifuges and piping. Thus, gas centrifuge enrichment of uranium uses only about 1/20th to 1/30th of the electricity per unit of separative work required by the gaseous diffusion process. The reliability of modern centrifuges allows them to operate nonstop for more than 15 years with a failure rate of only a few tenths of 1% per year. Once put into operation, gas centrifuges require no special preventive maintenance through their service lives, and the machines' separation characteristics remain practically constant over time. Provided that random failures occur, the fragments of the crashed machines can be left in the cascade because they will not have much effect on the overall cascade efficiency.

# Separation of Nonuranium Isotopes

## New Scientific Problems

More recently, interest has been shown in using gas centrifuges for other kinds of separations like stable isotopes or the isotopes of spent reactor uranium. For the stable isotopes, demand is growing in medicine and fundamental physics research, and the use of the gas centrifuge process makes it possible to produce isotopes economically when large (kilograms) quantities are needed. This is the case for the isotopes of xenon, krypton, tungsten, molybdenum, iron, tin, tellurium, sulfur, silicon, germanium, chromium and many others. The separation of stable isotopes by gas centrifuge has been underway for more than three decades in Russia. Currently, cascades of thousands of centrifuges are producing tens of kilograms of various isotopes. In another application, some countries have considered gas centrifuge technology as part of the reprocessing cycle of the re-enrichment of the spent uranium from power reactors which contains five isotopes: $U^{232}$, $U^{234}$, $U^{235}$, $U^{236}$ and $U^{238}$.

A centrifuge designed for binary separations with uranium hexafluoride cannot be efficiently used for nonuranium isotope separation with different chemical compounds. Therefore, centrifuges with specific characteristics must be designed for particular separation problems. For some isotopes, specially synthesized gaseous chemical compounds must be prepared for use as the process gas. The basic condition for the applicability of the process gas in these newly designed centrifuges is that the gas vapour pressure is not less than 5–10 mmHg under normal operating temperatures. In addition to the condition that this substance should not corrode the structural material of the centrifuge, it has to be sufficiently resistant to temperature dissociation, and preferably must possess the maximum possible content of the desired element in the molecule. The list of such substances includes fluorides and oxyfluorides of metals and nonmetals, metal–organic and complex compounds, phosphorus hydrides, boron hydrides, Freons and some others. Significant differences in the chemical and physical properties of process gases leads to the necessity to create a set of gas centrifuge designs for various ranges of molecular masses. The internal circulating flow must be optimized for each of these designs.

The synthesis of volatile compounds suitable for process gases was not the only scientific problem the researchers faced when separating nonuranium isotopes. In contrast to natural uranium, most chemical elements are polyisotopic. This property leads to additional difficulties in enriching the intermediate components of the isotope mixture.

As perhaps the greatest achievement in the development of centrifuge technology for enrichment of nonuranium isotopes, one may consider the complex isotope separation for iron, carbon and oxygen as a pentacarbonyl of iron – $Fe(CO)_5$. This separation has been realized on an industrial scale at the electrochemical plant in the Krasnoyarsk region of Russia. All chemical elements included in the molecule structure are polyisotopic: iron contains four, carbon two, and oxygen three isotopes. A natural isotope abundance of $Fe(CO)_5$ represents a mixture of 284 types of molecules with different isotope distributions. These types are distributed through 20 components with molecular masses from 194 up to 213. Almost every one of them contains several isotopes for each of the chemical elements. This isotopic overlapping of iron isotopes with intermediate masses and heavy isotopes of carbon and oxygen limits the direct enrichment by a centrifuge cascade. However, this limitation has been removed by introducing into the separation process the isotope exchange between the molecules of pentacarbonyl of iron in so-called photoreactors. The combination of isotope separation in gas centrifuges with isotope exchange has allowed the achievement in a single process of a complex mix of highly enriched isotopes: $^{57}Fe$ with concentration more than 99%, $^{13}C$ and $^{18}O$ with concentrations up to 80%.

## Advantages

The accumulated experience in separating isotopes of both light (boron, carbon, nitrogen and oxygen) and heavier chemical elements has shown gas centrifuge technology to be extremely promising. At the moment, isotopes of more than 20 chemical elements have been separated by gas centrifuges. The centrifuge cascade used for the separation of nonuranium isotopes usually has very low energy consumption and tens of times higher productivity than that of the electromagnetic installation and with a comparable output (**Figure 3**). Nonuranium isotope separation has been transformed into an independent area for development of gas centrifuge technology. It includes solving problems in the design of different types of centrifuges, in the theory of multicomponent isotope mixture separation, in chemical synthesis of process gases, in transformation of process gases with enriched isotopes required for use in chemical compounds, etc.

In many applications of isotope enrichment, rigid requirements on the isotope purity must be met, and undesired gas impurities must be reduced to a specified level. Gas centrifugation has also been recommended as an excellent practical tool to clean small

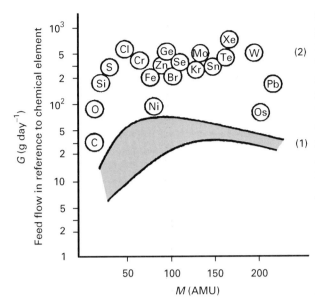

**Figure 3** Illustration to compare the separation of nonuranium isotopes by electromagentic separation (1) and by gas centrifuge (2). Production, $G$ (g day$^{-1}$: feed flow into the installation), versus molecular mass, $M$, of the separating chemical element. Reproduced with permission from Goldstein *et al.* (1993).

gas impurities. An extremely important benefit of gas centrifuges is the opening of an era of economical large scale production of many stable isotopes that have applications in medicine, industry and fundamental science. For example, experiments on neutrino physics with large detectors that require tens of kg of enriched isotope have been made possible by the cost reductions achieved through the large scale production by gas centrifugation.

## Future

The gas centrifuge is now a mature technology, but nevertheless the development potential has not been exhausted. The separation efficiency of existing gas centrifuges for separation of uranium and nonuranium isotopes can be further improved. Additionally, the technology can be applied in the near future to the re-enrichment of uranium from spent nuclear fuel as well as to enriching tails (depleted uranium of separating plants) to the natural isotopic concentration or higher. In yet another area, experiments have been performed which show promising results for the application of gas centrifuge technology to purification of process gas from aerosol particles that can be used, for example, in the semiconductor industry.

### See Colour Plate 102.

## Further Reading

Avery DG and Davies E (1973) *Uranium Enrichment by Gas Centrifuge.* London: Mills & Boon.

Beams JW and Haynes FB (1936) The separation of isotopes by centrifuging. *Physical Review* 50: 491.

Benedict M, Pigford LH and Levi HW (1981) *Nuclear Chemical Engineering.* New York: McGraw Hill.

Cohen K (ed.) (1951) *The Theory of Isotope Separation as Applied to the Large-scale Production of $U^{235}$.* New York: McGraw-Hill.

Goldstein S, Louvet P and Soulié E (eds) (1993) *Les Isotopes Stables Application–Production.* Paris: Centre d'Études de Saclay.

Lindemann FA and Aston FW (1919) The possibility of separating isotopes. *Philosophical Magazine* 37: 523.

London H (ed.) (1961) *Separation of Isotopes.* London: George Newnes.

Olander DR (1972) Technical basis of the gas centrifuge. *Advances in Nuclear Science and Technology* 6: 105.

Villani S (ed.) (1979) *Topics in Applied Physics: Uranium Enrichment.* New York: Springer-Verlag.

Whitley S (1984) Review of the gas centrifuge until 1962. Parts 1 and 2. *Reviews of Modern Physics* 56: 43.

# Liquid Chromatography

**L. Lešetický**, Charles University, Prague, Czech Republic

This article is reproduced from *Encyclopedia of Analytical Science*, Copyright © 1995 Academic Press

## Introduction

The application of isotopes in science and especially in analytical chemistry is based on two rather contradictory assumptions. Tracer experiments require that the isotopically modified and unmodified compounds behave in a very similar manner. A study of isotope effects must recognize, measure and interpret minute differences in the chemical and physical properties of compounds (isotopomers) that differ in their isotopic composition. These differences lead to different behaviour of isotopomers in all chromatographic processes, as has been demonstrated experimentally.

As early as 1938, Urey and coworkers published a paper on enrichment of $^6$Li by ion exchange chromatography and work in this field has been continuing to date. Gas–solid chromatography has been used for separation of the hydrogen isotopes, noble gases and some other gaseous elements and simple compounds. Low-molecular-mass volatile compounds labelled mainly with hydrogen isotopes have been separated by gas-liquid chromatography. During the late 1950s and the 1960s, papers were published concerning small but observable isotopic fractionation with liquid chromatography of isotopically labelled organic compounds. Over the last 20 years during which liquid chromatography (LC) has become a common analytical technique, many papers have been published dealing with separation of labelled compounds from their unlabelled counterparts. A review of these works, with a possible explanation of the mechanisms of the separation process, is presented.

## Isotope Effects

Chromatography can be considered as a process in which the compounds to be separated interact with a stationary and a mobile phase. These mutual interactions differ in magnitude even for different isotopes of the same element. The magnitude of the interaction energy for an isotopic pair depends on many parameters that are discussed below.

It is possible to draw the following general conclusions from the experimental data on isotope effects on the physical properties of molecules.

1. The $C-{}^2H$ bond is shorter than the $C-{}^1H$ bond (e.g. in ethane the C–H bond length is 111.2 pm; in hexadeuteroethane the $C-{}^2H$ bond length is 110.7 pm) and exhibits a higher electron density than the ${}^1H$ bond. The deuterium atom appears to be smaller than the hydrogen atom. There are small, but measurable differences in the dipole moments, e.g. $\mu(CH_3-{}^2H) = 3.7 \times 10^{-32}$ C m, $\mu({}^1HCl)-\mu({}^2HCl) = 1.7 \times 10^{-32}$ C m.
2. The $C-{}^2H$ bond has a polarizability lower than that of the $C-{}^1H$ bond.
3. The $C-{}^2H$ stretching vibrational frequency (around $2200$ cm$^{-1}$) is lower than the corresponding $C-{}^1H$ frequency (around $3000$ cm$^{-1}$).
4. Heavier isotopes have smaller atomic volumes; deuterated compounds have smaller molar volumes than the corresponding unlabelled compounds. The molar volume differences between, e.g. deuterated and unlabelled benzene are due to a molecular size effect caused by differences in the zero-point intermolecular motion of the molecules.
5. Deuterium ($^2H$) is more electropositive than protium ($^1H$) and thus some isotope effects can

**Table 1** Isotope effect on dissociation constants[a]

| Compound[b] | $log(K_H/K_D)$ |
| --- | --- |
| $^2HCOOH$ | $0.035 \pm 0.002$ |
| $C^2H_3COOH$ | $0.014 \pm 0.002$ |
| $(C^2H_3)_3CCOOH$ | $0.018 \pm 0.001$ |
| $C_6{}^2H_5COOH$ | $0.010 \pm 0.002$ |
| $(2,6-{}^2H_2)$benzoic acid | $0.003 \pm 0.001$ |
| $C^2H_3\overset{+}{N}H_3$ | $0.051$ |
| $(C^2H_3)_2\overset{+}{N}H_2$ | $0.117$ |
| $(C^2H_3)_3\overset{+}{H}N$ | $0.207$ |

[a]Willi (1983), pp. 58–61.
[b]Only the isotope-modified compound is specified in the table.

be discussed in terms of inductive effects. This has been clearly demonstrated in the measurement of the secondary isotope effect on ionization equilibria of some deuterated carboxylic acids and protonated amines, as shown in **Table 1**.

6. Deuterated compounds are less lipophilic than the corresponding unlabelled compounds

All these experimentally demonstrated phenomena may influence chromatographic separation processes. Isotope effects on the chromatographic behaviour of compounds labelled with isotopes of heavier elements (carbon, nitrogen, oxygen, etc.) are so small that they can only be detected for simple compounds of low relative molecular mass. The following discussion is limited to deuterated and tritiated compounds and only a few examples are given of separations of heavier isotopes.

It should be pointed out that discussion of isotope effects in terms of inductive or other electronic effects is sometimes helpful but represents a gross simplification. A more exact description of the origin of the isotope effects uses the zero-point energy and the vibrational frequencies of the molecules in question.

## Separation Processes

Two basic types of separation process can be distinguished. In systems with a polar stationary phase and nonpolar mobile phase (classical adsorption chromatography, normal-phase LC and chemically bonded polar stationary phase LC) the 'ordinary' (unlabelled) compounds are usually eluted first, i.e. the separation factor is less than unity ($\alpha = k_1/k_2 < 1$; subscripts 1, 2, 3 refer to protium, deuterium and tritium ($^3H$), respectively). As mentioned above, deuterated or tritiated compounds are more polar and thus are more strongly bound to polar stationary phases.

**Table 2** LC separation factors of some hydrocarbons versus their perdeuterated analogues

| Compound | $\alpha = k_1/k_2$ | Conditions[a,b] |
|---|---|---|
| [$^2H_{14}$]Hexane[a] | 1.036 | 51 |
|  | 1.049 | 40 |
| [$^2H_{18}$]Octane[a] | 1.054 | 51 |
| [$^2H_{12}$]Cyclohexane[a] | 1.044 | 33 |
| [$^2H_6$]Benzene[a] | 1.043 | 23 |
|  | 1.048 | 18 |
| [$^2H_6$]Benzene[b] | 1.049 | 30 |
| [$^2H_8$]Toluene[a] | 1.046 | 33 |
|  | 1.057 | 23 |
| [$^2H_8$]Toluene[b] | 1.052 | 40 |
| [$^2H_{10}$]Phenanthrene[b] | 1.067 | 70 |
| [$^2H_{14}$]Durene[b] | 1.071 | 70 |
| [$^2H_{10}$]Biphenyl[b] | 1.062 | 55 |

[a]$\mu$-Bondapak C$_{18}$ column; methanol in water (mol%); Tanaka (1977).
[b]Ultrasphere C$_{18}$ column; acetonitrile in water (v/v); Baweja (1987).

Partition chromatography and reversed-phase LC usually lead to the opposite result, i.e. a heavier isotopomer is eluted first ($\alpha > 1$). It has been suggested that a major contribution to the isotope effect is then a hydrophobic interaction. The fact that the separation factors are higher (see **Table 2**) when the mobile phase contains more water demonstrates that they are affected by more restricted motion of the C–H bonds (in solute), caused by a tighter solvation of the C–H bonds within the aqueous mobile phase relative to the hydrophobic stationary phase. On the other hand a less restricted motion of the C–H bonds in the stationary phase would tend to favour protium over deuterium, and this could contribute to the observed isotope effect.

Another important factor is the position of the label in a molecule. As these effects are essentially primary isotope effects, the best situation occurs when an isotopic atom is pushed out from the rest of a molecule, so that it is readily accessible for interaction. When the atom in question is in the shadow of a bulky group (e.g. *ortho* hydrogen in benzoic acid) or in an inconvenient configuration, hydrophobic interactions and adsorption can not take place or are greatly suppressed and no separation is attained. The other contributions, such as differences in solubility, molar volume and adsorption on the residual nonderivatized sites on the silica particles, seem to be of a far less importance for the magnitude of the isotope effect.

The above discussion deals with a direct interaction of an isotopic atom or bond with the stationary and mobile phases (primary isotope effect). However, there also exist chromatographic separations based on secondary isotope effects. In molecules deuterated or tritiated in the position adjacent to an ionizable functional group, the magnitude of the isotope effect is often greatly enhanced. As follows from Table 1 $\alpha$-deuterated amines are stronger bases than unlabelled amines, and $\alpha$-deuterated acids are weaker acids than their protiated counterparts. In an elution system with a pH value near the p$K_a$ values of the deuterated/unlabelled amine pair, this isotope-induced base strengthening would alter the ratio of protonated and unprotonated forms, leading to differences in the chromatographic mobility. However, at a pH much higher or much lower than the p$K_a$ value where both labelled and unlabelled amines are completely protonated or unprotonated, the p$K_a$ effect of isotopic substitution on the chromatographic separation is greatly suppressed. A similar effect was observed for labelled carboxylic acids.

## Survey of Separations

### Hydrocarbons and their Simple Derivatives

A separation of perdeuterated aliphatic and aromatic hydrocarbons has been performed on reversed-phase columns with similar results (**Figure 1**). A higher content of water in the mobile phase leads to higher separation factors (Table 2).

Chromatography of various isotopomers of benzoic acid shows a moderate effect of the number of deuterium atoms on the separation factor; however, an important effect is exerted by the position of the label (**Table 3**). Isotopomers containing deuterium in the *ortho* positions display lower isotopic separation than isotopomers with deuterium in

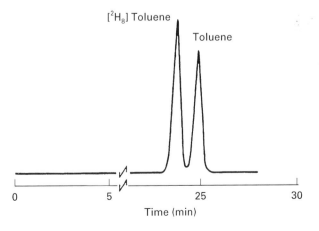

**Figure 1** Reversed-phase LC separation of [$^2H_8$]toluene and unlabelled toluene. Mobile phase: water–acetonitrile 60 : 40 (v/v). (From Baweja, 1987).

**Table 3** LC separations factors of unlabelled versus deuterated carboxylic acids

| Labelled compound | $\alpha = k_1/k_2$ |
|---|---|
| [$^2$H$_{31}$]Palmitic acid | 1.076[a,b] |
| [$^2$H$_{23}$]Lauric acid | 1.066[a,c] |
| [$^2$H$_5$]Benzoic acid | 1.040[a,d] |
| | 1.038[e] |
| [3,4,5-$^2$H$_3$]Benzoic acid | 1.029[e] |
| [2,3,5-$^2$H$_3$]Benzoic acid | 1.023[e] |
| [3,4-$^2$H$_2$]Benzoic acid | 1.019[e] |
| [3,5-$^3$H$_2$]Benzoic acid | 1.019[e] |
| [2,4-$^2$H$_2$]Benzoic acid | 1.013[e] |
| [2,5-$^2$H$_2$]Benzoic acid | 1.010[e] |
| [2,6-$^2$H$_2$]Benzoic acid | < 1.010[e] |

[a]$\mu$-Bondapak C$_{18}$ column; Tanaka *et al.* (1977).
[b]64 mol%; CH$_3$OH–H$_2$O.
[c]51 mol%; CH$_3$OH–H$_2$O.
[d]H$_2$O, pH 2.51.
[e]Hypersil C$_{18}$ column; CH$_3$OH–H$_2$O (3 : 7, v/v) + 1% HCOOH, Lockley *et al.* (1989).

the *meta* or *para* positions. The isotope effects associated with the *ortho* positions are minimal, probably because interactions between the *ortho* C–$^1$H or C–$^2$H bonds and the stationary phase are minimized by steric shielding of the carboxyl group. [9, 10, 12, 13-$^3$H]Linoleic acid methyl ester and [9, 10, 12, 13-$^3$H]oleic acid methyl ester were chromatographed on silica impregnated with silver nitrate, attaining partial separation from the appropriate [1-$^{14}$C]methyl esters. The tritium atom bound on the olefinic carbon atoms affects the equilibrium constant of the formation of the silver–olefin complex.

## Natural Substances

Tritium-labelled steroids have often been used for the elucidation of biochemical and physiological trans-

formations. During the chromatographic purification of such compounds, isotopic fractionation has often been observed. Good, sometimes baseline, separations of tritiated aldosterone, cortisone, oestrone, testosterone, prednisolone and oestradiol from their unlabelled or $^{14}$C-labelled analogues were obtained in reversed-phase LC, tritiated compounds being eluted first. Other techniques such as column partition chromatography and paper chromatography have also been successful.

Vitamin D$_3$ (cholecalciferol, [I]) is hydroxylated in living organisms to the 1,25-dihydroxy- or 24,25-

[I]

dihydroxy-derivatives. A significant chromatographic isotope effect was observed with both the derivatives labelled with tritium in positions 26 and 27, whereas labelling in positions 23 and 24 causes a substantially smaller effect. A similar behaviour was found in chromatography of the corresponding trimethylsilyl ethers (**Table 4**). As indicated in the table, the heavier isotopomers were eluted first on the reversed phase, and later than lighter isotopomers on the normal phase.

Catalytic hydrogenation with protium, deuterium and tritium of echinocandin B, a macrocyclic peptide

**Table 4** LC separation factors of vitamin D$_3$ metabolites versus their tritiated counterparts

| Vitamin D$_3$ metabolite | $\alpha = k_1/k_3$ | Eluent (v/v) |
|---|---|---|
| 1,25-Dihydroxy-[23,24-$^3$H]-TMS[a] | 0.977[b,c] | Hexane–CH$_2$Cl$_2$ (85 : 15) |
| | 0.989[b,d] | Hexane–CH$_2$Cl$_2$–CH$_3$CN (90 : 10 : 0.035) |
| 1,25-Dihydroxy-[23,24-$^3$H] | 0.991[b,d] | Hexane–2-propanol |
| 25(R),26-dihydroxy-[23,24-$^3$H]-TMS[a] | 0.965[b,d] | Hexane–CH$_2$Cl$_2$ (85 : 15) |
| 1,25-Dihydroxy-[26,27-$^3$H$_6$] | 0.983[e,f] | Hexane–ethanol (94 : 6) |
| 24,25-Dihydroxy-[26,27-$^3$H$_6$] | 0.965[e,f] | |
| 25,26-Dihydroxy-[23,24-$^3$H] | 1.008[e,g] | Methanol–water (3 : 1) |
| 1,25-Dihydroxy-[26,27-$^3$H$_6$] | 1.023[e,g] | Methanol–water (1 : 1) |

[a]TMS = trimethylsilyl derivative.
[b]Halloran *et al.* (1984).
[c]Zorbax-sil column.
[d]$\mu$-Porasil column.
[e]Worth and Retallack (1988).
[f]Econosphere silica column.
[g]Radial-pak column.

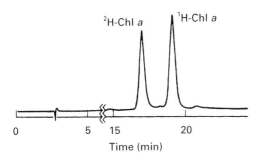

**Figure 2** Reversed-phase separation of a deuterated chlorophyll *a* and undeuterated chlorophyll *a* on a 25 cm × 4.6 mm i.d. C$_{18}$-Ultrasphere ODS column. Mobile phase: water–methanol–acetonitrile–tetrahydrofuran 5 : 28 : 38 : 23 (v/v/v/v). UV detector 663 nm. (From Baweja 1986.)

[III]

[II]  X=$^1$H, $^2$H or $^3$H

possessing antibiotic and antifungal properties, leads to the corresponding tetrahydroderivatives [II]. The isotope effect on the reversed-phase LC mobility was surprisingly large: $\alpha_2 = k_1/k_2 = 1.0190$ for the deuterated compound; $\alpha_3 = k_1/k_3 = 1.0233$ for the

tritiated compound. A Partisil 5 C$_{18}$ column and a mixture of 0.1% phosphoric acid (45%) with CH$_3$CN/THF/H$_3$PO$_4$ (90 : 20 : 0.1, v/v/v) (55%) as the eluent were used. Even though the chromatographic conditions described do not allow resolution of labelled and unlabelled species, such a possibility exists, e.g. when a column recycling system is used.

Reversed-phase LC of deuterated chlorophylls [III] obtained from green algae *Rhodospirillum rubrum* and *Rhodospirillum spheroides* grown in media containing 50%, 80%, 90% and 99.7% $^2$H$_2$O, exhibited baseline separation from isotopically unmodified chlorophyll. As usual, the labelled species were eluted first and the greater is the percentage of deuterium in the compound the faster it moved along the column (**Figure 2, Table 5**).

## Drugs

[$^2$H$_{10}$]Diphenylhydantoin (phenytoin [IV]) and [$^2$H$_{10}$]–5H–dibenz[*b,f*]azepine–5-carboxamide (carbamazepine [V]) were separated from their unlabelled parent compounds by reversed-phase chromatography on a C$_{18}$ column with H$_2$O/CH$_3$CN/THF (80 : 16 : 4, v/v/v) as the eluent.

**Table 5** LC separation factors of chlorophylls versus deuterated chlorophylls[a]

| Labelled compound | $\alpha = k_1/k_2$ | Eluent (A/B)[b] (v/v) |
|---|---|---|
| [$^2$H]Chlorophyll *a* | 1.132 | 5 : 95 |
| [$^2$H]Chlorophyll *a'* | 1.158 | 5 : 95 |
| [$^2$H]Chlorophyll *b* | 1.156 | 5 : 95 |
| [$^2$H]Bacteriochlorophyll *a*[c] | 1.149 | 10 : 90 |
| [$^2$H]Bacteriochlorophyll *a*[d] | 1.118 | 7 : 93 |
| [$^2$H]Pyrochlorophyll *a* | 1.167 | 0 : 100 |

[a]Baweja (1986).
[b]A = H$_2$O; B = methanol–acetonitrile–THF (30 : 40.5 : 24.5, by vol).
[c]Contains geranylgeranyl side-chain.
[d]Contains phytyl side-chain.

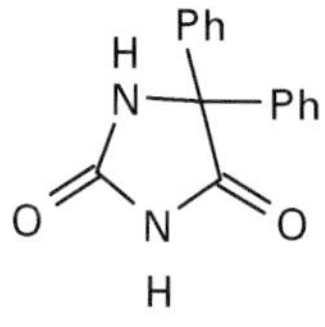

[IV]

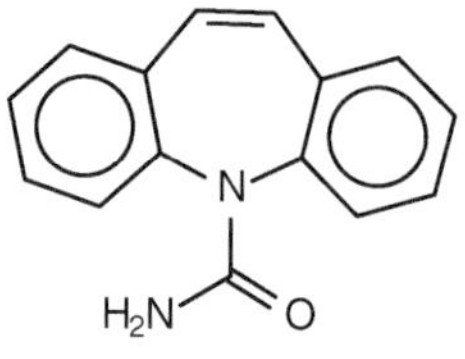

[V]

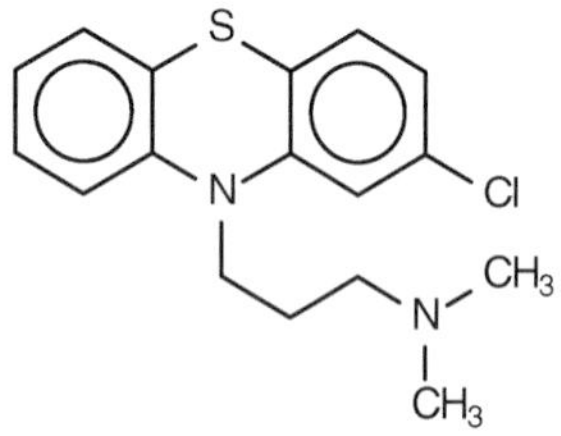

[VI]

[VII]

[VIII]

Baseline separation was attained even when using extracted serum samples. The calculated resolution for the isotopomer pairs was 1.3. (Resolution $R_s = (t_2 - t_1)/[0.5(w_1 + w_2)]$.)

[CH$_3$–$^3$H]Chlorpromazine [VI] and its metabolite [CH$_3$–$^3$H]-7-hydroxychlorpromazine were separated on a chemically-bonded stationary phase (Spherisorb CN) from their unlabelled counterparts. Although the specific activity of tritiated compound was about 40 Ci mmol$^{-1}$ indicating approximately 1.5 tritium atoms per methyl group, the measured separation factors were rather high (**Table 6**). The separation factors depend strongly on the pH of the mobile phase, so that the isotope effect on the basicity of the dimethylamino group is probably significant. The labelled species was eluted later.

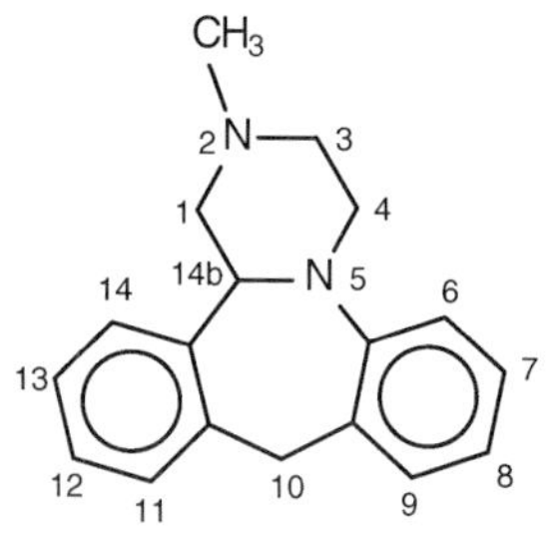

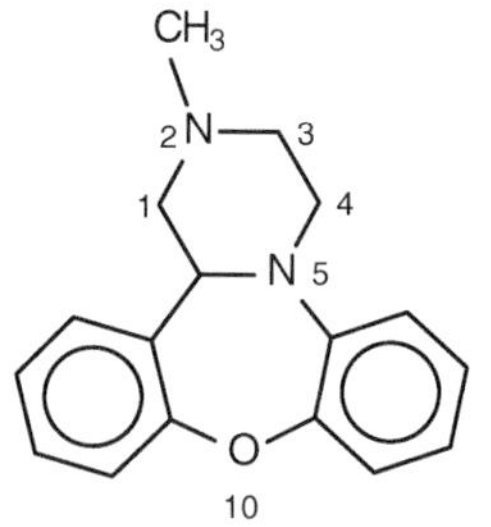

An almost complete resolution was attained between unlabelled and di-, tri- or tetradeuterated drugs active on the central nervous system – 1,2,3,4,10,14b-hexa-hydro-2-methyldibenzo[*c,f*]pyrazino[1,2-a]azepine (mianserin [VII]) and 1,3,4,14b-tetrahydro-2,7-dimethyl-2H-dibenzo[*b,f*] pyrazino[1,2-a]-1,4-oxa-zepine (Org GC 94 [VIII]). [3,3,4,4-$^2$H$_4$]Org GC 94 was chromatographed on a $\mu$-Porasil column and n-hexane–2-propanol (90 : 10, v/v) to which 4% of ethanol and 0.1% of ammonia solution were added as the mobile phase. Various isotopomers of deuterated mianserin were chromatographed on LiChrosorb Si 60 and Spherisorb C$_{18}$ columns (**Table 7**). The deuterated compounds were always moving more slowly than the unlabelled ones. The greatest isotope effects have been found for compounds labelled on the piperazine ring (with the exception of position 14b) and on the methyl group,

**Table 6** LC separation factors of chlorpromazine versus tritiated chlorpromazine[a]

| Labelled compound | $\alpha = k_1/k_3$[b] | pH of eluent[c] |
|---|---|---|
| [CH$_3$–$^3$H]Chlorpromazine | 0.855 | 7 |
| | 0.901 | 6 |
| | 1.00 | 5 |
| [CH$_3$–$^3$H]-7-Hydroxychlorpromazine | 0.847 | 7 |
| | 0.910 | 6 |
| | 0.952 | 5 |

[a]Yeung *et al.* (1984).
[b]Spherisorb CN column.
[c]10% 0.05 mol L$^{-1}$ sodium acetate buffer in methanol.

**Table 7** LC separation factors of mianserin versus isotopomers of deuterated mianserin[a]

| Labelled compound[b] | $\alpha_2$ (LiChrosorb)[c] | $\alpha_2$ ($C_{18}$)[d] |
|---|---|---|
| [3,3,4,4-$^2$H$_4$]Mianserin | 0.893 | 0.935 |
| [N-C$^2$H$_3$]Mianserin | 0.893 | 0.935 |
| [3,3-$^2$H$_2$]Mianserin | 0.910 | 0.935 |
| [1,1-$^2$H$_2$]Mianserin | 0.926 | |
| [4,4-$^2$H$_2$]Mianserin | 0.971 | |

[a]Kaspersen *et al.* (1984).
[b]For other isotopomers, namely [8-$^2$H], [6,8-$^2$H$_2$], [10,10-$^2$H$_2$], [12-$^2$H], [13-$^2$H] and [14b-$^2$H] no fractionation occurred.
[c]n-Hexane–2-propanol (9 : 1, v/v) + 0.1% NH$_3$, aq.
[d]0.1 mol L$^{-1}$ sodium acetate-acetonitrile (1 : 4, v/v).

indicating that the basicity change plays the most important role.

2-(*N*-Propyl-*N*-2-thienylethyl)-5-hydroxytet ra-linamine (N-0437 [**IX**]), an effective drug against Parkinson's disease and glaucoma, can be separated from its counterparts deuterated and tritiated at the propyl group. The corresponding diastereoisomeric glucuronides were prepared by enzymatic synthesis and separated in a reversed-phase LC system (Nova-Pak C$_{18}$). The separation factors were pH-dependent and relatively large, considering the remote positions of the labels from the nitrogen atom, e.g. $\alpha_2 = 1.033$; $\alpha_3 = 1.037$.

The above-mentioned effect of isotope substitution on the basicity of amines was demonstrated by thin-layer chromatography (TLC) of the popular antidepressant imipramine [**X**]. [$^2$H$_{10}$]Imipramine labelled in both methyl groups and in aromatic rings and its parent compound were resolved on silica plates using different eluents, in which the pres-

[**XI**]

ence of ammonia was necessary; e.g. benzene–acetone–NH$_3$ (300 : 60 : 1, v/v/v); $R_F$[$^1$H] = 0.24; $R_F$[$^2$H] = 0.18. Labelling on the benzene ring exerted no effect on the chromatographic mobility.

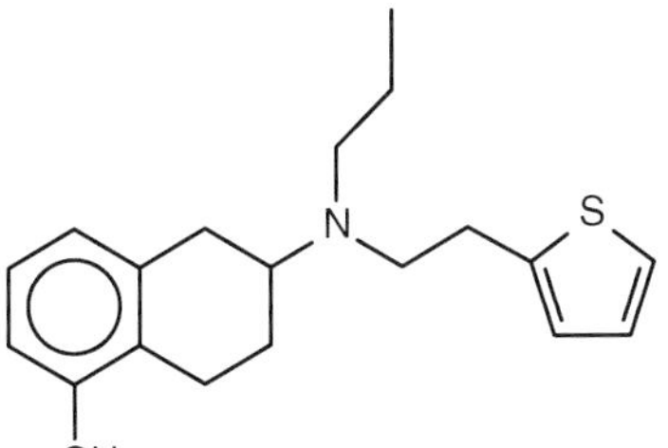

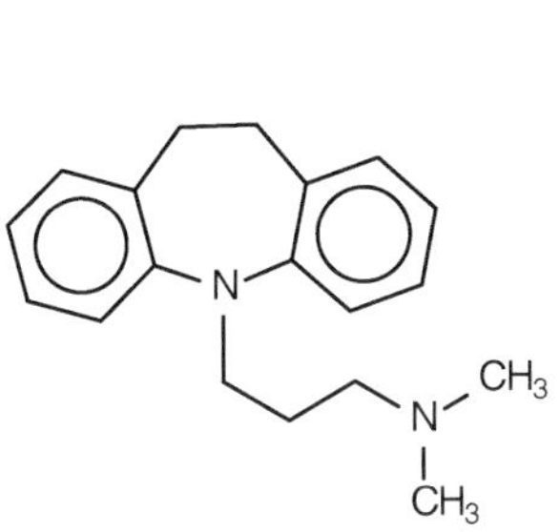

Similar TLC behaviour has been observed with (*R*, *S*)-6-[CH$_3$-$^2$H$_6$]dimethylamino-4,4-diphenyl-heptan-3-one.HCl ([$^2$H$_6$]methadone [**XI**]) (silica gel Merck; benzene–methanol–NH$_3$; 85 : 15 : 0.65, v/v/v); $R_F$[$^2$H] = 0.58; $R_F$[$^1$H] = 0.79. Reversed-phase LC showed baseline resolution, whereas gas–liquid chromatography afforded no separation.

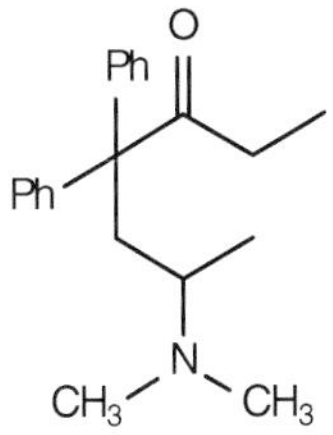

### Isotopes of Heavier Elements

Liquid chromatographic separations of the isotopes of elements other than hydrogen have been rather rare. A high-efficiency liquid–liquid chromatography system consisting of porous silica microspheres covered with 25% (w/w) bis(2-ethylhexyl)phosphoric acid in dodecane as the stationary phase and nitric acid as the mobile phase obtained a certain enrichment of heavier isotopes of calcium in the front of the elution curve. Separation factors calculated by Glueckauf (1961) for $^{42}$C, $^{45}$Ca, $^{48}$Ca versus $^{40}$Ca were 1.0012–1.0029.

Better resolution was obtained when isotopic pairs of $^{144}$Sm–$^{154}$Sm, $^{140}$Ce–$^{142}$Ce, and $^{151}$Eu–$^{153}$Eu were chromatographed on LiChrosorb 60 with di-2-propylether-THF-65% HNO$_3$ (100 : 20 : 5, v/v/v). Heavier isotopes were always eluted first, the separation factors being two orders of magnitude higher than in ion exchange chromatography (1.030–1.085).

[**IX**]

[**X**]

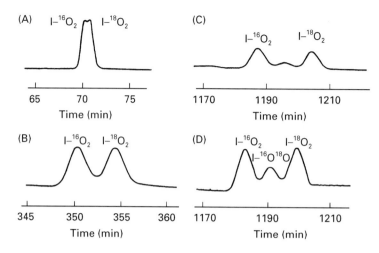

**Figure 3**    Separation of benzoic acid (I) isotopomers by recycle chromatography. Column: Cosmosil 5-C$_{18}$-P × 4, 15 cm × 4.6 mm i.d. Mobile phase: methanol-0.05 mol L$^{-1}$ acetate buffer (pH 4.83) 20 : 80 (v/v). Cycles: (A) 1, (B) 5, (C) and (D) 17. (From Tanaka, 1986.)

There are very few references in the literature to liquid chromatographic separations of organic compounds labelled with isotopes heavier than hydrogen. For example, Tanaka *et al.* (1986) demonstrated the oxygen isotope effect on the reversed-phase LC separation of $^{18}$O-labelled benzoic acid (**Figure 3**). They demonstrated that successful separation was due to differences in the dissociation constants between the labelled and parent compounds. The isotope effect was pH-dependent and attained a maximal value around pH 5 (**Table 8**). An equation was derived to determine the $^{18}$O isotope effect on the dissociation constants from LC data. This isotope effect was calculated to be $^{16}K_a/^{18}K_a = 1.020 \pm 0.002$. A very good, nearly baseline, resolution was attained in a recycle system for the pair benzoic acid – [$^{18}$O$_2$] benzoic acid after five cycles, and for the mixture benzoic acid – [$^{18}$O]benzoic acid – [$^{18}$O$_2$]benzoic acid after 17 cycles. Similar results were obtained in chromatography of [$^{18}$O$_2$]-4-chlorobenzoic acid and [OH–$^{18}$O]-4-nitrophenol with their unlabelled counterparts. The $^{18}$O–H bond is stronger than the $^{16}$O–H bond so that $^{18}$O-labelled acids are weaker acids than unlabelled ones. Therefore, a longer retention time in reversed-phase LC was expected and found for the labelled carboxylic acid compared to the unlabelled acid.

The same procedure as above was successfully applied to the separation of aniline and [$^{15}$N]aniline on the same column with 0.05 mol L$^{-1}$ acetate buffer containing 5% methanol and 0.01% triethylamine as the eluent. At pH 4.24 the separation factor was found to be $\alpha = k_{14}/k_{15} = 1.010$. The complete separation was accomplished after 20 cycles.

*See also:* **II/Chromatography; Liquid:** Mechanisms: Normal Phase; Mechanisms: Reversed Phases. **III/Pharmaceuticals:** Thin-Layer (Planar) Chromatography.

## Further Reading

Baweja R (1986) HPLC separation of deuterated photosynthetic pigments from their protio analogues. *Journal of Chromatography* 369: 125–131.

Baweja R (1987) Application of reversed-phase HPLC for the separation of deuterium and hydrogen analogues of aromatic hydrocarbons. *Analytica Chimica Acta* 192: 345–348.

Filer CN (1999) Isotopic fractionation of organic compounds in chromatography. *Journal of Labelled Compounds and Radiopharmaceuticals* 42: 169–197.

Glueckauf E (1961) Isotope separation by chromatographic methods. In: London H (ed.) *Separation of Isotopes*, pp. 209–248. London: G. Newnes Ltd.

**Table 8**    Dependence on eluent pH of LC$^a$ separation factors of benzoic acid versus [$^{18}$O$_2$]benzoic acid$^b$

| | Eluent pH | | | | | | | |
|---|---|---|---|---|---|---|---|---|
| | 2.44 | 3.13 | 4.07 | 4.40 | 4.90 | 5.29 | 5.86 | 6.27 |
| $\alpha$ | 1.000 | 1.000 | 0.995 | 0.991 | 0.988 | 0.988 | 0.992 | 0.995 |

$^a$Cosmosil C$_{18}$-P; methanol-0.05 mol L$^{-1}$ acetate buffer (20 : 80); 30°C.
$^b$Tanaka *et al.* (1986).

Halloran BP. Bikle DD and Whitney JD (1984) Separation of isotopically labelled vitamin D metabolites by HPLC. *Journal of Chromatography* 303: 229–233.

Kaspersen FM, van Acquoy J, van de Laar GLM, Wagenaars GN and Funke CW (1984) Deuterium isotope effects in mianserin. *Recueil Travaux Chimique de Pay-Bas* 103: 32–36.

Klein PD (1966) Occurence and significance of isotope fractionation during analytical separation of large molecules. *Advances in Chromatography* 3: 3–65.

Lešetický L (1985) Isotope separation by chromatographic methods [in Czech]. *Radioisotopy* 26: 113–126.

Lockley WJS (1989) Regiochemical differences in the isotopic fractionation of deuterated benzoic acid isotopomers by reversed phase HPLC. *Journal of Chromatography* 483: 413–418.

Simon H and Palm D (1966) Isotope effects in organic chemistry and biochemistry [in German]. *Angewandte Chemie* 78: 993–1007.

Tanaka N, Araki M and Kimata K (1986) Separation of oxygen isotopic compounds by reversed-phase liquid chromatography. *Journal of Chromatography* 352: 307–314.

Tanaka N and Thornton ER (1977) Structural and isotopic effects in hydrophobic binding measured by HPLC. A stable and highly precise model for hydrophobic interaction in biomembranes. *Journal of American Chemical Society* 99: 7300–7306.

Willi AV (1983) *Isotope Effects in Chemical Reactions* [in German]. Stuttgart: G. Thieme.

Worth GK and Retallack RW (1988) Tritium isotope effect in high-pressure liquid chromatography. *Analytical Biochemistry* 174: 137–141.

Yeung PKF, Hubbard JW, Baker BW, Looker MR and Midha KK (1984) Isotopic fractionation of *N*-([³H]methyl)chlorpromazine and *N*-([³H]methyl)-7-hydroxychlorpromazine by reversed-phase high performance liquid chromatography. *Journal of Chromatography* 303: 412–416.

# LEAD AND ZINC ORES: FLOTATION

**M. Barbaro**, Instituto Trattamento Minerali CNR, Rome, Italy

## Outline of the Problem

Today, owing to the limitation of sources and supplies of mineral raw materials and the need to treat ores of increasingly lower grades, as well as those which are fine, complex mineralogically, and refractory, new flotation technology must be developed.

Flotation is in fact the most common process in metallic mineral separation and is the main way for recovering such valuable metals as lead (Pb) and zinc (Zn), or copper (Cu) from ores.

It is known that separation by flotation of useful minerals from gangue in an aqueous pulp happens when particles with polar, hydrophilic or wettable surfaces remain in the liquid phase, whilst particles with apolar hydrophobic or not wettable surfaces adhere to air bubbles. Collectors or depressing/activating reagents modify surface characteristics of minerals thus influencing affinity towards water. Thus, research into new separation technologies is mainly concerned with the search for new flotation reagents.

In fact the value of a flotation concentrate containing a given mineral, from which a desired metal is extracted by a metallurgical process, decreases with an increase in the presence of minerals containing metals other than the one of prime interest. It is thus necessary to design new specific collectors to separate the desired mineral from the gangue. Collectors generally employed in flotation are surfactants that form, for instance, electrostatic bonds with the solids (Leja 1982). Difficulty therefore arises when a particular metallic mineral, such as Pb or Zn mineral, has to be separated from an ore of complex composition or low grade (complex sulfide ores) or when the surface properties of the mineral (oxidized Pb and Zn ores) make the response to flotation extremely poor. To overcome this basic drawback in metal ore flotation, the possibility of using new compounds endowed with a strong affinity for metals themselves has been investigated. The search for new reagents for mineral flotation therefore aims to discover collectors (or depressants) capable of linking more selectively with a given element present in the ore.

The present work deals with the recovery of Pb and Zn by flotation and reviews the main problems faced by research engineers in this field.

## Complex Pb-Zn Sulfide Ores

Complex sulfide ores have been defined as those ores for which it is difficult to recover one or more selective product of acceptable quality and economic value with minimal losses and at reasonable costs. Complex sulfide ores are fine-grained, intimate associations of chalcopyrite ($CuFeS_2$), sphalerite (ZnS) and galena (PbS), disseminated in dominant pyrite, and containing valuable amounts of minor elements.

Generally, collectors employed for sulfide recovery are of the thiol type, and the most commonly used are xanthates. A great number of studies have been carried out on xanthates, examining their adsorption mechanism by spectroscopic techniques such as infrared spectroscopy (Giesekke 1983, Kongolo et al. 1984, Little et al. 1961, Marabini et al. 1983), IR-ATR techniques (Mielczarski et al. 1987, Mielczarski et al. 1981), X-ray photoelectron spectroscopy (XPS) (Laajalehto et al. 1988, Page et al. 1989), and also by calorimetric techniques (Partyka et al. 1987, Arnaud et al. 1989).

However xanthates are active towards the whole class of sulfide minerals, rather than towards one individual mineral. Thus, in order to float a given mineral from a mixture of minerals belonging to the same sulfide class, modifiers are used in order to render the action of the collector more specific, and to improve separation efficiency (Finkelstein et al., 1976).

However, there are many problems in this procedure and the desired results are not always obtained, especially in the case of minerals of a complex composition as in the case treated.

Hence the importance of seeking out collectors capable of linking selectively with one single given mineral rather than with the whole class. Selective linkage is possible if the collector structure incorporates active groups having specific affinity for certain cation characteristic of the mineral surface.

Thus, the search for new, more selective reagents for sulfide mineral flotation is mainly concerned with chelate-forming reagents. In fact, chelating reagents are particular complexing reagents consisting of large organic molecules capable of linking to the metal ion via two or more functional groups, with the formation of one or more rings, thus forming a very stable bond. The stability of metal chelates is influenced by many factors which govern the selectivity and specificity of the chelation reaction.

Examples of the use of chelating reagents in flotation were known from studies on traditional thiol collectors. In fact, both dixanthogen and thionocar-bamate act as chelating reagents (Ackerman et al., 1987).

There are many studies on chelating agents as collectors more selective than xanthates (Marabini et al., Rinelli et al., Usoni et al., Barbaro et al., Somasundaran) and a review on this topic has been published by Pradip (Pradip 1988). Use of reagents of chelate type offers the possibility of improving selectivity in the flotation separation of complex sulfide minerals (Marbini et al. 1990 and 1991).

## Oxidized Zn and Pb Minerals

It is well known that there is a difference between sulfurized and oxidized minerals as regards their separation by flotation. It is easy to recover Zn from sphalerite and Pb from galena even using xanthate collectors but it is not the same for Zn from smithsonite (Zn $CO_3$) and Pb from cerussite ($PbCO_3$).

Flotation of Zn and Pb oxidized minerals is difficult because there are no known direct-acting collectors capable of producing single metal concentrates. The need for new specific collectors is felt particularly in the case of oxidized lead and zinc minerals because their surface – unlike that of the sulfide variety – is not easily rendered hydrophobic by the collectors generally used, to achieve efficient flotation. Furthermore, the solubility of these oxide minerals is high. Consequently the collector also interacts with metal cations which have gone into solution, thus greatly increasing the amount of reagent required for flotation. It is therefore common practice to sulfurize such minerals prior to flotation so as to prepare their surface to receive xanthates, the collectors generally adopted for concentrating sulfides. Generally the collectors normally used in beneficiation plants act only if the ore has been subjected to a preliminary sulfidization phase which is extremely delicate and critical. In fact, sulfurization calls for careful dosage to avoid rendering the mineral surface inert.

Thus classical collectors have an affinity towards given mineralogical classes, whilst chelating reagents – when chemically adsorbed on the mineral surface – have specificity towards given cations, independently from the mineralogical form of the solid.

However this approach also has two main disadvantages, firstly, excessive consumption (Marabini 1973, Marabini et al. 1983), and secondly, lack of an aliphatic chain which renders the mineral surface hydrophobic.

In fact the chelating reagents commercially available are almost all aromatic molecules without a long hydrocarbon chain; thus, although the chelated min-

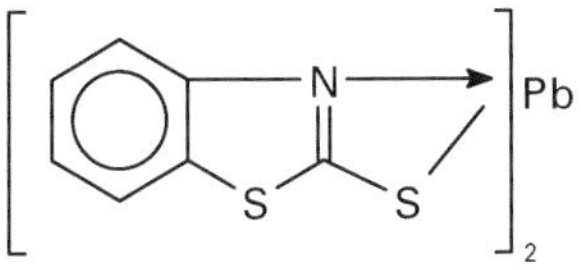

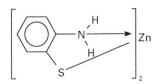

Figure 1    Structure of MBT-Pb chelate.

eral particle is fairly hydrophobic, it is not sufficiently aerophilic to ensure flotation. Studies on oxidized minerals (Usoni *et al.* 1971, Rinelli *et al.* 1973, Marabini 1975, Rinelli *et al.* 1976) were performed rendering particles hydrophobic by making contemporary available long-chain organic groups (as fuel-oil or oily frother) and chelating agents.

The first application of this concept is from 1973. A chelating reagent, namely 8-hydroxyquinoline (**Figure 1**) with fuel oil was used to float mixed oxide-sulfide minerals of Zn and Pb (Rinelli *et al.* 1973). Good recoveries have been attained on an ore containing 7.3% Zn with 1.4% as sphalerite, and 0.9% Pb with 1.4% as galena.

On the basis of the points made so far it is apparent that known chelating compounds form a class of reagents which can be used for the flotation of metallic ores, providing artificially the long chain organic portion by introducing a neutral oil (fuel oil). But the introduction of a new liquid phase into flotation pulp is damaging to the system as a whole and is not available on an industrial scale.

Studies, therefore, have been oriented towards the synthesis of new organic molecules containing both selective functional aromatic chelating groups and hydrophobic long alkyl chain portions. This is done by modifying known chelating collectors.

Indeed, much research was performed on the design of selective chelating collectors; this resulted in numerous structures being proposed and synthesized for laboratory-scale testing on lead/zinc ores prior to the performance of pilot-scale and plant-scale trials.

On the basis of a thermodynamic calculations for the selection of complexing collectors theoretically selective towards a cation (Marabini *et al.* 1983), two classes of reagents have been proposed by Marabini *et al.* (Marabini *et al.* 1988 and 1989, Nowak *et al.* 1991) for the flotation of oxidized Zn and Pb in a pilot plant. Much has been written on the role of the aliphatic chain in conventional collectors (Cases 1968, Predali 1968, Somasundaran 1964) but the work concerns new chelate-type reagents, of the mercaptobenzothiazole (MBT) and aminothiophenol (ATP) types having a mixed aromatic–aliphatic structure. The aromatic part contains specific functional chelating groups that are selective towards the zinc or the lead of oxidized minerals (MBT is selective towards lead and ATP towards zinc) while the aliphatic part consists of a hydrocarbon chain which renders the surface-complex hydrophobic.

The collecting action of MBT is thus attributable to the formation of a surface film selectively chemisorbed on the mineral surface rendered hydrophobic by the aliphatic chain.

In fact in the case of hydrophilic oxidized minerals, the aromatic-heterocyclic portion of the MBT alone does not suffice to render the surface sufficiently hydrophobic to ensure flotation. Hence an aliphatic chain has to be introduced in the molecular structure. The aliphatic chain is necessary to ensure a hydrophobic condition and hence collecting power for the aromatic chelating (MBT or ATP) reagent.

It has been demonstrated that three carbon atoms is the minimum chain length needed to ensure collecting power that improves with aliphatic chain length. Performance is enhanced slightly by the presence of an ether oxygen atom.

Where reagents of the ATP type are concerned, these (as the Schiff bases derived therefrom) exert chelating action towards Zn (Barbaro *et al.* 1997). Chelation occurs through weak bonds with nitrogen and –SH as shown (**Figure 2**).

The formation of a chemisorbed surface film is sufficiently stable to account for the collecting action. The selectivity of molecules containing ATP and different aliphatic chains has been studied by flotation tests.

In this case the role of the aliphatic chain and of the ether oxygen is of more decisive importance than for MBT in assuring the stability of the adsorbed phase and thus floatability. The selectivity increases with the number of carbon atoms in the chain.

In particular, the presence of the oxygen in the chain enhances selectivity, whilst in MBT class reagents only chain length is effective. This difference can be explained by the different chemical structure of the two reagents. In the case of MBT, the effect of the aliphatic substituent is due mainly to its hydrophobicizing effect, and thus to its

Figure 2    Structure of ATP-Zn chelate.

length which favours reciprocal attraction of the chains of the adsorbed layer.

By contrast, in the case of ATP, the effect of the aliphatic substituent is due not only to its hydrophobicizing effect, but also to its effect on reactivity of the aromatic polar head of the molecule. In fact the ATP chelating functional group has a weaker reactivity in comparison with MBT, and therefore is more sensitive to the effect of the substituent on its unique benzenic ring (whilst MBT has two aromatic structures). For this reason, in the case of ATP it is possible to observe that the presence of the oxygen in the chain greatly enhances selectivity. The positive effect of the R-O group in the para position *vis-à-vis* the nitrogen of ATP can be explained with the electron-releasing effect due to resonance of the oxygen with the benzene ring, which increases reactivity with the nitrogen group (Morrison 1973).

In the case of ATP, which forms a less stable bond with the mineral cation and which consists of a single benzene ring, the conjugative effect of the ether oxygen and the hyperconjugative effect of the alkyl groups are more evident than for MBT. Selectivity is improved by the insertion of oxygen in the chain and also by an increase in chain length. Here the effect of the alkyl chain on the aromatic functional group is more marked, permitting modulation of selectivity.

This research based on the design and synthesis of new flotation reagents opens new possibilities in the field of metallic Pb and Zn mineral recovery by flotation.

*See also:* **II/Flotation:** Hydrophobic Surface State Flotation.

## Further Reading

Ackerman PK, Harris GH, Klimpel R and Aplan FF (1987) Evaluation of flotation collectors for copper sulphides and pyrite. I. Common sulphydryl collectors, II. Non-sulphydryl collectors, III. Effect of xanthate chain length and branching. *Int. J. Min. Proc.* 21: 105–156.

Arnaud M, Partyka S and Cases JM (1989) Ethylxanthate adsorption onto galena and sphalerite. *Coll. Surf.* 37: 235–244.

Barbaro M, Herrera Urbina R, Cozza C *et al.* (1997) Flotation of oxidized minerals of copper using synthetic chelating reagents as collectors. *Int. J. Min. Proc.* 50: 275–287.

Finkelstein NP and Allison SA (1976) The chemistry of activation, deactivation and depression in the flotation of zinc sulphides: a review. In Fuerstenau MC (ed.) *Flotation*, Gaudin AM memorial volume. New York: AIME.

Giesekke EW (1983) A review of spectroscopic techniques applied to the study of interactions between minerals and reagents in flotation systems. *Int. J. Min. Proc.* 11: 19–56.

Kongolo M, Cases JM, Burreau A and Predali JJ (1984) Spectroscopic study of potassium amylxanthate adsorption on finely ground galena. In Jones and Oblatt (eds) *Reagents in Mineral Industry*, pp. 79–87. Rome: IMM.

Leja J (1982) *Surface Chemistry of Froth Flotation*. New York: Plenum.

Marabini A, Barbaro M and Ciriachi M (1983) A calculation method for selection of complexing collectors having selective action on a cation. *Trans. IMM*, Sect. C 92: 20–26.

Marabini A and Cozza C (1983) Determination of lead ethylxanthate on mineral surface by IR spectroscopy. *Spectrochimica Acta* 388: 215.

Marabini AM and Rinelli G (1986) Flotation of lead-zinc ores. In Advances in mineral processing. *Proc Symp. honoring N. Arbiter*, N. Orleans, March 3–5, pp. 269–288.

Marabini AM, Alesse V and Barbaro M (1988) New synthetic collectors for selective flotation of zinc and lead oxidised minerals. In Forssberg (ed.) *XVI Int. Min. Proc. Congr.*, Amsterdam: Elsevier, pp. 1197–1208.

Marabini A, Barbaro M and Passariello B (1989) Flotation of cerussite with a synthetic chelating collector. *Int. J. Min. Proc.* 25: 20.

Marabini A and Barbaro M (1990) Chelating reagents for flotation of sulphide minerals. In: *Sulphide Deposits – Their Origin and Processing*. London: Institution of Mining and Metallurgy.

Marabini AM, Barbaro M and Alesse V (1991) New reagents in sulphide mineral flotation. *Int. J. Min. Proc.* 33: 291–306.

Mielczarski J and Leppinen J (1987) Infrared reflection-absorption spectroscopy study of adsorption of xanthates on copper. *Surface Science* 187: 526–538.

Nowak P, Barbaro M and Marabini A (1991) Flotation of oxidised lead minerals with derivatives of 2-mercaptobenzothiazole. Part 1: Chemical Equilibria in the System 6-methyl-2-mercaptobenzothiazole-lead salts. *Int. J. Min. Proc.* 32: 23–43.

Page PW and Hazell LB (1989) X-ray photoelectron spectroscopy studies of potassium amylxanthate adsorption on precipitated PbS related to galena flotation. *Int. J. Min. Proc.* 25: 87–100.

Partyka S, Arnaud M and Lindheimer M (1987) Adsorption of ethylxanthate onto galena at low surface coverage. *Coll. Surf.* 26: 141–153.

Pradip (1988) Application of chelating agents in mineral processing. *Min. Met. Proc.* 80.

Predali JJ (1968) Flotation of carbonates with salts of fatty acids. *IMM* 77: 140–147.

Rinelli G and Marabini A (1973) Flotation of zinc and lead oxide–sulphide ores with chelating agents. *10th Int. Min. Proc. Congr.* London: IMM.

Rinelli G, Marabini AM and Alesse V (1976) Flotation of cassiterite with salicylaldehide as a collector, Flotation, Gaudin AM Memorial Volume, Fuerstenau MC (ed.) vol. 1, p. 549.

Somasundaran P and Nagaraj PR (1984) Chemistry and applications of chelating agents in flotation and floccula-tion. In Jones M and Oblatt R (eds) *Reagents in Mineral Industry*. London: IMM.

Usoni, L, Rinelli G and Marabini AM (1971) Chelating agents and fuel oil: a new way to flotation. *AIME Centennial Annual Meeting*. New York, Feb 26–March 3.

# LIPIDS

# Gas Chromatography

**A. Kuksis**, University of Toronto, Charles H Best Institute, Toronto, Canada

## Introduction

Natural lipids consist of complex mixtures of molecular species, which are found in association with cell membranes, lipoproteins and other subcellular structures. The composition differs among different cell and tissue types, reflecting the function of lipids in these body structures. Industrial and food products may be of plant, animal or synthetic origin. The gas chromatography (GC) of fatty acids by James and Martin in 1956 provided the first success in dealing with the complexity of the hydrolysis products of fats and oils. A few years later Kuksis and McCarthy developed methods for the resolution of intact triacylglycerols by high temperature GC.

Improved design of GC equipment, culminating in the development of reliable capillary columns, together with enzymic and chemical derivatization of samples now permits the separation of molecular species of all lipids on a routine basis. In modern analyses, conventional or high temperature GC serves as the final step in the multi-method resolution and quantification of individual components of a total lipid extract. The high molecular weight and low volatility, however, require constant vigilance in quantitative GC analysis of natural lipids.

### Nonpolar Capillary Columns

Nonpolar capillary columns provide resolution based on the overall molecular weight of the lipid molecules. The nonpolar phases are typically polymethyl-siloxane polymers with 5% phenyl groups. They are stable to 350°C or higher temperatures. Other nonpolar phases for GC are provided by certain hydrocarbons, which are limited to much lower temperatures and are operated isothermally.

### Polarizable Capillary Columns

The methylsiloxane liquid phases containing 50–65% phenyl groups become polar as the temperature increases above 290°C, as indicated by the longer retention of the unsaturated compared to saturated triacylglycerols. Below this temperature the liquid phase is nonpolar, as indicated by the earlier elution of the unsaturated compared to saturated fatty acid trimethylsilyl (TMS) esters. The polarizable liquid phases are stable up to 360°C. Since these liquid phases possess high temperature stability, they are well suited for the resolution of molecular species of seed oil and milk fat triacylglycerols.

However, columns at least 25 m long are needed to provide sufficient resolving power for the separation of the saturated and unsaturated species of natural diacyl and triacylglycerols, and ceramides.

### Polar Capillary Columns

Capillary columns containing polar and very polar liquid phases are utilized mainly for the separation of saturated and unsaturated fatty acids as the methyl esters. The more stable polar capillary columns can be programmed to 280°C and used for the separation of the TMS ethers of diacylglycerols and ceramides as well as low molecular weight triacylglycerols.

### Detection and Quantification of GC Peaks

The principal detector for the GC of lipids is the flame ionization detector (FID), but electron-capture detectors are also used (e.g. for pentafluorobenzyl esters). A number of authors have evaluated the quantitative GC analysis of fatty acids and triacylglycerols. GC with online electron impact (EI) mass spectrometry (GC-EI-MS) yields largely qualitative information from which the structure of unknown molecules can nevertheless be deduced. However, both total ion current and single ion monitoring have been utilized for quantitative analyses using appropriate calibration curves.

## Resolution of Neutral Lipids

Neutral lipids, including common triacylglycerols, are readily resolved according to their molecular weight or number of carbon atoms by high temperature GC. Mixtures of high and low molecular weight neutral lipids are best dealt with by temperature programming. These methods are also suitable for the GC of certain polar lipids provided the polar head groups have been removed or masked. The neutral lipid separations are usually performed on nonpolar liquid phases, but polarizable liquid phases of high temperature stability have also been employed for the separation of intact neutral lipids. In both instances some prefractionation of the lipid mixture is necessary for optimum analysis.

### Isolation and Preparation of Derivatives

Usually this includes the removal of the polar lipids from the neutral compounds by thin-layer chromato-graphy (TLC), high performance liquid chromatography (HPLC) or simple adsorbent cartridges. Before GC, the free functional groups of the lipids must be protected in order to avoid dehydration and to increase volatility, as well as to improve other GC properties. This is readily accomplished by preparing TMS derivatives; in special instances other derivatives may be prepared. Free carboxyl groups may be methylated with diazomethane and alcohol groups may be acetylated with acetic anhydride in the presence of other ester bonds. **Table 1** lists the more common reagents for derivatization of neutral lipids along with the reaction conditions. Polar lipids, such as glycerophospholipids, can be converted into neutral lipids by dephosphorylation. This can be readily accomplished by hydrolysis with phospholipase C (*Bacillus cereus*), which releases the phosphocholine moiety from phosphatidylcholine, lysophosphatidylcholine and sphingomyelin, and the phosphoethanolamine moiety of both diacyl and alkenylacylglycerophosphoethanolamines. The *B. cereus* enzyme also attacks the plasma inositol phosphatides to yield diacylglycerol moieties. Alternatively, the plasma phospholipids may be dephosphorylated by pyrolysis, acetolysis and silolysis, but the latter procedures lead to isomerization of the acylglycerols, incomplete conversion and partial destruction of the lipid samples.

**Table 1**  Preparation of derivatives for GC analysis of neutral lipids, fatty acids and prostanoids[a]

| Reagent | Ratio | Temperature | Time | Application |
|---|---|---|---|---|
| *Neutral lipids* | | | | |
| Pyridine/HMDS/TMCS | 12/5/2 | Ambient | 0.5–1 h | Trimethylsilyl ethers of acylglycerols and sterols |
| *t*-BDMCS/imidazole/ dimethylformamide | 1 : 2.5 in DMF | 80°C | 20 min | *tert*-Butyldimethylsilyl ethers of acylglycerols and sterols |
| Ac₂O/pyridine | 1 : 10 | 80°C | 1 h | Acetates of acylglycerols and sterols |
| PFB₂O/pyridine | 1 : 10 | 80°C | 1–2 h | Pentafluorobenzoates of acylglycerols and sterols |
| *Fatty acids* | | | | |
| BF₃/MeOH | 15% | Ambient | 20 min | Methyl esters of free fatty acids |
| BF₃/MeOH | 6–15% | 60°C/reflux | 2–10 min | Methyl esters of fats and oils |
| H₂SO₄/MeOH | 6% | 80°C/reflux | 2 h | Methyl esters and dimethylacetals of fatty acids |
| HCl/MeOH | 5% | 60°C | 0.5–2 h | Methyl esters of free and bound fatty acids |
| CH₂N₂/ether | Dilute | Ambient | 5 min | Methyl esters of free fatty acids |
| KOH/MeOH/benzene | 0.2–2 N | Ambient | 0.5 min | Methyl esters of glyceryl esters |
| NaOH/MeOH | 0.5–2 N | Refluxing | 0.5–1 h | Methyl esters of steryl esters |
| 2-NH₂-2-MePr (DMOX) | | 170°C | 18 h | 4,4-Dimethyloxazolines of unsaturated fatty acids |
| DEADMS/3-pyridyl carbinol | | 60°C | 10 min, 10 min | Picolinyldimethylsilyl (PICSI) esters |
| (TFA)₂O/3-pyridyl carbinol | 2 steps | 50°C | 1 h | Picolinyl esters of fatty acids |
| Oxalyl Cl/pyrrolidide | 2 steps | Ambient | 30 min | *N*-acyl pyrrolidides of fatty acids |
| *Prostanoids* | | | | |
| PFBBr/*N*,*N*-DIPEtn/ Meoxamine/BSTFA/pyridine | 3 steps | Ambient, 60°C, and 60°C, resp. | 10 min, 16 h, 15 min | PFB/MO/TMS derivatives of prostanoids, thromboxanes, hepoxilins |

[a]Abbreviations and experimental details are found in the text, in legends to figures and in references cited.

The dephosphorylated lipids are recovered by TLC or adsorbent cartridges and the exposed hydroxyl groups masked by acetylation or preparation of TMS derivatives. The diacyl, alkylacyl and alkenylacyl subclasses released from glycerophospholipids by phospholipase C can be readily resolved by normal-phase HPLC or TLC prior to the GC resolution of the molecular species of the sn-1,2-diradylglycerol moieties. Likewise, the method is suitable for the resolution of the molecular species of TMS ethers of the sn-1,2- and sn-2,3-diacylglycerol moieties of triacylglycerols derived by Grignard degradation, chiral-phase HPLC resolution as the dinitrophenylurethanes, and silolysis.

## Total Neutral Lipid Profiling

Total neutral lipids are easily resolved into component lipid classes by GC on short (8–15 m) nonpolar capillary columns. Longer (25 m) polarizable capillary columns provide separations of molecular species. The components ranging from free fatty acids to triacylglycerols can be effectively quantified by tridecanoylglycerol added as an internal standard to the mixture prior to the GC analysis. **Table 2** lists the more common liquid phases for GC of neutral lipids along with the column conditions and selected applications.

**Nonpolar capillary GLC**  **Figure 1** shows a GC separation of total lipids in plasma, following dephosphorylation and preparation of TMS derivatives, on a nonpolar capillary column along with that of a mixture of ceramides and monoacylglycerols released from the combined sphingomyelin and lysophosphatidylcholine fraction isolated by preliminary TLC. The separations on the nonpolar capillary column are limited to resolution by carbon number, although the unsaturated species are eluted slightly ahead of the saturates of the same carbon number. The ceramides (peaks 32–42) are eluted over the same temperature range as the diacylglycerols (peaks 34–40). GC separations similar to those obtained for the total plasma lipids are readily obtained for individual plasma lipoproteins and lymph chylomicrons, as well as for other total lipid extracts from natural sources that can be converted to neutral lipids. GC-MS of plasma total lipids provides confirmation of the peak identity. Mass chromatograms of characteristic fragment ions, retrieved from the total ion current by a computer, permit identification and quantification of overlapping molecular species of free fatty acids, their esters and amides, as well as free sterols and steryl esters.

Partial separation of saturated and unsaturated plasma cholesteryl esters has been reported on a nonpolar capillary column and identification confirmed by negative ammonia ionization mass spectrometry. GC-MS analysis of synthetic steryl esters by nonpolar capillary GLC and EI and chemical ionization has been reported. The GC separation of $C_{27}$ sterols has

**Table 2**  Commonly used liquid phases and columns for GC of neutral lipids[a]

| Chemical composition | Commercial names | Column dimensions | Type of separation (temperature programme) | Applications |
|---|---|---|---|---|
| Methylsilicone | BP-1 (OV-1, SE-30, SP-2100) | 12 m × 0.22 mm i.d. | Carbon number (200–350°C) | Cholesteryl esters |
| 5% Phenyl, 95% methylsilicone | SE-54 (DB-5, HP-5, BP-5, OV-5[b]) | 8 m × 0.32 mm i.d. | Carbon number (200–350°C) | Triacylglycerols, steryl esters |
| 65% Phenyl, 35% methylsilicone | RSL-300 (OV-22, Rtx-65-TG) | 25 m × 0.25 mm i.d. | Carbon and double bond number (40–360°C) | Triacylglycerols, steryl esters, ceramides |
| Methylsilicone | SE-52 (DB-, SE-54) | 26 m × 0.3 mm i.d. | Carbon number (200–350°C) | Triacylglycerols, steryl esters |
| 100% Dimethylsilicone | OV-1 (SP-2100, BP-1, DB-1) | 5 m × 0.32 mm i.d. | Carbon number (100–350°C) | Triacylglycerol core aldehydes |
| 100% Dimethylsilicone | OV-1 (SE-30, BP-1, SP-2100) | 15 m × 0.3 mm i.d. | Carbon number (200–350°C) | Triacyglycerol core aldehydes |
| 68% Cyanopropyl, 32% dimethylsiloxane | Rtx-2330 (SP-2330, SP-2560) | 15 m × 0.32 mm i.d. | Carbon and double bond number (250°C, isothermal) | Diradylglycerol TMS and TBDMS ethers |
| 100% Dimethylsilicone | SE-30 (SP-2100, DB-1, BP-1) | 12 m × 0.25 mm i.d. | Carbon number (40–350°C) | Diradylglycerol TMS and TBDMS ethers |

[a]Abbreviations and experimental details are found in the text, in legends to figures and in references cited.
[b]Comparable liquid phases are given in brackets.
(Modified with permission from Restek Corporation (1998) *Chromatography Products*, International Version, pp. 544–545.)

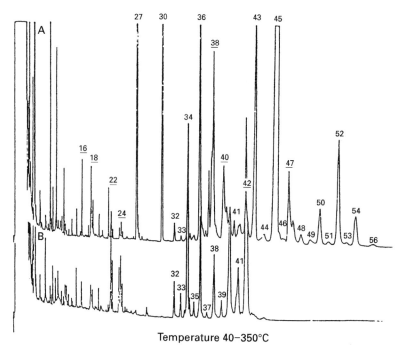

Temperature 40–350°C

**Figure 1** GC of total lipids of normal human plasma on a nonpolar capillary column. A, Total neutral lipids as obtained by dephosphorylation with phospholipase C; B, ceramides and monoacylglycerols released from sphingomyelin and lysophosphatidylcholine. Simplified peak identification: 16 and 18, free fatty acids; 22–24, monoacylglycerols; 27, free cholesterol; 30, tridecanoylglycerol (internal standard); 32–42, diacylglycerols and ceramides; 43–47, cholesteryl esters; 48–56, triacylglycerols. GC conditions: column, 5 m × 0.25 mm i.d., 100% dimethylsilicone (SP-2100, Supelco); temperature, programmed, 170–350°C at 8°C min$^{-1}$ with hydrogen as carrier gas. (Reproduced with permission from Myher JJ and Kuksis A (1984) Determination of plasma total lipid profiles by capillary gas liquid chromatography. *Journal of Biochemical and Biophysical Methods* 10: 13–23.)

been reviewed and the retention times tabulated on DB-5 and CP-WAX columns for a large number of compounds as the TMS derivatives in relation to 5α-cholestane.

**Polarizable capillary GLC   Figure 2** compares the plasma total lipid profiles as obtained by GC on (A) nonpolar and (B) polarizable capillary columns. The nonpolar column yields prominent peaks for free cholesterol, diacylglycerols and ceramides, cholesteryl esters and triacylglycerols. The lipid ester classes are resolved according to the total number of carbons. The polarizable column permits a separation of the glycerolipids and cholesteryl esters on the basis of both total carbon and double bond number. There is an extensive overlap among the molecular species of the diacylglycerols and ceramides but the cholesteryl esters are well resolved from each other and from the triacylglycerols of both higher and lower molecular weight. Characteristic lipid profiles are also obtained for the individual plasma lipoprotein classes. Cholesteryl arachidonate suffered some degradation and is incompletely recovered. The plasma triacylglycerols appear to be fully recovered, except for the more highly unsaturated long chain (56:4–66:18) species, which are only partially recovered.

## Molecular Species of Diradylglycerols and Ceramides

The early GC analyses of molecular species of diacylglycerols were performed on nonpolar packed or capillary columns following a preliminary resolution based on unsaturation by argentation TLC. Polarizable capillary columns provide an improved resolution of the molecular species of diacylglycerols and especially of ceramides. **Figure 3** compares the order of peak elution of diacylglycerols (partial Grignard deacylation products of lard triacylglycerols) and the ceramide moieties of plasma sphingomyelin. The diacylglycerol and ceramide peaks are eluted in order of increasing number of double bonds within a carbon number of each lipid class. Thus, the diacylglycerol 16:0–18:1 is eluted ahead of 16:0–18:2 and 18:0–18:1 is eluted ahead of 18:0–18:2, but 18:0–18:2 tends to overlap with 16:0–20:4. The ceramide d18:1–16:0 elutes earlier than the diacylglycerol 16:0/16:0 on the polarizable column, whereas on nonpolar columns they overlap.

The molecular species of diacylglycerols are best resolved by GC on polar capillary columns similar to those used for separation of fatty acid methyl esters. These columns provide especially detailed analyses of

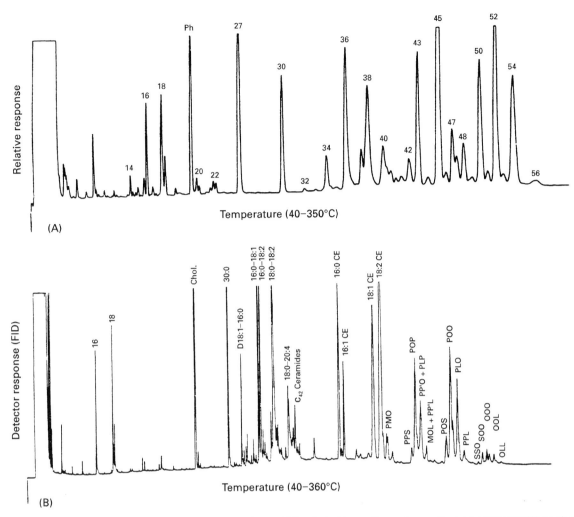

**Figure 2** GC profiles of plasma total lipids on (A) nonpolar and (B) polarizable capillary columns. Peak identification: (A) as given in Figure 1; Ph, TMS ester of phthalic acid; (B) as given in the figure; L, linoleic; M, myristic; O, oleic; P, palmitic; S, stearic acids. GC conditions: (A) column, 5 m × 0.25 mm i.d., dimethylsilicone (SP-2100, Supelco); temperature, programmed, 40–350°C at 8°C min$^{-1}$ with hydrogen as carrier; (B) column, 25 m × 0.25 mm i.d., 65% phenylmethylsilicone (OV-22, Quadrex); temperature, programmed, 40–360°C at 50°C min$^{-1}$ to 150°C, then 10°C min$^{-1}$ to 310°C, and 2°C min$^{-1}$ to 360°C with hydrogen as carrier gas. (Reproduced with permission from Kuksis A, Myher JJ and Geher K (1993) Quantitation of plasma lipids by gas liquid chromatography on high temperature polarizable capillary columns. *Journal of Lipid Research* 34: 1029–1038.)

the TMS ethers of the diradylglycerol moieties of glycerophospholipids and triacylglycerols. **Figure 4** shows the resolution of the molecular species of the diacyl, alkylacyl and alkenylacyl subclasses of human plasma ethanolamine glycerophospholipids as obtained by polar capillary GC of the TMS ethers of the derived diradylglycerols. There are marked differences in the distribution of the chain length and unsaturation among the three subclasses of the plasma ethanolamine glycerophospholipids as anticipated from the fatty acid and diradylglycerol carbon number distribution. Polar capillary GC resolution of molecular species of the *sn*-1,2- and *sn*-2,3-diacylglycerols has been extensively utilized in structural

analyses of triacylglycerols. Tables have been compiled of GC retention factors of diradylglycerol TMS ethers on polar capillary columns.

## Molecular Species of Triacylglycerols

GC is well suited for the separation of the molecular species of triacylglycerols. Nonpolar columns provided essentially carbon number or molecular weight resolution, while the polarizable liquid phases give effective separations of molecular species of saturated and unsaturated triacylglycerols. Polar capillary columns also provide separations based on both carbon and double bond number, but are limited to low molecular weight triacylglycerol mixtures.

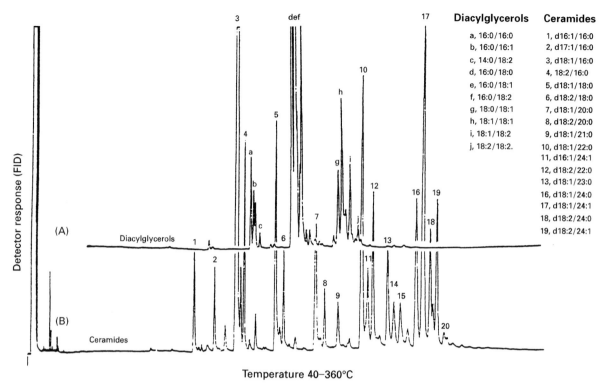

**Figure 3** GC profiles of (A) common diacylglycerols and (B) the ceramides of plasma sphingomyelin as obtained on a polarizable capillary column. Peak identification is as given in figure. GC conditions are as given in Figure 2B. (Reproduced with permission from Kuksis A, Myher JJ and Geher K (1993) Quantitation of plasma lipids by gas liquid chromatography on high temperature polarizable capillary columns. *Journal of Lipid Research* 34: 1029–1038.)

**Nonpolar liquid phases** The first GC resolutions of molecular species of triacylglycerols were obtained on nonpolar packed columns of short length. **Figure 5** compares the separation of butteroil triacylglycerols obtained on a nonpolar packed column and on a capillary column. Both columns yield essentially carbon number resolution, although the capillary column shows some peak splitting due to separation of saturated and unsaturated species within a carbon number. A combination argentation TLC or HPLC with nonpolar GC separation is required for a more extensive separation of molecular species.

Nonpolar capillary GC columns have been effectively utilized for resolution of the reduction products of ozonized lard, rapeseed and palm oil triacylglycerols, as well as the reduction products of ozonized partially hydrogenated soybean oil. **Figure 6** shows the elution profile on a nonpolar column of the triacylglycerols of soyabean oil after reductive ozonolysis. The peaks are identified by the triplets of the fatty acids: palmitic (P), stearic (S) and the $C_9$ aldehyde (U) of the unsaturated fatty acid.

**Polarizable liquid phases** The usefulness of the polarizable liquid phases for the separation of saturated and unsaturated triacylglycerols was discovered by

Geeraert and Sandra in 1984. **Figure 7** shows the resolution of the molecular species of butteroil triacylglycerols. Peak identification presents a problem because of extensive resolution of isobaric species. However, preliminary resolution by argentation TLC in combination with GC-MS enabled the pattern of elution to be established. At the present time, the polarizable capillary column provides the most complete GC resolution of the molecular species of triacylglycerols.

The resolution of triacylglycerols seen in the GC-MS profiles is somewhat reduced in comparison to GC-FID. The reduced resolution complicates postcolumn processing of the GC-MS data due to lack of discrete peaks in parts of the total ion chromatogram. GC-MS quantification of milk fat triacylglycerols is improved by employing the Biller–Biemann enhancement technique to produce mass-resolved total ion chromatograms. The Biller–Biemann enhancement algorithm examines each mass that appears in successive scans in the TLC to detect masses that are rising and falling in intensity.

**Polar liquid phases** The polar capillary columns commonly employed for fatty acid methyl ester separation can resolve low molecular weight

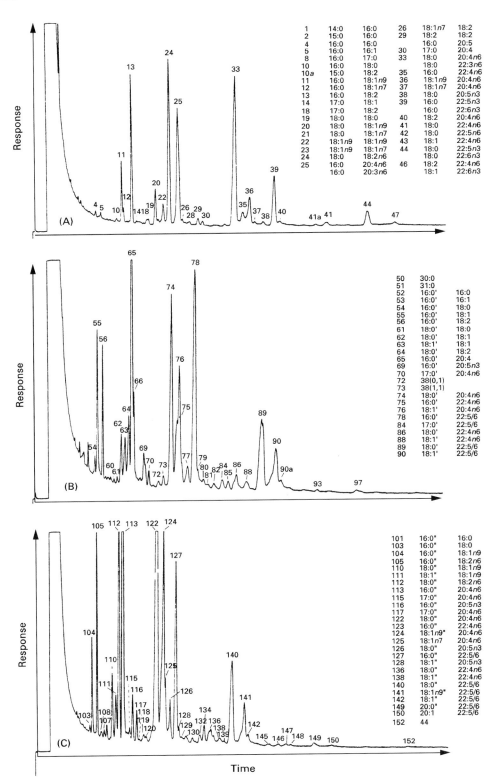

**Figure 4**  GC profiles of the TMS ethers diradylglycerol moieties of human plasma diradylglycerophosphoethanolamines (GPE). Peaks are identified in the figures. (A) Alkylacyl GPE; (B) alkenylacyl GPE; (C) diacyl GPE. Peak identification is given in the figures. GC conditions: column, 15 m × 0.32 mm, i.d. 68% cyanopropyl/32% dimethylsiloxane (RTx 2330, Restek); temperature, isothermal, 25°C with hydrogen as carrier gas. (Reproduced with permission from Myher JJ, Kuksis A and Pind S (1989) Molecular species of glycerophospholipids and sphingomyelins of human plasma: comparison to red blood cells. *Lipids* 24: 408–418).

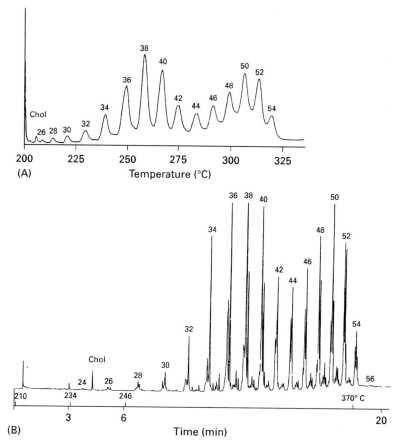

**Figure 5** GC profiles of butteroil triacylglycerols as obtained on nonpolar columns. (A) Packed column; (B) capillary column. Peaks are identified by total acyl carbon as shown in the figures. GC conditions: (A) packed column, 50 × 0.25 cm, i.d. 2.5% SE-30, Applied Science; temperature, linearly programmed from 200 to 325°C with nitrogen as a carrier gas (Reproduced with permission from Kuksis A and McCarthy MJ (1962) *Canadian Journal of Biochemistry and Physiology* 40: 679–685.) (B) Capillary column, 25 m × 0.25 mm i.d., methyl silicone (OV-1, Ohio Valley); temperature, programmed from 210 to 370°C in 20 min with hydrogen as a carrier gas. (Reproduced with permission from Geeraert, 1987.)

triacylglycerols and the diacetates of monoacyl-glycerols containing saturated and unsaturated fatty acids. Conventional polar capillary columns are not sufficiently stable at the high temperatures needed to elute long chain triacylglycerols.

## Resolution of Fatty Acids

Fatty acids constitute the most extensively investigated class of lipids for GC separation, identification and quantification. Although the routines for the common fatty acids are well established, determination of minor fatty acids, including branched-chain and positional and configurational isomers, requires specialized approaches. Other problems arise from the presence of oxygenated functional groups in the fatty chains. Many of these fatty acids have been successfully identified by preparing nitrogenous derivatives in combination with GC-MS.

### Isolation and Preparation of Derivatives

Proper isolation and preparation of derivatives are essential for both identification and quantification of fatty acids. The common procedures for isolating fatty acids consist of extracting lipids from biological materials, saponification and esterification of the recovered acids. Underivatized volatile fatty acids are difficult to quantify by GC because these highly polar compounds form hydrogen bonds that interact with the active sites on the column coatings and cause peak tailing and ghosting due to dimerization. These undesirable effects are avoided by converting the fatty acids into methyl esters or other volatile derivatives. Specially developed columns, however, may avoid both tailing and ghosting.

**Preparation of methyl esters** Table 1 also lists the more popular methods of derivatization of the common fatty acids for GC analysis. Formation of fatty

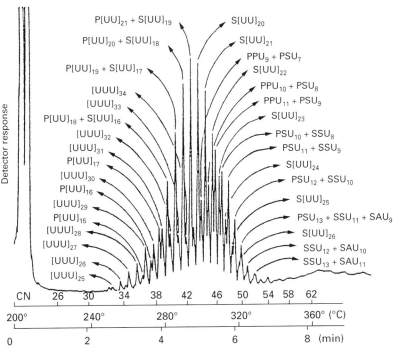

**Figure 6** Non-polar capillary GC profile of the triacylglycerols of soybean oil after reductive ozonolysis. Peak identification is as given in figure: P, palmitic; S, stearic; $U_n$, unsaturated fatty acid residues. GC conditions: column 15 m × 0.32 mm i.d. 100% methylsilicone (OV-1, Ohio Valley); temperature, 200°C (0.15 min) isothermal then 20°C min⁻¹ to 360°C with hydrogen as carrier gas. (Reproduced with permission from Geeraert E (1985) In: Sandra P (ed.) *Sample Introduction in Capillary Gas Chromatography*, vol. 1, pp. 133–158. Heidelberg: Alfred Hüetig Verlag.

acid methyl esters is usually accomplished in the presence of acid or alkaline catalysts. Acidic catalysts transesterify glycerolipids and other complex lipids as well as esterify any free fatty acids in the presence of methanol. Inclusion of butylated hydroxytoluene as an antioxidant in the transmethylation solution results in the formation of methoxy butylated hydroxytoluene, which emerges in the 14:0–16:0 fatty

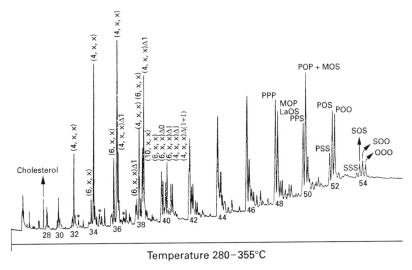

**Figure 7** Polarizable capillary GC profile of butteroil triacylglycerols. Peak identification is as given in figure: P, palmitic; O, oleic; S, stearic; M, myristic acids; 28–54, total acyl carbon numbers; short chain triacylglycerols are shown as combinations with major long chain acids (x) of different degrees of unsaturatation (Δ0 − Δ1 + 1) as identified elsewhere (Myher JJ, Kuksis A, Marai L and Sandra P (1988) *Journal of Chromatography* 452: 93–118.) GC conditions: column, 25 m × 0.25 mm i.d., 50% phenylmethylsilicone (RSL-300, supplied by Sandra P); temperature, linearly programmed from 280 to 355°C over 30 min with hydrogen as the carrier gas. *Major acetate peaks. (Reproduced with permission from Geeraert E and Sandra P (1987) Capillary GC of triglycerides in fats and oils using a high temperature phenylmethylsilicone stationary phase. Part II. The analysis of chocolate fats. *Journal of the American Oil Chemist's Society* 64: 100–105.)

acid elution region. Phthalate esters have also been detected among plasma lipids, even when prepared in the absence of external contamination. The precise point of elution relative to fatty acid derivatives is dependent on the nature of the phthalate ester and the stationary phase, but typically it is in the same range as the $C_{18}$–$C_{22}$ fatty acids. In many instances it is convenient to perform the acid transmethylation *in situ*, e.g. in the presence of silica gel scrapings from a TLC plate.

Acid-catalysed transmethylation also leads to significant dehydration of the sterol moiety and degradation of conjugated fatty acids. The peaks formed during acid-catalysed methylation were identified as positional allylic methoxy isomers of 18:1 by GC-MS. Re-evaluation of the $H_2SO_4$–isopropanol reaction showed just as extensive isomerization of conjugated dienes as the HCl–MeOH method.

Alkaline catalysts transesterify neutral lipid esters in anhydrous methanol much faster (a few seconds to a few minutes), but they are unable to esterify free fatty acids, which constitutes a serious shortcoming. Furthermore, N-acyl lipids are not methylated. The presence of water leads to saponification. Alkaline reagents dissolve silica gel and cannot be employed with gel scrapings from TLC plates. Diazomethane can be prepared in ether solution by the action of alkali on a nitrosamide (e.g. *N*-methyl-*N*-nitroso-*p*-toluenesulfonamide) in the presence of alcohol. This reagent is commercially available as Diazald (Aldrich Chemical Co). Diazomethane is used for preparation of fatty acid methyl esters from lipids that have been first saponified. It has the disadvantages that it is highly toxic, explosive and likely to cause specific sensitivity.

**Preparation of other derivatives** Table 1 also lists a selection of reagents for derivatization of oxygenated fatty acids and prostanoids. Monohydroperoxy fatty acids are reduced to hydroxy acids after reaction for 1 h at room temperature with triphenylphosphine in diethyl ether prior to GC analysis of the methyl ester TMS ethers. The monohydroxy fatty acid methyl esters are hydrogenated for 5 min at room temperature in the presence of rhodium on alumina in methanol. They are converted to their TMS ether derivatives by reaction with *N*-methyl-*N*-trimethylsilyltrifluoroacetamide for 30 min at room temperature. *tert*-Butyldimethylsilyl ethers of secondary hydroxy fatty acid methyl esters are prepared by dissolving the sample in a mixture of dry toluene, dimethylformamide and pyridine, and adding *N*-*tert*-butyldimethylsilylimidazole and *N*-methyl-*tert*-butyl-dimethylsilyltrifluoroacetamide containing 1% *tert*-butyldimethylsilylchloride (TBDMS).

Of the many different derivatives described in the literature for GC analysis of prostaglandins, the pentafluorobenzoate–methoxime–TMS (PFB–MO–TMS) derivatives yield the most satisfactory results in terms of GC properties, electron-capture detection and MS response and stability. These derivatives show single well-shaped peaks for each compound, except for prostaglandin $E_2$–PFB–MO–TMS, whose *syn-anti* isomers can be seen as well-separated peaks.

Normal oxygenated and nonoxygenated fatty acids have been converted into nitrogenous derivatives for structural analyses by GC-MS. The picolinyl esters are prepared from the fatty acids by first converting them into acid chlorides by dissolving in thionyl chloride. The acid chlorides are then treated with a dilute solution of 3-pyridylcarbinol in acetonitrile. The picolinyldimethylsilyl derivatives of fatty alcohols are prepared by dissolving the alcohol in dry pyridine, adding diethylaminodimethylsilyl-3-pyridylcarbinol and heating at 60°C for 10 min. The picolinyl esters of the epoxides are prepared by dissolving the epoxide in dry methylene chloride and adding a freshly prepared solution of 1,1'-carbonyl-diimidazole followed by 3-pyridylcarbinol solution of triethylamine. The preparation of picolinyl esters via fatty acid chlorides cannot be used for derivatization of the epoxides because of their instability in acidic solutions.

The pyrrolidides are prepared from the lipid ester. The sample is heated with an excess of pyrrolidine in the presence of acetic acid for 30 min at 100°C. 4,4-Dimethyloxazoline derivatives also have useful properties for locating functional groups by MS, but quantitative preparation requires heating the methyl esters with 2-amino-2-methylpropanol at 180°C overnight.

### Separation of Saturated and Unsaturated Fatty Acids

Table 3 lists the more popular liquid phases for the separation of saturated and unsaturated fatty acids and prostanoids. The Carbowax phases based on polyethylene glycol appear to provide the best general-purpose columns.

**Normal chain fatty acids and dimethylacetals** For equivalent chain length (ECL) measurements, the separations of saturated and unsaturated fatty acids are performed isothermally. Isothermal analyses give good ECL values and these are remarkably consistent for the bonded polyglycol columns. Extensive tables of ECL values have been compiled for the methyl ester derivatives of natural fatty acids on silicone,

**Table 3** Common liquid phases for capillary GC of fatty acids and their oxygenated derivatives[a]

| Chemical composition | Commercial names | Column dimensions | Type of separation (temperature programme) | Applications |
|---|---|---|---|---|
| 90% bis-Cyanopropyl/ 10% cyanopropylphenylsilicone | SP-2380 (OV-275/CP-Sil-88) | 15 m × 0.32 mm i.d. | Carbon and double bond number (260–270°C) | Short chain triacylglycerols of butteroil distillates |
| Cross-bonded polyethylene glycol | Supelcowax-10 (Carbowax) | 30 m × 0.25 mm i.d. | Saturates and polyunsaturates (185–220°C) | Fish oil fatty acids |
| Cross-bonded polyethylene glycol | Stabilwax (Carbowax CP 51) | 60 m × 0.25 mm i.d. | Saturated and unsaturated esters (212°C, isothermal) | Seed oils and fats, marine oils |
| 90% bis-cyanopropyl/10% cyanopropylphenyl | Rtx-2330 (CP-Sil-88, BPX-70)[b] | 100 m × 0.25 mm i.d. | $\Delta^5$-$\Delta^{16}$-isomers of trans-18:1 (160°C, isothermal) | Trans-18:1 of beef tallow and human milk |
| Biscyanopropyl, 68%/ dimethylsilicone, 32% | SP-2380 (BPX-70/CP-Sil-84) | 50 m × 0.25 mm | Cyclic fatty acid methyl esters (180°C, isothermal) | Hydrogenated cyclic fatty acid monomers from vegetable oils |
| 14% Cyanopropylphenyl/ 86% polysiloxane | CP Sil 19 (DB-1701) | 25 m × 0.25 mm | Methyl ethers of isomeric hydroxy fatty acids (100–250°C) | Hydrogenated hydroperoxides of plasma lipids |
| 2% Dimethylsilicone | OV-101 | 0.4 m × 3 mm, packed | Carbon number (315–330°C, isothermal) | TMS ethers of mycolic acids |
| 95% Dimethyl/5% phenyl silicone | SE-54 (HP-5/ DB5/CP-Sil-8) | 18 m × 0.28 mm | Double bond location (70–240°C) | Oxazoline derivatives of fish oil fatty acids |
| Dimethylpolysiloxane | DB-1 (SE-30/OV-1) | 30 m × 0.25 mm | Epoxide location (200–250°C) | 3-Pyridinylmethyl esters of epoxides |
| Dimethylpolysiloxane | DB-1 (SE-30/OV-1) | 20 m × 0.25 mm | Methoxy PFB esters (100–310°C) | Urinary prostanoids |
| 86% Dimethyl/14% cyano propylphenyl polysiloxane | DB-1701 (DB-5, CP-Sil 19CB) | 15 m × 0.2 mm | $F_2$-isoprostane TMS ethers and PFB esters (190–300°C) | Arachidonic acid peroxidation products |
| 86% Dimethyl/14% cyano propylphenyl polysiloxane | DB-1701 (DB-5, CP-Sil 19CB) | 30 m × 0.25 mm | Methoxy PFB esters (18–300°C) | Eicosanoids in blood |

[a]Abbreviations and details of applications are given in the text, in legends to figures and in references cited.
[b]Comparable liquid phases are given in brackets.
(Modified with permission from Restek Corporation (1998) *Chromatography Products*, International Version, pp. 544–545).

Carbowax, Silar 5CP and CP-Sil 84 columns. **Figure 8** shows the resolution of the fatty acid methyl esters of menhaden oil on a 30 m polyethylene glycol (Famewax, Restek) column using temperature programming (190–225°C). This liquid phase is stable to 250°C. The elution order of the complex polyunsaturated fatty acid methyl esters is comparable to that obtained on other Carbowax columns.

Acid-catalysed transmethylation leads to conversion of vinyl ethers (when present) into dimethylacetals, which are eluted ahead of the fatty acids of corresponding carbon number. **Figure 9** shows the resolution of the fatty acid methyl esters and dimethylacetals derived from human erythrocyte membranes on a deactivated cyanopropylsiloxane column. Although the separation of many of the dimethylacetals and the methyl esters is incomplete, it is still possible to obtain a good indication of the

relative amounts of plasmalogens that may be present in the sample. Resolution of the major dimethylacetals and methyl esters on cyanopropylsiloxane columns has been obtained of one-third the length used by earlier workers. Prior to GC, the fatty acid methyl esters and dimethylacetals are resolved by normal-phase HPLC.

**Identification of very long chain fatty acids** An extensive series of long chain polyunsaturated fatty acids has been shown to occur in phosphatidylcholines of the vertebrate retina, where they make up a homologous series of even carbon polyenes having up to 36 carbon atoms. A detailed study has been made of these acids in bovine retina, where they were shown to belong to the same families as well-known fatty acids of the *n*-6 and *n*-3 series like arachidonate (20:4*n*-6), docosapentaenoate (22:5*n*-3) and docosa-

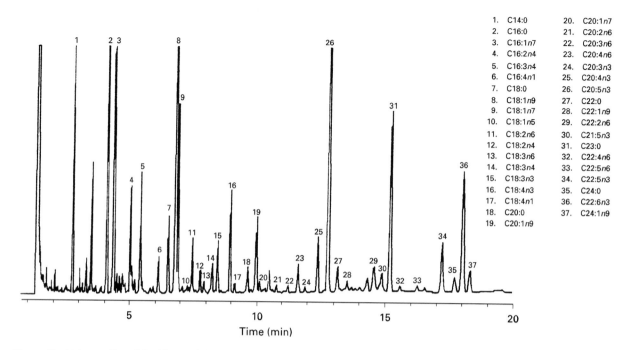

| 1. | C14:0 | 20. | C20:1n7 |
| 2. | C16:0 | 21. | C20:2n6 |
| 3. | C16:1n7 | 22. | C20:3n6 |
| 4. | C16:2n4 | 23. | C20:4n6 |
| 5. | C16:3n4 | 24. | C20:3n3 |
| 6. | C16:4n1 | 25. | C20:4n3 |
| 7. | C18:0 | 26. | C20:5n3 |
| 8. | C18:1n9 | 27. | C22:0 |
| 9. | C18:1n7 | 28. | C22:1n9 |
| 10. | C18:1n5 | 29. | C22:2n6 |
| 11. | C18:2n6 | 30. | C21:5n3 |
| 12. | C18:2n4 | 31. | C23:0 |
| 13. | C18:3n6 | 32. | C22:4n6 |
| 14. | C18:3n4 | 33. | C22:5n6 |
| 15. | C18:3n3 | 34. | C22:5n3 |
| 16. | C18:4n3 | 35. | C24:0 |
| 17. | C18:4n1 | 36. | C22:6n3 |
| 18. | C20:0 | 37. | C24:1n9 |
| 19. | C20:1n9 | | |

**Figure 8**  Polar capillary GC of fatty acid methyl ethers of menhaden oil. Peaks are identified in the figure. GC conditions: column, 30 m × 0.25 mm i.d., 0.25 μm cross-bonded polyethylene glycol (Famewax, Restek); temperature, 190°C (hold 4 min) then 4°C min$^{-1}$ to 225°C; carrier gas, helium; split injection 1:50 (Reproduced with permission from Restek Corporation (1998) *Chromatography Products*, International Version, p. 447).

hexaenoate (22:6n-3). The fatty acids were analysed by temperature-programmed GC using glass columns packed with 15% OV-225. The position of the double bonds in the long chain polyenes was determined by means of oxidative ozonolysis.

**Separation of short chain fatty acids**  Because of their high volatility, the short chain fatty acids create difficulties in GC separation and quantification when present in mixtures with long chain acids. The volatility can be effectively reduced by preparing higher molecular weight esters (propionyl or butyl esters) which also provide a more comparable response in the FID. It also permits a clear resolution of the short chain fatty acids from the solvent front and identification and quantification of the short chain acids in the presence of long chain acids. Recently developed high speed capillary columns (20 m × 0.10 mm i.d., 0.10 μm film thickness DB-WAX or DB-225, Restek) have been found to give an effective separation and quantification of both short and long chain fatty acids as methyl esters. A total of 37 fatty acids can be analysed in less than 15 min.

### Separation of Oxygenated Fatty Acids and Prostanoids

Oxygenated fatty acids occur in nature as a result of enzymatic and nonenzymatic oxidation, and because

of their physiological activity there is much interest in their analysis.

Fatty acid hydroperoxides are labile key intermediates in plant and mammalian lipid metabolism, acting as precursors of a variety of lipid-derived mediators such as prostaglandins and leukotrienes. Until recently, these analyses were usually done by GC, but now reversed-phase HPLC has been shown to have advantages for the analysis of the thermo-labile oxygenated derivatives of the fatty acids.

**Hydroxides, epoxides, hydroperoxides and isoprostanes**  A 30 m nonpolar methylsiloxane column was used to measure the ECLs of hydrogenated derivatives of isomeric hydroxydocosahexaenoates. Mass spectra of hydrogenated compounds indicated the presence of hydroxyl groups at carbons 20, 16, 17, 13, 14, 10, 11, 7, 8 and 4. The isomers were apparently racemic mixtures.

GC-MS was used to identify the arachidonate epoxides/diols. The epoxide regioisomers of arachidonic acid as the methyl esters overlapped on nonpolar capillary columns. The isomeric epoxides were identified as their hydrolysis products, the dihydroxyeicosatrienoic (DHET) acids, in the form of the PFB ester derivatives. The four regioisomers were not resolved by GC as PFB, TMS, TBDMS or Me esters. However, after being hydrolysed to the dihydroxyeicosatrienoic acids, three of the four regio-

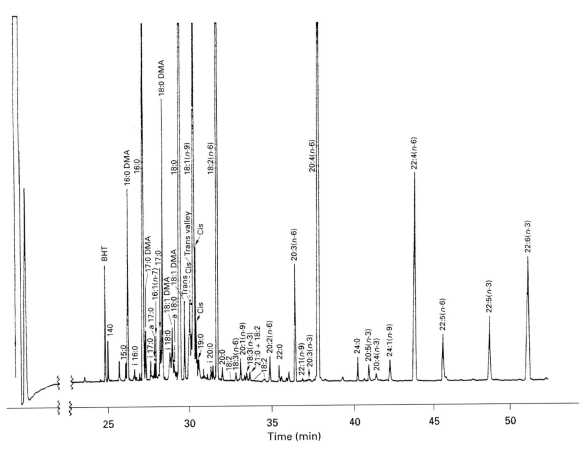

**Figure 9** Polar capillary GC of fatty acid methyl esters and dimethylacetals of human erythrocyte membranes. Peaks are identified by short-hand notation as indicated in figure. GC conditions: column, 100 m × 0.25 mm i.d., 0.2 μm deactivated cyanopropylsiloxane (SP 2560, Supelco); temperature, 80°C (hold 2 min) then 8°C min$^{-1}$ to 220°C and hold for 32 min; carrier gas, helium; split injection 1 : 30. (Reproduced with permission from Alexander LR, Justice JB and Madden J (1985) Fatty acid composition of human erythrocyte membranes by capillary gas chromatography mass spectrometry. *Journal of Chromatography and Biochemical Applications* 342: 1–12.)

isomers were resolved as TMS ethers, PFB esters. The fourth regioisomer, 5,6-DHET, was resolved after being converted to a δ-lactone. The regioisomers of DHET were resolved as the (bis)-TBDMS, PFB esters on a 60 m nonpolar column operated isothermally at 300°C. The 8,9-DHET was followed by 11, 12-DHET, which was followed by 14,15-DHET.

TBDMS ethers have been used for GC-MS separation and identification of synthetic secondary hydroxy fatty acid isomers with carbon chain lengths of 16–20. With the exception of the spontaneous formation of γ- and δ-lactones of C$_4$-OH and C$_5$-OH fatty acids, the TBDMS ethers of all hydroxy fatty acid methyl esters in a mixture are readily identified by GC-MS, yielding information on chain length, location of the hydroxyl group and degree of unsaturation. The method has been applied to the identification of a complex mixture of bovine skim milk hydroxy fatty acids, of which 19 were newly identified. The separations were performed on a 30 m column coated with 100% methylsilicone. Hydroperox-

ides are not sufficiently stable for GC analysis. Therefore, the fatty acid ester hydroperoxides are reduced to the corresponding alcohols with triphenylphosphine and transmethylated with sodium methoxide before GC analysis. A 20 m × 0.32 mm i.d. DB-1 column has been used for GC-MS analysis of linoleate and arachidonate peroxidation products. The hydroxy fatty acid methyl esters were isolated by TLC and converted into TMS ethers. By a combination of normal-phase HPLC and GC-MS, the linoleate oxidation products were identified as 9- and 13-OH derivatives and those arising from arachidonate oxidation as the 5-, 8-, 9-, 12- and 15-hydroxyeicosatetraenoates.

Among the free radical catalysed peroxidation products of arachidonic acid, prostaglandin-like compounds named isoprostanes have been recognized which arise independent of the cyclooxygenase enzyme. The evidence has been reviewed for the natural occurrence of these compounds, their mechanism of formation and methods of determination. They have

GC properties similar to those of prostaglandin $F_{2\alpha}$, with which they may be confused. GC-MS has been recently employed for the measurement of plasma isoprostanes.

**Prostaglandins**   The prostaglandins and thromboxanes are cyclo-oxygenase metabolites of arachidonic acid, which are involved in numerous pathophysiological pathways. High resolution GC in combination with mass spectrometry is generally recognized as the most specific and reliable method for qualitative and quantitative analysis of prostanoids. In fact, the use of high resolution GC is mandatory when the stable cyclo-oxygenase metabolites of arachidonic acid have to be analysed as a group. The application of GC-MS techniques to the analysis of prostaglandins and related substances has been reviewed, as well as the application of GC-MS to the analysis of other oxygenated fatty acids. Electron capture GC-MS analysis of the PFB-TMS derivatives has provided the most sensitive method for molecular mass determination. This method, however, does not provide structural information and even low energy collision-induced dissociation (CID) of the carboxylate anions from these derivatives does not yield structurally significant ions.

**Figure 10A** shows by means of authentic standards that a complete separation of all major metabolites of arachidonic acid via the cyclo-oxygenase pathway can be obtained in a relatively short time (30 min) using a 30 m capillary column of OV-101–OV-17 (8 : 2). The selected ion monitoring profiles of the prostaglandins and thromboxanes were obtained with the PFB–MO–TMS derivatives at 220°C with helium as carrier gas. Figure 10B shows a chromatogram of the same substances detected by electron capture. A critical aspect of the application of high resolution GC in routine analyses of prostaglandins is that it requires greater purification of biological material.

### Structural Identification

The location of double bonds and other modifications of the fatty chains is required for a complete characterization of the structure of the fatty acids. GC-MS techniques approach this problem by chemical modification prior to analysis.

**Double-bond isomers**   Derivatization at the double bond to oxygenated compounds (carbonyl compounds and vicinal diols) leads to distinctive fragmentation patterns that allow determination of the points of unsaturation or epoxidation but this method leads to the formation of highly polar derivatives and an increase in molecular weight, which makes it unsuitable for the analysis of polyunsaturated fatty acids.

An alternative technique is based on derivatization of the terminal carboxylic group (remote site modification), whereby fatty acids are converted to N-acyl

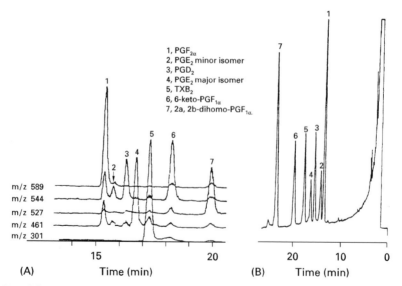

1, $PGF_{2\alpha}$
2, $PGE_2$ minor isomer
3, $PGD_2$
4, $PGE_2$ major isomer
5, $TXB_2$
6, 6-keto-$PGF_{1\alpha}$
7, 2a, 2b-dihomo-$PGF_{1\alpha}$

**Figure 10**   Polar capillary GC and selected ion monitoring (SIM) of authentic prostaglandins and major metabolites of arachidonic acid as PFB-MO-TMS derivatives. (A) SIM; (B) high resolution GC with electron capture detection. Peak identification is given in the figure. GC conditions: column, 30 m × 0.3 mm i.d., dimethylsiloxane/50% phenylmethylsiloxane 8 : 2 (OV-1/OV-17, Ohio Valley); temperature, 220°C isothermal; carrier gas, helium. (Reproduced with permission from Chiabrando C, Noseda A and Fanelli R (1982) Separation of prostaglandins and thromboxane $B_2$ by high resolution gas chromatography coupled to mass spectrometry or electron-capture detection. *Journal of Chromatography* 250: 100–108.)

pyrrolidides or picolinyl (3-hydroxymethylpyridinyl) esters that easily stabilize the ions containing the double bonds or substituents of the fatty chain. The pyrrolidides have been studied most often for the analysis of natural samples, but recently have been overtaken by the picolinyl esters or 4,4-dimethyloxazoline (DMOX) derivatives. Although distinctive modes of fragmentation can be obtained, the interpretation of spectra becomes substantially more difficult if the number of double bonds is greater than four. The picolinyl esters of isomeric unsaturated fatty acids give more abundant diagnostic ions than the pyrrolidides. **Figure 11** shows the capillary GC separations obtained with the picolinyl and pyrrolidide derivatives of pig testis fatty acids on OV-101 and BP-20 liquid phases, respectively. On the nonpolar methylsilicone (OV-101) column, the more unsaturated picolinyl esters emerge ahead of the less unsaturated and saturated fatty acids of the same acyl carbon number. On the polar BP 20, the pyrrolidide derivatives emerge in order of increasing unsaturation. The picolinyl esters require column temperatures about 50°C higher than methyl esters. The picolinyl esters are resolved according to number of double bonds on polar phases of high temperature stability, such as BPX-70 or Supelcowax 10, but

fattty acids of more than 20 acyl carbons may cause difficulty.

It has been shown that the DMOX derivatives of fatty acids are only slightly less volatile than the corresponding methyl esters and that they can be subjected to GC analysis on polar capillary columns to give resolutions based on degree of unsaturation. The GC separation has been carried out with the DMOX derivatives of fatty acids of rat testis and fish oil on a nonpolar methylsilicone (SE-54) column, from which the more unsaturated derivatives emerged ahead of the less unsaturated ones of the same acyl carbon number.

The picolinyl esters and DMOX derivatives can be separated by reversed-phase or argentation HPLC before GC-MS analysis, which is important for the detection of minor components in natural samples. These derivatives have high volatility and their mass spectra show easily recognizable diagnostic peaks for determination of the position of unsaturation. Tables have been prepared of characteristic ions in EI mass spectra of DMOX derivatives of 22 unsaturated fatty acids. The DMOX derivatives have been employed for successful identification of $\Delta$5-unsaturated poly-methylene-interrupted fatty acids from conifer seed oils. Most analysts now prefer either picolinyl esters

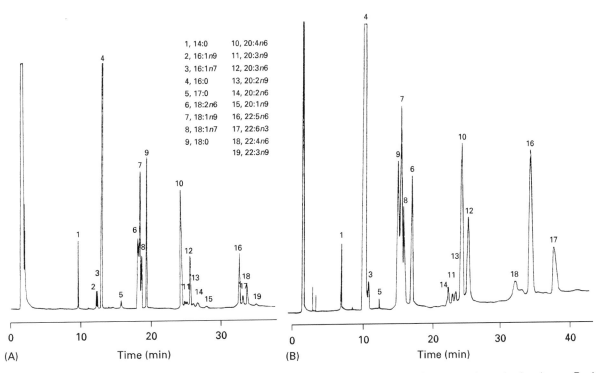

**Figure 11**  Capillary GC of (A) picolinyl and (B) pyrrolidide derivatives of pig testis fatty acids on nonpolar and polar phases. Peak identification is given in the figure. GC conditions: (A) column, 50 m × 0.22 mm i.d., methylsilicone (OV-101, Ohio Valley); temperature, programmed from 195°C (hold 3 min) then 1°C min⁻¹ to 235°C; carrier gas, helium; (B) column, 12 m × 0.22 i.d., polyethylene glycol (BP 20, SGE); temperature, programmed from 190°C (hold 3 min) then 1°C min⁻¹ to 230°C; carrier gas, helium. (Reproduced with permission from Christie WW, Brechany EY, Johnson SB and Holman RT (1986) A comparison of pyrrolidide and picolinyl ester derivatives for the identification of fatty acids in natural samples by gas chromatography-mass spectrometry. *Lipids* 21: 657–661.)

or DMOX derivatives for structural studies on fatty acids, but the pyrrolidides also provide informative mass spectra in many instances.

*cis/trans*-**Isomers and conjugated acids**  Identification of *cis/trans* isomers of unsaturated fatty acids cannot usually be achieved by GC-MS without reference substances. Satisfactory separations of the geometric isomers from margarine fatty acids and 44 synthetic $C_{18}$ unsaturated fatty acids have been obtained by GC on a 60 m capillary column coated with 100% cyanoethylsilicone (SP-2340). Overlaps occur between different double-bond systems and between some mono- and diethylenic, as well as between di- and triethylenic fatty acids. Because of these overlaps, the isomers cannot be determined by GC alone on the SP-2340 or on any other cyanosilicone phase. Prefractionation by silver ion TLC or HPLC is necessary. The GC separation of the *cis*- and *trans*-acids has been reported in milk samples using a 100 m capillary column coated with bis-cyano-propyl-32% dimethylsiloxane (SP-2560). The authors were able to calculate the total *trans* acid content by adding the relative percentage of all *trans* monounsaturated fatty acids and *cis/trans* linoleate and total conjugate dienes in the milk chromatograms. SP-2560 capillary columns were used to conduct an extensive investigation of the effect of various methods of methylation upon the estimation of the conjugated fatty acid content of bovine milk. Acid-catalysed methods caused extensive isomerization of conjugated dienes and formed allylic methoxy artefacts and are therefore not recommended for this purpose. Base-catalysed methods caused no isomerization of conjugated dienes and the formation of artefacts. GC–Fourier transform infrared–MS (GC/FTIR/MS) has been used to identify fatty acid methyl esters and differentiate between *cis/trans* isomers. In the FTIR spectra *cis/trans* isomers are identified by analysis of bands arising from C–H out-of-plane bending: for both fatty acid methyl esters and DMOX derivatives *cis*-1,2-disubstituted double bonds give a strong band near $720 \text{ cm}^{-1}$ and the corresponding *trans* isomers near $967 \text{ cm}^{-1}$. A greatly improved resolution of individual *trans*-18:1 isomers has been reported by capillary GC on a 100 m cyanopropyl polysiloxane (CP-Sil 88) column.

**Epoxides**  The picolinyl esters are also adequate for the location of epoxide functions in linoleic, arachidonic and docosahexaenoic acids. The electron impact (EI) mass spectra of these derivatives show a molecular ion and a sequence of peaks, with two characteristic abundant ions that result from formal cleavage of the carbon–carbon bonds at the oxirane ring. Both these ions retain the ester group. This fragmentation pattern allows the unequivocal identification of the separate epoxide isomers. The picolinyl esters of the epoxides are resolved on a 30 m methylsiloxane (DB-1) column, temperature-programmed from 220 to 300°C. These picolinyl esters provide characteristic GC-MS profiles that allow differentiation of the isomeric epoxides at nanogram levels.

**Branched-chain and cyclic acids**  The methyl esters of branched and cyclic fatty acids are usually eluted ahead of their normal-chain counterparts. The methyl branched acids may be confused with straight-chain odd-carbon number isomers. Long (50 m) capillary columns with nonpolar (5% phenylmethylsilicone) stationary phases are preferred and cyclopropane acids are resolved from their monounsaturated counterparts. The position of the methyl branching and cyclopropane ring is not readily determined by GC-MS with EI even after preparation of pyrrolidide and picolinyl derivatives, although exceptions are known. Only small differences are seen among the DMOX derivatives, but they may be sufficient for distinction between close isomers.

It has been shown that low-energy CID of the molecular ions of fatty acid methyl esters obtained by electron ionization (70 eV) in the tandem quadrupole mass spectrometer yield a regular homologous series of carbomethoxy ions. This method can be used to determine methyl (or alkyl) branching positions, as shown by enhanced radical site cleavage at the alkyl branching positions of several methyl esters, including phytanic acid, isomethyl and anteiso-methyl branched acids and tuberculostearic acid. Analyses of various stable isotope variants support the hypothesis of alkyl radical migration to the carboxy carbonyl oxygen atom, with subsequent radical site directed cleavage, either with or without a cyclization event.

# GC-MS Analyses of Stable Isotope-Labelled Lipids

In addition to identification of structures, GC-MS also provides the distribution and content of the stable isotope, both natural and enriched. Highly sensitive and reproducible estimates have been obtained for the isotopomer distribution by GC combined with isotope ratio mass spectrometry (isotopomer analysis). Furthermore, the utilization of appropriately labelled isotopic homologues as internal standards for the solute to be measured has greatly improved quantification by GC.

## Isotopomer Analyses

Mass isotopomer distribution analysis (MIDA) uses stable isotopes with quantitative mass spectrometry. It is based on the principle that the isotope distribution or labelling pattern of a fatty acid (e.g. palmitic acid) synthesized from an isotopically perturbed monomeric precursor pool conforms to a binominal expansion. The proportion of unlabelled, single-labelled, double-labelled, and so on, molecular species of a fatty acid is a function of the probability that each precursor subunit is isotopically labelled in the fraction of newly synthesized fatty acid present. The advantages and disadvantages of this approach in studies of lipid metabolism have been discussed in the literature. Analogous combinatorial probability methods have long been used to measure endogenous synthesis of fatty acids. The analytical data for the biosynthesis of palmitic acid from [$^2$H-ethanol], [$^2$H$_2$O] or [$^{13}$C-acetate], for example, are provided by GC-MS analysis of the palmitic acid methyl ester and the methyl esters of any other fatty acids of interest. In this method every newly synthesized molecule is counted and the total synthesis calculated by summation of the newly formed molecules.

## Definitive Methods

One of the major applications of GC-MS is in quantitative analysis with stable isotope dilution and selected ion monitoring. A known amount of a stable isotope-labelled variant of the molecule to be measured is added to the biological sample as an internal standard, followed by extraction, purification and chemical derivatization. The ions specific for the substance and the internal standard are then monitored. The ratio of the two ion abundances provides the quantitative measurement, which is independent of absolute sensitivity. The superiority of this approach has been demonstrated in the quantitative GC-MS analysis of prostaglandin F$_{2\alpha}$ in a biological extract using a deuterated derivative. Examples of this approach are provided by the highly sensitive measurement of thromoboxane B$_2$ (TXB$_2$). Another group based their method on the use of low-blank ($^1$H < 0.2%) tetradeuterated 18,18,19,19-$^2$H$_4$ TXB$_2$, which they synthesized as the internal standard. After purification and HPLC, the samples were derivatized to give an open chain derivative of TXB$_2$, a PFB–MO–TMS ether derivative, most suitable for negative ion chemical ionization mass spectrometry. In the selected ion monitoring mode, the detection limits per injection for pure standards and biological samples were 10 and 30 pg, respectively. GC was carried out on a 50 m × 0.25 mm i.d. bonded SE-30 column. An antibody-mediated extraction method

for GC-MS analysis of thromoboxane A$_2$ (TXA$_2$) urinary metabolites has been reported. An antibody (Ab) raised against TXB$_2$ (35% cross-reacting with 2,3-dinor-TXB$_2$) was coupled to CNBr-activated Sepharose 4B (Se) and used as a stationary phase for simultaneous extraction of both compounds from urine. Quantification was performed by GC-negative chemical ionization-MS, monitoring the carboxylate ion. The GC was performed on a 26 m cyanopropyl siloxane (CP-Sil 5CB) column.

The vasoconstrictor and platelet activator TXA$_2$, the predominant product of arachidonic acid in the platelet, is a very reactive substance. It is rapidly converted by nonenzymatic hydrolysis to the stable TXB$_2$. Detailed GC-MS analyses of the PFB–MO–TMS ether derivatives of the thromboxanes and their major metabolites have been described.

GC-MS with selected ion monitoring has been used to demonstrate the global changes in the prostaglandins and thromboxanes in rat circulation after administration of arachidonic acid. **Figure 12** illustrates the measurements of the prostaglandins with multiple deuterium isotope dilution involving separation of pentafluorobenzyl esters, O-methyl oximes and TMS ether derivatives by high resolution GC and specific detection by NICI-MS in the selected ion mode. The selected ion monitoring profiles show the detection of reference prostaglandins ($d_0$ and $d_4$) and their 15 K and 15 KD metabolites: top chromatogram, total ion current; other chromatograms, selected ion mass chromatograms.

### GC-MS-MS analyses

In GC-MS-MS, a specific ion is selected from the initial mass spectrum. The selected parent ion is allowed to enter a field-free reaction region, where it may undergo unimolecular or CID. The resulting fragments are analysed by a second mass spectrometer, which provides the MS-MS spectrum. Although commonly referred to as MS-MS, tandem mass spectrometric analysis is usually MS-CID-MS. The application of GC-MS-MS for the analysis of long chain carboxylic acids and their esters has proved enormously successful.

## Conclusion

The complex nature and large number of molecular species with similar physicochemical properties make GC the preferred method of lipid analysis. GC provides the highest resolution and the shortest analysis time compared to other chromatographic techniques. GC with FID is highly sensitive and readily automated. Its applicability is limited by high temperature

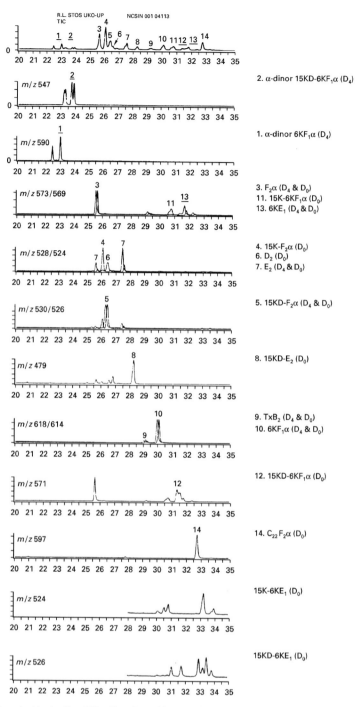

**Figure 12**  GC-negative chemical ionization-MS with selected ion monitoring of reference prostaglandins ($D_0$ and $D_4$) and their 15 K and 15 KD metabolites. Peak identification is given in the figure. GC conditions: column, 60 m × 0.2 i.d., 0.25 µm methyl silicone (DB-1, J & W); temperature, programmed from 100°C (hold for 2 min) to 280°C at 30°C min$^{-1}$; carrier gas, hydrogen. (Reproduced with permission from Pace-Asciak, 1987.)

stability and low volatility of the analytes, which may be overcome by enzymic and chemical modification of the samples. With appropriate strategy, automated GC provides a precise tool for the resolution and quantification of lipids ranging from fatty acids to triacylglycerols. Although appropriate GC strategy also provides identification of unknowns in relation to standards, this must remain tentative unless combined with mass spectrometry. GC-MS of appropriate derivatives of fatty acids provides details of the molecular structure of the fatty acids and their esters. With continued development of liquid phases of increased thermal stability and lipid derivatives of increased volatility and informative mass spectra, GC and GC-MS is likely to remain the method of choice for analyses of molecular species of most lipid classes for the immediate future.

**See Colour Plate 103.**

*See also:* **II/Chromatography: Gas:** Column Technology; Derivatization. **III/Lipids:** Liquid Chromatography; Thin-Layer (Planar) Chromatography.

## Further Reading

Ackman RG (1991) Application of gas-liquid chromatography to lipid separation and analysis: qualitative and quantitative analysis. In: Perkins EG (ed.) *Analyses of Fats, Oils and Lipoproteins*, pp. 270–300. Champaign, IL: American Oil Chemists' Society Press.

Christie WW (1998) Gas chromatography–mass spectrometry methods for structural analysis of fatty acids. Review. *Lipids* 33: 343–353.

Dobson G (1998) Cyclic fatty acids: qualitative and quantitative analysis. In: Hamilton RJ (ed.) *Lipid Analysis in Oils and Fats*, pp. 136–180. London: Blackie Academic & Professional.

Eder K (1995) Review. Gas chromatographic analysis of fatty acid methyl esters. *Journal of Chromatography* B 671: 113–131.

Geeraert E (1987) Polar capillary GLC of intact natural diacyl and triacylglycerols. In: Kuksis A (ed.) *Chromatography of Lipids in Biomedical Research and Clinical Diagnosis*, pp. 48–75. Amsterdam: Elsevier.

Gerst N, Ruan B, Pang J *et al.* (1997) An updated look at the analysis of unsaturated $C_{27}$ sterols by gas chromatogrphy and mass spectrometry. *Journal of Lipid Research* 38: 1685–1701.

Harvey DJ (1992) Mass spectrometry of picolinyl and other nitrogen-containing derivatives of lipids. In: Christie WW (ed.) *Advances in Lipid Methodology*, vol. 1, pp. 19–80. Dundee, Scotland: Oily Press.

Kuksis A (1994) GLC and HPLC of neutral glycerolipids. In: Shibamoto T (ed.) *Lipid Chromatographic Analysis*, pp. 177–222. New York: Marcel Dekker.

Myher JJ and Kuksis A (1989) Relative gas-liquid chromatographic retention factors of trimethylsilyl ethers of diradylglycerols on polar capillary columns. *Journal of Chromatography* 471: 187–204.

Myher JJ and Kuksis A (1995) General strategies in chromatographic analysis of lipids. *Journal of Chromatography* B 671: 3–33.

Pace-Asciak CR (1987) Application of GC-MS techniques to the analysis of prostaglandins and related substances. In: Kuksis A (ed.) *Chromatography of Lipids in Biomedical Research and Clinical Diagnosis*. Journal of Chromatography Library, vol. 37, pp. 107–127. Amsterdam: Elsevier.

Ruiz-Gutierrez V and Barron LJR (1995) Methods for the analysis of triacylglycerols. *Journal of Chromatography* B 671: 133–168.

Spitzer V (1997) Structure analysis of fatty acids by gas chromatography–low resolution electron impact mass spectrometry of their 4,4-dimethyloxazoline derivatives – a review. *Progress in Lipid Research* 35: 387–408.

Thompson RH (1996) Simplifying fatty acid analysis in multicomponent foods with a standard set of isothermal GLC conditions coupled with ECL determinations. *Journal of Chromatographic Science* 34: 495–504.

Tvrzicka E and Mares P (1994) Gas-liquid chromatography of neutral lipids. In: Shibamoto T (ed.) *Lipid Chromatographic Analysis*, pp. 103–176. New York: Marcel Dekker.

# Liquid Chromatography

**A. Kuksis**, University of Toronto, Charles H Best Institute, Toronto, Canada

## Introduction

Natural lipids consist of complex mixtures of molecular species, which are found in association with cell membranes, lipoproteins and other subcellular structures. The composition differs among different cell and tissue types, reflecting the function of lipids in these body structures. Much experimental effort and imagination has been expended in determining the exact composition of lipid species during dietary alterations and physiological activity.

The complex nature of natural lipids and their high solubility in organic solvents makes the separation and isolation of individual lipid classes and molecular species by most physical methods difficult or impossible. Only chromatographic methods have proven suitable for this. The earlier thin-layer and gas chromatographic routines have been complemented in recent years by high performance liquid chromatography (HPLC), which has proven to have nearly universal applicability.

## Principles of Liquid Chromatography

### Normal-phase Columns and Solvents

Normal-phase HPLC provides resolution based on the overall polarity of the lipid molecules. The stationary phase is typically silica gel. The best column and mobile phase for a specific application are selected by trial and error or based on previous experience using thin-layer chromatography (TLC). More refined solvent selection is based on chemometric methods. The solvents selected for the mobile phase in either isocratic or gradient separations should be excellent lipid solubilizers and should not degrade the stationary phase. Therefore, the addition of buffering salts should be minimized and strong acids should be avoided. Other restrictions on the selection of the mobile phase may be imposed by the choice or availability of the detection system, e.g. ultraviolet (UV) and refractive index (RI) detectors.

Silver ion HPLC provides resolutions based on the number, position and configuration of the double bonds present in the lipid molecule. It complements other methods of separation. For this purpose silica gel columns are impregnated with silver nitrate or silver ions are immobilized on columns containing benzenesulfonic acid groups. Chlorinated solvents as the mobile phase, with acetone or acetonitrile as a polar modifier, afford especially good separations.

### Reversed-phase Columns and Solvents

Reversed-phase HPLC is believed to separate lipid molecules based on their partition properties in a biphasic liquid–liquid system, although the exact mechanism is unknown. The most widely used and most important reversed phases for lipid analysis are silicas with relatively long hydrocarbon chains chemically bonded to the surface. Octyl- and octadecyl-bonded chains have been found to provide the best resolution of fatty acids and triacylglycerols.

### Chiral-phase Columns and Solvents

HPLC on columns containing a stationary phase with chiral moieties bonded chemically to a silica matrix has proven well suited for the resolution of chiral glycerolipids. The approach has been to prepare 3,5-dinitrophenyl urethane (DNPU) derivatives of mono- and diacyl-sn-glycerols and related compounds, where the hydrogen atom on nitrogen in the urethane group is available for hydrogen bonding with the stationary phase. The 3,5-dinitrophenyl moieties of the urethanes contribute to charge–transfer interactions with functional groups having $\pi$ electrons on the stationary phase; they are also advantageous for detection by UV absorption. A column with a polymer of (R)-( + )-1-(1-naphthyl)ethylamine moieties chemically bonded to silica gel (YMC-Pack A-KO3™) has been applied to the resolution of diacyl-sn-glycerol

derivatives in a similar manner. The best solvents so far have been the hexane/dichloromethane/ethanol mixtures, which, for the purpose of online mass spectrometry, have been replaced by mixtures of hexane/dichloroethane/acetonitrile or isooctane/tert-butyl methyl ether/acetonitrile/isopropanol.

There are no foolproof recommendations for the selection of solvents. Previous success appears to be the best guide. However, the polarity of the solvents must be chosen so as not to destroy the hydrogen bonding responsible for the differential affinity between the enantiomers and the stationary phase.

### Detectors and Quantification of Solutes

In early work, RI detectors were used in normal-phase HPLC. However, this detector is only applicable to isocratic elution, because RI detection is very sensitive to change in solvent composition. Short wavelength UV (205–210 nm) is also frequently employed for lipid detection but again suffers from sensitivity towards changes in solvent composition. However, UV absorption at longer wavelengths can be effectively used for the specific detection of UV-absorbing solutes. Furthermore, lipids with functional groups can be converted into UV-absorbing or fluorescing derivatives (e.g. free fatty acids, aminophospholipids), which have been widely exploited in reversed-phase HPLC. Many of these derivatives are sufficiently sensitive to permit detection of femtomole levels of the solute (see below).

In HPLC with flame ionization detection (FID), the analyte after solvent removal is burned in a flame and the ions formed are collected by applying an electric field. This nearly universal detector yields a linear relationship between the mass of the solute and peak area over a wide concentration range. More recently, the evaporative light-scattering detector (ELSD) has become the detector of choice in most analytical lipid separations. Like the FID, it is destructive, and must be employed with a stream splitter for peak collection.

It is generally agreed that all detectors require preliminary calibration with reference species to examine the response–structure relationship, and to determine the character of the calibration graph constructed in response–quantity dimensions to be able to work in the linear range. In principle, the quantification is based on measurements of peak area, normalization of the values and calculation of the relative percentage of each component.

Another detector for HPLC of universal application is provided by the mass spectrometer using a variety of ionization techniques to generate total or single ion current response. The use of this detector is discussed along with the use of online mass spectrometry for peak identification (see below).

# Liquid Chromatography–Mass Spectrometry/(LC-MS)

## LC-MS

Online mass spectrometry allows one to obtain direct evidence about the nature of chromatographic peak, e.g. purity, molecular weight and characteristic fragment ions. This information, together with the knowledge of the relative retention time, is usually sufficient for peak identification. More complete identification may be obtained by MS-MS, which is based on the specific mass spectrometric fragmentation of primary ions. Recent reviews of LC-MS applications to lipid analyses are available (see Kuksis and Myher, 1995 and Kuksis, 1997 in Further Reading, below).

The online LC-MS analysis of lipids is accomplished using interfaces which eliminate the HPLC solvent and effect a reliable and efficient transfer of the solute to the ion source. An early method of interfacing HPLC and a mass spectrometer utilized a direct liquid inlet. Several successful applications to lipid analyses with chemical ionization mode were reported. In the positive ion mode, this method produced mass spectra similar to those recorded in electron impact mass spectrometry. Thus, for a triacylglycerol species, a pseudomolecular ion along with characteristic diacylglycerol-like ions were obtained. These ions are frequently sufficient to identify the molecular weight and degree of unsaturation of the component fatty acids. Furthermore, the regiodistribution of the fatty acids in acylglycerols can also be obtained by this ionization method. More recently, the softer ionization techniques, thermospray (TS), electrospray (ES) and atmospheric pressure chemical ionization (APCI) have been utilized for online monitoring of triacylglycerols resolved by HPLC. The latter techniques also allow direct LC-MS of the molecular species of intact glycerophospholipids, which yield largely or exclusively the pseudomolecular ions $[M + 1]^+$ and $[M - 1]^-$ respectively in the positive and negative ionization modes. This information is sufficient for tentative identification of molecular species when combined with knowledge of HPLC retention times and the overall composition of the fatty acids.

## LC-MS-MS

Simple LC-MS is not sufficient to establish the exact composition of all molecular species, which requires the identification of the component fatty acids

in each parent acylglycerol molecule. The fast atom bombardment (FAB) and especially the ES ionization techniques are compatible with LC-MS-MS approaches and have been extensively utilized in polar lipid analysis. In many instances, flow ES-MS-MS has also proven adequate for identification of molecular species.

## Pseudo MS-MS

LC-ES-MS ionization can be used to produce collision-induced dissociation (CID) spectra of singly charged species with greater sensitivity than can be achieved with flow ES-MS-MS systems. The HPLC effluent is carried into the ES source via a stainless-steel or fixed silica needle at flow rates of $1-40 \, \mu L \, min^{-1}$. When analytes are present in the sprayed solution, molecular adduct ions from these analytes, typically protonated ions $[M + H]^+$, are formed. If a low voltage of 50–120 V is applied to the capillary exit, the molecular ion remains intact and the molecular weight of the analyte is obtained. If higher voltages are applied (e.g. 200–300 V) to the capillary exit, extensive and reproducible fragmentation of the molecular adduct ion is realized (pseudo MS-MS).

# Isolation of Natural Lipids

In order to determine the composition of the lipid phase associated with a particular function it is necessary to isolate the appropriate subcellular structure and to determine the component lipid classes and molecular species. For the purpose of discussion the analyses are considered separately as those of neutral lipids, glycerophospholipids and sphingolipids, including glycosphingolipids.

## Preparation of Lipid Extracts

The neutral lipids can be extracted from natural sources by means of benzene, chloroform and other nonpolar solvents. More complete lipid isolation is obtained by extraction with more polar solvents, such as mixtures of chloroform and methanol. However, use of chloroform/methanol results in low recoveries of acidic phospholipids, lysophospholipids and nonesterified fatty acids, while acidified solvents generate lysophospholipid artefacts from tissues containing plasmalogens.

## Purification and Preliminary Separation

Recently, organic solvent extraction has been replaced by solid-phase extraction. It is a simple, rapid technique and can be up to 12 times faster than liquid extraction when executed with commercially

**Table 1** Separation scheme for fractionation of lipid classes from wheat flour using combined silica and aminopropyl solid-phase extraction columns

| Absorbent | Solvents (volume ratio) | Volume (mL) | Lipid class eluted |
|---|---|---|---|
| Silica | Hexane/Et$_2$O (200 : 3) | 15 | SE |
| Silica | Hexane/Et$_2$O (96 : 4) | 20 | TG |
| Silica | Hexane/HOAc (100 : 0.2) | 20 | |
| Silica | Hexane/Et$_2$O/HOAc (100 : 2 : 0.2) | 20 | FFA |
| Silica | Hexane/ethyl acetate (95 : 5) | 15 | |
| Silica | Hexane/ethyl acetate (85 : 15) | 15 | 1,2-DG, 1,2-DG |
| Silica | Et$_2$O/HOAc (100 : 0.2) | 15 | $\alpha$-MG, $\beta$-MG |
| Silica | Et$_2$O/acetone (50 : 50) | 20 | MGDG, MGMG |
| Silica | Acetone | 20 | DGDG, DGDG |
| Silica | THF/ACN/isopropanol (40 : 35 : 25) | 5 | Trace GL |
| Silica | THF/ACN/isopropanol (30 : 35 : 35) | 5 | |
| Silica | THF/ACN/isopropanol (20 : 35 : 45) | 5 | NAPE |
| Silica | THF/ACN/MeOH (15 : 45 : 40) | 5 | NAPE, NAPE |
| Silica | THF/ACN/MeOH (15 : 35 : 50) | 5 | |
| Silica | THF/ACN/MeOH (10 : 35 : 55) | 5 | PC |
| Silica | THF/ACN/MeOH (5 : 35 : 60) | 5 | |
| Silica | CAN/MeOH (35 : 65) | 5 | lyso-PC |
| Aminopropyl | CHCl$_3$/MeOH/ammonium hydroxide (85 : 15 : 0.1) | 25 | |
| Aminopropyl | CHCl$_3$/MeOH/ammonium hydroxide (80 : 20 : 0.1) | 20 | NAPE |
| Aminopropyl | CHCl$_3$/MeOH/ammonium hydroxide (75 : 25 : 0.1) | 20 | |
| Aminopropyl | CHCl$_3$/MeOH/ammonium hydroxide (50 : 50 : 0.1) | 20 | NALPE |
| Aminopropyl | CHCl$_3$/MeOH/ammonium hydroxide (0 : 100 : 0.1) | 20 | |

Et$_2$O, Diethyl ether; HOAc, acetic acid; THF, tetrahydrofuran; ACN, acetonitrile; MeOH, methanol; CHCl$_3$, chloroform. SE, steryl esters; TG, triacylglycerols; FFA, free fatty acids; DG, diacylglycerols; MG, monoacylglycerols; DGDG, diacylglycerol digalactoside; DGDG, diacylglycerol diglucoside; GL, glycolipd; NAPE, *N*-acyl phosphatidylethanolamine; NALPE, *N*-acyllyso phosphatidyl-ethanolamine; lyso-PC, lysophosphatidylcholine. Modified with permission from Prieto JA *et al.*, 1992.

prepared cartridges. These cartridges have proven adequate for rapid removal of nonlipid components from total lipid extracts and for preliminary separation of both neutral and polar lipid classes. For this purpose, the cartridges are eluted by passing through them measured volumes of solvents of appropriate polarity. **Table 1** summarizes a separation scheme successfully applied to the fractionation of lipid classes from wheat flour that is also applicable to the resolution of most lipid classes from animal tissues. A major disadvantage of the solid-phase method is the difficulty of monitoring the separations, so that TLC must be frequently used to assess the results of such isolations.

The currently available techniques concerning extraction and characterization of the different lipids from biological specimens are designed for particular families and do not address consecutive isolation of lipid constituents in their totality. It must be pointed out that conventional TLC remains a convenient, rapid and reliable technique for lipid class isolation which permits efficient extraction of lipid components including gangliosides, without preferential loss of any one group and without the uncertainty of working blindly. The protocol is applicable to biological samples of limited availability.

**Derivatization**

In order to improve resolution, detection and recovery, natural lipids may be subjected to derivatization prior to HPLC. Since neutral lipids are more readily resolved than polar lipids, the polar functional groups of natural lipids may be removed or masked prior to separation. Thus, the phospholipids may be subjected to dephosphorylation by phospholipase C and the resulting diradylglycerols silylated, acetylated or benzoylated before normal-phase or reversed-phase HPLC. This procedure allows improved resolution of the diradylglycerol classes by normal-phase HPLC and of molecular species by reversed-phase HPLC. The preparation of the benzoates also improves the UV detection of the molecular species by reversed-phase HPLC. The preparation of the benzoates also improves the UV detection of the molecular species, while a preparation of the pentafluorobenzoates improves the sensitivity of mass spectrometric detection of the lipid classes and molecular species. The diacylglycerols generated randomly from natural triacylglycerols by Grignard degradation may be converted into the naphthylethyl urethane derivatives by an enantiomeric reagent prior to normal-phase separation of the resulting diastereomers. Using another approach to stereospecific analysis of triacylglycerols, the racemic diacylglycerols resulting from the Grignard degradation are converted into the dinitrophenyl urethanes of the diacylglycerols prior to separation of the enantiomers. In other instances, free fatty acids, free acylglycerols and aminophospholipids may be converted into UV-absorbing or fluorescent derivatives prior to reversed-phase HPLC. The total lipid extracts or any fraction of them may be hydrogenated, reduced with borohydride, peroxidized or ozonized prior to HPLC to provide reference materials or to improve the chromatographic behaviour of the solutes.

## Separation of Neutral Lipids and Free Fatty Acids

Neutral lipids and free fatty acids are made up of monoacylglycerols, diacylglycerols, triacylglycerols, unesterified and esterified sterols, unesterified fatty acids and various other minor components of natural lipids, which migrate with neutral lipids just listed, e.g. tocopherols, alcohols, hydrocarbons, ketones, aldehydes and simple esters. This definition also includes neutral lipids derived from polar lipids by enzymic or chemical transformation (e.g. acylglycerols and ceramides), as well as the peroxidation products of lipids and prostanoids. These lipids possess excellent chromatographic properties.

### Normal-phase Separations

Total lipid profiling on adsorption columns was practised in various forms long before the existence of any other comparable method. Reproducibility and quantification were the major problems which have now been resolved by the combination of HPLC and light scattering or mass spectrometric detection. **Figure 1** shows the separation of selected neutral and polar lipid standards and rat liver lipids by automated normal-phase HPLC using a light-scattering detector. Both neutral and phospholipid classes are well separated and this approach is suitable for use with ES-MS. Normal-phase HPLC is also well suited for the separation of natural and peroxidized fatty acid esters. **Figure 2** shows the separation of a mixture of standard oxosterols along with various other mixtures of oxosterols. The peaks were detected by UV at 205 nm.

Normal-phase HPLC can be employed for separation of the diastereomeric diacylglycerol naph-

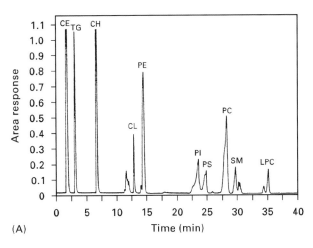

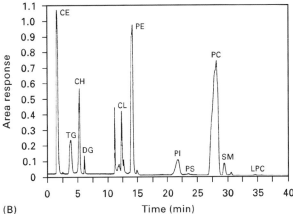

**Figure 1** Separation of (A) standard and (B) rat liver lipids by automated normal-phase HPLC with light-scattering detection. Peak identification: CE, cholesteryl esters; TG, triacylglycerols; CH, unesterified cholesterol; DG, diacylglycerols; CL, cardiolipin; PE, phosphatidylethanolamine; PI, phosphatidylinositol; PS, phosphatidylserine, PC, phosphatidylcholine; SM, sphingomyelin; LPC, lysophosphatidyl choline. HPLC conditions: column, 5 µm Ultrasphere Si (25 cm × 4.5 mm); solvent, a binary gradient of three different solvent mixtures made up of hexane/tetrahydrofuran (99 : 1, v/v), isopropanol/chloroform (4 : 1, v/v) and isopropanol/water (1 : 1, v/v). (Reprinted with permission from Redden and Huang, 1991.)

thylethyl urethanes for the purpose of stereospecific analysis of the positional distribution of the fatty acids in natural triacylglycerols. **Figure 3** shows the separation of the diastereomeric diacylglycerols as the naphthylethyl urethanes. The sn-1,2- and sn-2,3-enantiomers are resolved with the elution order depending on the S- or R-configuration of the reagent. The diastereomeric naphthylethyl urethanes are prepared by reacting the free sn-1,2(2,3)-diacylglycerols with either the S- or R-isocyanate. The partial overlap between molecular species of enantiomeric diacylglycerols from natural sources does not compromise the identification and quantification of the species by ES-LC-MS.

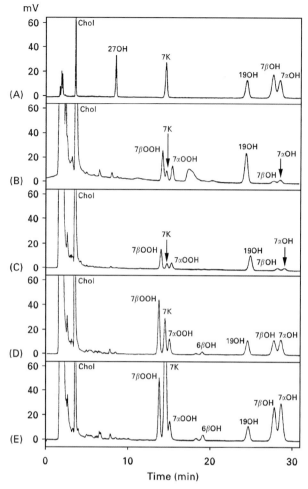

**Figure 2** Normal-phase HPLC of (A) standard oxosterols and (B–E) oxidation products of cholesterol. B, Liposomal cholesterol oxidized with azoamidopropane at 37°C for 20 h; C–E, LDL cholesterol oxidized with $Cu^{2+}$ for 4 h, 8 h and 24 h, respectively. Peak identification: Chol, cholesterol; 27OH, 27-hydroxycholesterol; 7K, 7-ketocholesterol; 19OH, 19-hydroxycholesterol (internal standard); 7βOH, 7β-hydroxycholesterol; 7αOH, 7α-hydroxycholesterol; 7βOOH, 7β-hydroperoxycholesterol; 7αOOH, 7α-hydroperoxycholesterol; 6βOH, 6β-hydroxycholesterol. HPLC conditions: two columns, 3 μm Ultramex (10 × 0.46 cm) in series with a 3 cm guard column; solvent, hexane/isopropanol/acetonitrile (95.8 : 3.90 : 0.30, by vol) at 1.5 mL min$^{-1}$. (Reprinted with permission from Brown AJ, Leong S-L, Dean RT and Jessup W (1997) 7-Hydroperoxycholesterol and its products in oxidized low density lipoprotein and human atherosclerotic plaque. *Journal of Lipid Research* 38: 1730–1745.)

times but can be distinguished from each other by their UV spectra. Normal-phase HPLC also separates the hydroxyeicosatetraenes (HETEs) and oxo-eicosatetraenes (oxo-ETES) with a mobile phase of hexane/isopropanol/acetic acid (90.05 : 0.45 : 0.5 by vol), as shown in the same figure. As expected from its polarity, 15-oxo-ETE had a retention time considerably shorter than that of 15-HETE. The retention time of 12-oxo-ETE is also shorter than that of 12-HETE, but only marginally so.

Normal-phase HPLC on columns containing immobilized silver ions has been used for the fractionation of simple fatty acid esters and triacylglycerols according to the number and configuration of double bonds. **Figure 5** shows an $Ag^+$-HPLC separation of a commercial mixture of conjugated linoleic acids (CLA). Using 0.1% acetonitrile in hexane, 12 peaks were obtained, which emerged into three groups of four peaks each. Evidence for the identity of the individual isomers was obtained by comparison with standards and by complementary chromatographic and MS techniques.

### Reversed-phase Separations

**Figure 6** shows the separation of the molecular species of randomized butterfat by reversed-phase HPLC with light-scattering detection. Butterfat constitutes one of the most difficult mixtures for any chromatographic separation. The analysis of a randomized sample has the advantage that all the

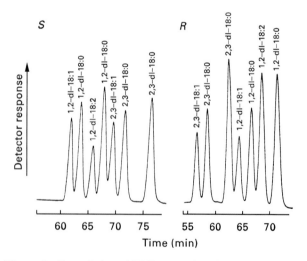

**Figure 3** Normal-phase HPLC separation of 1,2- and 2,3-diacyl-*sn*-glycerols in the form of 1-(1-naphthyl)ethyl urethane derivatives. *S*: (*S*)-( + )-1-(1-naphthyl)ethyl urethanes; *R*: (*R*)-(1)-1-(1-naphthyl)ethyl urethanes. HPLC conditions: two columns, 3 μm Hypersil™ (250 × 4.6 mm i.d.) in series; solvents: hexane/isopropanol (99.5 : 0.5, v/v) at 0.8 mL min$^{-1}$ and UV detection at 280 nm. (Reproduced with permission from Laakso and Christie, 1990.)

Normal-phase HPLC is used for the separation of the leukotriene $B_4$ ($LTB_4$) metabolites, with a mobile phase consisting of hexane/isopropanol/acetic acid (96 : 4 : 0.1 by vol; **Figure 4**). The retention times of these compounds are closely related to their polarities, with $LTB_4$ having the longest and 10,11-dihydro-12-oxo-$LTB_4$ the shortest time. The 12-oxo-$LTB_4$ and 10,11-dihydro-$LTB_4$ have similar retention

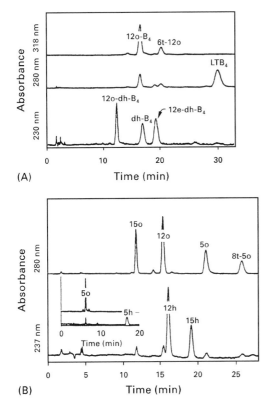

(A)

(B)

**Figure 4** Normal-phase HPLC resolution of (A) LTB$_4$, 12-oxo-LTB$_4$ (12o-B$_4$), 6-*trans*-12-oxo-LTB$_4$ (6t-12o), 10,11-dihydro-LTB$_4$ (dh-B$_4$), 12-oxo-10,11-dihydro-LTB$_4$ (12o-dh-B$_4$) and 12-epi-10,11-dihydro-LTB$_4$ (12e-dh-B$_4$) with hexane/isopropanol/acetic acid (96 : 4 : 0.1, v/v) and (B) HETEs and oxo-ETEs with hexane/isopropanol/acetic acid (99.05/0.45/0.5, v/v) as the mobile phase. The inset shows the normal-phase HPLC separation of 5-oxo-ETE and 5-HETE using a stronger mobile phase (hexane/isopropanol/acetic acid, 97.2 : 2 : 0.1, v/v). (Reproduced with permission from Powell *et al.* (1997) High-pressure liquid chromatography of oxo-eicosanoids derived from arachidonic acid. *Journal of Biochemistry* 247: 17–24.)

molecular species can be calculated on the basis of random distribution and sorted by number of acyl carbons and double bonds. Although there is nearly complete separation of the various triacylglycerol subclasses, numerous molecular species overlap. It is therefore necessary to employ MS to distinguish among the molecular species within each chromatographic peak. There is no resolution of enantiomers or regio-isomers.

Reversed-phase HPLC provides impressive separation of the complex fish oil triacylglycerols using either light-scattering or MS detection. **Figure 7** shows the separation of a triacylglycerol mixture containing 43% docosahexaenoic acid by LC-ESI-MS. Some 33 major components are detected. An examination of the ions generated from each triacylglycerol peak revealed the presence of both

molecular and diacylglycerol-like fragment ions from which the triacylglycerol composition of each peak was determined. Figure 7B and 7C illustrates the identification of peaks 20 and 26 in the chromatographic profile. This extremely powerful method allows identification of species that have the same elution times but different elution times.

The triacylglycerol structures in the HPLC effluent can be determined by atmospheric pressure chemical ionization (APCI) using a corona discharge to ionize vaporized molecules to form both molecular and diacylglycerol-like fragment ions. **Figure 8** compares the elution patterns recorded for a mixture of 35 triacylglycerols by reversed-phase HPLC with FID or APCI detection. Although different gradients of acetonitrile and dichloromethane were used in the two systems, the chromatographic patterns are similar.

**Figure 9** relates the retention times of various synthetic oxotriacylglycerols to their theoretical carbon numbers used to aid identification of peroxidized

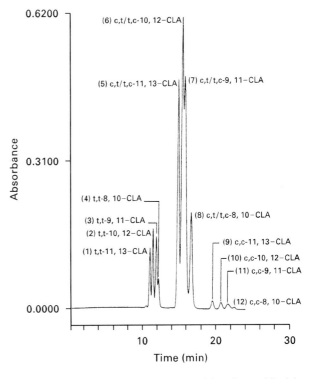

**Figure 5** Silver ion HPLC of a commercial conjugated linoleic acid standard. Peak identification is as given in figure: CLA, conjugated linoleic acid. HPLC: column, 5 μm ChromSper, AgNO$_3$ (250 × 4.6 mm i.d.); solvent, isocratic 0.1% acetonitrile in hexane. (Reproduced with permission from Sehat N, Yurawecz MP, Roach JAG *et al.* (1998) Silver ion HPLC separation and identification of conjugated linoleic acid isomers. *Lipids* 33: 217–221.)

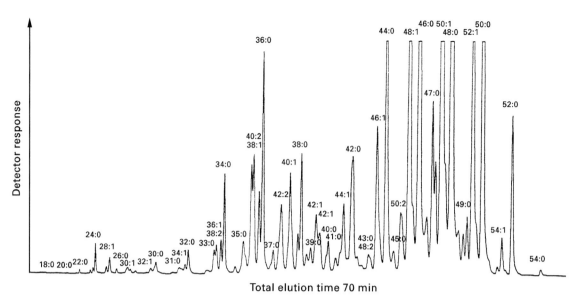

**Figure 6** Reversed-phase HPLC elution profile of randomized butterfat triacylglycerols as monitored by light-scattering detection. Peak identification is given in the figure on the basis of total acyl carbon : double bond number. HPLC conditions : column, 5 μm Supelcosil $C_{18}$ (250 × 4.6 mm i.d.) solvent gradient: 10–90% isopropanol in acetonitrile in 90 min. (Reproduced with permission from Marai *et al.*, 1994.)

natural triacylglycerols by reversed-phase HPLC with ES-MS. The theoretical carbon numbers and correction factors for the oxidized and unsaturated triacylglycerols were calculated using the curve for the saturated triacylglycerols as a reference.

Tocopherols exist in nature as a complex mixture of 2-methyl-6-chromanol homologues and aromatic ring position isomers, each having a three-terpene-unit side chain at the C-2-position. These components of closely related structures can be separated by normal-phase HPLC on silica-based columns. However, the most extensive separations have been obtained by reversed-phase HPLC on a column of octadecyl polyvinyl alcohol sorbent. **Figure 10** shows the separation obtained for the α-, β-, γ-, δ- and $\varepsilon_2$-tocopherols on such a column with acetonitrile/water or methanol/water as the mobile phase.

Cholesterol, sitosterol and their metabolic precursors are also separated by reversed-phase HPLC. Conversion to UV-absorbing derivatives greatly facilitates their detection and quantification. Likewise, reversed-phase HPLC is suitable for the separation of cholesteryl esters. **Figure 11** demonstrates the separation of the cholesteryl esters in a lipid extract from cholesterol-loaded J774 macrophages showing esters ranging from eicosapentaenoate (docosahexaenoate) of cholesterol to stearate in order of their partition number. There are several minor peak overlaps. Normally, the cholesteryl esters in a total lipid extract would overlap with the triacylglycerols

also present in the mixture. This problem can be solved by a mild alkaline hydrolysis, which destroys the triacylglycerols without affecting the cholesteryl esters.

Similarly, reversed-phase HPLC can be employed for the separation of retinyl esters. **Figure 12** illustrates the separation of 15 synthetic retinyl esters with minimal overlap, except for retinyl linolenate, laurate and arachidonate, which are unlikely to occur together in a natural mixture.

Reversed-phase HPLC has been most extensively employed for the separation of the molecular species of diradylglycerols derived from glycerophospholipids by hydrolysis with phospholipase C and from triacylglycerols by hydrolysis with lipase or Grignard degradation. Both UV-absorbing and fluorescent derivatives are prepared to facilitate detection and quantification (**Table 2**). Reversed-phase HPLC is also excellent for the separation of the molecular species of the diacylglycerol DNPU derivatives recovered from chiral HPLC.

Finally, reversed-phase HPLC is suitable for the separation of fatty acids as UV-absorbing or fluorescent derivatives (**Table 3**) and may rival gas chromatography for specific applications. Thus, excellent separation and sensitive detection of the 9-anthrylmethyl esters and the 1-pyrenyldiazomethane derivatives of free fatty acids has been obtained (**Figure 13**). The method has been applied to the determination of endogenous fatty acids released from a cell culture upon stimulation.

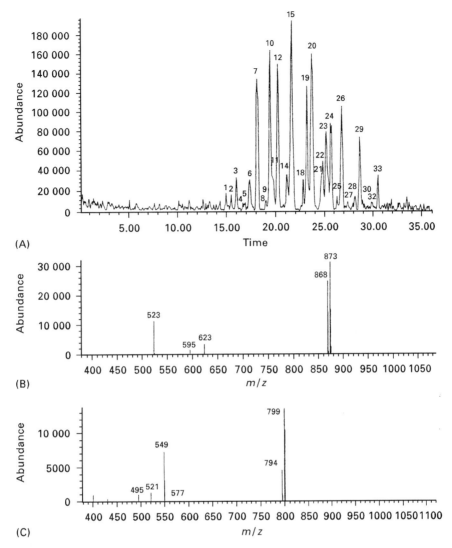

**Figure 7** Reversed-phase HPLC of docosahexaenoic acid-rich oil with online ESI/MS. (A) Total positive ion current profile; (B) ESI-CID-MS of peak 20 (14 : 0/16 : 0/22 : 6): $m/z$ 868, [M + 18]$^+$; $m/z$ 873, [M + 23]$^+$; $m/z$ 523, [M-RCOO]$^+$ (30 : 0 DG); $m/z$ 595, [M-RCOO]$^+$ (36 : 6); $m/z$ 623, [M-RCOO]$^+$ (38 : 6 DG); (C) ESI-CID-MS of peak 26 (14 : 0/14 : 0/18 : 1; 14 : 0/16 : 0/16 : 1): $m/z$ 794, [M + 18]$^+$; $m/z$ 799, [M + 23]$^+$; $m/z$ 495, [M-RCOO]$^+$ (28 : 0 DG); $m/z$ 521, [M-RCOO]$^+$ (30 : 1 DG); $m/z$ 549, [M-RCOO]$^+$ (32 : 1 DG); $m/z$ 577, [M-RCOO]$^+$ (34 : 1 DG). HPLC conditions: column, 5 μm Supelcosil LC-18 (250 × 4.6 mm i.d.); solvent, linear gradient of 20–80% isopropanol in acetonitrile in 30 min; ESI-CID-MS conditions, capillary exit voltage 215 V. (Reproduced with permission from Myher *et al.*, 1997.)

## Chiral-phase HPLC

Chiral HPLC permits the separation of enantiomeric diacylglycerols derived from Grignard degradation or lipase hydrolysis. **Figure 14** shows the separation of the *sn*-1,2(2,3)-diacylglycerols derived from Grignard degradation of a complex triacylglycerol mixture containing 43% docosahexaenoic acid; there is excellent separation of the enantiomers. With this chiral phase, the *sn*-2,3-enantiomers emerge last. There is considerable resolution of molecular species, especially within the longer-retained *sn*-2,3-enantiomers. The *X*-1,3-isomers not removed by borate TLC

emerge just ahead of, or overlap with, the *sn*-1,2-enantiomers. A chiral-phase LC-MS analysis of the 3,5-DNPU derivatives of the *sn*-1,2- and *sn*-2,3-diacylglycerols revealed the presence of a high proportion of species containing two long chain fatty acids per acylglycerol molecule, including 20 : 2–20 : 4 and 22 : 6–22 : 6.

## Separation of Glycerophospholipids and Sphingomyelins

HPLC analysis of glycerophospholipids and sphingomyelins is usually performed with the total phos-

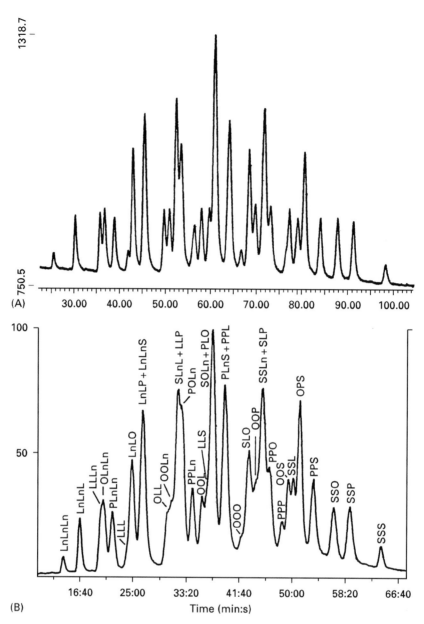

**Figure 8** Reversed-phase HPLC separation of synthetic mixture of 36 triacylglycerols containing five randomly distributed fatty acids. Peaks are identified by component fatty acids: Ln, linolenic; L, linoleic; O, oleic, P, palmitic; S, stearic. HPLC conditions: column, 5 μm Adsorbosphere $C_{18}$ (250 × 4.6 mm, i.d.) in series with 10 μm Adsorbosphere UHS $C_{18}$ (250 × 4.6 mm); solvent, linear gradient of acetonitrile/dichloromethane 70 : 30 to 40 : 60, by vol, over 120 min; detector, FID. APCI conditions: initial acetonitrile/dichloromethane 65 : 35, v/v, followed by a 20–25 min linear gradient acetonitrile/dichloromethane 60 : 40, v/v, and held until 85 min. (Reproduced with permission from Byrdwell WC, Emken EA, Neff WE and Adlof RO (1996) Quantitative analysis of triglycerides using atmospheric pressure chemical ionization-mass-spectrometry. *Lipids* 31: 919–935.)

pholipid fraction recovered from the preliminary isolation of the lipid classes, unless it already involved the separation of the individual phospholipid classes. The purified phospholipid classes can be separated further into molecular species by HPLC using the original molecules or their enzymatic or chemical transformation products.

## Normal-phase HPLC

Normal-phase HPLC is well suited for the separation of the phospholipids. Various silica gel columns yield excellent separations of the major phospholipid classes which, in many instances, also provide a readily discernible separation of the minor components.

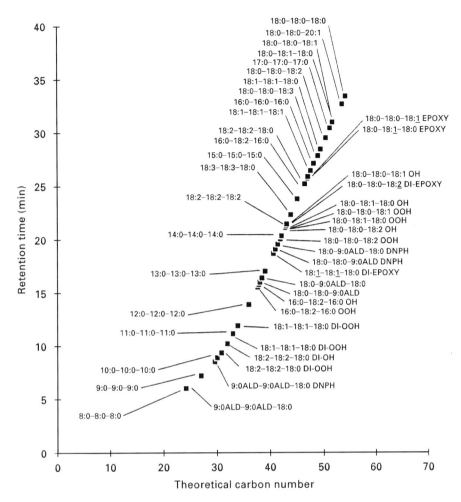

**Figure 9** Plot of theoretical carbon numbers (TCN) versus retention times of reference oxo-triacylglycerols along with a series of saturated monoacid triacylglycerols. TCN and correction factors were calculated for oxotriacylglycerols and unsaturated triacylglycerols using the saturated triacylglycerols as a reference curve. (Reproduced with permission from Sjovall O *et al.*, 1997.)

**Figure 15** shows the separation obtained with a silica gel column for the ethanolamine, choline, inositol and glycerol glycerophospholipids and sphingomyelin isolated from a subfraction of human high density lipoprotein preparation with online MS detection. In the positive ion mode only the choline-containing phospholipids are readily detected, although the ethanolamine glycerophospholipids can also be seen at low intensity. The acidic glycerol, inositol and serine phosphatides, along with any ethanolamine phospholipids, are best detected in the negative ion mode. The negative ion mode also registers the choline phospholipids as the chloride adducts. There is a complete baseline separation for all phospholipids without significant resolution of molecular species, except for SM, which is separated into long chain and short chain species.

Normal-phase HPLC with online MS can be used to assess the molecular species present in the individual phospholipid classes. It is possible to obtain single ion chromatograms retrieved from the total positive ion current spectra for the major molecular species of the choline and ethanolamine phosphatides. In this normal-phase system the newly identified glycated diradylglycerophosphoethanolamine migrates with the front of the phosphatidylcholine peak. The single ion chromatograms retrieved by the computer from the total negative ion current permit accurate quantification of the major molecular species of the acidic glycerophospholipids.

Normal-phase HPLC can be used for the separation of the alkylacyl, alkenylacyl and diacyl subclasses of the ethanolamine glycerophospholipids as the trinitrophenyl derivatives. The diradyl subclasses of the choline glycerophospholipids cannot be separated by chromatography of the intact parent molecules. For this purpose, diradylglycerophosphocholines must be dephosphorylated and the re-

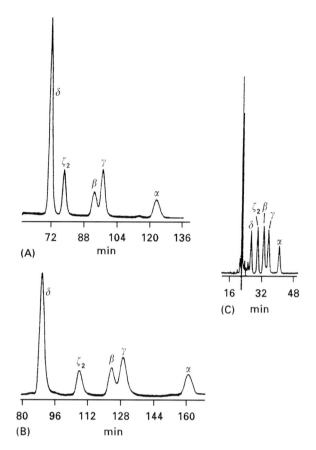

**Figure 10** Reversed-phase HPLC separation of tocopherols. HPLC conditions: columns (A and B), 5 μm Asahipak containing octadecyl polyvinyl alcohol phase; column C, 5 μm Phenomenex Curosil-PFP phase (250 × 4.6 mm i.d.); solvents: (A) acetonitrile/water) (85 : 15, v/v); (B and C) methanol/water (87.5 : 12.5, by vol). (Reproduced with permission from Abidi and Mounts, 1997.)

sulting diradylglycerols converted into UV-absorbing or fluorescent derivatives (Table 2) prior to HPLC separation unless an ELSD system is used. The molecular species separation of the diradylglycerols is carried out by reversed-phase HPLC, as described for the diacylglycerols derived from triacylglycerols by Grignard degradation.

### Reversed-phase HPLC

The total phospholipid mixture can also be separated on a reversed-phase column. Although this leads to extensive separation of the molecular species, there is little overlap among the different phospholipid classes. Both phospholipid classes and molecular species are readily identified and quantitated by online MS with ES ionization. Using 0.5% ammonium hydroxide in a water/methanol/hexane mixture on a $C_{18}$ column, complex mixtures of phospholipid classes and molecular species were identified mainly as protonated or natriated molecules.

The molecular species of the underivatized phospholipids can be separated by reversed-phase HPLC with a mixture of organic solvents and a counterion. The molecular species of intact aminophospholipids have previously been resolved as the UV-absorbing trinitrophenyl derivatives. The reversed-phase systems are also capable of separating the hydroxylated and hydroperoxidized glycerophospholipids from their unoxidized parent species.

### Chiral-phase HPLC

The stereochemical configuration of phosphatidylglycerols has been assessed with chiral phases. Although natural phosphatidylglycerols possess two chiral carbons and are diastereoisomers, they are not readily separable by normal-phase columns. However, the bis-3,5-dinitrophenylurethanes can be separated by chiral-phase HPLC (**Figure 16**). The molecular species of all synthetic phosphatidylglycerol derivatives examined can be separated into diastereomeric peaks in a short time using a mobile phase of hexane/dichloromethane/methanol containing a small amount of trifluorocetic acid.

## Separation of Sphingolipids and Gangliosides

The neutral glycosphingolipids or cerebrosides are ceramide monohexosides, lactosides and higher sugar glycosides. The great complexity and number of new glycosphingolipid components being reported challenge the best contemporary methods of characterization. These lipids have been frequently investigated by FAB, chemical ionization and electron ionization MS with prior chromatographic separation. Likewise, the sulfatides and the sialic acid-containing glycosphingolipids (gangliosides) have been separated by HPLC prior to MS.

### Normal-phase HPLC

A highly sensitive analytical method that allows the separation of ganglioside mixtures and quantification of individual nonderivatized gangliosides has been reported using Spherisorb-$NH_2$. Gangliosides in the 2 pmol to 1 nmol range are separated on a 1 mm i.d. column with a gradient of acetonitrile/phosphate buffer. **Figure 17** shows the resolution of standard gangliosides and gangliosides from the serum of a healthy human female and from human oligodendroglioma. Complete separations are obtained for GM3, GM2, GM1, GD3, GD1a, GD1b, GT1b and GQ1b. The

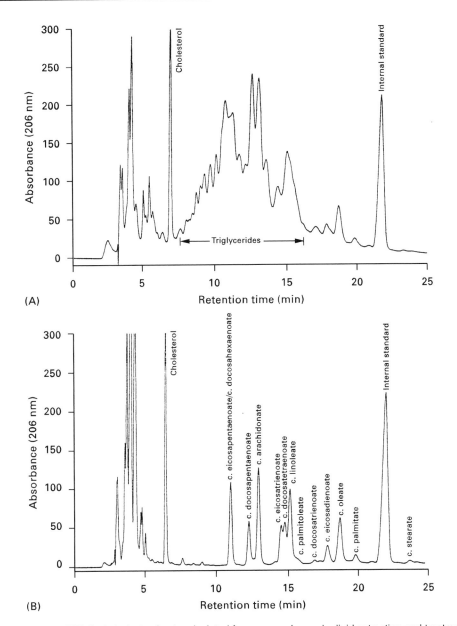

**Figure 11**  Reversed-phase HPLC of cholesteryl esters isolated from macrophages by lipid extraction and treatment of the extract with a dilute solution of ethanolic potassium hydroxide. Peak identification is given in the figure. HPLC conditions: column, 3 μm Spherisorb ODS2 (250 × 4 mm, i.d.); isocratic solvent, isopropanol/heptane/acetonitrile (35/12/52, by volume) and detected by UV absorption at 206 nm. (Reproduced with permission from Cullen P, Fobker M, Teglkamp K *et al.* (1997). An improved method for quantification of cholesterol and cholesteryl esters in human monocyte-derived macrophages by high performance liquid chromatography with identification of unassigned cholesteryl ester species by means of secondary ion mass spectrometry. *Journal of Lipid Research* 38: 401–409.)

new method of separation bypasses the earlier difficulties regarding baseline stability of the 195 nm absorption by using a high purity phosphate buffer.

Other workers have employed both FAB-MS and ES-MS to characterize monosialogangliosides of human myelogenous leukaemia HL60 cells and normal human leukocytes. The gangliosides were extracted and subjected to extensive segregation and examina-

tion of the selectin-binding ability of each fraction. Fractions were resolved on an Iatrobead column pre-equilibrated with isopropyl alcohol/hexane/water (55:40:5) and subjected to a linear gradient of isopropyl alcohol/hexane/water 55:40:5 to 55:25:20 with a flow rate of 1 mL min$^{-1}$. They were also reanalysed on a semipreparative Iatrobead column with a linear gradient of isopropyl alcohol/hexane/water 55:40:5 to 55:25:20 over

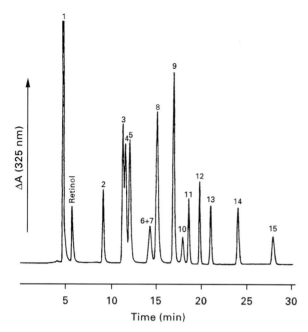

**Figure 12** Reversed-phase HPLC of retinyl esters. Peak identification: 1, acetate; 2, caprate; 3, linolenate; 4, laurate; 5, arachidonate; 6/7, palmitoleate/linoleate; 8, myristate; 9, pentadecanoate; 10, oleate; 11, palmitate; 12, heptadecanoate; 13, stearate; 14, arachidate; 15, behenate. HPLC conditions: column, 5 μm Suplex pKb 100 (250 × 4.6 mm) with a 20 mm guard column; solvent, isocratic elution (14 min) with solvent A (acetonitrile/methanol/dichloromethane/hexane, 88 : 4 : 4 : 4, by volume) at 1 mL min$^{-1}$ followed by a linear gradient of 100% B (acetonitrile/methanol/dicholoromethane/hexane, 70 : 10 : 10 : 10, by volume) over a 2 min period; isocratic elution with the final solvent composition continued for 14 min at 1.5 mL min$^{-1}$. Detection was at 325 nm. (Reproduced with permission from Wingerath T, Kirsch D, Spengler B *et al.* (1997) High performance liquid chromatography and laser desorption/ionization mass spectrometry of retinyl esters. *Analytical Chemistry* 69: 3855–3860.)

200 min with a flow rate of 0.5 mL min$^{-1}$. The final fractions from the normal-phase HPLC were analysed by TLC and the pure components subjected to negative and positive ion FAB-MS.

**Table 2** Selected derivatives of diacyl and monoacylglycerols for UV and fluorescent detection

| UV absorption | Fluorescent detection |
| --- | --- |
| Anthroyl derivatives | |
| Benzoates | |
| Dinitrobenzoates | |
| 3,5-Dinitrophenylurethanes | Phenylurethanes |
| Naphthylethylurethanes | Naphthylurethanes |
| *p*-Nitrobenzoates | Anthroylurethanes |
| Pentafluorobenzoates | |

Modified with permission from Bell, 1997.

**Table 3** Selected ester derivatives for UV and fluorescent detection of fatty acids

| UV detection | Fluorescent detection |
| --- | --- |
| Anthrylmethyl | 9-Anthrylmethyl |
| Benzyl | 9-Aminophenanthrene |
| *p*-Bromophenacyl | 4-Bromomethyl-7-acetoxycoumarin |
| *p*-Chlorophenacyl | 9-Anthryldiazomethane |
| 2-Naphthacyl | Dansyl-ethanolamine |
| *p*-Nitroanilides | 4-Methyl-7-methoxycoumarin |
| *p*-Nitrobenzyl | 4-Methyl-6,7-dimethoxycoumarin |
| *p*-Nitrophenacyl | 4-Methyl-7-acetoxycoumarin |
| Pentafluorobenzyl | 2-Naphthacyl |
| Phenacyl | |
| *p*-Phenphenacyl | |

Modified with permission from Purdon, 1991.

## Reversed-phase HPLC

Of the sphingolipids, the ceramides, cerebrosides and sphingomyelins have been most extensively studied by reversed-phase HPLC. Sphingomyelins obtained from bovine brain, chicken egg yolk and bovine milk fat were separated using a binary solvent system consisting of *n*-butanol/water isopropanol/isooctane on a C$_{18}$ column. The positive ion mass spectra exhibit prominent ions related to the amine base structure and fragments which can be utilized for identification of molecular species.

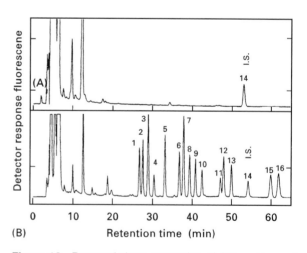

**Figure 13** Reversed-phase separation of fatty acids as the 1-fluorescent pyrenyldiazomethane derivatives. Peak identification; 1, 20 : 5$n$ − 3; 2, 14 : 1$n$ − 9; 3, 18 : 3$n$ − 3/18 : 3$n$ − 6; 4, 22 : 6$n$ − 3; 5, 20 : 4$n$ − 6; 6, 14 : 0/16 : 2$n$ − 9; 7, 18 : 2$n$ − 6; 8, 20 : 3$n$ − 6; 9, 22 : 4$n$ − 6; 10, 24 : 5$n$ − 6; 11, 18 : 1$n$ − 9; 12, 16 : 0; 13, 24 : 4$n$ − 6; 14, internal standard; 15, 20 : 1$n$ − 9; 16, 18 : 0. HPLC conditions: column, 5 μm LC-18 Supelcosil (250 × 4.6 mm i.d.) with a Pelliguard precolumn (4.6 × 20 mm) from Supelco; solvent: a gradient between water (solvent I) and acetonitrile (solvent II) was used as follows: 0–40 min, 90–100% solvent II and 40–70 min, isocratic 100% II at a flow rate of 1 mL min$^{-1}$. (Reproduced with permission from Brekke *et al.*, 1997.)

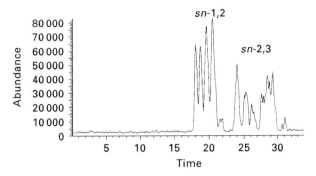

**Figure 14** Chiral-phase HPLC of the dinitrophenyl urethane derivatives of the diacylglycerols from an oil rich in docosahexaenoic acid. *Sn*-1-2-, *sn*-1,2-diacylglycerols; *sn*-2,3-, *sn*-2,3-diacylglycerols. HPLC conditions: chiral column, 25 cm × 4.6 mm i.d. tube containing R-( + )-1-(1-naphthyl)-ethylamine polymeric phase chemically bonded to 30 nm wide-pore spherical silica (YMC-pack A-KO₃); solvent: isocratic hexane/dichloromethane/ethanol 40 : 10 : 1 by volume, at 0.5 mL mim⁻¹; UV detector at 254 nm. (Reproduced with permission from Myher JJ, Kuksis A and Park PW (1996) Stereospecific analysis of docosahexaenoic acid-rich triacylglycerols by chiral-phase HPLC with on-line electrospray mass spectrometry. In: McDonald RE and Mossoba MM (eds) *New Techniques and Applications in Lipid Analysis*, pp. 100–120. Champaign, IL: American Oil Chemists' Society.)

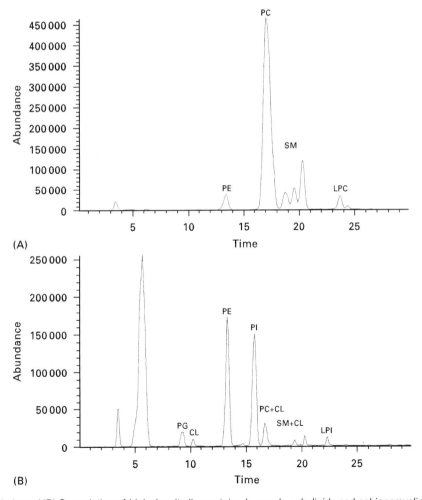

**Figure 15** Normal-phase HPLC resolution of high density lipoprotein glycerophospholipids and sphingomyelins in (A) the positive and (B) negative ion mode as recorded by online electrospray mass spectrometry. Peak identification is as given in Figure 1; PAF, platelet-activating factor. HPLC conditions: column, 5 μm Spherisorb (250 × 4.6 mm i.d.); solvent, a linear gradient of 100% A (chloroform/methanol/30% ammonium hydroxide 80 : 19.5 : 0.5, by volume) to 100% B (chloroform/methanol/water/30% ammonium hydroxide 60 : 34 : 5.5 : 0.5, by volume) in 30 min. (Unpublished results of Kuksis A and Ravandi A, 1997.)

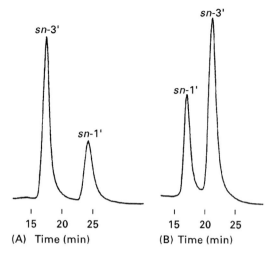

**Figure 16** Chiral-phase HPLC resolution of the bis-3,5-dinitrophenylurethane derivatives of the diastereomeric 1,2-dilinoleoyl-*sn*-3-phospho-1'-*sn*-glycerol (*sn*-1') and 1,2-dilinoleoyl-*sn*-glycero-3-phospho-3'-*sn*-glycerol (*sn*-3) on liquid phases of opposite configuration. (A) (*R*)-( + )-1-(1-naphthyl)-ethylamine column (YMC A-KO3); (B) (*S*)-( − )-1-(1-naphthyl)ethylamine column (YMC A-LO3); solvent, hexane/dichloromethane/methanol/trifluoroacetic acid (60/20/20/0.2, by volume) at 1.0 mL min⁻¹; column temperature 10°C. (Reproduced with permission from Itabashi and Kuksis, 1997.)

*See also:* **II/Chromatography: Liquid:** Derivatization; Detectors: Mass Spectrometry; Detectors: Ultraviolet and Visible Detection; Mechanisms: Chiral; Mechanisms: Normal Phase. **III/Lipids:** Thin-Layer (Planar) Chromatography. **Silver Ion:** Liquid Chromatography; Thin-Layer Planar Chromatography.

## Further Reading

Abidi SL and Mounts TL (1997) Reversed-phase high-performance liquid chromatographic separations of tocopherols. *Journal of Chromatography A* 782: 25–32.

Bell MV (1997) Separations of molecular species of phospholipids by high-performance liquid chromatography. In: Christie WW (ed.) *Advances in Lipid Methodology – Four*, pp. 45–82. Dundee: Oily Press.

Bligh EG and Dyer WJ (1959) A rapid method of total lipid extraction and purification. *Canadian Journal of Biochemistry and Physiology* 37: 911–917.

Brekke O-L, Sagen E and Bjerve KS (1997) Tumor necrosis factor-induced releases of endogenous fatty acids analyzed by a highly sensitive high-performance liquid chromatography method. *Journal of Lipid Research* 38: 1913–1922.

Dreyfus H, Guerold B, Freysz L and Hicks D (1997) Successive isolation and separation of the major lipid fractions including gangliosides from single biological samples. *Analytical Biochemistry* 249: 67–78.

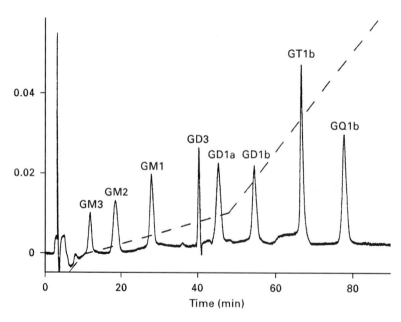

**Figure 17** Normal-phase HPLC of standard gangliosides. Peak identification: GM3, GM2, GM1, GD3, GD1a, GD1b, GT1b and GQ1b denote gangliosides according to Svennerholm. HPLC conditions: column, microbore 3 μm, Spherisorb-NH₂ (250 × 1 mm i.d.). A guard column (1 × 20 mm) was filled with the same material; solvent gradient: as indicated by dashed lines; it starts with 100% A, 0% B, and ends with 0% A, 100% B. Solvent A, acetonitrile/5 mmol L⁻¹ phosphate buffer, pH 5.6 (83 : 17, v/v); solvent B: acetonitrile/20 mmol L⁻¹ phosphate buffer, pH 5.6 (1 : 1, v/v) at 88 μL min⁻¹ at 20°C. (Reproduced with permission from Wagener R, Kobbe B and Stoffel W (1996) Quantification of gangliosides by microbore high performance liquid chromatography. *Journal of Lipid Research* 37: 1823–1829.)

**Plate 85 Herbicides: Gas Chromatography.** Scientist spraying part of an unripe grain field during a herbicide experiment. Residues of herbicides will persist in the plant or in the soil for a variable time, depending on their physiochemical properties and on the environmental conditions. Analysis of herbicide residues in these matrices is important, not only from the point of view of the efficacy of the application, but also to know the distribution and persistence of these compounds in food and the environment. (Reproduced with permission from Rosenfeld Images Ltd/Science Photo Library.)

**Plate 86 (left) Heroin: Liquid Chromatography and Capillary Electrophoresis.** Poppy field in the Far East. The separation and quantitative determination of opiates is required for a wide variety of purposes and applications. (Reproduced with permission from Eye Ubiquitous.)

**Plate 87 (below) Humic Substances: Liquid Chromatography.** Molecular model of the lowest energy conformations of humic acid building blocks linked to form a hexamer. Carbon atoms are green, oxygen atoms are red, nitrogen atoms are purple and hydrogen atoms are not shown. (Reproduced with permission from Davies and Ghabbour, 1999.)

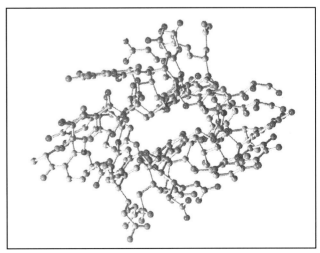

**Plate 88 Inks: Forensic Analysis by Thin-Layer (Planar) Chromatography.** Offset lithography photographed at 27X.

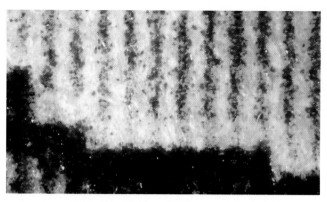

**Plate 89 Inks: Forensic Analysis by Thin-Layer (Planar) Chromatography.** Full colour toner photographed at 27X.

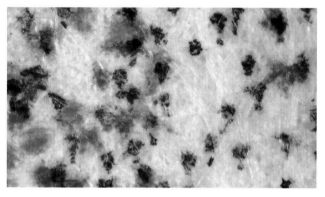

**Plate 90 Inks: Forensic Analysis by Thin-Layer (Planar) Chromatography.** Full colour ink jet photographed at 27X.

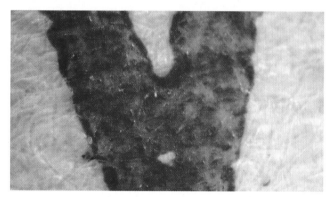

**Plate 91 Inks: Forensic Analysis by Thin-Layer (Planar) Chromatography.** Letterpress ink photographed at 27X.

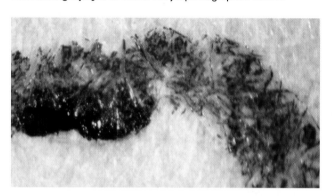

**Plate 92 Inks: Forensic Analysis by Thin-Layer (Planar) Chromatography.** Writing ink photographed at 27X.

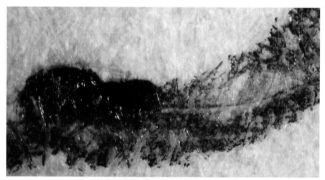

**Plate 93 Inks: Forensic Analysis by Thin-Layer (Planar) Chromatography.** Ballpoint writing ink photographed at 27X.

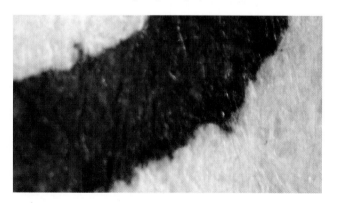

**Plate 94 Inks: Forensic Analysis by Thin-Layer (Planar) Chromatography.** Non-ballpoint writing ink (solvent-based) photographed at 27X.

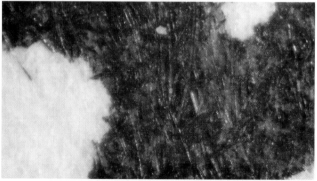

**Plate 95 Inks: Forensic Analysis by Thin-Layer (Planar) Chromatography.** Non-ballpoint writing ink (fountain pen) photographed at 27X.

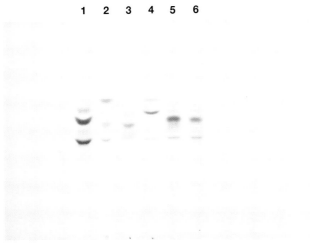

**Plate 96 Inks: Forensic Analysis by Thin-Layer (Planar) Chromatography.** Thin-layer chromatogram of full colour jet inks, solvent system I.

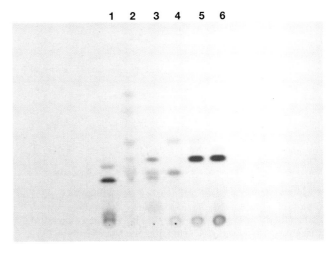

**Plate 97 Inks: Forensic Analysis by Thin-Layer (Planar) Chromatography.** Thin-layer chromatogram of full colour jet inks, solvent system V.

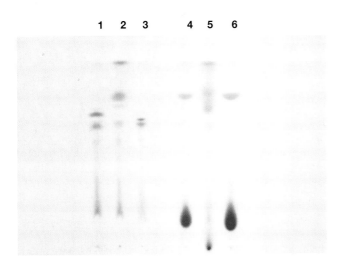

**Plate 98 Inks: Forensic Analysis by Thin-Layer (Planar) Chromatography.** Thin-layer chromatogram of ballpoint inks, solvent system I.

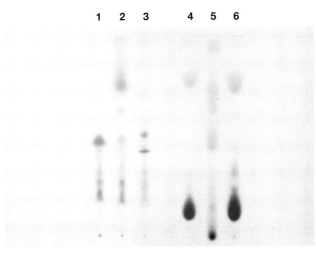

**Plate 99 Inks: Forensic Analysis by Thin-Layer (Planar) Chromatography.** Thin-layer chromatogram of ballpoint inks, solvent system II.

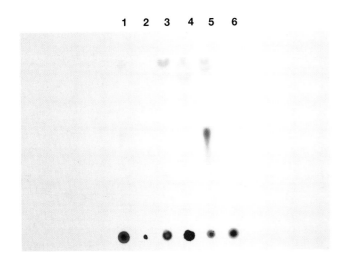

**Plate 100 Inks: Forensic Analysis by Thin-Layer (Planar) Chromatography.** Thin-layer chromatogram of full colour toner, solvent system I.

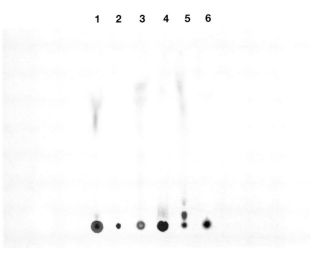

**Plate 101 Inks: Forensic Analysis by Thin-Layer (Planar) Chromatography.** Thin-layer chromatogram of full colour toner, solvent system IV.

**Plate 102 Isotope Separations: Gas Centrifugation.** Nuclear explosion. Aerial view of the nuclear explosion, code-named Seminole, at Enewetak Atoll in the Pacific Ocean on 6 June 1956. This atomic bomb was detonated at ground level and had the same explosive force of 13.7 thousand tonnes (kilotonnes) of TNT. It was detonated as part of Operation Redwing, an American programme, which tested systems for atomic bombs. (Reproduced with permission from Science Photo Library.)

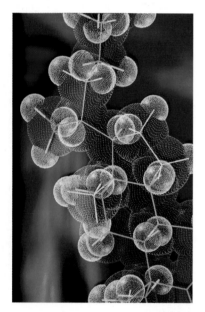

**Plate 103 Lipids.** Computer graphic of part of a cholesterol molecule. Pink spheres represent carbon atoms, yellow spheres represent hydrogen and the lines represent the bonds between them. Cholesterol is an important constituent of cells, playing a crucial role in the synthesis of hormones and bile salts. It also helps to transport fats in the bloodstream to tissues throughout the body. (Reproduced with permission from US Department of Energy/Science Photo Library.)

**Plate 104 Marine Toxins: Chromatography.** Shuttle photograph of a phytoplankton, or algal bloom (a colouring of water arising from a high concentration of plankton) in a north-west Coral Sea off the Queensland Coast, Australia. Plankton drift more or less passively with the water flow, hence orbital photographs of plankton colonies can

provide information on the prevailing currents. Here, the leading edge of a probable concentration of plankton is seen as a light irregular line and sheen in the sea at top right. Reports from ships in this area refer to floating mats of algae up to 2 metres thick. The discolouration of the water in and around the bay is due to the sediment load present in the rivers feeding into it. (Reproduced with permission from NASA/Science Photo Library.)

**Plate 105 Natural Products: Thin-Layer (Planar) Chromatography.** Thin Layer Chromatogram (TLC) of an extract of thylakoid membranes from the leaf of annual meadow grass *Poa annua.* TLC plastic sheets are coated with a 60 F254 silica gel that measures 0.2 millimetres thick. The extract was placed just above the solvent level at the bottom of the lane used to separate the sample. The picture shows the components of the extract separated according to their relative migration distance in the separation system. Six bands are seen; top (orange) is carotene; 2 (green) pheophytin; 3 (green) chlorophyll A; 4 (green) chlorophyll B; 5 (yellow) & 6 (mere trace) are carotenoids. The line across the top of the image is the solvent front. (Reproduced with permission from Science Photo Library.)

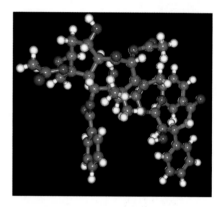

**Plate 106 Natural Products: Thin-Layer (Planar) Chromatography.** Taxol. Computer graphic of a molecule of taxol, an anti-cancer drug. The atoms and their bonds are colour-coded: carbon (C, dark blue); hydrogen (H, white); oxygen (O, red); nitrogen (N, light blue). Taxol has a chemical formula $C_{47}H_{51}NO_{14}$. It is found in the bark of the Pacific Yew tree, *Taxus brevifolia,* and can also be artificially synthesised. Taxol is particularly effective against advanced breast cancer and ovarian cancer. The drug works by inhibiting the protein *tubulin* to form microtubules, important structures involved in cell division (mitosis). The presence of taxol stabilises the microtubules, disrupting mitosis in the rapidly dividing cancer cells. (Reproduced with permission from Science Photo Library.)

**Plate 107 Neurotoxins: Chromatography.** Prairie rattlesnake's mouth. View of the open mouth and fangs of the northern Pacific prairie rattlesnake (*Crotalus viridis oreganus*). This is a subspecies of the prairie rattlesnake that lives in the north-western parts of North America. The fangs are seen enclosed by a membrane at the tip of the snake's upper jaw. These snakes feed mainly on small rodents, which they kill by injecting venom with these fangs. The prairie rattlesnake is widely distributed throughout western North America, with many subspecies existing. Their venom is dangerous and potentially fatal to humans. (Reproduced with permission from Science Photo Library.)

Folch J, Lees M and Sloane-Stanley GH (1957) A simple method for the isolation and purification of total lipids from animal tissues. *Journal of Biological Chemistry* 226: 497–509.

Itabashi Y and Kuksis A (1997) Reassessment of stereochemical configuration of natural phosphatidyl-glycerols by chiral-phase high-performance liquid chromatography and electrospray mass spectrometry. *Analytical Biochemistry* 254: 49–56.

Kaufmann P (1992) Chemometrics in lipid analysis. In: Christee WW (ed.) *Advances in Lipid Methodology – One*, pp. 149–180. Dundee: Oily Press.

Kuksis A (1996) Analysis of positional isomers of glycer-olipids by non-enzymatic methods. In: Christie WW (ed.) *Advances in Lipid Methodology – Three*, pp. 1–36. Dundee: Oily Press.

Kuksis A (1997) Mass spectrometry of complex lipids. In: Hamilton RJ (ed.) *Lipid Analysis of Oils and Fats*, pp. 181–249. London: Chapman & Hall.

Kuksis A and Myher JJ (1995) Application of tandem mass spectrometry for the analysis of long-chain carboxylic acids. *Journal of Chromatography B* 671: 35–70.

Laakso P and Christie WW (1990) Chromatographic res-olution of chiral diacylglycerol derivatives: potential in the stereospecific analysis of triacyl-sn-glycerols. *Lipids* 25: 349–353.

Marai L, Kuksis A and Myher JJ (1994) Reversed-phase liquid chromatography–mass spectrometry of the un-common triacylglycerol structures generated by ran-domization of butteroil. *Journal of Chromatography A* 672: 87–99.

Moreau A (1996) Quantitative analysis of lipids by HPLC with flame-ionization detector or an evaporative light-scattering detector. In: Shibamoto T (ed.) *Lipid Chromatographic Analysis*, pp. 251–272. New York: Marcel Dekker.

Myher JJ and Kuksis A (1995) General strategies in chromatographic analysis of lipids. *Journal of Chromatography B* 671: 3–33.

Myher JJ, Kuksis A and Park PW (1997) Stereo-specific analysis of docosahexaenoic acid-rich triacyl-glycerols by chiral-phase HPLC with on-line electrospray mass sepectrometry. In: McDonald RE and Mossoba MM (eds) *New Techniques and Applications in Lipid Analysis*, pp. 100–120. Champaign, IL: American Oil Chemists' Society Press.

Nikolova-Damyanova B (1997) Reversed-phase high-per-formance liquid chromatography: general principles and application to the analysis of fatty acids and triacyl-glycerols. In: Christie WW (ed.) *Advances in Lipid Methodology – Four*, vol. 8, pp. 193–251. Dundee, Scotland: Oily Press Lipid Library.

Patton GM, Fasulo JM and Robins SJ (1982) Separation of phospholipids and individual molecular species of phos-pholipids by high-performance liquid chromatography. *Journal of Lipid Research* 23: 190–196.

Powell WS, Wang L and Khanapure SP *et al.* (1997) High-pressure liquid chromatography of oxo-eicosanoids derived from arachidonic acid. *Analytical Biochemistry* 247: 17–24.

Prieto JA, Ebri A and Collar C (1992) Optimized separation of nonpolar and polar lipid classes from wheat flour by solid-phase extraction. *Journal of American Oil Chem-ists' Society* 69: 387–391.

Purdon MP (1991) Application of HPLC to lipid separation and analysis: mobile and stationary phase selection. In: Perkins EG (ed.) *Analyses of Fats, Oils and Lipo-proteins*, pp. 166–191. Champaign, IL: American Oil Chemists' Society.

Redden PR and Huang Y-S (1991) Automated separation and quantitation of lipid fractions by high-performance liquid chromatography and mass detection. *Journal of Chromatography Biomedical Applications* 567: 21–27.

Sjovall O, Kuksis A, Marai L and Myher JJ (1997) Elution factors of synthetic oxotriacylglycerols as an aid in iden-tification of peroxidized natural triacylglycerols by re-verse-phase high performance liquid chromatography with electrospray mass spectrometry. *Lipids* 32: 1211–1218.

Takagi T (1991) Chromatographic resolution of chiral lipid derivatives. *Progress in Lipid Research* 29: 277–298.

# Thin-Layer (Planar) Chromatography

**B. Fried**, Lafayette College, Easton, PA, USA

## Introduction

Thin-layer chromatography (TLC) is widely used for the separation and identification of lipid classes with silica gel as the most frequently used stationary phase. Numerous mobile phases (solvent systems) are avail-able for the separation of lipids and there are many nonspecific and specific detection reagents (visualiz-ation reagents) that are useful for detection.

There is no consensus as to the definition of a lipid. Kates considers lipids as compounds generally insol-uble in water but soluble in a variety of organic solvents. He recognized the following classes of lipids: hydrocarbons, alcohols, aldehydes, fatty acids and derivatives such as glycerides, wax esters, phos-pholipids, glycolipids and sulfolipids. Gunstone and

**Table 1** Lipids frequently separated by TLC

| Neutral lipids | Phospholipids | Glycolipids |
|---|---|---|
| Diacylglycerols | Diphosphatidylglycerol | Gangliosides, e.g. monosialoganglosides; disialoganglosides; trisialoganglosides; tetrasialoganglosides |
| Free fatty acids | Lysophosphatidylcholine | |
| Free sterols | Lysophosphatidylethanolamine | Plant and bacterial glycolipids, e.g. mono- and digalactosyldiacylglycerols |
| Monoacylglycerols | Phosphatidic acid | |
| Triacylglycerols | Phosphatidylcholine | Sphingolipids, e.g. ceramides; sphingomyelin; cerebrosides; globosides; sulfatides |
| Wax esters | Phosphatidylethanolamine | |
| | Phosphatidylglycerol | |
| | Phosphatidylinositol | |
| | Phosphatidylserine | |
| | Phosphonolipids | |

Herslöf consider that lipids are compounds based on fatty acids or closely related compounds such as the corresponding alcohols or sphingosine bases. Christie noted that a variety of diverse compounds usually soluble in organic solvents are classified as lipids and set up a convenient system of lipid classification that is followed here. His system considers the simple lipids (compounds that upon hydrolysis yield no more than two types of primary products per mole), also referred to as neutral or apolar lipids. According to Christie, the polar or complex lipids (compounds that upon hydrolysis yield three or more primary products per mole) are the glycerophospholipids (or simply phospholipids) and the glycolipids (also termed glyceroglycolipids or glycosphingolipids), including gangliosides. **Table 1** lists the major neutral lipids, phospholipids and glycolipids of interest in studies on the TLC of lipids.

## Functions

Lipids are involved in many functions of animals, plants and microorganisms and these functions are often studied using TLC. Lipids are important as storage depots for energy reserve. In mammals the storage depot is usually in the form of adipose tissue and TLC analysis shows that the major storage lipids are triacylglycerols, free fatty acids and mixed glycerides. Less information is available on lipid storage in invertebrates but TLC studies have shown that storage sites exist in invertebrates, including chlorogagen tissue in earthworms, the digestive glands in snails and specialized organs called trophosomes in some nematodes. As shown by TLC, triacylglycerols are major storage components in invertebrates. Lipids

(mainly sterols, phosphoglycerides, glycolipids and sphingolipids) are important in the structural integrity of cells and comprise the major components of membranes. Phosphoglycerides in the membranes of nervous tissue are involved in the transmission of electrical signals. Phosphoinosides are involved in cellular communication. Neutral lipids serve as pheromones or carrier of pheromones in both invertebrates and vertebrates. TLC has been used extensively for at least tentative identification of these pheromones.

Christie has documented numerous lipid functions, including their role in abnormal lipid metabolism associated with various disorders; accumulation of lipids associated with coronary blood vessel and cardiac diseases; the importance of lipids in human welfare, including nutrition and disease; the role of lipids as important dietary factors and suppliers of calories for humans and animals; and the importance of lipids to the palatability of foods.

Glycolipids play an important role in cellular metabolism and TLC has helped to elucidate this role. Glycolipids occur at the external surfaces of cell membranes and help regulate cell growth; they also serve as receptors for toxins and hormones and modulate immune responses.

## Sample Preparation

Lipid analysis should be done as soon as possible after samples have been obtained from plants and animals. If this is not possible, samples should be maintained at 4°C overnight or at −20°C for longer periods. Tissues that have been fixed in formalin, alcohol or other preservatives should not be used. Glass vessels

**Table 2** Frequently used methods for sample preparation of lipids

| Lipid extraction technique | Comments |
| --- | --- |
| *Vertebrate and invertebrate organ and tissue samples* <br> Chloroform–methanol (2 : 1); typically 1 part of tissue or fluid to 20 parts of the solvent | Most widely used method of lipid extraction; useful for TLC of neutral and complex lipids |
| Chloroform–methanol–$H_2O$ (1 : 2 : 0.8); following extraction, dilute the sample with 1 vol of chloroform and 1 vol of water to get a biphasic system | Particularly useful for extraction of more polar lipids such as gangliosides |
| Pre-extraction of brain tissue with 0.25% acetic acid followed by chloroform–methanol (2 : 1) | Nonlipid material first removed with the acetic acid; relatively pure lipid fraction then obtained by treatment with chloroform–methanol |
| Chloroform–methanol (1 : 2); typically 1 part of tissue to 3 parts of solvent mixture | Good for large amounts of tissue where complete recovery of lipid is not needed; does not use as much solvent as the previous extraction techniques |
| *Blood and amniotic fluids* <br> Chloroform–isopropanol–water (7 : 11 : 2) | Extracts lipids but not pigments; lipids are not contaminated with blood pigments; neutral and complex lipids are quantitatively extracted |
| Amniotic fluid or blood plasma (about 1 mL) is added directly to a Spice $C_{18}$ solid-phase extraction cartlidge (Analtech, Newark, DE); the analyte is eluted with chloroform–methanol | Good separation of phospholipids achieved, since most extraneous material is removed; technique also useful for separating neutral lipids from steryl esters and phospholipids |
| *Plant tissues* <br> Tissues first treated with isopropanol and then chloroform–isopropanol (1 : 1) prior to usual chloroform–methanol (2 : 1) extraction procedure | Isopropanol inhibits the action of plant lipases |

are recommended for lipid analysis, along with aluminium foil or Teflon-lined lids. Plastic vessels should be avoided because they may dissolve in the organic solvents used during the TLC process. Most samples are extracted in mixtures of chloroform–methanol (**Table 2**) to remove quantitatively the lipids prior to subsequent chromatographic techniques. The first procedure listed in Table 2, usually referred to as the Folch extraction procedure, is the one most frequently used. In brief, this procedure generally uses a 20 : 1 ratio of chloroform–methanol (2 : 1) to sample so that, for example, if 100 mg of tissue is being extracted, 2 mL of chloroform–methanol is suitable for total lipid extraction. The tissue is usually extracted in a glass homogenizer and the extract passed through a glass wool filter; the lipid-containing filtrate is collected and used for TLC following concentration of the sample under nitrogen gas. Many variants of this extraction procedure are available.

## Chromatographic Systems

The chromatographic system consists of the sample mixture (the analyte), the stationary phase (the sorbent) and the mobile phase (the development solvent). Along with the sample mixture, lipid standards (usually obtained from a commercial supplier) are run at the same time. Development of the plate in a suitable mobile phase, from the origin to the solvent front, constitutes the essential part of the chromatographic process. Following development, the plate is allowed to dry and the analytes are detected (see detection, next section) and compared to the standards on the plate.

Numerous stationary phases are available for TLC, but the one of choice for lipid work is silica gel. There are many different types of silica gel plates and sheets and most workers use commercially prepared plates. Silica gel plates can be modified for particular purposes. For example, silver nitrate plates can be prepared and used to separate *cis*-enoic compounds based on unsaturation. Other examples exist that show how commercial and home-made silica gel plates can be altered for specialized lipid applications. The 1990s have seen considerable use of high performance thin-layer chromatography (HPTLC) for lipid analysis. HPTLC plates are made of fine silica particles of narrow size distribution, and have excellent resolving power. The quantity of sample applied to such plates can be reduced markedly from that applied to conventional TLC layers. Many samples can be analysed on the same plate with minimal amounts of mobile phase. HPTLC plates are now being used frequently in densitometric studies on lipids.

**Table 3**  $R_F$ values of common neutral lipids separated in five frequently used solvent systems on silica gel

| Compound | $R_F \times 100$ | | | | |
|---|---|---|---|---|---|
| | $S_1$ | $S_2$ | $S_3$ | $S_4$ | $S_5$ |
| Cholesteryl esters | 97 | 97 | 85 | 94 | 90 |
| Triacylglycerols | 63 | 79 | 70 | 60 | 82 |
| Free fatty acids | 42 | 21 | 62 | 39 | 50 |
| Cholesterol | 28 | 42 | 38 | 19 | 30 |
| 1,3-Diacylglycerols | 24 | 66 | 46 | 21 | 40 |
| 1,2-Diacylglycerols | 21 | 53 | 41 | 15 | 25 |
| Monoacylglycerols | 8 | 11 | 10 | 2 | 5 |

$S_1$, hexane–diethyl ether–formic acid (80 : 20 : 2). $S_2$, toluene–diethyl ether–ethyl acetate–acetic acid (80 : 10 : 10 : 0.2). $S_3$, isopropyl ether–acetic acid (96 : 4) followed by petroleum ether–diethyl ether–acetic acid (90 : 10 : 1) in the same direction. $S_4$, petroleum ether–diethyl ether–acetic acid (80 : 20 : 1). $S_5$, heptane–isopropyl ether–acetic acid (60 : 40 : 4).

Numerous mobile phases are available for lipid TLC and most are used with a single development in the ascending mode. However, some systems have been designed for two or more developments in the same direction. A good example of this is the classical Skipski system (**Table 3**) that uses two developments

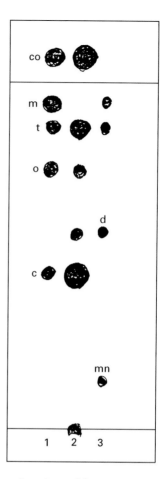

**Figure 2**  Separation of a snail liver extract on a silica gel sheet using the dual solvent system of Skipski (see text). Lane 1 contains a neutral lipid mix as described in Figure 1. Lane 2 shows the separation of snail neutral lipids; note the abundance of free sterols and cholesteryl esters in this tissue. Lane 3 contains a neutral lipid mix with equal parts of monolein (mn), diolein (d), triolein (t), and methyl oleate (mm). (Reproduced with permission from Fried and Sherma (1999).)

in the same direction. Some mobile phases have been designed with two-dimensional TLC in mind, in which the second development is done after the plate is turned through a 90° angle.

Of the many unidimensional solvent systems available to resolve neutral lipids, the Mangold system and its modifications are most frequently used (**Figure 1**). This system consists of different combinations of petroleum ether (or hexanes), diethyl ether and acetic acid with a typical ratio of 80 : 20 : 1 v/v (petroleum ether–diethyl ether–acetic acid; Table 3). Changes in the ratios will affect $R_F$ values of the neutral lipids being separated. Double development in the same direction as in the Skipski system is used to ensure good separation of glycerols from free fatty acids and free sterols (**Figure 2**). Although two-dimensional systems are infrequently used to separate neutral lipids, an example of such use is shown in

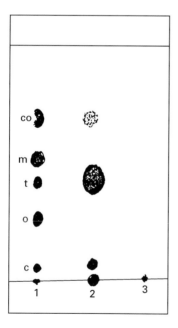

**Figure 1**  Separation of a hen's egg yolk–saline extract on a silica gel sheet. Lipids were developed 10 cm from the origin in petroleum ether–diethyl ether–acetic acid (80 : 20 : 1) and detected by spraying with 5% phosphomolybdic acid in ethanol. Lane 1 contains a neutral lipid mix consisting of equal parts of cholesterol (c), oleic acid (o), triolein (t), methyl oleate (m) and cholesteryl oleate (co). Lane 2 shows the presence of triacylglycerols and free sterols as the predominant neutral lipids in the yolk–saline extract. Lane 3 contains saline alone that is neutral lipid-negative. (Reproduced with permission from Fried and Sherma (1999).).

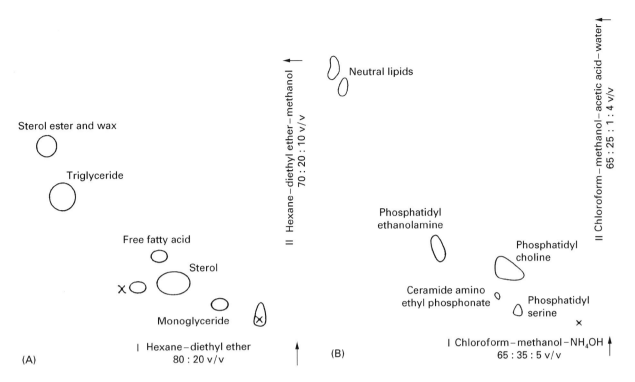

**Figure 3(A)** Chromatogram of the neutral lipids from an extract of the digestive gland–gonad (DGG) complex of the medically important snail *Biomphalaria glabrata*. The silica gel G plate was developed in the first direction in hexane–diethyl ether (80 : 20) and in the second direction in hexane–diethyl ether–methanol (70 : 20 : 10). Lipids were detected by spraying the plate with $H_2SO_4$. (Reproduced with permission from Thompson SN (1987) *Comparative Biochemistry and Physiology* 87B: 357–361.) **(B)** Chromatogram of phospholipids from a similar extract as in Figure 3(A). The silica gel plate was developed in the first direction in chloroform–methanol–$NH_4OH$ (65 : 35 : 5) and in the second direction in chloroform–methanol–water–acetic acid (65 : 25 : 4 : 1). Phospholipids were detected using various specific phospholipid detection reagents. (Reproduced with permission from Thompson SN (1987) *Comparative Biochemistry and Physiology* 87B: 357–361.)

Figure 3A, where a neutral lipid map is obtained. Two-dimensional development is used more frequently to resolve complex lipid mixtures and Figure 3B shows such a chromatogram resolving phospholipids in the medically important snail *Biom-* *phalaria glabrata*. **Table 4** shows the most widely used solvent systems for two-dimensional separations of complex lipids in animal and plant tissues.

Numerous one-dimensional systems for phospholipids are available. The commonly used ones for

**Table 4** Four recommended solvent systems for two-dimensional separations of phospholipids and glycolipids on silica gel

| System | Direction | Solvent composition and ratio | Comments |
|---|---|---|---|
| A | First<br>Second | Chloroform–methanol–water (65 : 25 : 4)<br>*n*-Butanol–acetic acid–water (60 : 20 : 20) | Systems A and B are good for separating polar lipids of animal tissue on silica gel H; 10–15 complex lipids are separated |
| B | First<br>Second | Chloroform–methanol–28% aq. $NH_3$ (65 : 35 : 5)<br>Chloroform–acetone–methanol–acetic acid–water (10 : 4 : 2 : 2 : 1) | |
| C | First<br>Second | Chloroform–methanol–7 mol L$^{-1}$ $NH_4OH$ (65 : 30 : 4)<br>Chloroform–methanol–acetic acid–water (170 : 25 : 25 : 6) | Good for separating bacterial polar lipids on silica gel G; 10–15 polar lipids are separated |
| D | First<br>Second | Chloroform–methanol–0.2% aq. $CaCl_2$ (60 : 35 : 8)<br>*n*-Propanol–water–28% aq. $NH_3$ (75 : 25 : 5) | Good for separating a wide variety of ganglioside species on HPTLC plates |

separating phospholipids on silica gel are shown in **Table 5**. The most widely used system is that of Wagner (see $S_5$ in Table 5), consisting of chloroform–methanol–water (65 : 25 : 4 v/v). In this system, the neutral lipids are moved as one or several bands near the solvent front and the common phospholipids of plant and animal tissues are clearly resolved. **Figure 4** shows a chromatogram of such a separation.

Glycolipid separation by unidimensional chromatography can be achieved in mobile phases consisting of various combinations of chloroform–methanol–water or chloroform–acetone–methanol–acetic acid–water. However, unidimensional separation of glycolipids is best done in the absence of phospholipids. Because glycolipids are difficult to separate completely by one-dimensional TLC, various two-dimensional procedures have been designed. A recommended two-dimensional system for glycolipids is chloroform–methanol–7 mol L$^{-1}$ ammonium hydroxide (65 : 30 : 4 v/v) in the first direction and chloroform–methanol–acetic acid–water (170 : 25 : 25 : 6 v/v) in the second direction.

One-dimensional TLC can be used to separate gangliosides with various combinations of chloroform–methanol–water or $n$-propanol–water as solvent systems. As noted for glycolipids, because of the diversity of oligosaccharides associated with gangliosides, it is difficult to achieve complete separation of these lipids using one-dimensional TLC. Thus, various maps have been prepared in which ganglioside species have been separated in two-dimensional systems. These maps follow the nomenclature of Svennerholm and use esoteric designations

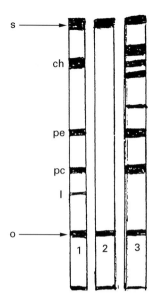

**Figure 4** Separation of phospholipids extracted from snail-conditioned water (SCW) on a silica gel plate in the Wagner solvent system (see text). Lane 1 contains a standard with equal amounts of lysophosphatidylcholine (l), phosphatidylcholine (pc), phosphatidylethanolamine (pe) and cholesterol (ch). Lane 2 is a water blank without snails and is lipid-negative. Lane 3 contains SCW and shows that snails release significant amounts of phosphatidylcholine (pc) and phosphatidylethanolamine (pe), along with an unidentified component that migrates ahead of the pe. The bands near the solvent front are neutral lipids. Note the origin (o) and the solvent front (s). (Reproduced with permission from Fried and Sherma (1999).)

to indicate the nature of the gangliosides being separated. Details of this work are beyond the scope of this article. Experience, patience, and trial and error are needed to effect good separations of glycolipids and gangliosides with one- and two-dimensional TLC.

## Detection

Following development, lipids are usually detected by a wide variety of detection (visualizing) agents either sprayed or dipped on to the plate. Detection reagents may be nondestructive and reversible such as iodine or destructive and nonreversible such as sulfuric acid. The afore-mentioned agents are considered general because they react with numerous different compound types. **Table 6** provides a list of such general detection reagents frequently used to localize lipids on TLC plates. More or less specific detection reagents are used widely in TLC lipid studies and they generally indicate a particular compound or functional group, e.g. ninhydrin is used to localize amine groups associated with phospholipids such as phosphatidylserine or phosphoethanolamine. A list of spe-

**Table 5** $R_F$ values of common phospholipids separated in five frequently used solvent systems on silica gel

| Compound | $R_F \times 100$ | | | | |
|---|---|---|---|---|---|
| | $S_1$ | $S_2$ | $S_3$ | $S_4$ | $S_5$ |
| Diphosphatidyl-glycerol | 91 | 94 | – | – | – |
| Phosphatidylglycerol | – | – | 90 | 78 | – |
| Phosphatidyl-ethanolamine | 65 | 56 | 81 | 59 | 40 |
| Phosphatidylserine | – | 47 | 51 | 47 | 13 |
| Phosphatidylinositol | – | 34 | 34 | 52 | 13 |
| Phosphatidylcholine | 24 | 21 | 90 | 30 | 20 |
| Lysophosphatidyl-choline | 6 | 6 | – | – | 6 |

$S_1$, chloroform–methanol–water (25 : 10 : 1). $S_2$, chloroform–methanol–acetic acid–water (25 : 15 : 4 : 2). $S_3$, chloroform–light petroleum–methanol–acetic acid (50 : 3 : 16 : 1). $S_4$, chloroform–ethanol–triethylamine–water (30 : 34 : 30 : 8). $S_5$, chloroform–methanol–water (65 : 25 : 4).

**Table 6** Five nonspecific reagents useful for the detection of lipids on TLC plates

| Reagents | Procedure | Results |
|---|---|---|
| Iodine | Spray as 1% alcoholic solution or place a few crystals in the bottom of a closed tank | Dark-brown spots on a pale yellow or tan background in a few minutes |
| 2′,7′-Dichlorofluorescein | Spray with a 0.2% solution in 95% ethanol. Observe in UV light | Saturated and unsaturated polar lipids give green spots on purple background |
| Phosphomolybdic acid | Spray with a 5% solution in ethanol; heat at 100°C for 5–10 min | Blue-black spots on yellow background |
| Sulfuric acid | Spray with 50% aq. $H_2SO_4$; heat as above | Black spots on a colourless background |
| Cupric acetate–phosphoric acid | Dissolve 3 g of cupric acetate in 100 ml of an 8% aq. phosphoric acid solution. Heat at 130–180°C for up to 30 min | Black spots on a colourless background |

cific detection reagents frequently used in lipid TLC is shown in **Table 7**.

## Quantification

There are many methods available for quantifying lipids using TLC. Some involve scraping and eluting lipids from the plate followed by spectrophotometric, gravimetric or chromatographic determination. The 1990s saw widespread use of direct *in situ* quantification, usually by densitometric methods, and an extensive literature is now available. Densitometry is performed in the reflection or transmittance mode with a specific brand of commercial densitometer. Any lipid that can be detected by ultraviolet or visible light is subject to densitometric analysis. Suitable standards, usually purchased from a commercial supplier, are needed and should match closely the compounds of interest. Quantification involves bracketing the analyte between two standards, one of slightly lower concentration, and the second of slightly higher concentration. Standards can be used to construct a calibration curve.

A list of recent TLC lipid applications by densitometry for the quantification of lipids is shown in **Table 8**.

## Concluding Remarks and Future Developments

The most extensive use of TLC is for the analysis of pharmaceuticals, followed by all aspects of lipid analysis. Silica gel TLC is an excellent tool for the separation and identification of neutral and complex lipid classes. Densitometry allows for the quantification of these compounds at least at the class level. Numerous specific detection reagents are helpful for identifying lipids. Moreover, techniques in which

**Table 7** Specific chemical detection reagents for various lipids

| Compound class | Reagent | Results |
|---|---|---|
| Cholesterol and cholesteryl esters | Ferric chloride | Cholesterol and cholesteryl esters appear as red-violet spots |
| Free fatty acids | 2′,7′-Dichlorofluorescein–aluminium chloride–ferric chloride | Free fatty acids give a rose colour |
| Lipids containing phosphorus | Molybdic oxide–molybdenum Zinzadze reagent | Phospholipids appear as blue spots on a white background within 10 min of spraying the plate |
| Choline-containing phospholipids (phosphatidylcholine and lysophosphatidyl-choline) | Potassium iodide–bismuth subnitrate Dragendorff reagent | Choline-containing lipids appear in a few minutes as orange-red spots |
| Free amino groups (phosphatidyl-ethanolamine and phosphatidylserine) | Ninhydrin | Lipids with free amino groups show as red-violet spots |
| Glycolipids | α-Naphthol–sulfuric acid | Glycolipids (cerebrosides, sulfatides, gangliosides, and others) appear as yellow spots |
| Gangliosides | Resorcinol | Gangliosides appear as a violet-blue colour; other glycolipids appear as yellow spots |

**Table 8**  Selected applications of densitometric TLC to the quantitative analysis of lipids

| Material | Comments |
|---|---|
| Neutral lipids in egg yolk | HPTLC silica gel; Mangold solvent system of petroleum ether–diethyl ether–acetic acid (80 : 20 : 2) for determination of cholesterol, triacylglycerols, and free fatty acids, and *n*-hexane–petroleum ether–diethyl ether–acetic acid (50 : 20 : 5 : 1) for cholesteryl esters. Lipid detection and quantification as described for neutral lipids in *Biomphalaria glabrata* snails in this table |
| Neutral lipids in *Biomphalaria glabrata* snails | HPTLC silica gel plates; petroleum ether–diethyl ether–acetic acid (80 : 20 : 2) mobile phase; detection by spraying with 5% ethanolic phosphomolybdic acid; lipid zones measured by scanning at 700 nm with a Shimadzu CS 930 TL densitometer operated in the single-beam reflectance mode |
| Phospholipids in *Biomphalaria glabrata* snails | HPTLC silica gel plates; multiple developments in a chloroform–methanol–isopropanol–0.25% and KCl–ethyl acetate (30 : 9 : 25 : 6 : 18) mobile phase; detection by spraying with 10% cupric sulfate–8% phosphoric acid solution; phospholipids measured by reflectance scanning at 400 nm with a Shimadzu CS-930 densitometer in the single-lane/single-beam mode |
| Lecithin and sphingomyelin from amniotic fluid | Silica gel plates developed in chloroform–methanol–water (75 : 25 : 4); sprayed with phosphomolybdic acid; scanned in a densitometer at 450 nm in double-beam transmission mode; detection of each lipid at 0.2 μg level |
| Sphingolipids in the parasitic protozoan, *Blastocystis hominis* | Sphingolipids along with neutral lipids and phospholipids were quantified on HPTLC plates; sphingolipids resolved on plates with chloroform–methanol–water (70 : 22 : 3) and detected with the orcinol reagent; plates scanned with a Shimadzu Flying Spot densitometer operated in the reflectance mode at 580 nm |
| Brain gangliosides | Complex sample preparation; use of HPTLC plates; chloroform–methanol–0.22% CaCl$_2$ (55 : 45 : 10) solvent system; detection by spraying with resorcinol–hydrochloric acid reagent; chromatogram scanned at 580 nm in transmission mode; separation and quantification of 4–8 brain gangliosides |

plates can be impregnated with special agents have allowed for the separation and identification of molecular species within classes, e.g. molecular species of triacylglycerols.

A new area of work has used multiphase TLC, in which components are separated in two directions according to different parameters, e.g. conventional silica gel in one direction and reversed-phase in the other. This technique has proved useful in the analysis of triacylglycerols. The use of HPTLC-densitometry has revolutionized our ability to quantify lipids by relatively simple procedures. There are attempts underway to automate various aspects of the TLC process. Certainly with more automated methodology, TLC will be used more widely in the future by both chemists and biologists interested in lipid separations.

*See also:* **II/Chromatography: Thin-Layer (Planar):** Densitometry and Image Analysis; Layers; Spray Reagents. **III/Lipids:** Gas Chromatography; Liquid Chromatography.

## Further Reading

Christie WW (1982) *Lipid Analysis*, 2nd edn. Oxford: Pergamon.

Christie WW (1987) *High Performance Liquid Chromatography and Lipids*. Oxford: Pergamon.

Fried B and Sherma J (eds) (1996) *Practical Thin-layer Chromatography – A Multidisciplinary Approach*. Boca Raton: CRC.

Fried B and Sherma J (1999) *Thin-layer Chromatography – Techniques and Applications*, 4th edn. New York: Marcel Dekker.

Gunstone FD and Herslöf BG (1992) *A Lipid Glossary*. Ayr: Oily.

Gunstone FD and Padley FB (eds) (1997) *Lipid Technologies and Applications*. New York: Marcel Dekker.

Gurr MI and Harwood JL (1991) *Lipid Biochemistry – An Introduction*, 4th edn. London: Chapman & Hall.

Hammond EW (ed.) (1993) *Chromatography for the Analysis of Lipids*. Boca Raton: CRC.

Kates M (1986) *Techniques of Lipidology, Isolation, Analysis, and Identification of Lipids*, 2nd edn. Amsterdam: Elsevier.

Mukherjee KD and Weber N (eds) (1993) *CRC Handbook of Chromatography – Analysis of Lipids*. Boca Raton: CRC.

Padley FB (ed.) (1996) *Advances in Applied Lipid Research: A Research Annual*, vol. 2. Greenwich: JAI.

Sherma J and Fried B (eds) (1996) *Handbook of Thin Layer Chromatography*, 2nd edn. New York: Marcel Dekker.

Shibamoto T (ed.) (1994) *Lipid Chromatographic Analysis*. Boca Raton: CRC.

# LIQUID CHROMATOGRAPHY-GAS CHROMATOGRAPHY

**L. Mondello, P. Dugo and G. Dugo**,
University of Messina, Messina, Italy
**K. D. Bartle and A. C. Lewis**,
University of Leeds, Leeds, UK

## Introduction

High resolution gas chromatography (HRGC) is the most suitable technique for the analysis of volatile compounds. If the sample is a complex matrix, such as natural products, food products or environmental pollutants, direct gas chromatographic (GC) analysis is not advisable for several reasons, and a sample pretreatment is necessary. In fact, peaks of different classes of compounds may overlap, rendering the qualitative and quantitative analysis of some compounds difficult. Moreover, if the compounds of interest are present only as trace amount, a preconcentration step is necessary before the GC analysis.

If the mixture is subjected to a preliminary separation by liquid chromatography (LC), the fraction so obtained can be analysed by GC. Offline coupling of LC and GC is laborious, involving numerous steps with the risk of contamination and possible greater loss of part of the sample than if online techniques were used. The online coupling of LC and GC offers a number of advantages compared to offline coupling: the amount of sample required is much lower; no sample work-up, evaporation or dilution is necessary and complex automated sample pretreatment is possible.

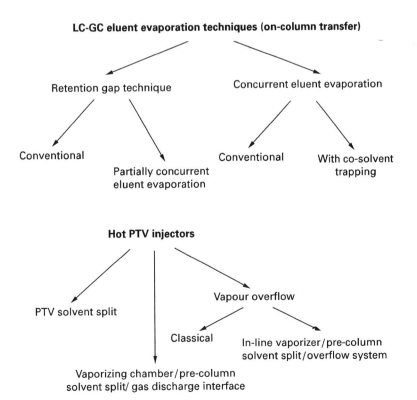

**Figure 1** LC-GC transfer techniques. Possible mechanisms for LC-GC eluent evaporation.

The introduction of large amounts of solvent into a GC column requires the use of special techniques to separate the solvent from the sample selectively. At present, the principal techniques of eluent evaporation that allow transfer of large LC fraction into GC are based on the use of a modified on-column injector or on the use of a programmed temperature vaporizing (PTV) injector. **Figure 1** summarizes the transfer techniques used for online LC-GC coupling.

## Concurrent Eluent Evaporation

Concurrent eluent evaporation is the most used technique, because of its simplicity and the possibility of transferring large amounts of solvent (up to 10 mL!). The technique involves complete evaporation of the eluent during its introduction into the GC system. This technique is suitable for the analysis of solutes with intermediate to high elution temperature, depending on the volatility of the eluent and on the volume of the LC fraction transferred.

When the sample solvent evaporates at the front end of the liquid, volatile compounds co-evaporate with the solvent and immediately start passing into the analytical column. The consequence is that, if the solvent vapours are vented through an early vapour exit, the volatile compounds are lost with the solvent vapour while, if venting is delayed, the most volatile compounds reach the detector even before the end of solvent evaporation. This causes a broadening of the initial band of solutes, which will give peaks that start with the solvent peak and spread out, masking some of the volatile peaks. Only high boiling substances, which migrate slowly during the transfer, will elute after the oven temperature increase, and will give sharp peaks. In practice, the first properly shaped peaks are eluted some 40–120°C above the transfer temperature.

The fraction to be analysed is contained in a loop, connected to a switching valve. The opening of the valve allows the sample in the loop to be driven by the carrier gas into the GC. Usually an early vapour exit is located after a few metres of deactivated precolumn and 3–4 m of retaining column (cut from the analytical column). The retaining pre-column provides a short zone where the solute can be focused in the stationary phase, so that volatile components will be trapped, avoiding their loss through the vapour exit. This is opened during solvent evaporation to reduce the amount of solvent that would reach the detector, and at the same time to increase the solvent evaporation rate.

**Figure 2** shows the scheme of the loop-type interface.

**Figure 3** shows an example of tocopherols, free sterols and esterified sterols determined in a single analytical run after acetylation. These compounds were pre-separated from the triacylglycerols by normal-phase LC.

## Retention Gap

The retention gap method represents the best approach in the case of qualitative and quantitative analysis of samples containing highly volatile compounds. In fact, the retention gap technique allows the analysis of substances eluting immediately after the solvent peak, due to the reconcentration of these components by the so-called solvent effects (primarily solvent trapping). **Figure 4** shows the scheme of the retention gap transfer technique.

The term retention gap means a column inlet of a retention power lower than that of the analytical column. In the retention gap technique, the sample is introduced into the GC at a temperature below the boiling point of the LC eluent (corrected for the current inlet pressure). In this way, the solvent

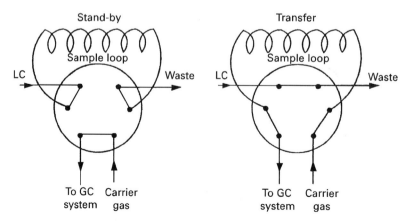

**Figure 2**  Loop interface scheme for concurrent eluent evaporation.

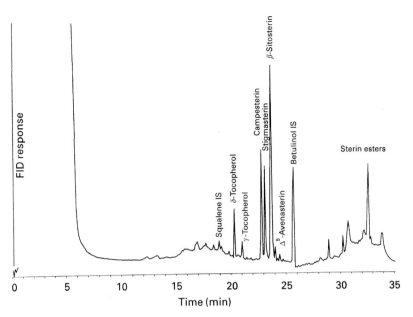

**Figure 3**  GC chromatogram of tocopherols, free sterols and esterified sterols transferred from the LC pre-separation system using the concurrent solvent evaporation technique. FID, Flame ionization detector. (Reproduced with permission from Lechner M and Lorber E (1998) *20th International Symposium on Capillary Chromatography*, Riva del Garda, Italy. Copyright P. Sandra and AJ Rackstraw.)

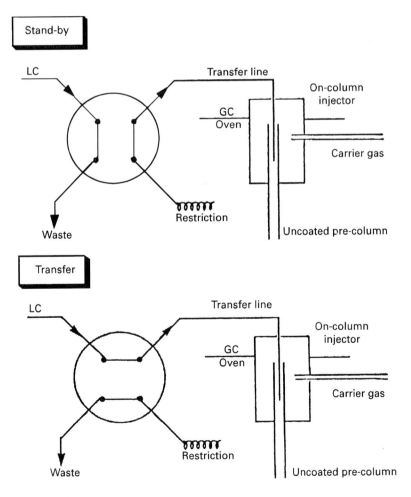

**Figure 4**  Retention gap scheme.

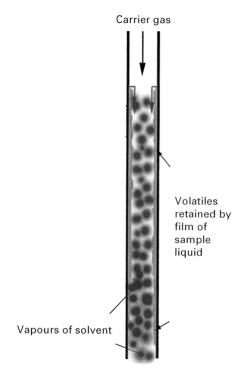

Carrier gas

Volatiles retained by film of sample liquid

Vapours of solvent

**Figure 5** Solvent evaporation in a retention gap. The evaporation occurs from the rear part of the solvent.

vapours replace only part of the carrier gas, which continues to flow through the column. In the carrier stream, solvent evaporates from the rear towards the front of the sample layer (**Figure 5**). Under these conditions, the volatile compounds that are liberated from the solvent envelope start moving, but they are immediately trapped again by the solvent present ahead of the evaporation site (solvent-trapping effect). In this way, the volatile components can start moving in the carrier gas only when the process of solvent evaporation is complete.

A second effect, called phase soaking, occurs in the retention gap technique. This effect is obtained because the carrier gas is saturated by solvent vapours, so it swells the stationary phase with this amount of solvent. The consequence is that the column shows an increased retention power that can be used to trap the volatile compounds.

During the evaporation process, band broadening in space spreads the high boiling compounds. Two retention gap effects can reconcentrate these solute bands. First, compounds migrate more rapidly through the retention gap zone than through the main column; reconcentration depends on the ratio between the retention power in the pre-column and in the main column (phase-ratio focusing). Second, the compounds arrive at the entrance of the main column at a temperature at which they are practically unable to migrate, so they accumulate until the temperature

is increased, and concentrate. The limitation of the method is that, due to the poor capacity of uncoated pre-columns to retain liquid, only modest volumes can be transferred (100–150 μL of eluent) and long uncoated pre-columns are needed.

Two additional techniques have been developed to overcome the drawbacks of the techniques described above, with the aim of obtaining sharp peaks at elution temperatures below 120–150°C, for concurrent eluent evaporation, and to transfer large LC fractions, for the retention gap technique: partially concurrent solvent evaporation and co-solvent trapping.

## Partially Concurrent Solvent Evaporation

Partially concurrent solvent evaporation allows working under conditions that still produce a zone flooded by the eluent (retention gap), providing solvent trapping. This causes a large amount of eluent to evaporate during its introduction (concurrently), so that shorter uncoated pre-columns or larger volumes of transferred fraction can be handled. In practice, an early vapour vent may be placed between the uncoated pre-column and the analytical column, but this makes closure of the vent critical for partial losses of early eluted peaks. A section of the analytical column may be installed after the uncoated pre-column but before the solvent vapour exit.

**Figure 6** shows GC chromatograms obtained after pre-separation by normal-phase LC, applying the partially concurrent solvent evaporation. This application consists of the measurement of the enantiomeric ratio of linalol contained in sweet and bitter orange essential oil. This ratio characterizes each oil, differentiating them, so that the presence of the less valuable sweet orange oil can be detected in the more valuable bitter orange oil.

## Co-solvent Trapping

Co-solvent trapping is used to obtain sharp solute peaks of correct size even for volatile compounds, with concurrent evaporation technique. A small amount of a higher boiling co-solvent is added to the main solvent (for example, n-heptane added to pentane), to prevent co-evaporation of the volatile compounds with the main solvent (**Figure 7**).

It is necessary to adjust the concentration of co-solvent so that some co-solvent is left behind as a liquid, while some is evaporated with the main solvent. In this way, the co-solvent remaining forms a layer of liquid film that evaporates from the rear to the front, as in the retention gap technique. The solutes are trapped by the solvent-trapping effect, and

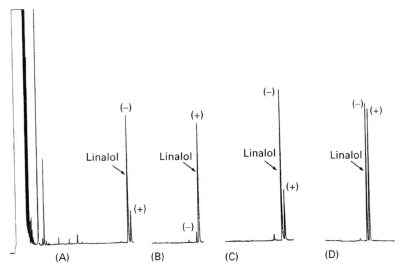

**Figure 6**  GC chiral separation of linalol from sweet orange, bitter orange and mixtures of the two essential oils, showing the different enantiomeric ratios. Linalol fraction was transferred from the LC system using the partially concurrent solvent evaporation technique. (A) Bitter orange; (B) sweet orange; (C) bitter orange 90%, sweet orange 10%; (D) bitter orange 70%, sweet orange 30%. (Reproduced with permission from Dugo G, Venzera A, Cotroneo A, Stagno d'Alcontres W, Mondello L and Bartle KD (1994) *Flavour and Fragrance Journal* 9: 99. Copyright John Wiley & Sons Limited.)

are released after the co-solvent has been evaporated. As can be seen, this technique shows some similarity to partially concurrent solvent evaporation. The main difference is that it is designed for the loop-type interface, while partially concurrent solvent evaporation works with an on-column interface. The optimization of the transfer conditions is not easy, so this technique has never been routinely applied to the transfer of normal-phase eluents, because the partially concurrent solvent evaporation technique is easier to use. It can be the technique of choice for the transfer of water-containing solvents, because in contrast to retention gap techniques, concurrent eluent evaporation does not need wettability, and co-solvent trapping retains the volatile solute. Under these conditions, transfer occurs at 110–120°C, so the first compounds should elute at this temperature. Unfortunately, the problem is the lack of an uncoated pre-column resistant to condensed water. Studies on improving pre-column deactivation have been carried out, but the scant interest in reversed-phase LC coupled to GC prevented re-evaluation of this promising technique with improved methods of deactivation.

## Vaporization with Hot Injectors

Together with the techniques described above, other techniques using hot injectors for the transfer from LC to GC, have been developed. PTV with solvent trapping in packed beds can be used successfully to transfer large volume fractions into the GC instead of into a capillary pre-column (uncoated, deactivated silica tubing). This technique shows some advantages compared to the techniques that use on-column sample introduction:

1. Packed beds retain more liquid per unit internal volume.
2. Wettability is no longer critical.
3. Packing material is more stable than deactivated silica tubing.
4. PTV is more easily heated than capillary column.

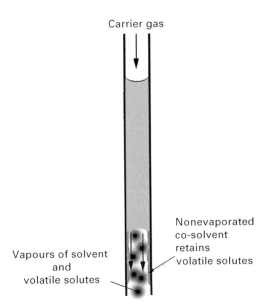

**Figure 7**  Concurrent eluent evaporation with co-solvent trapping effect. The volatile components are retained by the solvent film.

Because of the high retention power, the packed chambers have to be heated above the column temperature to release the solutes. This is a drawback in the analysis of thermally labile compounds, and high boiling compounds. In the PTV solvent split mode, the injection is made into a cool injector, and the solvent is largely removed via the split valve. Volatile compounds can evaporate with the solvent; many studies have been carried out to minimize the loss of volatiles, for example, reducing the temperature of the injector. On the other hand, high boiling compounds are difficult to release, due to the high retention power of the packing material. Introduction of large volumes in the PTV solvent-split mode was introduced in the 1970s and, since its introduction, many different approaches have been described.

The first configuration was a glass tube with a packed bed with a cool injector. The solvent vapours are discharged through the split valve. After solvent evaporation the split valve is closed and the injection chamber is heated to transfer the solutes into the column. Liners are usually packed with Tenax TA and Thermotrap TA. To increase trapping efficiency, liners with sintered porous glass beads were introduced.

Recently a new, fully automated online LC-GC coupling has been introduced. This interface consists of a flow cell, where the fraction is sampled by a large volume autosampler, and automatically injected into a PTV device using the solvent vent mode. This technique was successfully applied to determine pesticide residues in complex natural matrices such as essential oils (**Figure 8**).

## Vapour Overflow

The vapour overflow technique is intended for introducing samples into large volumes of solvent by syringe injection of dilute samples or by coupled LC-GC. The liquid is introduced into a packed (generally with Tenax) vaporizing chamber maintained above the solvent boiling point at a pressure which is near or below ambient. This technique is performed in the absence of carrier gas and vapours are discharged by expansion during evaporation (overflow). The vaporizing chamber is filled with packing material of a GC retention power for the volatile components. The carrier gas supply line is equipped with a switching valve, allowing the gas supply to be stopped during sample introduction. The sample is released by a syringe or a transfer line from the LC near the bottom of the voporizing chamber. Vapours expand and, driven by the expansion, leave the system through the septum purge (**Figure 9**).

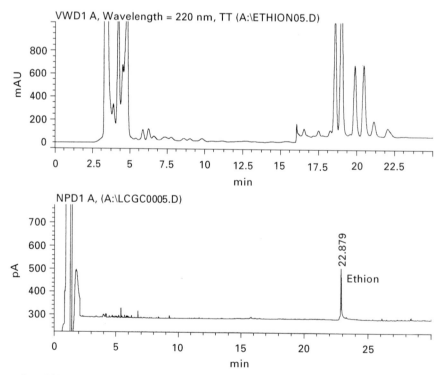

**Figure 8** LC separation of (A) orange essential oil and (B) GC/NPD chromatogram of the fraction containing the pesticide ethion, transferred from the LC system, using the PTV interface type in the solvent vent mode. NPD, nitorgen-phosphorous detector. (Reproduced with permission from David F, Correa RC and Sandra P (1998) *20th International Symposium on Capillary Chromatography*, Riva del Garda, Italy. Copyright P. Sandra and AJ Rackshaw.)

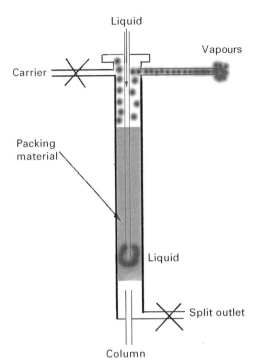

**Figure 9** Vapour overflow interface. The solvent vapours are eliminated through the septum purge of the PTV injector.

After solvent evaporation is completed, the septum purge is closed, the injector is warmed up (PTV) and the trapped analytes are released into the column. More recently, this system has been modified for use with a conventional hot injector.

## In-line Vaporizer/Pre-column Solvent-split/Overflow System

The in-line vaporizer/pre-column solvent-split/overflow system was developed by Grob for large volume liquid injection and for online LC-GC. This system is an overflow-based technique, but with an improvement for the retention of more volatile compounds. As can be seen from **Figure 10**, the vaporizer consisted of a transfer line (from the LC) of fused silica. Inside the vaporizer a 5 cm length of steel wire (untreated or deactivated) or of fused silica is inserted for complete evaporation of the liquid (250–350°C). Moreover the system is equipped with a retaining pre-column and an early vapour exit. The oven in this case is maintained at a low temperature (lower than with a loop interface) to improve retention of sample components by phase soaking. The improvement in the arrangement corresponds to about four extra carbon atoms retained (undecane instead of pentadecane).

## Vaporizing Chamber/Pre-column Solvent-split/Gas Discharge Interface

The sample is injected by an autosampler or by online transfer from a high performance liquid chromatography (HPLC) into a heated device. The vaporizer is packed and maintained at a temperature suitable for solute evaporation. Compounds that are sensitive to high temperatures are injected into a PTV injector in

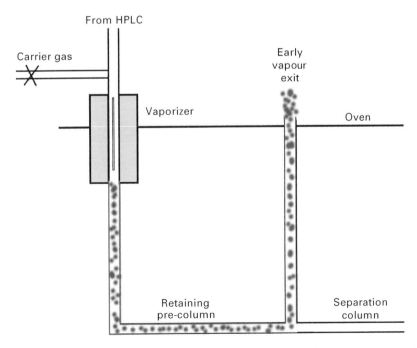

**Figure 10** In line vaporizer/pre-column solvent-split/overflow system. The more volatile components are retained by the retention gap, letting the solvent out from the early vapour exit.

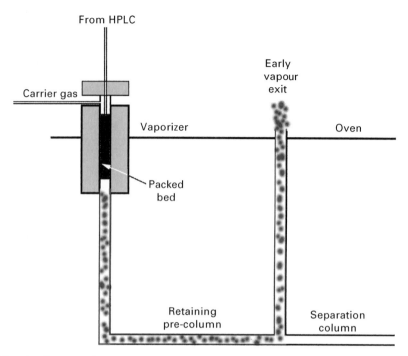

**Figure 11**  Vaporizing chamber/pre-column solvent-split/gas discharge interface. The vaporizer is packed and heated at a suitable temperature for solvent evaporation. The vapour exit can be positioned at the end of the retention gap.

order to evaporate the solvent first and then the solutes. As shown in **Figure 11**, the vapours are discharged through a pre-column and an early vapour exit. If solvent trapping is needed, an uncoated pre-column is installed after the injector and before the solvent vapour exit. Moreover, the use of a retaining pre-column is possible depending on how critical is the solvent vapour exit closure.

## Conclusions

Coupled LC-GC is an excellent online method for sample preparation and clean-up. For the analysis of nonaqueous media the transfer technique of choice is certainly concurrent eluent evaporation using a loop-type interface. The main drawback of this technique is the loss of volatile solutes due to the solute co-evaporation with the solvent. If volatile compounds are present, transfer from LC to GC is best achieved by a retention gap technique. This on-column method yields sharp peaks starting from the programming of the GC oven temperature during eluent transfer. However, due to the limited capacity of uncoated pre-columns for liquid retention, the technique is only suited for the transfer of relatively small fractions. Larger fractions can be transferred by partially concurrent eluent evaporation since this includes the retention gap procedure. Presently, PTV solvent-split injection is considered as the method of choice for

large volume injection of 'dirty' or water-containing samples. Finally, vaporizing chamber/pre-column solvent-split/gas discharge interface and in-line vaporizer/pre-column solvent-split/overflow systems seem to be promising techniques for LC-GC coupling in a wide range of applications.

*See also:* **I/Chromatography. II/Chromatography: Gas Column Technology: Multidimensional Gas Chromatography; Theory of Gas Chromatography. Chromatography: Liquid:** Multidimensional Chromatography; Theory of Liquid Chromatography.

## Further Reading

Grob K (1986) *On-column Injection in Capillary Gas Chromatography*. Heidelberg: Huethig.

Grob K (1991) *On-line Coupled LC-GC*. Heidelberg: Huethig.

Kelly GW and Bartle KD (1994) The use of combined LC-GC for the analysis of fuel products: a review. *Journal of High Resolution Chomatography* 17: 390.

Mondello L, Dugo G and Bartle KD (1996) On-line microbore high performance liquid chromatography-capillary gas chromatography for food and water analyses. A review. *Journal of Microcolumn Separations* 8: 275.

Vreuls JJ, de Jong GJ, Ghijsen RT and Brinkman UATh (1994) LC coupled on-line with GC: state of the art. *Journal of the Association of Official Analytical Chemist International* 77: 306.

# MARINE TOXINS: CHROMATOGRAPHY

**A. Gago-Martínez and J. A. Rodríguez-Vázquez,**
Universidad de Vigo, Vigo, Spain

## Introduction

The marine environment may be seriously affected by contamination due to the massive proliferation of toxic phytoplanktonic species for example toxic algae or 'algal blooms', which appear at certain times of the year under the influence of various environmental factors. These phytoplanktonic microorganisms are essential food for filter-feeding bivalve shellfish as well as other types of marine seafood. This phenomenon is commonly known as a 'red tide'.

These toxic algal blooms cause important socio-economic damage to those regions that depend on aquaculture or fisheries industry, owing to their environmental and human health impacts.

It is believed that the first reference to a harmful algal bloom appears in the Bible:'... all the waters in the river turned to blood, the fish died'(Exodus 7:20–21). One of the first recorded fatal cases of human poisoning happened in 1793 in Poison Cove (British Columbia), caused by a group of alkaloids now called 'paralytic shellfish poisoning' toxins (PSP). Since this toxic event, a number of different toxic episodes have been reported in several places worldwide.

There are different type of shellfish poisoning: the main causative organisms and important toxicological effects are summarized in **Table 1**. In this article we will focus on the paralytic, diarrhoeic and amnesic shellfish poisoning toxing (PSP, DSP and ASP) because they are responsible for most of the toxic events on the European Atlantic Coast and are also of general occurrence in many other places worldwide.

These toxins cause important cause important human contamination and their impact has increased considerably over the last few years. For this reason, strict control of these toxins is necessary to prevent serious damage to health. The conventional mouse bioassay (still considered as official and routine methodology in most countries) is useful for determining most of the implicated toxins, but alternative analytical methods are desired, especially for research

**Table 1**  Main types of harmful algal blooms

| Type of poisoning | Causative organism | Symptoms |
|---|---|---|
| Paralytic shellfish poisoning (PSP) | *Alexandrium catenella* <br> *A. minutam* <br> *A. tamarense* <br> *Gymnodinium catenatum* <br> *Pyrodinium bahamense* | Typical neurological symptoms. The severity of these symptoms is related to the ingested dose. Also symptoms like dizziness, nausea, vomiting, diarrhoea, muscular paralysis, respiratory difficulties, death through respiratory paralysis (in exteme cases). |
| Diarrhoeic shellfish poisoning (DSP) | *Dinophysis acuminata* <br> *D. acuta* <br> *D. fortii* <br> *D. norvegica* <br> *Prorocentrum lima* | Vomiting and diarrhoea typical of gastrointestinal disorders. Chronic exposure may cause tumour promotion, especially in the digestive system (stomach intestine and colon). |
| Amnesic shellfish poisoning (ASP) | *Pseudo-nitzschia multiseries* <br> *Pseudo-nitzschia* <br> *Pseudo-nitzschia austrials* | Typical gastrointestinal disorders (nausea, vomiting, diarrhoea). Neurological symptoms, especially in older people and people with chronic illnesses – confusion, short-term memory loss. |
| Neurotoxic shellfish poisoning (NSP) | *Gymnodinium breve* | Headache, diarrhoea, muscle weakness, nausea and vomiting. Paraesthesia, respiratory difficulties, vision alterations, speech difficulties. |
| Ciguatera poisoning | *Gambierdiscus toxicus* <br> *Prorocentrum lima* | Gastrointestinal disorders (diarrhoea, abdominal pain, nausea, vomiting), difficulty in balance, heart problems (low heart rate and blood pressure), death through respiratory failure (in extreme cases). |

purposes where high sensitivity and specificity are required.

Owing to the complexity of the sample matrix, separation techniques are required to remove interferences and to increase the selectivity in the analytical response. During the 1990s considerable research has been focused on the development of chromatographic approaches for the analysis of PSP, DSP and ASP toxins. Liquid chromatography (LC) has been shown to be one of the most successful alternatives for the sensitive determination of these compounds using different detection modes. Capillary electrophoresis (CE) is another analytical alternative that has been recently developed for the analysis of PSP and ASP toxins and, as we will discuss later, this is a promising technique for the analysis of such compounds.

## Separation Techniques for the Analysis of Marine Biotoxins

### PSP Toxins

Paralytic shellfish poisoning (PSP) is a worldwide problem caused by consumption of shellfish that have accumulated potent neurotoxins produced by toxicogenic dinoflagellates. The PSP toxins include saxitoxin (STX) and several of its derivatives formed by addition of sulfo, hydrosulfate and N-1-hydroxyl groups (**Figure 1**).

The most common chemical method used for the analysis of PSP toxins is the combination of LC with online post-column oxidation and fluorescence detection, using two different isocratic and gradient elution modes. This evolved from earlier work, which showed that STX could be easily oxidized to a fluorescence derivative by hydrogen peroxide under alkaline conditions. Hydrogen peroxide is not able to oxidize the N-1-hydroxylated derivatives, which are better oxidized by using periodate. This reagent is therefore commonly used in post-column oxidation systems in order to detect all PSP toxins. LC combined with post-column oxidation has resulted in a successful approach but unfortunately is not without difficulties, especially concerning the operation of the equipment. This has made it necessary to optimize parameters such as stationary phase, mobile phase, etc. An alternative LC method employing pre-chromatographic oxidation has been reported. This method results in improved separation and quantitation of most PSP analogues. Modification of the periodate oxidation reaction for the N-hydroxy-containing toxins has led to improved sensitivity and stability of the products, enabling overnight analysis.

All the high performance liquid chromatography (HPLC) methods mentioned have resulted in valid approach for the control of these toxic compounds, but although these methods offer good sensitivity and dynamic range, the sensitivity is dependent on parameters such as reagent concentration, reaction times, pH and temperature of the oxidation reaction. In addition to the elaborate procedure required to achieve reliable and reproducible results, the main drawback of the alkaline oxidation reaction is the reliance on fluorescence response factors for the different PSP toxins based on the only commercially available standard. **Figure 2** shows an example of the

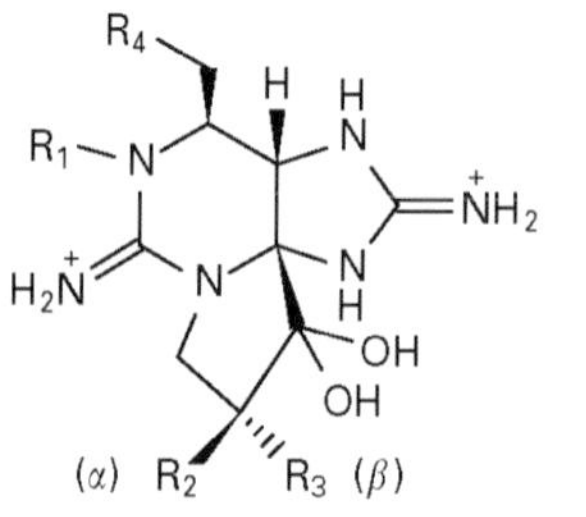

| $R_1$ | $R_2$ | $R_3$ | Carbamoyl toxins $R_4=OCONH_2$ | N-Sulfocarbamoyl toxins $R_4=OCONHSO_3^-$ | Decarbamoyl toxins $R_4=OH$ |
|---|---|---|---|---|---|
| H | H | H | STX | GTX5 | dcSTX |
| H | H | $OSO_3^-$ | GTX2 | C1 | dcGTX2 |
| H | $OSO_3^-$ | H | GTX3 | C2 | dcGTX3 |
| OH | H | H | neoSTX | GTX6 | dcNEO |
| OH | H | $OSO_3^-$ | GTX1 | C3 | dcGTX1 |
| OH | $OSO_3^-$ | H | GTX4 | C4 | dcGTX4 |

**Figure 1** Chemical structure of PSP toxins.

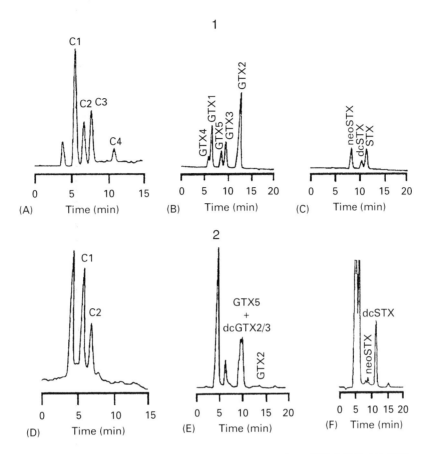

**Figure 2** Chromatogram obtained for the PSP toxins profile by using post-column HPLC-FLD. (A) STX group; (B) GTX group; (C) C toxin group. 1, standard of PSP toxins; 2, mussel samples.

application of the LC technique with fluorescence detection for the analysis of PSP toxins in contaminated extracts of mussels for Galicia (northwest Spain). The conditions used to carry out this isocratic HPLC analysis with post-column oxidation are described in **Table 2**. Under these conditions good resolution was achieved for most of PSP toxins, with the exception of the gonyautoxin (GTX) group, owing to the presence of other GTX components with similar

retention times (this was confirmed using mass spectrometric detection).

The successful separation of these toxins by both ion exchange chromatography and cellulose acetate electrophoresis prompted investigations of the application of CE to their analysis. With the exception of the Ciguatera (C) toxins, which have neutral overall charge, all PSP toxins have positive charge under acidic conditions, and can be separated by elec-

**Table 2** Conditions for the post-column HPLC-FLD analysis of PSP toxins

| | |
|---|---|
| HPLC instrument | Perkin-Elmer series 10-LC |
| | Column: reversed-phase, prodigy 5 µm $C_8$ Phenomenex 4.6 mm × 15 cm |
| Mobile phase | Flow rate, 0.8 mL min$^{-1}$ |
| Mobil phase A (*for C toxin group*) | 2 mmol L$^{-1}$ tetrabutylammonium phosphate, pH 5.8 |
| Mobile phase B (*for GTX toxin group*) | 2 mmol L$^{-1}$ sodium 1-heptanesulfonate in 10 mmol L$^{-1}$ ammonium phosphate, pH 7.3 |
| Mobile phase C (*for STX group*) | 2 mmol L$^{-1}$ sodium 1-heptanesulfonate in 30 mmol L$^{-1}$ ammonium phosphate, pH 7.1, 5% v/v acetonitrile |
| Oxidizing reagent | 7 mmol$^{-1}$ potassium periodate in 50 mmol L$^{-1}$ potassium phosphate buffer, pH 9.0; flow rate: 0.4 mL min$^{-1}$ |
| Reaction system | In 10 m Teflon tubing (0.5 mm i.d.) at 65°C in water bath |
| Acid solution | 0.5 mol L$^{-1}$ acetic acid; flow rate: 0.4 mL min$^{-1}$ |
| Detection | Hitachi F1000 fluorescence detector, double monochromator. Excitation wavelength 330 nm; emission wavelength 390 nm |

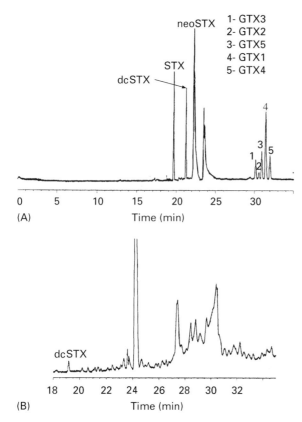

**Figure 3** Electropherogram obtained for the PSP toxins profile by using CE-UV-DAD. (A) standard of PSP toxins; (B) mussel sample.

trophoresis. Capillary electrophoresis should provide an efficient separation of most PSP components; however, some of the inherent difficulties in analysing these compounds are the lack of a chromophore absorbing in the usual UV range as well as the lack of standards to confirm the electrophoretic peak identity.

To overcome situations where the electrophoretic peak identity presents some uncertainties, or standards are not available to correlate with the peaks of interest, it is necessary to use a complementary technique, such as mass spectrometric detection using electrospray ionization. This technique has shown excellent sensitivity for PSP as well as for other marine toxins. Mass spectrometric detection coupled with CE has been used for the rapid and efficient determination of PSP toxins. This technique has also been applied for the analysis of real mussel samples and an efficient separation for most of PSP toxins was achieved.

An example of the application of CE to the analysis of PSP toxins in mussels is shown in **Figure 3**. Under the conditions described in **Table 3**, several contaminated Galician mussel samples were analysed by this technique. Clean-up of the samples was required to remove interferences, but after this clean-up good resolution was obtained for most PSP toxins with the exception of the C toxins, which are not ionized in acidic media. The potential of this technique in terms of sensitivity was clearly increased by using isotacophoresis. Resolution in terms of efficiency, by means of theoretical plates, was clearly higher than that achieved by using HPLC.

### DSP Toxins

Since PSP toxins are potent neurotoxins, the term 'diarrhoeic shellfish poisoning' (DSP) has been associated with a number of different groups of toxic compounds; these include polyether compounds such as okadaic acid (OA), dinophysistoxins (DTX1, DTX2, DTX3), pectenotoxins and the fused polyether yessotoxin. Toxicological studies have shown that okadaic acid and dinophysis compounds are potent phosphatase inhibitors; their common symptomatology is related with the occurrence of diarrhoea. The mechanism of action of the other compounds has not been fully established, but it seems that they do not cause diarrhoea; instead they are described as hepatotoxins. The main reason for including these toxins in the DSP group is probably their polyether structure (**Figure 4**).

DSP toxins are also produced by certain toxic dinoflagellates; the chemical structure of these toxins emerged following the isolation of new polyether toxin named okadaic acid, from the sponge *Hali-*

**Table 3** Conditions for the CE-UV-DAD analysis of PSP toxins

| CE-UVD System | HP $^{3D}$CE (Hewlett-Packard); voltage 20 kV |
|---|---|
| Capillary | Polyvinyl alcohol (PVA) capillary (75 μm i.d. and 104 cm length) |
| Background buffer | 50 mmol L$^{-1}$ morpholine in water adjusted at pH 5 with formic acid |
| Injection | Sample injection into capillary was 20% of capillary volume. Pressure, 50 mbar; time, 120 s |
| ICTP | Voltage, 20 kV |
| | Leading buffer, 10 mmol L$^{-1}$ formic acid |
| | Terminating buffer, 50 mmol L$^{-1}$ morpholine in water adjusted at pH 5.0 with formic acid |
| | Time, 90 s |
| UVD detection | Wavelength 200 nm |

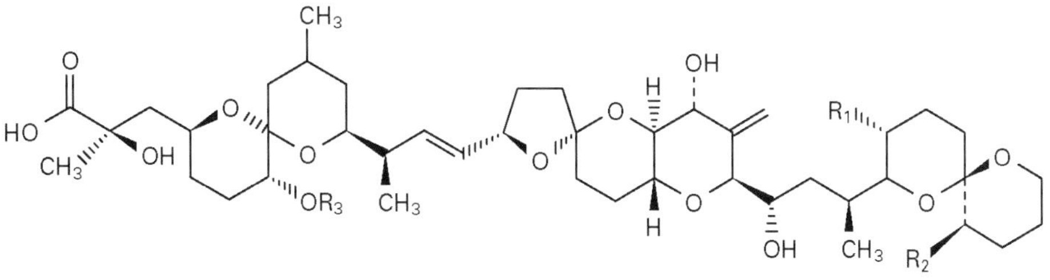

**Figure 4**  Chemical structure of DSP toxins.

| R₁ | R₂ | R₃ | |
|----|----|----|---|
| CH₃ | H | H | Okadaic acid (OA) |
| CH₃ | CH₃ | H | Dinophysistoxin-1 (DTX1) |
| H | CH₃ | H | Dinophysistoxin-2 (DTX2) |
| H / CH₃ | H / CH₃ | Acil | Dinophysistoxin-3 (DTX3) |

*chondria okadai.* The similarities between okadaic acid and DTXs were quickly recognized. Several places worldwide have been affected by such toxic outbreaks since the first toxic event, which took place in the Netherlands in 1960. Although the toxic effects of okadaic acid are related to gastrointestinal disorders, these toxins have been shown to have the potential to bind to protein phosphatases. Consequently the toxicological activity of DSP toxins is also associated with the promotion of tumours, especially in the stomach, intestine and colon.

As for PSP toxins, mouse bioassay is the method commonly used for the analysis of DSP compounds, but the problems associated with this bioassay, such as long assay time, poor reproducibility and false positives, have instigated the search for alternative techniques. Since DSP toxins are lipid-soluble, organic solvents are required for their extraction. However, such lipid-soluble extracts are considerably more complex than aqueous extracts of the same organism and for this reason additional clean-up steps are required before analysis. DSP toxins can be detected by thin-layer chromatography (TLC), although the presence of interferences causes difficulties when using this technique.

DSP toxins may be separated by reversed-phase HPLC using an octadecylsilica (ODS) stationary phase and an acidified aqueous acetonitrile or methanol mobile phase. Detection can be accomplished with UV absorbance at 205–215 nm or with a refractive index detector, to give a detection level of about 10 μg mL$^{-1}$ in solution, but the low selectivity of these detectors requires a high degree of clean-up prior to analysis. DSP toxins have a carboxyl group that is easily converted into a fluorescent ester derivative, which allows HPLC analysis with fluorescence detection. Pioneering work using this fluorescence

detection method uses 9-anthryldiazomethane (ADAM) as the derivatization reagent. Several other reagents have also been used for derivatization to try to overcome the problems of instability of ADAM; however, none has proven as selective and sensitive as ADAM. An HPLC method has been developed with fluorescence detection, using ADAM as derivatization reagent, which was synthesized *in situ* and used immediately. This method offers a reformulation of the previous *in situ* method. The ADAM method is very sensitive for DSP toxins, being able to detect 10 pg of the okadaic acid (OA) derivative injected on-column; the practical quantitation limit is about 10 ng g$^{-1}$ tissue. **Figure 5** shows an example of the application of the *in situ* ADAM-HPLC analysis of standards and a real Galician mussel sample under the chromatographic conditions described in **Table 4**. This ADAM method is not suitable for the analysis of DTX-3 compounds owing to their high molecular weight and lipophilicity. They must first be converted back to OA, DTX-1 or DTX-2 via alkaline hydrolysis; nevertheless, these compounds can be directly analysed by mass spectrometric techniques.

Mass spectrometry is a powerful tool for the analysis of marine toxins. This technique can provide structural information, as well as offering high sensitivity and selectivity; this structural information is useful not only for the confirmation of toxin identity, but also for the identification of new toxins. The combination of HPLC with electrospray mass spectrometry (LC-ESMS) appears to be one of the most sensitive and rapid methods of analysis for DSP toxins. The detection limit found for this technique is about 1 ng g$^{-1}$ in whole edible shellfish tissue. This mass spectrometric detection has been also applied for the analysis of Galician samples, allowing the first confirmation of DTX-2 in these mussels.

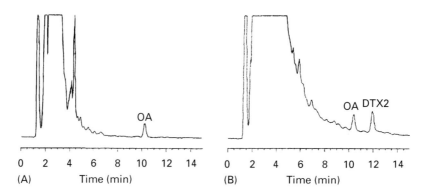

**Figure 5**   Chromatogram obtained for the DSP toxin profile by using ADAM-HPLC-FLD.

## ASP Toxins

This new type of seafood toxicity was first described after a contamination that took place in Prince Edward Island, Canada, in 1987. None of the known shellfish toxins was implicated in this incident and eventually domoic acid was identified as the toxic agent. Amnesic shellfish poisoning was originally isolated from a red macroalga, *Chondria armata*, by Japanese researchers studying insecticidal properties of algal extracts. Most of the people affected by this intoxication experience gastroenteritis but many older people develop neurological symptoms including memory loss. Intraperitoneal injections of acidic aqueous extracts of mussels contaminated with domoic acid into mice cause death with unusual neurotoxic symptoms very different from those of paralytic shellfish poison and other known toxins.

Domoic acid (DA) is a known neurotoxin that is absorbed through the gastrointestinal system, causing damage in the central nervous system. The source of this toxin is a diatom, *Nitzschia pungens multiseries*, which is ingested by shellfish such as mussels during normal filter feeding. The chemical structure of domoic acid is shown in **Figure 6**. This rare naturally occurring amino acid is a member of a group of potent neurotoxic amino acids that act as an agonist to glutamate, a neurotransmitter in the central nervous system. A number of DA isomers that show varying degrees of toxicity have also been identified. Isomerization of DA can occur photochemically or thermally, the latter being significant in cooked seafood.

Like the toxins previously described, reliable methods for the analysis of DA and isomers in seafood products are extremely important for protection of public health. Domoic acid can be analysed semiquantitatively by TLC, but instrumental methods of analysis are most commonly used. HPLC or ion exchange chromatography using ultraviolet absorbance detection are the methods of choice. HPLC has been used since 1987 by Canadian regulatory agencies to prevent other incidents of shellfish poisoning, and is also the official method of analysis for domoic acid in most countries. Domoic acid may be extracted from shellfish tissues by boiling with water or by blending with aqueous methanol. The latter is most commonly used because it is better suited for trace analysis and combines well with a highly selective clean-up based on strong anion exchange. The detection of domoic acid is facilitated by its strong absorbance at 242 nm. The main problem of using this technique is associated with the presence of interferences, which can given false positives with crude extracts. This is the case of tryptophan and some of its derivatives. Since these compounds are often present in shellfish and elute close to domoic acid, SAX-SPE clean-up prior to HPLC-UV analysis avoids the problem caused by these

**Table 4**   Conditions for the ADAM-HPLC-FLD analysis of DSP toxins

| | |
|---|---|
| HPLC system | Liquid chromatograph, HP-1050 |
| Column | Reversed-phase column, HP-Hypersil ODS (4 mm i.d. × 25 cm, 5 μm) |
| Mobile phase | MeCN : H$_2$O (85 : 15) |
| Flow rate | 1.0 mL min$^{-1}$ |
| Detection | Fluorescence detector, HP-1046A: excitation wavelength 254 nm emission wavelength 412 nm |

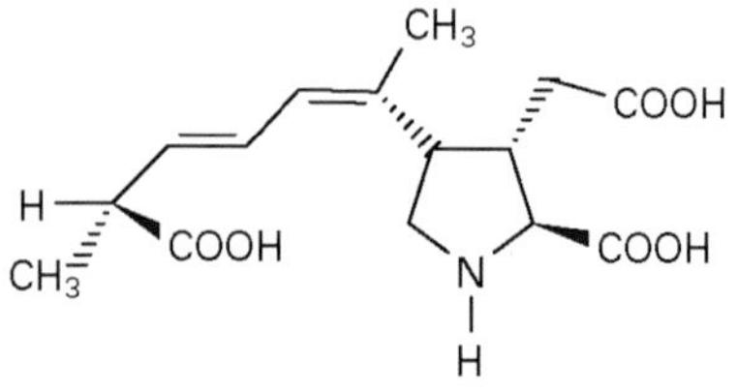

**Figure 6**   Chemical structure of domoic acid, main toxin responsible for ASP toxicity.

**Table 5** Conditions for the HPLC-UV analysis of ASP toxins

| HPLC system | Liquid chromatograph Jasco PU-980 pump |
|---|---|
| Column | Prodigy-ODS 5 µm column (Phenomenex) 4.6 mm × 25 cm |
| Mobile phase | 10% v/v aqueous acetonitrile with 0.2 mol L$^{-1}$ formic acid; flow rate, 1 mL min$^{-1}$ |
| UV detector | Perkin-Elmer LC-95 UV/Vis; wavelength 242 nm |

interferences. An example of the application of the HPLC-UV technique for the analysis of domoic acid in standards, mussel tissue reference material and real contaminated bivalves under the chromatograhic conditions described in **Table 5** in shown in **Figure 7**.

A very sensitive alternative HPLC method using fluorescent detection has been proposed. This method uses 9-fluoronylmethyl chloroformate (FMOC) as derivatization reagent and the FMOC derivative is analysed with fluorescence detection.

Capillary electrophoresis offers the potential of fast high resolution separation of DA and its isomers and the possibility of trace analysis with very small amounts of sample. This method is also attractive in terms of being inexpensive and complementary to HPLC.

Extraction and clean-up procedures are priority steps in order to achieve the best chromatographic and electrophoretic resolution; when combined with a selective extraction and clean-up procedure, CE with UV detection is an excellent method for the separation and quantitative analysis of domoic acid and its isomers in shellfish tissues. **Figure 8** shows the

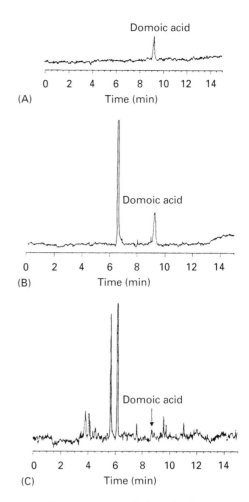

**Figure 8** Electropherogram obtained for domoic acid by CE-UV-DAD. (A) Standard of domoic acid; (B) mussel tissue reference material (MUS-1); and (C) Galician mussel sample. (Reproduced with permission from James KJ et al. Journal of Chromatography 798: 147.)

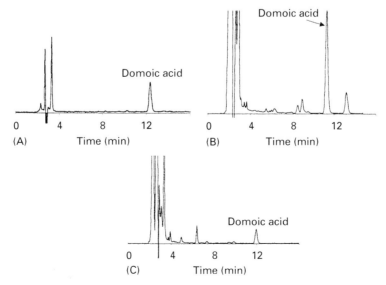

**Figure 7** Chromatogram obtained for domoic acid by HPLC-UV. (A) Standard of domoic acid; (B) mussel tissue reference material (MUS-1); and (C) Galician mussel sample.

**Table 6**  Conditions for the CE-UV-DAD analysis of domoic acid

| | |
|---|---|
| CE-DAD system | HP $^{3D}$CE (Hewlett-Packard); voltage, 30 kV |
| Capillary | Bare fused silica capillary |
| | (50 µm i.d. × 66 cm length) |
| Background buffer | 25 mmol L$^{-1}$ borate at pH 9.2 |
| DAD detector | Wavelength 242 nm |
| Injection | Pressure 50 mbar; time, 12 s |

application of CE to the analysis of standard solution of domoic acid, mussel tissue reference material (MUS-1) and Galician mussel samples. The conditions for this CE analysis are summarized in **Table 6**.

The use of cyclodextrins allows an increase on the separation efficiency when compared with LC-UVD. The better mass detection limit provided by this CE technique could be very useful for situations where limited sample size is available.

## Future Trends

The techniques of HPLC and CE for the analysis of marine toxins provide the separation efficiency required for complicated matrices such as marine samples as well as the high resolution required to determine the toxins present in contaminated samples with very low detection limits. Extraction and clean-up steps prior to the chromatographic analysis are essential in order to obtain accurate quantitative results and also to prevent interferences that can cause false positives. New, fast and on-line automated procedure could give shorter and more accurate analyses. The possibility to combine this automation with simple methods is also desirable, especially for routine control purposes.

Taking into account the lack of standards for all toxins and also the appearance of new, unknown toxic compounds, the development of confirmatory techniques for their detection is an important research field. The development and optimization of coupling techniques, such as LC-MS or CE-MS, and the development of adequate interfaces as well as efficient ionization modes is a high priority.

**See Colour Plate 104.**

*See also:* **II/Chromatography: Liquid:** Detectors: Mass Spectrometry; Instrumentation; Mechanisms: Reversed Phases. **Electrophoresis:** Capillary Electrophoresis; Capillary Electrophoresis-Mass Spectrometry.

## Further Reading

Bates HA and Rapoport H (1975) Chemical assay for saxitoxin, the paralytic shellfish poison. *Journal of Agriculture and Food Chemistry* 23: 237.

Bialojan C and Takai A (1988) Inhibitory effect of a marine-sponge toxin, okadaic acid on protein phosphatases specificity and kinetics. *Biochemical Journal* 256: 283–290.

Fujiki H, Suganuma M, Suguri H, Yoshizawa S, Takagi K, Uda N, Wakamatsu K, Yamada K and Murata M (1988) Diarrhetic shellfish toxin dinophysistoxin-1 is a potent tumour promoter on mouse skin. *Japanese Journal of Cancer Research* 79: 1089–1093.

Gago-Martínez A, Rodríguez-Vázquez JA, Quilliam MA and Thibault P (1996) Simultaneous occurrence of diarrhetic and paralytic shellfish poisoning toxins in Spanish mussels in 1993. *Natural Toxins* 4: 72–79.

Hallegraeffe GM, Anderson DM and Cambella AD (eds) (1995) *Manual on Harmful Marine Microalgae.* IOC Manuals and Guides No. 33, UNESCO.

Lawrence JF, Charbonneau CF, Menard C, Quilliam MA and Sim PG (1989) Liquid chromatographic determination of domoic acid in shellfish products using the paralytic shellfish poison extraction procedure of the Association of Official Analytical Chemists (AOAC). *Journal of Chromatography A* 462: 349–356.

Lawrence JF, Menard C and Cleroux Ch (1995) Evaluation of prechromatographic oxidation for liquid chromatographic determination of paralytic shellfish poisons in shellfish. *Journal of AOAC International* 78(2): 514–520.

Lee JS, Yanagi T, Kenma R and Yasumoto T (1987) Fluorimetric determination of diarrhetic shellfish toxins by high performance liquid chromatography. *Agricultural and Biological Chemistry* 51: 877–881.

Locke SJ and Thibault P (1994) Improvement in detection limits for the determination of paralytic shellfish poisoning toxins in shellfish tissues using capillary electrophoresis/electrospray mass spectrometry and discontinuous buffer systems. *Analytical Chemistry* 20: 3436–3446.

Oshima Y, Machida M, Sasaki K, Tamaoki Y and Yasumoto R (1984) Liquid chromatographic-fluorometric analysis of paralytic shellfish toxins. *Agricultural and Biological Chemistry* 48: 1707–1711.

Pocklington R, Milley JE, Bates SS, Bird CJ, De Freitas ASW, Quilliam MA (1990) Trace determination of domoic acid in seawater and plankton by high-performance liquid chromatography of the fluorenylmethoxycarbonyl (FMOC) derivative. *International Journal of Environmental and Analytical Chemistry* 38: 351–368.

Quilliam MA, Gago-Martínez A, Rodríguez-Vázquez JA (1998) Improved method for preparation and use of 9-anthryldiazomethane for derivatization of hydroxycarboxylic acids. Application to diarrhetic shellfish poisoning toxins. *Journal of Chromatography A* 807: 229–239.

Quilliam MA, Sim PG, McCulloch AW and McInnes AG (1989) High-performance liquid chromatography of domoic acid, a marine neurotoxin, with application to shellfish and plankton. *International Journal of Environmental and Analytical Chemistry* 36: 139–154.

Sullivan JJ (1990) High-performance liquid chromatographic method applied to paralytic shellfish poisoning research. In Hall S and Strichartz G (eds) *Marine Toxins,* pp. 66–77. ACS Symposium Series 418.

Sullivan JJ and Wekell MM (1987) In: Kramer DE and Liston J (eds) *Seafood Quality Determination*, p. 357. New York: Elsevier, North Holland.

Thibault P, Pleasance S and Laycock MV (1991) Analysis of paralytic shellfish poisons by capillary electrophoresis. *Journal of Chromatography A* 542: 483–501.

Wright JLC, Boyd RK, De Freitas ASW *et al.* (1989) Identification of domoic acid, a neuroexcitatory amino acid, in toxic mussels from eastern Prince Edward Island. *Canadian Journal of Chemistry* 67: 481–490.

Zhao JY, Thibault P and Quilliam MA (1997) Analysis of domoic acid and isomers in seafood by capillary electrophoresis. *Electrophoresis* 18: 268–276.

# MECHANICAL TECHNIQUES: PARTICLE SIZE SEPARATION

**A. I. A. Salama,** Natural Resources Canada, Devon, Alberta, Canada

## Introduction

Particles of many kinds and various sizes have played an important role in man's interaction with his physical environment. They abound in the soil and earth below; they are also present in water, air, chemical products, and many other sources. If particles were spherical or cubical, it would be easy to characterize them. Unfortunately, most of the particles present in our environment are of irregular size and shape. Therefore, it is desirable to try to develop methodologies and techniques to characterize particles of irregular size and shape, and this is the main objective of particle size analysis. Moreover, particle size analysis is important in studying particle behaviour in a medium as in many analytical sciences and industrial applications.

Particle size analysis in physical, chemical, and biological processes involves many concepts and techniques; however, this article focuses on the methods of particle size analysis utilizing mechanical techniques.

This article will first introduce some basic principles used in particle size analysis. This will be followed by a summary of the applicable particle size ranges for the different methods and the size ranges of most common particles found in industrial, chemical, environmental, and clinical applications. The most common mechanical techniques and methods used in particle size analysis will be briefly presented.

## Particle Properties

Particle size analysis plays an important role in many analytical sciences and industrial applications. To assist in developing useful methodologies and tech-

niques it is essential to identify the main factors that control the behaviour of particles in a medium. Such factors include particle density, shape, size, size distribution, concentration, and surface characteristics, and the carrier medium dynamics (**Figure 1**). This article focuses on particle size analysis using mechanical techniques in relation to clinical, industrial, and environmental applications: therefore, the particles under consideration could be solid or liquid and the medium could be liquid or gas. In aerosol systems the medium is gas (air).

### Density

Particles originating from a solid will have the same density as that of the parent material. However, if the material undergoes hydration or surface oxidization or if it agglomerates in clusters, its specific gravity will change. The particle density plays an important role in the separation of solids as in centrifugal and gravitational sedimentation, for example.

### Particle Shape

**Shape factors**  The method of formation influences the resultant particle shape. In comminution, attrition or disintegration, the generated particle resembles the parent material. On the other hand, if the method of formation is condensation from vapour, the smallest unitary particle may be spherical or cubical. In many cases condensation is followed immediately by solidi-

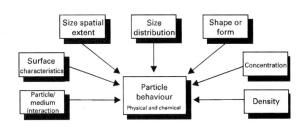

**Figure 1**  Main factors affecting particle behaviour in a medium.

fication and formation of chain-like aggregate (e.g. iron oxide fumes, carbon black).

Based on experimental data it has been found that for a collection of groups of particles having an average diameter $D_{pi}$ for group $i$, the total surface area can be expressed as:

$$A_p = \alpha_s(\sum n_i D_{pi}^2) \qquad [1]$$

where $\alpha_s$ is defined as the surface shape factor. The total volume can be expressed as:

$$V_p = \alpha_v(\sum n_i D_{pi}^3) \qquad [2]$$

where $\alpha_v$ is defined as the volume shape factor. The surface and volume shape factors may be related to a combined shape factor $k_p$ as:

$$k_p = \frac{\alpha_s}{\alpha_v} \qquad [3]$$

In the case of spherical or cubical particles it can be shown that the shape factor is 6. For irregular particles, values vary from 6 to 10. Some other relationships can be obtained by using eqns [1]–[3]:

$$A_v = \frac{A_p}{V_p} = \frac{k_p}{D_p^*} \qquad [4]$$

$$D_p^* = \frac{\sum n_i D_{pi}^3}{\sum n_i D_{pi}^2} \qquad [5]$$

$$A_m = \frac{k_p}{\rho_p D_p^*} \qquad [6]$$

where $A_v$ = specific surface area (surface area per unit volume): $D_p^*$ = specific surface diameter: $A_m$ = surface area per unit mass: $\rho_p$ = density of the particle material. The surface and volume information are used in estimating the equivalent spherical particle diameter which is used in studying particle behaviour in a medium.

**Fractal geometry** Mandelbrot introduced the basic concepts and theories of a new type of geometry called fractal geometry, in order to describe rugged structures. The main idea put forward by Mandelbrot is that the boundary of a rugged system can be described in its embedding space by a fractal dimension which describes its space filling effect. In the case of a fractal surface the estimate of surface area ($A_\lambda$) tends to increase without limit as the step size (resolution) $\lambda$ of an elemental square decreases. This can be expressed as:

$$A_\lambda = k_a \lambda^{-(\delta-2)} \qquad [7]$$

where $k_a$ is a constant and the fractal dimension $\delta$ is greater than 2. Hence a plot of $\ln(A_\lambda)$ versus $\ln(\lambda)$ will have a slope of $[-(\delta-2)]$, where 'ln' designates a natural logarithm.

Kaye applied fractal geometry in his studies of the profiles of carbon black agglomerates. In subsequent studies he demonstrated the usefulness of fractal geometry in studying boundary and mass fractal dimensions of aerosol systems, fractal structures of fine particle systems, fragmentation, description of porous bodies and gas adsorption.

## Surface Characteristics and Interfacial Phenomena

The surface characteristics of small particles include surface area, rate of evaporation and condensation, electrostatic charge, adsorption, adhesion and light scatter. In certain circumstances, changes in the environment of a particle during sampling and particle size analysis may change its size or state of aggregation or its surface characteristics. Such changes must be considered in the selection of a suitable sampling device or method for particle size analysis.

**Surface area** One of the important characteristics of small particles is the rapid increase in exposed surface area per unit mass as size decreases, which leads to increased chemical reaction rate. Fine powders of organic and inorganic oxidable materials (such as coal, iron, flour, sugar, and starch) burn vigorously or explode violently when in the form of an aerosol. Moreover, an increase in surface area increases the toxicity of some granular materials.

**Evaporation and condensation** Evaporation and condensation are diffusion mass transfer processes which proceed at rates proportional to the surface area exposed. The temperature and partial pressure in the vicinity of the surface control the time required for small particles (e.g. water) to evaporate into still air. The evaporation time is given by:

$$\tau = \frac{RT}{8M} \frac{\rho_p D_p^2}{D\Delta_p} \qquad [8]$$

where $\tau$ = evaporation time (s); $\rho_p$ = density of particle material (kg m$^{-3}$); $D_p$ = particle diameter (m); $D$ = diffusion coefficient of vapour from particle (m$^2$ s$^{-1}$); $\Delta_p$ = difference between the particle pressure at the particle surface and in the surrounding fluid (N m$^{-2}$); $R$ = gas constant (8.3144 J mol$^{-1}$ K$^{-1}$); $T$ = absolute temperature (K); $M$ = molecular weight of evaporating particulate material. Finer particles can act as centres for

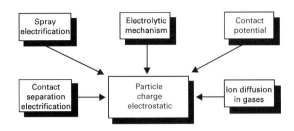

**Figure 2**  Mechanisms producing natural charge on particles.

condensation of moisture, leading to an increase in their size.

**Electrostatic charge**  Electrostatic charge represents an excess or deficiency of electrons on the particle surface. This charge may be assumed to reside on the particle surface in an absorbed gas or moisture film. Mechanisms which produce natural charge on particle surfaces are shown in **Figure 2**. The electrostatic charge generated on a particle is proportional to the particle surface area, which is the principle used in the design of electrostatic classifiers and precipitators. Furthermore, the presence of electrostatic charge on particle surfaces controls the behaviour of particles in an electric field (see II/PARTICLE SIZE SEPARATION/Electrostatic Precipitation).

**Scattering properties**  Scattering of radiation arises from inhomogeneities, such as dispersed dust or water drops, in the fluid medium. Scattering is often accompanied by absorption, and both scattering and absorption remove energy from the incident beams. The quantitative response of the intensity of transmitted and/or scattered beams can be used to characterize the size of a particle.

## Kinetic Behaviour of Particles

The equivalent spherical particle diameter of an aggregate of irregularly-shaped particles can be determined by studying the inertial motion of particles in a medium. Such inertial motion can be found in many applications such as pulmonary deposition, design of industrial ventilation, particle collectors, and electrostatic precipitation. The various processes that affect particle motion in a field are shown in **Figure 3**.

| Field forces | Fluid mechanics | Stochastic |
|---|---|---|
| Gravitational | Inertial | Diffusion |
| Electrical | Drag | (concentration gradient- |
|   coulombic | Centrifugal | particle phase) |
|   image | Turbulent | Diffusiophoresis |
|   dielectrophoretic | Sheer gradient | (concentration gradient- |
|   induced | Coriolis | molecular species in |
|   space charge | | gas) |
| Magnetic | | |

**Figure 3**  Processes affecting particle motion in a medium.

### Medium Resistance

For a small spherical particle moving in a medium at low velocities (i.e. laminar flow), the drag (medium resistance) force acting on the particle is given by Stokes' Law as:

$$F_R = 3\pi\mu_m D_p V_{pm} = C_D \frac{\rho_m A V_{pm}^2}{2} \qquad [9]$$

where $F_R$ = medium resistance (N); $D_p$ = particle diameter (m); $\rho_m$ = medium density (kg m$^{-3}$); $\mu_m$ = medium viscosity (kg m$^{-1}$ s$^{-1}$); $V_{pm}$ = relative velocity between particle and medium (m s$^{-1}$); $A$ = projected area of particle normal to its motion (m$^2$).

It is useful to relate the magnitudes of the inertial and viscous forces in the form of a dimensionless Reynolds' Number as:

$$Re_p = \frac{\rho_m D_p V_{pm}}{\mu_m} \qquad [10]$$

The relationship between the drag coefficient $C_D$ and the Reynolds' Number can be found in any fluid mechanics textbook. However, for spheres with $Re_p < 1$, $C_D = 24/Re_p$.

In finite containers a particle experiences an increase in the drag force due to two effects. First the fluid streamlines around the particle impinge on the container walls and are reflected back causing increased drag on the particle. The second effect occurs because the fluid is stationary at a finite distance from the particle, and there is a distortion of the flow pattern which reacts back on the particle. Taking into consideration these two effects, the drag force may be modified as:

$$F_R = 3\pi\mu_m D_p V_{pm}\left(1 + \frac{kD_p}{L}\right) \qquad [11]$$

where $L$ represents the distance from centre of particle to the container walls and $k = 0.563$ for a single wall or container bottom and $k = 2.104$ for a cylindrical container.

### Particle Motion

**Particle motion in a gravitational/drag field**  The linear motion of a particle in a direction $X$ relative to time $t$ and under the influence of the drag, gravitational, and buoyancy forces is governed by:

$$\frac{\pi}{6}\rho_p D_p^3\left(\frac{d^2X}{dt^2}\right) = \frac{\pi}{6}(\rho_p - \rho_m)D_p^3 g - 3\pi\mu_m D_p \frac{dX}{dt}$$

$$[12]$$

where $g$ denotes gravitational acceleration. If the particle starts from zero velocity it will accelerate until it reaches a terminal velocity given by:

$$\left(\frac{dX}{dt}\right)_{t\to\infty} = V_{gt} = \left[\frac{D_p^2(\rho_p - \rho_m)}{18\mu_m}\right] \cdot g \qquad [13]$$

from which the particle diameter can be expressed as:

$$D_p = \left[\frac{18\mu_m}{(\rho_p - \rho_m)g} \cdot V_{gt}\right]^{1/2} \qquad [14]$$

Note that eqns [12] to [14] are applicable for $Re_p < 1$.

For a particle settling in air, it is usual to neglect the buoyancy correction, since $\rho_p$ is of the order of unity, the air density is of the order of $10^{-3}$ g cm$^{-3}$, and viscosity of ambient air is $1.8 \times 10^{-5}$ kg m$^{-1}$ s$^{-1}$. Equation [13] reduces to:

$$V_{gt} = 3.03 \times 10^4 \, \rho_p D_p^2 \qquad [15]$$

where $V_{gt}$, $\rho_p$, and $D_p$ are expressed in m s$^{-1}$, kg m$^{-3}$, and m, respectively.

Particles with diameters less than or close to the mean free path of the fluid molecules begin to slip between molecules and settle at a higher velocity than that predicted by eqn [13] for settling velocity. Cunningham considered this so-called slip effect and introduced a correction term to the terminal settling velocity as:

$$V_{gtc} = C_c V_{gt} \qquad [16]$$

$$C_c = \left(1 + \frac{2\alpha\lambda}{D_p}\right) \qquad [17]$$

where $C_c$ denotes the Cunningham slip correction factor, $\lambda$ is the mean free path of medium molecules, and $\alpha$ is a constant of approximately one.

**Particle motion in a rotational (centrifugal) field** In a centrifugal field the radial motion of a particle at a distance $R$ from the centre of rotation is governed by:

$$\frac{\pi}{6}\rho_p D_p^3\left(\frac{d^2R}{dt^2}\right) = \frac{\pi}{6}(\rho_p - \rho_m)D_p^3\omega^2 R - 3\pi\mu_m D_p\frac{dR}{dt}$$

$$[18]$$

where $\omega$ is the angular velocity of the particle. Assuming that the particle movement outward is resisted by viscous drag, Stokes' law provides a reasonable approximation for the drag. Therefore, the terminal radial velocity at equilibrium is given by:

$$\left(\frac{dR}{dt}\right)_{t\to\infty} = V_{rt} = \left[\frac{D_p^2(\rho_p - \rho_m)}{18\mu_m}\right] \cdot \omega^2 R \qquad [19]$$

which is equivalent to eqn [13].

To evaluate the performance of a centrifugal separation process, a separation factor 'SF', defined as the ratio of centrifugal acceleration to gravitational acceleration:

$$SF \text{ or } g\text{-force} = \frac{\omega^2 R}{g} \qquad [20]$$

is used. For dust-collecting cyclones (particles $> 100$ µm), $SF = 200$, while conventional centrifuges used in precipitation of submicron particles and large molecules in liquid suspension have $SF = 5000$. See II/PARTICLE SIZE SEPARATION/Hydrocyclones for Particle Size Separation.

**Particle motion in an electrostatic field** When particles larger than 1 µm are passed through a corona discharge as the result of bombardment charging, they acquire charges from electrons and adsorbed gas ions proportional to the surface area of the particle. The saturation charge acquired is given by:

$$Q_{pb} = ne = \pi\varepsilon_0\varepsilon_1\kappa D_p^2 E \qquad [21]$$

$$\kappa = \frac{3\varepsilon_2}{\varepsilon_2 + 2\varepsilon_1} \qquad [22]$$

where $Q_{pb}$ = saturation bombardment charge acquired (C); $n$ = number of electron charges acquired; $e$ = electron charge ($1.6022 \times 10^{-19}$ C); $\varepsilon_0$ = permittivity of vacuum ($8.8542 \times 10^{-12}$ F m$^{-1}$); $\varepsilon_1$ = relative permittivity of medium (gas); $\varepsilon_2$ = relative permittivity of particle material; $E$ = external electric field strength (V m$^{-1}$). For particles less than 0.2 µm, diffusion charging predominates and the charges acquired at time $t$ are given approximately by:

$$Q_{pb} = ne = \frac{D_p kT}{2e}\ln\left(1 + \frac{\pi D_p V_i N_0 e^2}{2kT}t\right) \qquad [23]$$

where $Q_{pd}$ = diffusion charge acquired (C); $k$ = Boltzmann constant ($1.3807 \times 10^{-23}$ J K$^{-1}$); $T$ = absolute temperature (K); $N_0$ = ion density (ions m$^{-3}$); $V_i$ = ion velocity (root mean square, m s$^{-1}$). Based on the acquired charge on the particle and assuming that air resistance is approximated by Stokes' Law, the particle terminal velocity in an electric field is given by:

$$V_{et} = C_c\left(\frac{EQ_p}{3\pi\mu_m D_p}\right) \qquad [24]$$

where $C_c$ = Cunningham slip correction factor; $\mu_m$ = medium (gas) viscosity; $Q_p$ = acquired (bombardment or diffusion) charge on the particle.

**Particle motion in a thermal gradient field** It has been observed that particles in suspension move from hotter to colder regions. Later it has been shown that in a thermal gradient field and at atmospheric pressure, the thermal force acting on a particle is given by:

$$F_t = (-9\pi)\left(\frac{D_p}{2}\right)\left(\frac{\mu_m^2}{\rho_m T}\right)\left(2 + \frac{\kappa_p}{\kappa_m}\right)^{-1}\left(\frac{dT}{dx}\right) \quad [25]$$

where $\kappa_p$ = thermal conductivity of the particle material (J m$^{-1}$ s$^{-1}$ K$^{-1}$); $\kappa_m$ = thermal conductivity of the air (J m$^{-1}$ s$^{-1}$ K$^{-1}$); $T$ = absolute temperature (K); d$T$/d$x$ = temperature gradient in the air (K m$^{-1}$). The negative sign in the equation indicates that the force is in the direction of negative thermal gradient. By setting the thermal force equal to the resistive force of the medium, the terminal velocity of a particle is given as:

$$V_{tt} = (-1.5)\left(\frac{C_c \mu_m}{\rho_m T}\right)\left(2 + \frac{\kappa_p}{\kappa_m}\right)^{-1}\left(\frac{dT}{dx}\right) \quad [26]$$

The particle velocity and the flow velocity and geometrical configuration are used in the design of thermal precipitators.

## Common Methods of Particle Size Analysis and Particle Size Ranges

To assist in covering the scope of the different techniques used in particle size analysis, an attempt was made to survey the current literature on particle size analysis techniques. The results of this investigation are summarized in **Figure 4**. It should be emphasized that this summary is not exhaustive. There are other sophisticated techniques which are outside the scope of this article. Of more importance is the identification of the size ranges of most common particles found in industrial, environmental, chemical and clinical applications (**Figure 5**). From an industrial point of view, the different types of gas particulate collecting equipment are summarized in **Figure 6**. Some basic definitions of particle size ranges of gas and atmospheric dispersoids and soil are included in **Figure 7**.

## Mechanical Techniques for Particle Size Analysis

Based on the information given in Figures 4–7, it can be seen that the most common mechanical techniques for particle size analysis relevant to analytical sciences and industrial applications are: sieving, sedimentation (gravitational or centrifugal), elutriation, electrostatic precipitation, thermal precipitation, hydrodynamic chromatography and impaction. Each technique will be briefly presented; the details of each technique can be found in the references listed in the bibliography.

### Sieving

Sieving is an obvious and most widely used technique for particle size analysis. The particles are classified based on their size, independent of any other particle characteristics such as density and surface properties.

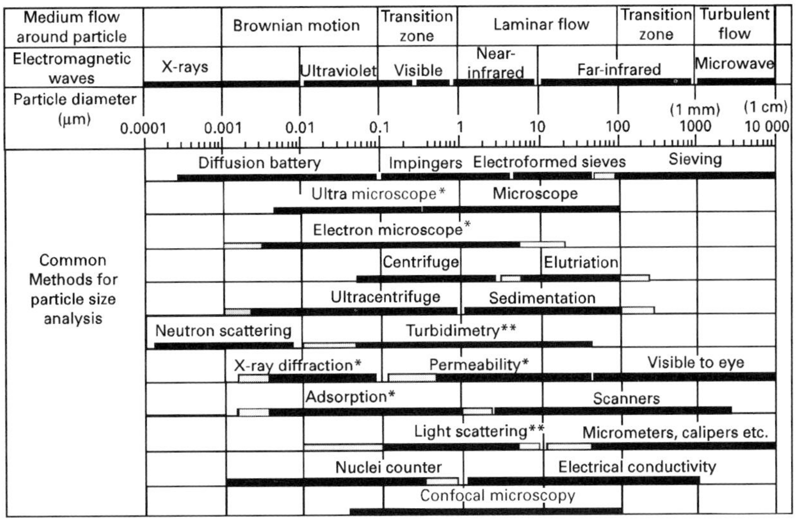

**Figure 4** Common methods of particle size analysis. *Average particle diameter but not size distribution. **Size distribution may be obtained by special calibration.

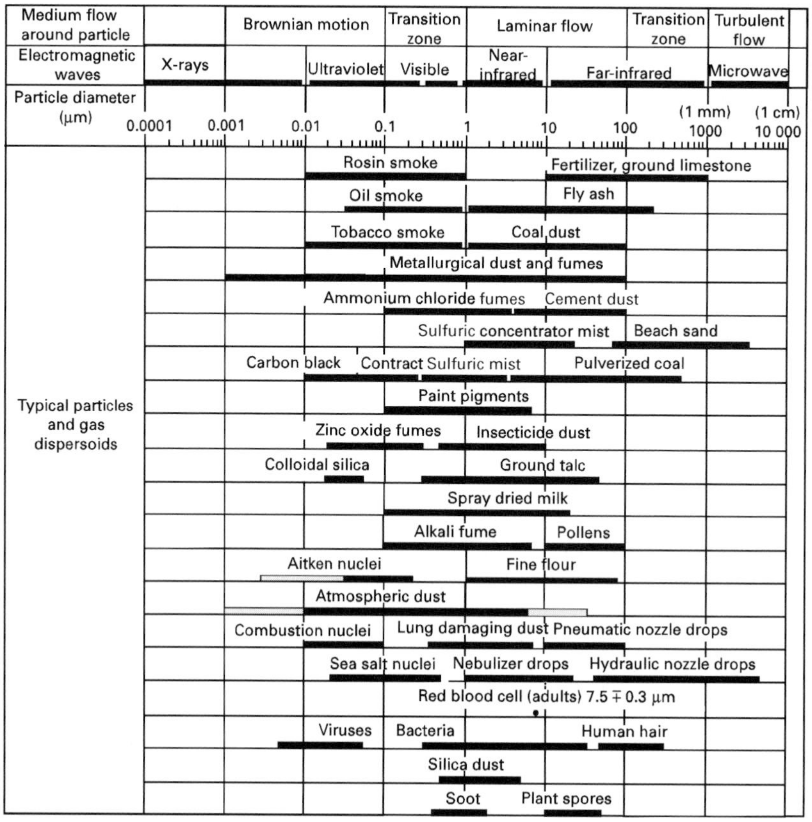

**Figure 5**   Typical particles and gas dispersoids.

Micromesh sieves are used to classify particles of size range 5–20 μm, while particles of size range 20–125 μm are classified in the standard woven wire sieves. Coarse particles (> 125 μm) are classified in punched plate sieves. Punched plate sieves are commonly used in industrial applications where the openings are circular or rectangular; the sieves can take different configurations.

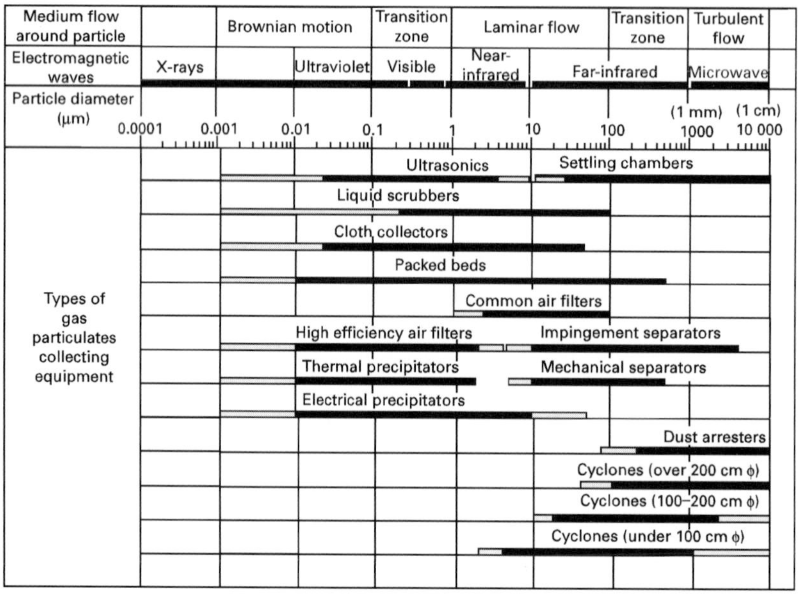

**Figure 6**   Different types of gas particulates collecting equipment ($\phi$, diameter).

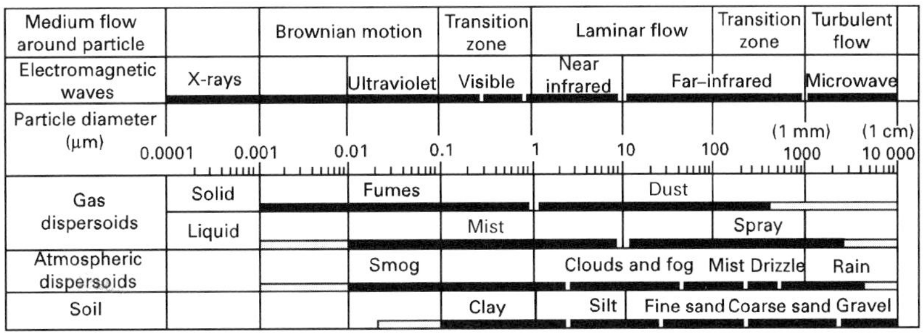

| Medium flow around particle | | | Brownian motion | | Transition zone | Laminar flow | | Transition zone | Turbulent flow |
|---|---|---|---|---|---|---|---|---|---|
| Electromagnetic waves | X-rays | | Ultraviolet | Visible | Near infrared | Far-infrared | | | Microwave |
| Particle diameter (µm) | 0.0001 | 0.001 | 0.01 | 0.1 | 1 | 10 | 100 | (1 mm) 1000 | (1 cm) 10 000 |
| Gas dispersoids | Solid | | Fumes | | | Dust | | | |
| | Liquid | | | Mist | | | Spray | | |
| Atmospheric dispersoids | | | Smog | | Clouds and fog | | Mist Drizzle | Rain | |
| Soil | | | Clay | | Silt | Fine sand | Coarse sand | Gravel | |

**Figure 7** Some size range definitions.

The sieving test is conducted using up to 11 sieves stacked with progressively larger aperture openings towards the top, and placing the powder on the top sieve. A closed pan (receiver) is placed at the bottom. There are several schemes for shaking the sieves by mechanical or ultrasonic means. The residues in each sieve are recorded and expressed in percentage as cumulative values against the nominal sieve aperture values.

The common methods for fine sieving are machine, wet, hand and air-jet sieving. Wet-sieving is recommended for material originally suspended in a liquid and is necessary for powders which form aggregates when dry-sieved. In such tests the stack of sieves is filled with liquid and the sample is fed to the top sieve. Sieving is accomplished by rinsing, vibration, reciprocating action, vacuum, ultrasonication or a combination of these.

Table 1 presents the different international sieve standards and the corresponding sieve types. There are several sieve aperture progression ratios commonly available depending on the different international standards. In the USA, a progression ratio of $2^{1/2}$ is used. This ratio corresponds to successive particle groups of 2 : 1 particle surface ratio. The progression rate of $2^{1/3}$ ($10^{0.1}$) which has been adopted by the French corresponds to successive particle groups of 2 : 1 particle volume ratio. The progression ratios of $10^{0.1}$ and $10^{0.05}$ are recommended for narrow size distributions.

**Table 1** International sieve standards

| Country | Standard | Sieve type |
|---|---|---|
| Great Britain | BS 410 | Woven wire |
| USA | ASTM E11 | Woven wire |
| | ASTM E161-607 | Micromesh (electroformed) |
| Germany | DIN 4188 | Woven wire |
| | DIN 4187 | Perforated plate |
| France | AFNOR NFX 11-501 | Woven wire |
| International | ISO R565 1972(E) | Woven wire |
| | | Perforated plate |

The probability of a particle passing through sieve apertures depends on the particles size distribution, the number of particles on the sieve (sieve loading), the method of sieve shaking, the dimension and shape of the particle, and the ratio of open area of sieve to total area. In addition, the sieving operation can be affected by the friability and cohesiveness of the powder.

**Sedimentation**

Sedimentation of fine powders in a suspension is an important tool in the determination of particle size distribution. Table 2 presents a classification of the methods and techniques used in sedimentation (gravitational or centrifugal). To conduct particle size analysis using sedimentation, the suspension can be prepared using the line start (two-layer) or the homogeneous suspension techniques. In the two-layer technique, the powder is introduced at the top of a column of clear liquid. In the homogeneous suspension technique, the powder is uniformly dispersed in the liquid. As the particles start to settle the change in solids concentration at a particular fixed height with time or the sedimentation time rate is measured. The solids concentration or density measurements is used in the incremental methods, while the settling rate measurement is used in the cumulative methods. Incremental methods may be divided as fixed time and fixed depth methods, the latter being more popular, although a combination is sometimes used.

A powder is made up of three types of particles: primary particles, aggregates and agglomerates. The

**Table 2** Sedimentation methods and techniques

| Incremental methods | Cumulative methods |
|---|---|
| *Solids concentration variation* | *Sedimentation rate* |
| Line start | Line start |
| Homogeneous suspension | Homogeneous suspension |
| *Suspension density variation* | |
| List start | |
| Homogeneous suspension | |

primary particles are crystalline or organic structures bound together by molecular bonding, while the aggregates are primary particles tightly bound together at their point of contact by atomic or molecular bonding. The force required to break these bonds is considerable. In case of agglomerates, the primary particles are loosely bound together with weak van der Waals forces. It is often necessary to disperse the powder in a liquid prior to analysis. Dispersion is affected by the use of wetting agents which break down the agglomerates to their constituent parts. This process is facilitated by mechanical or ultrasonic agitation.

**Gravitational sedimentation**   In gravitational sedimentation there are four main techniques: volume sample, mass sample, manometry and sedimentation vessel-wall pressure sensing as shown schematically in **Figure 8**.

**Incremental methods**   Based on the results of motion of a particle in a gravitational/drag field, it can be shown that the solids concentration of settling suspension at depth $h$ can be related to the cumulative undersize mass distribution as:

$$\frac{C(h, t)}{C(h, 0)} = \frac{\int_{D_{\min}}^{D_t} f(D)\,\mathrm{d}D}{\int_{D_{\min}}^{D_{\max}} f(D)\,\mathrm{d}D} = \Psi \qquad [27]$$

$$D_t = \left[ \frac{18\mu_m}{(\rho_p - \rho_m)g} \cdot \frac{h}{t} \right]^{1/2} \qquad [28]$$

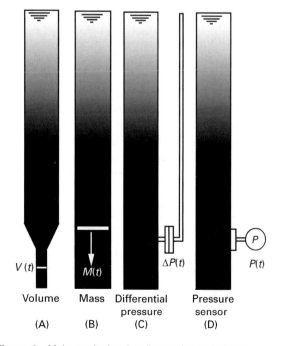

$V(t)$   $M(t)$   $\Delta P(t)$   $P(t)$

Volume   Mass   Differential pressure   Pressure sensor

(A)   (B)   (C)   (D)

**Figure 8**   Main gravitational sedimentation techniques.

where $\mathrm{d}m(D) = f(D)\mathrm{d}D$ represents the mass fraction having particle size between $D$ and $D + \mathrm{d}D$. In eqns [27] and [28], it is assumed that the volume of particles (powder) in suspension is very small compared with the total volume of suspension. By plotting $100\Psi$ against the free-falling particle diameter $D_t$ the resulting curve shows the cumulative undersize percentage curve by mass.

Similarly, it can be shown that:

$$\frac{C(h, t)}{C(h, 0)} = \frac{\phi(h, t) - \rho_m}{\phi(h, 0) - \rho_m} = \Phi \qquad [29]$$

where $\phi(h, 0)$ and $\phi(h, t)$ are the suspension density at a height $h$ at $t = 0$ and $t = t$, respectively. By plotting $100\Phi$ against $D_t$, the resulting curve shows the cumulative undersize percentage size curve.

*The pipette method*   In this method, the changes in concentration occurring within a settling suspension are determined by drawing off definite volumes at a set of discrete intervals of time by means of a pipette. The solids concentration in the suspension is required to be between 0.2 and 1.0 vol%. If the concentration exceeds 1% the hindered settling adversely affects the results of the analysis. Initially the powder is made into a paste; this is followed by slow addition of the dispersing liquid, using a spatula plus mixing, to form a slurry. Further dispersion may be carried out in an ultrasonic bath. The suspension is washed into a sedimentation vessel. The analysis starts with violent agitation of the vessel avoiding the use of a stirrer. It is recommended that the container be continually inverted by hand for 1 min. Since initially the particles are not at rest, it is advisable to wait 1 min before withdrawing samples. The use of a time scale progression of 2 : 1 which produces a $2^{1/2}$ particle size progression is recommended. The collected samples are prepared to determine the solids concentration for particle size cumulative mass determination.

*Hydrometers and divers*   The variation in density of settling suspension may be monitored with hydrometers, a method used widely in the cement industry. The method starts with a fully dispersed suspension and the densities at known depths are recorded as the solid phase settles out. The hydrometer technique is useful for quality control but not as an absolute method.

Divers are an extension of the hydrometer technique. They act as miniature hydrometers where each diver is calibrated to a particular density. Several divers of different densities are added to the fully dispersed suspension and each will settle at a height where its density is equal to the suspension around it.

Sealed in each diver is a copper ring which enables an external search coil to monitor and determine the location of the diver using high-frequency alternating current.

*The specific gravity balance*   The specific gravity balance may be used to monitor the change within a settling suspension. Such a balance comprises two bobs, one in clear fluid and the other in the suspension being studied. The bobs are connected to the two arms of a beam balance. The depth of immersion of the bobs is adjustable. The change in buoyancy is counterbalanced by means of solenoids which are connected to a pen recorder. From the trace of the pen recorder, the particle size distribution can be calculated.

**Cumulative methods**   Cumulative methods have an advantage over incremental methods in that the amount of sample required is small (about 0.5 g), which reduces the interaction between particles. This is a useful feature when only a small quantity of powder is available or when dealing with toxic materials.

*Line start method*   In this case the size distribution may be directly determined by plotting the fractional weight settled against the free-falling diameter of particles. Special care needs to be exercised to eliminate the streaming problem, especially when the suspension at the top has higher density than the liquid.

*Homogeneous suspension method*   Let us consider a powder with a mass distribution such that $dM = f(D)dD$, where $dM$ represents the fractional mass of particles having a diameter between $D$ and $D + dD$. Let us assume that the powder is completely dispersed in a liquid, and consider a suspension chamber of height $h$. It can be reasoned that mass per cent $P$ which has settled out at time $t$ is made up of two parts:

(1)   all the particles with a free-falling speed greater than that of $D_t$ as given by Stokes' Law or some related law, where $D_t$ is the size of particle which has a velocity of fall $h/t$:
(2)   particles smaller than $D_t$ which started off at some intermediate position in the chamber. The falling velocity of one of these smaller particles is $v$, the fraction of particles of this size that have fallen out at time $t$ is $(vt/h)$.

This mechanism can be represented mathematically as:

$$P = \int_{D_t}^{D_{max}} f(D)\, dD + \int_{D_{min}}^{D_t} \frac{vt}{h} f(D)\, dD \qquad [30]$$

which after some manipulation can be rewritten as:

$$P = M + t\frac{dP}{dt} \qquad [31]$$

eqn [31] may be written in a different form as:

$$M = P - \frac{dP}{d\ln(t)} \qquad [32]$$

Both eqns [31] and [32] can be used to determine $M$. The most obvious way is to tabulate $t$ and $P$, and hence derive $dP$, $dt$, and finally $M$ (cumulative percentage oversize) versus $D_t$. Equation [32] is recommended in cases of wide size distribution.

**Centrifugal sedimentation**   Gravitational sedimentation for particle size analysis has limited flexibility. Firstly, the only means of varying the particle velocity is by selecting a medium with different density or viscosity. Secondly, gravitational sedimentation cannot handle particles smaller than 5 μm. Thirdly, most sedimentation devices suffer from the effects of convection, diffusion and Brownian motion. These difficulties can be reduced by speeding up the settling process by centrifuging the suspension. Furthermore, by the use of a centrifugal field, a substantially lower size limit and reduced analysis time can be achieved. As with gravitational methods the data may be cumulative or incremental and the sample may be homogeneous or two-layer.

Calculations of size distribution from centrifugal data are more difficult than calculations from gravitational data, since particle velocities increase as they move away from the axis of rotation (i.e. the particle velocity depends on its radial position). One way to overcome this difficulty is to use a relatively small settling radial zone at a far distance from the centre of rotation (i.e. the centrifugal force acting on all particles is approximately the same). Another solution is to use the line start technique. The most common techniques used in centrifugal sedimentation are schematically presented in **Figure 9** where $S$ and $D$ designate source and detector, respectively.

*Line start method*   Rewriting eqn [19] as:

$$V_{rt} = \frac{dR}{dt} = \left[ \frac{D_t^2(\rho_p - \rho_m)}{18\mu_m} \right] \cdot \omega^2 R \qquad [33]$$

together with separation of variables and integration of eqn [33], yields:

$$D_t = \left[ \frac{18\mu_m}{(\rho_p - \rho_m)\omega^2 t} \cdot \ln\left(\frac{R}{S}\right) \right]^{1/2} \qquad [34]$$

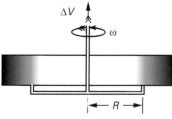

Pipette centrifugation

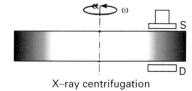

X–ray centrifugation

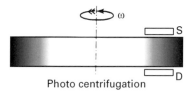

Photo centrifugation

**Figure 9**  Main centrifugal sedimentation techniques.

where $t$ is the time for a particle of size $D_t$ to settle from the surface of the fill (at distance $S$ from the centre of rotation) to a radial distance $R$. Hence, at time $t$, all particles at $R$ will be of size $D_t$. Monitoring the per cent solids or density of the suspension at specified intervals of time will produce the particle size distribution which can be represented using cumulative values at different values of $D_t$.

*Homogeneous suspension method*  Equation [34] still applies: however, at time $t$, all particles of size greater than $D_t$ will have settled out radially to a distance $R$. Conversion of the sedimentation curve into a cumulative curve is not as simple in this case as for that of gravitational sedimentation. Difficulties involved in evaluating the sedimentation curve may be overcome in the case of a centrifugal field by assuming a constant centrifugal field, i.e. for $(R - S)$ interval is small enough to allow the approximation:

$$\ln\left(\frac{R}{S}\right) \approx \frac{(R - S)}{R} \quad [35]$$

when the value $(R - S)$ is one twentieth of $R$, the cumulative curve can be obtained directly by the pipette technique with an error of 1%.

## Elutriation

In fluid classification, the effects of different forces on the movement of suspended particles control the separation of dispersed particles. As discussed in the kinetic behaviour of particles, the field forces are gravitation, as with elutriators, or centrifugal or Coriolis force in classifiers. The medium is usually water or air. In general all fluid classifiers can be divided into two classes, counterflow equilibrium (elutriation) and inverse flow separation. The elutriation is presented briefly below.

In the elutriation technique, the field and the drag forces act in opposite directions and particles leave the separation zone in one of two directions, depending on their size. Particles of a certain size stay in equilibrium in the separation zone. The grading is carried out in a series of vessels (cylindro-conical form) of successively increasing diameter. Hence, the fluid velocity decreases in each stage, the coarse particles being retained in the smallest vessel and the relatively finer particles in the following vessels. For air elutriation, the analysis is considered complete if the rate of change of weight in residues is less than 0.2% of the initial weight in half an hour, and for water elutriation the analysis ends when there is no sign of further classification.

Using Stokes' Law, the particle size retained in an intermediate vessel can be predicted approximately as

$$D_p = \left[\frac{18\mu_m}{(\rho_p - \rho_m)g} \cdot V_{max}\right]^{1/2} \quad [36]$$

$$V_{max} = 2V = 2\frac{Q}{A} \quad [37]$$

where $Q$ = volumetric flow rate, $A$ = vessel cross-sectional area at the equilibrium zone. In eqn [37] it is assumed that the flow profile in the vessel is parabolic. It is clear that elutriation is only suitable for rough dispersions. With small particles, sedimentation may be speeded up by using a centrifuge. This technique is utilized when classifying aerosols using a stream of air which flows in the direction opposite to the centrifugal force.

## Electrostatic Precipitation

The electrostatic precipitator consists of an ionizing cathode at high potential surrounded by a collecting anode; typically, these anodes consist of concentric cylinders, the inner one often being a single wire. The gas suspension passes between the cylinders, picks up the charge, and travels to the anode where the charge is deposited. The transfer of electrons from one anode to the other constitutes an electric current. The magnitude of this current is proportional to the number of particles deposited. Particle sizes can be determined by varying the flow rate or applied potential.

Classification in an electrostatic field by differences in charge is related explicitly to particle size. This type of instrument has been used for collecting aerosol bacteria. The instrument consists of a glass cylinder with a central electrode. The inner surface of the cylindrical glass is coated with a suitable material to act as the other electrode and to collect samples. The principal advantages of this type of instrument are high collection efficiency over a wide size range, low resistance and high flow-rate capacity.

## Thermal Precipitation

Particles in a thermal gradient medium move in the direction of negative gradient, i.e. from hotter to colder regions. Based on this principle, the instrument typically consists of two parallel round microscopic plates and a heated wire in between as shown in **Figure 10**. The sample is drawn between the plates and the particles deposit on the glass plates and are collected for further analysis.

Normally a sample flow of 1–2 cm³ s⁻¹ is recommended and the collection efficiency is high for particles smaller than 5 μm. The collecting device may be modified so that the sample is collected directly on an electron microscope grid. Modifications of the basic design include means of centering the wire in position, substitution of the wire by a ribbon to give more uniform deposits, and using inlet elutriators to exclude coarse particles. The practical application of thermal precipitation in gas cleaning plants has only rarely been attempted.

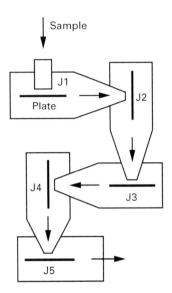

**Figure 11**  Cascade impactor.

## Impaction

**Impactor**  Inertial impaction devices cause an air sample to be drawn into a round or rectangular nozzle where the gas velocity is substantially increased. The jet from the nozzle is discharged against an adjacent flat surface, causing the air to diverge sharply. Particles in the air have more inertia than the air, and tend to continue forward as the air passes off to the sides, causing some of the particles to impact onto the surface. To prevent the particles from re-entrainment, a viscous material such as silicone fluid (or substrate) is used to coat the plate. The efficiency of impaction may be defined in terms of the dimensionless impaction factor as:

$$I = \left( \frac{C_c \rho_p D_{pa}^2 V_j}{18 \mu_m D_j} \right)^{1/2} \qquad [38]$$

where $D_{pa}$ = particle aerodynamic diameter: $D_j$ = jet diameter or width: $V_j$ = average air velocity at the jet outlet.

Classification of a particle cloud into discrete sizes using cascade impaction may be interpreted as measuring aerodynamic (equivalent spherical particle) diameter. Several impaction stages (cascade impactor) are used in the classification of a polydisperse cloud (see **Figure 11**). The stages are arranged to permit jet velocity to increase with each succeeding stage (by successive reduction in jet diameter or width), and to therefore cause particles of progressively smaller sizes to be impacted. In effect, the cascade impactor classifies particles according to their aerodynamic size. The aerodynamic diameter can be expressed in terms of Stokes'

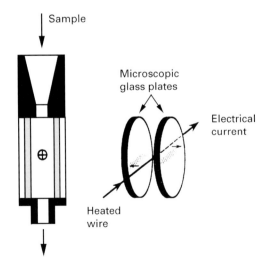

**Figure 10**  Thermal precipitation.

diameter as:

$$D_{pa} = D_{\text{Stokes}} \rho_p^{1/2} \qquad [39]$$

where $D_{\text{Stokes}}$ is the measured diameter. The aerodynamic size is important because it controls the motion of a particle in an air stream. Therefore, it is significant for studies concerning lung inhalation, spray effectiveness, and gaseous cleaning devices. A special duty impactor (five-stage cascade impactor) has been developed for sampling and grading acid mists in the size range 0.3–3 μm.

**Impinger**    Another sampling instrument, referred to as an impinger, also utilizes inertial impaction: however, deposition occurs at the bottom of a liquid-containing vessel. The downward-directed air-jet displaces the liquid and uncovers the bottom of the vessel. The particles that impinge against the wet surface are subsequently washed off by the liquid. The undeposited particles may be caught as air bubbles rise through the liquid. The particles are usually examined in the liquid suspension. Water is the most commonly used liquid.

### Hydrodynamic Chromatography

Size information about colloidally suspended particles (0.01–1 μm) can be obtained by employing hydrodynamic chromatography (HDC). A medium (an aqueous solution) is pumped through a column packed with impermeable spheres. A pulse of colloidal suspension (0.2 cm³) containing about 0.01 wt% polymer is injected into the flowing stream of the column entrance. The mobile phase from the column effluent is passed through a suitable detection system, such as a flow through spectrophotometer of the type used in liquid chromatography, and the detector response of the colloid is determined as a function of elution time. An extra step is needed to determine the concentration of solids in the eluted solution.

It has been observed that the larger particles elute faster than the small ones. It has also been found that the smaller the packing diameter the better the separation. Other factors which affect the rate of transportation through the bed are the size of bed particles, the ionic strength, and flow velocities, as well as the particle size of eluting particles.

In general, HDC has been successfully applied to the size characterization of a number of polymer lattices. The method is applicable to the size separation of particles between 0.02 and 1 μm, if they are rigid. Moreover, it is expected that HDC has wide applicability to sub-μm particles such as in lattices, carbon black, colloidal silica, paint, and photo-graphic pigments, dyes, food colours, natural and artificial blood, and metallic fumes.

Another extension of HDC is to replace the packed bed with a long capillary. Capillary particle chromatography (CPC) requires 30 kPa pressure and has a separating range of 0.2–200 μm. In such techniques the particle transit time is a logarithmic function of particle size.

### Liquid Particle Size Measurement Techniques

In the areas of combustion and chemical processes and in the pharmaceutical and agriculture industries, measurement of droplet size and velocity distributions is important. In one respect the measurement of droplets is easier than that of solid particles because droplets are usually spherical and smooth. In another respect the measurement of droplets is more difficult because they are not easy to collect and stabilize, and they may be volatile. Thus, an *in situ* measurement is usually preferable. The applications commonly used can be grouped as spray diagnostics (using ensemble scattering techniques and optical single-particle analysis) and spray measurement.

## Acknowledgement

This work was supported in part by the Federal Panel on Energy Research and Development (PERD).

*See also:* **II / Chromatography:** Hydrodynamic Chromatography. **Flotation:** Cyclones for Oil/Water Separations; Historical Development; Oil and Water Separation. **Particle Size Separation:** Electrostatic Precipitation; Field Flow Fractionation: Thermal; Hydrocyclones for Particle Size Separations; Sieving/Screening.

## Further Reading

Allen T (1997) *Particle Size Measurement*, 5th edn. New York: Chapman and Hall.

Barth HG (ed.) (1984) Modern methods of particle size analysis. In: *Chemical Analysis* (A Series of Monographs on Analytical Chemistry and its Application) pp. 1–75, vol. 73. New York: John Wiley.

Barth HG and Sun ST (1985) Particle size analysis. *Analytical Chemistry* 57, 151R–175R.

Barth HG, Sun ST and Nikol RM (1987) Particle size analysis. *Analytical Chemistry* 59: 142R–162R.

Böhm J (1982) *Electrostatic Precipitators*. New York: Elsevier.

Hirleman DE, Bachalo WD and Felton FG (eds) (1990) *Liquid Particle Size Measurement Techniques*, vol. 2. American Society for Testing and Materials, STP; 1083.

Kaye BH (1981) *Direct Characterization of Particles*. New York: John Wiley.

Kaye BH (1989) *A Random Walk Through Fractal Dimensions*. Weinheim, Germany: VCH Publishers.

Mandelbrot BB (1983) *The Fractal Geometry of Nature.* San Francisco: W. Freeman Publishers.

Miller BV and Lines R (1988) Recent advances in particle size measurement: a critical review. *Critical Reviews in Analytical Chemistry* 20(2): 75–116.

Oglesby S Jr and Nichols GB (1978) *Electrostatic Precipitation.* New York: Marcel Dekker Inc.

Provder T (ed.) (1998) Particle Size Distribution III, Assessment and Characterization. *ACS Symposium Series* 693, ACS.

Salama AIA and Mikula RJ (1996) Particle and suspension characterization. In *Suspensions: Fundamentals and Applications in the Petroleum Industry.* Advances in Chemistry Series 251, ACS.

Silverman L, Bellings CE and First MW (1971) *Particle Size Analysis in Industrial Hygiene.* New York: Academic Press.

Syvitski James PM (1991) *Principles, Methods and Applications of Particle Size Analysis.* Cambridge: Cambridge University Press.

# MEDICINAL HERB COMPOUNDS: HIGH-SPEED COUNTERCURRENT CHROMATOGRAPHY

**T. Zhang**, Beijing Institute of New Technology Application, Beijing, China

Copyright © 2000 Academic Press

## Introduction

Medicinal herbs are an important source of natural products for medicine. They include various chemical components ranging from fat-soluble to water-soluble compounds. The isolation of the biologically active components is the starting point of further research in chemistry and pharmacology as well as in the utilization of these compounds.

Traditional Chinese medicine is an extremely rich source of the experience acquired over a long period of time. In order to make greater use of traditional Chinese medicine, modern scientific methods are used to find the bioactive compounds in the traditional drugs and to use them as leading compounds for new drug design. New drugs developed in this way include anisodamine and the antimalarial agent Qinghaosu (artemisine).

For separating and purifying bioactive compounds from medicinal herbs, modern chromatographic techniques, such as gas chromatography, high performance liquid chromatography, thin-layer chromatography and electrophoresis have significantly raised the technical level and have shortened the time required for research projects.

High speed countercurrent chromatography (HSCCC) has been recognized as an effective means for separation and purification of a wide variety of bioactive components. It is a liquid–liquid partition chromatography system based on a coil planet centrifuge system without the use of any solid support. This technique has developed rapidly during the past decade. It has been demonstrated to have preparative capabilities and unique properties for fractionating a variety of natural products and medicinal herbs.

Here some applications of the separation of bioactive compounds, such as alkaloids and flavonoids, in medicinal herbs by HSCCC are described.

## Separations of Alkaloids

### Separations of Alkaloids Extracted from *Stephania tetrandra* S. Moore

Dried roots of *Stephania tetrandra* S. Moor (Menispermaceas) or Fenfangji in Chinese is a traditional Chinese drug used for rheumatism and arthritis. The total active alkaloid content in the natural products is 2.3%. Three major alkaloids have been identified as tetrandrine (I, 1%), fangchinoline (II, 0.5%) and cyclanoline (III, 0.2%). I and II are inseparable by conventional methods, while III is well separated from the other two. As illustrated in **Figure 1**, I and II are both bisbenzylisoquinoline alkaloids, whereas III is a water-soluble quaternary protoberberine-type alkaloid.

A sample solution was prepared as a mixture of I and II with purified III to obtain a 10 : 5 : 2 weight ratio to simulate their composition in the natural drug. 3 mg of this sample was dissolved in 0.5 mL of the upper stationary phase of the selected solvent system. The solvent system was composed of n-hexane/ethyl acetate/methanol/water at two different volume ratios of 3 : 7 : 5 : 5 in the first experiment and 1 : 1 : 1 : 1 in the second. In both cases the lower phase was used as the mobile phase at a flow rate of $60 \text{ mL h}^{-1}$ in the normal elution mode. The apparatus used in these experiments was a Pharma-Tech Model CCC-2000 analytical countercurrent

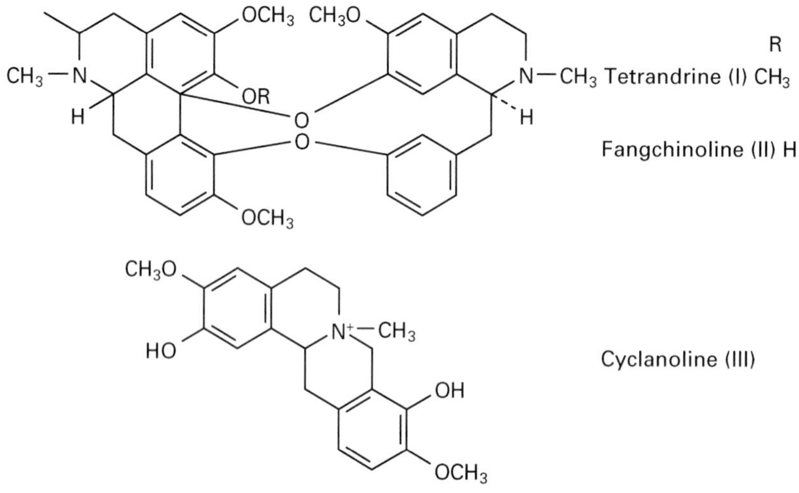

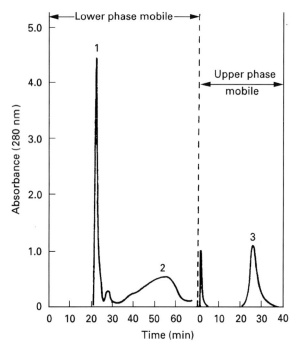

**Figure 1**   The chemical structures of tetrandrine(I), fangchinoline(II) and cyclanoline(III).

chromatograph made by Pharma-Tech Research Corp. It is equipped with a column holder at a 6.4 cm revolution radius. A multilayer coil prepared from a 70 m length of heavy wall 0.85 mm i.d. PTFE tubing is coaxially mounted on the holder. The total capacity of the column is 43 mL. The maximum speed of this centrifuge is 2000 rpm. The apparatus is equipped with a metering pump, a speed controller with a digital rpm display, and a pressure gauge.

**Figure 2** shows the chromatogram obtained from the first experiment. In the normal elution mode,

peaks 1 and 2 were completely resolved and collected in 70 min. This was followed by a reversed elution mode without interrupting the centrifuge run to collect the third peak in an additional 30 min. As shown in the chromatogram, a very small amount of impurity present between peaks 1 and 2 was also resolved.

The chromatogram obtained from the second experiment is shown in **Figure 3**. It demonstrates an alternative approach where the solvent composition was adjusted to modify the partition coefficients of the compounds to shorten the separation time without the use of a reversed elution mode.

A Finnigan MAT mass spectrometer was used to analyse the peak fractions to identify the compounds in peaks 2 and 3 in Figure 3 as purified fangchinoline and cyclanoline respectively.

### Semipreparative Separation of Alkaloids from *Cephalotaxus fortunei* Hook f.

The alkaloids, isoharringtonine (I), homoharringtonine (II) and harringtonine (III) isolated from *Cephalotaxus fortunei* Hook f., possess anticancer potency. Among those compounds, (I) and (III) are isomers, and (II) and (III) differ only by a –CH$_2$ group (**Figure 4**).

A crude alkaloid powder was prepared from the leaves and branches of *C. fortunei* Hook f. 20 mg of the crude powder was dissolved in 1 mL of the solvent mixture (upper phase and lower phase) as the sample solution for each separation. The two-phase solvent system was composed of chloroform/0.07 M sodium phosphate buffer solution at the volume ratio of 1 : 1 (pH 5.0).

The apparatus used in these experiments was a GS-10A HSCCC made by Beijing Institute of New Technology Application. A pair of column holders

**Figure 2**   Chromatogram of the sample mixture of tetrandrine (1)–fangchinoline (2)–cyclanoline (3) (10 : 5 : 2). Solvent system n-hexane–ethyl acetate–methanol–water (3 : 7 : 5 : 5).

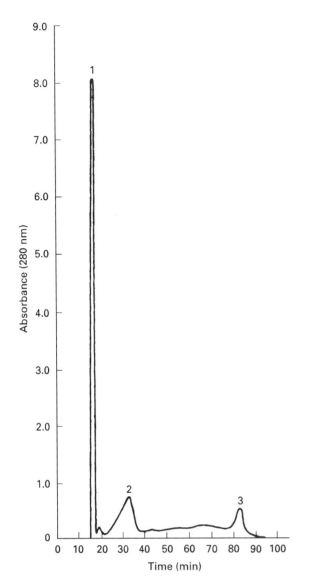

**Figure 3** Chromatogram of the sample mixture of tetrandrine (1)–fangchinoline (2)–cyclanoline (3) (10 : 5 : 2). Solvent system n-hexane–ethyl acetate–methanol–water (1 : 1 : 1 : 1).

are held symmetrically on the rotary frame at a distance of 8 cm from the central axis of the centrifuge. The multilayer coil separation column was prepared by winding a 130 m length of 1.5 mm i.d. PTFE tubing directly on to the holder hub. The total capacity of the 14 layer column was 230 mL. The speed of this machine was regulated with a speed controller in a range between 0 and 1000 rpm, while 800 rpm was used as the optimum speed for the conventional two-phase solvent system.

In each separation, the column was first entirely filled with the stationary aqueous phase followed by injection of the sample solution. The apparatus was rotated at 800 rpm while the nonaqueous mobile phase was pumped into the column at a flow rate of 2 mL min$^{-1}$. Effluent from the outlet of the column was continuously monitored with a UV monitor at 254 nm and collected with a fraction collector at 2 min intervals. After a group of nonpolar compounds was eluted, the centrifuge was stopped while pumping was resumed to collect a polar compound still retained on the column. **Figure 5** shows the chromatogram obtained. Each fraction was analysed by TLC and the fractions corresponding to peaks 2, 4 and 6 were identified as isoharringtonine, homoharringtonine and harringtonine.

In order to increase the yield of the component homoharringtonine, 300 mg of crude extract was applied to produce 70 mg of the pure product. The method is useful for semipreparative separation of alkaloids and other natural products with similar polarity.

## Separation of Flavonoids

### Separation of Flavonoids in Crude Extract from Sea Buckthorn

The results of separations of a crude ethanol extract from dried fruits of sea buckthorn (*Hippophae rhamnoides*) are described. Five flavonoid constituents were successfully separated by P.C. Inc. A two-phase solvent system composed of chloroform–methanol–water at a 4 : 3 : 2 volume ratio was used. The sample solution was prepared by dissolving the crude ethanol extract in the solvent mixture of upper and lower phases at a concentration of 2.2 g %.

Separation was performed as follows: The coiled column was first entirely filled with the upper phase as the stationary phase followed by injection of sample solution containing 100 mg crude mixture through the sample port. The apparatus was rotated at the optimum speed of 800 rpm while the lower phase was pumped into the head end of the column as the mobile phase at 200 mL h$^{-1}$ flow rate. Effluent from the outlet end of the column was continuously monitored at 278 nm and fractionated into test tubes. An aliquot of each fraction was diluted with methanol and the absorbance was determined at 260 nm with a spectrophotometer. **Figure 6** shows the chromatogram obtained. Five flavonoid components were completely resolved from each other as symmetrical peaks and eluted in 2.5 h. Partition efficiencies computed from the equation $N = (4R/W)^2$ (where $N$ is the efficiency expressed in terms of theoretical plates (TP), $R$ is retention time of the peak maximum, $W$ is the peak width expressed in the same unit as $R$) range from 800 TP (2nd peak) to 530 TP (5th peak). By calculating the partition coefficients of each peak and comparing the values

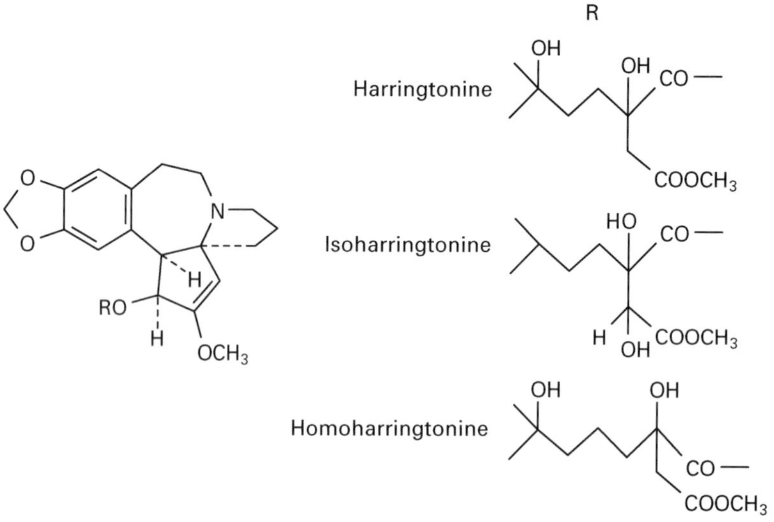

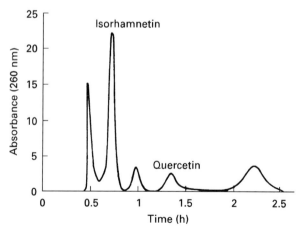

**Figure 4**   Chemical structures of harringtonine, isoharringtonine and homoharringtonine.

obtained with those of the pure compounds, quercetin and isorhamnetin peaks were identified as labelled in the chromatogram.

The same crude sample of ethanol extract from dried fruits of sea buckthorn can be separated with an analytical HSCCC for a series of rapid separations. The apparatus was a Pharma-Tech Model CCC-2000 made by Pharma-Tech Research Corp. The multilayer coiled column was prepared from a single piece of 0.85 mm i.d. heavy wall PTFE tubing. The total capacity of the column is 43 mL including 3 mL in the flow tubes. The revolution speed of the centrifuge can be continuously adjusted up to 2000 rpm. A LDC/Milton Roy Pump, a speed controllor with digital rpm display and a pressure gauge are also included.

In these separations, 3 mg of sample of the flavonoid mixture were dissolved in 0.5 mL of solvent mixture and loaded for each experiment. The centrifuge was rotated at the optimum speed of 1800 rpm. The effluent from the outlet of the column was continuously monitored with an LKB Uvicord S at 278 nm and then collected as 1 mL fractions with an LKB fraction collector. Each fraction was diluted with 2 mL of methanol and the absorbance was determined at 260 nm with a Zeiss spectrophotometer Model PM6.

**Figure 7** shows the result of separation obtained at a flow rate of $60 \text{ mL h}^{-1}$. The high efficiency of the separation is evidenced by a minor peak present between the first and the second major peaks, which

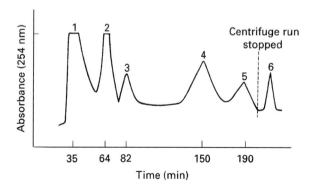

**Figure 5**   High speed CCC separation of harringtonine, isoharringtonine and homoharringtonine from a crude extract of *Cephalotaxus fortunei* Hook f. Peak 2, isoharringtonine; 4, homoharringtonine and 6, harringtonine.

**Figure 6**   Countercurrent chromatogram of flavonoids in crude extract from dried fruits of sea buckthorn by a multilayer coil planet centrifuge. Conditions for CCC: sample size, 100 mg; solvent system, chloroform–methanol–water (4 : 3 : 2); mobile phase, lower phase; flow rate, 200 mL h⁻¹; speed, 800 rpm; fraction volume, 6 mL.

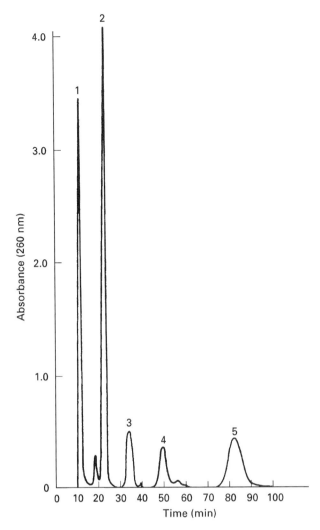

**Figure 7** Chromatogram obtained from 3 mg extract by analytical countercurrent chromatography, flow rate of 60 mL min⁻¹.

was not detected in the semipreparative separation shown in Figure 6. Flow rates as high as 300 mL h⁻¹ can be used to give a separation time of only 15 min, which is quite comparable with that of analytical HPLC.

The analytical HSCCC can be used to interface with MS to provide a new analytical methodology HSCCC-MS, combining the versatility and high resolution of HSCCC with the identification capability and low detection limit of mass spectrometry.

### Preparative Separation of Kaempferol, Isorhamnetin and Quercetin from Extracts of *Ginkgo biloba* L.

Extracts of the leaves of *Ginkgo biloba* L. are used to increase peripheral and cerebral blood flow. *Ginkgo* extracts contain active compounds such as flavonoids and terpene lactones (ginkgolides and bilobalide).

They show effects on vascular and cerebral metabolic processes and inhibit platelet-activating factor. The standard compounds used to control the quality of *Ginkgo* extracts and its preparations, such as isorhamnetin, kaempferol and quercetin, the three major flavone aglycones in the leaves of *Ginkgo biloba* L., are expensive and difficult to obtain. Some commercial quercetin standards, usually used in quantitative analysis of total flavonoids, are not pure and contain isorhamnetin, kaempferol and impurities. The preparative separation and purification of flavonoids from plant materials by classical methods are tedious and usually require multiple chromatographic steps. HSCCC as a form of liquid–liquid partition chromatography without using a solid support matrix is very suitable for the separation of flavonoids.

HSCCC was performed using a Model GS-10A multilayer coil planet centrifuge made by the Beijing Institute of New Technology Application. A NS1007 constant-flow pump, a model 8823A UV monitor operating at 254 nm and a manual sample injection valve with a 35 mL loop, were used in the experiment. A Rainin Model SD-200 HPLC was used for analysis.

In these studies, a two-phase solvent system composed of chloroform–methanol–water at a volume ratio of 4 : 3 : 2 was used. The crude *Ginkgo* flavone aglycones were prepared by several steps, from the extract of *Ginkgo* leaves. A second sample was a commercial quercetin standard. The sample solutions were prepared by dissolving these two samples in equal volumes of upper and lower phases.

In each separation, the column was first entirely filled with the upper aqueous phase as stationary phase, and then the apparatus was rotated at 800 rpm, while the lower chloroform phase was pumped into the column at a flow rate of 2 mL min⁻¹. After the mobile phase front emerged and the two phases had established hydrodynamic equilibrium in the coil column, the sample solution containing 200 mg of crude *Ginkgo* flavone aglycones or commercial quercetin standard was introduced through the injection value. HPLC analysis of the crude *Ginkgo* flavone aglycones is shown in **Figure 8A**. According to MS analysis, quercetin, kaempferol, isorhamnetin and some impurities were present. HPLC analysis of the commercial quercetin standard is shown in Figure 8B indicating that a larger amount of unknown impurity was present.

After repurification of the isorhamnetin peak by HSCCC, the purities of the three flavone aglycones were more than 99% according to the results of HPLC analysis. The identities of the three flavone aglycones were confirmed by MS analysis.

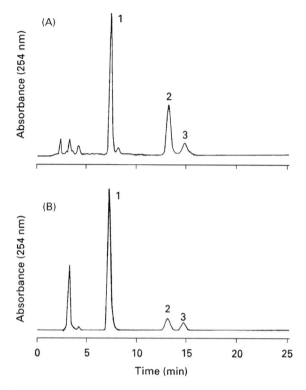

**Figure 8**  Chromatograms of the HPLC analyses of the crude *Ginkgo* flavone aglycones from the Ginkgo extracts (A) and the commercial quercetin standard (B). Mobile phase: methanol-0.04% $H_3PO_4$ (50:50, v/v); flow rate: 1.0 mL min$^{-1}$; detection: 254 nm. Peak 1: quercetin: Peak 2: kaempferol; Peak 3: isorhamnetin.

## Semipreparative Separation of Taxol Analogues

Taxol is a promising new anticancer drug. Currently, taxol intended for human consumption is mainly obtained from the bark of *Taxus*. Although taxol is the compound most commonly used in clinical treatment, numerous analogues have also been identified. Some of them are of similar biological activities to taxol, and some can be used as natural precursors to semisynthesize taxol by simply modifying the C13 side chains. Meanwhile, the study of these compounds may lead to the discovery of new alternatives with improved pharmaceutical properties or economic benefits. The separation and purification of taxol analogues by classical methods such as preparative TLC, open column chromatography and HPLC in general are complicated and have low recoveries. HSCCC can be used for the semipreparative separation of some taxol analogues such as cephlomanine and 7-epi-10-deacetyltaxol.

HSCCC was performed with a model GS-10A multilayer coil planet centrifuge made by the Beijing Institute of New Technology Application. The chromatography system was the same as that described previously.

A crude extract containing 10% taxol from *Taxus yunnannesis* was pre-separated through a $C_{18}$ column eluted by an acetonitrile–water gradient. After recovering the majority of the taxol and removing some other polar components, the fraction mainly containing cephalomannine, 7-epi-10-deacetyltaxol, residual taxol and an unknown compound was collected, concentrated and dried. The sample solution was prepared by dissolving the sample in the solvent mixture of upper and lower phases.

The two-phase solvent system, a quaternary system of n-hexane–ethyl acetate–ethanol–water was selected. In those studies a two-step separation was chosen. Firstly, the solvent system in 1:1:1:1 proportion was employed to separate the sample into two groups, A and B. Secondly, the system in 3:3:2:3 and 4:4:3:4 proportions was employed to separate the two compounds in A and B. In each separation, the coiled column was first entirely filled with the upper phase as the stationary phase. After the apparatus was rotated at 800 rpm, the lower phase was pumped into the column at a flow rate of 2 mL min$^{-1}$, while the sample solution was introduced via the injection valve. The effluent from the outlet of the column was continuously monitored at 254 nm. Twelve consecutive injections of 10–16 mg of sample mixture were performed without changing the stationary phase in the coiled column.

The four compounds were first separated into two groups as shown in **Figure 9**. Peak A contains mainly cephalomannine and taxol, while peak B contains mainly 7-epi-10-deacetyltaxol and the unknown compound. After 12 consecutive injections, 84.4 mg of A and 42.1 mg of B were obtained. Cephalomannine and taxol in peak A were well separated in the second step with the 3:3:2:3 solvent system

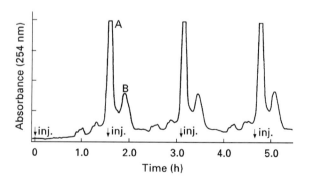

**Figure 9**  First step separation of taxol and its analogues by HSCCC with three consecutive injections. Solvent system: n-hexane–ethyl acetate–ethanol–water (1:1:1:1, v/v/v/v). A, cephalomannine + taxol; B, unknown + 7-epi-10-deacetyltaxol.

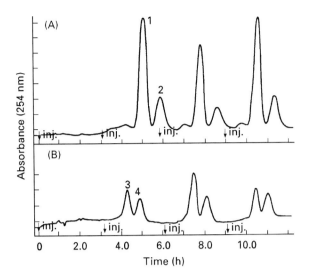

**Figure 10** Second step separations of peak fractions (A) and (B) in Figure 9 by HSCCC with three consecutive injections. Solvent system: n-hexane–ethyl acetate–ethanol–water. (A) 3 : 3 : 2 : 3, (B) 4 : 4 : 3 : 4. 1, Cephalomannine; 2, taxol: 3, unknown; 4, 7-epi-10-deacetyltaxol.

as shown in **Figure 10**. 49.3 mg of cephalomannine and 14.6 mg of taxol were obtained after six consecutive injections. The purities were about 99% and 87% respectively according to the results of HPLC analysis. Similarly with peak B, 7-epi-10-deacetyl-taxol and the unknown compound were separated with the 4 : 4 : 3 : 4 solvent system as shown in Figure 10B: 12 mg of 7-epi-10-deacetyltaxol and 10 mg of the unknown compound were obtained after three consecutive injections. The purities were about 85% according to the results of HPLC analysis.

## Conclusions

Many applications of HSCCC such as the separation of alkaloids, flavonoids, hydroxyanthraquinone derivatives, etc., have been performed successfully. The main advantages of HSCCC are as follows:

1. Since no solid supports are used, all compounds in the sample solution are recovered. The crude sample can be injected directly into the column, which simplifies sample preparation.
2. Numerous two-phase solvent systems with a broad spectrum of polarity can be used with either the aqueous or organic phase as the mobile phase. Some examples are shown in **Table 1**.
3. The quantity of purified compound may range from milligrams to grams. Consecutive injection can be applied for large scale separation.
4. HSCCC will become widely used in the research and production of natural drugs, since the instrument is inexpensive, convenient to operate, solvent saving and has preparative capacity.

**Table 1** Solvent systems for separation of some substances by high speed countercurrent chromatography

| Substances separated | Solvent system |
|---|---|
| Alkaloids | $nC_6H_{14}$–EtOAc–MeOH–$H_2O$ (1 : 1 : 1 : 1) |
| | $nC_6H_{14}$–EtOH–$H_2O$ (6 : 5 : 5) |
| | $CHCl_3$–MeOH–$H_2O$ (4 : 3 : 2) |
| | $CHCl_3$–MeOH–0.5%HBr–$H_2O$ (5 : 5 : 3) |
| | $Bu^nOH$–0.1 M NaCl (1 : 1) |
| Flavonoids | $CHCl_3$–MeOH–$H_2O$ (7 : 13 : 8), (4 : 3 : 2) |
| Flavonoid glycosides | $CHCl_3$–EtOAc–MeOH–$H_2O$ (2 : 4 : 1 : 4) |
| | EtOAc–$Bu^nOH$–$H_2O$ (2 : 1 : 2) |
| Indole | $nC_6H_{14}$–EtOAc–MeOH–$H_2O$ (3 : 7 : 5 : 5), (1 : 1 : 1 : 1) |
| Xanthone glycosides | $CHCl_3$–MeOH–$H_2O$ (4 : 4 : 3) |
| Lignans | $nC_6H_{14}$–$CH_3CN$–EtOAc–$H_2O$ (8 : 7 : 5 : 1) |
| Lignan glycosides | $CHCl_3$–MeOH–$H_2O$ (5 : 5 : 3) |
| | $nC_6H_{14}$–$CH_2Cl_2$–MeOH–$H_2O$ (2 : 4 : 5 : 2) |
| Phenolic glycosides | $CHCl_3$–MeOH–$H_2O$ (7 : 13 : 8) |
| | $C_6H_{12}$–$Me_2CO$–EtOH–$H_2O$ (7 : 6 : 1 : 3) |
| Polyphenols | $CHCl_3$–MeOH–$H_2O$ (7 : 13 : 8) |
| Tannins | $Bu^nOH$–0.1 M NaCl (1 : 1) |
| Coumarins | $nC_6H_{14}$–EtOAc–MeOH–$H_2O$ (3 : 7 : 5 : 5) |
| Saponins | $CHCl_3$–MeOH–$Bu^iOH$–$H_2O$ (7 : 6 : 3 : 4) |
| | $CHCl_3$–MeOH–$Pr^iOH$–$H_2O$ (5 : 6 : 1 : 4) |
| | $CHCl_3$–MeOH–$H_2O$ (7 : 13 : 8) |
| Hydroxyanthraquinone derivatives | $nC_6H_{14}$–EtOAc–MeOH–$H_2O$ (9 : 1 : 5 : 5) |
| 25-Hydroxycholecalciferol | $nC_6H_{14}$–EtOAc–MeOH–$H_2O$ (5 : 1 : 5 : 1) |

See also: II/Chromatography: Countercurrent Chromatography and High-Speed Countercurrent Chromatography: Instrumentation. Chromatography: Liquid: Countercurrent Liquid Chromatography. III/Alkaloids: Gas Chromatography; High-Speed Countercurrent Chromatography; Liquid Chromatography; Thin-Layer (Planar) Chromatography.

## Further Reading

Conway WD and Petroski RJ (1995) *Modern Countercurrent Chromatography*. American Chemical Society Symposium Series 593. Washington DC.

Ito Y (1986) *CRC Critical Reviews in Analytical Chemistry* 17: 65.

Ito Y and Conway WD (1996) *High-Speed Countercurrent Chromatography*. New York: John Wiley.

Mandava NB and Ito Y (1988) *Countercurrent Chromatography*. New York and Basel: Marcel Dekker.

Zhang TY (1991) *Countercurrent Chromatography*. Beijing: Beijing Science and Technology Press.

Zhang TY (1996) HSCCC on medicinal herbs. In: *High-Speed Countercurrent Chromatography*. New York: John Wiley.

# MEDIUM-PRESSURE LIQUID CHROMATOGRAPHY

**K. Hostettmann and C. Terreaux**,
University of Lausanne, Lausanne, Switzerland

## Introduction

Medium-pressure liquid chromatography (MPLC) is one of the various preparative column chromatography techniques. Separation under pressure renders the use of smaller particle size supports possible and increases the diversity of usable stationary phases. MPLC was introduced in the 1970s as an efficient technique for preparative separation of organic compounds. MPLC overcame one major drawback of low pressure liquid chromatography (LPLC), i.e. the limited sample loading. This separation method is now routinely used beside or in combination with the other common preparative tools: open-column chromatography, flash chromatography, LPLC or preparative high performance liquid chromatography (HPLC). The distinction between low pressure, medium pressure and high pressure LC is based on the pressure ranges applied in these techniques and the overlap is often considerable. MPLC allows purification of large compound quantities and, unlike open-column chromatography and flash chromatography, faster and improved separations are obtained. Packing of material with lower particle size under pressure enhances separation quality and moreover the solid phase can be reused. **Table 1** provides a comparative description of these different methods. Simplicity and availability of the instrumentation, together with recycling of packing materials and low maintenance costs, contribute to the attractiveness of this technique. More details about experimental conditions are given below.

## Instrumentation

A schematic representation of a simple MPLC setup is shown in **Figure 1**. The instrumentation is made up of

**Table 1** Comparative description of various preparative column chromatographic techniques

| Technique | Stationary phase particle size (µm) | Pressure (bar) | Flow rate (mL min⁻¹) | Sample amount (g) | Solvents | General |
|---|---|---|---|---|---|---|
| Open-column chromatography | 63–200 | Atmospheric | 1–5 | 0.01–100 | General solvent used | Frequent packing; RP not possible |
| Flash chromatography | 40–63 | 1–2 | 2–10 | 0.01–100 | General solvent used | Frequent packing |
| Low pressure LC | 40–63 | 1–5 | 1–4 | 1–5 | More solvent required | Prefilled Lobar columns |
| Medium pressure LC | 15–40 | 5–20 | 3–16 | 0.05–100 | More solvent required | Infrequent packing |
| Preparative HPLC | 5–30 | >20 | 2–20 | 0.01–1 | High purity solvent required | Higher resolution |

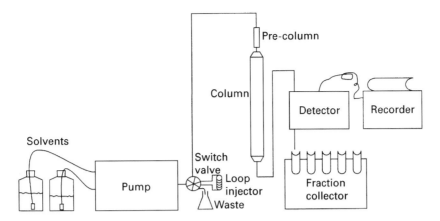

**Figure 1** Typical MPLC instrumental setup.

a pump for solvent delivery, a sample injection system, and a self-packed column. Product separation can be followed either manually by monitoring with thin-layer chromatography (TLC) or automatically with a detector and a recorder connected to the column outlet. Separated compounds are collected by means of a fraction collector.

## Pump

Chromatographic separation of 0.1–100 g sample within a few hours requires flow rates ranging from about 5 to 200 mL min$^{-1}$, with a maximal pressure of 40 bar. Several companies provide pumps suitable for MPLC. Criteria for selecting an MPLC pump include: flow rate range; presence or absence of a pulse-damper, which provides regular flow rates and pressures during separation and increases the reproducibility of the separation; presence of a pressure cut-off device; and presence of a gradient former. Some manufacturers provide pumps with exchangeable piston heads, thus allowing flow rates from 0.5 to 160 mL min$^{-1}$ with pressures up to 40 bar.

## Column

The column is the central point when optimizing a preparative chromatographic separation and criteria such as amount of sample to be purified, amount of packing material and column length versus column diameter, have to be carefully considered. MPLC columns are generally made of thick glass coated with protective plastic and can withstand pressures up to 50 bar; however, some columns from some manufacturers cannot withstand pressures exceeding 20 bar. The columns vary in length as well as in internal diameter (i.d.) and sizes are expressed as filling volumes. Filling volumes range from 63 mL (9 mm i.d. × 100 mm) to 15 000 mL (105 mm i.d. ×

1760 mm) for the larger columns available. Selection of the column dimensions depends on the sample amount to be separated, ranging from 0.1 g with the smallest columns up to 100 g with large columns. Selectivity ($\alpha$) and retention factor ($k$) are the prominent factors influencing resolution. Sample loading can greatly affect resolution. Therefore, when separations are 'easy' ($\alpha > 1.2$; high resolution between the eluted compounds), larger sample amounts can be loaded. Increasing the column diameter allows injection of a larger sample mass (higher throughput), but also makes use of smaller particle size material possible. On the other hand, increasing the column length results in higher resolution but has little or no effect on sample throughput. The back pressure increase with longer columns often implies the use of larger particle size material. The influence of column dimensions on resolution has been studied through the separation of standard mixtures. The correlation between resolution and amount of packing material was shown to be linear either when testing columns with identical internal diameter and different lengths or when varying the internal diameter in a set of columns of the same length. However, a lesser increase in resolution was observed with the use of larger internal diameter and a constant length. Consequently, longer columns are preferred in order to improve the separation of a given sample. The column system supplied by Büchi (Flawil, Switzerland) gives the possibility to couple columns together vary simply by means of a Teflon sealing joint, resulting in an increased resolving power.

### Detector and Recorder

Monitoring of a MPLC separation can be performed by TLC of the collected fractions. Online detection is also routinely used with single-wavelength UV/Vis detectors. Most available UV detectors are designed for analytical purposes and are of little use for preparative separations. Accommodation of high flow rates is a prerequisite for a preparative chromatographic detector. This results in a loss of sensitivity, which is compensated by the usual high concentration of the eluate. In fact, these concentrations are often so large when coming through the detection cell that the detector is overloaded. This problem can be solved by the use of detectors with a splitting system before the UV cell: one part of the sample goes directly from the column outlet to the fraction collector, while another part is diverted through the detector. Detectors with a pathlength < 0.1 mm are also very useful. Selection of a detection wavelength where absorption of the products is low can also be an alternative to avoid detection overload. The Gow-Mac 80-800 LC-UV detector is a specially designed detector for preparative separations: flow rates up to $500 \, mL \, min^{-1}$ are possible and the eluate arrives through a needle and passes as a thin film on a 6.5 cm wide quartz cell. Connection to a recorder allows visualization of the chromatographic separation.

### Fraction Collector

Automatic collection of fractions can be performed by connecting a fraction collector to the column or detector outlet. The volume of the collected fractions is of course strongly dependent on the internal diameter of the column and the flow rate; it is in most cases time-monitored. Presence of a built-in peak detector or connection to an external one allows peak-monitored fraction collection. In its standard MPLC setup, the Büchi system provides a fraction collector with a total capacity of $240 \times 20 \, mL$ tubes, $120 \times 50 \, mL$ tubes or $48 \times 250 \, mL$ tubes. This type of fraction collector has proved to be particularly suitable for MPLC.

## Column Packing

### Packing Material

Selection of the stationary phase is probably the most crucial parameter affecting separation quality. Several types of packing material are commonly used in MPLC and various factors have to be considered when choosing the packing material:

- particle size
- column length
- operating pressure
- type of sample
- cost.

With regard to cost-effectiveness, the most frequently utilized stationary phase is silica gel. Beside its economic advantage, silica gel possesses other advantages such as a wide range of possible solvents as eluents, easy evaporation of the fractions and elution with high flow rates. The risk of irreversible adsorption is a possible major drawback of this support.

A wide range of particle sizes is commercially available. The smallest average particle size (5–10 μm) is used for analytical HPLC, while for preparative LC, stationary phases with sizes starting at 15 μm are the most convenient. Optimal separations are generally obtained with sizes around 20 μm. The influence of particle size on resolution has been investigated. A large decrease in resolution was observed when the average particle size changed from 15 μm to 30 μm. Using particle sizes of 52 μm or 130 μm resulted in a slower resolution decrease but the retention times increased significantly: 3 h more with an average particle size of 130 μm than with one of 15 μm.

The use of modified silica gel phase (bonded phases) has become more common. These inherent advantages, such as a lower risk of sample decomposition and less irreversible adsorption, allowing an easier recycling of the column sorbent. Reversed-phase (such as RP-18, RP-8 or RP-4) or dihydroxypropylene-bonded (Diol, Merck, Germany) silica gels are frequently used for MPLC separations. Moreover, it is possible to use other commercially available bonded phases for MPLC separations.

### Column Packing Methods

Different filling methods are described for packing MPLC columns. Filled columns should possess an optimal homogeneity and a good density. Two methods are most frequently used: dry filling and the slurry method.

**Dry filling** Dry filling is generally applied for silica gel. This method usually gives a 20% better packing density than the slurry method. The 'tap-and full' technique can be used with particle sizes larger than 20–30 μm; however, when applying eluent pressure up to 40 bar, the packing density obtained may not be sufficient. Packing under nitrogen pressure allows use of 15 μm particle size silica gel and provides a high packing density. Filling is carried out manually by connecting a reservoir to the top of the column, which is then filled with dry stationary phase until it contains approximately enough phase to fill another 10% of the column (**Figure 2**). The system is then connected to a nitrogen cylinder and a 10 bar pressure is applied (with the column outlet open) until the level of packing material remains constant. The nitrogen valve can be closed and the pressure slowly goes down to atmospheric. Vacuum at the column exit can

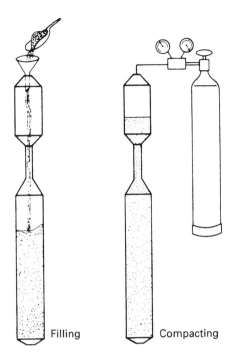

**Figure 2** Dry filling of Büchi MPLC glass columns. (Reproduced with permission from Hostettmann *et al.*, 1997.)

be used as an alternative to nitrogen pressure. An automatic mechanism has been suggested to slowly and homogeneously fill the column (3–4 g min⁻¹). Passing mobile phase through the column induces compression of the stationary phase, which is compensated by the further addition of dry silica gel.

**Slurry method**   The slurry method is an alternative method to pack silica gel, with the inconvenience of a lower packing density. However, slurry filling is the preferred method for packing bonded phases. The slurry is prepared by suspending homogeneously an appropriate amount of stationary phase in the eluent. The mixture is then poured into the column and the eluent is passed through the column until the stationary phase level is constant.

**Column preparation and regeneration**   Before sample introduction, it is recommended that a separation test is performed with a standard mixture of compounds. A mixture of phthalic acid dimethyl-, diethyl- and dibutyl esters is convenient for testing silica gel columns, whereas the separation of benzene and naphthalene can be used for reversed-phase columns.

Usually, packing material is regenerated after each chromatographic separation. For silica gel supports, this can be performed by washing successively with methanol, ethyl acetate and *n*-hexane. However, after a certain time, the stationary phase should be

changed and the column repacked. Regular elution of a test mixture is a good method to determine column quality. Bonded-phase columns are generally easier to clean (for example with a mixture of methanol/tetrahydrofuran (1 : 1)) and thus have a longer working life. In order to prolong column lifetime, a pre-column can be used. Contaminating material remaining at the top of the column is thus eliminated after each MPLC separation.

## Solvent Selection

Selection of the eluent system is also a crucial point in development and optimization of a MPLC separation. The ideal case would be successive direct testing of various solvent mixtures on the MPLC column. However, in routine practice such an approach is obviously impossible because of the waste of time due to column equilibration, together with loss of sample, etc. Two methods are mainly used for solvent selection: optimization by TLC or transposition of analytical HPLC conditions on MPLC.

Preliminary TLC allows rapid screening of numerous possible solvents and it is now well established how TLC results on silica gel plates can be transposed to silica gel columns. Solvent testing on silylated TLC plates can be used for reversed-phase columns. One important factor that has to be considered is that the surface areas of silica gel used in TLC is about twice that of the column packing material. Therefore, it is recommended that sample constituents display a retention factor ($R_F$) lower than 0.3 on the TLC plate. The major drawback of this method is the lower separation and resolution observed when reducing the solvent strength to obtain an $R_F \leqslant 0.3$. An alternative has been suggested to circumvent this problem: the use of overpressured-layer chromatography (OPLC) as a pilot method for MPLC. In a first step, a suitable multicomponent eluent with a good selectivity is searched for by means of TLC. Adjustment of the solvent strength and fine tuning are performed with OPLC. Unlike TLC, OPLC is a closed and equilibrated system and can be viewed as a 'planar column'. Because of these properties, direct transposition from OPLC to MPLC is an accurate and efficient method. Such an approach is also applicable to the other preparative pressure chromatography techniques using normal silica gel as stationary phase.

Because of the similarities of the phases used in analytical HPLC and preparative packing materials, separation optimization on an analytical HPLC column very often provides excellent results and transposition to MPLC is straightforward and direct. This is particularly evident for separations on reversed-phase sorbents, where studies with TLC are more difficult. Due to the wider range of solvents available for normal-phase chromatography, preliminary tests on TLC are of major use prior to analytical HPLC optimization. Examples of transposition of analytical HPLC conditions to MPLC are given below.

Once the ideal conditions have been selected, a compromise has to be found between speed of separation and sample loading: decreasing the solvent strength (for example, by adding water to the solvent system in reversed-phase separations) will increase the separation between the different components and afford higher sample loading, but will require a considerable longer separation time. The influence of solvent strength on the resolution of a standard mixture has been studied and a linear decrease of resolution was observed when increasing the solvent strength. Running a gradient is also possible with MPLC, provided a suitable solvent delivery system is used. Peak sharpening can be obtained by a simple stepwise change of mobile phase composition.

Evaporation of large quantities of solvent takes place after fraction collection in order to concentrate the purified compounds. This procedure can cause the accumulation of considerable amounts of nonvolatile impurities from the solvent. As high purity solvents are very expensive, preliminary distillation of ordinary grade solvents to prepare the eluent can be a good compromise between solvent purity and quantity employed. Use of such lower grade solvents often implies an additional purification step by gel filtration, for example.

## Sample Introduction

Several criteria have to be considered before sample injection on a MPLC column:

- sample preparation
- sample mass and volume
- solubility characteristics of the sample
- type of injection used.

If solubility is not a problem, the eluent should be chosen to dissolve the sample. However, even in such a case, care has to be taken to adjust the sample volume: a sample that is too dilute (injection of a large volume) results in decreased separation efficiency, while precipitation at the top of the column may be observed by injection of samples that are too concentrated. High concentrations of the sample may alter the viscosity of the solution, which is then very different from that of the mobile phase. High viscosity leads to severe tailing, while fronting may result from lower sample viscosity compared to the

mobile phase. Despite these inconveniences, injection of a small volume of a concentrated solution is usually preferred. Sample solubilization in a solvent different from the mobile phase is also possible, but special care has to be taken with such an approach: solubility after mixing the sample solution with mobile phase has to be checked in order to avoid sample precipitation on the top of the column. Samples can be injected either directly on the column through a septum, or by means of a sample loop. In both cases, injection success depends on the quality of the column packing to ensure a homogeneous distribution of the sample at the top of the column.

The various problems mentioned above are more frequently encountered with separations on reversed-phase columns. The mobile phase usually contains a large proportion of water and organic compounds are often encountered that are insoluble in water. Solid injection or solid introduction is an alternative to circumvent low sample solubility. The introduction mixture is prepared by mixing dry powdered sample with a suitable amount of column packing material. The sample can also be preadsorbed on stationary phase by removing the volatile solvent (e.g. dichloromethane, ethyl acetate, acetone, etc.) in which it was solubilized from the suspension containing the stationary phase. Homogeneity of the injection powder is a prerequisite for efficient separations. The proportions of the introduction mixture are generally one part sample mixed with two to five parts stationary phase. The prepared sample is then placed directly onto the column inside a small precolumn and the eluent is passed through the system for separation.

## Applications: MPLC in Natural Product Isolation

MPLC has recently become widely used in the pharmaceutical, chemical and food industries, and many applications are found in natural product isolation. Both applications given below have been selected as examples of the transposition of analytical HPLC conditions to MPLC.

The methanol extract of *Halenia corniculata*, a Gentianaceae plant from Mongolia, was first passed through a Sephadex LH-20 gel column and the glycoside-rich fraction (300 mg) was then purified by MPLC on a reversed-phase RP-18 column, yielding six xanthone glycosides (1–6). The search for optimal conditions was performed by analytical HPLC (**Figure 3A**) and was followed by direct transposition to MPLC separation (**Figure 3B**).

The dichloromethane extract from the roots of *Tinospora crispa* (Menispermaceae) was first

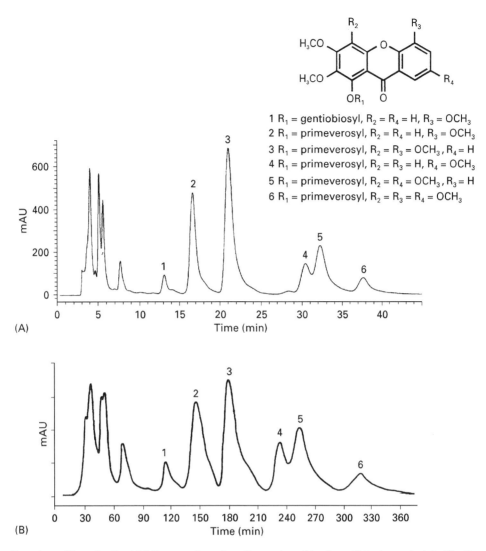

1 $R_1$ = gentiobiosyl, $R_2$ = $R_4$ = H, $R_3$ = $OCH_3$
2 $R_1$ = primeverosyl, $R_2$ = $R_4$ = H, $R_3$ = $OCH_3$
3 $R_1$ = primeverosyl, $R_2$ = $R_3$ = $OCH_3$, $R_4$ = H
4 $R_1$ = primeverosyl, $R_2$ = $R_3$ = H, $R_4$ = $OCH_3$
5 $R_1$ = primeverosyl, $R_2$ = $R_4$ = $OCH_3$, $R_3$ = H
6 $R_1$ = primeverosyl, $R_2$ = $R_3$ = $R_4$ = $OCH_3$

**Figure 3** Transposition of conditions for the MPLC separation of xanthone glycosides from *Halenia corniculata* (Gentianaceae). (A) Analytical HPLC on a Lichrosorb 7μm RP-18 (250 mm × 4 mm) column with MeOH/$H_2O$ 40 : 60 (v/v); flow rate 1 mL min$^{-1}$; (B) MPLC on Lichrosorb RP-18 (15–25 μm) with MeOH/$H_2O$ 40 : 60 (v/v); flow rate 3 mL min$^{-1}$; column dimensions 460 mm × 12 mm. (Reproduced with permission from Hostettmann *et al.*, 1997.)

fractionated by centrifugal partition chromatography and one fraction was submitted to analytical HPLC with an acetonitrile gradient (**Figure 4A**). Owing to the lower convenience of acetonitrile for preparative purposes (cost, toxicity), conditions were found with methanol on analytical HPLC (Figure 4B). The selected isocratic eluent system was transposed directly to MPLC and 6-h separation led to the isolation of three phenylpropane derivatives (7–9) (Figure 4C).

## Conclusion

Since the early 1980s MPLC has been confirmed as an excellent preparative chromatographic tool that is now routinely used in many laboratories. The extended use of various bonded phases in MPLC no longer restricts the use of this technique to the isolation of lipophilic substances with silica gel. For reasons of economy, recycling the stationary phase by simple washing or repacking of the column is of great interest in MPLC. Furthermore, a working experimental setup can be easily and rapidly assembled. The wide range of sample amounts that can be separated with this technique, together with the use of TLC and analytical HPLC in the search for optimal conditions, are also major benefits of this chromatographic method. However, good column packing and adequate sample preparation are prerequisites for successful separations. Further developments in MPLC

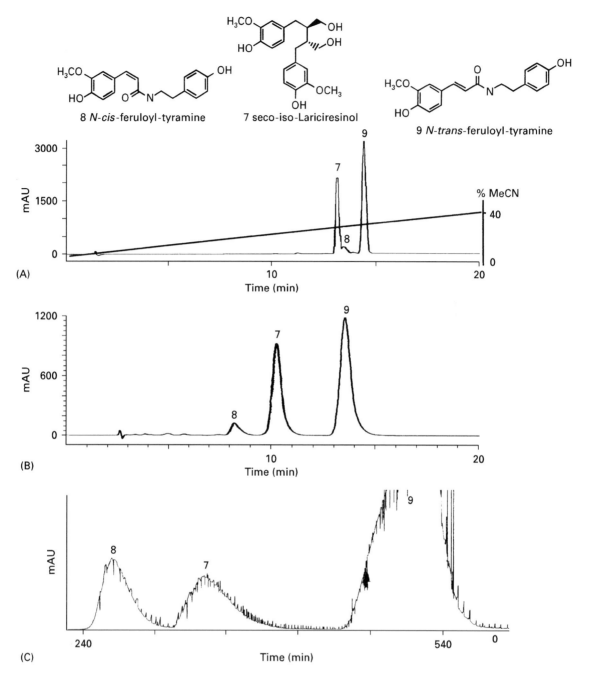

**Figure 4** Transposition of analytical HPLC conditions for MPLC separation of phenylpropane derivatives from *Tinospora crispa* (Menispermaceae). (A) Analytical HPLC on a Lichrosorb 7 μm RP-18 (250 mm × 4 mm) column with a MeCN/water gradient 0 : 100 to 40 : 60 (v/v) in 20 min; flow rate 1 mL min⁻¹; (B) analytical HPLC on a Lichrosorb 7 μm RP-18 (250 mm × 4 mm) column with MeOH/water 40 : 60 in 20 min; flow rate 1 mL min⁻¹; (C) MPLC on Lichrosorb RP-18 (15–25 μm) with MeOH/water 30 : 70; flow rate 4 mL min⁻¹; column dimensions 460 mm × 12 mm.

will mainly concern detection problems with the optimization of detectors that can accommodate high sample loads.

*See also:* **II/Chromatography: Liquid:** Large-Scale Liquid Chromatography. **III/Flash Chromatography. Natural Products:** Liquid Chromatography; Thin-Layer (Planar) Chromatography.

## Further Reading

Cavin A, Hostettmann K, Dyatmyko W and Potterat O (1998) Antioxidant and lipophilic constituents of *Tinospora crispa. Planta Medica* 64: 393–396.

Hostettmann K, Marston A and Hostettmann M (1997) *Preparative Chromatography Techniques – Applications in Natural Product Isolation*, 2nd edn. Berlin: Springer-Verlag.

Leutert T and Von Arx E (1984) Präparative Mitteldruck-Flüssigkeitschromatographie. *Journal of Chromatography* 292: 333–344.

Nyiredy S, Dallenbach-Toelke K, Zogg GC and Sticher O (1990) Strategies of mobile phase transfer from thin-layer to medium-pressure liquid chromatography with silica as the stationary phase. *Journal of Chromatography* 499: 453–462.

Porsch B (1994) Some specific problems in the practice of preparative high-performance liquid chromatography. *Journal of Chromatography A* 658: 179–194.

Rodriguez S, Wolfender J-L, Odontuya G, Purev O and Hostettmann K (1995) Xanthones, secoiridoids and flavonoids from *Halenia corniculata*. *Phytochemistry* 40: 1265–1272.

Verzele M and Geeraert E (1980) Preparative liquid chromatography. *Journal of Chromatographic Science* 18, 559–570.

Zogg GC, Nyiredy S and Sticher O (1989a) Operating conditions in preparative medium pressure liquid chromatography (MPLC). II. Influence of solvent strength and flow rate of the mobile phase, capacity and dimensions of the column. *Journal of Liquid Chromatography* 12, 2049–2065.

Zogg GC, Nyiredy S and Sticher O (1989b) Operating conditions in preparative medium pressure liquid chromatography (MPLC). I. Influence of column preparation and particle size of silica. *Journal of Liquid Chromatography* 12, 2031–2048.

# MEMBRANE CONTACTORS: MEMBRANE SEPARATIONS

**J. G. Crespo, I. M. Coelhoso and R. M. C. Viegas,**
Universidade Nova de Lisboa, Monte de Caparica,
Portugal

Membrane-based processes are receiving recognition for their flexibility and efficiency. Processes like reverse osmosis, ultrafiltration and dialysis are already well developed and recently, membrane application to other separation processes, such as absorption and liquid–liquid extraction, have been gaining considerable attention.

In these latter processes, the porous membrane acts as contacting media for gas–liquid or liquid–liquid phases with comparable advantages to the traditional continuous contact equipment. While in conventional two-phase processes, dispersion of one phase into another immiscible phase is used in order to promote an efficient contact and increase the transport rate, membrane extraction is accomplished without dispersion of the two phases.

Consider a liquid–liquid extraction process and a microporous hydrophobic membrane with the aqueous–organic interface stabilized inside the membrane pores (**Figure 1**). Since the membrane is hydrophobic, the organic phase spontaneously wets the membrane and may permeate through the pores to the aqueous phase. This breakthrough problem can be controlled by applying a higher pressure on the phase that does not wet the pores. This higher pressure must not exceed a critical value, $\Delta p_{cr}$, otherwise the nonwetting fluid will

penetrate the pores and contaminate the other fluid phase.

If a hydrophilic membrane is used, the procedure is analogous, but in this case it is necessary to impose an organic phase pressure which is higher than that of the aqueous phase.

Although extraction can be conducted using a number of different membrane configurations, including flat-sheet, spiral-wound, rotating annular and hollow fibres, hollow fibres have received the most attention due to their high packing density: typical interfacial areas of contact per unit volume range from 1500 to 7000 $m^2 m^{-3}$.

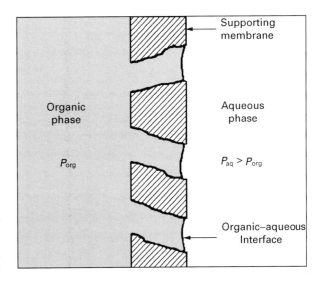

**Figure 1** Organic–aqueous interface immobilized in a micro-porous hydrophobic membrane. $P_{aq}$, aqueous-phase pressure; $P_{org}$, organic-phase pressure.

The associated advantages of this configuration, which will be discussed in detail in the next section, have led to the development of an enormous range of processes, either with liquid–liquid phases or with gas–liquid phases.

# Comparison between Membrane Contactors and Conventional Equipment

## Advantages of Membrane Contactors

**High contact area per unit volume**  Using a suitable module configuration such as a hollow fibre, membrane contactors can provide a contact area per unit volume which is 20–100 times higher than conventional equipment. The higher the interfacial area provided, the more efficient the contactor becomes and the smaller its size required for a given separation.

**No loading and flooding constraints**  As the fluids to be contacted flow on the opposite sides of the membrane, both flow rates can be set independently. The available contact area remains constant even at very low or very high flow rates. This feature is particularly useful in applications where the required solvent/feed ratio is very low or very high, in contrast to conventional equipment, which is subjected to flooding at high flow rates and unloading at low ones.

**Reduction of phase back-mixing**  When using membrane contactors, the mass transfer between the two phases occurs at the fluid–fluid interface immobilized at the mouth of the pores. This nondispersive contact minimizes emulsion formation and the occurrence of back-mixing is also reduced. By reducing back-mixing, a higher number of transfer units (NTU) can be achieved.

**No need for density between the phases**  Unlike conventional dispersed-phase contactors, no density difference is required between fluids in liquid–liquid extraction because coalescence and separation of the dispersed phase are not necessary when using membrane contactors.

**Reduced solvent hold-up**  Solvent hold-up is rather low when using membrane contactors; this may be important when using expensive solvents.

**Modular design – direct scale-up**  The modular character of this equipment allows an easy straightforward scale-up procedure. Membrane operations usually scale linearly and thus, when an application requires several contactors in series or parallel, this modular design allows a given process to be easily tested on a reduced scale.

**Easy process integration**  As membrane contactors do not involve the dispersion of the two fluid streams, it is easy to combine them with other operation units. These hybrid processes can be highly advantageous from a technical and economic point of view. For example, a combined extraction/stripping process can be designed by coupling two membrane contactors in series, without the need for an intermediate coalescence step.

## Limitations of Membrane Contactors

**Additional membrane resistance**  Besides the mass transfer resistances associated with the boundary layers of the two fluid phases, the membrane provides a third resistance. While this additional resistance is often negligible it may, under some conditions, contribute significantly to the overall mass transfer resistance. It will be discussed later in this chapter a few heuristic rules to minimize this effect.

**Fluid distribution in the shell side**  In hollow-fibre contactors the spatial distribution of the fibres is not perfectly uniform. This uneven distribution can induce fluid flow maldistribution in the shell side and, eventually, bypassing, especially when high flow rates are used. This problem may become particularly important when large scale modules are used.

**Transmembrane pressure constraints**  The transmembrane pressure can become quite important in porous membrane contactors, because it may induce flow across the membrane (breakthrough), causing unwanted froth, foam and dispersion between the two phases. For this reason, the option between operation with co- or counter-current mode and the setting of the fluid flow velocity has to take into consideration the pressure drop profile developed along the module. This explains why laminar flow conditions are usually used.

**Construction materials**  Membrane contactors employ polymeric membranes and potting adhesive resins to bond the fibre bundle to the module casing. These materials may have a limited compatibility with certain organic solvents, especially with aromatic compounds.

**Linear up-scaling factor**  It was mentioned as an advantage that the scaling-up of membrane contactors is rather simple and linear. On the other hand, this linearity is a serious drawback when comparing with traditional contacting equipment where the factor for scale-up cost is typically 0.6. This means that, if we want to double the capacity of a membrane contactor, the cost will be twice the original one, while for conventional equipment the cost will be $2^{0.6}$ of the original cost.

## Membranes and Modules

### Membrane Selection

Unlike most membrane operations, in membrane contactors the chemistry of the membrane is relatively unimportant, as it imparts no selectivity to the separation. The goal is to choose a membrane whose effect is not negative, i.e. that has no influence on mass transfer. Thus, the success of membrane contactors greatly depends on minimizing the membrane resistance to mass transfer. As a general rule, choose the membrane that is wet by the fluid to which the solute has more affinity (higher partition): if the solute partitions favourably to the solvent (organic), a hydrophobic membrane should be used; if it partitions favourably to the aqueous phase, then a hydrophilic membrane would be the best choice.

For gas–liquid contact two modes of operation are possible: wetted mode and dry mode. The wetted mode occurs when the pores are filled with liquid and the dry is when the pores are filled with gas: a hydrophilic membrane operates in wetted mode if the liquid phase is aqueous and in dry mode if it is organic, whilst a hydrophobic fibre will operate the inverse way. The dry mode is usually preferred to take advantage of the higher diffusivity of the solute in the gas phase, except in systems with an instantaneous interfacial reaction where the gas-phase resistance controls.

Still concerning the maximization of mass transfer, microporous membranes, typically with pore sizes between 0.05 and 1.0 μm and 20–100 μm thick have been used, in order to hinder solute diffusions as little as possible.

Hydrophobic membranes present the following advantages:

1. Higher pH and chemical stability
2. Reduced fouling with whole cells
3. Easier sterilizability

On the other hand, hydrophilic membranes are advantageous in the following conditions:

1. Systems with lysed cells or proteins, as they are likely to foul the membrane's surface to a lesser extent

2. Systems with very low interfacial tension. As hydrophilic membranes are commercially available with smaller pore sizes than the hydrophobic ones it can be easier to stabilize the interface (higher $\Delta p_{cr}$).

### Module and Operating Mode Selection

Due to its high packing density (providing interfacial areas of contact up to 7000 $m^2\ m^{-3}$), the hollow-fibre modules are the most attractive configuration.

Any design should be preceded by some preliminary considerations regarding the operating mode:

1. Should the feed stream flow in the tube or in the shell side of the module?
2. Should the extraction be carried out co- or counter-currently?
3. Should the operation be carried out in unsteady-state batch mode or in continuous mode?

As a general rule, the feed stream should circulate in the tube side whilst the extract should flow in the shell side. This observation stems from the fact that commercially available hollow-fibre modules still present deficient mass transfer in the shell side due to uneven distribution of the fibres, that can produce effects of channelling, bypassing and back-mixing on the shell side. With the aim of minimizing these problems, new modules have recently been marketed with baffles and better distribution of the fibres.

However, exception may be considered when dealing with feeds containing solids or a high degree of particles. Although cost considerations should be taken into consideration, a prior filtration is suggested.

Regarding co- or counter-current operation mode, although an attractive higher driving force could be attained with the latter, the stability of interface must be taken into consideration: when operating counter-currently, the transmembrane pressure difference along the module presents a higher variation which can interfere with the interface stability and even lead to breakthrough. The breakthrough pressure, $\Delta p_{cr}$, is determined by the pore size of the membrane, the interfacial tension between the two fluids and the contact angle, according to the Laplace equation:

$$\Delta p_{cr} = \frac{2\gamma \cos \theta}{r_p} \qquad [1]$$

As to operating in batch recirculation or continuous contact mode, attention must be paid mainly towards the degree of extraction needed: the batch mode cannot achieve an extraction beyond the final equilibrium concentrations of both phases, while the single-pass continuous counter-current operation can reach further extraction, depending on the fibre length and flow rates. For not such a high degree of extraction, the design equations must be looked at so as to choose the mode that minimizes the costs inherent in the desired extraction.

### Commercially Available Membrane Contactors

The best known module is the Liqui-Cel® Extra-Flow, marketed by Celgard LLC (**Figure 2**). This module uses Celgard microporous polypropylene fibres, up to 22 500, that are woven into a fabric and wrapped around a central tube feeder that supplies the shell side fluid. It also contains a central shell side baffle that improves efficiency by minimizing shell side bypassing and provides a component of normal velocity to the membrane surface, which results in higher mass transfer coefficients than those achieved with strictly parallel flow. The larger modules can operate with liquid flow rates up to several thousand litres per minute.

Also, commercial hollow-fibre microfiltration and ultrafiltration modules with hydrophilic membranes, generally with parallel flow, can also be used as membrane contactors.

For bubble-free gas–liquid mass transfer applications, Membrane Corporation and W.L. Gore market modules with different nonporous membrane arrangements: the first with the fibres potted at one end only and individually sealed at the other end, so that all entering gas permeates the membrane; the second has the fibres arranged as a helix, offering higher shell side mass transfer coefficients than the parallel configuration.

## Equipment Design

A hollow-fibre membrane contactor is a continuous contact equipment and so the well-known concept of mass transfer unit is also applied here:

$$L = HTU*NTU \qquad [2]$$

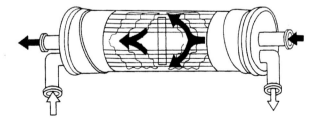

**Figure 2** Schematic representation of a Liqui-Cel Extra-Flow membrane contactor.

where $HTU$ is the height of the transfer unit and $NTU$ is the number of transfer units necessary for a given separation. This equation allows the evaluation of the height or length of the contactor necessary to obtain the required extent of mass transfer.

The mathematical description of this design equation can be illustrated for an extraction process where the aqueous phase circulates in the tube side. In this case, a differential mass balance to the hollow-fibre module can be used to determine the change in solute concentration during a single pass:

$$-Q_{aq} \cdot dC_t = K_t \cdot dA_m \cdot (C_t - C_t^*) \qquad [3]$$

where $K_t$ is the overall mass transfer coefficient, $Q_{aq}$ represents the aqueous-phase flow rate, $A_m$ the membrane transfer area, equal to $\pi.d_i.L.n_f$, where $n_f$ is the number of fibres, and $C_t$ is the solute concentration in the tube side phase. The superscript * refers to the solute concentration in the tube side (aqueous phase) in equilibrium with the solute concentration in the shell side (organic phase).

Eqn [3] can be integrated for the fibre length. For the module inlet ($z = 0$) $C_t = C_t^{in}$ and for the outlet ($z = L$) $C_t = C_t^{out}$, where $C_t^{in}$ and $C_t^{out}$ are the solute concentrations entering and exiting the module, respectively:

$$L = HTU*NTU = \frac{v_{aq}}{K_t a_i} \int_{C_t^{in}}^{C_t^{out}} \frac{dC_t}{C_t^* - C_t} \qquad [4]$$

where $v_{aq}$ is the fluid velocity circulating in the tube side, $K_t$ is the module-averaged overall mass transfer coefficient and $a_i$ is the interfacial area per unit module volume.

If a constant partition coefficient, $P$, can be assumed during the extraction process, integration of eqn [4] using $C_t^* = C_s/P$ where $C_s$, the solute concentration in the shell side phase, is obtained by mass balance, yields an analytical expression for the contactor length. The $NTU$ expressions for a hydrophobic membrane with aqueous phase in fibre lumen and organic phase in the shell side, respectively for co-current flow and counter-current flow are the following:

$$NTU = \frac{1}{1 + \frac{Q_{aq}}{Q_{org}P}} \ln \frac{C_t^{out} - C_s^{out}/P}{C_t^{in} - C_s^{in}/P} \qquad [5]$$

$$NTU = \frac{1}{1 - \frac{Q_{aq}}{Q_{org}P}} \ln \frac{C_t^{in} - C_s^{out}/P}{C_t^{out} - C_s^{in}/P} \qquad [6]$$

For gas–liquid separations the partition coefficient may be replaced by $H$, Henry's law constant, and the aqueous and organic flow rates replaced by liquid and gas flow rates.

For systems with a variable partition coefficient it is necessary to introduce the equilibrium relation between $C_t^*$ and $C_s$ and a numerical integration is required.

## Evaluation of Mass Transfer Coefficients

Three individual mass transfer resistances may be considered in membrane contactor extraction processes:

1. the inside tube boundary layer resistance
2. the membrane resistance to the solute diffusion through the pores
3. the shell side boundary layer resistance

The resistances are inversely proportional to the local mass transfer coefficients and a function of the system's geometry. Thus, for a hollow-fibre system when the membrane is wetted by the shell side phase, we obtain:

$$\frac{1}{K_t \cdot A_i} = \frac{1}{k_t \cdot A_i} + \frac{1}{P \cdot k_m \cdot A_{lm}} + \frac{1}{P \cdot k_s \cdot A_o} \qquad [7]$$

where $K_t$ represents the overall mass transfer coefficient (based on the tube side phase), $k_t$, $k_m$ and $k_s$ are

the local mass transfer coefficients on the tube side, membrane and shell side, respectively, and $A_i$, $A_o$ and $A_{lm}$ are the fibres' internal, external and logarithmic mean areas, respectively.

As the membrane may be hydrophobic or hydrophilic and the aqueous phase may circulate either in the fibre lumen or in the shell side, four different expressions for the overall mass transfer resistance for liquid–liquid extraction can be determined. Figure 3 shows the concentration profiles and the overall mass transfer resistances.

The tube side and the shell side mass transfer coefficients can be obtained experimentally and several correlations may be found in the literature.

## Mass Transfer Correlations

Since laminar flux is predominant in hollow-fibre membrane contactors, a Lévêque type equation can be used to correlate both the tube side and the shell side mass transfer coefficients:

$$Sh_t = \alpha \cdot Sc_t^{bt} \cdot Re_t^{ct} \cdot \left(\frac{d_i}{1}\right)^{1/3} \qquad [8]$$

$$Sh_s = \beta \cdot Sc_s^{bs} \cdot Re_s^{cs} \cdot \left(\frac{d_h}{1}\right) \qquad [9]$$

where the subscripts $t$ and $s$ refer to the tube and shell sides, respectively, and $\alpha$ and $\beta$ are constants. The

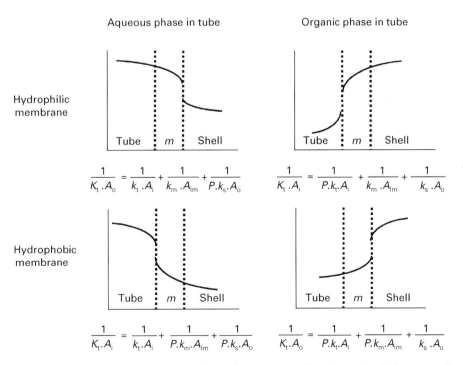

**Figure 3** Concentration profiles and overall mass transfer resistance expressions for a chemical system with a solute partition coefficient favourable to the aqueous phase ($P < 1$).

**Table 1** Mass transfer correlations for the tube side

| Equation | Characteristics | Reference |
|---|---|---|
| $Sh = 1.62\ Gz^{0.33}\ Gz > 25$ | Theoretical; laminar flux | Lévèque MC (1928) Les lois de transmission de chaleur par convection. *Annal. Mines* 13: 201 |
| $Sh = 1.86\ Gz^{0.33}\ Gz > 100$ | Empirical; laminar flux, heat transfer | Sieder EN and Tate GE (1936) Heat transfer and pressure drop of liquids in tubes. *Ind. Eng. Chem.* 28: 1429 |
| $Sh = 1.64\ Gz^{0.33}\ 30 < Gz < 2000$ | Hollow-fibre module; gas–liquid extraction | Yang MC and Cussler EL (1986) Designing hollow-fiber contactors. *AIChE J.* 32: 1910 |
| $Sh = 1.5\ Gz^{0.33}$ | Hollow-fibre module; liquid–liquid extraction | Dahuron L and Cussler EL (1988) Protein extraction with hollow fibers. *AIChE J.* 34: 130 |
| $Sh = 1.4\ Gz^{0.33}\ 50 < Gz < 1000$ | Hydrophobic fibre; liquid–liquid extraction | Takeuchi H, Tahamashi K and Nakano EM (1990) Mass transfer in single oil containing microporous hollow fiber contactors. *Ind. Eng. Chem. Res.* 29: 1471 |
| $Sh = 0.2\ Re\ Sc^{0.33}(d_i/L)^{0.33} Gz < 65$ | Hollow-fibre module; liquid–liquid extraction | Viegas RMC, Rodríguez M, Luque S, Alvarez JR, Coelhoso IM and Crespo JPSG (1998) Mass transfer correlations in membrane extraction: analysis of Wilson–Plot methodology. *J. Memb. Sci.* 145: 129 |

$Gz$: Graetz number $= Re\ Sc(d_i/L)$.

exponents of the Schmidt numbers $(Sc)$, $b_t$ and $b_s$, are usually 0.33; however, the exponents of the Reynolds numbers $(Re)$, $c_t$ and $c_s$, may be different from that value.

**Tables 1** and **2** show some correlations collected from the literature for both tube side and shell side mass transfer coefficients in hollow-fibre membrane contactors.

**Table 2** Mass transfer correlations for the shell side

| Equation | Characteristics | Reference |
|---|---|---|
| $Sh = 1.25(Re d_h/L)^{0.93} Sc^{0.33}$ | Hollow-fibre module; gas–liquid extraction | Yang MC and Cussler EL (1986) Designing hollow-fiber contactors. *AIChE J.* 32: 1910 |
| $Sh = 8.8(d_h/L) Re Sc^{0.33}\ Re < 100$ | Hollow-fibre module; liquid–liquid extraction | Dahuron L and Cussler EL (1988) Protein extraction with hollow fibers. *AIChE L.* 34: 130 |
| $Sh = 5.85(1 - \phi)d_h/L) Re^{0.6} Sc^{0.33}$ $\phi < 0.2\ Re < 500$ | Hollow-fibre module; liquid–liquid extraction | Prasad R and Sirkar KK (1988) Dispersion-free solvent extraction with microporous hollow fiber modules. *AIChE J.* 34: 177 |
| $Sh = 0.85(d_h/L)^{0.25}$ $(d_e/d_s)^{0.45} Re^{0.33} Sc^{0.33}\ Re < 700$ | Hydrophobic fibre; liquid–liquid extraction | Takeuchi H, Tahamashi K and Nakano EM (1990) Mass transfer in single oil containing microporous hollow fiber contactors. *Ind. Eng. Chem. Res.* 29: 1471 |
| $Sh = 0.017$ $(d_e/d_s)^{0.57} Re^{0.8} Sc^{0.33}\ 700 < Re < 2000$ | Hydrophobic fibre; liquid–liquid extraction | Takeuchi *et al.* (1990) |
| $Sh = (0.53 - 0.58?) Re^{0.53} Sc^{0.33}$ | Hollow-fibre module; gas–liquid extraction | Costello MJ, Fane AG, Hogan PA and Schofield RW (1993) The effect of shell side hydrodynamics on the performance of axial flow in hollow fibre modules. *J. Memb. Sci.* 80: 1 |
| $Sh = 8.7\ Re^{0.74}(d_h/L)\ Sc^{0.33}\ Re < 50$ | Hollow-fibre module; liquid–liquid extraction | Viegas RMC, Rodríguez M, Luque S, Alvarez JR, Coelhoso IM and Crespo JPSG (1998) Mass transfer correlations in membrane extraction: analysis of Wilson–Plot methodology. *J. Memb. Sci.* 145: 129 |

$d_e$, external fibre diameter; $d_s$ shell diameter; $d_h$ hydraulic diameter; $\phi$ packing factor $= n_t d_e^2/d_s^2$.

Concerning the tube side, in most published works an exponent of $c_t = 1/3$ is usually obtained for a higher tube side Reynolds numbers range. However, using tube side Reynolds numbers in a low range ($Re < 50$ and $Gz < 100$), values of $c_t = 1$ were reported.

For the shell side, the values of the exponent of the Reynolds number are $0.5 < c_s < 1$. Values of 0.5 for laminar flow and 0.6 for turbulent flow on the shell side of a shell and tube heat exchanger are reported. Deviation from these values may be due to the nonuniform distribution of the fibres and their deformation by action of organic solvents, both inducing an irregular flow due to the formation of stagnant zones, preferential pathways and deficient mixing.

## Applications

### Liquid–Liquid Extraction

Liquid–liquid extraction cover quite a broad range of applications, including metal extraction, wastewater treatment and extraction of pharmaceutical and other products of biotechnological interest, such as organic acids and proteins.

Recovery of metals from industrial process wastewater is important, not only because metals are valuable, but also because of environmental legislation restrictions. Several examples of metal extraction (Cu, Zn, Ni, Cr(VI), Cd) using membrane contactors, have been reported. Reactive extraction is usually employed and extractants such as organophosphorous compounds (TOPO, $D_2$EHPA), tertiary amines (tri-n-octylamine) and liquid ion exchangers (Aliquat 336) are used.

Extraction of pollutants from wastewater such as phenol, toluene and volatile organic compounds (VOCs) using methyl isobutyl ketone (MIBK), hexane and kerosene as solvents and also reactive extraction with several extractants were reported. A pilot-scale plant for extraction of chlorinated and aromatic compounds from industrial wastewaters in the Netherlands was in operation for periods up to 3 months, reducing contaminant levels to 10 μg L$^{-1}$.

Reactive extraction of organic acids produced by fermentation, such as acetic, citric, lactic and succinic acid, were also reported. Extraction of amino acids using reversed micelles was also studied.

Protein extraction can be accomplished either by two-phase aqueous extraction or by reversed micelles. Problems of interface stabilization caused by emulsions due to the adsorption of surfactant to the membrane surface, thus lowering the interfacial tension, were reported. Careful control of the pressure difference across the membrane was required for a stable operation.

### Gas–Liquid Contactors

In gas–liquid extraction with membrane contactors, most efforts have been conducted in the areas of gas absorption/stripping and of wastewater treatment. Other fields like dense gas extraction and semiconductor cleaning water have been the object of study more recently.

Commercial applications include the Pepsi bottling plant in West Virginia that has been operating a bubble-free membrane-based carbonation line since 1993, showing reduced foaming, improved yield, lower $CO_2$ pressures and increased filling speed at high temperatures. Also, several beer production plants are using this technology for $CO_2$ removal to obtain a dense foam head, while others remove oxygen from beer to preserve its flavour; the stripping of oxygen from water, which is then used to dilute beer, is also being applied. Other commercial applications include the treatment of boiler feedwater, stripping of $CO_2$ from anion exchange feed streams and ultrapure water production for semiconductor manufacture.

In wastewater treatment, air stripping of VOC has been studied using polypropylene hollow fibres to remove chloroform, tetrachloroethylene, carbon tetrachloride, 1,1,2-trichloroethane and trichloroethylene from aqueous streams. Also, several studies have reported the use of membrane contactors for bubble-free aeration in wastewater treatment. Advantages include the absence of foaming, higher aeration rates, lower power input and ability to handle solids.

A list of applications is summarized in **Table 3**.

### Membrane Distillation

In membrane distillation a nonwetted hydrophobic microporous membrane separates two phases of different water chemical potential. The difference in the water chemical potential may be induced by a temperature difference of the aqueous

**Table 3** Applications of membrane contactors on gas absorption/stripping processes

| Application |
| --- |
| $SO_2$, $CO_2$, CO and $NO_x$ removal from flue gases |
| $CO_2$ and $H_2S$ removal from natural gas |
| $CO_2$ removal from biogas |
| VOC removal from offgas |
| $NH_3$ removal from air in intensive farmery |
| Recovery of volatile bioproducts |
| $O_2$ transfer in blood oxygenation and in aerobic fermentation |
| Ultrapure water production for semiconductor manufacturing |
| Dense gas extraction |
| Separation of saturated/unsaturated (ethane/ethylene) |

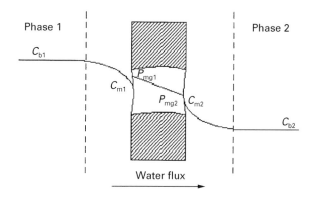

**Figure 4** Schematic representation of water transport in osmotic distillation.

solutions corresponding to a difference in vapour pressure at both ends of the membrane. It may also be due to the different nature and concentration of the solute components of both phases, thus causing a different osmotic pressure of the two liquids. Since the osmotic pressure of an electrolyte solution is about 10 times higher than an equimolar solution of electrically uncharged particles, salt solutions (NaCl, $CaCl_2$, $KH_2PO_4$) are efficient and relatively cheap systems to create high osmotic pressure differences. In both cases, water evaporates in the solution of higher chemical potential and the water vapour formed is transported across the membrane pores before being condensed in the solution of the lower water potential (**Figure 4**).

Solutions whose vapour pressure is relatively unaffected by the presence of the solute are ideal candidates for concentration using osmotic distillation. This includes solutions consisting primarily of sugars, such as fruit juices.

Two pilot-plant facilities located in Melbourne and Mildura (Australia) are successfully operating for concentration of fruit juices. Colour, flavour and aroma retention is good due to the lower operating temperature and stresses.

Grape juice concentrates used for production of high quality wines may also be concentrated by membrane distillation. Since these concentrates are stable for long periods of storage they can be shipped over long distances and high priced wines can be produced in regions where these grapes are not available or are too expensive.

Opportunities also exist for the concentration of pharmaceutical products, which are susceptible to thermal degradation.

### Biphasic Membrane Bioreactors

Multilayer membrane bioreactors can readily be constructed from several membrane films to which different biocatalysts have been attached or from combinations of permselective and catalytic membranes.

The use of permselective membranes in conjunction with catalytic films (or catalytic compartments) makes possible a high degree of control over the fluxes of the reaction participants, and hence over the course of reaction, that is impossible to achieve with catalyst particles. By using an adequate permselective membrane the fluids on either side of a membrane can be segregated, thus providing an additional degree of freedom in reactor design.

Most research work has been oriented to the development of biphasic membrane bioreactors where a microporous (hydrophobic or hydrophilic) membrane is used to separate an aqueous from an organic compartment. In this way, two immiscible liquid phases can be contacted across a membrane without one of the phases having to be dispersed in the other, as is required in most conventional multiphase reaction systems. In this type of reactor the biocatalyst may be linked to the membrane (**Figure 5**) or dispersed in one of the bulk phases.

The opportunity for development of biphasic membrane bioreactors is quite clear: the demand for selective removal of defined pollutants and the need for enantioselective transport and reaction for chiral synthesis in the pharmaceutical and food industries require new approaches in this field. Special attention has been devoted to the development of biphasic membrane bioreactors for enzymatic esterification and hydrolysis reactions.

## Future Developments

Membrane contactors are unique equipment for promoting mass transfer while avoiding the dispersion of the fluid phases involved. This article has briefly reviewed the potential of membrane contactors in different areas of application and the problems which are still to be solved.

The industrial future of membrane contactors for liquid–liquid extraction processes, and in some defined situations for gas absorption, will depend very strongly on the ability to synthesize specific carriers or receptors with the potential to achieve recognition of individual solutes. Therefore, the trend will be the development of very selective carriers, in some cases with the ability for chiral recognition.

Membrane stability, in the sense of avoiding contamination between the two contacting fluids, is of major importance for the penetration of membrane contactors in some industrial markets. In particular, in the food and the pharmaceutical industries, trace contamination between the two fluid phases is a sufficiently strong reason to reject this type of process.

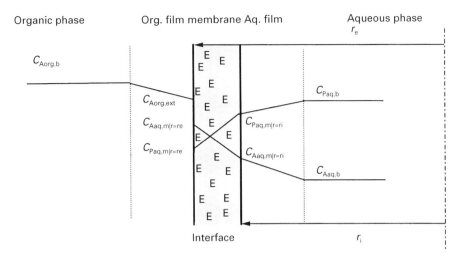

**Figure 5**  Representation of the enzymatic conversion of a substrate (subscript A) soluble in the organic phase to a product (subscript P) soluble in the aqueous phase. The enzyme is entrapped inside the porous structure of a hydrophilic membrane.

Development of organic solvents which are insoluble in water, nonvolatile and without the tendency to form emulsion would be highly desirable. Recently, the development of ionic liquids has been reported in the literature. These are entirely comprised of ions (complete absence of water), and are nonvolatile and insoluble in water. This type of solvent opens a world of new opportunities for liquid–liquid extraction and gas absorption using membrane contactors, without the risk of fluid cross-contamination. Also the use of dense gases and supercritical fluids has been suggested in membrane contactors. Again, the problem of contamination with the solvent phase could be eliminated.

Finally, new module design and new manufacturing materials will be welcome for certain type of applications, especially when viscous fluids and aggressive solvents are used. Further increases in membrane contactor performance are expected with the use of hollow-fibre fabrics and baffled modules.

## Further Reading

Cussler EL (1994) Hollow fiber contactors. In: Crespo JG and Boddeker KW (eds) *Membrane Processes in Separation and Purification*. Dordrecht: Kluwer Academic Publishers.

Gabelman A and Hwang S-T (1999) Hollow fiber membrane contactors. *Journal of Membrane Science* 159: 61–106.

Hogan PA, Canning RP, Petersen PA *et al.* (1998) A new option: osmotic distillation. *Chemical Engineering Progress* 49–61.

Kunz W, Behabiles A and Ben-Aim R (1996) Osmotic evaporation through macroporous hydrophobic membranes: a survey of current research and applications. *Journal of Membrane Science* 121: 25–36.

Matson SL and Quinn JA (1992) Membrane reactors. In: Ho WSW and Sirkar KK (eds) *Membrane Handbook*. New York: Chapman & Hall.

Mulder M (1996) *Basic Principles of Membrane Technology*, 2nd edn. Dordrecht: Kluwer Academic Publishers.

Prasad K and Sirkar KK (1992) Membrane-based solvent extraction. In: Ho WSW and Sirkar KK (eds) *Membrane Handbook*. New York: Chapman & Hall.

Reed BW, Semmens MJ and Cussler EL (1995) Membrane contactors. In: Noble RD and Stern SA (eds) *Membrane Separations Technology. Principles and Applications*. Dordrecht: Kluwer Academic Publishers.

Viegas RMC, Rodriguez M, Luque S *et al.* (1998) Mass transfer correlations in membrane extraction: analysis of Wilson-plot methodology. *Journal of Membrane Science* 145: 129.

# MEMBRANE PREPARATION

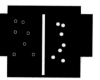

## Hollow-Fibre Membranes

**M. van Bruijnsvoort and P. J. Schoenmakers**,
University of Amsterdam, Amsterdam,
The Netherlands

## Introduction

Hollow-fibre membranes are an attractive alternative to conventional flat membranes in a growing number of important applications. Arguably, they have proliferated most in the biomedical and biochemical fields, thanks to their simple geometry and small dimensions in combination with their inherently high surface-to-volume ratio. Dialysis is the technique in which hollow-fibre membranes are most commonly employed.

There exist a number of extensive reviews on the technology, application and fabrication of membranes. A true classic is the book by Mulder, which contains a detailed overview of developments in the preparation of membranes and their use in purification. McKinney has provided a useful review of the preparation of organic hollow-fibre membranes. Tsapatsis has reviewed the preparation of inorganic membranes.

We have a particular interest in hollow fibres for polymer separations, with hollow-fibre flow field–flow fractionation (HF$_5$) as the analytical principle. Inorganic fibres are of great interest for polymers that require (strong) organic solvents. This article is biased due to this specific interest, our limited personal experience with different types of fibres and their preparation, and by our background as analytical chemists. The latter also provides us with an original perspective. To us there is an obvious analogy between the preparation of hollow-fibre membranes and open-tubular columns for chromatography. Both areas may benefit from such a comparison, presented in **Table 1**.

### Types of Membranes

Basically, three types of membranes are distinguished:

- porous membranes;
- nonporous membranes; and
- liquid membranes.

All three types can be used in the hollow-fibre geometry. Porous membranes allow the passage of relatively large molecules. Depending on the size of the pores, one speaks of microfiltration (pore size > 100 nm), ultrafiltration (pore size < 100 nm), or nanofiltration (molecular weight cut-off > ca. 1000 Da). Nonporous membranes are permeable to very small molecules (gases). Liquid membranes are of interest because of the selectivity and flexibility they provide. A broad review of liquid membranes has been provided by Sastre and co-workers.

Hollow-fibre membranes are often the preferred geometry, offering distinct advantages over flat or tubular (diameter larger than 1 mm) membranes, for a number of reasons:

- high surface-to-volume ratio;
- conceptual simplicity;
- easy incorporation in flow streams;
- broad availability.

In preparing hollow-fibre membranes, we must try and capitalize on these advantages. For example, if the surface-to-volume ratio is a key parameter, then narrow-bore fibres are most interesting.

Membranes are used in many different ways in analytical chemistry. Most of their applications are in the areas of sampling and sample preparation, thanks to the fundamental ability of semipermeable

**Table 1** Analogy between the properties of membranes and chromatographic columns

| Type of membrane selectivity | Chromatographic equivalent(s) |
| --- | --- |
| Preferential interaction and adsorption from a gas, liquid or supercritical fluid | Gas chromatography with solid or polymeric stationary phases (GC) Normal-phase and reversed-phase liquid chromatography (LC) Supercritical-fluid chromatography (SFC) |
| Size selectivity (sieving effect) | Molar sieve columns (GC) Size exclusion chromatography (LC) |
| Charge selectivity | Ion exchange chromatography Ion exclusion chromatography |
| Affinity membranes | Affinity chromatography |

membranes to differentiate between different materials (viz. matrix and analytes). This selectivity of the membrane can be based on molecular size, affinity, charge, or a combination of these properties. There are only a limited number of situations in which the actual analysis relies on the application of hollow-fibre membrane interfaces. These include the following:

- *Affinity chromatography*. Porous membranes combine a large surface area with a high permeability. By bonding specific groups to the surface, targeted species can be bound very strongly to the membrane. After the entire sample has been passed through the membrane, the analyte(s) can be removed. Affinity chromatography is of particular interest in the biochemical and biomedical areas. Since the membranes have a high permeability, the danger of degradation of vulnerable proteins is reduced. Moreover, the high fluxes possible allow more mass to be purified within the same period of time. Two extensive reviews (Roper and Josic) have been dedicated to this important field.
- *Extractions*. The intrinsic ability of hollow-fibre membranes to separate two distinct phases has found multiple applications in the field of extractions. Recently, Gabelman and Hwang published an extensive review on the subject of membrane contactors, i.e. membranes used to separate two immiscible phases.
- *Chiral separations*. The field of preparative chiral separations using hollow-fibre membranes is too interesting to remain unmentioned. In particular, in the pharmaceutical industry there is a great demand for the separation of racemic mixtures on a preparative scale. For example, a hollow-fibre supported liquid membrane can be used to separate two phases, with a chiral selector present in one of these. Alternatively, a chiral selector can be chemically bonded to the membrane surface, prepared in a manner similar to the membrane used for affinity chromatography.
- *Hollow-fibre flow field–flow fractionation* (HF$_5$). A technique in which the simplicity of hollow-fibre membranes is very advantageous is flow field–flow fractionation (flow FFF). The concept of flow FFF was brilliantly conceived by Giddings. In flow FFF, macromolecules or particles are injected into a channel with a porous wall and displaced by a flow perpendicular to its direction of movement. Based on the difference in average velocity lines that are occupied by particles with different diffusion characteristics (related to particle size or molecular mass), a fractionation is obtained. Ideally (i.e. for spherical particles), the re-

tention time is directly related to the diffusion coefficient of the analyte.

HF$_5$ was introduced by Carlshaf and Jönsson in 1988 as an instrumentally simple and cost-effective alternative to flow FFF channels with a flat configuration. However, during the 1990s only a limited number of papers have been dedicated to this promising technique. Flow FFF poses very high demands on the quality of the hollow-fibre membrane and, especially, on the fibre-to-fibre repeatability. This is definitely the most important reason for slow progress on the subject. Recently, Lee *et al.* showed an impressive fractionation of latex particles (**Figure 1**). For this type of fractionation their HF$_5$ system performed at least as good as alternative techniques, such as the more-established flat-channel flow FFF.

- *Isoelectric focusing*. In 1998, Korlach published an original article on pH-regulated electro-retention chromatography (ERC). This technique is quite similar to (electrical) FFF. A voltage is applied

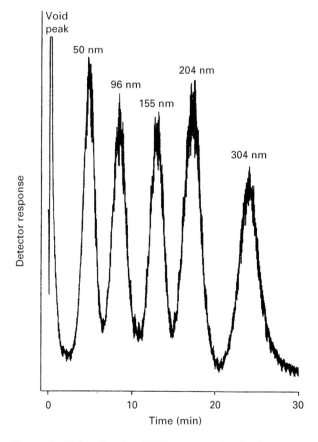

**Figure 1** Hollow-fibre flow FFF fractogram showing the separation of polystyrene latex beads. (Reprinted with permission from Lee WJ, Min BR and Moon MH (1999) Improvement in particle separation by hollow fiber flow field-flow fractionation and the potential use in obtaining particle size distribution. *Analytical Chemistry* 81: 3446–3452.)

across the diameter of a hollow-fibre membrane. Subsequently, a pH gradient is applied in a fluid reservoir which surrounds the fibre. The gradient gradually diffuses into the hollow-fibre membrane. At some point (when the pH crosses the isoelectric point or p*I* value), the charge on the analyte reverses and the ions will start to migrate to the middle of the fibre, from where they will be rapidly eluted owing to the higher axial flow velocity. The bottleneck in the development of this technique, which is mainly applied to the separation of proteins, is adsorption of the analytes on the membrane.

Semipermeable membranes allow us to manipulate processes that take place at the interface between two (miscible or nonmiscible) phases. They can be used in the gas phase for sampling purposes (e.g. membrane-assisted headspace injection in GC, membrane inlet mass spectrometry, MIMS) or in the liquid phase for sample preparation (e.g. dialysis) or sample concentration. Membranes used for gas sampling usually consist of simple fibres made of polymeric materials, such as polysiloxanes. More sophisticated separations, such as affinity chromatography, require more sophisticated membranes with dedicated selectivities. Also when in contact with liquid phases, most of the fibres used in analytical chemistry are based on organic (polymeric) materials.

The most important characteristic of a membrane is its separation selectivity. Differences in permeability for different materials can be based on a sieving effect (size selectivity) or on the chemical structure of the membrane. Hydrophobic membranes are more permeable to (nonpolar) organic molecules, while hydrophilic membranes are more permeable to water. Membranes will also show adsorption effects. Certain materials (analytes or matrix components) may be preferentially contained in and on the membrane. In some cases this is a desired effect (affinity membranes are the obvious example). Where fibres are used for sampling or sample clean-up in analytical chemistry, adsorption effects are often undesirable.

## Preparation of Hollow-Fibre Membranes

The art of hollow-fibre-membrane preparation has taken a high flight in recent years. Nowadays, hollow-fibre membranes are attainable in a fantastic variety of membrane and support materials, pore sizes, diameters, thicknesses, etc. An overview of the techniques involved in the preparation of hollow-fibre membranes is given below.

The various processes for preparing fibres tend to be rather complex, and certainly are laborious and time-consuming. The most difficult step is the optimization of the process, viz. ensuring repeatable (in-house) and ultimately reproducible (transferable) results. Hence, for small-scale applications of hollow-fibre membranes, including all practical applications within analytical chemistry, the most sensible approach to the technology is to obtain suitable fibres from a commercial source.

### Preparing Fibres

Most commonly, polymer tubing is prepared by a process referred to as spinning. This can be seen as an extrusion process. A viscous polymer solution (or melted polymer) is pressed through a small hole, while a fluid (bore liquid) is pumped through the centre. This construction is known as a spinneret. It allows precise control of the fibre dimension, thickness, etc. On exiting the spinneret, the fibre is drawn through a coagulation bath, after which additional treatment steps may take place.

The coagulation bath may merely be used to solidify the polymer and stabilize the fibre, but it may also be used to deposit a membrane (see next section). Many different polymers can be processed this way and the porosity of the resulting microfiltration membrane is affected by a large number of parameters. A summary is provided in **Table 2**. In the case of porous fibres, the type of material used is usually of little relevance to the properties of the membrane. The latter are almost totally determined by the parameters of the pores (size distribution, shape) and the membrane thickness. The material is selected based on other criteria, such as compatibility with the materials (fluids) encountered in the application, mechanical strength and cost. A typical example of a hollow-fibre membrane is shown in **Figure 2**.

### Membrane Films

Polymeric membrane films are usually formed by transferring a polymer from the liquid phase (solution, melt or suspension) to the solid phase. Such a process is known as *phase inversion*. Alternatively, membranes can be formed by *in situ* polymerization. Several processes will be considered below in some detail.

Liquid membranes from a special class of membrane films. They are prepared simply by immersing a microporous hollow fibre in a liquid. The liquid membrane, supported by the hollow fibre, separates two phases; by adding a selective carrier to the membrane (carrier-mediated transport), an increased selectivity can be obtained.

**Table 2** Summary of variables affecting fibre properties

| Variable | Values | Effects |
| --- | --- | --- |
| Consumption of casting solution | Suspension of polymer in non-solvent ('dry spinning') <br> Melted polymer ('melt spinning') <br> Polymer solution ('wet spinning') | Great effect on fibre porosity (and other properties) |
| Spinning parameters | Extrusion rate (mass flux of solution) <br> Tearing rate (speed of drawing the fibre) <br> Bore fluid rate <br> Distance between spinneret and coagulation solution <br> Composition of coagulation bath <br> Temperature at the various stages | Determines fibre dimensions (internal and external diameters) <br> Significant effect on fibre porosity |
| Fibre treatments | Washing <br> Chemical modifications | Removes contamination <br> Determines selectivity |

**Film deposition techniques** In order to create a film of a polymeric material on the inside of the fibre, the complete fibre can be filled with a solution of a polymer. Upon evaporation of the solvent, a film will be formed (*static coating*). The critical step is the evaporation of the solvent. This must take place slowly and regularly, in order to obtain a membrane with constant properties throughout the fibre. After a film has been deposited, it may be stabilized by heat treatment or by *in situ* cross-linking.

Instead of an evaporation step, a solvent displacement step may be introduced. In this case, a nonviscous liquid is pumped through the column containing the (viscous) film of polymer solution. The newly introduced liquid may either be a solvent or a non-solvent for the polymer, but it must dissolve the initial solvent. The solvent is then either extracted from the film, or replaced by a non-solvent. In either case, phase inversion will occur and a solid polymeric layer is obtained.

A second method for depositing a film is to press a plug of a polymer solution or a liquid polymer through the fibre. Behind this plug, a polymeric film will be left on the wall (*dynamic coating*).

Important parameters determining the properties of the membrane include the solution thermodynamics of the specific polymer–solvent combination, the concentration of the polymer in the solution, the temperatures at various stages of the process, the rate of solvent evaporation (or the rate of replacement of the solvent by a non-solvent) and the presence of additives in the solution.

***In situ* polymerization** Using the processes of static or dynamic coating described above, it is also possible to deposit a film of a solution containing monomers or polymer precursors (pre-polymers). This has the considerable advantage of a low solution viscosity, allowing the formation of relatively thick films without the need to apply high pressures. The residual solvent needs to be evaporated slowly and carefully after the polymerization to avoid cracks in the film or radial differences in film thickness.

**Sol–gel deposition** A promising class of membranes is that of the inorganic membranes, which are resistant to a wide variety of solvents, including potent organic solvents such as tetrahydrofuran or hexafluoro-isopropanol. The most common technique of preparation is sol–gel deposition on a ceramic support. This procedure was developed in the early 1980s by Burggraaf's group. A sol of nanometre-sized γ-alumina particles is deposited on the wall of the fibre from an aqueous solution containing a small percentage of polyvinyl alcohol. The deposited sol

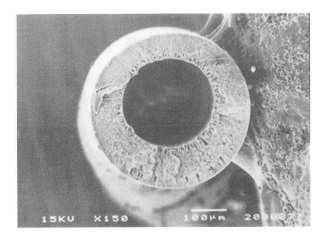

**Figure 2** Typical scanning electron micrograph of an asymmetrical composite hollow-fibre membrane. (Reprinted with permission from Chung TS, Teoh SK and Hu X (1999) Formation of ultra-thin high-performance polyethersulfone hollow-fiber membranes. *Journal of Membrane Science* 133: 161–175.)

can be converted to a defect-free membrane by sintering.

Okubo pointed out in 1991 that for the preparation of membranes inside a narrow-bore ceramic hollow fibre (1.4 mm inner and 2.0 mm outer diameter) dynamic coating procedures (see Film Deposition Techniques, above) are required. By forcing the sol through the hollow fibre, a stable and thick defect-free layer can be obtained. Reducing the diameter of hollow-fibre membranes is important, as it results in an increase of the surface-to-volume ratio, so that very small volumes can be separated or purified. Many applications in medicine, biology and analytical chemistry stand to benefit.

### Chemical Modification of Membranes

Polymers or reactive groups can be chemically bonded to suitable groups on the solid surface inside membrane pores. A large number of surface modification reagents are readily available and specific functional groups, to create the desired membrane selectivity, can be readily attached to such molecules. The process of chemically modifying membrane surfaces may also involve several reaction steps, which, however, may result in a less well-defined product.

Polymers can be attached to the surface ('grafted') through reactive functional groups, or through a pre-deposition radiation treatment. This process allows the creation of a great variety of selective membranes, which can be tailored for specific separations. These tailor-made phases allow affinity-type separations to be performed, in which extremely selective interactions are realized between the membrane surface and a selected analyte. Such interactions can be very strong, but desorption is possible by an appropriate change of conditions (displacing solvent or buffer).

Polymeric films can also be modified by introducing ionic groups, which drastically alters the properties of the membrane. Highly inert, nonpolar polymeric membranes, such as polyethylene or polytetrafluoroethylene (Teflon®), can be modified to yield highly polar, ionic interfaces.

## Hollow Fibres in Analytical Chemistry

As stated earlier, in most applications of hollow-fibre membranes in analytical science commercially available fibres are used. In only a limited number of cases (i.e. liquid membranes and affinity membranes) have analytical scientists reverted to in-house preparation. A nonexhaustive list of major suppliers of hollow-fibre membranes is provided in **Table 3**. Also, two suppliers of ceramic tubular membranes are listed in this table. Ceramic tubular membranes are not yet commercially available in diameters small enough ($d < 1$ mm) to be called fibres.

In **Table 4** we present a selection of recently developed applications of hollow-fibre membranes in analytical chemistry, with emphasis on the types of fibres used and their preparation or commercial availability. Here, we concentrate on the use of hollow-fibre membranes directly coupled with analytical techniques. The case in which the membrane solely determines the separation process has been covered above. Omitted from this table are instruments featuring on-line couplings of either dialysis of filtration units to standard high performance liquid chromatography (HPLC) instruments. These techniques are already well established and commercially available. A very useful review on the state of the art in the on-line coupling of dialysis to HPLC and capillary electrophoresis (CE) has been provided by van de Merbel.

In ion chromatography, hollow-fibre membranes can be used either before the analytical separation column (pre-column) for sample preparation, or post-column to suppress the conductivity of the effluent (mobile phase) prior to conductivity detection. In an interesting article, Kaufmann described the insertion of a hollow-fibre membrane between an HPLC

**Table 3**  Some major suppliers of hollow-fibre membranes

| Producer | Location | Web address (April 2000) |
|---|---|---|
| Dow-Corning | Midland, MI, USA | www.dowcorning.com |
| Hoechst Celgard | Wiesbaden, Germany | www.celgard.de |
| Minn Tech | Minneapolis, MN, USA | www.minntech.com |
| Millipore | Bedford, MA, USA | www.millipore.com |
| Sepracor | Marlborough, MA, USA | www.sepracor.com |
| A/G Technology | Needham, MA, USA | www.agtech.com |
| Tech-Sep[a] | Lyon, France | – |
| US Filter/Schumacher[a] | Asheville, NC, USA | schumacher-usa.com |

[a]Supplier of (tubular) ceramic membranes.

**Table 4** A selection of applications of hollow-fibre membranes in analytical chemistry, connecting specific applications (techniques and analytes) with types and sources of membranes, and useful reviews*

|  | Method | Compound | Author | Membrane type | Manufacturer |
|---|---|---|---|---|---|
| Gas phase extractions | GC (MESI) | Volatile organic carbohydrates | Mitra (1996)<br>Yang (1994) | Silicon<br>Silicon | Dow-Corning<br>Dow-Corning |
|  | MIMS | Volatile organic carbohydrates | Ketola (1998)<br>Cisper (1995)<br>Srnivisan (1997)*<br>Degn (1992)* | Silicon<br>Silicon | Dow-Corning<br>Dow-Corning |
| Liquid phase pre-column | $\mu$-LC | Bambuterol | Thordarsson (1996) | Polypropylene, 0.03 $\mu$m | Hoechst, Celanese |
|  | CE | Bambuterol | Palmarsdottir (1997) | Polypropylene, 0.2 $\mu$m | AKZO Nobel |
|  | CE | Organochlorides | Bao (1998) | Celgard X-10 | AKZO Nobel |
|  | CE | Proteins | Zhang (1997) | Cupruphan, 10 kDa | Hoechst Celanese |
|  | CE | Methamphetamine | Pedersen (1999) | Polypropene, 0.2 $\mu$m | AKZO Nobel |
|  | cIEF |  | Wu (1999)* |  |  |
| Post-column | Reaction | Barbiturates<br>Bromate | Haginaka (1987)<br>Inoue (1997) | AFS-2<br>Nafion | Dionex<br>Dupont |
|  | Ion exchange | Small anions | Hanaoka (1982) | Nafion | Dupont |
|  | Buffer exchange | Polyethylene glycol, proteins | Kaufmann (1993) | Cuprophan C1 | Akzo Nobel |
|  | Dialysis–electrospray MS | Proteins | Lutz (1999) | Regenerated cellulose, 13 kDa | Spectrum Medical Instruments |

Abbreviations: LC, liquid chromatography; CE, capillary electrophoresis; cIEF, capillary isoelectric focusing.

column and a detector for continuously exchanging buffer ions.

Following Davis, Haginaka's group has developed a hollow-fibre membrane-based post-column reactor. By immersing the fibre in, for example, an alkaline solution, the pH of the eluent can be altered to enable the detection of penicillins, amino acids, barbiturates, etc. However, since the process is diffusion-limited, a long fibre is usually required, leading to additional band broadening. Nonetheless, this is an elegant method to change the pH without the need to introduce an additional reagent stream and a post-column mixing coil.

Two important applications of hollow-fibre membranes are as sample preparation devices to extract specific analytes from a gaseous matrix. When the extracted components are directly fed into the ion source of a mass spectrometer, we speak of membrane-inlet mass sepectometry or MIMS. When the membrane serves as the inlet for a gas chromatograph various acronyms are used, of which MESI (membrane extraction with a sorbent interface) is the most common. The main objective of using membranes in these cases is to prevent large amounts of water (vapour) from entering the analytical instrument.

The membrane in MIMS can be used in different configurations. One of these involves a flat-disc membrane at the tip of a tubular probe, which is inserted or immersed in the sample or sample stream. In most cases, however, a tubular membrane is used. Polysiloxane tubes (of surgical quality) are the most popular. The sample can either flow through this tube or be on the outside. In the first case, the membrane tube will be inside the mass spectrometer. In the second case, the membrane forms the interface between the mass spectrometer and the outside (chemical) world. A purge gas may pass through the inside of the tube, or it may just be connected to the mass spectrometer vacuum system.

The membrane tubes used for mass spectrometry are usually 1–2 mm in diameter. Using narrower tubes or fibres will not lead to lower detection limits, as the response of a mass spectrometric system increases with the amount (mass) of sample introduced per unit time. In principle, the mass flow of sample is proportional to the tube diameter, so that larger tube diameters are more favourable in this respect. A very large area can also be obtained by using flat, folded membranes. The use of hollow-fibre membranes for MIMS has the advantage that very small samples or sample streams suffice. An efficient parameter by which to affect the sensitivity of the system is the thickness of the membrane (tube wall). The selectivity of the system may be influenced by varying the tube materials. MIMS is most commonly applied to liquid samples, although the concept is

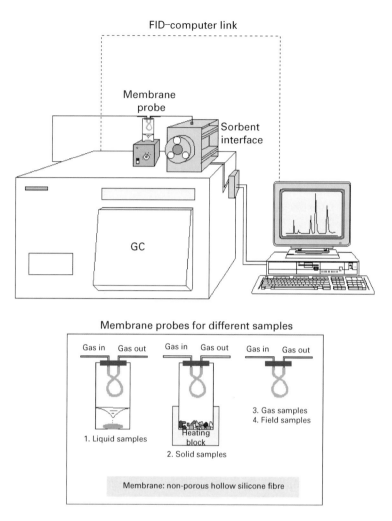

**Figure 3**    Principle of the MESI set-up. (With permission from the Web page ⟨http://sciborg.uwaterloo.ca/chemistry/pawliszyn/⟩.)

equally valid for gaseous samples. The concept is very attractive for introducing components into the mass spectrometer from aqueous samples, such as those encountered in biotechnology (e.g. measuring the amounts of gases in fermentation broths) or in waste management, but can also be a practical tool in the laboratory. There are at this time relatively few known applications of MIMS for process monitoring, although this is one of the most promising areas.

The principle of MESI is illustrated in **Figure 3**. Analytes, following selective passage through the membrane, are trapped onto a sorbent interface. After a sufficient amount of the analytes has been accumulated, these components are desorbed. In GC this can be done by rapidly increasing the temperature (thermal desorption). Finally, the analytes are separated on the GC column. Most commonly, the membrane probe is used to sample a gaseous phase, either a gaseous sample or sample stream, or the

headspace of a liquid or solid sample. However, there is no fundamental reason why a liquid (e.g. aqueous) phase cannot be sampled directly. The technique can elegantly be used for the field analysis of air.

An interesting trend is the online coupling of hollow-fibre membranes to modern miniaturized separation techniques, where the intrinsic small volumes of hollow-fibre membranes come fully to their right. The hollow-fibre membrane introduces selectivity between analytes and matrix components (sample preparation), and can be used to concentrate the analytes prior to analysis. A liquid membrane device for sample preparation, developed by Mathiasson and Jönsson in 1996, is shown in **Figure 4**. The fibre is positioned in a small channel ($d < 1$ mm) that serves as the donor compartment. From the receptor compartment, i.e. the lumen of the fibre, small volumes can be manipulated towards the attached separation devices through narrow-bore capillaries. The device has been coupled to both $\mu$-LC and CE, and has been

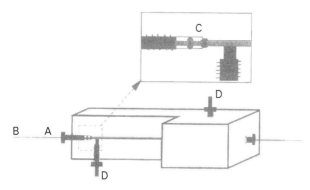

**Figure 4** A liquid membrane device for sample preparation. A, hollow fibre (reaching through a hole drilled through the whole block); B, fused silica capillaries inserted in the ends of the fibre; C, O-rings for fixing the fibre and capillaries; D, connectors for the donor channel. (Reprinted with permission from Thordarson E, Palmarsdottir S, Mathiasson L and Jonsson JA (1996) Sample preparation using a miniaturized supported liquid membrane device connected on-line to packed capillary liquid chromatography. *Analytical Chemistry* 68: 2559–2563.)

employed for the analysis of drugs in a matrix of blood plasma.

*See also:* **III/Membrane Preparation:** Interfacial Composite Membranes; Phase Inversion Membranes.

## Further Reading

Degn H (1992) Membrane inlet mass spectrometry in pure and applied microbiology. *Journal of Microbiological Methods* 15: 185.

Gabelman A and Hwang ST (1999) Hollow fibre membrane contactors. *Journal of Membrane Science* 159: 61.

Giddings JC (1991) *Unified Separation Science*. New York: Wiley.

McKinney R (1987) A practical approach to the preparation of hollow fibre membranes. *Desalination* 62: 37.

Mulder M (1991) *Basic Principles of Membrane Technology*. Dordrecht: Kluwer.

Roper DK and Lightfoot EN (1995) Separation of biomolecules using adsorptive membranes. *Journal of Chromatography A* 702: 3.

Sastre AM, Kumar A, Shukla JP and Singh RK (1998) Improved techniques in liquid membrane separations: an overview. *Separation and Purification Methods* 27: 213.

Tsapatsis M and Gavalas GR (1999) Synthesis of porous inorganic membranes. *MRS Bulletin* 24: 30.

van de Merbel NC (1999) Membrane-based sample preparation coupled on-line to chromatography or electrophoresis. *Journal of Chromatography A* 856: 55.

# Interfacial Composite Membranes

**J. E. Tomaschke**, Hydranautics Oceanside, CA, USA

## Introduction

The development of asymmetric cellulose acetate membranes in the 1960s was a breakthrough in membrane technology. These membranes consisted of a thin surface skin layer on a microporous support. The skin layer performed the separation required and because it was very thin fluxes were high. The microporous support provides the mechanical strength required. Following these developments Rozell *et al.* in 1967 described the preparation of the first interfacial (IFC) composite membranes. These membranes have since become the standard for reverse osmosis (RO) and nanofiltration (NF) applications.

IFC membranes have the same asymmetric status of the first-generation cellulose acetate membranes but are made by a very different procedure, shown schematically in **Figure 1**. In a first step a microporous polysulfone support membrane is impregnated with an aqueous solution containing a multifunctional amine. The impregnated membrane

is then contacted with a hexane solution containing a multifunctional acid chloride. Because the two solutions are immiscible the reactants can only combine at the membrane interface and so a thin polymer film layer forms at the surface. This layer performs the separation required.

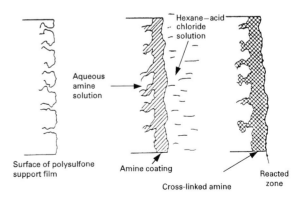

**Figure 1** Schematic of the interfacial polymerization procedure. (From Cadotte JE and Petersen RJ (1981) Thin-film composite reverse osmosis membranes: origin, development and recent advances. In Turbak AF (ed.) *ACS Symposium Series 153*, Washington, DC, pp. 305–326.)

IFC membrane development can be divided into two time periods. The earlier development from 1967 to approximately 1980 was characterized by work funded through the US Department of the Interior, whereas the majority of the development since 1980 has been industry funded. The early membrane preparations experienced a transition from the use of polymeric to monomeric amine reactants resulting in more durable products. Today the state-of-the-art IFC membrane chemistry consists of cross-linked aromatic polyamides derived from monomeric reactants. A review by Cadotte of composite RO membranes gives an account of the evolution of developments leading up to the commercialization of high performance membranes.

The basic interfacial method of membrane preparation using porous support has changed little since its inception though improvements in reactant chemistry and processing conditions have been made. The laboratory-scale preparation method remains a valuable intiation step in the development of IFC membranes because of its efficiency and simplicity. This method will be given a detailed discussion in the sections which follow.

## Interfacial Polycondensation

### Early Thin Film Synthesis

The origins of interfacial polycondensation reactions can be traced to Morgan of Du Pont who studied the interfacial polymerization of numerous polyamides and polyesters. He found the Schotten–Bauman reaction of diamines with acid chlorides to be an effective laboratory process, which was termed interfacial polycondensation. In this process, the irreversible polymerization of two highly reactive monomers takes place near the interface of the two phases of nonmiscible liquids, as demonstrated by the model system hexamethylenediamine sodium hydroxide–water/sebacoyl chloride–hydrocarbon solvent to produce Nylon 610. This model system is the basis for the discussion which follows.

When the two liquid phases containing diamine and acid chloride are brought together and the hydrocarbon or halogenated solvent, etc.) solvent is a nonsolvent for the final polymer, a thin film of the polymer will be formed rapidly at the liquid interface. Generally this polymer is found to be tough and of high relative molecular mass. In a very short time interval equivalent amounts of reactants combine nearly quantitatively, with elimination of hydrogen chloride, and produce a thin film. In Nylon 610 polymerization, the optimal molar ratio of diamine to diacid chloride was found to be about 6.5, indicating the rate-limiting

**Table 1**  Interfacial polycondensation variables

1. Reactivity of amine and acid chloride
2. Partition coefficient of amine water: organic solvent
3. Diffusion rate of amine into organic solvent
4. Concentration of reactants
5. Concentration ratio of reactants
6. Polymer film growth rate
7. Acid chloride hydrolysis rate
8. Polymer film permeability
9. Interfacial tension
10. Acid acceptor type
11. Surfactant type

feature of diamine diffusion across the interface and through the growing polymer film. Also noteworthy is the observation that polymer film growth occurs exclusively in the organic solvent phase owing to the extremely low solubility of acid chloride reactants in the aqueous phase. In general it is found that the mass transfer of the diamine is the rate-controlling step at all concentrations of reactants. The variables affecting interfacial polycondenzation determined from experimentation are listed in **Table 1**.

### Mechanism of Interfacial Polycondensation

The mechanism of membrane formation has been studied using the reaction between diamines and diacid chloride and can be generalized in eqn [1]:

$$H_2N-R-NH_2 \quad + \quad Cl-\underset{O}{\underset{\|}{C}}-R'-\underset{O}{\underset{\|}{C}}-Cl$$

$$\longrightarrow \left( \underset{H}{\overset{H}{\underset{|}{N}}}-R-\underset{H}{\overset{H}{\underset{|}{N}}}-\underset{O}{\underset{\|}{C}}-R'-\underset{O}{\underset{\|}{C}} \right)_n \quad + \quad 2HCl \qquad [1]$$

Normally in the interfacial polymerization, sodium hydroxide or other suitable base is added to the aqueous phase as an acid acceptor to neutralize the hydrogen chloride formed and drive the reaction to completion. In some systems excess diamine reactant can serve as the acid acceptor since amine hydrochlorides are highly water soluble and at the same time insoluble in hydrocarbon solvents. In addition to the simple difunctional reactants shown in eqn [1], trifunctional and combinations of di- and trifunctional reactants may be used to achieve the desired degree of polymer cross-linking.

Initially the polymer film grows rapidly; growth then slows and finally a constant film thickness is reached. This is due to the inability of the amine reactant to diffuse through the polymer film to react with the acyl halide. Enkelmann and Wegner described this process in eqn [2]:

$$\frac{dx}{dt} = K\frac{c}{x} - k'x \qquad [2]$$

where $x$ is the membrane thickness; $c$ is the concentration of diamine; $K$ is the diffusion coefficient of diamine through the membrane; and $k'$ is the rate constant of the inhibiting reaction ( $\propto$ acid chloride hydrolysis).

When the limiting thickness $x_\infty$ of the film is reached, $dx/dt = 0$ and eqn [2] simplifies to $x_\infty = \sqrt{Kc/k'}$. The limiting film thickness is therefore proportional to the square root of the diamine concentration. Subsequently, Enkelmann and Wegner established as a solution of eqn [2] the rate law of membrane growth:

$$x' = \frac{x}{x_\infty} = [1 - \exp(-2k't)]^{1/2} \qquad [3]$$

where $x'$ is the reduced membrane thickness. Solving in terms of $t$ (seconds) gives:

$$t = \log\frac{(1-x')^2}{-2k'} \qquad [4]$$

From this equation one can obtain values for early film growth from a period of seconds to over 10 min for more complete growth. It was found from this work that the limiting film thickness depended on both the absolute concentration and concentration ratio of the diamine and diacyl chloride reactants. Enkelmann also showed by X-ray diffraction techniques that in Nylon 610 membranes the polymer chains are ordered perpendicular to the interface. It was also concluded that membrane properties could be regulated by selecting particular reactive monomer ratio and concentrations, solvents and reaction times.

## Early IFC Membranes

### The IFC Membrane Structure

The development of IFC membranes is a logical outcome following the earlier development of asymmetric cellulose acetate (CA) membranes, as well as the previously discussed interfacial polycondensation work. The CA membrane is comprised of a soluble polymer or blend of polymers of varied cross-sectional morphology with the uppermost surface (skin) forming a permselective barrier. The IFC membrane, which is now the state-of-the art product, contains a microporous support layer of one polymer and a separate permselective skin or thin film of another polymer. The advantage of the IFC membrane is that the chemistry of the all-important permselective thin-film layer can be chosen independently from the underlying porous support material. Asymmetric membranes require polymers that are soluble in solvents necessary for the phase inversion process and this limits the number and type of polymers that can be utilized. Many useful crystalline, semicrystalline, and all cross-linked polymers are thus excluded from asymmetric membrane manufacture. The thin films of IFC membranes are in the range of 20–300 nm thick and when coupled with microporous supports of low hydrodynamic resistance provide membranes with unmatched productivity and solute retention.

### NS-100 and PA-300 Membranes

In the discussions of IFC membranes that follow, technical milestones are highlighted with emphasis on commercially significant developments. The early period of membrane development shown in **Table 2** began in 1967 with the investigation of various aqueous diamine and hexane–diacyl chloride interfacial solutions upon polysulfone porous supports by Rozelle *et al.* at North Star Research Institute. These first IFC membranes had low salt rejections, probably due to lack of film integrity since the resultant polymers were not cross-linked. This pioneering work, however, is significant in that the essential elements for the preparation of IFC membranes were demonstrated. Shortly thereafter, in 1970, the first high salt-rejecting IFC membrane, NS-100, was also developed at North Star Research. This membrane was made from polyethylenimine (PEI) in the aqueous solution and toluene diisocyanate (TDI) in the hexane solution. The coated and drained polysulfone support was subsequently dried at 110°C to yield a dry composite membrane with greater than 99% salt rejection on a synthetic seawater feed at 1000 psig (6.9 MPa). A later related membrane, designated NS-101, substituted isophthaloyl chloride (IPC) for TDI as the cross-linker and provided similar results. The selective layers in these membranes consisted of cross-linked polyurea and polyamide films, respectively. The membranes demonstrated high permselectivity but were mechanically delicate and highly vulnerable to attack by chlorine disinfectant.

The sensitivity of early interfacial membranes to chlorine attack was a serious problem that has still not been completely solved. Chlorine is routinely added to water to prevent bacterial growth on the membrane surface. However, exposure to even p.p.m. levels of chlorine destroyed the permselective layer of IFC membranes within a few hours.

**Table 2**  Early interfacial composite membrane developments

| Date | Development |
|------|-------------|
| 1967 | First IFC membranes investigated at North Star Research and Development Institute. |
| 1970 | NS-100 membrane |

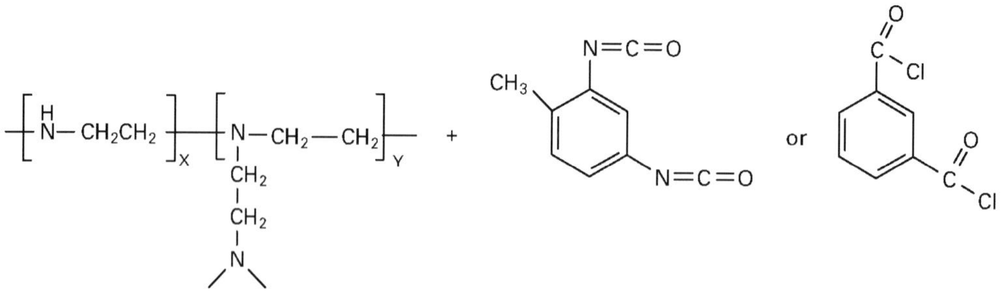

PEI    +    TDI    or    IPC

1975    PA-300 membrane

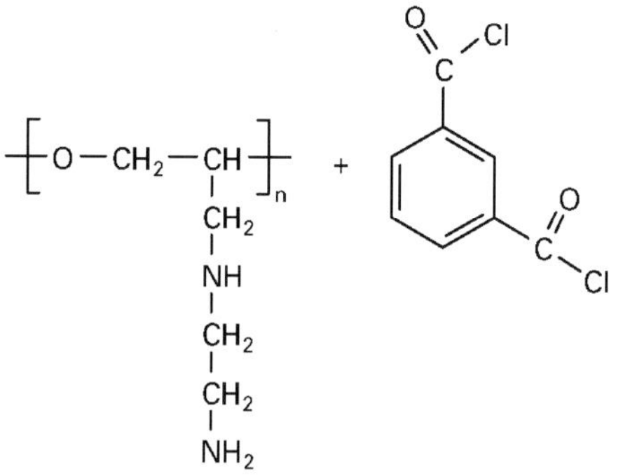

Polyepiamine        IPC

NS-300 membrane

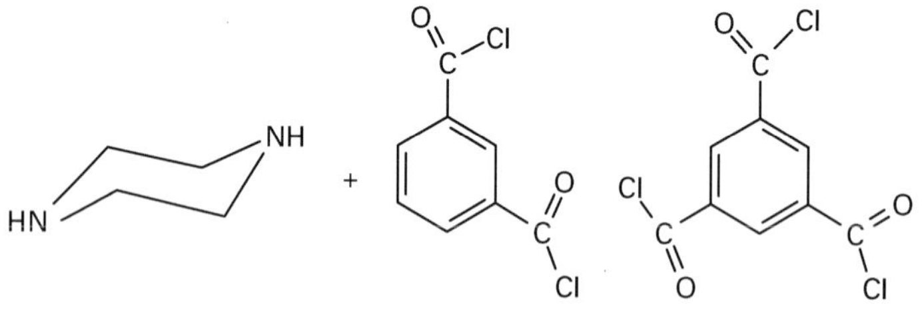

Piperazine        IPC        TMC

Another early membrane developed from a polyamine reactant was the PA-300 membrane by Riley *et al.* at UOP Fluid Systems Division in 1975. The advantage of this polyamide membrane prepared from IPC cross-linker was the lack of residual amines or amide functional groups in the polymer backbone, which exhibited improved chlorine tolerance. The performance of the PA-300 membrane was similar to that of the NS-100 and was the first IFC membrane to be utilized in a large-scale commercial desalination facility located in Jeddah, Saudi Arabia.

**NS-300 Membrane**

The last example of the earlier generation IFC membranes – the NS-300 – differed from its predecessors in that it was prepared from a difunctional *monomeric* amine, piperazine, and a trifunctional acyl chloride, TMC. This cross-linked polyamide membrane, developed at North Star division of Midwest Research Institute in 1975 by Cadotte *et al.*, demonstrated improved tolerance to chlorine compared to its predecessors due to absence of the vulnerable amidic hydrogen. Later variants of this membrane included addition of the difunctional IPC acyl chloride, which resulted in increased salt rejection and decreased flux. As might be expected, this is probably due to the decrease in residual carboxylic acid functionality resulting from decrease of the trifunctional TMC cross-linker. Another interesting structural aspect of this polyamide is the nearly 90° out-of-plane orientation of the piperazine ring relative to the aromatic ring. This rigid polymer structure containing a high volume geometry may in part account for the high permeability of this membrane.

## Contemporary IFC Membranes

### Performance Goals

The goal of further membrane development was to maximize solvent passage while at the same time minimizing solute passage. In a typical reverse osmosis desalination application, this means developing membranes with high water permeability yet low salt passage. This effort applies to nanofiltration membranes as well, except in this case passage of monovalent salts and organics of low relative molecular mass is preferred. Since the two performance properties of solvent flux and solute retention are competing, it is found in practice that one generally observes a trade-off in these values with membrane optimization. Both the thin film chemistry and morphology determine its transport properties. Additional goals of recent IFC membrane development include durability, chlorine and other oxidant stability, and fouling resistance.

### MPD-Based Membranes

**Table 3** lists recent significant IFC membrane developments. Beginning with the wholly aromatic polyamide FT-30 membrane developed by Cadotte at Film-Tec in 1978, it is seen that all of the subsequent membrane examples rely on the aromatic diamine monomer m-phenylene diamine (MPD). With the exception of the A-15 membrane, all of the MPD-based membranes provide very high salt rejection and similar water fluxes. Consistent with the general trade-off principle, the A-15 yields higher water flux with commensurately lower salt rejection, making it what is commonly called in the industry a 'loose RO' membrane. The cross-linked aromatic polyamide remains the-state-of-the-art in IFC membrane chemistry. Membranes of this kind are durable, hydrolytically stable, temperature stable, and exhibit high transport properties. A range of commercially successful membranes encompassing nanofiltration, brackish RO and seawater RO applications have

been achieved with the basic MPD/TMC reactants. These and other modern IFC membranes are made essentially by the same techniques of interfacial polymerization onto porous polysulfone substrates as were their predecessors.

## IFC Membrane Preparation

This section provides general information on how IFC membranes have been prepared and discusses guidelines for others to follow in preparing their own such membranes. The laboratory-scale preparations are discussed in an ordered sequence below with emphasis on techniques commonly practised in the desalination membrane industry for flatsheet IFC membranes. The basic principles of these interfacial techniques are also applicable to the less commercially significant hollow fibre IFC membranes or other composite membrane formats.

### Porous Support Preparation

The preferred polymer for use in porous support preparation is polysulfone, a moderately priced material with many desirable chemical and mechanical properties. In addition to strength and temperature stability, it is resistant to hydrolysis and oxidative attack. Its disadvantages, though relatively minor, are its hydrophobicity and lack of solvent resistance. The former property necessitates inclusion of surfactants or wetting agents for some aqueous coating methods used in IFC membrane manufacuture and the latter property limits its applications to ones which are predominantly aqueous or contain nonaggressive solvents such as alcohols and aliphatic hydrocarbons. Nevertheless, polysulfone has been and remains the polymer of choice for the porous support of RO and NF IFC membranes.

Preparation of the polysulfone microporous support may be carried out using laboratory, pilot, or full-scale production equipment. Regardless of scale, all of these procedures involve conversion of a polymer in solvent solution to a porous solid layer in what is called the phase inversion process. As the water in the gelation or solidification bath replaces the solvent, the clear polysulfone solution, or casting solution, is transformed to an opaque plastic layer on to the surface coated. With laboratory preparation this is normally carried out by applying a 14–18% polysulfone solution in N,N-dimethylformamide (DMF) onto a flat glass plate using a Gardner blade or other suitable device set with a blade gap of 0.13–0.26 mm, then immersing the plate into a small tank of water. For better strength and ease of later processing, it is advisable to do the solution coating onto a calendered polyester fabric or related material attached to the

**Table 3** Recent interfacial composite membrane developments

| Date | Development |
| --- | --- |
| 1978 | FT-30 membrane (US 4 277 344) |

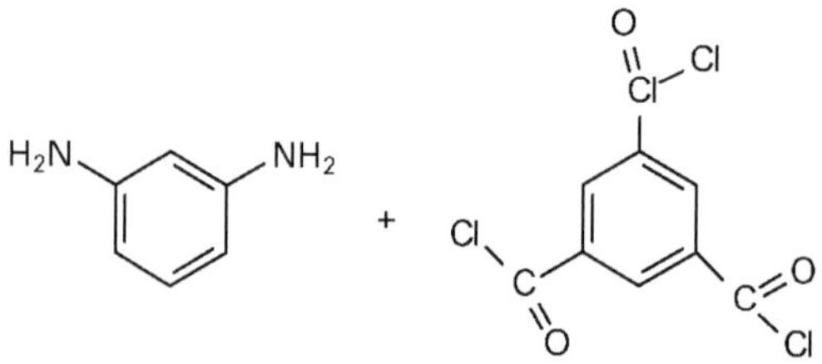

MPD                TMC

| 1984 | A-15 membrane (US 4 520 044) |

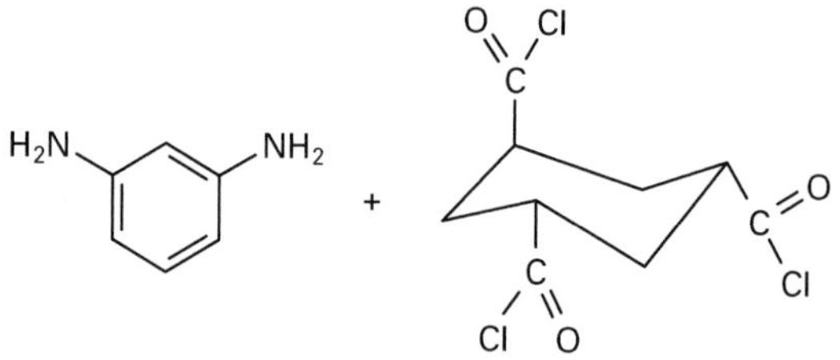

MPD                CHTC

| 1986 | SU-700 membrane (US 4 761 234) |

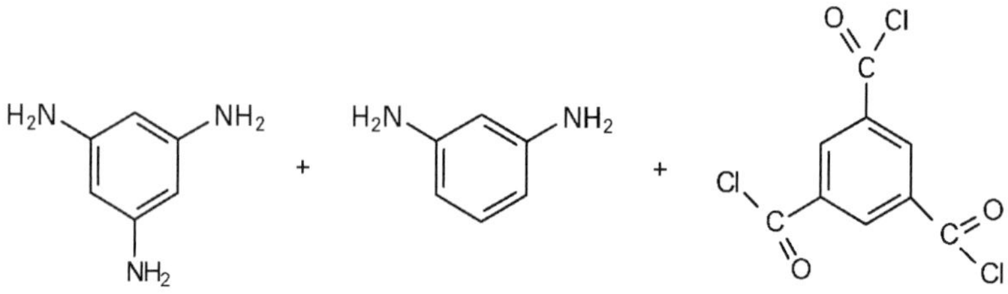

TAB          MPD          TMC

| 1990 | X-20 membrane (US 5 019 264) |

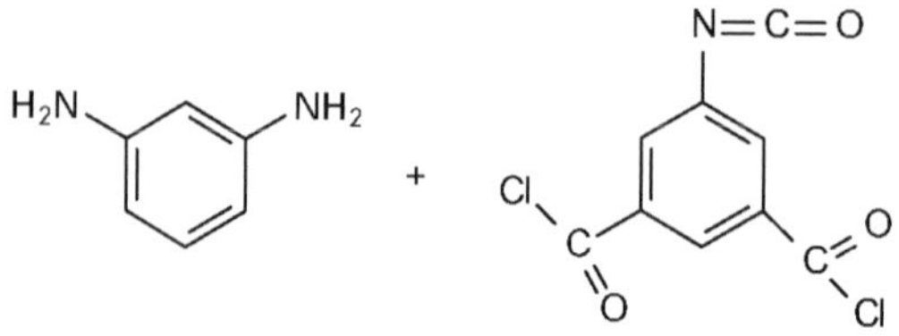

MPD                5IIPC

| 1991 | NCM membrane (US 5 254 261) |

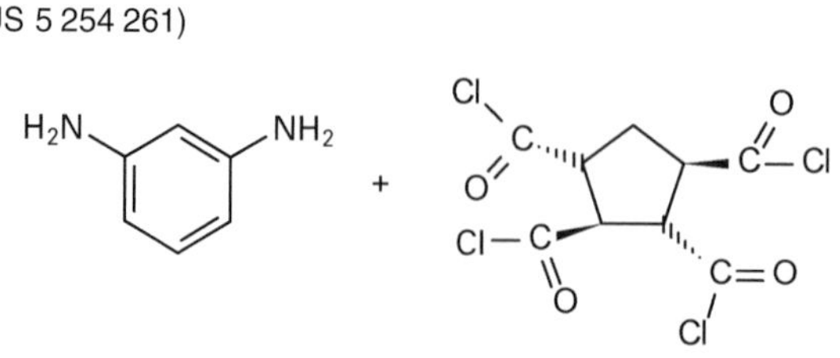

MPD                *trans*-CPTC

glass plate. After several minutes, immersion time in the water bath to remove all solvent, the newly formed porous substrate is immersed again in a fresh water bath as a final rinse. This batchwise process can be scaled up and carried out as continuous processes employing pilot 1 foot (30 cm) wide or production 40-in wide ( ∼ 100 cm) equipment. The advantages of utilizing the continuous process include not only the obvious efficiency but also better reproducibility in resultant porous support properties. However, it is sometimes necessary to pursue the laboratory batchwise process when experimenting with small quantities of costly new polymers or processing conditions that are not easily implemented on the larger-scale continuous equipment.

A typical polysulfone microporous support used in RO or NF IFC membrane fabrication is, by its pore size designation, an ultrafilter (UF) with surface pore sizes ranging from 0.005 to 0.05 μm. Since this size is an order of magnitude smaller than the interfacial film thickness, it easily supports the film even under operating pressures as high as 1000 psi (6.9 MPa). The thickness of the PS support must be sufficient to cover completely the carrier fabric surface plus irregularities caused by improper calendaring, debris and lack of flatness during the casting operation. In practice the net thickness of the PS layer ranges from 25 to 75 μm and that of the carrier fabric upon which it lies ranges between 75 and 150 μm. Scanning electron micrographs (SEM) of a typical PS porous support cross-section and top view are shown in **Figure 2**A and B, respectively. The anisotropic structure is plainly evident with the finest and most supportive pores residing in the upper surface of the support. The finished PS support is normally stored fully immersed in water or at least damp and protected from dust, debris and biological growth. In some cases it is necessary to include a biocide in the storage water, particularly if it is to be stored a long time. A simplified drawing of a continuous casting machine designed to manufacture PS porous supports in which a carrier fabric is used is given in **Figure 3**. A few additional comments regarding the porous support should be noted. In addition to polysulfone, other similar aromatic polyethers may be used such as polyether sulfone. However, these and other variants are significantly more expensive and, except for certain specialized applications, are generally not warranted. The final PS support should be rinsed free of the casting dope solvent otherwise this residual may fuse the porous structure when the IFC membrane is dried.

### Aqueous Amine Reactant Application

The two basic formulations used in the RO IFC membrane industry contain the diamines piperazine

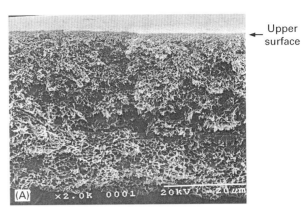

**Figure 2**  (A) SEM polysulfone porous support cross-section. (B) SEM polysulfone porous support (top view).

(Pip) or *m*-phenylenediamine (MPD). Because both the reactivity and solubility (partition coefficients) of these two monomers are different, it is necessary to utilize each at different absolute concentrations as well as different concentration ratios with the cross-linker. When using the Pip formulation it is usually necessary to include an acid acceptor such as sodium hydroxide (NaOH) to neutralize the hydrochloric acid by-product of the polyamidization reaction. This is not necessary when using MPD since it is a much weaker base than Pip and used in a higher excess concentration so that, excess MPD serves as its own acid acceptor. It is generally preferred also to include a surfactant in the aqueous amine formula to acid in the wetting and thus even coverage of the PS support. An anionic or neutral surfactant type is preferred.

There are many acceptable techniques for applying the aqueous amine solution to the PS support. Examples of these include dipping, pouring on, spraying, kiss coating, cloth coating, reverse roll coating, etc. A simple yet effective laboratory-scale method involves sandwiching a 6-in. ( ∼ 15 cm) square piece of PS support between two plastic frames using metal clips to hold the two pieces together. An excess of amine solution is then poured onto the top surface of

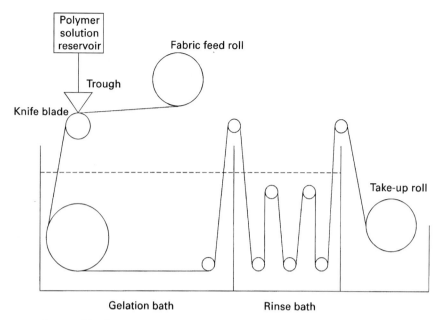

**Figure 3**  Continuous casting machine for porous support.

the PS support and after a brief time interval is drained off leaving an even, wet layer. Depending on the particular formulation the excess amine may be further removed by rubber roller, squeegee, or air knife. It is important that some degree of wetness remains prior to the contact with the cross-linker solution, otherwise the amine cannot transfer effectively via the water–solvent interface.

### Cross-Linker Reactant Application

The choice of solvent for the acyl chloride reactant is dictated by the following requirements:

1. It must completely dissolve the acyl chloride (or other cross-linker) but not react with it.
2. It must be insoluble or virtually insoluble in water.
3. It must not dissolve or swell porous support.
4. It must have a sufficient volatility such that membrane-degrading temperatures are not required for its evaporation.

In practice, the only solvents meeting all of the above requirements are aliphatic hydrocarbons and chlorofluorocarbons (CFCs). There may also be some examples of hydrogenated chlorofluorocarbons (HCFCs) that are acceptable and at the same time are more environmentally friendly than the CFCs. For manufacturing purposes, further restrictions may include preferences for flash point above 100°F and (37.8°C) and low level toxicity. As mentioned previously, the concentration of cross-linker required will be different for the two types of diamines. The concentration of acyl chloride needed for the Pip

formulation is approximately five times that needed for the MPD formulation.

The method of cross-link solution application is generally limited to those which do not disrupt the biphasic nature of the interfacial reaction. If excessive disturbance to this step occurs, the growing polymer may be disrupted, leading to thin film discontinuity and ultimately to high salt passage through the defect regions. Dipping, pouring on gently, kiss coating, etc., are effective methods. For the laboratory-scale techniques, excess acyl chloride cross-link solution is gently poured onto the amine solution-coated PS support contained in the frame and kept horizontal for a brief period. The cross-link solution is then drained off vertically, leaving behind the delicate IFC film residing between thin aqueous and solvent layers. The final step involves some form of evaporation of these two solvents as described below.

### Drying the IFC Membrane

Since the freshly polymerized thin film resides on a thin layer of water, this layer must be removed for the film to strongly adhere to the PS support surface. It is also desirable to remove the cross-link solvent so that the finished IFC membrane can be safely and conveniently handled in a dry state.

Depending on the volatility of the cross-link solvent and amount of moisture present under the IFC film layer, a temperature range of from ambient to 150°C is required. The higher temperature is a practical upper limit owing to tendency for discoloration and degradation of the IFC membrane. Though many

**Table 4** IFC membrane optimization variables: simple approach

| Aqueous solution | |
|---|---|
| 1. | Amine monomer concentration |
| 2. | Acid acceptor concentration |
| 3. | Amount of solution applied to porous substrate |
| | |
| Organic solvent | |
| 4. | Cross-link monomer concentration |
| 5. | Cross-link solution contact time with amine solution |
| 6. | Organic solvent volatility |
| 7. | Drying temperature |

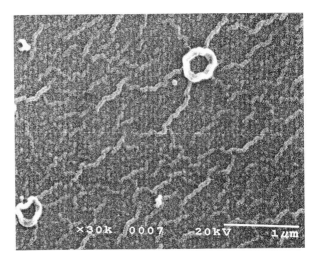

**Figure 4** SEM piperazine IFC membrane (top view).

forms of heating are possible forced air can be used with less heat because of the efficient mechanical effect it has on liquid evaporation. This can be an advantage for IFC membranes that are vulnerable to excessive dehydration. In general, it is found that lower boiling solvents combined with lower drying temperatures often result in membranes with higher fluxes and lower salt rejections than their higher boiling, higher drying temperature counterparts. Of course, longer time periods of drying can be employed in a similar manner with lower drying temperatures to achieve a similar effect, but the choice is ultimately dictated by mechanical and space requirements of the manufacturing equipment. For laboratory scale preparation, it is convenient to use forced hot air devices such as hair-dryers and/or laboratory convection ovens to dry the IFC membrane after the cross-link solution is drained off the frame. Because the thin film is not yet adhered to the PS support excessive air velocity is to be avoided.

The laboratory scale membrane preparation method described in previous sections is now recalled, combining the various steps together: Six inch ( ∼ 15-cm) square pieces of a polysulfone ultrafilter support are clamped between two Teflon® frames and coated on the upper surface with an aqueous solution of the amine monomer for several seconds; the excess solution is removed by any of the various methods previously mentioned. This freshly coated surface is immediately contacted with the cross-linker–solvent solution horizontally for a period of several seconds then drained vertically for several seconds and finally dried by either forced air and/or convection oven for several minutes. The precise conditions for each of the above steps will depend on the particular type of IFC membrane product that is desired, i.e. NF or RO application, and high productivity or high solute retention, etc., according to the development–performance relationships discussed in the previous sections. A listing of the major IFC membrane prepara-

tion variables is given in **Table 4**. SEM pictures of Pip- and MPD-based membranes are given in **Figures 4** and **5**, in which difference in surface roughness is seen. **Figure 6** gives a simplified diagram of the continuous IFC membrane manufacturing process.

## Testing and Optimization

### Test Criteria

The membrane performance throughout the optimization process is obtained by testing on various saline or other solute feeds as appropriate for the particular type of membrane being developed. Examples of test feeds commonly used for RO and NF membrane evaluation are shown in **Table 5**. Because these feeds are in some respects arbitrary, one can easily substitute other concentrations of solutes, types

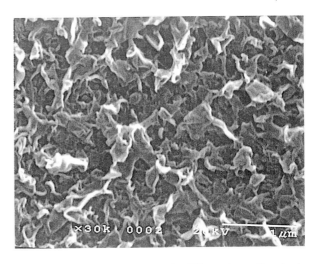

**Figure 5** SEM *m*-phenylenediamine IFC membrane (top view).

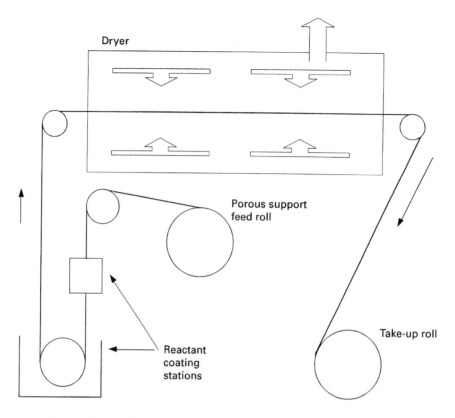

**Figure 6**   Continuous coating machine for IFC membrane.

of solutes, or test pressures to suit the desired application for the membrane.

The permeate quantity and quality are measured for each membrane sample tested and utilized as performance criteria. In the desalination industry, the former is termed membrane flux, with units of gallons/foot$^2$-day (gfd) ($\times 40.8$ L/m$^2$-day) and the latter as salt rejection (%). Flux measurements are made by collecting a volume of permeate under a controlled temperature and time interval. Knowing the active area of the membrane sample and utilizing a temperature correction factor for the viscosity of water, one can calculate the flux normalized to 25°C. Salt rejection is calculated from electroconductivity measurements of the permeates with correction for specific conductance as a function of sodium chloride concen-

tration, or via specific ion probe measurement. Salt rejection is finally calculated as

$$\left(1 - \frac{\text{permeate p.p.m.}}{\text{feed p.p.m.}}\right) \times 100$$

In the case of organic solutes, measurements of permeate and feed are done with a total organic carbon (TOC) analyser with rejection calculated in the same manner as before.

During the membrane development process it is often necessary to rank membranes based upon an objective evaluation. This is difficult because membrane flux and rejection both change. For example, how can a 20 gfd (815 L/m$^2$-day) 99.0% membrane be ranked against a 15 gfd (611 L/m$^2$-day)

**Table 5**   Test feeds for IFC RO and NF membranes

| RO membranes | | | |
|---|---|---|---|
| 35 000 ppm NaCl | 1500 ppm NaCl | 1500 ppm NaCl | 1500 ppm isopropanol |
| 800 psig | 150–225 psig | 1000 ppm CaCl$_2$ | 150–225 psig |
| | | 150–225 psig | |
| **NF membranes** | | | |
| 500 ppm NaCl | 500 ppm MgSO$_4$ | 500 ppm NaCl | 500 ppm sucrose |
| 75 psig | 75 psig | 300 ppm MgSO$_4$ | 75 psig |
| | | 75 psig | |

99.3% membrane, that is, one with a lower flux but higher rejection. A simple ranking method, if the salt rejection of the membrane is 75% or higher, is to take the ratio of flux/salt passage (F/SP) in which salt passage is simply 100 − salt rejection. This value correlates well with the more sophisticated ranking calculation of $A^2/B$, in which $A$ is the pure water permeability constant (g mol cm$^{-2}$ s$^{-1}$ atm$^{-1}$) and $B$ is the salt transport coefficient (cm s$^{-1}$). In the discussion below, both a simple and a more sophisticated performance optimization example are presented for membrane development.

### Simple Approach Optimization

If one has some development experience with a particular amine and cross-linker reactant system such that the workable range of reactant concentrations and processing conditions are at least roughly definable, or if highly optimized membrane performance is not essential, a simplified approach may be pursued. A listing of the recommended minimum number of optimization variables has been given in Table 4. The membrane optimization plan should be carried out in the order shown in this table since it is ordered from highest to the lowest criticality. This approach relies on selecting previously known conditions or educated estimates of some of the variables to be surveyed. The amine monomer concentration experiment would begin by comparing amine concentrations ranging, for example, from $x$ to $3x$ with increments in-between while holding all other variables in Table 4 constant. This requires some discretion in selecting median values of the held constant variables based on prior knowledge. After determining the 'optimum' amine concentration one would proceed in order to the next variable and carry out the next experiment, holding all other variables constant except the acid acceptor concentration that is to be varied. This procedure continues until all the variables have been individually optimized. It is strongly recommended after a once-through optimization to reiterate this process at least once more since new values of many of the variables are likely to have been established. The second time through is likely to result in refinements of both the optimization variable values and the membrane performance.

It should be pointed out that this simple single-variable optimization approach can suffer errors due to interactive variables that can only be optimized in concert. For example, it is likely that, consistent with general principles of chemical reactants, when the amine concentration increases the need for cross-linker increases but so does that for the acid acceptor. In this example there is a three-variable interaction, not merely the two-variable one that the simple method examines. Thus it is often desirable to consider a more sophisticated approach to optimization in which the best combination of variables is found. A multivariable optimization process is offered below.

### Self-Directing Optimization Approach (SDO)

In the SDO process, a regular simplex in $K$ variables is constructed. The experiment can be intiated in $K$ variables with $K + 1$ experiments and upon completion of the $K + 1$ experiments, the results are ranked from best to worst. In the case of IFC desalination membranes, ranking is performed using either F/SP or $A^2/B$ calculations made from the flux and rejection results from a specified test feed type and operating pressure. An example of a Placket–Burmann SDO plan containing 11 variables A–K in 12 experiments is presented in Table 6. Corresponding to the + and − symbols are high and low levels, respectively, to be selected for each variable A–K. The experiments 1–12 are carried out as one series in a random order. After completion of the first series of 12 experiments, the best eight cases, for example, will have the averages of each of the variables calculated. These average values are then multiplied by 2, then from these are subtracted each of the conditions of the four worst cases. The four new experiments created are then run and ranked against the previous eight best cases as before with subsequent elimination of the four worst cases. This process is repeated several times, each time eliminating the worst cases and creating new ones to be compared with the previous best. Eventually the variables will be found to converge such that the optimization is complete.

## Future Developments

Within the polyamide family of chemistry used in IFC membrane preparation dramatic differences in transport properties can be obtained. It is believed that the thin film polymer chemistry and macrostructure play critical roles in determining these performance differences, thus it is expected that future development will rely heavily on the understanding of polymer structure–property relationships. Though there is relatively little such information available to date concerning membrane polymers, recent computer molecular modelling studies are beginning to show promise. Studies such as these and ones involving the mechanism of solvent/solute transport in permselective polymers will lead to future intelligent design of polymer membranes for specific separation processes. In addition to transport performance, there is still need for improvement in chlorine tolerance and fouling resistance by both RO and NF membranes.

**Table 6** Plackett–Burman optimization plan

| Exp. | Amine conc. | Acid accept. conc. | Amine applic. Etc. | A.C. conc. | A.C. time | Sol. B.P. | Etc. | Etc. | Drying time | Etc. | Flux | Rej | A/B | Rank |
|---|---|---|---|---|---|---|---|---|---|---|---|---|---|---|
| 1 | + | + | − | + | + | − | − | − | + | − | | | | |
| 2 | − | + | + | + | + | + | + | + | − | + | | | | |
| 3 | + | − | + | − | − | + | + | + | − | − | | | | |
| 4 | − | + | − | + | + | + | + | + | − | − | | | | |
| 5 | − | − | + | + | + | + | + | + | + | − | | | | |
| 6 | − | + | − | + | + | + | + | + | + | + | | | | |
| 7 | + | + | − | + | − | − | − | − | + | + | | | | |
| 8 | + | + | − | − | − | + | + | + | − | + | | | | |
| 9 | + | + | + | − | − | − | − | − | + | − | | | | |
| 10 | − | − | + | − | − | + | + | + | + | + | | | | |
| 11 | + | + | + | + | + | − | − | − | − | + | | | | |
| 12 | − | − | − | − | − | − | − | − | − | − | | | | |

There are performance gaps in presently available membrane products for the NF area of separations involving species with relative molecular masses ranging from 100 to 3000. It is forseeble that markets will expand for NF membrane applications in high value separations for biotech, chemical, food, and pharmaceutical industries if well-defined relative molecular mass cutoffs can be achieved.

With respect to commercial IFC membrane manufacture, there is a need for improved uniformity and quality of carrier fabrics upon which the porous support is cast. Lack of control here can result in defects that are translated right through the completed composite membrane product. An additional future goal is the development of real-time membrane film integrity and/or performance measurement so that corrections to the process can be made during the course of the manufacture.

*See also:* **II/Membrane Separations:** Membrane Preparation; Reverse Osmosis; Ultrafiltration.

## Further Reading

Al-Gholaikah A, El Ramly N, Janyoon I and Seaton R (1978) The world's first large seawater reverse osmosis desalination plant, at Jeddah, Kingdom of Saudi Arabia. *Desalination* 27: 215–231.

Cadotte JE (1984) In: Lloyd DR (ed.) *Evolution of Composite Reverse Osmosis Membranes, Materials Science of Synthetic Membranes*, p. 273. Washington, DC: ACS Symposium Series.

Cadotte JE, Cobian KE, Forester RH and Petersen RJ (1976) *Continued Evaluation of Insitu-Formed Condensation Polymers for Reverse Osmosis Membranes.* NTIS Report No. PB 253193. Springfield: US Department of Interior.

Cadotte JE, Petersen RJ, Larson RE and Erickson EE (1980) A new thin film composite membrane for seawater desalting applications. *Desalination* 32: 25–31. Amsterdam: Elsevier Science B.V.

Enkelmann V and Wegner G (1976) Mechanism of interfacial polycondensation and the direct synthesis of stable polyamide membranes. *Makromolekulare Chemie* 177: 3177–3189.

Enkelmann V and Wegner G (1972) *Makromolekulare Chemie* 157: 303.

Hirose M, Minamizaki Y and Kamiyama Y (1997) The relationship between polymer molecular structure of RO membrane skin layers and their RO performances. *Journal of Membrane Science* 123: 153–163.

Morgan PW, Kwolek S L and Wittbecker EL (1959) Interfacial polycondensation I and II. *Journal of Polymer Science* XL: 289–326.

Riley RL, Fox RL, Lyons CE et al. (1976) Spiral-wound poly(ether amide) thin-film composite membrane systems. *Desalination* 19: 113–127.

Rozelle LT, Cadotte JE, Corneliussen RD and Erickson EE (1967) *Development of New Reverse Osmosis Membranes for Desalination*, Report No. PB-206329. Springfield, IL: VA National Technical Information Service.

Rozelle LT, Cadotte JE, Cobian KE and Kopp CV Jr (1977) In: Souriragan S (ed.) *Nonpolysacharide Membranes for Reverse Osmosis. NS-100 Membranes for Reverse Osmosis and Synthetic Membranes*, p. 249. Ottawa, Canada: National Research Council Canada.

Souriragan S (1970) *Reverse Osmosis*. New York: Academic Press.

# Phase Inversion Membranes

**M. Mulder**, University of Twente, Enschede, The Netherlands

## Introduction

Phase inversion is the most versatile technique with which to prepare polymeric membranes. A variety of morphologies can be obtained that are suitable for different applications, from microfiltration membranes with very porous structures, to more dense reverse osmosis membranes, to gas separation and pervaporation membranes, with a complete defect-free structure. **Table 1** gives an overview of the techniques that are commonly applied for the preparation of synthetic polymeric membranes.

Most commercially available membranes are prepared by phase inversion. This is a process by which a polymer is transformed from a liquid or soluble state to a solid state. The concept of phase inversion covers a range of different techniques such as immersion precipitation or 'diffusion-induced phase separation', thermal-induced phase separation, 'vapour-phase' precipitation and precipitation by controlled evaporation. The technique of phase inversion has been known for quite some time; the first paper on the preparation of porous nitrocellulose membranes by phase inversion appeared in 1907.

**Table 1** Frequently used techniques for the preparation of synthetic polymeric membranes

| Process | Techniques |
| --- | --- |
| Microfiltration | Phase inversion, stretching, track-etching |
| Ultrafiltration | Phase inversion |
| Nanofiltration | Phase inversion, interfacial polymerization [a] |
| Reverse osmosis | Phase inversion, interfacial polymerization [a] |
| Pervaporation | Dipcoating [a], plasma polymerization [a] |
| Gas separation | Phase inversion, dipcoating [a], plasma polymerization [a] |
| Vapour permeation | Dipcoating [a] |

[a] Support layer prepared by phase inversion.

After World War I the number of publications on membrane preparation and characterization increased significantly and led to the development of the first methods for producing porous nitrocellulose membranes in a reproducible way. The 'Membranfiltergesellschaft Sartorius-Werke' in Göttingen was the first company to produce microfiltration membranes on a commercial scale, based on the work of Zsigmondy. This early work on preparation and characterization was reviewed by Ferry in 1936.

Until World War II most membrane research was performed in Germany, but after the war the technology was transferred to USA. In 1960 Goetz developed a new method for the production of porous membranes. Some years later the Millipore Corporation was founded, which commercialized this production method. The membranes were typically microfiltration membranes and were still based on cellulosic materials. It was more than two decades before ultrafiltration membranes were developed. Alan Michaels, founder of the Amicon Corporation, promoted the development of ultrafiltration membranes. Until that time the research was still focused on cellulosics as material but it became clear that due to the limited thermal and chemical stability other materials were required. This resulted in the development of various ultrafiltration membranes from polyacrylonitrile, polysulfone and polyvinylidene fluoride. Today polymeric materials are still the most commonly employed materials both in ultrafiltration and microfiltration. The early companies such as Sartorius and Schleicher and Schuell still exist and have expanded their membrane business to the technical market. Recently the market for the production of drinking water and industrial water from surface water has become important. Here both microfiltration/ultrafiltration and nanofiltration/reverse osmosis are either used as a single separation unit or in combination with each other or with another technique. The nanofiltration and reverse osmosis membranes are either thin film composites or, less commonly, asymmetric phase inversion membranes. In the case of composite membranes a phase inversion membrane is usually used as support (see Table 1). For gas separation, vapour permeation and pervaporation composite membranes are generally applied with a porous support membrane prepared by phase inversion. Some gas separation membranes, such as polyphenylene oxide, are prepared by immersion precipitation, which results in completely defect-free asymmetric membrane.

## Phase Inversion Membranes

Phase inversion is a process whereby a polymer is transformed in a controlled way from a solution state to a solid state. The concept of phase inversion covers a range of different techniques such as precipitation by controlled evaporation, thermal precipitation from the vapour phase and immersion precipitation. The majority of phase inversion membranes are prepared by immersion precipitation.

### Precipitation from the Vapour Phase

This method was used as early as 1918 by Zsigmondy. A cast film, consisting of a polymer and a solvent, is placed in a vapour atmosphere where the vapour phase consists of a nonsolvent saturated with the solvent. The high solvent concentration in the vapour phase prevents the evaporation of solvent from the cast film. Membrane formation occurs because of the penetration (diffusion) of nonsolvent into the cast film. This results in a porous membrane without a top layer. With immersion precipitation an evaporation step in air is sometimes introduced and, if the solvent is miscible with water, precipitation from the vapour will start at this stage. An evaporation stage is often introduced in the case of hollow fibre preparation by immersion precipitation ('wet–dry spinning') exchange between the solvent and nonsolvent from the vapour phase, leading to precipitation.

### Precipitation by Controlled Evaporation

In this method the polymer is dissolved in a mixture of solvent and nonsolvent where the solvent is more volatile than the nonsolvent. The composition shifts during evaporation to a higher nonsolvent and polymer content. This eventually leads to polymer precipitation, resulting in the formation of a skinned membrane.

### Thermally Induced Phase Separation

A solution of polymer in a mixed or single solvent is cooled to enable phase separation to occur. Evaporation of the solvent often allows the formation of a skinned membrane. This method is frequently used to prepare microfiltration membranes, as will be discussed later.

### Immersion Precipitation

Most commercially available membranes are prepared by immersion precipitation: a polymer solution (polymer plus solvent) is cast on a suitable support and immersed in a coagulation bath containing a nonsolvent. Precipitation occurs because of the exchange of solvent and nonsolvent. The membrane structure ultimately obtained results from a combination of mass transfer and phase separation.

All phase inversion processes are based on the same thermodynamic principles, as will be described in the next section.

## Phase Separation

The change in Gibbs free enthalpy of mixing (dG) for a two-component system $i$ and $j$, where the numbers of moles are $n_i$ and $n_j$, respectively, is given by:

$$dG = V\,dP - S\,dT + \left(\frac{dG}{dn_i}\right)_{T,P,n_i} dn_i + \left(\frac{dG}{dn_j}\right)_{T,P,n_i} dn_j$$

[1]

Here $V$ is the volume, $S$ the entropy, $P$ the pressure and $T$ the temperature ($K$). The chemical potential of a component $i$, which is the partial molar free enthalpy, is defined as:

$$\mu_i = \left(\frac{\partial G}{\partial n_i}\right)_{P,T,n_j,n_k,\ldots}$$

[2]

where $\mu_i$ is equal to the change in free enthalpy of a system containing $n_i$ moles when the pressure, temperature and the number of moles of all the other components are held constant. For a multicomponent system eqn [1] becomes:

$$dG = V\,dP - S\,dT + \sum \mu_i\,dn_i$$

[3]

The chemical potential $\mu_i$ is defined at temperature $T$, pressure $P$, and composition $x_i$. For the pure component ($x_i = 1$), the chemical potential may be written as $\mu_i^o$.

The free enthalpy $G_m$ of a mixture consisting of two components is given by the sum of the chemical potentials (the partial free enthalpy). If $G_m$ is expressed per mole, then:

$$G_m = x_1\mu_1 + x_2\mu_2$$

[4]

The dependence of the free enthalpy on the composition of the mixture is shown schematically in **Figure 1**. The value of the $G_m$ at the $y$-axis represent

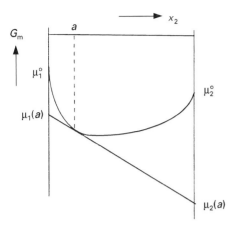

**Figure 1** Schematic drawing of the free enthalpy of a mixture at temperature $T$ as a function of the composition.

the chemical potential of the pure components. $\mu_1^\circ$ and $\mu_2^\circ$, respectively.

For ideal solutions the free enthalpy of mixing per mole is given by:

$$\Delta G_m = RT(x_1 \ln x_1 + x_2 \ln x_2) \qquad [5]$$

The solubility behaviour of polymer solutions differs completely from that of a solution containing components of low relative molecular mass because the entropy of mixing of the long polymeric chains is much lower. Flory and Huggins used a lattice model to describe the entropy of mixing of polymer solutions. In general for a binary system $\Delta G_m$ is given by:

$$\Delta G_m = RT(n_1 \ln \phi_1 + n_2 \ln \phi_2 + n_1 \phi_2 \chi) \qquad [6]$$

where an additional term has been added that was originally derived as an enthalpic contribution and which contains the Flory–Huggins interaction param-

eter $\chi$. In the original Flory theory $\chi$ was considered to be constant but for many systems it has been proven that this is not the case. In addition, $\chi$ is considered to be an excess free energy parameter containing all nonideality (including excess entropy). Differentiation of eqn [6] with respect to $n_1$ and $n_2$, respectively, gives the partial molar free enthalpy difference of component 1 ($\Delta\mu_1$) and ($\Delta\mu_2$) upon mixing:

$$\Delta\mu_1 = \mu_1 - \mu_1^\circ = \left(\frac{\partial \Delta G_m}{\partial n_1}\right)_{P,T,n_2}$$

$$= RT\left(\ln \phi_1 - \left(1 - \frac{V_1}{V_2}\right)\phi_2 + \chi\phi_2^2\right) \qquad [7]$$

and:

$$\Delta\mu_2 = \mu_2 - \mu_2^\circ = \left(\frac{\partial \Delta G_m}{\partial n_2}\right)_{P,T,n_1}$$

$$= RT\left(\ln \phi_2 - \left(1 - \frac{V_2}{V_1}\right)\phi_1 + \chi\frac{V_2}{V_1}\phi_1^2\right) \qquad [8]$$

In the case of polymer solutions the entropy term is very small and a positive enthalpy of mixing will cause demixing. Decreasing the temperature often causes an increase in the enthalpy of mixing.

**Figure 2** shows two plots of $\Delta G_m$ versus $\phi$ for two different temperatures. At temperature $T_1$ (Figure 2A), the system is completely miscible over the whole composition range. This is indicated by the tangent to the $\Delta G_m$ curve, which can be drawn at any composition. For example, at composition $a$ the intercept at $\phi_2 = 0$ gives $\mu_1(a)$ (the chemical potential of component 1 in the mixture of composition $a$) and the intercept at $\phi_2 = 1$ gives $\mu_2(a)$. This means that the chemical potentials of both components 1 and 2 de-

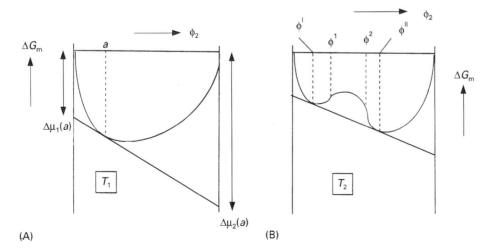

(A)    (B)

**Figure 2** Free energy of mixing as a function of composition for a binary mixture. $T_2 < T_1$ ($H_m > 0$).

crease (or $\Delta\mu_i < 0$). At temperature $T_2$ (Figure 2B), the curve of $\Delta G_m$ exhibits an upward bend between $\phi^I$ and $\phi^{II}$. These two points lie on the same tangent and are thus in equilibrium with each other. All the points on the tangent have the same derivative $(= \partial\Delta G_m/\partial n_i = \Delta\mu_i)$, i.e. the chemical potentials are the same. In general, increasing the temperature leads to an increase in miscibility, which means that the enthalpy term becomes smaller. The two points on the tangent will approach each other and eventually they will coincide at the so-called critical point. This critical point is characterized by $(\partial^2\Delta G_m/\partial\phi_i^2) = 0$ and $(\partial^3\Delta G_m/\partial\phi_i^3) = 0$. Two points of inflection are also observed in Figure 2B, i.e. $\phi^1$ and $\phi^2$. A point of inflection is the point at which a curve changes from being concave to convex, or vice versa. These points are characterized by $(\partial^2\Delta G_m/\partial\phi_i^2 = 0)$. Plotting the locus of the minima in a $\Delta G_m$ versus $\phi$ diagram leads to the binodal curve. The locus of the inflection points is called the spinodal. A typical temperature–composition diagram is depicted in **Figure 3**.

The location of the miscibility gap for a given binary polymer–solvent system depends principally on the chain length of the polymer (see **Figure 4**). As the chain length increases the miscibility gap shifts towards the solvent axis as well as to higher temperatures. The critical point shifts towards the solvent

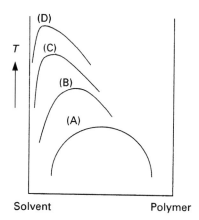

**Figure 4** Schematic drawing of a binary mixture with a region of immiscibility. Binodal (A): mixture of two components of low relative molecular mass; binodals (B), (C), (D): mixtures of a solvent with low relative molecular mass and a polymer with increasing relative molecular mass.

axis, while the asymmetry of the binodal curve increases. The interaction between polymer and solvent is another important parameter, and this is expressed by the Flory–Huggins interaction parameter.

The location of the phase diagram in a binary or ternary system can be determined experimentally, e.g. by cloud point measurements, or theoretically by applying Flory–Huggins thermodynamics with suitable values for the interaction parameters.

## Demixing Processes

### Liquid–Liquid Demixing (Binary Systems)

To understand the mechanism of liquid–liquid demixing more easily, a binary system consisting of a polymer and a solvent will be considered. The starting point for preparing phase inversion membranes is a thermodynamically stable solution (see **Figure 5**), for example one with the composition $A$ at a temperature $T_1$ (with $T_1 > T_c$). All compositions with a temperature $T > T_c$ are thermodynamically stable. As the temperature decreases demixing of the solution will occur when the binodal is reached. The solution demixes into two liquid phases and this is referred to as liquid–liquid demixing.

Suppose that the temperature is decreased from $T_1$ to $T_2$. The composition $A$ at temperature $T_2$ lies inside the demixing gap and is not stable thermodynamically. The curve of $\Delta G_m$ at temperature $T_2$ is also given in Figure 5. At temperature $T_2$ all compositions between $\phi^I$ and $\phi^{II}$ can reduce their free enthalpies of mixing by demixing into two phases with compositions $\phi^I$ and $\phi^{II}$, respectively (see Figure 3). These two phases are in equilibrium with each other since they lie on the same tangent to the $\Delta G_m$ curve, i.e. the

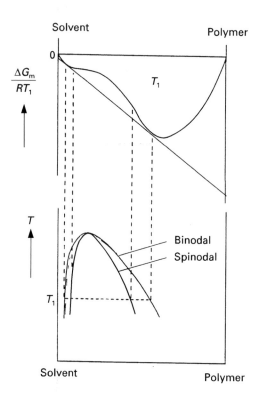

**Figure 3** Temperature–composition phase diagram for a binary polymer–solvent system.

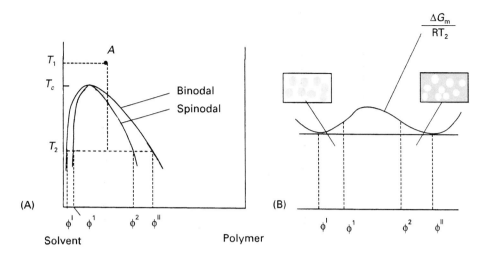

**Figure 5** Demixing of a binary polymer solution by deceasing the temperature. $T_C$ is the critical temperature.

chemical potential in phase $\phi^I$ must be equal to that of phase $\phi^{II}$.

**Figure 6** gives the curve of $\Delta G_m$ plotted against composition at a given temperature (e.g. $T_2$), together with the first and second derivative. Two regions can clearly be observed from the second derivative

(the lowest figure). Over the interval $\phi^1 < \phi < \phi^2$ the second derivative of $\Delta G_m$ with respect to $\phi$ is negative, implying that the solution is thermodynamically unstable and will demix spontaneously into very small interconnected regions of composition $\phi^1$ and $\phi^{II}$:

$$\frac{\partial^2 \Delta G_m}{\partial \phi^2} < 0 \qquad (\phi^1 < \phi < \phi^2) \qquad [9]$$

The amplitude of small fluctuations in the local concentration increases in time, as shown schematically in **Figure 7**. In this way a lacy structured membrane is obtained, and the type of demixing observed is called spinodal demixing. Over the intervals $\phi^I < \phi < \phi^1$ and $\phi^2 < \phi < \phi^{II}$, the second derivative of $\Delta G_m$ with respect to $\phi$ is positive and the solution is metastable. This means that there is no driving force for spontaneous demixing and the solution is stable towards small fluctuations in composition. Demixing

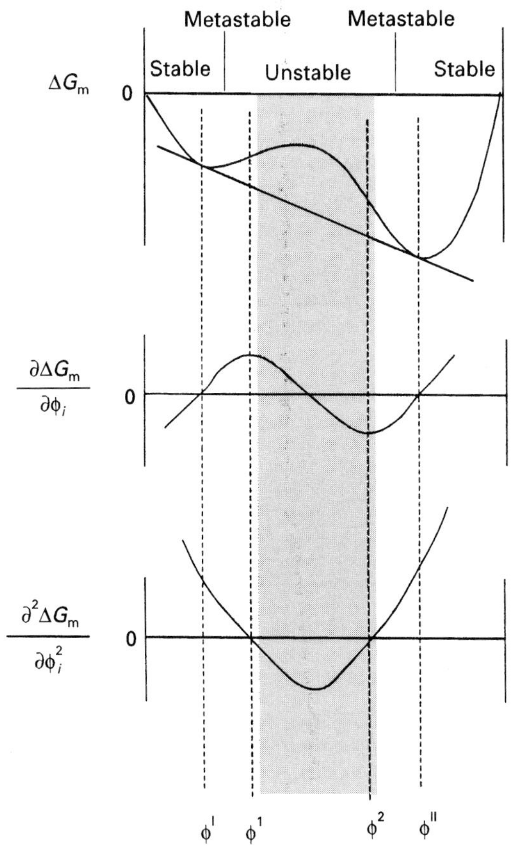

**Figure 6** Plots of $\Delta G_m$, the first derivative of $\Delta G_m$ and the second derivative of $\Delta G_m$ against $\phi$.

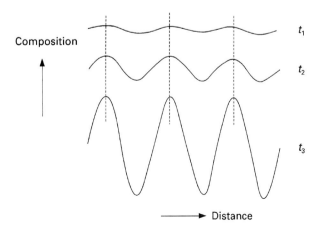

**Figure 7** Spinodal demixing: increase in amplitude with increasing time ($t_3 > t_2 > t_1$).

can commence only when a stable nucleus has been formed. A nucleus is stable when it lowers the free enthalpy of the system; hence over the interval $\phi^I < \phi < \phi^1$ the nucleus must have a composition near $\phi^{II}$, and over the interval $\phi^2 < \phi < \phi^{II}$ it must have a composition near $\phi^I$:

$$\frac{\partial^2 \Delta G_m}{\partial \phi^2} > 0 \quad (\phi^I < \phi < \phi^1) \quad \text{and} \quad (\phi^2 < \phi < \phi^{II})$$

[10]

After nucleation, these nuclei grow further in size by downhill diffusion whereas the composition of the continuous phase moves gradually towards that of the other equilibrium phase. The type of structure obtained after liquid–liquid demixing by nucleation and growth depends on the initial concentration.

Starting with a very dilute polymer solution (see Figure 5), the critical point will be passed on the left-hand side of the diagram and liquid–liquid demixing will start when the binodal curve is reached and a nucleus is formed with a composition near $\phi^{II}$. The nucleus formed will grow further until thermodynamic equilibrium is reached (nucleation and growth of the polymer-rich phase). A two-phase system is formed consisting of concentrated polymer droplets of composition $\phi^{II}$ dispersed in a dilute polymer solution with composition $\phi^I$. In this way a latex type of structure is obtained, which has little mechanical strength. When the starting point is a more concentrated solution (composition $A$ in Figure 5), demixing will occur by nucleation and growth of the polymer-lean phase (composition $\phi^I$). Droplets with a very low polymer concentration will now continue to grow until equilibrium has been reached.

As can be seen from Figure 5, the location of the critical point is close to the solvent axis. Hence the binodal curve for a polymer–solvent system will be reached on the right-hand side of the critical point, indicating that liquid–liquid demixing will occur by nucleation of the polymer-lean phase. These tiny droplets will grow further until the polymer-rich phase solidifies. If these droplets have the opportunity to coalesce before the polymer-rich phase has solidified, an open porous system will result.

**Liquid–Liquid Demixing (Ternary Systems)**

In addition to temperature changes, changes in composition brought about by the addition of a third component, a nonsolvent, can also cause demixing. Under these circumstances we have a ternary system consisting of a solvent, a nonsolvent and a polymer. The liquid–liquid demixing area must now be represented as a three-dimensional surface. The free enthalpy of mixing is a function of the composition, as can be seen from **Figure 8**, where the $\Delta G_m$ surface is depicted at a certain temperature. All pairs of compositions with a common tangent plane to the $\Delta G_m$ surface constitute the solid line projected in the phase diagram, the binodal. **Figure 9** shows a schematic illustration of the temperature dependency of such a three-dimensional liquid–liquid demixing surface for a ternary system. The demixing area takes the form of a part of a beehive. As the temperature increases the demixing area decreases, and if the temperature is sufficiently high the components are miscible in all proportions. From this figure an isothermal cross-section can be obtained at any temperature as shown in **Figure 10**.

The corners of the triangle in Figure 10 represent the pure components polymer, solvent and

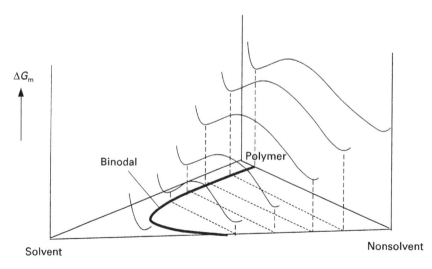

**Figure 8** Schematic drawing of the free enthalpy of mixing ($\Delta G_m$) as a function of the composition for a ternary system consisting of polymer, solvent and nonsolvent.

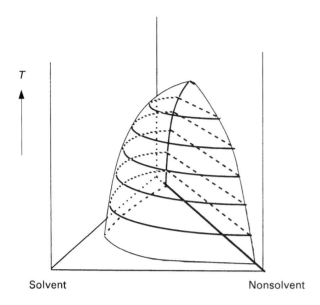

$T$

Solvent                                Nonsolvent

**Figure 9**  Three-dimensional representation of the binodal surface at various temperatures for a ternary system consisting of polymer, solvent and a nonsolvent.

have the same chemical potential. By minimizing the following function the compositions of the end points may be obtained:

$$F = \sum f_i^2 \qquad [11]$$

with $f_i = (\Delta\mu_i' - \Delta\mu_i'')$ and $i = 1, 2, 3$. The polymer-lean phase is indicated by a single prime (') and the polymer-rich phase is indicated by a double prime (").

The initial procedure for membrane formation from such ternary systems is always to prepare a homogeneous (thermodynamically stable) polymer solution. This will often correspond to a point on the polymer–solvent axis. However, it is also possible to add nonsolvent as long as all the components are still miscible. Demixing will occur by the addition of such an amount of nonsolvent that the solution becomes thermodynamically unstable.

When the binodal curve is reached liquid–liquid demixing will occur. As in the binary system, the side from which the critical point is approached is important. In general, the critical point is situated at low to very low polymer concentrations (see Figure 10). When the metastable miscibility gap is entered at compositions above the critical point, nucleation of the polymer-lean phase occurs. The tiny droplets formed consist of a mixture of solvent and nonsolvent with very little polymer dispersed in the polymer-rich phase, as described in the binary example (see Figure 5). These droplets can grow further until the surrounding continuous phase solidifies via crystallization, gelation or when the glass transition temperature has been passed (in the case of glassy polymers). Coalescence of the droplets before solidification leads to the formation of an open porous structure.

nonsolvent. A point located on one of the sides of the triangle represents a mixture consisting of the two corner components. Any point within the triangle represents a mixture of the three components. In this region a spinodal curve and binodal curve can be observed. The tie lines connect points on the binodal curve that are in equilibrium. A composition within this two-phase region always lies on a tie line and splits into two phases represented by the two intersections between the tie line and the binodal curve. As in the binary system, one end point of the tie line is rich in polymer and the other end point is poor in polymer. The binodal curve may be calculated numerically. The tie lines connect the two coexisting phases that are in equilibrium with each other, and these

### Solid–Liquid Demixing (Crystallization)

Many polymers are partially crystalline. They consist of an amorphous phase without any ordering and an ordered crystalline phase. Crystallization may occur if the temperature of the solution is below the melting point of the polymer. **Figure 11** shows the free enthalpy of mixing ($\Delta G_m$) for a binary system of polymer and solvent (or diluent) that shows no liquid–liquid demixing. However, below the melting point the chemical potential of the polymer in the solid state will be smaller than that in the solution. Therefore, the solution can lower its free enthalpy by phase separation into a pure crystalline solid state ($\phi_2$) and a liquid state ($\phi_a$ in Figure 11) that are in equilibrium with each other ($\Delta\mu_{2,L} = \Delta\mu_{2,S}$). The corresponding melting temperature for this mixture $\phi_a$ is $T_1$. This is shown schematically in Figure 11B. $T_m^o$ is the melting point of the pure polymer and the melting

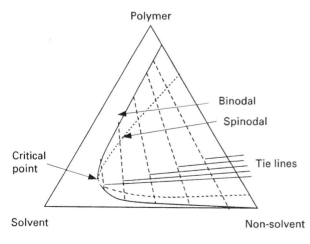

Polymer

Binodal

Spinodal

Critical
point

Tie lines

Solvent                                Non-solvent

**Figure 10**  Schematic representation of a ternary system with a liquid–liquid demixing gap.

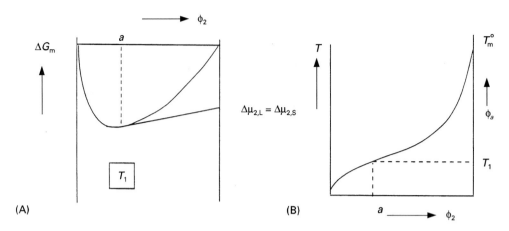

**Figure 11** Schematic drawing of the free enthalpy of mixing for a binary system in which component 2 is able to crystallize (A) and the melting point curve as a function of the composition (B). $\phi_a$ is the volume fraction at point *a*.

point depression for a binary polymer–solvent system, as derived by Flory, is given below:

$$\frac{1}{T_m} - \frac{1}{T_m^o} = \frac{R}{\Delta H_f} \frac{V_2}{V_1} (\phi_1 - \chi \phi_1^2) \qquad [12]$$

Here $\phi_1$ is the volume fraction of solvent and $\chi$ is the polymer–solvent interaction parameter; $T_m$ is the melting temperature of the diluted polymer; $\Delta H_f$ is the heat of fusion per mole of repeating units; and $V_1$ and $V_2$ are the molar volume of the solvent and of the polymer repeating unit, respectively.

For a ternary system with a semicrystalline polymer a similar ternary diagram can be constructed, as shown in **Figure 12**. However, it is somewhat more complex since solid–liquid demixing occurs in addition to liquid–liquid demixing. Except for the homogeneous region (I) where all components are miscible

with each other and a region where liquid–liquid demixing occurs (II), other phases can be observed. The curve PQ is the crystallization curve and a composition somewhere in the region of P–Q–polymer will contain crystalline pure polymer that is in equilibrium with a composition somewhere on the crystallization line PQ. A possible morphology of a semicrystalline polymer is shown schematically in **Figure 13**. Spherulitic structures are frequently observed in semicrystalline polymers.

Many morphologies are possible ranging from a completely crystalline to a completely amorphous conformation. The formation of crystalline regions in a given polymer depends on the time allowed for crystallization from the solution. In very dilute solutions the polymer chains can form single crystals of the lamellar type, whereas in medium and concentrated solutions more complex morphologies occur, e.g. dendrites and spherulites.

Membrane formation is generally a fast process and only polymers that are capable of crystallizing rapidly (e.g. polyethylene, polypropylene, aliphatic polyamides) will exhibit an appreciable amount of

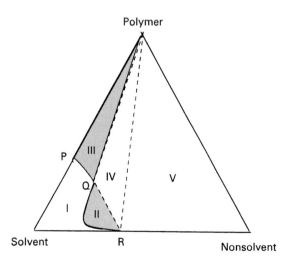

**Figure 12** Ternary system of a semicrystalline polymer, solvent and nonsolvent system. I, homogeneous, II, liquid–liquid; III, Solid–liquid; IV, solid–liquid–liquid; V, solid–liquid.

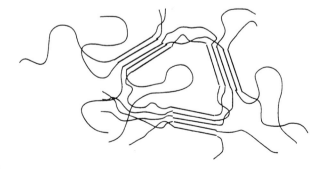

**Figure 13** Morphology of a semicrystalline polymer (fringed micelle structure).

crystallinity. Other semicrystalline polymers contain a low to very low crystalline content after membrane formation. For example, PPO (2,6-dimethylphenylene oxide) shows a broad melting endotherm at 245°C. Ultrafiltration membranes derived from this polymer, prepared by phase inversion, hardly contain any crystalline material, indicating that membrane formation was too rapid to allow crystallization.

### Gelation

Gelation is a phenomenon of considerable importance during membrane formation, especially for the formation of the top layer. It was mentioned in the previous section that a large number of semicrystalline polymers exhibit a low crystalline content in the final membrane because membrane formation is too fast. However, these polymers generally undergo another solidification process, i.e. gelation. Gelation can be defined as the formation of a three-dimensional network by chemical or physical cross-linking. Chemical cross-linking, the covalent bonding of polymer chains by means of a chemical reaction, will not be considered here.

When gelation occurs, a dilute or more viscous polymer solution is converted into a system of infinite viscosity, i.e. a gel. A gel may be considered as a highly elastic, rubber-like solid. A gelled solution does not demonstrate any flow when a tube containing the solution is tilted. Gelation is not a phase separation process, and it may also take place in a homogeneous system consisting of a polymer and a solvent. Many polymers used as membrane materials exhibit gelation behaviour, e.g. cellulose acetate, poly(phenylene oxide), polyacrylonitrile, poly(methyl) methacrylate, poly(vinyl chloride) and poly(vinyl alcohol). Physical gelation may occur by various mechanisms dependent on the type of polymer and solvent or solvent–nonsolvent mixture used. In the case of semicrystalline polymers especially, gelation is often initiated by the formation of microcrystallites. These microcrystallites, which are small ordered regions, are in fact the nuclei for the crystallization process but without the ability to grow further. However, if these microcrystallites can connect various polymeric chains together, a three-dimensional network will be formed. Because of their crystalline nature these gels are thermo-reversible, i.e. upon heating the crystallites melt and the solution can flow. Upon cooling, the solution again gels. The formation of helices often occurs during the gelation process. Gelation may also occur by other mechanisms, e.g. the addition of complexing ions ($Cr^{3+}$) or by hydrogen bonding.

Gelation is also possible in completely amorphous polymers (e.g. atactic polystyrene). In a number of systems the involvement of gelation in the membrane

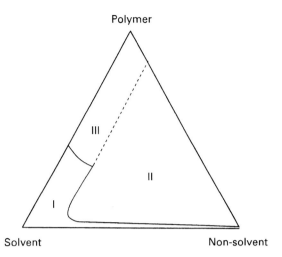

**Figure 14** Isothermal cross-section of a ternary system containing a one-phase region (I), a two-phase region (II) and a gel region (III).

formation process often involves a sol–gel transition. This is shown schematically in **Figure 14**. As can be seen from this figure, a sol–gel transition occurs where the solution gels. The addition of a nonsolvent induces the formation of polymer–polymer bonds and gelation occurs at a lower polymer concentration. These sol–gel transitions have been observed in a number of systems, e.g. cellulose acetate/acetone/water, cellulose acetate/dioxane/water, poly(phenylene oxide)/trichloroethylene/octanol and poly(phenylene oxide)/trichloroethylene/methanol.

### Vitrification

There are polymers that show neither crystallization nor gelation behaviour. Nevertheless, these polymers finally solidify during a phase inversion process. This solidification process may be defined as vitrification, which is the stage where the polymer chains are frozen in a glassy state, i.e. it is a phase where the glass transition temperature has been passed and the mobility of the polymer chains has been reduced drastically. In the absence of gelation or crystallization, vitrification is the mechanism of solidification in any membrane-forming system with an amorphous glassy polymer.

The glass transition of a polymer is reduced by the presence of an additive, i.e. a solvent or nonsolvent. This glass transition depression can been described by various theories, the Kelley–Bueche theory being widely used. A schematic phase diagram of the system PPO/trichloroethylene/methanol is shown in **Figure 15**. Four regions can be observed:

1. a one-phase region where all the components are miscible with each other;

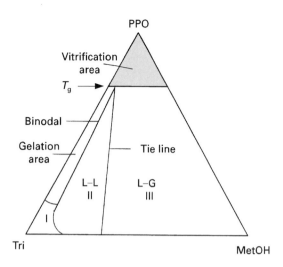

**Figure 15** Schematic phase diagram of the quasi ternary system PPO/trichloroethylene/methanol.

2. a gel region where the polymer is able to form a three-dimensional network, providing that certain conditions have been established (a sol–gel transition for the system PPO/DMAc has been determined, however a minimum time of 1 h is necessary for gel formation whereas in immersion precipitation the timescale is much shorter);

3. a glassy region or vitrification region where the glass transition or the polymer has been passed. During immersion precipitation the diffusion of solvent and nonsolvent proceeds according to their corresponding driving forces independent of whether gelation occurs. The final solidification may be a combined gelation/vitrification process, or in absence of gelation vitrification will be the dominant process.

4. a two-phase region where liquid–liquid demixing occurs. In the figure only one tie line is given in which the polymer-rich phase has entered the vitrification area. On the left side of this tie line (II) the (equilibrium) system is still a liquid, whereas on the right side (III) vitrification of the polymer-rich phase had occurred.

## Membrane Formation

### Thermally Induced Phase Separation (TIPS)

Before describing immersion precipitation in detail, a short description of thermal precipitation or 'thermally induced phase separation' (TIPS) is given.

This process allows the ready preparation of porous membranes from a binary system consisting of a polymer and a solvent. Generally, the solvent has a high boiling point, e.g. sulfolane (tetramethylene sulfone, bp $287°C$) or oil (e.g. nujol). The starting

point is a homogeneous solution, for example composition $A$ at temperature $T_1$ (see Figure 5).

This solution is cooled slowly to the temperature $T_2$. When the binodal curve is attained liquid–liquid demixing occurs and the solution separates into two phases, one rich in polymer and the other poor in polymer. When the temperature is decreased further to $T_2$, the composition of the two phases follow the binodal curve and eventually the compositions $\phi^I$ and $\phi^{II}$ are obtained. At a certain temperature the polymer-rich phase solidifies by crystallization (polyethylene), gelation (cellulose acetate) or on passing the glass transition temperature (atactic polymethylacrylate). Frequently, semicrystalline polymers are used (polyethylene, polypropylene, aliphatic polyamides) which crystallize relatively fast, and hence a solid–liquid phase transition should be included.

**Figure 16** shows how the liquid–liquid (L–L) demixing area and the solid–liquid (S–L) for a binary system. In the case of glassy amorphous polymers the melting line may be replaced by a vitrification line. This concept may be applied to various systems, **Table 2** provides some examples of this thermally induced phase separation (TIPS) process.

### Immersion Precipitation

An interesting question remains after all of these theoretical considerations: what factors are important in order to obtain the desired (asymmetric) morphology after immersion of a polymer–solvent mixture in a nonsolvent coagulation bath? Another interesting question is: why is a more open (porous) top layer obtained in some cases whereas in other cases a very dense (nonporous) top layer supported by an (open) sponge-like structure develops? To answer these questions and to promote an understanding of the basic principles leading to membrane formation

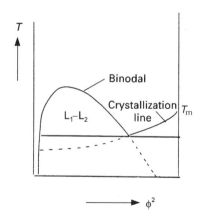

**Figure 16** Construction of a $T–\phi$ diagram for a binary system polymer–solvent. The solidification line is the glass transition temperature line.

**Table 2** Some examples of thermally-induced phase separation systems

| Polymer | Solvent |
|---------|---------|
| Polypropylene | Mineral oil (nujol) |
| Polyethylene | Mineral oil (nujol) |
| Polyethylene | Dihydroxy tallow amine |
| Poly(methyl) methacrylate | Sulfolane |
| Cellulose acetate/PEG | Sulfolane |
| Cellulose acetate/PEG | Dioctyl phthalate |
| Nylon-6 | Triethylene glycol |
| Nylon-12 | Triethylene glycol |
| Poly(4-methyl pentene) | Mineral oil (nujol) |

via immersion precipitation, a qualitative description will be given. For the sake of simplicity, the concept of membrane formation will be described in terms of three components: nonsolvent (1), solvent (2) and polymer (3). The effect of additives such as second polymer or material of low relative molecular mass will not be considered because the number of possibilities would then becomes so large and every (quarternary) or multicomponent system has its own complex thermodynamic and kinetic descriptions.

Immersion precipitation membranes in their most simple form are prepared in the following way. A polymer solution consisting of a polymer (3) and a solvent (2) is cast as a thin film upon a support (e.g. a glass plate) and then immersed in a nonsolvent (1) bath. The solvent diffuses into the coagulation bath ($J_2$) while the nonsolvent will diffuse into the cast film ($J_1$). After a given period of time the exchange of solvent and nonsolvent has proceeded so far that the solution becomes thermodynamically unstable and demixing takes place. Finally a solid polymeric film is obtained with an asymmetric structure. A schematic representation of the film–bath interface during immersion is shown in **Figure 17**.

The local composition at any point in the cast film depends on time. However, it is not possible to measure composition changes very accurately with time because the thickness of the film is only of the order of a few microns. Furthermore, membrane

formation can sometimes occur instantaneously, i.e. all the compositional changes must be measured as a function of place and time within a very small time interval. Nevertheless, these composition changes can be calculated. Such calculations provide a good insight into the influence of various parameters upon membrane structure and performance.

Different factors have a major effect upon membrane structure. These are:

- choice of polymer;
- choice of solvent and nonsolvent;
- composition of casting solution;
- composition of coagulation bath;
- gelation, vitrification and crystallization behaviour of the polymer;
- location of the liquid–liquid demixing gap;
- temperature of the casting solution and the coagulation bath; and
- evaporation time.

By varying one or more of these parameters, which are not independent of each other, the membrane structure can be changed from a very open porous form to a very dense nonporous variety.

Take polysulfone as an example. This is a polymer that is frequently used as a membrane material, both for microfiltration/ultrafiltration as well as a sublayer in composite membranes. These applications require an open porous structure, but in addition asymmetric membranes with a dense nonporous top layer can also be obtained that are useful for prevaporation or gas separation applications. Some examples are given in **Table 3** that clearly demonstrate the influence of various parameters on the membrane structure when the same system, DMAc/polysulfone (PSf), is employed in each case. To understand how it is possible to obtain such different structures with one and the same system, it is necessary to consider how each of the variables

**Table 3** Influence of preparation procedure on membrane structure

Evaporation PSf/DMAc ⇒ pervaporation/gas separation
Precipitation of 15% PSf/DMAc/THF in water ⇒ gas separation [a]
Precipitation of 35% PSf/DMAc in water ⇒ pervaporation/gas separation [b]
Precipitation of 15% PSf/DMAc in water ⇒ ultrafiltration
Precipitation of 15% PSf/DMAc in water/DMAc ⇒ microfiltration [c]

[a]After an initial evaporation step.
[b]It will be shown later that integrally skinned asymmetric membranes can be prepared with completely defect-free top layers.
[c]To obtain an open (interconnected) porous membrane, an additive, e.g. poly(vinyl pyrrolidone) must be added to the polymer solution.

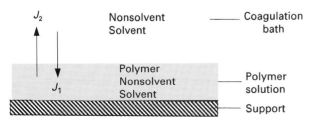

**Figure 17** Schematic representation of a film–bath interface. Components: nonsolvent (1), solvent (2) and polymer (3). $J_1$ is the nonsolvent flux and $J_2$ the solvent flux.

affects the phase inversion process. The ultimate structure arises through two mechanisms: (1) a diffusion processes involving solvent and nonsolvent occurring during membrane formation; and (2) demixing processes.

## Diffusional Aspects

Membrane formation by phase inversion techniques, e.g. immersion precipitation, is a nonequilibrium process that cannot be described by thermodynamics alone since kinetics have also to be considered. The composition of any point in the cast film is a function of place and time. To know what type of demixing process occurs and how it occurs, it is necessary to know the exact local composition at a given instant. However, this composition cannot be determined very accurately experimentally because the change in composition occurs extremely quickly (in often less than 1 s) and the film is very thin (less than 200 μm). However, it can be described theoretically. Cohen *et al.* were the first to describe mass transport in an immersion precipitation process. Since then many modified models have been published to describe better this highly nonideal complex multicomponent mass transfer system.

The change in composition may be considered to be determined by the diffusion of the solvent ($J_2$) and the nonsolvent ($J_1$) (see Figure 17) in a polymer fixed frame of reference. The fluxes $J_1$ and $J_2$ at any point in the cast film can be represented by a phenomenological relationship:

$$J_i = - \sum_{j=1}^{2} L_{ij}(\phi_i, \phi_j) \frac{\partial \mu_j}{\partial x} \qquad (i = 1, 2) \qquad [13]$$

where $-\partial\mu/\partial x$, the gradient in the chemical potential, is the driving force for mass transfer of component $i$ at any point in the film and $L_{ij}$ is the permeability coefficient. From eqn [13] the following relations may be obtained for the nonsolvent flux ($J_1$) and the solvent flux ($J_2$):

$$J_1 = - L_{11} \frac{d\mu_1}{dx} - L_{12} \frac{d\mu_2}{dx} \qquad [14]$$

$$J_2 = - L_{21} \frac{d\mu_1}{dx} - L_{22} \frac{d\mu_2}{dx} \qquad [15]$$

As can be seen from the above equations, the fluxes in a given polymer/solvent/nonsolvent system are determined by the gradient in the chemical potential as driving force, while they also appear in the phenomenological coefficients. This implies that a knowledge of the chemical potentials, or better the

factors that determine the chemical potential, is of great importance. An expression for the free enthalpy of mixing has been given by Flory and Huggins. For a three-component system (polymer/solvent/nonsolvent), the Gibbs free energy of mixing ($\Delta G_\mathrm{m}$) is given by:

$$\Delta G_\mathrm{m} = RT(n_1 \ln \phi_1 + n_2 \ln \phi_2 + n_3 \ln \phi_3 + \chi_{12} n_1 \phi_2$$
$$+ \chi_{13} n_1 \phi_3 + \chi_{23} n_2 \phi_3) \qquad [16]$$

where $R$ is the gas constant and $T$ is the absolute temperature. The subscripts refer to nonsolvent (1), solvent (2) and polymer (3). The number of moles and the volume fraction of component $i$ are $n_i$ and $\phi_i$, respectively. $\chi_{ij}$ is called the Flory–Huggins interaction parameter. In a ternary system there are three interaction parameters: $\chi_{13}$ (nonsolvent/polymer), $\chi_{23}$ (solvent/polymer) and $\chi_{12}$ (solvent/nonsolvent). $\chi_{12}$ can be obtained from data on excess free energy of mixing that have been compiled or from vapour–liquid equilibria. $\chi_{13}$ can be obtained from swelling measurements and $\chi_{23}$ can be obtained from vapour pressure or membrane osmometry. The interaction parameters account for the nonideality of the system and they contain an enthalpic as well as an entropic contribution. In the original Flory–Huggins theory they are assumed to be concentration independent, but several experiments have shown that these parameters generally depend on the composition. To account for such dependence the symbol $\chi$ is often replaced by another symbol, $g$, indicating concentration dependency.

From eqn [16] it is possible to derive the expressions for the chemical potentials of the components since:

$$\left( \frac{\partial \Delta G_\mathrm{m}}{\partial n_i} \right)_{P,T,n_i} = \Delta \mu_i = \mu_i - \mu_i^0 \qquad [17]$$

The eventual concentration dependency of the $\chi$ parameter must be taken into account in the differentiation procedure. The influence of the different interaction parameters $\chi$ (present in the driving forces) on the solvent flux and nonsolvent flux, and thus on the membrane structures obtained, will be described later.

The other terms present in the flux equations (eqns [16] and [17]) are phenomenological coefficients, and these must also be considered with respect to membrane formation. These coefficients are also mostly concentration dependent. There are two ways of expressing the phenomenological coefficients when the relationships for the chemical poten-

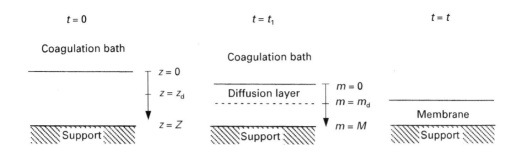

**Figure 18** Schematic drawing of the immersion process at different times.

tials are known: (1) in diffusion coefficients; and (2) in friction coefficients.

In most cases there is a (large) difference between the casting thickness and the ultimate membrane thickness. This implies that during the formation process the boundary between the nonsolvent bath and the casting solution moves, as is shown in **Figure 18**. For this reason, it is necessary to introduce a position coordinate to correct for this moving boundary.

The immersion process starts at time $t = 0$. At all times $t > 0$, solvent will diffuse out of the film and nonsolvent will diffuse in. If there is a net volume outflow (solvent flux larger than nonsolvent flux) then the film–bath interface is shifted from $z = 0$, i.e. the actual thickness is reduced. This process will continue until equilibrium is reached (at time $t = t$) and the membrane has been formed. In order to describe diffusion processes involving a moving boundary adequately, a position coordinate $m$ must be introduced (eqn [18]).

The film–bath interface is now always at position $m = 0$, independent of time. The position of the film–support interface is also independent of time (see Figure 18):

$$m(x, t) = \int_0^x \phi_3(x, t)\, \mathrm{d}x \qquad [18]$$

$$(\mathrm{d}m)_t = \phi_3\, (\mathrm{d}x)_t \qquad [19]$$

In the $m$-coordinate:

$$\frac{\partial(\phi_i/\phi_3)}{\partial t} = \frac{\partial J_i}{\partial m} \quad i = 1, 2 \qquad [20]$$

Combination of eqns [18]–[20] yields:

$$\frac{\partial(\phi_1/\phi_3)}{\partial t} = \frac{\partial}{\partial m}\left[ v_1 \phi_3 L_{11} \frac{\partial \mu_1}{\partial m} \right] + \frac{\partial}{\partial m}\left[ v_1 \phi_3 L_{12} \frac{\partial \mu_2}{\partial m} \right]$$

$$[21]$$

$$\frac{\partial(\phi_2/\phi_3)}{\partial t} = \frac{\partial}{\partial m}\left[ v_2 \phi_3 L_{21} \frac{\partial \mu_1}{\partial m} \right] + \frac{\partial}{\partial m}\left[ v_2 \phi_3 L_{22} \frac{\partial \mu_2}{\partial m} \right]$$

$$[22]$$

The main factor determining the type of demixing process is the local concentration in the film. Using eqns [21] and [22] it is possible to calculate these concentrations ($\phi_1$, $\phi_2$, $\phi_3$) as a function of time. Thus at any time and any place in a cast film the demixing process occurring can be calculated; in fact the concentrations are calculated as a function of place and elapsed time and the type of demixing process is deduced from these values. However, one should note that a number of assumptions and simplifications are involved in this model. Thus heat effects, occurrence of crystallization and relative molecular mass distributions are not taken into account. Nevertheless, it will be shown in the next section that the model allows the type of demixing to be established on a qualitative basis and is therefore useful as a first estimate. Furthermore, it allows an understanding of the fundamentals of membrane formation by phase inversion.

### Mechanism of Membrane Formation

It is shown in this section that two types of demixing process resulting in two different types of membrane morphology can be distinguished:

- instantaneous liquid–liquid demixing, where the membrane is formed immediately;
- delayed onset of liquid–liquid demixing, where the membrane takes some time to form.

The occurrence of these two distinctly different mechanisms of membrane formation can be demonstrated in a number of ways: by calculating the concentration profiles; by light transmission measurements; and visually.

The best physical explanation is given by a calculation of the concentration profiles. To calculate the concentration profiles in the polymer film during the delayed demixed type of phase inversion process, some assumptions and considerations must be made:

- diffusion in the polymer solution;
- diffusion in the coagulation bath – no convection occurs in the coagulation bath;
- thermodynamic equilibrium is established at the film–bath interface:
  $$\mu_i\,(\text{film}) = \mu_i\,(\text{bath}) \qquad i = 1, 2, 3;$$
- volume fluxes at the film–bath interface are equal, i.e.
  $$J_i\,(\text{film}) = J_i\,(\text{bath}) \qquad i = 1, 2.$$

In addition, the thermodynamic binary interaction parameters (the $\chi$ parameters or the concentration-dependent $g$ parameters) that appear in the expressions for the chemical potentials must be determined experimentally.

- $g_{12}$, from calorimetric measurements yielding values of the excess free energy of mixing, from literature compilations of $G^E$ and activity coefficients, from vapour–liquid equilibria and from Van Laar, Wilson, or Margules equations or from UNIFAC;
- $g_{13}$, from equilibrium swelling experiments or from inverse gas chromatography,
- $g_{23}$, from membrane osmometry or vapour pressure osmometry.

Two types of demixing process will now be distinguished that lead to different types of membrane structure. These two different types of demixing process may be characterized by the instant when liquid–liquid demixing sets in. **Figure 19** shows the composition path of a polymer film schematically at the very instant of immersion in a nonsolvent bath (at $t < 1$ s). The composition path gives the concentration at any point in the film at a particular time. For any other time, another compositional path will exist.

Because diffusion processes start at the film–bath interface, the change in composition is first noticed in the upper part of the film. This change can also be observed from the composition paths given in Figure 19. Point $t$ gives the composition at the top of the film while point $b$ gives the bottom composition. Point $t$ is determined by the equilibrium relationship at the film–bath interface $\mu_i\,(\text{film}) = \mu_i\,(\text{bath})$. The composition at the bottom is still the initial concentration in both examples. In Figure 19A places in the film beneath the top layer $t$ have crossed the binodal curve, indicating that liquid–liquid demixing starts immediately after immersion. In contrast, Figure 19B indicates that all compositions directly beneath the top layer still lie in the one-phase region and are still miscible. This means that no demixing occurs immediately after immersion. After a longer time interval, compositions beneath the top layer will cross the binodal curve and liquid–liquid demixing will start in this case also. Thus two distinctly different demixing processes can be distinguished and the resulting membrane morphologies are also completely different.

When liquid–liquid demixing occurs instantaneously, membranes with a relatively porous top layer are obtained. This demixing mechanism results in the formation of a porous membrane (microfiltration/ultrafiltration type). However, when liquid–liquid demixing sets in after a finite period of time, membranes with a relatively dense top layer are obtained. This demixing process results in the formation of dense membranes used for gas separation/pervaporation. In both cases the thickness of the top

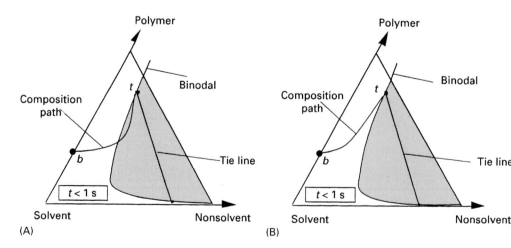

**Figure 19** Schematic composition path of the cast film immediately after immersion; $t$ is the top of the film and $b$ is the bottom. Part (A) shows instantaneous liquid–liquid demixing whereas (B) shows the mechanism for the delayed onset of liquid–liquid demixing.

**Table 4** Polymers that are frequently used as phase inversion membranes

| Polymer | Process |
|---------|---------|
| Cellulose acetate | MF, UF, NF, RO, GS |
| Nitrocellulose | MF |
| Polysulfone | MF, UF |
| Polyethersulfone | MF, UF |
| Polyacrylonitrile | MF, UF |
| Polyvinylidenefluoride | MF, UF |
| Polyimide | UF, GS |
| Aliphatic polyamide | MF, UF |
| Aromatic polyamide | NF, RO |
| Polyphenylene oxide | GS |

MF, microfiltration; UF, ultrafiltration; NF, nanofiltration; RO, reverse osmosis; GS, gas separation.

layer is dependent on a variety of membrane formation parameters (i.e. polymer concentration, coagulation procedure, additives, etc.).

### Polymers for Phase Inversion Membranes

The preparation of phase inversion membranes has only been described briefly. It may be evident that many polymers can be applied as long as they are soluble in a suitable organic solvent. This is the only limitation. Nevertheless, owing to their mechanical, thermal and chemical properties some polymers are more frequently applied. Two of these so-called 'engineering polymers', polysulfone and polyethersulfone, are used as micro- and ultrafiltration membranes or as support material in composite membranes for nanofiltration, reverse osmosis and gas separation. These two polymers have very good film-forming properties and are soluble in a range of solvents. Beside these two polymers a wide variety of other polymers are applied; these are listed in Table 4.

## Future Developments

Phase inversion will remain the most important technique for the preparation of polymeric membranes for microfiltration and ultrafiltration membranes and for use as support membranes in composite membranes for nanofiltration, reverse osmosis, gas separation, vapour permeation and pervaporation.

*See also:* **III/Membrane Preparation:** Hollow Fibre Membranes; Interfacial Composite Membranes

## Further Reading

Arnauts J and Berghmans H (1987) Atactic gels of atactic polystyrene. *Polymer Communications* 28: 66–68.

Altena FW and Smolders CA (1982) Calculation of liquid–liquid phase separation in a ternary system of a polymer in a mixture of a solvent and a nonsolvent. *Macromolecules* 15: 1491–1497.

Altena FW, Smid J, Van den Berg JWA, Wijmans JG and Smolders CA (1985) Diffusion from solvent from a cast CA solution. *Polymer* 26: 1531.

Boom RM, Wienk IM, Boomgaard van den T and Smolders CA (1992) Microstructures in phase inversion membranes. Part 2, the role of the polymer additive. *Journal of Membrane Science* 73: 277–292.

Boom RM, Boomgaard van den T and Smolders CA (1994) Mass transfer and thermodynamics during immersion precipitation for a two-polymer system: evaluation with the systems PES-PVP-NMP-water. *Journal of Membrane Science* 90: 231–249.

Bulte AMW, Folkers B, Mulder MHV and Smolders CA (1993) Membranes of semi-crystalline aliphatic polyamide Nylon 4-6. Formation by diffusion induced phase separation. *Journal of Applied Polymer Science* 50: 13–26.

Caneba GT and Soong DS (1985) Polymer membrane formation through the thermal inversion process. 1. Experimental study of membrane formation. *Macromolecules* 18: 2538–2545.

Caneba GT and Soong DS (1985) Polymer membrane formation through the thermal inversion process. 2. Mathematical modeling of membrane structure formation. *Macromolecules* 18: 2545–2555.

Cheng L-P, Soh YS, Dwan A-H and Gryte CC (1994) An improved model for mass transfer during the formation of polymer membranes by the immersion precipitation process. *Journal of Polymer Science Part B, Polymer Physics* 32: 1413–1425.

Cohen C, Tanny GB and Prager EM (1979) Diffusion-controlled formation of porous structures in ternary polymer systems, *Journal of Polymer Science, Polymer Physics Edition* 17: 477–489.

Eykamp W (1995) Microfiltration and ultrafiltration. In: Noble RD and Stein A (eds) *Membrane Separation Technology, Principles and Applications*, pp. 1–25. Amsterdam: Elsevier.

Ferry D (1936) Ultrafilter membranes and ultrafiltration. *Chemical Reviews* 18: 373–455.

Flory PJ (1953) *Principles of Polymer Chemistry*. Ithaca, NY: Cornell University Press.

Gaides GE and McHugh AJ (1989) Gelation in an amorphous polymer: a discussion of its relation to membrane formation. *Polymer* 30: 2118–2123.

Guillotin M, Lemoyne C, Noel C and Monnerie L (1977) Physicochemical processes occurring during the formation of cellulose diacetate membranes. Research of criteria for optimizing membrane performance. IV. cellulose diacetate-acetone-organic additive casting solutions. *Desalination* 21: 165–170.

Kamide K and Matsuda S (1984) Phase equilibria of quasi ternary systems consisting of multi-component polymers in a binary solvent mixture. II Role of initial concentration and relative amount of polymers. *Polymer Journal* 7: 515–530.

Koenhen DM, Mulder MHV and Smolders CA (1977) Phase separation phenomena during the formation of asymmetric membranes. *Journal of Applied Polymer Science* 21: 199–215.

Lloyd DR and Kinzer KE (1991) Microporous membrane formation via thermally induced phase separation. II Liquid–liquid phase separation. *Journal of Membrane Science* 64: 1–11.

McHugh AJ and Yilmaz L (1985) The diffusion equation for polymer membrane formation in ternary systems. *Journal of Polymer Science, Polymer Physics Edition* 23: 1271–1274.

McHugh AJ and Tsay CS (1992) Dynamics of the phase inversion process. *Journal of Applied Polymer Science* 46: 2011–2021.

Mulder MHV, Oude Hendrikman J, Wijmans JG and Smolder CA (1985) A rationale for the preparation of asymmetric pervaporation membranes. *Journal of Applied Polymer Science* 30: 2805–2830.

Mulder MHV, Franken ACM and Smolders CA (1985) Preferential sorption versus preferential permeation. *Journal of Membrane Science* 22: 155–173.

Mulder MHA (1996) *Basic Principles of Membrane Technology*. Dordrecht: Kluwer.

Radovanovic P, Thiel SW and Hwang S-T (1992) Formation of asymmetric polysulfone membranes by immersion precipitation. Part II. The effects of casting solution and gelation bath compositions on membrane structure and skin formation. *Journal of Membrane Science* 65: 231.

Radovanovic P, Thiel SW and Hwang S-T (1992) Formation of asymmetric polysulfone membranes by immersion precipitation. Part I. Modelling of mass transport during gelation. *Journal of Membrane Science* 65: 213–229.

Reuvers AJ, Altena FW and Smolders CA (1986) Demixing and gelation behaviour of ternary cellulose acetate solutions studied by differential scanning calorimetry. *Journal of Polymer Science, Polymer Physics Edition* 24: 793–804.

Reuvers AJ, Berg van den JWA and Smolders CA (1987) Formation of membranes by means of immersion precipitation. Part I. A model to describe mass transfer during immersion precipitation. *Journal of Membrane Science* 34: 45–65.

Reuvers AJ and Smolders CA (1987) Formation of membranes by means of immersion precipitation. Part II. The mechanism of membranes prepared from the system cellulose acetate–acetone–water. *Journal of Membrane Science* 34: 67–86.

Strathmann H, Koch K, Amar P and Baker RW (1975) The formation mechanism of asymmetric membranes. *Desalination* 16: 179–203.

Tan HM, Moet A, Hiltner A and Baer E (1983) Thermoreversible gelation of atactic polystyrene solutions. *Macromolecules* 16: 28–34.

Tsai FJ and Torkelson JM (1990) Roles of phase separation mechanism and coarsening in the formation of PMMA asymmetric membranes. *Macromolecules* 23: 775–784.

Tsay CS and McHugh AJ (1990) Mass transfer modelling of asymmetric membrane formation by phase inversion. *Journal of Polymer Science* 28: 1327–1365.

Tsay CS and McHugh AJ (1991) Mass transfer dynamics of the evaporation step in membrane formation by phase inversion. *Journal of Membrane Science* 64: 81–92.

Tsay CS and McHugh AJ (1991) The combined effects of evaporation and quench steps on asymmetric membrane formation by phase inversion. *Journal of Polymer Science Part B, Polymer Physics* 29: 1261–1270.

Vadalia HC, Lee HK, Myerson HS and Levon K (1994) Thermally induced phase separation in ternary crystallizable polymer solutions. *Journal of Membrane Science* 89: 37–50.

Wisniak J and Tamir A (1978) *Mixing and Excess Thermodynamic Properties*. Amsterdam: Elsevier.

Wijmans JG, Kant J, Mulder MHV and Smolders CA (1985) Phase separation phenomena in solutions of polysulfone in mixtures of a solvent and a nonsolvent: relationship with membrane formation. *Polymer* 26: 1539–1545.

Wijmans JG, Rutten HJJ and Smolders CA (1985) Phase separation phenomena in solutions of PPO in mixtures of trichloroethylene, octanol and methanol: relation to membrane formation. *Journal of Polymer Science, Polymer Physics* 23: 1941–1955.

Yilmaz L and McHugh AJ (1986) Analysis of solvent–nonsolvent–polymer phase diagrams and their relevance to membrane formation modelling. *Journal of Applied Polymer Science* 31: 997–1018.

Yilmaz L and McHugh AJ (1986) Modelling of asymmetric membrane formation. I. Critique of evaporation models and development of diffusion equation formalism for the quench period. *Journal of Membrane Science* 28: 287–310.

Zeman L and Tkacik G (1988) Thermodynamic analysis of a membrane forming system water–N-methyl-2-pyrrolidone/polyethersulfone. *Journal of Membrane Science* 36: 119–140.

# METABOLITES

*See* **III / DRUGS AND METABOLITES: Liquid Chromatography-Mass Spectrometry; Liquid Chromatography-Nuclear Magnetic Resonance-Mass Spectrometry**

# METAL ANALYSIS: GAS AND LIQUID CHROMATOGRAPHY

**P. C. Uden**, University of Massachusetts, Amherst, MA, USA

## Introduction

Since its inception as a separatory technique, gas–liquid chromatography (GLC) has had great impact in the quantitative resolution of mixtures of volatile compounds. Although the sister discipline of gas–solid chromatography (GSC) is used for inorganic gas analysis along with metalloid halides and hydrides, the potential of both GLC and GSC for inorganic compounds, metal complexes and organometallics has been less realized. The relative obscurity of inorganic gas chromatography derives from expectations that inorganic compounds are inherently of very low volatility and/or thermally unstable under GC conditions, are incompatible with column substrates, show undesirable reactivity making GC difficult, or are incompatible with conventional detectors. Some of these strictures apply, but there are many examples of GC of metallic and metalloid compounds including organometallics incorporating sigma or pi carbon to metal bonds, and metal complexes with coordinate bonds between oxygen, sulfur, nitrogen, phosphorus, or halogen, and the metal atom. Some metal oxides and halides may be eluted at high temperatures. Chemical derivatization methods may also enable successful elution. There is often considered to be a minimal group of chemical properties to be possessed by a compound before gas chromatography can be successful: volatility, thermal stability, monomeric form, neutrality, relatively low molecular weight, coordinative saturation and shielding of the metal atom(s) by bulky and inert organic functional groups. The compound should appear to the GC column as a simple organic species and free metal atoms should not be exposed to reactive sites. Inertness of column materials, injection and detection pathways are particularly important. Fused silica columns allow GC of previously inapplicable species. Detectors that are more compatible, selective or specific for inorganic compounds have opened the way for quantitative and sensitive analysis.

Liquid chromatography (LC) has been used extensively in thin-layer and ion exchange for metal ion and compound separations. However high performance column chromatography (HPLC) with different column packings, instrumentation and detectors has given rise to many new capabilities. The literature of LC of inorganics and organometallics up to 1970 was presented by Michal with 700 references while that between 1970 and 1979 was surveyed by Schwedt with over 450 references. MacDonald's (1985) text *Inorganic Chromatographic Analysis* covers all LC and GC methods in a comprehensive review.

The various modes of HPLC allow a wider range of analytes to be chromatographed than in GC. Metal ions may be resolved by ion chromatography, or separated as ion pairs in a reversed-phase regime. Positively or negatively charged metal complexes may be similarly separated. Neutral metal complexes, chelates and organometallics may be chromatographed in reversed- or normal-phase systems. Metal-containing macromolecules such as metalloproteins may be separated by size exclusion. Chemical properties desirable in an inorganic compound for viable liquid partition chromatography include solvolytic stability, minimal adsorptive interactions and shielding of the metal by inert functional groups. As with GC, the analyte should appear to the column as an organic compound and free metal atoms should never be accessible to reactive sites. New detectors that are more compatible or specific for inorganic compounds have also made an impact in HPLC. Diode array UV, electrochemical detection and elemental and molecular mass spectroscopy have proven valuable. The inorganic chromatographer is often faced with a choice between GC and HPLC and each may be valuable in a complementary fashion. Supercritical fluid chromatography (SFC) has been little applied for metal compounds, successful applications following mostly from HPLC methodology.

The primary application of analytical chromatography in metals analysis is clearly in 'speciation' wherein the requirement is resolution and quantitation of specific chemical species incorporating the target metal. Situations occur (i) in which the chemical species is already amenable to the chosen chromatographic technique, and (ii) wherein it may be converted to such by physical or chemical means, as in pyrolysis, derivatization, etc. While total element analysis is less suited to chromatographic methods, sometimes suitable derivatization techniques enable this as well.

# Gas Chromatography

The classes of metallic substances for which GC is viable, either directly or by derivatization, are: binary metal and metalloid compounds such as halides, hydrides and oxides; sigma-bonded organometallic and organometalloid compounds such as alkyls and aryls; pi-bonded organometallics such as metal carbonyls and metallocenes; chelated metal complexes having nitrogen, oxygen, sulfur, phosphorus, etc., as ligand atoms. The text by Guiochon and Pommier (1973) *Gas Chromatography in Inorganics and Organometallics* provides an excellent summary of this topic augmented by Schwedt's coverage.

## Binary Metal and Metalloid Compounds

Binary metal compounds with adequate vapour pressures and thermal stabilities at normal GC temperatures include main group hydrides and halides. At 1000°C or higher, some metal oxides have adequate properties.

Among hydrides, those of boron, silicon, germanium, tin, arsenic, antimony, bismuth, selenium and tellurium are viable having boiling points ranging from 112°C to −2°C. Low column temperatures and inert systems are needed. Element-selective spectral detectors are also advantageous. These separations are important for trace-level determinations in electronic grade organometallics. Organo-hydrides are also readily chromatographed. Analysis of environmentally significant arsenic and antimony compounds with microwave plasma emission detection involves reduction of alkylarsonic acids with sodium borohydride to alkylarsines.

Certain metal halides are sufficiently volatile for GC, but difficulty lies in their high reactivity in the vapour and condensed phases, necessitating precautions to ensure maximum inertness of the system. Among chlorides that have been gas chromatographed are those of titanium, aluminium, mercury, tin, iron, antimony, germanium, gallium, vanadium, silicon and arsenic. Problems arise from reaction with even methyl silicone oils and thus inert fluorocarbon packings have been favoured for reactive chlorides and oxychlorides including $VOCl_3$, $VCl_4$, $PCl_3$ and $AsCl_3$. The less volatile metal bromides are very challenging and high temperature stationary phases such as alkali bromide salts coated on silica are needed. Some metal fluorides have low boiling points, e.g. tungsten (17.5°C), molybdenum (35°C), tellurium (35.5°C), rhenium (47.6°C) and uranium (56.2°C), for which low column temperatures are feasible. The determination of alloys and metal oxides, carbides, etc., after conversion to fluorides by fluorination appears feasible.

One of the most extreme modifications of GC has been in the area of very high temperature GC for metal oxides, hydroxides and oxychlorides. Bachmann employed temperatures as high as 1500°C for the separation of oxides and hydroxides of technetium, rhenium, osmium and iridium. Quartz granules were used as the substrate and oxygen and oxygen/water mixtures as carrier gases with the necessary equipment modifications.

## Organometallic Compounds

Crompton in *Gas Chromatography of Organometallic Compounds*, published in 1982, notes that during the preceding decade more than 1000 papers were published on this topic, relating to compounds of more than 50 elements. At least as many papers have been published since that time.

**Sigma-bonded compounds** GC is feasible for compounds of Al, Ga and In from Group III; Ge, Sn and Pb from Group IV; P, As, Sb and Bi from Group V; and Se and Te from Group VI. Also included are Hg and possibly Zn and Cd. The elements that have attracted most analytical interest have been silicon and lead, tin and mercury.

The organic functionalities are typically alkyl, aryl and substituted aryl groups; perfluoroalkyl and aryl groups, which typically impart enhanced volatility; and mixed alkyl-chloro and aryl-chloro systems. The two reported separations of aluminium alkyls leave doubts as to possible on-column decomposition. Trimethylgallium has been eluted along with $(CH_3)_2$ $GaCl$ and $CH_3GaCl_2$, but no GC of organometallic indium or thallium compounds has been reported. The GC characteristics of alkyl germanium, tin and lead compounds resemble those of silicon but with decreasing thermal stability. For alkylstannanes, on-column oxidation, hydrolysis or thermal degradation must be avoided. Selective detection has been widely used for organotin compounds, notably for environmental samples such as bis(tributyltin)oxide (TBTO) in marine paints, triphenylhydroxystannane pesticide after derivatization to triphenylmethylstannane, and tricyclohexylhydroxystannane in apple leaves after derivatization to tricyclohexylbromostannane.

GC of tetraalkyllead compounds is extensive and covers 30 years of developments. Crompton devotes 90 pages to discussion of organolead GC. The electron-capture detector is highly selective for tetraalkylleads, but is prone to contamination. Atomic spectral detection has been widely applied for lead-specific detection, e.g. flame photometric detection with an oxygen–hydrogen flame at 405.8 nm and graphite furnace electrothermal atomic absorption detection at 283.3 nm. Chau quantitated methylethylleads in water, sediment and fish in an investigation of bioalkylation processes, lead detection limits being between 0.01 and 0.025 $\mu g\,g^{-1}$ for solid samples. Plasma atomic emission detection for alkyl leads has been widely used; trimethyl- and triethyllead chlorides have been determined in water in the range of 10 ppb to 10 ppm and also derivatized by butyl Grignard reagent to form the trialkylbutylleads.

The alkyl and aryl derivatives of arsenic and antimony are more labile than those of Group IV elements, and require stringent GC conditions for successful elution. Bismuth compounds have not been chromatographed. Talmi determined $As^{3+}$ and $Sb^{3+}$ in environmental samples as triphenylarsine and triphenylstilbene, plasma emission detection giving limits of 20 pg and 50 pg, respectively.

GC of the environmentally significant organomercurials has attracted much attention, biomethylation of inorganic mercury being important. $(CH_3)_2Hg$ and $CH_3HgCl$ have been separated on packed glass columns using plasma emission detection. Non-flame 'cold vapour', atomic absorption spectrophotometry detection has proved useful after catalytic conversion of organomercurials to elemental mercury.

Although they are not formally organometallics, many alkoxides share similar GC characteristics with alkyl compounds. Germanium, tin, titanium, zirconium and hafnium form stable volatile alkoxides, as do some Group III elements, notably aluminium. There have been GC separations reported of oxy-carboxylate salts of beryllium and zinc.

**Pi-bonded compounds**  The most gas chromatographed transition metal organometallics are those containing carbonyl, arene and cyclopentadienyl ligand moieties. Their elution characteristics are very favourable with only rare on-column degradation or adsorption reported. Effective separation of $Fe(CO)_5$, $Cr(CO)_6$, $Mo(CO)_6$ and $W(CO)_6$ is possible and methods may be employed for the highly toxic $Ni(CO)_4$ since its high volatility permits ready elution. Many results have been reported for arene, cyclopentadienyl and related derivatives of metal carbonyls. Arenechromiumtricarbonyls and molybdenumtricarbonyls are easily gas chromatographed. Cyclopentadienylmanganesetricarbonyl $(C_5H_5Mn(CO)_3)$ and its derivatives are well suited to quantitative GC. Methylcyclopentadienylmanganesetricarbonyl, $(CH_3C_5H_4Mn(CO)_3)$-MMT) has been determined in gasoline using various detectors. Capillary GC is effective for compounds of this class; spe-

cific element detection simplifies such separations and provides qualitative and quantitative proof of elution. **Figure 1** shows such a capillary column separation with atomic emission detection of organometallics with differing metals and functionalities. Many of these organometallics are of interest as polymerization catalysts, etc.

Among GC of metallocenes, ferrocene, bis(cyclopentadienyl)iron, is the most familiar example. These compounds have favourable GC properties, ferrocene proving an ideal organometallic probe for determining column and system efficiency. Ferrocene derivatives chromatographed have included alkyl, vinyl, dialkyl, acetyl, diacetyl and hydroxymethyl compounds. GC behaviour is determined largely by the substituents; ruthenocene and osmocene have also been separated.

## Metal Chelates

Neutral metal complexes deriving from a number of anionic ligands with oxygen, nitrogen, sulfur or phosphorus donor atoms, have been widely studied. The range of organic ligands that has been shown to be suitable for GC analysis has been limited, but a considerable amount of development and application has been done.

**Beta-diketonates**  Beta-diketonates are readily formed with stability arising from multiple chelate rings. Ions with coordination numbers twice their oxidation state such as Al(III), Be(II) and Cr(III) form coordinatively saturated neutral complexes which are good for GC but the non-fluorinated beta-diketonates are generally of marginal thermal and chromatographic stability; they usually require column temperatures too high for thermal degradation to be completely absent. The major breakthrough in metal chelate GC involved fluorinated beta-diketone ligands, giving complexes of greater volatility and thermal stability. Moshier and Sievers gave the major impetus to this development, and a major portion of their monograph summarizes the analytical progress made to that time. Trifluoroacetylacetone (1,1,1-trifluoro-2,4-pentanedione-HTFA) and hexafluoroacetylacetone (1,1,1,5,5,5-hexafluoro-2,4-pentanedione-HHFA) have been the most widely studied and analytically developed of the fluorinated beta-diketonates. HTFA extended the range of metals that may be quantitated to include Ga(III), In(III), Sc(III), Rh(III) and V(IV). HTFA chelates of trivalent hexacoordinate metals such as Cr(III), Co(III), Al(III) and Fe(III) exhibit geometrical isomerism with facial (*cis*) and meridonal (*trans*) forms present and interconverting. Numerous analytical applications of HTFA chelates are reviewed in detail by Moshier and

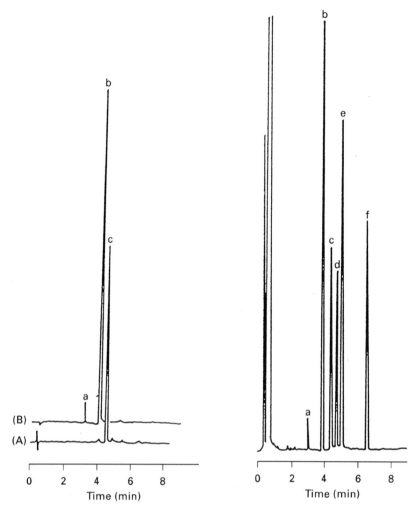

**Figure 1** Microwave plasma atomic emission capillary GC detection of organometallics. Left: (A) chromium monitored at 267.7 nm (peak c) and (B) manganese monitored at 257.6 nm (peaks a and b). Right: carbon monitored at 247.9 nm for organometallic mixture. The six peaks afford 'universal' carbon detection. The identities of the eluted peaks are: (a) $C_5H_5Mn(CO)_3$, (b) $CH_3C_5H_4Mn(CO)_3$, (c) $C_5H_5Cr(NO)(CO)_2$ and $(C_5H_5)_2Ni$ (unresolved), (d) $C_5H_5V(CO)_4$, (e) $(C_5H_5)_2Fe$ and (f) $C_5H_5(CH_3)_5Co(CO)_2$. (Reproduced with permission from Estes SA *et al.* (1980). *Journal of Chromatography and Chromatographic Communications* 3: 471. Copyright John Wiley & Sons Ltd.)

Sievers, Uden and Henderson and in the texts of Guiochon and Pommier and Schwedt.

Since modifications of the beta-diketone structure may be made readily, various such ligands have been evaluated for GC. The two major adaptations have been the replacement of the methyl group by higher branched alkyl groups, notably *t*-butyl, and the incorporation of longer chain perfluoroalkyl groups in the ligand. Sievers used 1,1,1,2,2,3,3-heptafluoro-7,7-dimethyl-4,6-octanedione (heptafluoropropanoyl-pivalylmethane (HFOD or HHPM)) for lanthanide separations.

**Alternative beta-difunctional chelates** The major classes of such ligands are summarized in the schematic diagram in **Figure 2**. This indicates the mode of formation of the main alternative bidentate ligands

with sulfur donors and bidentate and tetradentate ligands with nitrogen donors.

**Beta-thioketonates** There are a number of advantages of these chelates. The metals showing favourable GC properties are those whose diketonates are generally unsatisfactory, such as the divalent metals nickel, palladium, platinum, zinc and cobalt. Nickel has been subjected to a complete quantitative analysis, as the monothiotrifluoroacetylacetonate.

**Beta-ketoaminates** The presence of the nitrogen atom in the ligand dictates an intermediate place between beta-diketonates and monothioketonates in terms of the metals which are readily complexed. The nickel group of metals is favourably complexed, but in addition stable chelates of copper(II) are formed,

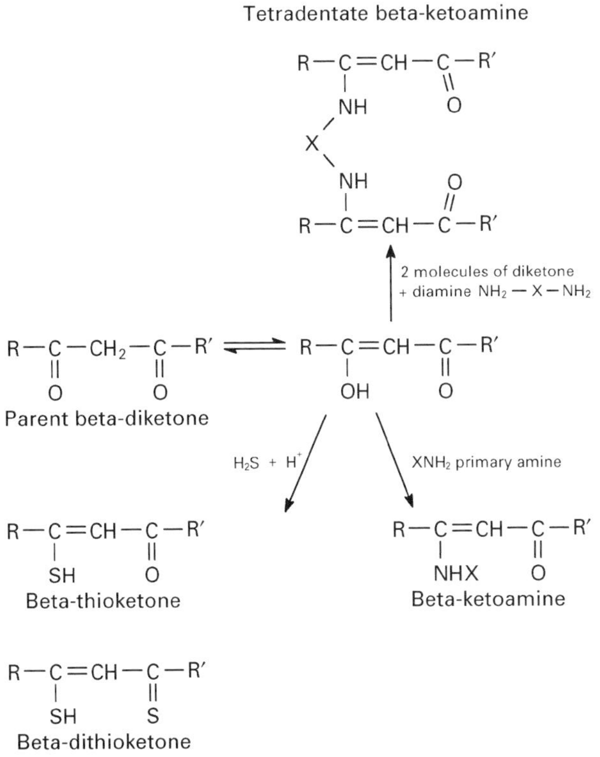

Tetradentate beta-ketoamine

$$R—C=CH—C—R'$$

Parent beta-diketone

Beta-thioketone

Beta-dithioketone

**Figure 2** Formation of beta-difunctional ligands for metal chelate gas chromatography. (Reproduced with permission from Estes SA *et al.* (1980). *Journal of Chromatography and Chromatographic Communications* 3: 471. Copyright John Wiley & Sons Ltd.)

as are those of vanadyl (V(IV)O). The bidentate ketoamines are of only marginal GC stability to submicrogram elution level, but tetradentate beta-ketoamine ligands form a useful group for GC of

**Table 1** Representative tetradentate beta-ketoamine ligands used for chelate formation and gas chromatography of divalent transition metals

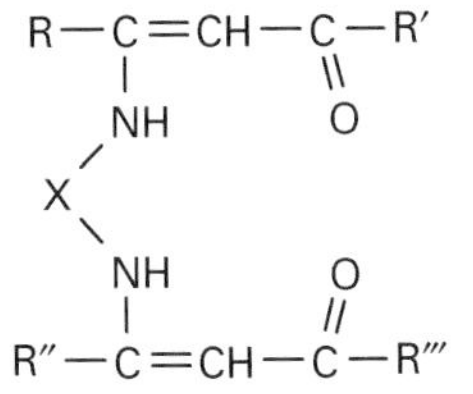

| R | R' | R'' | R''' | X | Ligand symbol |
|---|---|---|---|---|---|
| $CH_3$ | $CH_3$ | $CH_3$ | $CH_3$ | en | $H_2(enAA_2)$ |
| $CH_3$ | $CH_3$ | $CH_3$ | $CH_3$ | pn | $H_2(pnAA_2)$ |
| $CH_3$ | $CH_3$ | $CH_3$ | $CH_3$ | bn | $H_2(bnAA_2)$ |
| $CF_3$ | $CH_3$ | $CH_3$ | $CF_3$ | en | $H_2(enTFA_2)$ |
| $CF_3$ | $CH_3$ | $CH_3$ | $CF_3$ | pn | $H_2(pnTFA_2)$ |
| $CF_3$ | $CH_3$ | $CH_3$ | $CF_3$ | bn | $H_2(bnTFA_2)$ |
| $C(CH_3)_3$ | $CH_3$ | $CH_3$ | $C(CH_3)_3$ | en | $H_2(enAPM_2)$ |
| $C(CH_3)_3$ | $CH_3$ | $CH_3$ | $C(CH_3)_3$ | pn | $H_2(pnAPM_2)$ |
| $CF_3$ | $C(CH_3)_3$ | $C(CH_3)_3$ | $CF_3$ | en | $H_2(enTPM_2)$ |

en = $CH_2$–$CH_2$; pn = $CH(CH_3)$–$CH_2$; bn = $CH(CH_3)$–$CH(CH_3)$.

divalent transition metals. The addition of the extra five-membered ring stabilizes the complexes, which more than offsets their lowered volatility. Table 1 lists the tetradentate beta-ketoamine ligands that have been evaluated for GC of copper, nickel, palladium and vanadyl complexes. The fluorinated chelates have very great electron capturing abilities affording picogram level detection.

**Dialkyldithiocarbamates and dialkyldithiophosphinates** The metals complexed by these ligands are parallel to those chelated by the tetradentate beta-ketoamines for GC applications. Zinc, copper and nickel have been determined in marine bottom sediments and in sea sands and muds. In parallel to the development of fluorinated beta-diketone ligands, fluorinated dialkyldithiocarbamates show analytical promise due to their increased volatility over nonfluorinated analogues. Tavlaridis and Neeb first investigated di(trifluoroethyl)dithiocarbamates, eluting zinc, nickel, cadmium, lead, antimony and bismuth complexes at 185°C. Sucre and Jennings reported effective capillary separation of these complexes of nickel and cobalt(III) on 5 metre fused silica columns. It appears likely that further refinement of high resolution columns will broaden the application of these versatile complexes for analytical GC.

Cardwell and McDonagh separated zinc, nickel, palladium and platinum 0,0'-dialkyldithiophosphinates. These complexes are suitable for selective detection by flame photometry with monitoring of either the $S_2$ or HPO emission modes.

**Metalloporphyrins** One of the most important demonstrations of the expanded range of sample applications for inorganic GC brought about by high resolution fused silica capillary was of transition metal porphyrin complexes. Marriott achieved capillary GC elution of these closed macrocylic ring copper, nickel, vanadyl and cobalt aetioporphyrin I, and octaethyl-porphyrin chelates with Kováts indices in the range 5200–5600.

## Chromatographic Detection

Of particular value for metal compound chromatography are detectors giving 'selective' or 'specific' information on the eluates. Spectral property detectors such as the mass spectrometer (MS), the infrared spectrophotometer (IRS) and the atomic emission spectrometer (AES) fall into this class. The latter 'element selective' detectors are widely used: the microwave-induced plasma in particular for GC and the inductively coupled plasma mass spectrometer (ICPMS) for HPLC. These detector systems typically offer nanogram level detection or lower for metal content of analytes. Metals may be determined directly or through derivatization procedures which render them more readily separable and detectable.

# High Performance Liquid Chromatography

## Size-Exclusion Chromatography (SEC)

SEC separates chemical species by molecular size and shape in a column containing particles of a rigid packing with a defined pore structure. Many applications to inorganic and organometallic species have been reported ranging in size from a few hundred to $10^5$ Da or greater; from labile nickel complexes of alkyl substituted phosphorus esters to molecular clusters of inorganic colloids in the 1–50 μm range, aluminosilicate sols, humic acid metal complexes, metallopeptides and metalloproteins. Ferritin, an iron-containing protein which exists in a number of discrete forms, has been analysed using aqueous SEC, good repeatability being found for iron at the nanogram level. Trace levels of Cd, Zn and Cu metalloproteins in marine mussels were determined using SEC with sequenced UV absorption detection before the ICP. In the petroleum field, effective separation and determination of vanadium and nickel metalloporphyrins has been accomplished. Columns of pore size 10 nm in particular are a viable choice for many separation problems of inorganic chromatography.

## Reversed-Phase High Performance Liquid Chromatography (RPHPLC)

RPHPLC of inorganic and organometallics may be characterized by the detector used and some examples are noted.

Atomic absorption (AA)    The interfacing of column effluent and an AA spectrometer was attained in 1973, tetraalkyllead compounds being separated on $C_{18}$ μ-Bondapak with 70% acetonitrile and 30% water. The superiority of AA detection over UV detection was shown, since gasoline samples have components that mask the tetraalkyl lead compounds.

Inductively coupled plasma (ICP) and direct current plasma (DCP)    Plasma spectral detection for HPLC has emphasized the ICP and to some extent the DCP, in contrast to the dominance of the microwave-induced plasmas as element-selective GC detectors. Metal-specific detection is predominant and will probably remain so until more eluate-selective interface systems can be devised. The ICP became

commercially available in 1974 replacing atomic absorption spectroscopy as the method of choice for metal analysis. The technique and DCP has been coupled to HPLC for trace analysis and speciation of real-world samples such as arsenite, dimethyl arsenate and arsenate. A 130 ng mL$^{-1}$ detection limit for arsenic in organoarsenic acids was found for 100 μL samples after hydride formation. Separation of mercury cations employed an alkyl sulfonate anion as ion pair reagent and, after hydride formation, gave detection limits of 50–100 μg L$^{-1}$. A possible solution to overcome the difficulties in quantitative transfer of HPLC eluate involves a total injection microconcentric nebulizer (DIN) which can achieve almost 100% nebulization and transport efficiency. Detection limits down to 4 mg L$^{-1}$ for zinc have been reported. The DCP gave detection for Cr(III) and Cr(VI) linear over at least three orders of magnitude with a detection limit of 10 μg L$^{-1}$; applications include biological samples from ocean floor drillings, chemical dump sites, surface well water and waste water samples.

Visible and UV detection    Spectrophotometric complexing reagents are widely used for visible spectral determination of many metals at low concentrations, but these complexes are seldom suited to HPLC. Pyridylazonaphthol complexes of copper, nickel and cobalt are separable by reversed-phase LC using acetonitrile/water/citrate buffer at pH 5 (80 : 18 : 2), 0.01 mol L$^{-1}$ ammonium thiocyanate, and other such stable chelates are also amenable. In addition, non-absorbing species may be detected after conversion to chromophores by complexing, ion pairing or other chemical reactions. Neutral metal chelates such as beta-diketonates, beta-ketoaminates, dialkyldithiocarbamates and dialkyldithiophosphinates separable by GC may also be chromatographed by reversed-phase HPLC and less volatile complexes are also amenable to the technique.

Amperometric and differential pulse detection    Dithiocarbamates of Cu(II), Ni(II), Co(II), Cr(III) and Cr(VI) have been detected amperometrically after HPLC separation. Although both the oxidation and reduction reactions of these organometallics are well defined, the ubiquitous presence of reducible oxygen dissolved in the polar solvent results in the oxidative process being more desirable. Knowledge of solvent, electrodes and electrochemistry is essential before a correct HPLC-EC protocol can be stated. Cancer chemotherapeutic drugs cisplatin, mitocynin C and mitoxanthrone are separable on $C_{18}$ columns and are easily detected by oxidation, if the cisplatin oxidation potential is shifted by $0.10 \text{ mol L}^{-1}$ chloride to 0.80 V.

Detection selectivity may be improved by differential pulse detection as in the reversed-phase determination of alkyl- and aryl-mercurials, tri-*n*-butyl tin, triethyl tin and triphenyl tin.

### Ion Pair Chromatography (IPC)

Combination of ion exchange and partition mechanisms and the performance of 5 and 10 μm $C_{18}$ silica bonded phases give high performance ion separation in the direct and ion-paired modes. Pairing ions may be present in the mobile phase or retained on the surface. A $C_{18}$ column coated with $C_{20}$ alkyl sulfate gave baseline separation of Cu(II), Co(II) and Mn(II). The eluted metal ions were detected by absorption spectrophotometry at 530–540 nm after post-column reaction with 4-(2-pyridylazo)-resorcinol (PAR). Paired-ion HPLC separation of iron(II), nickel(II) and ruthenium as cationic 1,10-phenanthroline complexes with alkylsulfonic acids was achieved. Crown ethers can form stable complexes with metal cations, reversed-phase HPLC retention being dependent upon the relative size of the cation and the ring of crown ether. A typical elution order is $Li^+ < Na^+ < Cs^+ < Rb^+ < K^+$, and $Mg^{2+} Ca^{2+} < Sr^{2+} < Ba^{2+}$.

Overall, the choice in inorganic column liquid chromatography is now between reversed-phase, including paired-ion, and ion chromatography, the former typically affording the better resolution.

### HPLC-ICP-Mass Spectrometry (HPLC-ICP-MS)

The most extensively developed plasma mass spectral analytical technique is that of ICP-MS. The argon ICP acts as a mass spectral ion source; for a sample solution, after aerosol formation in a nebulizer and spray chamber, analyte is injected into the plasma where it undergoes desolvation, vaporization, atomization and ionization. A portion of the ions is sampled from the centre of the plasma and directed, through a low pressure interface, into the mass spectrometer. ICP-MS combines advantages of ICP-AES such as multi-element analysis, wide dynamic range and speed, with mass spectral acquisition, enhanced detection limits (typically $0.01–0.1 \text{ mg L}^{-1}$) and capability for isotopic analysis. In HPLC-ICP-MS detection limits as low as 100 pg/peak have been obtained for many elements. Ion exchange and ion pair chromatography were used for speciation of triorganotin species and arsenic speciation has been examined in a number of studies. The technique shows excellent prospects in biomedical and clinical studies in which analyte levels are usually below the capabilities of ICP emission

detection. Interfaced aqueous SEC with ICP-MS was used for element and isotope ratio detection of lead and copper in protein fractions of serum and blood cell haemolysate with molecular weight ranges from 11 kDa to > 600 kDa and detection limits in the 1–10 μg $L^{-1}$ range for metallodrugs and their metabolites, measured in samples from patients undergoing gold drug therapy for arthritis. The elements mercury, arsenic, lead and tin have attracted most interest from the trace element environmental point of view.

## Conclusions

High performance gas and liquid chromatography of metal compounds have always presented considerable challenges, but have in return provided many analytical and characterization insights, both qualitative and quantitative. In practical terms, the wider adoption of element-specific spectral detection depends on the continual development of commercial instrumentation to permit inter-laboratory comparisons of data and the development of 'recommended' methods of analysis which can be widely used. Many areas of analysis are subject to restrictions designed to ensure high levels of accuracy and precision. Fully integrated units which remove the need for analysts to interface their own GC, emission device and spectrometer may become as familiar in the future as GC-MS and GC-FTIR systems are today. Integrated HPLC and SFC systems will be longer delayed, but their eventual adoption is inevitable in view of the broad scope of these separation methods.

*See also:* **II / Chromatography: Gas:** Detectors: General (Flame Ionization Detectors and Thermal Conductivity Detectors); Detectors: Mass Spectrometry; Gas Chromatography-Infrared; Gas-Solid Gas Chromatography; Historical Development. **Chromatography: Liquid:** Detectors: Mass Spectrometry.

## Further Reading

Crompton TR (1982) *Gas Chromatography of Organometallic Compounds*. New York: Plenum Press.

Guiochon G and Pommier C (1973) *Gas Chromatography in Inorganics and Organometallics*. Michigan: Ann Arbor.

Krull IS (ed.) (1991) *Trace Metal Analysis and Speciation*. Journal of Chromatography Library, vol. 47. Amsterdam: Elsevier.

Lederer M (ed.) (1984) *Separation Methods in Inorganic Chemistry*, Chromatographic Reviews, vol. 29. *Journal of Chromatography* 313: 1.

McDonald JC (1985) *Inorganic Chromatographic Analysis*. New York: John Wiley.

Michal J (1973) *Inorganic Chromatographic Analysis.* New York: Van Nostrand.

Moshier RW and Sievers RE (1965) *Gas Chromatography of Metal Chelates.* Oxford, London: Pergamon.

Schwedt G (1981) *Chromatographic Methods in Inorganic Analysis.* Heidelberg, New York: Huthig Verlag.

Uden PC (1995) Element specific chromatographic detection by atomic absorption, plasma atomic emission and plasma mass spectrometry. *Journal of Chromatography* 703: 393.

Uden PC and Henderson DE (1977) Determination of metals by gas chromatography of metal complexes – a review. *Analyst* 102: 889.

# METAL COMPLEXES

## Ion Chromatography

**D. J. Pietrzyk**, University of Iowa, Iowa City, IA, USA

### Introduction

Many different and clever methods have been reported in the literature for the separation of metal ions. Most involve chemical reactions in which complexes are formed between the hydrated metal ions and an inorganic or organic ligand. If an organic ligand is employed coordination is usually through oxygen, sulfur or nitrogen atoms, individually or in combination.

In a simplified representation the stepwise and overall formation of metal–ligand complexes can be represented by the following:

$$M^{2+} + HL = ML^+ + H^+ \qquad [1]$$

$$ML^+ + HL = ML_2 + H^+ \qquad [2]$$

$$M^{2+} + nHL = ML_n^{+2-n} + nH^+ \qquad [3]$$

where $M^{2+}$ is a divalent hydrated cation (acceptor) and HL is a monoprotic weak acid ligand (donor) that replaces the waters of coordination in a series of steps. The formation and dissociation of the metal–ligand (donor–acceptor) complexes proceeds, often rapidly, by a series of equilibrium reactions, each of which is defined by a formation constant. Because of the metal ion coordination number, the stepwise equilibria, the formation constant, and concentrations of the metal ion and ligand, a series of complexes (see the expression of the total equilibrium steps in eqn [3]) may coexist in the solution. When the ligand is neutral the charge of the resulting complexes are positive. However, the complexes may associate with solution anions to produce neutral species. Ligands can also be anions, eqns [1]–[3], and thus the complexes that form can be cations, neutral or anionic, depending on the coordination number of the metal ion, the number of ligands bound to the metal ion, the formation constants for the complexes that form, and the metal ion and anionic ligand concentrations.

Conversion of a metal analyte into a complex is important in separations for two major reasons. First, the chemical and physical properties of the metal–ligand complex are sharply different than those of the metal ion hydrate. Solubility, ionic or polar character, and even volatility can be different. Formation constants for the metal–ligand complexes will differ and pH is often an important variable. Thus, the number of separation strategies that are applicable to the separation of metal ions are broadened and the number of variables that can be altered to bring about resolution in the separation of a complex mixture of metal ions is increased. Second, some metal–ligand complexes can be more easily detected, which improves quantitative estimation and even identification, than the metal ion hydrate. For example, metal–ligand complexes can be highly coloured, while some will fluoresce and/or have electrochemical properties that can be monitored.

Using complex formation between a ligand and metal ion to facilitate metal ion separations was an important strategy in the development of low efficiency liquid column chromatographic separations of metal ions, particularly in separations by ion exchange and partition column chromatography. Similarly, complex formation is a key elution parameter in high efficiency liquid column chromatographic separations. Each of these column strategies is briefly described in the following.

### Ion Exchange

In ion exchange column chromatography the column is packed with either a cation exchanger, which will exchange cations, or an anion exchanger, which will exchange anions. The exchangers are like insoluble electrolytes and exchange ions rapidly, reversibly and stoichiometrically. Cation exchange is represented in eqn [4] while eqn [5] illustrates anion exchange.

$$R^-C^+ + M^+ = R^-M^+ + C^+ \qquad [4]$$

$$R^+A^- + X^- = R^+X^- + A^- \qquad [5]$$

Here R is the ion exchanger matrix containing either an anionic or cationic ionogenic group, $C^+$ and $A^-$ are co-cation and co-anion, respectively, and $M^+$ and $X^-$ are the analyte cation and anion, respectively. The direction of the equilibrium in eqns [4] and [5] is determined by the selectivity coefficient for the exchanger towards the two competing ions. Thus, in the absence of mass action effects the exchanger will prefer the ion with the highest selectivity coefficient. For a mixture of analytes, for example a mixture of metal ions and a cation exchanger, the metal ion with the smallest selectivity coefficient would elute first and the metal ion with the largest selectivity coefficient would elute last when using a mobile phase containing an electrolyte that provides a cation of appropriate selectivity and concentration. To increase elution of the metal ion the electrolyte concentration in the mobile phase is increased, or a different electrolyte that provides a cation of higher selectivity is used.

While low efficiency metal ion separations are possible on cation exchangers, resolution is improved considerably when a ligand is included in the mobile phase and the separation occurs because of the properties of the metal complex. For example, elution of $M^+$ from the cation exchanger is enhanced with a mobile phase containing the ligand, HL, because of the formation of metal ion–ligand complexes. As shown in eqn [4], $M^+$ is retained on the cation exchanger through competition with the mobile phase cation $C^+$. When the ligand is in the mobile phase, the ligand will form a complex with the metal ion, causing the equilibrium in eqn [4] to shift to the left with the formation of the metal ion–ligand complex. The overall effect of the ligand can be represented by eqn [6]:

$$R^-M^+ + 2HL + C^+ = R^-C^+ + ML_2^- + 2H^+$$
$$[6]$$

From eqn [6] the best mobile phase ligand, assuming formation constants and solubility are favourable, will be one that forms anionic complexes with the metal ion.

On the other hand, if the metal ion–ligand complex that forms is anionic, the complex can be retained by an anion exchanger, or

$$M^+ + 2HL = ML_2^- + 2H^+ \qquad [7]$$

$$R^+C^- + ML_2^- = R^+ML_2^- + C^- \qquad [8]$$

In this case the metal ion is subsequently removed from the anion exchanger by reversing the equilibrium in eqn [7], which is done by reducing the concentration of the ligand in the mobile phase. This causes the equilibrium in eqn [8] to shift to the left, thus removing the metal ion from the anion exchanger.

## Low Efficiency Ion Exchange Separation of Metal Ions

In the 1940s ion exchange was recognized as an excellent strategy for the separation of metal ions. The advantages of including ligands in the mobile phase were soon realized and many different types of inorganic and organic ligands were evaluated to aid metal ion separations. Chloride ion was one inorganic ligand that was studied extensively. Many metal ions complex with $Cl^-$ stepwise and the equilibria describing the stepwise formation of metal ion–chloride complexes are summarized by eqn [9]:

$$M^{2+} + nCl^- = MCl^-, MCl_2, MCl_3^-, MCl_n^{2-n}$$
$$[9]$$

where the complexes that are present depend on the formation constants for the individual steps, the coordination number of the metal ion, and the $Cl^-$ concentration. Depending on these factors, the metal analyte may be a cation, a neutral species or an anion. Thus, the charge of the metal-containing species may not only be altered but actually reversed and separation of the metals can become one of cations from anions.

Separation of metal ions is possible on either a strong acid cation exchanger or a strong base anion exchanger using a HCl mobile phase. With the cation exchanger increasing the HCl concentration shifts the equilibria represented by eqn [9] to the right, thus causing the metal analytes to elute from the cation exchange column as the metal chloro complexes (see eqns [4] and [6]). Those metals that form chloro complexes more readily (more favourable formation constants) will elute first.

The better resolution for metal ion separations is actually obtained when using a strong base type anion exchanger with Cl⁻ as the mobile phase ligand. For the separation on an anion exchanger the mixture of metal ions is placed on the anion exchanger from a concentrated HCl solution. This converts the metal ions into chloro complexes (see eqn [9]) and these complexes are retained by the anion exchanger. The HCl concentration is then progressively reduced, the equilibria in eqn [9] shifts to the left, and the metal ions are removed from the column (see eqns [7] and [8]). In this case the more stable metal chloro complexes stay on the anion exchanger the longest and the least stable metal chloro complexes come off the earliest.

The retention of all the metal ions, some at several oxidation states, was determined on a strong base anion exchanger as a function of HCl concentration from  < 0.001 mol L⁻¹ to 12 mol L⁻¹ HCl. These data allow one to predict elution order and conditions for the separation. **Table 1** summarizes some of these results by listing the elution order for several metal ions on the anion exchanger from 12 mol L⁻¹ HCl to dilute HCl. As HCl concentration in Table 1 is reduced, those metal ions above the HCl concentration listed are eluted while those below are retained on the exchanger. Many metal ion separations, including complex mixtures, are possible with the HCl eluent. **Figure 1** shows the separation of the transition elements on a strong base anion exchanger by a successive decrease in mobile phase HCl concentration. The column was 26 cm × 0.29 cm² and about 6 mg of each cation was separated. Each metal ion was collected and its quantity was determined by chemical or instrumental methodology. Mn²⁺ was determined spectrographically, Fe³⁺ and Zn²⁺ were radioactive tracers and were determined by radioactivity, and Co²⁺, Ni²⁺, and Cu²⁺ were determined spectrophotometrically.

**Table 1** Elution order for common metal ions on a strong base anion exchanger with an HCl mobile phase

| Mobile phase HCl concentration (mol L⁻¹) | Metal ion |
|---|---|
| 12 | Not retained: Ni²⁺, Al³⁺, lanthanoids, Th⁴⁺  Slight retention: Sc³⁺, As⁵⁺, Cr³⁺, Mn²⁺ |
| 9.5 | Ti⁴⁺ |
| 8 | Hf⁴⁺ |
| 7.5 | Zr⁴⁺ |
| 7 | Fe²⁺ |
| 6 | U⁴⁺ |
| 4.5 | Co²⁺, As³⁺ |
| 3 | Cu²⁺ |
| 1 | UO₂²⁺, Fe³⁺ |
| 0.02 | Zn²⁺ |
| 0.001 | Cd²⁺ |

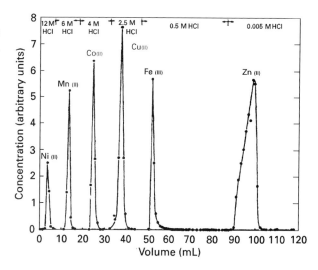

**Figure 1** Separation of transition elements on a strong base anion exchanger from an HCl mobile phase. M = mol L⁻¹. (Reproduced from Kraus and Moore, 1953. Reprinted by permission of American Chemical Society.)

Success with chloride ion as the eluent ligand led to studies with many other inorganic ligands. Like Cl⁻, F⁻ is also an important mobile phase ligand and was employed in a mobile phase that was 1 mol L⁻¹ HF as a function of HCl concentration. The presence of the F⁻ is particularly useful because it forms complexes with Group IV and V elements such as Zr⁴⁺, Hf⁴⁺, Nb⁵⁺ and Ta⁵⁺ and aids their separation. Other inorganic ligands have also been shown to be useful but typically do not have the broad scope of applications characteristic of the HCl and HCl/HF elution systems. These ligands include NO₃⁻, SO₄²⁻, Br⁻, I⁻, SCN⁻, CO₃⁻ and CN⁻. While the analytical applications of these ligands may be limited, several have been shown to be particularly useful in commercial recovery and purification applications, for example the isolation of uranium and thorium from low grade ores. **Figure 2** illustrates an elution diagram that was used for the separation of a multicomponent high temperature alloy on a strong base anion exchanger using a combination of the HCl and HCl/HF elution scheme. Note that the column dimensions and sample size of 1 g are large and by today's standards this would be called a prep column and separation. Because each metal analyte was separated in such a large quantity each metal ion was determined by chemical methodology, namely by ethylenediaminetetraacetic acid (EDTA) titration or by gravimetry.

Organic ligands have been used in eluents to improve resolution in low efficiency ion exchange metal ion separations. Iminodiacetic acid, nitrilotriacetic acid and EDTA are examples of mobile phase ligands that have been used that form very stable complexes with many metal ions, while citric

Ion exchange column: strongly basic anion exchange resin (200–400 mesh); diameter: 25 mm; length: 20 cm; weight of sample: 1 g

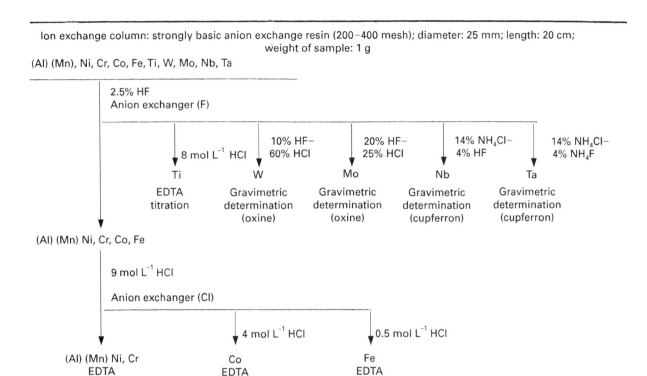

**Figure 2** A flow diagram for the separation of a high temperature alloy on a strong base anion exchanger. The anion exchanger is a 40–75 μm, 2.5 cm × 20 cm strong base anion exchange column and the sample size is about 1 g. (Reproduced from Wilkins, 1959. Reprinted by permission of Pergamon Press Ltd.)

acid, tartaric acid, malonic acid, oxalic acid, diglycolic acid and α-hydroxyisobutyric acid (HIBA) are examples of useful ligands that form metal ligand complexes of modest stability. In addition to formation constants, pH is also an important mobile phase variable that can be optimized to enhance resolution because all of these ligands are weak acids. Another mobile phase parameter that can be altered to influence elution is to use a mixed solvent for the mobile phase. The presence of the organic solvent in the mixture has a pronounced effect on the formation constants for the complexes as well as the exchange equilibrium with the exchanger. The type of organic solvent and its concentration then become variables in influencing elution time and resolution.

The unique complexing properties of organic ligands have been used to isolate and concentrate metals from ores and other geological samples. For example, one of the more successful applications of this kind was the isolation and separation of the lanthanoids on cation exchangers from mobile phases containing an organic ligand. Prior to this development individually pure lanthanoids were not readily available. Similar success was subsequently achieved in the isolation and purification of uranium and other actinoids, hafnium and zirconium, and many other less familiar metals from natural occurring ores and minerals.

**Table 2** lists several elution conditions that have been used for the low efficiency separation of lanthanoids on anion or cation exchangers using a ligand in the mobile phase. The separation of the lanthanoids on a strong acid cation exchanger with lactate as the mobile phase ligand and a small pH increase during the elution is shown in **Figure 3**. Each lanthanoid was collected in a series of fractions and the lanthanoid in each fraction was determined spectrophotometrically by complex formation with 1-(2-pyridylazo)-2-naphthol (PAN). The pH and the formation constants for the lactate complexes determine the elution order. When HIBA is used as the ligand instead of lactate, the same elution order is obtained for the lanthanoids. In another example cation exchange was used to separate 35 metal ions into six separate groups by using mobile phase ligands citric acid, N-hydroxyethyl(ethylenedinitrilo)triacetic acid, EDTA, HCl and pH control in a predetermined elution programme.

## Ligands in the Separation of Metal Ions by Partition Chromatography

Column liquid–liquid partition chromatography of metal ions can be one of two types. In one case the stationary phase is a nonaqueous immiscible solvent

**Table 2** Mobile phase ligand conditions used for the separation of the lanthanoids on anion and cation exchangers

| Mobile phase conditions | Exchanger | Application |
|---|---|---|
| HIBA at pH = 5.2 | Cation | Early lanthanoids |
| HIBA at pH = 4.7 | Cation | Lanthanoid radionuclides |
| HIBA, pH gradient from 3.4 to 4.0 | Cation | Lanthanoids and fission products |
| 0.24 mol L$^{-1}$ lactate at 87°C | Cation | Lanthanoids |
| 1.1–1.25 mol L$^{-1}$ lactate and 3.1 to 3.25 pH gradient at 87°C | Cation | Lanthanoids |
| Ammonium lactate gradient at pH 5 and 95°C | Cation | Radiochemical separation of lanthanoids from fission products |
| 26 mmol L$^{-1}$ EDTA at pH 3.62 and 87°C | Cation | Ce, Pr, Nd, Sm, La |
| 5–6 mol L$^{-1}$ HNO$_3$ | Anion | La, Ce, Pr group separation |
| EDTA | Anion | Lanthanoids |
| Citrate | Anion | Lanthanoids |

held up by an inert support and the mobile phase is usually an aqueous solution containing electrolyte, buffer, and a ligand. This type of column chromatography is called liquid–liquid reversed-phase partition chromatography (RPPC). The other strategy is the opposite – the stationary phase liquid is a polar or aqueous solvent held up by the inert support and the mobile phase is an immiscible organic solvent or solvent mixture of less polarity. In both cases the additives cause the metal ions to distribute between the two phases as the metal ions pass through the column. Separation occurs because of differences in the equilibrium positions imposed by the additives, namely the ligand, the resulting formation constants, the pH, and the properties of the two solvents that make up the two phases.

These techniques, which may also be carried out with solvent combinations that are partially miscible suffer from gradual changes in the stationary phase liquid over prolonged column use. This means it is difficult to maintain or reproduce a uniform,

constant stationary liquid phase, sometimes even in a single elution run and often over repeated runs. Even solvents that are thought to be completely immiscible will have a low level equilibrium distribution of one solvent into the other when they are brought together. Thus, the stationary liquid phase will be slowly removed (solvent bleed) from the inert support.

Because of solvent bleeding high efficiency liquid–liquid partition chromatography is rarely used and as a separation strategy it has been replaced by high efficiency reversed-phase and normal-phase column chromatography. Nevertheless, many difficult separations of metal ion mixtures are possible by RPPC and RPPC can be readily applied to cases where the chromatographer needs to separate large quantities of metal ions and is not faced with requirements of short analysis times, the best detection limits, and/or regulatory method controls. Several examples of RPPC described below illustrate the scope of using ligands in RPPC separations of metal ions.

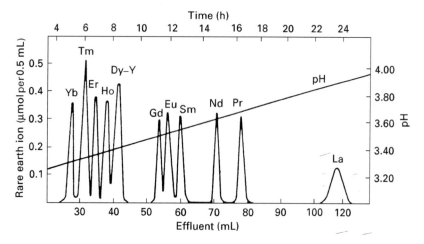

**Figure 3** Separation of the lanthanoids on a strong acid cation exchanger. The cation exchanger column is a 40–75 μm, 0.5 cm × 100 cm column at 80°C and the mobile phase is a lactate buffer solution and pH gradient at about 0.7 mL min$^{-1}$. (Reproduced from Inczédy, 1966, p. 166. Reprinted by permission of Pergamon Press Ltd.)

**Table 3** Selected applications of porous organic copolymers as support materials for the partition chromatographic separation of metal ions[a]

| Metals | Porous polymer | Extraction system |
|---|---|---|
| 27 metal ions | XAD-2 | Isopropyl ether/HCl |
| | | Isobutylmethyl ketone/HCl |
| | | Trioctylphosphine oxide/HCl |
| Ga, In, Th | XAD-2 | Isobutylmethyl ketone/HBr |
| $U^{6+}$ | XAD-2 | Dioctyl sulfoxide/1,2-dichloroethane |
| $Mo^{6+}$, $W^{6+}$, $V^{6+}$ | XAD-2 | Aliquat 336 liquid anion exchanger/ toluene/$H_2SO_4$ |
| Cu | XAD-2 | Aliphatic $\alpha$-hydroxyoxime/toluene |
| $Mo^{6+}$ | XAD-2 | 5,8-Diethyl-7-hydroxydecane-6-one oxine/toluene |
| Au | XAD-2 | HCl/$H_2O$ |
| Fe, Cu | XAD-7 | Kelex 100/$H_2O$ |
| Zn, Cd, Hg | Macroporous polystyrene–divinylbenzene | Dithiozone/dibutyl phthalate |
| Ni, Fe, Co | Ethylstyrene–divinylbenzene | Monothiobenzoylmethane/heptane |

[a]Reproduced from Pietrzyk, 1989, p. 144. Reprinted by permission of Wiley-Interscience.

RPPC shares many features with solvent extraction and a wealth of information is available in the literature on the solvent extraction of metal–ligand complexes from one phase to another. In most cases a successful extraction depends on the properties of the complex that forms between the ligand and the metal ion. Complex formation constants, metal coordination, solubility of the ligand in the aqueous phase and the nonaqueous phase, solubility of the complex in each of the two phases, and pH are the main factors that determine the percent extraction and the selectivity in the extraction, and consequently the ability to resolve one metal ion from another. These same factors influence metal ion resolution in RPPC and must be optimized to obtain a successful separation.

Two of the more successful inert stationary phases for RPPC are polystyrene–divinylbenzene and acrylic acid-based macroporous copolymers. These are capable of holding up appreciable quantities of nonpolar organic solvents, many of which are desirable for partitioning procedures. **Table 3** lists several RPPC quantitative metal ion separations that have been carried out where an inorganic or organic ligand is employed in the two-phase system to complex metal ions and bring about resolution on either the styrene or acrylic acid-type copolymers as the inert stationary phase.

## Chelating Ion Exchanger

Instead of adding a ligand to the mobile phase, a chelating group can be chemically bound to a solid, inert matrix. In this case coordination occurs between the metal ion and the bound chelating group. Resolution is possible because of differences in formation constants for the differential metal complexes that form with the stationary phase-bound chelating group, the mobile phase pH, and the ionic strength. Employing a second chelating group or ligand in the mobile phase to establish a competition between the bound chelating group and the mobile phase ligand towards the metal ion analytes is also an excellent elution procedure to effect resolution of the metal ion mixture.

Chelating ion exchangers often have several disadvantages, even though metal ion retention can be high and selectivity can be favourable. First, the kinetics for the exchange or dissociation of the metal ion from the chelating group bound to the stationary phase can be slow, and this causes poor elution behaviour. Second, few chelating ion exchangers are commercially available. One notable exception is a chelating ion exchanger that has the iminodiacetic acid group, $-CH_2N(CH_2CO_2H)_2$, attached to a polystyrene–divinylbenzene copolymer. And third, because most chelating ion exchangers have to be synthesized in the laboratory, reproducibility from column to column is only fair. Furthermore, column efficiency in analytical separations is rarely exceptional because the kinetics are usually not favourable.

The major advantage of a chelating ion exchanger, which is very important to environmental and ultratrace metal analysis, is that the exchanger can be used to isolate trace levels of metal ions from samples and/or to concentrate the trace metal ions prior to their determination. This has been proven to be particularly valuable for the isolation/concentration of metal ions other than the Group I metal ions, particularly transition metal ions, the lanthanoids and the actinoids. Often this can be done from solutions that may contain substantial quantities of

**Table 4** Chelating ion exchangers

| Chelating group | Chelating group |
| --- | --- |
| Anthranilic acid | N-substituted hydroxylamine |
| Arsonic acid | 8-Hydroxyquinoline |
| Crown ether | Iminodiacetic acid |
| $\beta$-Diketone | Isothiuronium |
| Dimethylglyoxime | Nitrilotriacetic acid |
| Dithiocarbamate | Rescorinol |
| Dithioizone | Salicylic acid |
| Ethylenediaminetetraacetic acid | Thioglycolate |
| Hydroxyamic acid | |

monovalent electrolyte, for example brine solution, boiler water and sea water. **Table 4** lists several chelating ion exchangers that have been synthesized in the laboratory and some typical applications of these chelating ion exchangers.

# High Efficiency Column Chromatographic Separation of Metal Ions and the Influence of Ligands on their Resolution

The analytical chemistry and role of the ligand in high efficiency or high performance liquid column chromatographic (HPLC) separations of metal ions is much the same as in low efficiency liquid column chromatographic separations. Mobile phase concentration of reagents including the ligand and metal ion analyte concentration are much lower due to more favourable chromatographic properties of high efficiency columns. Selectivity, retention, and particularly detection limits are improved. Formation constants, stepwise equilibria, coordination number, and formation of positive, anionic or neutral complexes are still crucial factors that influence separation. In addition, the rate of complex formation and dissociation is a major contributing factor because of the significant increase in linear velocity of the analyte in the high efficiency column.

Cation exchangers, anion exchangers, chelating exchangers, and reversed stationary phases are the major column stationary phases that are employed in high efficiency metal ion analyte separations. These modern, high efficiency stationary phase particles are available as small (5 and 10 μm), uniform, and spherical-sized particles of considerable physical strength that can be packed uniformly and reproducibly into columns to yield very favourable mass transfer, and thus high efficiency. All of these are properties that the column must possess in order to exhibit high efficiency. Columns that satisfy these criteria are commercially available.

Examples illustrating the scope of modern applications of ligands to improve the resolution in the separation of metal ions on high efficiency columns of cation and anion exchangers, reversed stationary phases and chelating exchangers are outlined in the following sections.

## High Efficiency Cation Exchangers

Metal ions, particularly multivalent cations, are highly retained on high efficiency strong acid-type cation exchangers, even though exchange capacities may be very low. One approach to elute metal ions, particularly multivalent cations, from the cation exchanger is to use a mobile phase cation derived from ethylenediamine, $H_2NCH_2CH_2NH_2$, (En). In an acidic solution En, which is basic, will exist as the dication, $^+H_3NCH_2CH_2NH_3^+$ ($H_2En^{2+}$), providing the pH is low enough. Thus, the eluent strength of the mobile phase is determined by the mobile phase $H_2En^{2+}$ concentration and metal ion analyte elution on a typical high efficient cation exchanger follows ordinary cation exchange. A second approach is to use a ligand in the mobile phase. A successful ligand is one that will form neutral or anionic complexes with the analyte metal ions that are stable (formation constants will influence elution order) and form and dissociate rapidly. It should be noted that although En is a ligand for many multivalent cations, in the diprotonated form its complexing ability is sharply reduced. When a ligand, for example citric acid, which forms complexes with metal ions, is included in the buffered mobile phase along with the $H_2En^{2+}$, resolution now depends on pH. This influences dissociation of citric acid (and also the eluent cation $H_2En^{2+}$ and its potential complexing ability), the formation constants for the metal ion citrate complexes (and metal ion En complexes if they form), and the concentration of the citric acid ligand. The more stable the metal ion citrate complex and the higher the concentration of the citrate, the more quickly the metal ions are eluted. **Figure 4** illustrates the high efficiency separation of several divalent metal ions on a strong acid high efficient cation exchanger using the combined effect of citrate as the ligand and $H_2En^{2+}$ as a divalent mobile phase cation. A similar separation is possible using tartrate or oxalate rather than citrate as the mobile phase ligand.

Because complex formation constants differ, small differences in selectivity and elution order are obtained when the elution behaviour of citrate, tartrate and oxalate as ligands are compared. Consequently, metal ion elution order and selectivity for given metal ions, for example for transition metals, can be changed through the selection of the ligand.

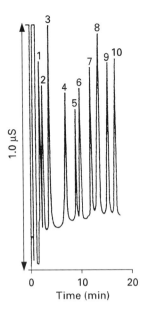

**Figure 4** Separation of alkali, alkaline earth, and transition element metal ions on a cation exchanger. The column is a TSK IC cation SW column and the mobile phase is 3.5 mmol $L^{-1}$ $H_2EN^{2+}$, 10 mmol $L^{-1}$ citric acid, pH 2.8, at a flow rate of 1.0 mL $min^{-1}$ with conductivity detection. Peaks are: 1, $Na^+$; 2, $K^+$; 3, $Cu^{2+}$; 4, $Ni^{2+}$; 5, $Co^{2+}$; 6, $Zn^{2+}$; 7, $Fe^{3+}$; 8, $Mn^{2+}$; 9, $Cd^{2+}$; 10, $Ca^{2+}$. (Reproduced from Timberbaev and Bonn (1993). Reprinted by permission of Elsevier Science Publishing.)

**Figure 5** illustrates the separation of several transition metals using a mobile phase that contains both oxalate and citrate as ligands and gives an elution order slightly different from the order obtained in Figure 4. Metal ion complex formation is also very important in detection in high efficiency ion exchange separations of metal ions. In Figure 5 the metal ions are converted into complexes postcolumn online by reaction of the metal ion with 4-(2-pyridylazo)resorcinol (PAR) and the complex is detected by absorbance at 520 nm.

HIBA at pH 4.6 is a good ligand to use to separate lanthanoids on a cation exchanger. This separation is shown in **Figure 6**, where a gradient of 0.018–0.070 mol $L^{-1}$ HIBA is used for the elution of the lanthanoids on a Nucleosil SCX column, a sulfonated bonded phase-type cation exchanger. In Figure 6 detection was by a postcolumn reaction, which produces the highly coloured complex between the ligand, 3-(2-arsonophenylazo)-4,5-dihydroxy-2,7-naphthalene disulfonic acid trisodium salt (Arsenazo 1) and each of the lanthanoids as they are separated. If $H_2En^{2+}$ is also included in the mobile phase with HIBA, separation of the first seven lanthanoids is obtained without employing an HIBA mobile phase gradient.

The properties of the cation exchanger used in the column will also influence the selectivity, even caus-

ing a reversal. For example, Figure 5 is a separation on a surface-sulfonated cation exchanger, while the cation exchanger used in Figure 4 is a surface/interior-sulfonated cation exchanger.

The latex-based exchanger affects elution order when ligands are used in the mobile phase because of the exchanger's composition. The latex-based cation exchanger, which is commercially available from the Dionex Corporation, for example Ion Pac CS5, is composed of a surface-sulfonated substrate as its central core, uniformly coated with a thin layer of aminated latex particles. This basic surface is then coated by a thin, uniform layer of sulfonated latex particles. The cation exchanger groups ($-SO_3H$) are on the surface while the aminated or anion exchange groups ($-NR_4^+$) are in the interior

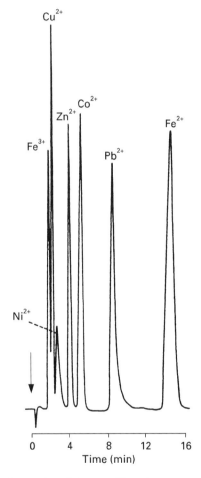

**Figure 5** Separation of the transition element metal ions on a surface-sulfonated cation exchanger. The column is an Ion Pac CS2 column and the mobile phase is 0.01 mol $L^{-1}$ oxalic acid, 0.0075 mol $L^{-1}$ citric acid, pH 4.2, at a flow rate of 1.0 mL $min^{-1}$. Detection is postcolumn absorbance at 520 nm after reaction with PAR. Sample injection is 50 μL containing 5 ppm $Fe^{3+}$, 0.5 ppm $Cu^{2+}$, 0.5 ppm $Ni^{2+}$, 0.5 ppm $Zn^{2+}$, 1 ppm $Co^{2+}$, 10 ppm $Pb^{2+}$, and 5 ppm $Fe^{2+}$. (Reproduced from Weiss, 1995, p. 197. Reprinted by permission of VCH Publishers.)

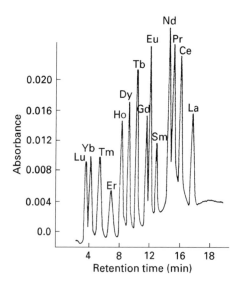

**Figure 6** Separation of the lanthanoids on a sulfonated silica-bonded phase cation exchanger. The column is a 10 μm, 0.4 cm × 10 cm, Nucleosil 10 SA column and the mobile phase is a linear gradient of 0.018 mol L$^{-1}$ to 0.070 mol L$^{-1}$ HIBA at pH 4.6. Sample injection is 10 μL containing about 10 μg mL$^{-1}$ of each lanthanoid. (Reproduced from Elchuk and Cassidy, 1979. Reprinted by permission of American Chemical Society.)

of the particle. Thus, the observed elution for metal ions on the Ion Pac CS5 will be influenced by both cation and anion exchange and the equilibria that favour cationic or anionic metal–complex formation. For example, the transition metal elution order in **Figure 7** on the Ion Pac CS5 column with oxalate as a mobile phase ligand is quite different from the order found in either Figure 4 or 5 and is due to retention of anionic oxalate complexes that form between several of the transition metal ions and oxalate. When the Ion Pac CS5 column is used with a mobile phase containing pyridine-2,6-dicarboxylic acid (PDCA) as the ligand at pH 4.8, the elution order that is obtained is similar to that obtained by the sulfonated, nonlatex-type cation exchangers used in Figures 4 and 5.

### High Efficiency Anion Exchangers

Some ligands will form very stable complexes with metal ions that are also anionic. It is often possible in these cases to separate the anionic metal–ligand complexes on high efficiency anion exchangers. For example, transition metal–PAR complexes are anionic and these complexes can be separated at high efficiency. Since the complexes are also highly coloured, detection of the separation is readily done at favourable detection limits by absorption. In another example the anionic complexes that form between multivalent cations and EDTA are anionic at the appropriate pH and can also be separated on

anion exchangers using an $HCO_3^-/CO_3^{2-}$ mobile phase that is common to ordinary ion chromatographic separation of anionic analytes. The basic requirement of the formation of a very stable anionic complex between the metal ion and the ligand is often a limiting factor and for this reason separation of metal ions in the presence of a ligand on a high efficiency cation exchanger is often a more versatile separation strategy.

The latex-based cation exchanger that possesses both cation and anion exchanger properties, such as the Ion Pac CS5, however, does provide a way to separate metal ion mixtures in the presence of mixed ligands where both cation and anion exchange are involved. This is illustrated in **Figure 8**, where both transition metals and lanthanoids are separated in one run on the high efficiency Ion Pac CS5 column using a complex gradient at the appropriate pH where the concentration of the ligands oxalic acid, diglycolic aci?d and PDCA are varied as a function of elution time. The PDCA causes the elution of the transition metals to be similar to that found on the

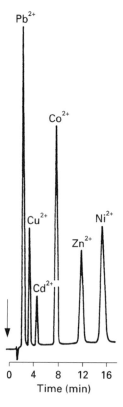

**Figure 7** Separation of several transition element metal ions with oxalic acid as the mobile phase ligand. An Ion Pac CS5 column and a 0.5 mol L$^{-1}$ oxalic acid, pH 4.8, mobile phase at 1 mL min$^{-1}$ was used. Injection volume was 50 μL and contained 4 ppm Pb$^{2+}$, 0.5 ppm Cu$^{2+}$, 4 ppm Cd$^{2+}$, 2 ppm Co$^{2+}$, 2 ppm Zn$^{2+}$, 4 ppm Ni$^{2+}$. (Reproduced from Weiss, 1995, p. 199. Reprinted by permission of VCH Publishers.)

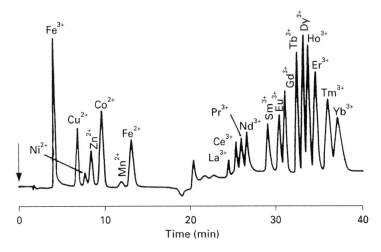

**Figure 8** Separation of the transition elements and lanthanoid metal ions using a three-ligand mobile phase. An Ion Pac CS5 column and a mobile phase gradient of PDCA, oxalic acid, diglycolic acid, and LiOH for pH adjustment at 1 mL min$^{-1}$ was used. Injection volume was 50 μL and contained 2 ppm Fe$^{3+}$, 1 ppm Cu$^{2+}$, 3 ppm Ni$^{2+}$, 4 ppm Zn$^{2+}$, 2 ppm Co$^{2+}$, 1 ppm Mn$^{2+}$, 3 ppm Fe$^{2+}$, 7 ppm of each lanthanoid. (Reproduced from Weiss, 1995, p. 205. Reprinted by permission from VCH Publishers.)

sulfonated cation exchanger (see Figure 5), while the lanthanoids remain on the column as trivalent anions. In the later stages of the gradient oxalic acid and diglycolic acid become more significant in concentration and cause the lanthanoids to be eluted in reverse order to that obtained by the high efficiency cation exchange separation of the lanthanoids (see Figure 6). If only the lanthanoids are to be separated, they can be separated in the order shown in Figure 8 on the Ion Pac CS5 column in about 25 min with the oxalic acid/diglycolic acid gradient. Detection is also made possible by the formation of metal complexes. In this example effluent from the column is combined with PAR to give the highly coloured metal ion–PAR complexes, which are readily detected by absorbance at 520 nm.

### Reversed Phase

Separation of metal ions as anionic complexes on high efficiency reversed stationary phases, such as a C$_{18}$ bonded phase silica or a polystyrene–divinyl-benzene copolymer, requires a mobile phase that must also include an ion interaction reagent, for example a quaternary ammonium salt (R$_4$N$^+$C$^-$) having lipophilic character. The metal ion–ligand anionic complex interacts with the R$_4$N$^+$C$^-$, which in turn interacts with the reversed stationary phase, A, as shown below:

$$A + R_4N^+C^- = A \cdots R_4N^+C^- \qquad [10]$$

$$A \cdots R_4N^+C^- + ML_2^- = A \cdots R_4N^+ML_2^- + C^-$$
$$[11]$$

where M$^{2+}$ is the metal ion analyte and ML$_2^-$ represents the anionic metal–ligand complex. The direction of the equilibrium and subsequently the elution of the metal–ligand complex is controlled by the selection of the reversed stationary phase and the R$_4$N$^+$C$^-$ salt, the concentration of the R$_4$N$^+$C$^-$ salt, the counteranion X$^-$ and its concentration via inert electrolyte, pH, solvent composition, the ligand, and the ligand concentration. This chromatographic strategy, or ion interaction chromatography (IIC), is known by several different terms and has been the subject of many studies to establish the nature of the interactions between the analyte, mobile phase components and stationary phase that are present. Clearly, several equilibria are involved and the success of the separation strategy in analytical applications requires careful control of all the equilibria.

Transition metal–PAR complexes are anionic and separations of these anionic complexes are possible on a reversed stationary phase. Other ligands can be used; the most useful ones are ligands that form kinetically stable anionic complexes with very high formation constants. Often the complexes are highly coloured and/or fluoresce and these properties allow sensitive detection.

Inorganic ligands that form very stable anionic complexes with metal ions can be used in a mobile phase that also contains a R$_4$N$^+$C$^-$ salt to separate the metal ions as complexes on a reversed stationary phase. **Figure 9** illustrates the separation of transition metals as anionic cyanide complexes on a C$_{18}$ reversed stationary phase column using a R$_4$N$^+$C$^-$ (Waters PICA additive) mobile phase. Detection in this separation is by absorbance at 214 nm.

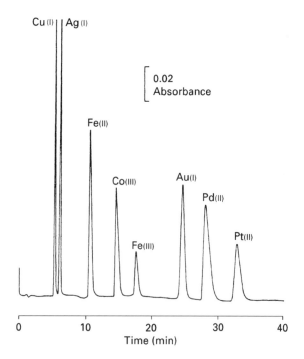

**Figure 9** Ion interaction chromatographic separation of metal ions as cyano complexes. A 5 µm, 0.39 cm × 15 cm, Waters Nova Pak $C_{18}$ column and an $H_2O/CH_3CN$ (23 : 77) 5 mmol $L^{-1}$ tetramethylammonium hydroxide mobile phase at 1.0 mL $min^{-1}$ was used. Sample injection was 10 µL and contained 0.15 µg $Cu^{1+}$, 1.5 µg $Ag^{1+}$, 0.02 µg $Fe^{2+}$, 0.2 µg $Co^{3+}$, 0.3 µg $Fe^{3+}$, 2.0 µg $Au^{1+}$, 0.4 µg $Pd^{2+}$, 0.2 µg $Pt^{2+}$ as cyano complexes. (Reproduced from Hilton and Haddad, 1986. Reprinted by permission from Elsevier Science Publishing.)

IIC with an anionic lipophilic reagent, such as an alkane sulfonic acid ($RSO_3^-H^+$), is also an important strategy for the high efficiency separation of metal ions. However, in this case the ligand in the mobile phase removes the metal ion from the reversed stationary phase according to the metal ion–ligand complex formation constants. The equilibria that are involved and which must be controlled for a successful separation are represented below (equilibria for complex formation are not shown – see eqns [1]–[3] where L is $CN^-$):

$$A + RSO_3^-H^+ = A \cdots RSO_3^-H^+ \qquad [12]$$

$$nA \cdots RSO_3^-H^+ + M^{n+} = (A \cdots RSO_3^-)_nM^{+n} + nH^+ \qquad [13]$$

$$(A \cdots RSO_3^-)_nM^{+n} + xL^{x-} + nH^+$$
$$= n(A \cdots RSO_3^-H^+) + ML_x^{+n-x} \qquad [14]$$

where A is the reversed stationary phase, $RSO_3^-H^+$ is the lipophilic ion interaction reagent, $M^{n+}$ is the metal analyte ion, and $L^{x-}$ is the mobile phase liquid.

Examples where this separation strategy has been successfully used are in the separation of transition metals or lanthanoids on a $C_{18}$ stationary phase and octanesulfonic acid as the IIC reagent in the buffered mobile phase. The former metal analytes are eluted with tartrate in the $RSO_3^-H^+$ mobile phase and the elution order is $Cu^{2+}$, $Pb^{2+}$, $Zn^{2+}$, $Ni^{2+}$ $Co^{2+}$, $Mn^{2+}$, which differs from the tartrate elution of these cations on the sulfonated cation exchange column and the latex-based exchanger (see Figures 4, 5 and 7). The lanthanoids are separated by including the ligand HIBA in the $RSO_3^-H^+$ mobile phase and the elution order is the same as that indicated in Figure 6.

## High Efficiency Chelating Exchangers

Few high efficiency chelating exchangers are commercially available. Even though various chelating groups have been chemically bonded to high efficiency silica or polystyrene–divinylbenzene copolymer particles, and metal ion retention is very high on these stationary phase particles, efficiencies generated on these columns are usually not comparable to efficiencies found for metal ion–ligand separations on ion exchangers or by reversed-phase ion interaction chromatography. In general, high efficiency is difficult to obtain, even when the core particle of the chelating exchanger meets high efficiency properties, because the complex between the metal ion and the bound chelating group often forms and dissociates slowly. Thus, metal ion analyte bands in elution can be broadened appreciably.

Using a ligand in the mobile phase to aid elution can sometimes reduce the analyte peak broadening. For example, the separation of transition metal ions on a high efficiency silica particle containing chemically bound iminodiacetic acid groups was achieved by using citrate, tartrate, nitrilotriacetic acid or PDCA in the buffered mobile phase. Since two complexing reactions are taking place, the metal ion elution order is dependent on the formation constant for the metal ion complex with the bound chelating group and with the mobile phase ligand in addition to pH. Other laboratory synthesized chelating exchangers where the chelating group (several of these are listed in Table 4) is bound to a high efficiency matrix, such as silica, have been evaluated for metal ion separations.

While metal ions can be separated on chelating ion exchangers, the more useful applications of chelating exchangers is still one of isolating and preconcentrating of metal ions from a complex sample matrix. Metal ions can be preconcentrated from biological samples, environmental samples and concentrated electrolyte solutions, such as seawater, brine and other industrial waters, and strong alkali solutions. It is possible to automate fully the preconcentration and separation/analysis procedure. In an example of this trace levels of $Mg^{2+}$, $Ca^{2+}$ and transition metals in biological and environmental samples are preconcentrated on two iminodiacetic acid-type chelating ion exchanger and one strong acid cation exchanger connected in series. The preconcentrated metal ions are then removed from the three columns and the concentrated metal ion mixture is separated on an analytical Ion Pac CS5 column by an elution programme that takes into account pH, ammonia/ammonium ion buffer concentration and PDCA concentration. Detection is by absorption after postcolumn reaction of the metal ions with PAR.

*See also:* **II/Chromatography: Liquid:** Mechanisms: Ion Chromatography. **Extraction:** Analytical Inorganic Extractions. **Ion Exchange:** Theory of Ion Exchange. **III/Ion Analysis:** Liquid Chromatography.

## Further Reading

Barkley DJ, Blanchette M, Cassidy RM and Elchuk S (1986) Dynamic chromatographic systems for the determination of rare earths and thorium in samples from uranium ore refining processes. *Analytical Chemistry* 58: 2222–2226.

Cagniant D (1992) *Complexation Chromatography.* New York: Dekker.

Elchuk S and Cassidy RM (1979) Separation of the lanthanides on high-efficiency bonded phases and conventional ion-exchange resins. *Analytical Chemistry* 5: 1434–1438.

Haddad PR and Jackson PE (1990) *Ion Chromatography Principles and Applications*, p. 232. Amsterdam: Elsevier.

Hilton DF and Haddad PR (1986) Determination of metal-cyano complexes by reversed phase ion-interaction high performance liquid chromatography and its application to the analysis of precious metals in gold processing solutions. *Journal of Chromatography* 361: 141–150.

Inczédy J (1966) *Analytical Applications of Ion Exchangers.* Oxford: Pergamon Press.

Kraus KA and Moore GE (1953) Anion exchange studies. VI. The divalent transition elements manganese to zinc in hydrochloric acid. *Journal of the American Chemical Society* 74: 1460–1462.

Pietrzyk DJ (1989) Macroporous polymers. In Brown PR and Hartwick RA (eds) *High Performance Liquid Chromatography*, p. 244, New York: Wiley-Interscience.

Siriraks A, Kingston HM and Riviello JM (1990) Chelation ion chromatography as a method for trace elemental analysis in complex environmental and biological samples. *Analytical Chemistry* 62: 1185–1193.

Timberbaev AR and Bonn GK (1993) Complexation in ion chromatography – an overview of developments and trends in metal analysis. *Journal of Chromatography* 640: 195–206.

Weiss J (1995) *Ion Chromatography*, 2nd edn, Weinheim: VCH.

Wilkins DH (1959) The separation and determination of nickel, chromium, cobalt, iron, titanium, tungsten, molybdenum, niobium, and tantalum in a high temperature alloy by anion exchange. *Talanta* 2: 355–360.

# Use in Gas Separation

*See* **III / GAS SEPARATION BY METAL COMPLEXES: MEMBRANE SEPARATIONS**

# METAL MEMBRANES: MEMBRANE SEPARATIONS

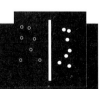

**Y. S. Lin and R. E. Buxbaum**, University of Cincinnati, Cincinnati, OH, USA

## Introduction

The use of metal membranes for hydrogen separation was first demonstrated over a century ago. Graham discovered in 1866 that palladium absorbs a suprising amount of hydrogen. Graham went on to show that palladium and palladium–silver membranes permeate only hydrogen, paving the way for the use of metal membranes for hydrogen extraction and purification. Much of the earlier knowledge in metal membranes is summarized in three reference books published in 1967–1968 (see Further Reading). Recent academic research is reviewed in seversl monographs on inorganic membranes and review articles. This article gives an overview of the general properties of metal membranes, followed with a description of methods that have been developed for membrane fabrication. Applications of the metal membranes in separation and chemical reaction processes and design of the metal membrane separation processes are descibed later in the article.

## General Properties of Metallic Membranes

A surprising amount of hydrogen can be absorbed reversibly by palladium alloys and by many transition metals over a large temperature range. The absorption proceeds via interstitial incorporation in the metal, generally leaving the crystalline structure intact. Thus, for example, face-centered cubic palladium retains its structure to the metal hydride phase, with the hydrogen progressively occupying tetrahedral and octahedral sites. The nature of the chemical bonding is still not well understood, but a widely used model assumes that the hydride is an alloy (in the usual metallic sense) of hydrogen and the host metal. Electrons from the hydrogen progressively occupy the d-bands of the transition metal, and the hydrogen exists essentially as slightly shielded protons in the host lattice. The other model is based on a predominantly covalent bond between the metal and hydrogen. In this model molecular hydrogen is dissociated to become atomic hydrogen, which subsequently forms chemical-type bonds to the host metal.

**Figure 1** shows hydrogen solubility relationships for several transition metals at different temperatures. At constant pressure, the hydrogen solubility of some metals increases with increasing temperature, while for some other metals it decreases. **Figure 2** shows typical relationships of hydrogen solubility (presented in the atomic ratio of hydrogen to metal) versus hydrogen partial pressures at different temperatures. The dashed curve in Figure 2 encloses the two-phase region of the palladium hydride system with $\alpha$-phase to its left and $\beta$-phase to its right. For all metals below their hydride phase transition, the solubility increases with hydrogen pressure. The most common relationship for this is

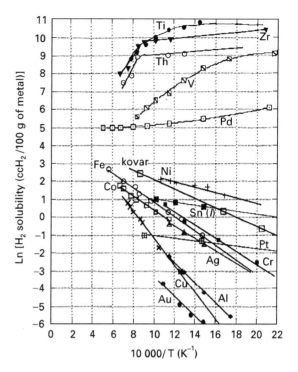

**Figure 1**  Hydrogen solubility of several transition metals at different temperatures ($P_{H_2} = 1$ atm). (Copyright REB Research & Consulting with permission.)

the Sievert's relation:

$$C = K_s P^{\frac{1}{2}} \qquad [1]$$

That is to say that the hydrogen concentration, expressed as the ratio of hydrogen : metal or as the volume of hydrogen per gram of metal, is proportional to a temperature dependent constant ($K_s$) times the hydrogen pressure to the power $\frac{1}{2}$. The power $\frac{1}{2}$ in eqn [1] comes from the entropic effect of every molecule of hydrogen dissociating into two hydrogen atoms in the metal.

Aside from solubility, the most important qualities of a metal for use in metallic membranes are the hydrogen diffusivity in metal and the rates of dissociation and recombination reactions on the membrane surfaces. At the upstream membrane surface the surface reaction step consists of adsorption of $H_2$, dissociation of this adsorbed $H_2$ into two hydrogen atoms (or charge-transfer reaction to form protons), and absorption of the atoms into the bulk metal. In the bulk there is a net transport of hydrogen atoms from high to low partial pressure sides along the concentraction gradient. At the downstream membrane surface there is another surface reaction step including the formation of hydrogen molecules from hydrogen atoms (or protons) and desorption of hydrogen molecules from the surface to the gas phase.

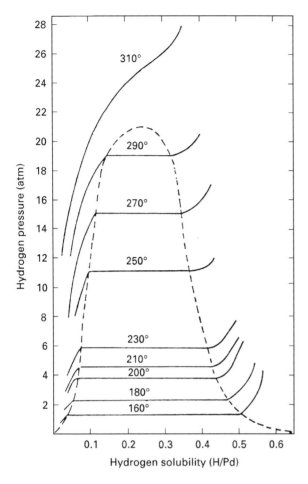

**Figure 2** Hydrogen solubility of palladium at different hydrogen pressures and temperatures (°C) (From Lewis FA (1967) *The Palladium Hydrogen System*. Fig. 2.3, p. 17. New York: Academic Press.)

Based on the Mass Action Law, the following simplified equations can be obtained to describe the combined effects on hydrogen permeation flux:

$$J = K_R(P_{UH} - P_I) \qquad [2]$$

$$J = (DK_s/2L)\,(P_I^{\frac{1}{2}} - P_{II}^{\frac{1}{2}}) \qquad [3]$$

$$J = K_R(P_{II} - P_{DH}) \qquad [4]$$

where $P_{UH}$ is the upstream hydrogen pressure, $P_{DH}$ is the downstream hydrogen pressure, $P_I$ is the imaginary pressure in equilibrium with the upstream surface of the metal, and $P_{II}$ is the imaginary presure in equilibrium with the downstream surface. $K_R$ is a lumped rate constant for the surface reaction step. In eqn [3], $DK_s$ is the product diffusivity and Sievert's constant in the bulk metal membrane and $L$ is the membrane thickness. Often $DK_s/2$ is called the hydrogen permeability and is plotted in **Figure 3** for several metals. It should be noted that

metal membranes are practically impermeable to other gases such as helium or nitrogen due to their low solubility and diffusivity in the metals. Although for all the metals of interest to separation applications, the hydrogen diffusivity increases and solubility decreases with increasing temperature, we find for some metals (e.g. palladium) the permeability increases with temperature, while for others (e.g. tantalum, niobium) it decreases. The determining factor is the relative temperature dependency of diffusivity and solubility. Because of slow reaction steps (eqn [2] and eqn [4]), permeation membranes of Nb, V and Ta must be coated with palladium to be used for separation applications.

For most metal membranes, bulk diffusion is rate-limiting. That is, $(DK_s/2L)$ is typically smaller than $K_R$. For such membranes eqn [3] describes the flux with $P_{II}$ and $P_I$ replaced by $P_{UH}$ and $P_{DH}$. In this case, the hydrogen permeance is inversely proportional to the membrane thickness with permeation flux exhibiting a $P_{H_2}^{\frac{1}{2}}$ dependence. This relationship is referred to as the Richardson's law. If the surface steps are rate-limiting, i.e. $(DK_s/2L)$ much larger than $K_R$, the hydrogen permeation flux is proportional to the transmembrane pressure drop. When resistances of both surface and bulk steps are equally important, the above equations are sometimes approximated with $P_{H_2}^{0.7}$ dependencies or similar simplified relations.

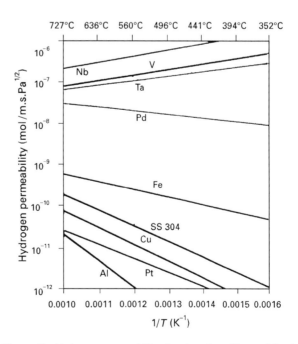

**Figure 3** Hydrogen permeability of various transition metals at different temperatures. (Copyright REB Research & Consulting with permission.)

Another critical behaviour for any metallic membrane is durability. Figure 2 shows that below a critical temperature (310°C) palladium hydride can exist either in the α-phase or β-phase, depending on hydrogen pressure and temperature. Above the critical temperature, palladium hydride remains in the α-phase regardless of the hydrogen partial pressure. Both the α-phase and β-phase have the same face-centered cubic lattice structure, but the lattice constant of the latter is as much as 3% larger than the former. Thus, a hydrogen pressure swing below the critical temperature or a temperature swing (above the critical pressure) will cause large mechanical strains during phase transformation of the palladium hydride between the two phases. These lattice strains would destroy the mechanical integrity of a membrane made of palladium alone. This phenomenon is referred to as 'hydrogen embrittlement'. For practical application this is avoided by alloying the palladium and by lowering the hydrogen pressure sharply before the membrane cools to room temperature. The same problems and solutions are also applicable to Pd-coated transition metals.

Alloying palladium with silver results in a phase diagram with a decreased critical temperature and pressure. If we accept the metallic alloy model for metal–hydride systems, the role of silver can be explained as an electron-donating behaviour where electrons from silver and hydrogen atoms compete for filling the 4d-band of palladium. Since 4d-band filling is associated with β-phase formation, 40% silver prevents β-phase formation entirely although most common alloys do not employ quite so much silver. Once the 4d-band is filled, electrons from further hydrogen absorption contributes to 5s-band filling and the net effect is decrease in both critical temperature and pressure in the phase diagram. Common palladium–silver (23–25% silver) membranes and palladium–copper membranes can be operated at much lower temperatures than pure palladium membranes without significant hydrogen-embrittlement problems. This is especially true if care is taken at start-up and shutdown to avoid exposure of the membranes to high-pressure hydrogen at room temperature.

Metal membranes are susceptible to surface poisoning. Prolonged exposure of many metal membranes to sulfur or even carbon containing gases under some conditions could deactivate the membrane surface or change the bulk structure of thin metal membrane, affecting adversely the membrane permeation and separation properties. The current approach is to avoid this problem by removing poisoning impurities from the gas stream prior to the contact with the membrane surface. Modification of the metal surface to improve its resistance to surface poisoning may provide an inherent solution to this problem. Research efforts in finding a suitable modification method have been limited, but are expected to increase in the future.

## Fabrication of Metal Membranes

Casting/rolling is the current method for commercial production of palladium-based metal membranes. This method involves several steps. Membrane raw materials are melted at a high temperature to form ingots which go subsequently through hot or cold forging or pressing to produce tube or sheet. The final membrane has a thickness typically in the range of 50 to 75 μm (0.002–0.003 in). The metal membranes prepared by this method are polycrystalline with large grains (submicron or micron size). Furthermore, cold rolling often generates lattice dislocation and it can enhance hydrogen solubility in palladium and some of its alloys due to the accumulation of hydrogen around the dislocation. As a result, these polycrystalline metal membranes have very high hydrogen solubility.

Based on the Richardson equation (eqn [3]), thin metal membranes are more desired than thick ones in terms of hydrogen permeance and metal cost savings. The problem has been in forming such membranes and keeping them stable during temperature and pressure cycling. To improve the durability, thin metal membranes have been formed by coating on porous metallic or ceramic support. Several methods have emerged in the past decade for this including liquid-phase electroless plating and chemical vapour deposition (CVD).

Electroless plating involves autocatalysed decomposition of a selected metastable metal presursor such as $Pd(NH_3)_4X_2$ where X can be $NO_3$, Cl or Br in alkaline aqueous solution at around 50°C. To avoid autodecomposition of the metal precursors a stabilizer such as ethylenediaminetetraacetic acid (EDTA) is often added. For nonmetallic supports, the surface (substrate) should be activated, e.g. by $SnCl_2/PdCl$, prior to electroless plating. Palladium or palladium–silver films with a thickness ranging from 5 to 30 μm have been prepared by this method on supports of porous alumina and stainless steel and on non-porous niobium, tantalum or vanadium alloy.

Thin metal membranes can also be prepared by CVD of metal salts and metal organic compounds, such as $PdCl_2$ and $Pd(O_2C_5H_7)_2$ and can also be produced by sputter coating of a solid metal. For CVD, the precursor vapour, with or without a reducing agent (hydrogen), is brought in contact with one surface of the porous support at a high temperature.

Reduction of the precursor occurs directly or by countercurrent exposure to the reducing agent at 200–400 °C. Palladium films as thin as 0.5 µm have been deposited on or inside porous alumina support this way. For sputter deposition a metal target of desired composition is placed apart from the substrate in a sputtering chamber filled with a working gas (argon) at low pressure (about $10^{-3}$ torr). A DC or radiofrequency AC electric-field gradient plasma is generated in the working gas and bombards the target. Fine metal particles or atoms dislodged from the target move towards the substrate and subsequently deposit. Thin palladium and palladium–silver membranes have been coated on porous γ-alumina and dense polymer substrates by this method.

Metal membranes prepared by the vapour methods are usually nanocrystalline with grain size in the range of a few tens of nanometers. Electroless-plated metal membranes appear to be more coarse-grained. For palladium on substrate membranes, the bulk diffusion step in the substrate is generally the rate-limiting step. Furthermore, fine-grained thin metal membranes appear to have a lower hydrogen permeability than coarse-grained thick metal membranes. Thus, while the metal cost for Pd-coated ceramics is typically much lower than for symmetric metal membranes, the flux is usually fairly similar and can even be lower than traditional metal membranes, and is typically proportional to the pressure dependence and not to $P_{H_2}^{\frac{1}{2}}$ as in eqn [3]. Similar membrances prepared by different methods can exhibit different dependency, and specific relationships should be obtained in order to design separation and reaction processes involving these metal membranes.

## Metal Membrane Applications

The major large-scale application for metal membranes today is hydrogen purification, with most of that hydrogen used for interated circuit manufacture. Different circuit technologies differ in the purity of hydrogen they require, and the current needs are between 5 nines (99.999%) and 8 nines (99.999999%) hydrogen, or 0.01 to 10 ppm impurities. The largest user, silicon-based semiconductor manufacture requires the lower levels of purity, and gallium arsenide the highest. These purity levels are expected to increase over time as transistor densities rise, doubling approximately every two years. Silicon applications should thus need 7 nines and gallium arsenide applications 9 nines hydrogen by 2005.

Metal membrane purification competes with adsorption technology for the market in electronic component manufacturing. The metal membrane market is valued at only $10 millions/year, but the ultrapure hydrogen delivery market is at least 100 times bigger suggesting significant room for growth in purifier sales. For electronic component manufacturing using cylinder hydrogen (the only choice in much of the Far East) there is no real alternative to membrane purification using tubular membranes of palladium–silver alloy (23–25% silver). The leading companies in this market are Johnson Matthey (75% market share) and Japan Pionics (25% market share.) Cylinder hydrogen is so impure that competing getter technology is not cost effective. Competing getter technology is often preferred with liquid hydrogen sources (avaliable in the USA and Europe) because the pressure drop is lower and since the delivered hydrogen is purer, the cost is lower as well. Even here, membranes are often used for high value-added parts since getters tend to release small amounts of oxide dust as they saturate with impurities. This oxide can interfere with semiconductors manufacture and presents a safety hazard, e.g. released vanadium oxide dust is highly toxic with an exposure limit of 0.05 mg m$^{-3}$. Membranes are also safety favourites since the main getter alloys are flammable in contact with air at operating temperatures as high as 600°C.

A second major application for metal membrane purified hydrogen is as a carrier gas for gas chromatography. This is true particularly in Europe and the Far East where helium is scarce, but also in the USA for fast-response, high-sensitivity applications. The purity need here could be met by ultrapure cylinders, especially backed by a getter trap, but it is cost effective to use a membrane. An 'A-size' cylinder, containing 65 ft$^3$ (STP) (1900 L (STP)) of ultrapure zero-grade hydrogen costs approximately $220 currently. Continuous use as a carrier and reference gas consumes about 150 cm$^3$ (STP) min$^{-1}$, emptying the cylinder in slightly over a week, and costing the customer $10 000 annually. A membrane purifier of very modest size will allow the customer to use a much cheaper, lower grade of hydrogen, returning the purchase cost in six months or less.

Palladium membranes are also used for isotope enrichment providing a separation factor of about 1.4 for protium over deuterium, and about 1.7 for protium over tritium. While the hydrogen volume need for this application is small, the low separation factor ensures that quite a lot of membrane is needed.

A rather new line of metal membrane applications is for membrane reactors, mainly to generate hydrogen for fuel cells, but also to promote thermodynamically unfavourable reactions like methane splitting and ethylene production. A membrane reactor combines, in one unit, a catalytic reactor bed and a membrane separator to remove a desired product (hydrogen) as it is formed. As a result membrane reactors drive the reaction towards completion and achieve advantages of enhanced catalyst use, higher feed space velocity and fewer side products.

Another benefit of membrane reactors, particularly for fuel cell applications that require high-purity hydrogen from source gases like methanol or gasoline, is that a membrane changes the way pressure affects extent of reaction and product recovery. Without the membrane in the reactor, a reforming reaction of this type would require low pressure operation to achieve high conversion, and would require a difficult hydrogen purification for the intended use. A membrane reactor allows the reaction to be driven by high pressures, greatly increasing the ease of hydrogen product recovery. At present, several membrane reactors are in trial and one at the pilot-plant scale for the production of hydrogen. Typical pressures of operation are at 17 atm, with operation temperatures dictated by the catalyst and material of construction. The size of the fuel cell market is currently small, but is believed to be rising fast.

## Designs of Metal Membrane Separation Processes

Figure 4 shows the basic design of a hydrogen purifier. Prominently shown is a heat exchanger set between the input hydrogen feed and the purified hydrogen product. Heat recovery of this type is critical because the membranes typically operate at 350–400°C in an environment where the source gas and product application are near room temperature. Heat exchange cools the output gas and reduces the electric heating costs. Heat exchange also promotes temperature uniformity within the purifier; this is an important consideration since for most of the hydrogen permeable alloys a 100°C temperature gradient in

the membrane would mean that part of the membrane would have to operate at significantly below their optimal temperature for flux, and part of the membrane would operate at significantly above the optimal temperature for long life.

Figure 4 also shows a bleed valve to control the removal of impurities from the purifier, and a system to provide inert gas (nitrogen) flushing. Setting these systems is something of an art, since with too little bleed, impurities would build up in the purifier reducing the hydrogen output. Too much bleed, by contrast, results in a loss of heat and pressure resulting in a similar reduction in output. Most larger purifiers provide a nitrogen purge system as shown, to help extend the life of the membrane by reducing hydrogen embrittlement of the palladium alloys. This purge system can also be a source of problems as purging can result in hydrogen being sucked back into the purifier. If the sucked back hydrogen is contaminated with other electronic gases (particularly arsine) this will result in an irreversibly poisoned membrane.

There are basically two types of tubular membrane based commercial purifier designs: pressure-outside and pressure-inside. Until recently, large-scale hydrogen purification ($>100$ L (STP) min$^{-1}$) was limited to pressure-outside design. Figure 5 shows schematic of a pressure-outside design. The inlet flow of fluid containing a mixed gas includes hydrogen flows over the outer surface of the metal membrane tubes housed inside a pressure vessel. The metal membrane tubes are operatively connected at one end to a header and the other end of the tubes are either capped or can be operatively attached to a floating head. Hydrogen permeates from outside into the inside of the metal membrane tubes, and leaves the membrane module through the outlet port. Pressure-outside designs (Johnson Matthey, Japan Pionics, and REB Research & Consulting) have longer life than the pressure-inside designs since the former do not require a braze seal at both ends of the membrane tube.

Pressure-inside designs, supplied by REB Research & Consulting, RSI and Power & Energy, are believed to output somewhat purer hydrogen because the impure gas flow past the membrane is more uniform than with pressure-outside designs. In the pressure-inside designs, the pressure drop and vibration effects in the tubes become excessive as tube lengths exceed 25–50 in. (63–127 cm). These effects could be effectively minimized by a scheme for placing several short lengths of Pd–Ag or similar tube in parallel while balancing the flow through each using matched, internally located flow restrictors. Forces of expansion and contraction by the membranes against the seals are believed to be a major

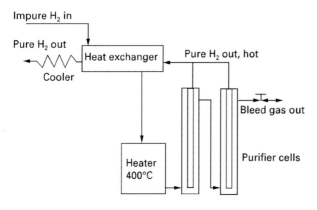

**Figure 4** Schematic diagram of a metal membrane hydrogen purifier process. (Courtesy of Johnson Matthey, plc.)

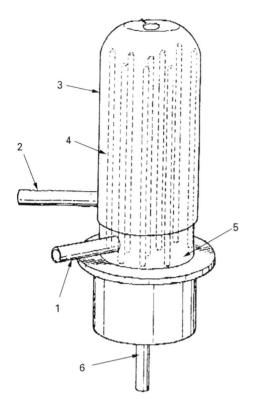

**Figure 5**  Schematic diagram of a pressure-outside design of metal membrane module: 1. Feed gas inlet. 2. Feed gas outlet. 3. Pressure vessel. 4. Metal membrane tubes. 5. Header. 6. Pure hydrogen outlet. (From US Patent 5,931,987, 1999.)

thickness appropriately, metal cost per unit area could be decreased significantly while flux could be increased somewhat. The early membranes did not prove durable though, and had to be removed from service. Results with all of these alternatives is encouraging but it is fair to say that the commercial metal membrane market is still dominated by unsupported palladium alloys. The main reason is purity and durability: both the Pd-coated refractory metals and the Pd coated porous substrates have shown less cycling durability than typically associated with the solid Pd alloys, and the purity of hydrogen purified through the newer membranes has still to reach the very high levels of solid membrane purifiers. This latter is particularly significant for the high value added applications that dominate the metal membrane market currently.

## Concluding Remarks

Metal membranes are the most hydrogen perm-selective inorganic membranes in the temperature range of 200–500°C. They are expected to remain so in the near future. Much progress has been made in understanding metal membrane properties and development of various methods for metal membrane fabrications. However, applications of metal membranes have been limited to hydrogen purification. In addition to the structural stability and cost problems, another major problem for metal membranes is their susceptibility to surface poisoning. This problem has to be solved in order for the metal membranes to gain wider acceptance in processes for separation of hydrogen containing gases of industrial importance.

## Further Reading

Armor JN (1998) Applications of catalytic inorganic membrane reactors to refinery products. *Journal of Membrane Science* 147: 217–233.

Bhave RR (1991) *Inorganic Membranes. Synthesis, Characteristics and Applications.* New York: van Nostrand Reinhold.

Burggraf AJ and Cot L (eds) (1995) *Fundamentals of Inorganic Membrane Science and Technology.* Amsterdam: Elsevier.

Buxbaum RE (1999) Membrane reactor advantages for methanol reforming and similar reactions. *Separation Science and Technology* 34: 2113–2123.

Buxbaum RE (1999) US Patent 5,888,273. *High Temperature Gas Purification System.* March 30.

Buxbaum RE and Kinney AB (1996) Hydrogen transport through tubular membranes of palladium-coated tantalum and niobium. *I&EC Research* 35: 530–537.

Hsieh HP (1996) *Inorganic Membranes for Separation and Reaction.* Amsterdam: Elsevier.

cause of fatigue failure in pressure-inside designs, as is burnthrough. Burnthrough is believed to be a bigger problem with pressure-inside designs since metals are inherently more stable in compression than in tension. That is a tube in compression will flow somewhat to heal a thin spot, but will flow away in tension turning a thin spot into a hole.

A third type of purifier using flat plates of Pd–Cu was recently demonstrated by Northwest Power. As flat plates can be made thinner than metal tubes the output flux with these membrane is exceptional. Unfortunately welded seals are much less reliable in this configuration than with tubes and instead graphite seals are used. The degree of purification from these devices is unacceptable for integrated circuit manufacture, but appears to be acceptable for use with fuel cells.

Almost all commercial metal membrane separation systems use the symmetric solid metal membranes fabricated by the conventional casting/rolling method. The only commercial attempt at using porous substrate supported metal membranes was in the 1950s and 1960s by Union Carbide who applied a thin layer of palladium–silver alloy on a porous ceramic substrate. By choosing the pore size and

Hwang ST and Kammermeyer K (1984) *Membranes in Separation*. Malabar, Florida: Rorbert E. Krieger.

Lewis FA (1967) *The Palladium Hydrogen System*. London: Academic Press.

Mueller WM, Blackledge JP and Libowitz GG (1968) *Metal Hydrides*. New York: Academic Press.

Shu J, Grandjean BPA, Van Neste, A and Kaliaguine S (1991) Catalytic palladium-based membrane reactors. A review. *Canadian Journal of Chemical Engineering* 69: 1036–1060.

Uemiya U (1999) State-of-the-art of supported metal membranes for gas separation. *Separation and Purification Methods* 28: 51–85.

Wise EM (1968) *Palladium Recovery, Properties and Uses*. London: Academic Press.

# METAL UPTAKE ON MICROORGANISMS AND BIOMATERIALS: ION EXCHANGE

**H. Eccles**, British Nuclear Fuels, Preston, UK

## Introduction

Microorganisms and biomaterials can be used for the removal/recovery of metals from process liquors and liquid wastes. The mechanisms involved in capturing metals from solution by microorganisms and related materials are now commonly referred to as biosorption. Biosorption has been defined as the removal of metals or metalloid species, compounds and particulates from solution by biological material. It is one of the fields in environmental biotechnology, which is itself a small, but growing, component of the biotechnology industry (**Figure 1**).

The use of microorganisms to treat waste liquors on a commercial scale dates back to the end of the nineteenth century when the first communal sewage plants in Berlin, Hamburg, Munich, Paris and other major cities came into operation. In the intervening 100 years the use of microorganisms and plants to protect the environment has developed into a multi-billion dollar (US$) industry. However, the foundations of this environmental biotechnology industry are still with the treatment of municipal/domestic effluents.

Although nature has demonstrated some subtle and intricate mechanisms for selectively controlling the mobility of pollutants in the environment, the conversion of this science to technology and to application has been very disappointing. In explaining this lack of application, it is important that these mechanisms

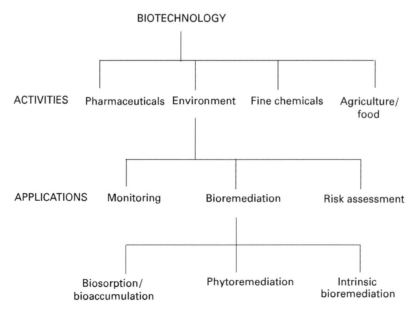

**Figure 1**  Biotechnology activities and applications. (This is an original developed by H Eccles.)

are appreciated; factors that influence their effectiveness need to be addressed to engineer robust and reliable processes.

## Microorganisms and Biomaterials for Metal Removal

The biological systems studied for metal removal/recovery range from living microorganisms and dead cell systems to other biomaterials such as peat, coconut shells and eggshell membrane. All these systems have at least one common characteristic, namely their absorptive surfaces, and (in the case of living microorganisms within the cell structure) have at least one chemical moeity (functional group) that has an affinity for metal(s). This affinity can be accomplished by active and passive mechanisms that may act singly or in combination. The mechanisms can be microbiologically dependent, for example active transport across the cell membrane, or physicochemical and chemical-dominated reactions such as adsorption, cation exchange, complexation, chelation and precipitation. Essentially, active transport is a function of living cells, such as bacteria, algae and fungi. The remainder of the mechanisms are passive and may occur with living or dead cells or with other biomass.

Passive uptake mechanisms generally occur at the cell wall level (see **Table 1**) and the biomasses involved are generally tolerant to their environment. In their interactions with metals, dead cell systems and biomaterials behave as surrogate ion exchange materials. Ion exchange is a proven commercial technology that is widely used for the removal of metals from various liquors. Where major differences occur between biosorption and conventional ion exchange is when living microorganisms are involved, which results in the participation of other metal-removal mechanisms.

The basis for these differences and also for variations in metal-accumulation abilities is partly the result of cell surface characteristics. Microbial cell walls are complex and are normally charged. Certain cell wall properties, such as the type of polar groups present and the charge within the cell wall macromolecules, may influence the metal capacity and selectivity. The types of macromolecules present in cell walls for bacteria, fungi and yeast are shown in **Table 2**.

Classification of bacteria, algae and fungi is based on aspects other than cell wall composition. Living organisms can be divided into two groups, prokaryotes and eukaryotes, based on cell structure. Bacteria belong to the prokaryote group, while algae and fungi belong to the eukaryotes. Further simple distinctions are that bacteria lack a nucleus and membrane-bound organelles and most bacteria have cell walls. Bacteria are without question the oldest and simplest organisms. Eukaryotic organisms have both a nucleus and membrane-bound organelles. Fungi are distinct from algae in that they have filamentous growth, lack chlorophyll and motile cells, and have chitin-rich cell walls (**Figure 2**).

Although these simple descriptions may be helpful in understanding the different behaviour of bacteria, fungi and algae with metals, they are by no means rigorous biological definitions, and the purist microbiologist may find them too superficial. It is probable, however, that the more rigorous and sophisticated explanations have contributed to the restrictive use of biotechnology in the environmental arena. This point will be discussed in more detail later.

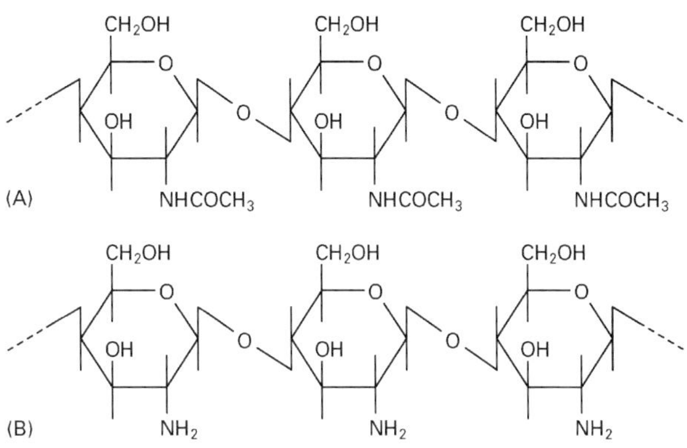

**Table 1** Cell wall metal functional groups

| Cell wall functional group | Metal affinity |
| --- | --- |
| Carboxyl | Ca, Mg, Cu, Zn |
| Imidazole | Cu, Pb |
| Sulfhydryl | Zn |
| Amino | Co, Ni, Cu |
| Phosphate | Ca, Mg, Fe, U |
| Sulfate | Ba, Ca, Sr |
| Thioether | Cu(I) |
| Amide | Cu, Co, Ni, Fe |
| Hydroxyl | Ca, Pb, Cu, Sr, Ba, Ni, Co, Zn |

**Table 2** Cell wall macromolecules

| Cell wall macromolecules | Microorganism |
| --- | --- |
| Polysaccharides, proteins, lipids, peptidoglycan | Bacteria |
| Mannan polysaccharides, chitin, galactosamine, proteins, lipids | Algae and fungi |

**Figure 2** Chemical structure of (A) chitin and (B) chitosan. (Hardman DJ, McEldowney S and Waite S (eds), *Pollution: Ecology and Biotreatment* (1993), p. 284, Addison Wesley Longman Ltd, UK.)

The complex nature of cell walls bestows some unique abilities on living microorganisms and to a lesser extent dead cell systems and biomaterials. Equally this complexity confers some serious drawbacks, as modelling of metal-removal mechanisms and predicting process efficiencies are extremely difficult when more than one metal functional group is available.

Notwithstanding these reservations, numerous microorganisms and biomaterials have been evaluated for a diversity of metals by many scientists worldwide. Some of the more definitive work is discussed in the following sections.

## Dead Cell Systems and Biomaterials

Dead cell systems can be derived from living cells by subjecting them to a physical or chemical method to terminate the living cell metabolic activity. As growth conditions can confer some metal affinity characteristics on living cells, it is prudent to explain briefly the growth phase and the various parameters that may be controlled to achieve the desired goal. The more important growth phase conditions that can confer metal adsorption characteristics are:

1. time of cell harvesting;
2. composition of nutrient medium.

The growth of bacterial populations is normally limited either by the exhaustion of available nutrients or by the accumulation of toxic products of metabolism. This is particularly true for batch growth conditions. As a consequence, the rate of growth declines after the exponential phase and growth eventually stops. At this point a culture is defined as being in the stationary phase (**Figure 3**). The transition between the exponential phase and the stationary phase involves a period of unbalanced growth during which the various cellular components are synthesized at unequal rates. Consequently, cells in the stationary phase have a chemical composition that is different from that of the cells in the exponential phase. In general, cells in the stationary phase are small relative to cells in the exponential phase and they are more resistant to adverse physical and chemical agents.

Bacterial cells held in a nongrowing state eventually die, largely due to the depletion of the cellular reserves of energy. When cells are transferred from a culture in the stationary phase to a fresh medium of the same composition they undergo a change of chemical composition before they are capable of initiating growth. This period of change is called the lag phase.

Culture age affects the biosorption properties of microorganisms. For example, younger cells (12 h growth) of *Saccharomyces cerevisiae* removed approximately five times more uranium than older cells (24 h-growth).

The surface charge of living cells can vary both their age and with the nature and composition of the growth medium. With respect to metal adsorption, surface charge will have a predominant, if not the prime, influence. The cell wall surface charge is itself strongly affected by the pH value of the growth medium. This pH influence is due to the different ionogenic groups at the cell surface being susceptible to protonation/deprotonation reactions. Experimental evidence involving bacterial cells shows that carboxyl groups are present in excess over amino groups, and thus they dominate electrokinetic behaviour.

Physical methods that have been used to kill living cells include vacuum and freeze drying, boiling, autoclaving and mechanical disruption such as ultrasonics. Chemical methods include contacting the cells with various organic and inorganic compounds. The main aim of controlling the growth conditions coupled with the appropriate killing stage is to produce a biomaterial that has metal affinity properties superior to those of the parent living cell. The advantages of dead cells over living cells are:

- the metal removal process is largely independent of toxicity limitations;
- there are no requirements for growth media and nutrients;
- biosorbed metals can be eluted and the dead cells re-used;
- there is an ample and ready supply of some biomasses (dead cells);
- pretreatment of the biomass can enhance the metal biosorptive characteristics;
- the process is simpler and akin to ion exchange;

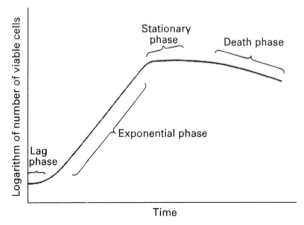

**Figure 3** Generalized growth curve of a bacterial culture. Stanier RY, Ingraham JL, Wheelis ML and Painter PR (eds) *General Microbiology* (1986), 5th edition, p. 185, Prentice-Hall Inc, NJ.)

**Table 3** Pretreatment methods to improve metal selectivity

| Biomass type | Pretreatment method | Metal studied |
|---|---|---|
| Penicillium | $0.1 \text{ mol L}^{-1}$ sodium hydroxide | Ni, Cu, Zn, Cd, Pb |
| digitalum | Contacted with dimethyl sulfoxide for 90 min | U |
| | Trichloroacetic acid | U |
| Saccharomyces cerevisiae | 10% (v/v) nitric acid, 30 min boiling | Cu |
| | 10% (w/v) formaldehyde | Cd, Zn |
| | Detergent | Th |
| Aspergillus niger | Freeze-dried | U |
| | 5% KOH | Cu, Cd, Zn, Co, Ni |

- disposal of spent and/or excess nutrient media or surplus cells does not present a problem;
- the shelf-life of the dead biomass is nearly infinite.

If the required metal affinity characteristics have not been achieved at the harvesting stage and by the process used to kill the living cells, the dead biomass can be treated to enhance the key metal biosorptive properties such as metal capacity, metal selectivity and rate of uptake.

In **Table 3** are reported some of the pretreatment methods employed and their effect on metal removal. Alkali treatment of fungal biomass has been shown to increase significantly the metal sorption capacity of *Aspergillus*, *Mucor* and *Penicillium*, and deacetylation of chitin in the cell wall to form chitosan–glucan complexes results in higher affinity for metal ions.

In what follows dead cell systems and biomaterials will be regarded as being equivalent, but any specific subtle differences will be highlighted. The biomasses that have attracted most attention are those that are readily available in significant quantities, with or without pretreatment to enhance their metal uptake capability. To date these materials have included: peat, lignates, seaweed waste (dealginated seaweed), yeast from brewing operations, algal and fungal biomasses and activated carbon from a variety of sources.

The last material is presently used commercially in the treatment of water for the removal of nonmetallic species; numerous papers have been published on the uses of activated carbon. In evaluating the above materials one feature regularly considered has been their particle size and the implications to scale-up. In their behaviour to metals these materials resemble ion exchange resins, which are invariably packed into columns. Thus laboratory studies have included the immobilization of biomaterials into particles of appropriate shape and size. Spherical particles 1–3 mm in diameter appear to be acceptable.

As the metal-removal processes are passive and of a chemical/physicochemical nature, the parameters considered are like those for ion exchange, namely pH, temperature, competing cations and metal speciation. These parameters are equally important to metal removal by living cell systems. Without doubt, pH is by far the most dominant influence on metal uptake. The optimal pH values for a variety of metals and biomaterials are presented in **Table 4**.

The efficiency of metal removal by the biomaterial as a function of pH will be related to: (1) the functional group involved and (2) metal speciation chemistry. At low pH values, i.e. about 3 or less,

**Table 4** Optimal pH values for metal uptake by various biomaterials/microorganisms

| Biomaterial/microorganism | Optimal pH value for metal removal | Metal(s) |
|---|---|---|
| Peat | 1.5 | Cr(VI) |
| | 5.0 | Cu(II) |
| | 7.0 | Ni(II) |
| Seaweed | 4.0 | Au |
| | 5.0–6.0 | Cu, Zn, Pb, Cd, Cr, Ni, Fe |
| S. cerevisiae | 1.5 | Mo(V) |
| | 3.0–4.0 | U(VI) |
| Penicillium digitatum | 7.0 | Cd |
| Bacillus subtilis | 4.1 | U |
| Chlorella salina | 4.0 | Tc(VII) |
| | 8.0 | Co |

metal(s) uptake will be comparatively low as the metal(s) will be in competition with the hydrogen ion and most functional groups with an exchangeable hydrogen ion will be undissociated. On increasing the pH, competition with the hydrogen ion diminishes, dissociation of the hydrogen of the functional group increases and the speciation of the metal(s) changes from being a hydrated cation to hydroxy metal species thus:

$$M(H_2O)_4^{2+} \xrightarrow[\text{increasing pH}]{} M(H_2O)_3(OH)^+ + H^+$$

This progressive displacement of the inner sphere of water molecules with increasing pH results in the metal ion becoming more amenable to adsorption by the biomaterial.

Temperature increases within a modest range, i.e. 20°C to 60°C, have little or no effect, either thermodynamically or kinetically, on metal removal for most biomaterial and metal systems. In general, optimal metal uptake will occur at 25°C for most dead cell systems and at a similar value (25–30°C) for biomaterials.

Metal affinity by many dead cell systems measured using single metal systems has been shown to comply with the Irving–Williams series. Most dead cell systems exhibit selectivity for heavier metals, such as Zn, Cu and Ni, over the lighter alkali metal ions. The selectivity of dead cell systems and biomaterials for a particular metal is highly dependent on the pH value of the solution under investigation and the functional group involved in the metal adsorption process. The determination of metal selectivity for dead cell systems and biomaterials is complicated in that more than one functional group may be involved in the biosorption process. These interactions (biosorption processes) may operate independently or may be interrelated. For example, Tsezas and Volesky indicated that three mechanisms are operative in uranyl ion biosorption by *Rhizopus arrihizus*. Two of the mechanisms occur simultaneously and rapidly (< 60 s to equilibrium). In the first process, uranyl ions coordinate with the amino nitrogen of the cell wall chitin, facilitating the second process in which the complexation sites acted as nucleation points for deposition of additional uranium. These two mechanisms accounted for 66% of the total uptake capacity (0.05 mmol U per g cell dry weight). The third process is considerably slower, only reaching equilibrium after 30 min, and involves the precipitation of uranyl hydroxide within the cell wall microcrystalline chitin.

Although most of the available literature is concerned with single metal experiments, a limited num-ber of studies have involved binary metal systems or multimetal solutions. These studies demonstrate that copper generally has the highest binding capacity and exerts the largest competing effect.

The presence of anions in the metal solution under consideration will have an impact on metal biosorption, depending on the concentration of the anion and its metal-complexing ability. The more common anions, such as nitrate, chloride and sulfate, rarely affect the biosorption of metals when in near stoichiometric concentrations to the metal. Increasing their concentration will affect the biosorption (lower it), but this will depend on the ability of the metal to form anionic species. When the anion species such as $EDTA^{2-}$, phosphate, citrate, etc., is capable of coordination/complexation with the metal, the metal capacity will be strongly reduced.

## Living Cell Systems

Living cells systems offer some unique and subtle metal-accumulation mechanisms not possible with dead cell systems, biomaterials or conventional ion exchange resins. These active metal-accumulation mechanisms, in addition to the passive ones, are generally coupled with the metabolic activity of the algal, bacterial, fungal living cells.

Many organisms have developed detoxification mechanisms to overcome the detrimental effects of metals. For the purpose of this article two mechanisms will be taken as predominant, namely bioaccumulation and bioreduction. Although these mechanisms are generally encompassed by the term biotransformation, this terminology is far more appropriate to the treatment of organic pollutants using microorganisms, as these pollutants can be biodegraded to bengin metabolic end products such as carbon dioxide and water. Metals are persistent and their toxicity is infinite. Microorganisms can only affect their physical and/or chemical state and transform them into more immobile forms that are less bioavailable to plants and other higher organisms. In achieving the biotransformation of metals, microorganisms are behaving as minute chemical factories generating chemicals, for example hydrogen sulfide, alkali and inorganic phosphate, or providing electrons from interconnected redox processes. These processes are summarized in **Table 5**. One process in particular will now be described in more detail.

In the natural environment the major mechanism for bacterial metal precipitation is through the formation of hydrogen sulfide and the immobilization of metal cations as metal sulfides. The bacteria involved in this process are the sulfate-reducing bacteria including members of the genera *Desulphovibrio*

**Table 5**  Metal bioaccumulation and bioreduction processes

| Process | Metal-removal mechanism | Microorganism |
|---|---|---|
| Bioaccumulation | Sulfide precipitation | Sulfate-reducing bacteria |
| | Phosphate precipitation | *Citrobacter* sp. |
| | Hydroxide precipitation | *Alcaligenes eutrophus* |
| Bioreduction | Hydroxide/oxide precipitation | *Shewanella alga* |
| | Sulfide precipitation | Sulfate-reducing bacteria |

and *Desulphotomaculum*. Sulfate-reducing bacteria (SRB) are widely distributed in anaerobic environments such as sediments and bogs. Sulfide production by sulfate-reducing bacteria is a consequence of their energy-generating processes. They couple the reduction of oxidized forms of sulfur, e.g. sulfates and sulfur, with the oxidation of reduced carbon in the form of simple organic molecules such as lactose or ethanol. The optimal chemical environment for effective sulfate reduction to sulfide is a pH value between 5.5 and 6.5 with a negative redox value of about 200–400 mV.

One of the underpinning features for the effectiveness of a SRB metal-removal process is the extremely low solubility product value for metal sulfides, as reported in **Table 6**. Other process advantages are:

- removal of metal anions such as chromate with initially the reduction of Cr(VI) to Cr(III) by the biogenic hydrogen sulfide;
- the ability of the microorganisms to nucleate the sulfide precipitation and thus assisting in the coagulation of metal sulfide particles;
- maintaining (slightly increasing) chemical neutrality of the process sulfate liquor due to the carbonate formed from oxidation of lactose or ethanol, or other suitable carbon sources, and sulfide from reduction of sulfate;
- the ability to remediate inorganic pollutants, for example toxic heavy metals and sulfate, as well as organic pollutants.

**Table 6**  Solubility product values for metal sulfides and hydroxides at 25°C

| Metal | Hydroxide | Sulfide |
|---|---|---|
| Ag | $2 \times 10^{-8}$ | $1.6 \times 10^{-49}$ |
| Cu(II) | $1 \times 10^{-20}$ | $8.5 \times 10^{-45}$ |
| Zn | $1 \times 10^{-17}$ | $1.2 \times 10^{-23}$ |
| Ni(II) | $1 \times 10^{-15}$ | $1.4 \times 10^{-24}$ |
| Co(II) | $1 \times 10^{-15}$ | $3.0 \times 10^{-26}$ |
| Fe(II) | $1 \times 10^{-15}$ | $3.7 \times 10^{-19}$ |
| Cd(II) | $1 \times 10^{-14}$ | $3.6 \times 10^{-19}$ |

These advantages may have been behind the development by Shell Research Ltd. of a biological process in preference to chemical ones for the removal of zinc, cadmium and sulfate from contaminated groundwater accumulating below a zinc smelter site. This process is the only biological metal-removal system employing specifically SRB operating on a commercial scale. The reasons for this unique situation is addressed in the next section.

The active metal biotransformation processes that rely on the precipitation of the offending metal(s) are not selective. The microorganisms require carefully controlled conditions both for their growth and to maintain enzyme activity, which are usually pH values ranging from 5.0 to 7.0, slightly aerobic 0 to + 200 mV or anaerobic − 200 mV to − 400 mV.

Metals capable of forming insoluble hydroxides/oxides, phosphates or sulfides will precipitate effectively under these conditions. Consequently, competing cations that form such insoluble materials are only a problem in that they consume the precipitant, which in turn generally requires more carbon substrate. Interfering anion conditions will apply, similar to those already discussed above.

Although dead and living cell systems have been shown to have some unique capabilities for metal removal, the transfer to commercial technology has been very limited. The reasons for this will now be discussed.

## Engineering and Process Considerations

Although dead cell systems, biomaterials and living cells have some distinct advantages compared with conventional ion exchange resins in the removal of metals from aqueous waste liquors, the number of recorded pilot-plant and commercial processes is extremely small, as illustrated in **Table 7**.

It would appear that even the rapid metal uptake by dead cell systems (less than 60 s for outer wall adsorption) combined with potentially cheap biosorbents and greater versatility of living cell systems such as SRB, capable of accommodating more than one

**Table 7**  Biological metal-removal pilot plants and commercial facilities

| Process | Biomaterial/microorganism | Target metal(s) | Process conditions |
|---|---|---|---|
| AMT-Bioclaim | *Bacillus subtilis* | Pb, Zn | Pilot plant (PP) <br> Capacity 20 BV h$^{-1}$ <br> Columns 18 L BV$^{-1}$ |
| Alga SORB | Algae | Cd, U, Pb, Hg | PP <br> Capacity 10 BV h$^{-1}$ <br> Columns 0.4 L BV$^{-1}$ |
| BIOFIX | Various biomasses | Ni, Zn, Mn, Cd | PP <br> Capacity 30 BV h$^{-1}$ <br> Columns 14.3 L BV$^{-1}$ |
| Shell Chemicals | Sulfate-reducing bacteria | Zn, Cd | PP <br> Capacity 3 m$^3$ h$^{-1}$ <br> USAB 12 m$^3$ <br> Commercial scale <br> Capacity 300 m$^3$ h$^{-1}$ <br> USAB 1800 m$^3$ |

BV, bed volume; USAB, upflow anaerobic sludge blanket reactor.

pollutant such as sulfate and metals, are generally insufficient to convince potential users to install, or even consider, a biological process in preference to a chemical/physicochemical process.

In considering process technologies for removal of metals from waste liquors, certain criteria need careful and critical appraisal. These process criteria and their implications to biological systems are briefly reviewed.

## Robustness

Dead cell systems and other biomaterials exhibit similar robust chemical properties to conventional ion exchange resins. They lack equality, however, because of their mechanical properties. Living organisms are significantly less robust as it is crucial to maintain cell growth and/or metabolic activity, which requires a carefully controlled environment. It is however possible, by judicious design, to arrange the biological stage of the process remote from and/or independent of the metal removal stage.

## Selectivity

The metal selectivity of dead cell systems and biomaterials can be significantly improved by a variety of chemical treatments, but obviously at a cost and in some instances to the detriment of other process considerations, e.g. chemical treatment may result in a lowering of the total metal affinity.

However, metal selectivity may not be a prime consideration as effluents often contain several offending metals, in which case living cell systems are more appropriate as they are less discriminatory.

## Compatibility

By definition effluent treatment processes have to be compatible with upstream, and in some instances downstream, operations. In many circumstances the compatibility requirement is exacerbated as effluent processes are retrofitted. Chemicals used in the biological metal-removal processes such as regeneration liquors for dead cell/biomaterial systems and excess sulfide and/or carbon substrate in SRB processes will require careful examination, in particular when water from the treated effluent is to be recycled.

## Reliability

Site effluent treatment systems operate continuously, virtually 365 days a year. Biological processes are equal in reliability to their chemical competitors but may require greater control to ensure that environments are maintained, in particular for living cell metal removal processes.

Reliability can be enhanced by the installation of duplicate facilities and/or buffer storage, but with a cost penalty.

## Simplicity

Modern automation has allowed effluent treatment facilities to be left unattended for significant periods. The greater the simplicity of metal removal, the lower the automation requirements. Dead cell/biomaterial systems are no more complicated than ion exchange resin processes, but this is not the case for living cells. Ensuring the activity of appropriate enzymes requires added process considerations.

## Predictability

At the outset, effluent treatment designers will have to be confident that appropriate scientific and engineering information is available to meet all eventualities, in particular maloperation of the facility and the environmental impact this may have.

When more than one metal-removal mechanism may be participating, either synergistically or antagonistically, such information may not be readily available. Even when this information is available, predictability becomes more complex and difficult. Dead cell/biomaterial systems may fall into this category.

## Efficiency

The biological process should have high metal-accumulation capacity and accumulation should be sufficiently rapid and efficient to compete with those of conventional technologies. In order to be competitive, the process should remove at least 99% of the target metals. There is clear evidence that biological systems can compete in efficiency with existing technologies, some reaching a sorption capacity of greater than 200 mg per g dry weight of dead cell systems.

The rate of heavy metal accumulation by biosorbents also compares favourably with existing separation techniques.

## Versatility

Since waste streams are highly variable, the efficiency of metal removal should ideally be unaffected by other waste stream constituents and should be relatively stable to variations in pH. This presents one of the biggest challenges in the development of liquid waste treatment. The effect of pH on metal removal is often significant and varies with the biomass and metals. The presence of inorganic and organic components other than the target metal or metals common in waste streams. Such components have often been found to alter the efficiency of sorption. There are several possible mechanisms for this effect, ranging from direct competition for the binding sites resulting in lower uptake of target metal(s), to organic pollutants forming soluble complexes with metal(s) and thus reducing removal efficiency.

## Economics

The dead cells/biomaterials and living cells should be cheap to grow and/or harvest. Clearly it is economically desirable to utilize waste biomass or material from other processes since the production cost will be

**Table 8**  Sources of waste biomass for use in heavy metal removal

| Waste biomass | Source |
| --- | --- |
| Activated sludge, digested anaerobic sludge | Wastewater treatment |
| *Saccharomyces cerevisiae* (yeast) | Brewing |
| *Bacillus subtilis* (Gram-positive bacterium) | Enzyme production |
| *Penicillium chrysogenum* (fungus) | Penicillin production |

reduced. Waste biomass is produced from a number of industrial processes (**Table 8**) but the use of waste biomass should not, however, be at the expense of process efficiency.

# Realizing the Potential of Biosorption/Bioaccumulation of Metals

The understanding of the interactions of microorganisms with metals for a variety of applications such as health care, environmental protection and process technology (biocatalysis) has been pursued for nearly half a century. In this time numerous microorganisms have been isolated, characterized and evaluated for a diversity of metals. Notwithstanding this colossal effort, to date there are few installations, possibly no more than five major ones, that use microorganisms (excluding activated sludge processes) to remove and/or recover metals from waste waters/liquid wastes. Why is this? First, we need to consider what microorganisms are capable of – they are no different to their chemical counterparts, in that they cannot, for example, convert lead to gold or 'eat' plutonium! The perception of microorganism capabilities for dealing with metal pollutants is for the nonbiologists clouded by the great successes, reported worldwide, that these minute chemical factories have secured in dealing with oil spillages and land contaminated with a variety of organic pollutants. Without a doubt the microbial degradation of such pollutants is truly the 'green ticket', assuming of course that these pollutants are ultimately and quickly degraded to carbon dioxide and water.

One major reason hampering potential is the approach taken by microbiologists. In the past, and in many instances even today, screening for new microorganisms has been a major preoccupation and many person-years effort are expended in the laboratory with little thought as to how these microorganisms can/will be engineered into a technological process. The interaction between microbiologists and workers in other scientific disciplines, in particular chemists, is now more strongly evident, largely because an array of scientific techniques is needed to characterize the microorganisms. The involvement of engineers is still lacking and consequently key questions are missed or omitted when considering the potential of microorganisms to treat complicated waste streams. Many engineers are surprised at the microbiologist's approach in tackling a seemingly new pollution problem. The technique of screening the polluted environment for thriving microorganisms is logical to the microbiologist, but curious to the engineer, who may not understand the subtleties of genera and strains.

It is this fusion of scientific and engineering approaches that is needed to enable bioremediation, and hence environmental biotechnology, to achieve its true potential.

Environmental legislation is now stringent and is likely to become even more so in the future. In this situation it is important not only that the process technology is understood, but also that the implications and consequences of perturbations to this technology can be accurately predicted. With environmental processes perturbations will undoubtedly arise, as to date there is no specification for effluents that is definitive.

Unfortunately, in the present commercial environment, the quest for scientific knowledge is too often perceived as no longer valuable or affordable. In this respect the success of biotechnology in other areas, e.g. pharamaceuticals, may well have a positive benefit to other markets, persuading nonscientists that knowledge and intellectual property is valuable and ignorance is unaffordable.

*See also:* **II/Ion Exchange:** Theory of Ion Exchange. **III/Biological Systems: Ion Exchange. Resins as Biosorbents: Ion Exchange.**

## Further Reading

Brierley JA (1990) In: Volesky B (ed.) *Biosorption of Heavy Metals*, pp. 305–311. Boca Raton, FL: CRC Press.

Darnall DW, Greene B, Hosea M, *et al.* (1986) In: Thompson R (ed.) *Trace Metal Removal from Aqueous Solutions*, pp. 1–24. Whitstable, Kent: Litho Ltd.

Eccles H (1995) *International Biodeterioration and Biodegradation*, Special Issue, *Biosorption and Bioremediation*, vol. 35, pp. 5–16.

Edyvean RGJ, Williams CJ, Wilson MM and Aderhold D (1997) In: Wase J and Forster C (eds.) *Biosorbents for Metal Ions*, pp. 165–182. London: Taylor & Francis.

Gadd GM (1988) In: Rehm H-J (ed.) *Biotechnology-Special Microbial Processes*, vol. 6B, pp. 401–433. Weinheim: VCR.

Hunt S (1986) In: Eccles H and Hunt S (eds) *Immobilisation of Ions by Biosorption*, pp. 15–46. Chichester: Ellis Horwood.

Kuyucak N and Volesky B (1990) In: Volesky B (ed.) *Biosorption of Heavy Metals*, pp. 173–198. Boca Raton, CRC Press.

Lovley DR, Phillips EJP, Gorby YA and Landa Y (1991) *Nature* 350: 413–416.

Macaskie LE (1991) *CRC Critical Reviews in Biotechnology* 11: 41–112.

McEldowney S (1990) *Applied Biochemistry and Biotechnology* 26(2): 159–180.

Scheeren PJH, Koch RO, Buisman CJN, Barnes LJ and Versteegh JH (1992) *Transactions of the Institution of Mineralogy and Metallurgy, Section C* 101 (Sept/Oct): 190–199.

Tsezos M (1997) In: Wase J and Forster C (eds) *Biosorbents for Metal Ions*, pp. 87–113. London: Taylor & Francis.

# METALLOPROTEINS: CHROMATOGRAPHY

**E. Parisi**, CNR Institute of Protein Biochemistry and Enzymology, Naples, Italy

## Classification and Characteristics of Metalloproteins

Metals are known to play essential roles in catalysis, macromolecular structure and membrane stabilization as well as hormonal and genetic regulation. Metals present at very low concentrations in tissues and biological fluids are termed oligoelements. Usually they do not occur in the biological matter as free ions, but as metal–protein complexes. The term metalloprotein is used to define a large group of proteins containing one or more atoms of metal bound to specific sites in the polypeptide chain. The binding sites on the protein are provided by histidine nitrogens, glutamate or aspartate oxygens and cysteine sulfurs; the metal ligand is usually represented by calcium, selenium, iron, zinc, copper and other heavy metals.

Metalloproteins can be divided into two groups: biologically active metalloproteins and proteins with no apparent biological activity. The first ones include metalloenzymes and zinc-containing DNA-binding proteins. Metalloenzymes are particularly suited to studies on metal–protein interaction aimed at a better understanding of the enzymatic mechanisms; their activity is measured by means of specific assays with an appropriate substrate. The metal can either stabilize the protein structure or be part of the active catalytic site; sometimes distinct atoms of the same metal may have structural or catalytic roles, depending on the site to which they are bound.

Metalloproteins that do not exhibit biological activity can be isolated by simply monitoring the metal. The most widely used technique for metal determination is atomic absorption spectrometry (AAS); however, alternative methods are also available, such as flame atomic emission spectrometry, plasma emission spectrometry, differential pulse polarography and the neutron activation analysis.

## Metalloenzymes

### Metalloenzyme Preparation

Metalloenzymes are widely represented in almost every group of enzymes. The methods used for the purification and separation of these enzymes from other components do not differ substantially from those usually employed for enzymes not containing metals as prosthetic groups. These methods include techniques such as gel permeation, ion exchange, affinity chromatography and high performance liquid chromatography (HPLC). The reader is referred to these specific topics for details on the chromatographic techniques.

The purification strategy varies from one enzyme to another; however, an effective procedure for some enzymes is affinity chromatography. Alcohol dehydrogenase, for example, can be extensively purified by affinity chromatography based on the interaction of the coenzyme with a Blue A column and subsequent elution of the enzyme with $NAD^+$.

Good recovery of the metalloenzymes from chromatographic columns may depend on the precautions adopted during the separation procedures. In particular, special care is required to preserve the integrity of the metal–protein complex during chromatography by avoiding conditions that may affect metal binding. A loss of metal can occur in the presence of chelating agents such as ethylenediaminetetraacetic acid (EDTA) or as a result of decreasing the pH of the medium. Buffer complexation with metals may also affect metalloenzyme stability. The use of inorganic buffers should be avoided as they may remove metals

essential to enzymatic activity. To avoid problems of metal chelation, the use of Good buffers (made of N-substituted taurine and glycine) is useful. Certain substances usually added to protein solvents may affect the metal binding; high concentrations of reducing agents such as dithiothreitol may remove metals with low binding affinity. In these cases, the best choice is to use 0.5% (v/v) mercaptoethanol as a reducing agent, as this substance does not remove metals.

### Separation of Protein and Metal Moieties

Metal-free enzymes are well-suited to the study of the interaction of protein and metal ions and the effect of such interaction on the structure and function of the enzyme. Separation of metal from protein is easily performed using chelating agents at a pH between 5.5 and 7.5. The most commonly used chelator for apoenzyme preparation is 1,10-phenanthroline, but it is a good rule to test several chelating agents for their efficacy. A list of chelating agents used to remove the metal from various metalloenzymes is given in **Table 1**. The use of EDTA as chelator is not recommended because it binds to protein and laboratory glassware; if its usage cannot be avoided, it is advisable to add a trace amount of $^{14}C$-EDTA to check its complete removal.

The procedure for apoenzyme preparation requires that the enzyme, at a concentration of $0.1–1$ mmol $L^{-1}$, is dialysed against several volumes of chelator solution with several changes. Dialysis tubings may contain heavy metals, hence they should be heated at 70–80°C for 2 h in metal-free water (see below) before use. Metal-free dialysis membranes (Spectra/por) are also commercially available. The chelator is removed by extensive dialysis against metal-free buffer (see below). If 1,10-phenanthroline or 8-hydroxyquinoline is used to chelate the metal, their removal can be monitored by following the optical densities of the dialysate, because these

**Table 1** Chelating agents commonly used for the preparation of metal-depleted enzymes

| Substance | Metal | Reference |
| --- | --- | --- |
| Ethylenediamine-tetraacetic acid | $Zn^{2+} Mg^{2+}$ | McConn et al. (1964) J. Biol Chel. 239: 3706 |
| 1,10-Phenanthroline | $Zn^{2+}$ | Prescott et al. (1983) Biochem. Biophys. Res. Commun. 114: 646 |
| Dipicolinic acid | $Zn^{2+}$ | Maret (1989) Biochemistry 28: 9944 |
| 8-Hydroxyquinoline-5-sulfonic acid | $Zn^{2+}$ | Jacob et al. (1998) Proc. Natl. Acad. Sci. USA 95: 3489 |

substances absorb light. The chelator can also be removed by gel filtration chromatography. A suitable chromatographic system is represented by a Bio-Gel P polyacrylamide pre-loaded column (Bio-Rad). For most proteins, an exclusion limit of 6000 may allow recovery of the apoenzyme in the void volume with little dilution of the sample.

An alternative procedure for apoenzyme preparation is provided by the use of Chelex 100. The sodium form of the resin, previously equilibrated in metal-free buffer, is mixed with the metalloenzyme solution at a ratio of about 20% in volume. If the metal is essential for catalytic activity, a time-dependent loss of enzyme activity may be observed during the treatment with Chelex 100. The advantage of this method is that the apoenzyme so prepared can be stored in the presence of the resin, thus avoiding its reactivation by adventitious metals.

In order to maintain the protein in a metal-free form, it is important to minimize the presence of contaminating metals. One of the first precautions for preventing an unwanted reassociation with metals is the use of metal-free water in all purification and separation procedures. Ultra pure water suitable for apoenzyme preparation can be obtained by repeated distillation or by using a Milli-Q apparatus (Millipore). The latter system can supply water with a metal content below the detection limits of sophisticated analytical methods such as AAS, provided the cartridge is changed frequently.

Although metal-free water is indispensable, sometimes its use is not sufficient to eliminate contamination unless polystyrene metal-free containers are used for its storage. Of course, even the use of high purity water will not avoid unwanted problems if reagents, glassware and buffers themselves are sources of contamination. The use of polystyrene labware must be preferred in place of polyethylene and polypropylene, because these may be sources of metal ions. The ubiquitous presence of metals such as zinc, mercury, iron and aluminium in many laboratory reagents is another factor that has to be controlled. It is advisable to remove these metals from buffers and solutions before using them in separation procedures. The most widely recommended methods for eliminating metals from aqueous media are dithizone extraction or treatment with a chelating resin.

Dithizone (diphenylthiocarbazone) is used as a complexing reagent to remove heavy metals from aqueous solutions by exploiting its high solubility in organic solvents compared to the low solubility in water. Dithizone may be recrystallized by dissolving 2 g of substance in 100 mL chloroform. The volume of this solution is then reduced to one-half by evaporation under a nitrogen stream. The crystals are collected by filtration, washed with carbon tetrachloride and dried under vacuum. All the manipulations with organic solvents must be carried out in a hood. The solution to be extracted is shaken for 5 min with 0.1 vol of a freshly prepared dithizone solution (0.02% in chloroform) in a separatory funnel equipped with a Teflon stopcock (do not grease the stopcock). The extraction procedure must be repeated several times with different aliquots of dithizone solution. At the end, any trace of organic solvent present in the aqueous phase is removed under reduced pressure. This procedure works very well for ions such as $Zn^{2+}$, $Cd^{2+}$, $Co^{2+}$, $Cu^{2+}$, $Fe^{2+}$ and $Ni^{2+}$, but is not as effective for $Mn^{2+}$. In addition, dithizone extraction cannot be used with buffers with a pH above 8, because at this pH dithizone solubility in water increases.

An alternative method uses a column of Chelex 100 as metal chelator. The resin is washed with 2 vol of 0.5 mol $L^{-1}$ HCl, 5–6 vol of water and 2 vol of 0.5 mol $L^{-1}$ NaOH. After a final wash with 5 vol of water, the resin is packed in a chromatographic column. The buffer to be demetallized is passed through the column at a flow rate of 10–20 mL $min^{-1}$ $cm^{-2}$, and collected after several bed volumes have been discarded.

## Metalloproteins with no Enzymatic Activity

### Detection of Metal-binding Proteins in Chromatographic Eluates

Most of the metalloproteins lacking catalytic activity are metal-binding proteins. In general, these molecules are characterized by a high metal-to-protein stoichiometry: usually from 4 to 7 atoms of heavy metal are bound to cysteinyl, glutamyl, aspartyl residues of the polypeptide chain. The best known among the metal-binding proteins are metallothioneins (MT), a family of cysteine-rich low molecular weight polypeptides present in all animal phyla, as well as in fungi, plants and cyanobacteria. Mammalian MT are single chain proteins made of 60 amino acids, including 20 cysteines arranged in two domains of metal–thiolate clusters containing 7 equivalents of metal (usually zinc, copper or cadmium).

Purification and separation of metal-binding proteins are performed by column chromatographic procedures, generally involving gel permeation and anion exchange methods and HPLC. As many metal-binding proteins do not exhibit evident biological activity, a widely used technique for monitoring chromatographic eluates is the determination of the metal by AAS. Flame AAS is mostly used for samples in solution, provided the analyte concentration is in the order of p.p.m. (1 p.p.m. = 1 mg L$^{-1}$) and enough volume of sample (at least 1 mL) is available for analysis. If the analyte concentration is in the range of p.p.b. (1 p.p.b. = 1 µg L$^{-1}$), or if one wishes to minimize the volume of sample to be employed in the analysis, the technique of choice is the furnace AAS.

This technique, initially proposed by L'vov, became commercially available in 1969. A small amount of sample (20 µL) is introduced in a graphite furnace that is rapidly brought to high temperature by electrical heating. The sample is converted in atomic vapour and part of a light produced by a lamp containing the element to be analysed is absorbed by the analyte. In some cases, the sample is placed on a small platform added to the furnace to delay vaporization until the temperature within the furnace has reached a stable plateau. The addition of a matrix modifier may help to stabilize the analyte to high temperatures. A drastic reduction of the background influence is achieved by the Zeeman correction. The experimental conditions for the determination of the elements most commonly found in metalloproteins are reported in **Table 2** for graphite furnace AAS.

In general, the quantification of the MT-bound metal in chromatographic eluates does not require pretreatment of the sample such as digestion with oxidizing agents or ashing. However, the extent of matrix interference should be evaluated by measuring the absorbance of a known standard in the presence of the sample (internal standard technique). If metal concentration is high, interference may be reduced by appropriate dilution of samples. The determination of volatile elements such as cadmium and mercury requires the addition of a modifier (Table 2).

## Separation of Metallothionein from High Molecular Weight Proteins

The procedure currently employed for MT isolation is gel permeation chromatography. In a typical preparation, 1 vol of an ethanol chloroform solution (1.05/0.08 v/v) is added dropwise to 1 vol of a tissue extract, previously centrifuged at 100 000 **g**. After the precipitate is centrifuged at 20 000 **g** for 15 min, the supernatant is mixed with 3 vol ethanol prechilled at − 20°C and maintained at the same temperature overnight. The resulting pellet, collected by centrifugation at 20 000 **g** for 20 min, is dissolved in 20 mmol L$^{-1}$ Tris/HCl buffer pH 8.6 and loaded on a Sephadex G-75 column equilibrated with 10 mmol L$^{-1}$ Tris/HCl pH 8.6. The column is eluted with the equilibration buffer and the eluate is monitored for metal content. Such a procedure is particularly suitable for low molecular weight proteins like MT, because these can be easily separated from the bulk of high molecular weight proteins. MT usually elutes at the level of standard cytochrome *c* (mol wt about 12 000) because of molecular asymmetry (**Figure 1**).

One should be aware, however, that the original metal composition of MT may be altered by the presence of adventitious metals. As zinc usually found associated with tissue MT can be exchanged for other heavy metals with higher affinity for thiol groups, such as cadmium and copper ions, that may be present as contaminants in chromatographic media or glassware, it is advisable to render these sources metal-free before use. The methods for removing heavy metals from buffers have been described above. It is advisable to wash the gel permeation column with 3 bed vol of 2 mmol L$^{-1}$ 1,10-phenanthroline to remove trace metals from the matrix. The column is then washed with metal-free buffer until disappearance of the absorbance at 320 nm.

## Separation of Metallothionein Isoforms

Genetic polymorphism is a typical feature of metallothionein. One or more MT isoforms have been found in most animal tissues; their intracellular levels may vary from tissue to tissue. These isoforms are often very similar, with only a few amino acid substitutions in their amino acid sequences. Because they are so similar it is sometimes difficult to separate these proteins; however, if two isoforms have different

**Table 2** Conditions for the determination of some elements by AAS with graphite furnace

| Element | Wavelength (nm) | Atomization temperature (°C) | Modifier | Graphite tube |
| --- | --- | --- | --- | --- |
| Cd | 288.8 | 1600 | 0.2 mg NH$_4$H$_2$PO$_4$ + 0.01 mg Mg(NO$_3$)$_2$ | Pyrolytic/platform |
| Cu | 324.8 | 2300 | None | Pyrolytic/platform |
| Co | 242.5 | 2500 | 0.05 mg Mg(NO$_3$)$_2$ | Pyrolytic/platform |
| V | 318.5 | 2650 | None | Pyrolytic/wall |
| Fe | 248.3 | 2400 | 0.05 mg Mg(NO$_3$)$_2$ | Pyrolytic/platform |
| Hg | 253.7 | 2000 | 10 mg Te in 1% HCl | Uncoated/wall |
| Zn | 213.9 | 1800 | 0.006 mg Mg(NO$_3$)$_2$ | Pyrolytic/platform |
| Se | 196.0 | 2100 | 0.05 mg Cu + 0.01 mg Mg(NO$_3$)$_2$ | Pyrolytic/platform |

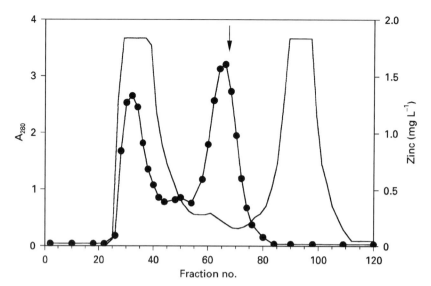

**Figure 1** Separation of metallothionein by gel permeation chromatography. About 8 mL of extract containing 200 mg protein from fish liver was loaded on a Sephadex G-75 column (2.6 × 35 cm) previously equilibrated with 10 mmol L$^{-1}$ Tris/HCl buffer pH 8.6. The column was eluted with the same buffer and fractions (2 mL) were monitored for absorbance at 280 nm (continuous line) and zinc content (circles). The arrow indicates the elution volume of standard rabbit liver metallothionein. Reproduced with permission from Scudiero R, Carginale V, Riggio M, Capasso C, Capasso A, Kille P, di Prisco G and Parisi E (1997) Difference in hepatic metallothionein content in Antarctic red-blooded and haemoglobinless fish: undetectable metallothionein levels in haemoglobinless fish is accompanied by accumulation of untranslated metallothionein mRNA. *Biochemical Journal* 322: 207–211.

net electric charges, they can be separated by anion-exchange chromatography. In **Figure 2** the elution profile from a diethylaminoethyl (DEAE)-cellulose column of two MT isoforms is shown. Metallothioneins were previously separated from high molecular weight proteins by solvent precipitation followed by gel permeation chromatography on a Sephadex G-75 column under conditions similar to those described above. Anion exchange chromatography was performed as usual, by loading the sample on the column; this was

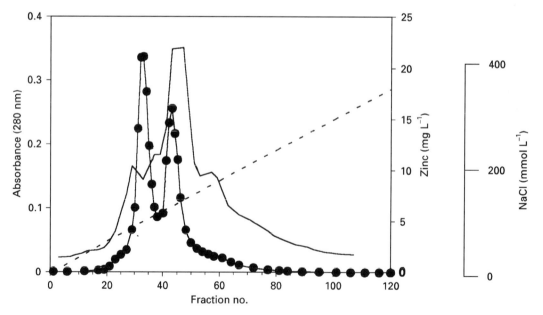

**Figure 2** Separation of sea urchin metallothionein isoforms by anion exchange chromatography. The metallothionein-containing fractions from a Sephadex G-75 column were pooled and loaded on a DEAE-cellulose column (1.6 × 25 cm) equilibrated with 20 mmol L$^{-1}$ Tris/HCl pH 8.6. The column was eluted with a linear gradient of NaCl (0–400 mmol L$^{-1}$) in equilibration buffer and fractions (3 mL) were monitored for absorbance at 280 nm (continuous line) and zinc content (circles). Reproduced with permission from Scudiero R, Capasso C, Carginale V, Riggio M, Capasso A, Ciaramella M, Filosa S and Parisi E (1997) PCR amplification and cloning of metallothionein cDNAs in temperate and Antarctic sea urchin characterized by a large difference in egg metallothionein content. *Cellular and Molecular Life Science* 53: 472–477.

**Table 3**  Chromatographic conditions for separation of metallothionein isoforms by HPLC

| Column type | Elution buffers | Running conditions | Reference |
|---|---|---|---|
| TSK G 3000 SW (Yokyo Soda) (600 × 7.5 mm i.d.) | 10 mmol L$^{-1}$ Tris/HCl pH 8.6 | 20 min at a flow rate of 1.0 mL min$^{-1}$ | Suzuki *et al.* (1984) *J. Chromatogr.* 303: 131 |
| LiChrosorb RP-18 (10 μm) (Bischoff Analysentechnik) (250 × 21 mm i.d.) | A: 25 mmol L$^{-1}$ Tris/HCl pH 7.5 B: 60% CH$_3$CN in buffer A | One-step linear gradient: 0–100 min, 0–40% B at a flow rate of 2 mL min$^{-1}$ | Hunziker *et al.* (1985) *Biochem. J.* 231: 375 |
| DEAE-5PW (Waters) (750 × 7.5 mm i.d.) | A: 10 mmol L$^{-1}$ Tris/HCl pH 7.4 B: 200 mmol L$^{-1}$ Tris/HCl pH 7.4 | One step linear gradient: 0–12 min, 0–40% B at a flow rate of 1 mL min$^{-1}$ | Lehman *et al.* (1986) *Anal. Biochem.* 153: 305 |
| Aquapore RP 300 (Brownlee Laboratories) (10 μm) (250 × 4.6 mm i.d.) | A: 10 mmol L$^{-1}$ Tris/HCl pH 7.5 B: 60% CH$_3$CN in buffer A | One-step linear gradient: 0–100 min, 0–40% B at a flow rate of 1 mL min$^{-1}$ | Ebadi *et al.* (1989) *Neurochemical Res.* 14: 69 |
| μBondapak C18 (Waters) (10 μm) (100 × 8 mm i.d.) | A: 10 mmol L$^{-1}$ NaH$_2$PO$_4$ pH 7.0 B: 40% CH$_3$CN in buffer A | Two-step linear gradient: 0–5 min, 0–10% B; 5–20 min, 10–25% B | Richards (1989) *J. Chromatogr.* 482: 87 |

washed with the low ionic strength buffer and eluted with a linear gradient of NaCl.

A suitable procedure for separation of MT isoforms is HPLC, and this can be performed with various types of column systems. Conditions most usually applied for the HPLC separation of MT are given in **Table 3**. Since MT are converted in the corresponding metal-free apoforms at acidic pH, the use of neutral buffer systems is recommended. A suitable system is a μBondapak C$_{18}$ column eluted with a two-step linear gradient of acetonitrile in phosphate buffer (**Figure 3**).

## Preparation of Metal-free Metallothionein for Metal Reconstitution

Metal substitution in MT is a common procedure to produce a complex with a defined metal composition for structural studies. It must be remembered that certain metals form complexes with different coordination numbers: while Zn and Cd in MT are tetrahedrally coordinated, Cu forms tridentate complexes. To perform metal substitution it is necessary first to prepare the metal-free MT. In doing so, it is recommended that adventitious metals in reagents and glassware are avoided by following the procedures described above.

Metal-free MT can be prepared by lowering the pH of the protein solution to 2. Under these conditions, the metal moiety dissociates from the protein; the apothionein is then separated from the metal by gel permeation chromatography on a Sephadex G-25 column. An acidified sample of about 5 mL containing approximately 15 mg of protein can be separated

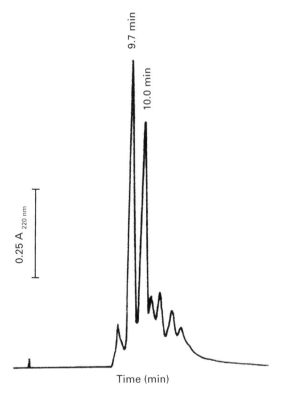

**Figure 3**  Separation of rabbit metallothionein isoforms by reversed-phase HPLC. The separation was performed on a μBondapak C$_{18}$ column using a two-step linear gradient of acetonitrile in 10 mmol L$^{-1}$ NaH$_2$PO$_4$ pH 7.0 at a flow rate of 3 mL min$^{-1}$ (0–6% from 0 to 5 min, 6–15% from 5 to 20 min). The amount of purified MT used in each run ranged from 50 to 100 μg. Modified with permission from Richards MP (1991) Purification and quantification of metallothioneins by reverse-phase high-performance liquid chromatography. *Methods in Enzymology* 205: 217.

from the metal on a 2 × 20 cm column in 0.1% trifluoroacetic acid, by recovering the apothionein in the void volume. The addition of a reducing agent is not required as the sulfhydryl groups of cysteines are maintained in the reduced form as long as the pH is acidic. The apothionein so prepared can be stored at − 80°C in an atmosphere of argon in a sealed ampoule. Prolonged storage should be avoided, as apothionein has a strong tendency to be oxidized. Anaerobic conditions are strictly required to prevent oxidation of the sulfhydryl groups with formation of large protein aggregates. The concentration of the demetallized protein can be determined by titrating the sulfhydryl groups with Ellman's reagent (5,5′-dithiobis(2-nitrobenzoic acid)), or by measuring the absorbance at 220 nm and assuming an extinction coefficient of 7.9 mg cm$^{-1}$ mL$^{-1}$ at pH 2. Metals such as zinc and cadmium are completely removed from the protein at low pH, but the procedure is not as effective for copper.

Removal of copper from Cu-thionein requires the use of chelating agents such as diethyldithiocarbamate (DTC). A solution of Cu-MT, brought to pH 5 by adding 30 μL of 3 mol L$^{-1}$ acetate buffer pH 5 per mL of solution, is mixed with solid DTC (1 mg mL$^{-1}$ of MT solution) and incubated at room temperature for 1 h. The colloidal precipitate of Cu-DTC is then removed by filtration on a 0.22 μm Millipore filter, and the filtrate containing the apothionein is desalted on a Sephadex G-25 column equilibrated and eluted with 0.1% trifluoroacetic acid.

Metal substitution is performed by mixing an argon-purged acidic solution containing 20 mg apothionein with 7–8 equivalents of metal (Zn, Cd, Hg, Bi or Pb). The solution is then rapidly titrated to pH 8.6 (7.6 for Pb-MT) with an oxygen-free solution of 0.5 mol L$^{-1}$ Tris base under fast stirring. Immediately after titration, the solution is mixed with a small aliquot of a Chelex 100 suspension in 20 mmol L$^{-1}$ Tris/HCl pH 8.6 (7.6 for Pb-MT) and stirred for 5 min. The resin is then removed by centrifugation and the solution is concentrated by ultrafiltration. MT is purified by gel permeation chromatography on a Sephadex G-50 column equilibrated and eluted with 20 mmol L$^{-1}$ Tris/HCl pH 8.6.

Reconstitution with Cu(I) is more troublesome as this ion has a marked tendency to oxidation. A stable form of Cu(I) can be prepared by dissolving Cu$_2$O in acetonitrile containing 2 mol L$^{-1}$ HClO$_4$ at 100°C. The solution is evaporated at room temperature to obtain crystals. Reconstitution is carried out by mixing apometallothionein with increasing amounts of a solution of Cu(I)-acetonitrile at pH 2 followed by titration to pH 7 with Tris/acetate buffer. The whole procedure must be carried out under anaerobic conditions.

## Conclusions

Developments in the field of metalloprotein research are strictly related to any future progress achieved in protein separation techniques. For metalloproteins, the main goals are reduction in the amount of the biomass required for their preparation, and decrease in separation time. Large amounts of protein are expected to be produced by expanding the use of recombinant DNA techniques combined with effective separation procedures such as affinity chromatography. Immunochemical methods can be applied for the detection of small protein quantities present in chromatographic eluates, and can be particularly useful for proteins lacking enzymatic activity. Techniques coupling HPLC and AAS may contribute to decreasing the time gap between sample separation and metal detection.

*See also:* **II/Centrifugation:** Analytical Centrifugation. **Chromatography: Liquid:** Mechanism: Size Exclusion Chromatography. **III/Peptides and Proteins:** Liquid Chromatography. **IV/Essential Guides for Isolation/ Purification of Enzymes and Proteins.**

## Further Reading

D'Auria S, La Cara F, Nazzaro F *et al.* (1996) A thermophilic alcohol dehydrogenase from *Bacillus acidocaldarius* not reactive towards ketones. *Journal of Biochemistry* 120: 498.

Deutscher MP (ed.) (1990) *Methods in Enzymology. Guide to Protein Purification*, vol. 182. New York: Academic Press.

Ellman GL (1959) Tissue sulfhydryl groups. *Archives of Biochemistry and Biophysics* 82: 70.

Klaassen CD (ed.) (1999) *Metallothionein IV*. Basel: Birkhäuser.

Riordan JF and Vallee BL (eds) (1988) *Methods in Enzymology. Metallobiochemistry*, vol. 158. New York: Academic Press.

Riordan JF and Vallee BL (eds) (1991) *Methods in Enzymology. Metallothionein and Related Molecules*, vol. 205. New York: Academic Press.

Suzuki KT, Sunaga H and Yajima T (1984) Separation of metallothionein into isoforms by column switching on gel permeation chromatography and ion-exchange columns with high-performance liquid chromatography-atomic-absorption spectrophotometry. *Journal of Chromatography* 303: 131.

Suzuki KT, Imura N and Kimura M (1993) *Metallothionein III*. Basel: Birkhäuser.

Vallee BL (1960) Metal and enzyme interactions: correlation of composition, function and structure. In: Boyer PD, Lardy H and Myrbäk K (eds) *The Enzymes*, Vol. 3. New York: Academic Press.

Vallee BL and Wacker EC (1970) Metalloproteins. In: H. Neurath (ed.), *The Proteins*, vol. 5. New York: Academic Press.

# MICROORGANISMS: BUOYANT DENSITY CENTRIFUGATION

*See* **III / FOOD MICROORGANISMS: BUOYANT DENSITY CENTRIFUGATION**

# MICROWAVE-ASSISTED EXTRACTION: ENVIRONMENTAL APPLICATIONS

**G. N. LeBlanc**, CEM Corporation, Matthews, NC, USA

Speed and efficiency are always prime considerations of an analytical technique. In addition, we are seeing environmental considerations – 'greening' of methods – becoming another important factor. A solvent extraction technique that reduces extraction time, improves extraction efficiency and reduces solvent consumption by a factor of 10 is an important one. The latter is critical since it is estimated that about 100 million litres of organic solvent are used annually in organic analytical laboratories worldwide. The use of microwave heating for the extraction of compounds from a variety of sample matrices has been performed since 1985. This review article provides information on the development of the microwave-assisted extraction (MAE) technique, microwave instrumentation and its capabilities, and a perspective for extraction applications.

## Development

Microwave heating was first introduced commercially in 1947. It took some time to catch on but it is now a standard item in most kitchens for cooking uses. Its use in the chemistry laboratory has significantly lagged behind domestic applications. Microwaves were first used in 1975 as a heating source for acid digestion under atmospheric conditions. The sample preparation step was reduced from 1–2 h to under 15 min, producing an overall reduction in analysis time. This work initiated the use of microwave energy as a heating source for the chemistry laboratory. Other applications include distillation, organic and inorganic synthesis, evaporation and solvent extraction.

The early applications for MAE were for the extraction of compounds with nutritional interest from plant and animal tissues. From 1986 to 1990, Ganzler and co-workers published a series of four papers exploring the use of microwave energy to partition various compounds from soils, seeds, food and animal feeds as a sample preparation method prior to chromatographic analyses. The extraction step was performed under atmospheric conditions. They extracted a variety of compounds that included antinutritives, crude fat and pesticides. They used solvent schemes similar to their traditional Soxhlet technique to allow a direct comparison of the recoveries between the two techniques. The heating programme consisted of multiple 30 s microwave heating cycles (up to seven cycles) followed by a cooling step. This approach allowed observation of the samples in the microwave cavity to prevent sample boil-over.

The microwave extracts gave recoveries that were 100–120% of the Soxhlet technique. However, the MAE times were a factor of 100 less than the traditional Soxhlet approach. The authors postulated that improved extraction efficiency was due to the polar nature of the extracted compounds or the water contained in the materials. They also noted a decrease in extraction efficiency for the recovery of nonpolar compounds when a nonpolar solvent was used.

The published work of Freitag and John in 1990 expanded the application and technique. These authors explored the use of microwave heating at elevated temperatures and pressure for the extraction of additives from polyolefins. The additives were antioxidants, Irganox 1010 and Irgafos 168, and the light stabilizer Chimassorb 81. The polyolefin matrices were polyethylene and polypropylene with a particle size of 20 mesh and a sample size of 1 g. The MAE was performed with 30 mL of 1,1,1-trichloroethane or a mixture of acetone and *n*-heptane. They obtained 90–100% recoveries of the additives with extraction times of 3–6 min without degradation of the analytes at the elevated temperatures of the extraction. This compared favourably with the conventional 16 h Soxhlet or 0.5–2 h reprecipitation techniques.

Pare *et al.* from the Ministry of Environment of Canada introduced the microwave-assisted process (MAP™) through a US patent in 1991. In this work, Pare used microwave energy to extract a variety of essential components from natural products and foods, such as biologicals and consumer products. The primary examples are the extraction of essential oils from peppermint, garlic and cedar. The major feature of this work is the release mechanism for the compounds of interest from the substrate. The microwave energy was used to disrupt the glandular and vascular system of the tissue without damaging the surrounding tissue. The solvent was used to trap and dissolve the compounds released from the tissue.

In general, the mechanism involves localized heating of the free water present in the sample. Once the water is at or above its boiling point, the water causes the cell membrane to rupture. This water, as steam, transports the target analyte from the solid to the nonabsorbing solvent. In this type of work, the sample is a good dielectric while the solvent is a poor dielectric. The microwave process gave higher yields than the traditional steam distillation process. The authors postulated that the improved efficiency was due to the lower bulk temperatures and shorter extraction times.

In 1992, Bichi *et al.* published work from the pharmaceutical area based on MAE using closed-vessel technology with temperature-feedback control. They extracted pyrrolizidine alkaloids from dried plants using 25–50 mL of methanol. The extractions were performed over a temperature range of 65–100°C for 20–30 min. The MAE technique gave qualitatively and quantitatively identical chromatographic results relative to the samples extracted by the Soxhlet procedure with a significant reduction in extraction time and solvent consumption. The temperature feedback control provided highly reproducible extractions.

In 1993, Onuska and Terry published the first data on the use of MAE for pollutants from environmental samples. They successfully extracted organochlorine pesticide residues from soils and sediments using a 1 : 1 mixture of isooctane–acetonitrile using sealed vials. The samples were irradiated for five 30 s intervals. This produced faster and more reliable results than the conventional methods. They also studied the use of a nonpolar solvent, iso-octane, for the extraction of wet sediment samples. Pesticide residue recoveries increased as the moisture content increased to a maximum of 15% and then levelled off. This shows the importance of a polar co-solvent when using a nonpolar extracting solvent for MAE tech-

niques that are temperature-dependent. In 1994 Lopez-Avila *et al.* published their work to expand the use of MAE to 187 volatile and semivolatile organic compounds from soils. The compounds included polyaromatic compounds (PNA), phenols, organochlorine pesticides and organophosphorus pesticides. This work was performed using a temperature feedback-controlled microwave heating system with closed vessels. This demonstrated the viability of the technique for the extraction of many compounds of interest to the US Environmental Protection Agency with relatively small volumes of solvent and extraction times of only 10 min. This culminated in 1998 with the approval of the microwave extraction technique by the SW-846 Organic Methods Workgroup of incorporation into SW-846 as the latest of the 'green' methods.

In 1997 there were a number of significant extensions of the technology. Incorvia-Mattina *et al.* investigated the use of MAE to extract taxanes, used in ovarian and breast cancer research, from Taxus biomass. MAE offered significant time savings versus the standard shaking technique and improved yields versus synthetic approaches. Stout *et al.* combined MAE with liquid chromatography–electrospray ionization–mass spectrometry for the determination of imidazolinones in soils at concentrations of less than 1 p.p.b. McNair *et al.* reported the combination of MAE with solid-phase microextraction for the analysis of flavour ingredients at concentrations of 2–10 p.p.b. in solid food samples. This technique showed good selectivity for the target analytes in a variety of foods.

## Instrumentation

MAE is the process of heating solid sample–solvent mixtures with microwave energy and the subsequent partitioning of the compounds of interest from the sample to the solvent. The most common approach is to perform the extraction in a sealed vessel that is microwave-transparent. This allows a temperature elevation significantly above the atmospheric boiling point of the solvent (**Table 1**) and hastens the extraction process.

The alternative approach is to perform the extraction in an open vessel at atmospheric conditions. This approach is common when the solvent is nonpolar or microwave-transparent and the sample is a biological or agriculture tissue that has a polar constituent, usually water. This is the basis for the patent issued to Environment Canada for the MAP™.

For the scope of this review article, we will limit the instrumentation discussion to the closed-vessel microwave heating system. In approaching MAE

**Table 1** Solvent boiling point – closed vessel temperature comparison

| Solvent | Boiling point (°C) | Closed vessel temperature (°C)[a] |
| --- | --- | --- |
| Acetone | 56.2[b] | 164 |
| Acetone : Cyclohexane 7 : 3 vol/vol | 52 | 160 |
| Acetonitrile | 81.6[b] | 194 |
| Dichloromethane | 39.8[b] | 140 |
| Hexane | 64.7[b] | 162[c] |
| Methanol | 68.7[b] | 151 |

[a]At 175 psig.
[b]*Lange's Handbook of Chemistry*, 14 edn. Dean JA (ed.) New York: McGraw-Hill, Inc., 1992: 11.10–11.12.
[c]Using carboflon heating insert.

applications, it is first necessary to understand the analysts' objectives. These objectives include:

1. Selection of the optimum solvent for the analytes of interest.
2. Minimization of any steps prior to the extraction step.
3. The use of a minimum amount of solvent for the extraction step.
4. Effective and reproducible extraction conditions.
5. High sample throughput.
6. Safe operation.

Considering these objectives, a microwave-assisted extraction system was designed that was a derivative of the successful microwave acid digestion system. The key components are:

1. Microwave instrumentation
2. Solvent safety features

3. Vessel technology
4. Temperature control system
5. Indirect heating source (for heating nonpolar solvents)
6. Stirring mechanism

Each component is reviewed in terms of its technical merit and how it assists in meeting the analysts' objectives for the extraction step.

**Microwave Instrumentation**

**Figure 1** shows the major components of the microwave system, including the magnetron, isolator, wave guide, cavity and mode stirrer. Microwave energy is generated by the magnetron, propagated down the wave guide and introduced into the cavity. The mode stirrer distributes the energy in various directions while the cavity acts as a containment housing for the energy until it is absorbed by the sample load within the cavity. The isolator protects the magnetron from

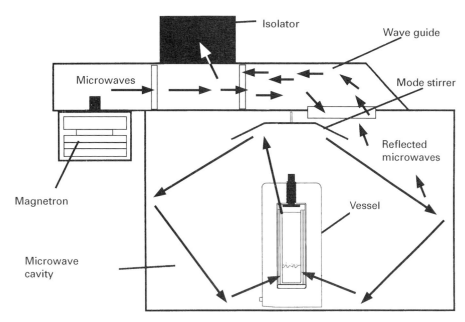

**Figure 1** Microwave system components.

reflected energy that would decrease its power output. A good analogy is a one-way mirror – it allows energy to go from the magnetron to the cavity but will not allow it to go from the cavity to the magnetron. A turntable can be used to rotate the sample load within the cavity to ensure even energy distribution.

Microwave heating is significantly different from conductive heating methods. Conductive heating is sample-independent. All samples placed inside a conduction heating oven will equilibrate to the programmed temperature. This can take quite some time. Microwave heating is sample-dependent. The temperature rise rate of samples will depend on their microwave-absorbing characteristics. The microwave design provides the ability to heat uniformly a large number of samples in a short period of time based on the sample load characteristics.

## Solvent Safety Features

Due to the flammable characteristics of many organic solvents, there is a major safety issue when heating a solvent in a microwave field. This safety issue is magnified when heating these solvents in sealed vessels at temperatures up to 100°C above their atmospheric boiling point. The microwave system should have redundant safety features, each acting as a back-up to prevent possible fire or explosion from occurring inside the cavity. Instruments should be designed to eliminate possible ignition sources, to detect solvent leaks and to remove leaking solvent. **Figure 2** is an illustration of the interior of a commercially available microwave extraction system. The safety aspects are an exhaust fan which evacuates the cavity air

volume approximately once per second. If the exhaust fan fails or there is a block downstream of the fan, the air flow switch shuts down the system. The solvent detector monitors the cavity for the presence of solvent. The detector shuts the system down if solvent concentrations reach one-tenth of the lower explosive limit for acetone, sets off an alarm and posts a message for the operator. The cavity is Teflon-coated to minimize the potential of high energy discharges. The system door is designed to withstand an event equivalent to the explosion of 1 g of TNT. It will partially open, allowing gases to escape, and then the compression springs will pull it closed. This will contain any of the contents associated with a vessel-related event inside the system's cavity.

## Vessel Technology

The closed vessels used for MAE are designed for temperatures up to 200°C and pressures of 200 psi (14 bar). The materials of construction for the components that are in contact with the sample–solvent mixture, either Teflon or glass, are inert to solvents. However, since these materials are relatively weak, an outer body is used that is much stronger – either a reinforced thermoplastic or a frame of polypropylene, or both. In addition, these materials of construction absorb a minimal amount of microwave energy. **Figure 3** illustrates a standard extraction vessel and a control extraction vessel. The vessels are composed of glass or Teflon liners, Teflon, PFA® seal cover, polyetherimide load disc and sealing screw, glass-filled polyetherimide sleeve and polypropylene support frame. The vessel has a built-in pressure relief

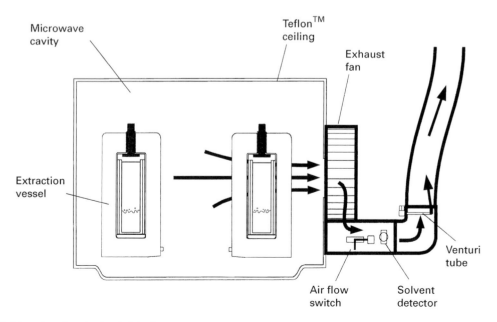

**Figure 2**  Safety exhaust and solvent detector.

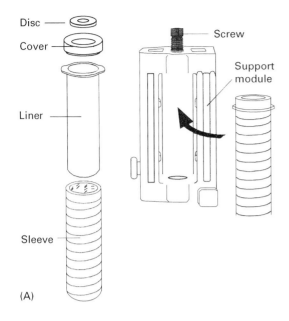

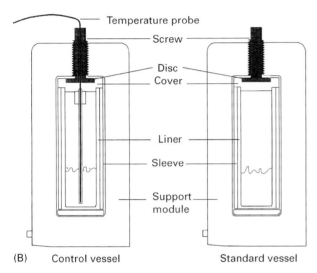

**Figure 3**  (A) Exploded view of a microwave extraction vessel. (B) Cut-away view of microwave extraction vessels.

of the target analytes, and to provide reproducible operating conditions. This is achieved with a temperature measurement system that is microwave-transparent so it does not cause any self-heating. The temperature probe is inserted directly into the control vessel to measure the temperature of the solvent–sample mixture. It is then used in a feedback control loop to regulate the microwave power output to achieve and maintain the operator-selected extraction temperature. This approach provides temperature control for one of the samples in the batch and assumes equivalent reaction conditions for all the other samples. This control technique is augmented with an indirect infrared temperature measurement system. The infrared sensor is located underneath the cavity floor. It monitors the temperature of each vessel as it passes over the sensor. This temperature reading is correlated to the direct temperature reading from the control vessel to provide temperature data for all of the samples in the batch.

### Indirect Heating Source

In some extraction applications, the solvent of choice for the target analytes is nonpolar and therefore does not heat when exposed to the microwave field. This would normally preclude the use of the MAE technique unless the analyst is willing to alter the solvent scheme to include a polar co-solvent. This is not desirable since it alters the extraction efficiency (or selectivity). This problem is overcome with the use of an insert that is a chemically inert fluoropolymer filled with carbon black, a strong microwave absorber. The insert is placed into the vessel with the solvent–sample mixture. The insert absorbs microwave energy and transfers the thermal energy it generates to the mixture. The performance characteristics of the heating inserts for heating with n-hexane in a microwave field are seen in **Table 2**. The use of the heating insert allows transfer of existing methods to the MAE technique without a change in the solvent scheme.

### Stirring Mechanism

Stirring increases the surface area contact between the sample and solvent. This offers the benefit of improved extraction efficiency and decreased solvent consumption. Stirring is achieved with the use of a rotating magnet below the cavity floor and the placement of a magnetic spinbar in the extraction vessel. The magnet creates a rotating magnetic field that couples with the spinbar in the vessel to create a stirring effect. The spinbars are either coated with an unfilled fluoropolymer for applications using polar solvents or a carbon black filled fluoropolymer for applications with nonpolar solvents.

mechanism for safety purposes. If the pressure inside the vessel exceeds the operating limits, the vessel will automatically vent. The control vessel is modified to accept a probe to monitor and control the extraction temperature. A turntable of 14 extraction vessels is placed into the instrument's cavity for batch processing. The ability to rapidly achieve elevated solvent temperatures under controlled conditions for a large batch of samples is a major advantage of the MAE technique.

### Temperature Control System

Temperature control is necessary to optimize the extraction efficiency, prevent thermal degradation

**Table 2**  MAE heating rate for *n*-hexane using heating insert

| Temperature set point (°C) | Time to temperature (min) | | |
|---|---|---|---|
| | 1 Vessel | 6 Vessels | 12 Vessels |
| 100 | 1:45 | 2:45 | 3:45 |
| 125 | 2:30 | 4:15 | 6:15 |
| 150 | 3:35 | 6:15 | 9:30 |

Notes: Each vessel contained 50 mL of *n*-hexane and 1 heating insert. The starting temperature was 25°C. As a point of reference, 12 vessels containing 50 mL of acetone will reach 125°C in 5 min under equivalent conditions.

These components create a system to perform MAE. This technique has the capability to reduce extraction time, reduce solvent consumption and improve extraction efficiency. However, it does not come without a cost. The extraction step is but one of many necessary steps to obtain the final analytical result. The analyst must take into account the differences in using the MAE technique versus their existing approach. The primary difference is the finished sample form. It is the same sample–solvent mixture originally placed into the vessel. The analyst needs to separate the sample from the solvent at the completion of the extraction step. If the analyst can work with an aliquot of the solvent for analysis, this objection can be overcome with the use of a syringe filtration technique. A secondary consideration is the vessel manipulations used with MAE. These manipulations will be new for the analyst.

## Applications

MAE has been applied to a wide variety of samples in which traditional Soxhlet extractions are performed. **Table 3** gives a performance comparison of MAE versus conventional extraction techniques for a variety of sample types. For the scope of this work, we will limit the application discussion to the extraction of plastics and polymers, pesticides and environmental samples.

### Plastics and Polymers

The additive package used in the production of polyolefins is designed to improve processing efficiency or to impart specific performance characteristics. The package can contain antioxidants, antistatic agents, slip agents, anti-block agents, UV stabilizers or antifogging agents. It is important for production efficiency and product quality that the appropriate amount of each additive is present. A fast and reliable method is needed to determine the additive concentration level. The conventional extraction approach is a reflux technique with an appropriate sol-

vent for 1–48 h followed by high performance liquid chromatography (HPLC) analysis. An alternative extraction technique is sonication for 30–60 min, but the gain in speed is offset by a loss in extraction efficiency. MAE has the ability to address the time and extraction deficiencies of the reflux and sonication techniques. Freitag and John first demonstrated the potential for MAE when they obtained excellent antioxidant recoveries from polypropylene and polyethylene.

### Pesticides and Herbicides

Pesticides and herbicides are used to protect a wide variety of agricultural commodities. There is an interest in pesticides, herbicides and their degradation product concentrations in plant and animal tissues and soil and sediment samples. The underlying assumption is that extractable compounds are labile in the environment and constitute a threat to the environment if they are hazardous. Specific examples are from the work of Fish and Revesz on chlorinated pesticides from soils and the work of Stout's group on imidazolinone herbicides in plant tissues. Fish and Revesz showed chlorinated pesticide recoveries from a Certified Reference Soil greater than or equal to those achieved using the standard Environmental Protection Agency Soxhlet technique, method 3540. This was achieved with extraction times of only 20 min and solvent volumes of 50 mL. Stout *et al.* incorporated the use of MAE with liquid chromatography–electrospray ionization–mass spectrometry to shorten the clean-up procedure and method development time of residue methodologies for determining the imidazolinones and their metabolites in crops. This application area is of concern not only to the traditional commercial testing laboratory, but to agro-chemical producers.

### Environmental

The organic side of the environmental laboratory market constitutes the majority of the analytical testing load. The extraction of priority pollutants, as well

**Table 3** MAE versus conventional extraction techniques

| Sample | Analyte | Microwave | | | Conventional | | |
|---|---|---|---|---|---|---|---|
| | | Solvent volume (mL) | Extraction time (min per sample)[a] | Concentration | Solvent volume (mL) | Extraction time (min per sample)[a] | Concentration |
| **Environmental** | | | | | | | |
| Method 3546 | Priority pollutants | 25 | 7 | – | 300 | 1080 | – |
| Soil | TPH | 30 | 7 | 943 mg kg$^{-1}$ | 300 | 60 | 773 mg kg$^{-1}$ |
| Soil | OCP | 50 | 7 | 92.3%[b] | 300 | 1080 | 83.4%[b] |
| Soil | PCBs | 25 | 7 | 47.7 µg g$^{-1}$ | 250 | 1080 | 44.0 µg g$^{-1}$ |
| Sediment | Methylmercury | 10 | 6 | 80 µg g$^{-1}$ | 200 | 150 | 81 µg g$^{-1}$ |
| Sediment | Dioxins | 30 | 8 | 565 pg g$^{-1}$ | 300 | 1440 | 542 pg g$^{-1}$ |
| Biomaterial | Organotin | 20 | 5 | 1.28 µg g$^{-1}$ | – | | 1.3 µg g$^{-1}$ |
| **Plastics** | | | | | | | |
| HDPE | Anti-oxidants | 30 | 6 | 157 µg g$^{-1}$ | 200 | 60 | 140 µg g$^{-1}$ |
| LDPE | Erucamide | 30 | 6 | 480 µg g$^{-1}$ | 200 | 60 | 491 µg g$^{-1}$ |
| PET | Oligomers | 40 | 8.5 | 1.16% | 190 | 1440 | 1.24% |
| Nylon | Plasticizer | 30 | 8 | 11.80% | 200 | 120 | 11.78% |
| Polyamide | % Extractables | 35 | 8 | 6.62% | 150 | 960 | 6.60% |
| Cellulose acetate | % Oil | 50 | 6 | 2.42% | 135 | 250 | 2.40% |
| **Agrochemical** | | | | | | | |
| Radish | PCNB | 25 | 6 | 0.42 p.p.b. | 300 | 60 | 0.36 p.p.b. |
| Soil | Imidazolinone | 20 | 5 | 11.2 p.p.b. | 400 | 120 | 10 p.p.b. |
| **Food** | | | | | | | |
| Corn | % Fat | 40 | 7 | 49.02% | 75 | 360 | 49.11% |
| Feed | % Fat | 35 | 6 | 8.73% | 250 | 120 | 8.75% |
| **Other** | | | | | | | |
| Fibre glass | % Extractables | 75 | 7 | 43.80% | 200 | 120 | 39.00% |
| Paper | % Wax | 50 | 5 | 0.88% | 150 | 60 | 0.71% |
| Carbon fibre | % Extractables | 60 | 7 | 0.55 | 250 | 1080 | 0.55% |

[a]Includes weighing, reagent addition, vessel manipulation, heating and cooling time.
[b]Value is % recovery. TPH, total petroleum hydrocarbons; OCP, organochlorine pesticides; PCB, polychlorinated biphenyls; HDPE, high density polyethylene; LDPE, low density polyethylene; PET, polyethylene terephthalate; PCNB, pentachloronitrobenzene.

as other organic molecular species, from solid samples is a primary concern. The workload in the environmental laboratory is expected to increase significantly and will thus require extraction techniques that offer increased throughput, reduced solvent consumption, improved efficiency and reproducibility. MAE has the potential to address these needs. McMillin of US Environmental Protection Agency-Region VI demonstrated this in a comparison of various soil extraction techniques for semivolatile analysis. He used an abbreviated MAE technique consisting of a small modification to regular MAE. He worked with only 10 mL of solvent versus the conventional 30 mL. This eliminated the subsequent concentration step and allowed the sample to be injected straight from the extraction vessel into a GC. The abbreviated MAE provided better extraction efficiencies and reproducibility than the three conventional techniques. The extraction time averaged 16 min per sample with a solvent use of 10 mL per sample.

One difficulty in the use of MAE for the environmental laboratory market is Environmental Protection Agency approval. The methodology has been approved by the SW-846 Organic Methods Workgroup for incorporation into SW-846. However, the method has not been promulgated and thus can only be used when regulations do not specifically require SW-846 methods.

## Future Developments

The instrumentation for MAE will continue to evolve, as will its potential applications. There will be developments in the vessel technology to address the separation issue of the sample and solvent after the extraction step. This will allow the technique to be a true replacement for the Soxhlet. There is also a need for larger vessel sizes. The current vessel has a working volume of 100 mL. There is a need to increase this to 250 mL and even higher for bulky samples and the inevitable push for lower detection limits. Finally, the microwave system's use should be extended to concentration of the sample after the extraction step. This will create a multi-tasking tool for the analytical laboratory.

MAE has focused on extraction applications from solid matrices. However, its speed and efficiency suggest that this technique will be used for isolating pharmaceutical compounds during the drug dis-

covery process. The recent addition of sample stirring suggests that it can be extended to liquid–liquid extraction applications. It could also be coupled with solid-phase microextraction to lower detection limits significantly. As MAE becomes more widely accepted and instrumentation evolves, we should see a significant increase in its applicability.

*See also:* **II/Extraction:** Microwave-Assisted Extraction. **Environmental Applications:** Soxhlet Extraction.

## Further Reading

Bichi C, Beliarab F and Rubiolo P (1992) Extraction of alkaloids from species of seneio. *Lab 2000* 6: 36–38.

Fish J and Revesz R (1996) Microwave solvent extraction of chlorinated pesticides from soil. *LC-GC* 14(3): 230–234.

Freitag W and John O (1990) Fast separation of stabilizers from polyolefins by microwave heating. *Die Angewandte Makromo Lekulare Chemie* 175: 181–185.

Incorvia Mattina M, Iannucci Berger W and Denson C (1997) Microwave-assisted extraction of taxanes from taxus biomass. *Journal of Agricultural and Food Chemistry* 45: 4691–4696.

Kingston H and Haswell S (eds) (1997) *Microwave-enhanced Chemistry*. Washington, DC: American Chemical Society.

Kingston H and Jassie L (eds) (1988) *Introduction to Microwave Sample Preparation*. Washington, DC: American Chemical Society.

Lesnik B (1998) Method 3546: microwave extraction for VOCs and SVOCs. *Environmental Testing and Analysis* 7(4): 20.

Lopez-Avila V, Young R and Beckert W (1994) Microwave-assisted extraction of organic compounds from standard reference soils and sediments. *Analytical Chemistry* 66(7): 1097–1106.

McNair R, Wang Y and Bonilla M (1997) Solid phase microextraction associated with microwave assisted extraction of food products. *Journal of High Resolution Chromatography* 20: 213–216.

Onuska F and Terry K (1993) Extraction of pesticides from sediments using microwave technique. *Chromatographia* 36: 191–194.

Stout S, Dacunhà A and Safarpour M (1997) Simplified determination of imidazolinone herbicides in soil at parts-per-billion level by liquid chromatography/electron ionization tendem mass spectrometry. *Journal of AOAC International* 80(2): 426–432.

# MOLECULAR IMPRINTS FOR SOLID-PHASE EXTRACTION

**P. A. G. Cormack**, University of Strathclyde, Glasgow, UK
**K. Haupt**, INTS-INSERM, Paris, France

Among the most important fields in analytical chemistry are medical, food and environmental analysis, where the target analytes are often present at very low concentrations in rather complex matrices. Methods currently used for analysis, such as liquid chromatography, gas chromatography or capillary electrophoresis, must therefore be preceded by a selective isolation and concentration step. Commonly used techniques for the extraction and clean up of analytes from environmental and biological samples are liquid–liquid extraction (LLE) and solid-phase extraction (SPE). Supercritical-fluid extraction (SFE) has also attracted interest, although it has not found widespread application thus far because of the specialized equipment required. The advantages of SPE compared with LLE are that it is faster, more reproducible, cleaner extracts are obtained, emulsion formation is not an issue, solvent consumption is reduced, and smaller sample sizes are required. It may also be cheaper than LLE when the costs for solvent disposal are taken into consideration. Moreover, SPE can be easily incorporated into automated analytical procedures. For a detailed introduction to SPE, the reader is directed to the relevant chapters in the Encyclopedia.

Solid phases currently employed for SPE include polystyrene divinylbenzene resins and chemically modified silica with either hydrophobic or ion-exchange groups. These materials generally yield satisfactory results if the conditions for extraction, washing and elution are carefully determined according to the chemical characteristics of the analytes and the other components in the sample. Nevertheless, due to the non-specificity of hydrophobic or ionic interactions, a large number of contaminants are often co-extracted, particularly if the sample matrix is very complex. A solution to this problem can be to use affinity sorbents, thus taking advantage of the specificity of biological recognition.

Theoretically, biomolecules like antibodies or receptors have the potential to meet the analytical demands for almost any target analyte since, with the advent of phage display antibody libraries and recombinant antibodies, a suitable recognition element can be found in many cases, even if a natural receptor does not exist. Unfortunately, biomolecules also have a major drawback, which is their poor stability, particularly in the presence of organic solvents. Artificial receptors have therefore been gaining in importance as a possible alternative to natural systems. Molecular imprinting is becoming increasingly recognized as a versatile technique for the preparation of synthetic polymers bearing tailor-made recognition sites. These molecularly imprinted polymers (MIPs) are obtained by copolymerizing functional and cross-linking monomers in the presence of the analyte, which acts as a molecular template. The molecular imprinting technique is reviewed in more detail elsewhere in the Encyclopedia (*see* Affinity Separation/Imprint Polymers). Molecularly imprinted polymers combine the advantages of biological antibodies or receptors (very specific binding) with those of synthetic polymers (high physical and chemical stability, compatibility with both aqueous and organic solvents). Their use in sample preconcentration and clean up by solid-phase extraction is therefore highly attractive.

## Analytical-Scale SPE

The applicability of imprinted polymers for SPE has been demonstrated on a number of compounds such as herbicides and drugs, which have been selectively extracted even from complex samples like beef liver extract, bile, blood serum and urine. **Figure 1** shows the principle of molecularly imprinted solid-phase extraction (MISPE). Typically, the sample is brought into contact with the imprinted polymer. This results in binding of the analyte and some impurities. Adsorption of the sample onto the polymer can be either from aqueous solution or from an organic or aqueous/organic solution after a solvent extraction step. The latter is often necessary to recover the analyte from solid or semi-solid samples, for example, soil or tissue. In both cases, the adsorption step is followed by a first elution step where impurities are washed away, whereas the specifically bound target analyte remains bound to the imprinted polymer. The last step is the elution of the analyte in concentrated and purified form using, for example, a competitor.

It has been shown that the target analyte can be recovered from samples at concentrations in the

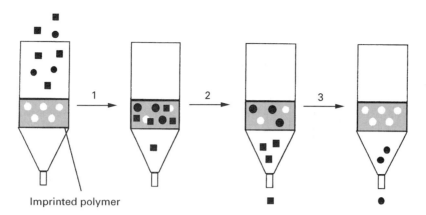

Imprinted polymer

**Figure 1**   Principle of solid-phase extraction with an imprinted polymer. The sample is loaded onto the imprinted polymer, resulting in binding of both the analyte and some contaminants (step 1). A first elution step removes contaminants and the analyte remains in the specific binding sites (step 2). The analyte is then eluted from the polymer (step 3). ●, analyte; ■, other sample components.

lower $\mu g \, dm^{-3}$ to the $mg \, dm^{-3}$ range. The lower limit is determined by the dissociation constant of the analyte binding to the polymer. The upper limit is set by the binding capacity of the polymer. If the quantity of polymer used for extraction is increased, more analyte will bind, although non-specific binding of contaminants will also increase considerably. It is therefore necessary to optimize the extraction protocol for a specific application in terms of the amount of polymer used, the sample load, and the adsorption, washing and elution conditions, etc.

The quantification of the herbicide atrazine in beef liver is a good example of the utility of imprinted polymers in SPE. In a first step, atrazine was extracted from liver tissue with chloroform. The imprinted polymer was then used to clean the chloroform extract and to further concentrate the analyte prior to quantification. In this specific example, the binding capacity of the polymer for atrazine in chloroform was found to be $19 \, \mu mol \, g^{-1}$. The analyte was eluted from the polymer with a suitable solvent (acetonitrile containing 10% acetic acid) and quantified, after drying and reconstitution in acetonitrile or buffer, by reversed-phase high performance liquid chromatography (RP-HPLC) or enzyme-linked immunosorbent assay (ELISA). When comparing the purified with the non-purified chloroform extracts in RP-HPLC, the SPE step with the imprinted polymer considerably improved the accuracy and precision of the HPLC method and lowered the detection limit from $20 \, \mu g \, dm^{-3}$ to $5 \, \mu g \, dm^{-3}$. This was achieved due to the removal of interfering components in the sample, resulting in baseline resolution of the atrazine peak. Furthermore, analyte recovery was increased from 60.9% to 88.7% (quantification by HPLC) and from 79.6 to 92.8% (quantification by ELISA).

## Preparative-Scale SPE

What has been described thus far is, by and large, the use of molecular imprints as solid-phase extraction media in *analytical* applications, where the inherent selectivity of the imprints enables *efficient* clean up and/or preconcentration of samples prior to analyte quantification. Imprinted polymers are well suited to this purpose, and their functional capacity (the mass of analyte that can be bound per unit mass of polymer) places no undue restrictions on their widespread usage. In contrast, capacity is an important consideration in preparative-scale SPE, even if the solid-phase can be regenerated many times. Bearing in mind that the functional capacities of imprinted polymers prepared thus far have been moderately low, it is relatively easy to appreciate why only a few reports describing their potential use in preparative-scale work have appeared. However, there are certain niche applications that could conceivably be serviced by state-of-the-art materials where the low capacity is more than compensated for by the additional benefits that imprints confer, e.g. stability, selectivity, etc. In the long term it is certain that numerous opportunities will exist for the use of imprinted polymers in preparative-scale SPE once the low capacity issue has been addressed satisfactorily. They are well suited to product recovery from fermentation broths, production waste streams, and during chemical and enzymatic syntheses.

The term *facilitated chemical synthesis* in the context of preparative-scale SPE refers to chemical reactions that are performed and/or worked up in the presence of imprinted polymers. In the simplest case, this involves the addition of a polymer imprinted against the product to a vessel containing the crude reaction mixture. The imprint selectively binds the product in preference to reactants, reagents, catalysts, etc., and is readily separated from the other components by virtue of its cross-linked insoluble character (the use of magnetic imprinted polymer particles or beads, which are available, would make this process even simpler). Alternatively, the crude reaction mixture can be passed through an appropriately sized SPE column, or the imprinted polymer can be placed in a product stream. Following selective adsorption, the product is washed out from the imprinted polymer and isolated. It can then be purified further if required. Another possibility for simple, inexpensive, and large-scale separations using imprinted-polymer particles is adsorptive bubble fractionation. The target compound is selectively adsorbed to the imprinted polymer in suspension within a cylindrical column containing a glass frit at the base of the column. Bubbles are formed when a gas is injected through the frit and they rise to the top of the column. Imprinted-polymer particles, which adhere to the gas bubbles, are transported to the top of the column where they accumulate and can then be easily recovered. It has been shown that enantiometric enrichment of L-phenylalanine anilide from a racemic solution can be achieved in this way, using an L-phenylalanine anilide-imprinted polymer. Overall then, imprinted polymers appear to offer an efficient general method for product isolation and purification, at least in principle.

Bearing in mind the functional capacity limitations associated with imprinted polymers at present, it should be clear why preparative-scale MISPE is currently not an attractive option. For the time being, it is more practical to use state-of-the-art imprinted materials in the selective extraction of impurities that are present in low or trace concentrations in crude reaction mixtures, where the functional capacities of imprinted polymers place considerably fewer restrictions on their application. In the following example, a recently reported model system focused on the chemical synthesis of the artificial sweetener α-aspartame (**Figure 2**). In the synthetic sequence, Z-protected L-aspartic acid anhydride (**2**) was reacted with L-phenylalanine methyl ester (**3**) to give Z-α-aspartame (**4**). The α-aspartame product (**1**) was obtained following removal of the Z-protecting group from the intermediate (**4**). The key feature about this reaction sequence was that a by-product, Z-β-aspartame (**5**), was formed during the first step, which then had to be removed. The typical composition of the crude

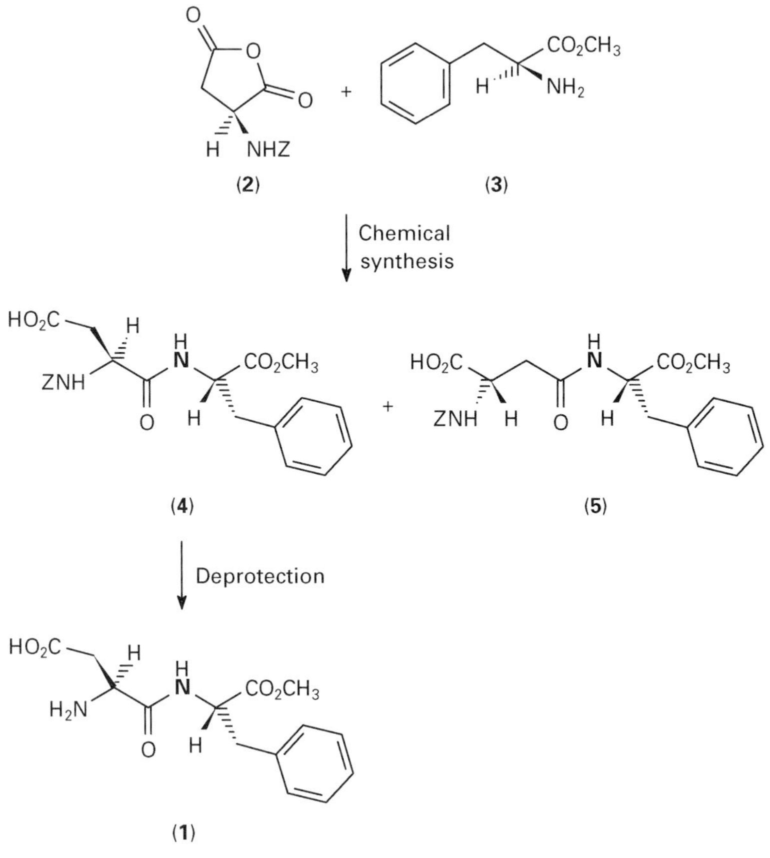

**Figure 2** The chemical synthesis of α-aspartame. (1) α-aspartame, (2) Z-protected L-aspartic acid anhydride, (3) L-phenylalanine methyl ester, (4) Z-α-aspartame, (5) Z-β-aspartame.

reaction mixture was as follows: Z-α-aspartame, 59%; Z-β-aspartame, 19%; Z-L-aspartic acid, 22%. A polymer was imprinted against the by-product Z-β-aspartame (5) and used in the SPE mode for the selective removal of the by-product from the crude reaction mixture. After five passes through a solid-phase extraction column, the product purity was increased from 59 to 96%. In a control experiment using a nonimprinted polymer, the final purity achieved was only 86%.

It is also possible to use imprinted polymers advantageously *during* chemical reactions, to drive chemical equilibria in particular directions and thus influence product distributions in a controlled fashion. To prove this concept, the enzymatic condensation of Z-protected L-aspartic acid (6) with L-phenylalanine methyl ester (3) using the enzyme thermolysin was studied (see **Figure 3**). The equilibrium for this reaction, which normally lies to the left thus favouring reactants, can be driven to the right, in the direction of products by working in a nonsolvent for Z-α-aspartame. Another way of achieving the same effect would be to use a product-trap (product-sink) in the reaction itself, for example, a polymer imprinted against the product Z-α-aspartame. Indeed, when the reaction was carried out in the presence of

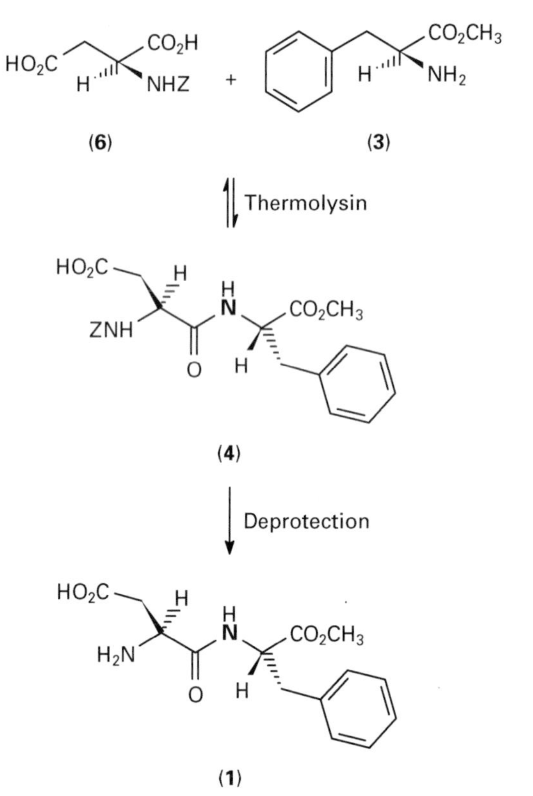

**Figure 3** The enzymatic synthesis of α-aspartame. (1) α-aspartame, (3) L-phenylalanine methyl ester, (4) Z-α-aspartame, (6) Z-protected L-aspartic acid.

a Z-α-aspartame-imprinted polymer, the reaction yield was found to increase from 15 to 63%.

## Present Limitations and Remedial Solutions

Any sorbent used in SPE must satisfy a range of performance criteria for a given application. Criteria that are of importance for imprinted polymers include: (1) Strong, selective, and reversible binding to the analyte; (2) fast mass-transfer kinetics; (3) high functional capacity; (4) minimal interaction of analyte with polymer backbone; (5) effective displacers available; (6) zero bleeding of template; (7) compatible with many solvents; (8) pressure resistant; (9) batch-to-batch reproducibility; (10) physical form of imprinted polymer; and (11) economics. Whilst imprints perform very well in many respects, there are certain limitations that need to be overcome in specific applications.

One of the most unsatisfactory features associated with the application of imprinted polymers as SPE sorbents in ultra-trace analysis is template leakage. Generally, once an imprinted polymer has been prepared, it is exhaustively extracted to remove the template from the polymer matrix. The difficulty in extracting 100% of the template molecule from an imprinted polymer has long been recognized, although until relatively recently it was widely believed that the few per cent of template remaining within the polymer was permanently entrapped. Recent work clearly demonstrates that this is not necessarily the case. What can and does occur is slow leakage of a portion of the remaining template from the polymer matrix over a period of time, even after exhaustive (solvent) extraction of the polymer beforehand. This template *leakage* or *bleeding* can have serious implications when the polymer is being used as an SPE sorbent in ultra-trace analysis, although is of much less concern in trace analysis.

Whilst a general solution to the bleeding problem is being sought by researchers across the globe, a possible method of circumventing the bleeding problem entirely is to use a template analogue during the imprinting step, rather than the template itself. One of the first demonstrations of this approach was described by Andersson *et al.* in a paper detailing the use of MISPE for the preconcentration of the drug sameridine (7) in human plasma, prior to its quantification via gas chromatography (GC). At the nanomolar concentration levels used in the study, leakage of template from the polymer matrix during sample handling was considerable and easily detectable via GC analysis. Bearing in mind the application and the concentration window, such leakage was completely

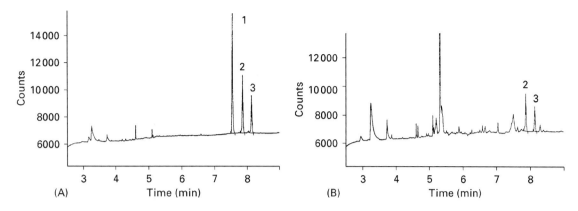

**Figure 4** GC traces of human plasma samples spiked with sameridine and an internal standard, and subjected to (A) solid-phase extraction with an imprinted polymer and (B) standard liquid–liquid extraction. The peaks are (1) the template molecule (a close structural analogue of sameridine), (2) the analyte sameridine and (3) the internal standard. (Reproduced with permission from Andersson, Papricia and Arvidsson, 1997).

unacceptable because it led to large errors in the precision of the analytical measurement. Rather than take steps to minimize or eradicate leakage, a close structural analogue of sameridine (8) was used as the template molecule in the imprinting step, which yielded an imprinted polymer that still displayed a strong affinity for sameridine.

Following solid-phase extraction of sameridine from human plasma using this polymer, leakage of the analogue from the polymer matrix did occur, but sameridine and the analogue were readily resolved using GC and the sameridine was subsequently quantified. The results obtained were as good as those obtained via a standard liquid–liquid extraction method, with the added advantage that the plasma sample contained fewer matrix contaminants with MISPE than for LLE, i.e. the sample for assay was much cleaner (**Figure 4**).

The template-analogue method does rely upon the availability of a close structural analogue of the analyte, and also a strong affinity between the polymer imprinted against the analogue and the analyte. One or both of these criteria may not always be fulfilled, therefore new approaches are required to tackle the bleeding issue.

In a solid-phase extraction operation, one ideally wants strong binding of the analyte to the sorbent during the loading and washing steps, and rapid stripping of the analyte from the sorbent during the elu-

tion step, ideally in as small a volume as possible. Efficient loading and washing is therefore favoured by a strong affinity between the polymer and the analyte, whereas efficient elution is obtained when the affinity is moderate to weak. In some cases, MIPs can actually bind analytes too strongly, which means that stronger displacers or larger volumes of eluting solvent are required than would otherwise be considered ideal. In some circumstances, therefore, having an MIP of lower affinity but with the same selectivity would be desirable. One way of achieving this goal might be via thermal pretreatment of the polymer at high temperature prior to use, which can have the effect of killing off a proportion of the high-energy binding sites.

As indicated already, the low functional capacity of imprinted polymers does not unduly limit their potential in analytical applications, although it does place restrictions on their immediate value in preparative-scale solid-phase extractions. With the advent of new developments in the imprinting area, and the advantageous knock-on effect this is likely to have on capacities, it is expected that preparative-scale applications will become more feasible in the future. Scale-up of the polymerization process to an industrial scale must also be addressed. In future developments, the elaboration of imprinting methodologies in polar environments is also expected; at present, it is a challenge to prepare good imprints in these 'competitive' environments. Although it is possible to imprint certain templates in polar environments, polar solvents (e.g. water) tend to interfere to an unacceptable level with the non-covalent interactions between template and functional monomer that are often relied upon in imprinting protocols. What is more common and easier to deal with, however, is the use of imprints effectively in buffer and/or polar environments. Finally, the imprinting of larger templates (e.g. proteins) is rather challenging at present due to their 'fragile' nature. New synthetic developments should lead to progress in this area also.

## Conclusions

Molecularly imprinted polymers constitute a new class of sorbents which combine the robust character of cross-linked polymers with the attractive properties of natural receptors. In sample clean up and concentration for trace and ultra-trace analysis, they offer distinct advantages over both liquid–liquid extraction and solid-phase extraction using classical sorbents and immunosorbents. Besides analytical applications, imprinted polymers are being increasingly considered for preparative-scale SPE applications, even though their present low functional capacity sets a limit on their widespread utility. However, concomitant with improvements in their capacity, it will become increasingly appealing to use imprinted polymers to remove products and/or by-products from reaction vessels/streams, and to influence directly the course of chemical reactions by 'equilibrium shifting'. Finally, it is worth noting that the area of molecular imprinting as a whole is undergoing rapid expansion at present. What this implies for MISPE is that one can expect tailored, high-performance imprinted polymers to become increasingly attractive and more widely available as methodologies improve and breakthroughs are made.

*See also:* **II/Extraction:** Solid-Phase Extraction. **III/Immunoaffinity Extraction. Selectivity of Imprinted Polymers: Affinity Separation:**

## Further Reading

Andersson LI, Papricia A and Arvidsson T (1997) A highly selective solid phase extraction sorbent for pre-concentration of sameridine made by molecular imprinting. *Chromatographia* 46: 57–62.

Amstrong DW, Schneiderheinze JM, Hwang YS and Sellergren B (1998) Bubble fractionation of enantiomers from solution using molecularly imprinted polymers as collectors. *Analytical Chemistry* 70: 3717–3719.

Bartsch RA and Maeda M (eds) (1998) Molecular and ionic recognition with imprinted polymers. *ACS Symposium Series* 703.

Katz SE and Siewierski M (1992) Drug residue analysis using immunoaffinity chromatography. *Journal of Chromatography* 624: 403–409.

Muldoon MT and Stanker LH (1997) Development and application of molecular imprinting technology of residue analysis. *ACS Symposium Series* 657: 314–330.

Muldoon MT and Stanker LH (1997) Molecular imprinted solid phase extraction of atrazine from beef liver extracts. *Analytical Chemistry* 69: 803–808.

Ramström O, Ye L, Krook M and Mosbach K (1998) Applications of molecularly imprinted materials as selective adsorbents: emphasis on enzymatic equilibrium shifting and library screening. *Chromatographia* 47: 465–469.

Reid E, Hill H and Wilson I (1998) *Drug Development Assay Approaches, Including Molecular Imprinting and Biomarkers.* Guilford: Guilford Academic Associates.

Sellergren B (1994) Direct drug determination by selective sample enrichment on an imprinted polymer. *Analytical Chemistry* 66: 1578–1582.

Ye L, Ramström O and Mosbach K (1998) Molecularly imprinted polymeric absorbents for by-product removal. *Analytical Chemistry* 70: 2789–2795.

# MOLECULAR RECOGNITION TECHNOLOGY IN INORGANIC EXTRACTION

**J. D. Glennon**, University College Cork, Cork, Ireland

Selective ion recognition, binding and transport are important processes in living systems. From the active sites of metalloproteins, such as amine oxidase, to lower molecular weight ligands like valinomycin, the $K^+$-selective macrocyclic antibiotic, the underlying principles of such selective ion binding and utilization have for some time been the subject of intensive investigation and modelling by many researchers.

**Table 1** Developments in macrocyclic host chemistry of major importance in ion recognition

| | |
|---|---|
| von Baeyers (1872) | Origin of calixarenes in phenol-formaldehyde reactions |
| Meadow and Reid (1934) | Preparation of hexathia-18-crown-6 in low yield |
| Zinke and Ziegler: Niederl and Vogel (1940s) | Cyclic oligomeric structures assigned to products from phenol-formaldehyde reactions |
| Pedersen (1967) | Pioneering work on crown ethers, using picrate extraction |
| Rosen and Busch (1969) | Macrocyclic thioether ligands |
| Lehn, Sauvage and Diedrich (1969) | First macrobicyclic ligands, the cryptands |
| Gutsche (1978) | Introduced name calixarene and suggested potential as molecular receptors |
| Cram (1979) | Development of spherands, with more fixed pre-organized cavities for cation selectivity |
| Izatt and Christensen (1983) | Selective extraction of cesium ions by *p-tert*-butylcalixarenes in liquid membranes |
| Pedersen, Lehn and Cram (1987) | Nobel Prize in chemistry |

Many new branches of chemistry have developed which aid the challenge, including supramolecular chemistry, the study of the formation and properties of larger molecular aggregates from the self-assembly of smaller complementary molecules through non-covalent intermolecular forces. Specifically, the creation of pre-organized cavities in synthetic host molecules for the selective reception of a neutral or ionic species is described as host–guest chemistry or molecular recognition, an approach gleaned from enzyme–substrate interactions. Macrocyclic ligands have received particular attention, as host molecules, since the pioneering work of Pedersen on the synthesis and metal extraction properties of crown ethers. This began a revolution in macrocyclic ligand and receptor design, acknowledged with the award of the Nobel Prize in chemistry in 1987 to three of the main contributing scientists, Pedersen, Lehn and Cram (**Table 1**), a revolution which continues today as new host molecules with unique selectivities of binding and mechanisms of release are produced.

Many macrocyclic ionophores are host molecules, which vary in the extent to which they are pre-organized for selective recognition of guest species and the fitting of guest to host in a complementary fashion. Some of the most important macrocyclic ligands are listed in **Table 2**. Cram recognized that the more highly organized the host was for binding and low

**Table 2** Macrocyclic ligands

| Molecular recognition ligands | Examples of metals complexed |
|---|---|
| Crown ethers | Na, K, Rb, Cs |
| Thiacrowns | Ag, Au, Pt, Hg |
| Azacrowns | Cu, Ni |
| Cryptands | Na, K, Rb, Cs |
| Spherands, hemispherands, cryptaspherands | Li, Na, K, Rb, Cs |
| Calixarenes | Cs, Fe, Cu |
| Calixarenes (functionalized) | Na, Cs, Ca, Pb, Ag, Fe, $UO_2^{2+}$ |
| Calixcrowns | K, Cs |

solvation, prior to complexation, the greater would be complex stability. Greater cation selectivity resulted, using fixed pre-organized host cavities of controlled size, provided by such macrocycles as cryptands and spherands. The chemical structures of representative macrocyclic host molecules are given in **Figure 1**, alongside the ions which have been shown to reside selectively in the guest cavities. The concepts that cavity size and shape could be tailor-made and fine-tuned to suit the selected cation diameter, and that donor atom choice determines the cation selectivity, have caught the imagination of many chemists over the last 30 years. In particular, separation scientists have been quick to demonstrate selective metal extractions, such as the separation of trace amounts of silver from mercury using 14-thiacrown-4 and the extraction of mercury and lead by 18-crown-6. Important studies were carried out on monocyclic aza crowns and cryptands, which are macrobicyclic compounds, where the complexed cation is completely enclosed by ligands, containing O and N, in a central cavity. More recently, attention has been focused on calixarenes, which are a class of functionalized meta-cyclophanes possessing convergent phenolic groups arranged around the periphery of a central aromatic cavity. These macrocycles have been described by Shinkai as the third host molecule after cyclodextrins and crowns as the first two major impact host molecules. The unique ionophoric properties of functionalized calixarenes have been clearly demonstrated using nuclear magnetic resonance (NMR) spectroscopy and liquid–liquid extraction studies. Functionalization, at the upper and lower rim, can lead to an enormous number of derivatives, water- or organic-soluble, with varied ionophoric selectivities.

In this review, representative examples of these molecular recognition reagents and their roles in selective metal ion extraction are presented. The emphasis is on illustrating through examples the influences that contribute to selectivity of complexation and extraction, and on the new materials and

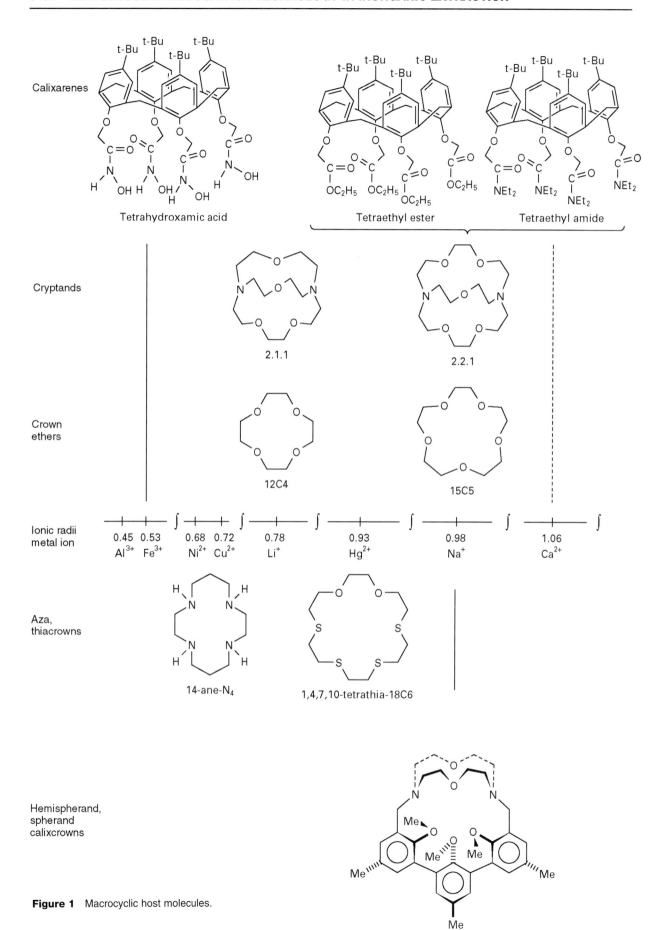

**Figure 1** Macrocyclic host molecules.

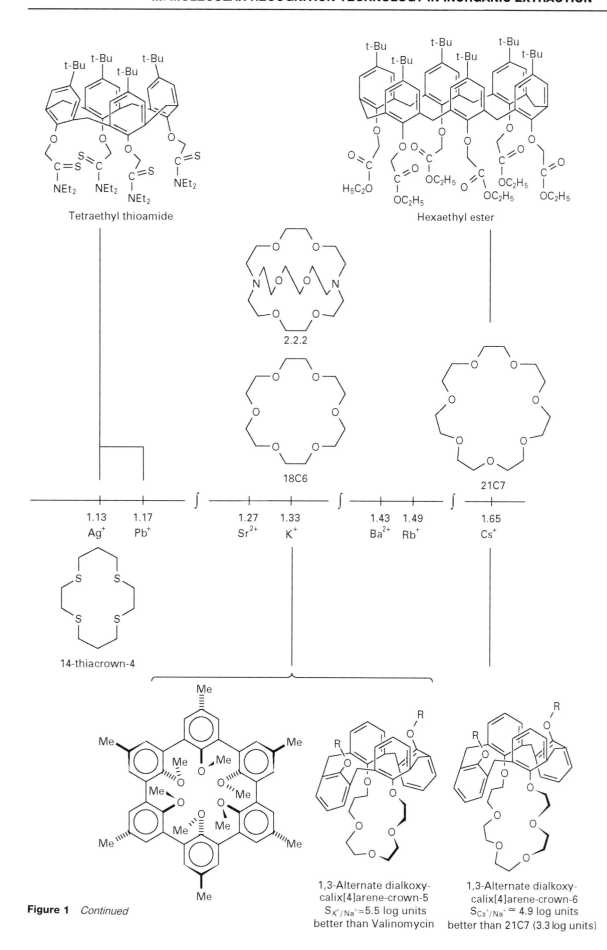

**Figure 1** *Continued*

experimental approaches that have been adapted to demonstrate analytical and preparative-scale selective metal ion extraction using molecular recognition reagents.

## Selectivity of Metal Ion Complexation

Determination of the selectivity of complexation is an important goal, since it gives further pointers to where the synthetic effort should focus, and it is a necessary step in assessing how far the synthetic design has progressed to meet the challenge presented by such natural ionophores as valinomycin.

The selectivity of molecular recognition ligands for metal ions is chiefly described in two ways, as the ratio of the measured stability constants for a pair of ions or from phase transfer and transport studies of selected cations from aqueous solution into organic solvents.

### Ratio of Stability Constants

The stoichiometric stability constant $K_s$ (sometimes symbol $\beta$) for the complexation of a metal ion, $M$, by a ligand, $L$, in solution at a constant ion strength, is defined as:

$$K_s = [ML]/[M][L]$$

in the simplest situation where a $1:1$ complex is formed and where the square brackets refer to molarities of complex, free metal and free ligand in solution. This is commonly the situation when macrocyclic ligands are involved. Thus, when comparing the relative stabilities of complexation of two metal ions, $M_1$ and $M_2$, with the same macrocyclic ligand, the ratio of the stoichiometric stability constants is a useful measure of selectivity:

$$\text{Complexation selectivity } S_{M1/M2} = K_s(M_1L)/K_s(M_2L)$$

### Extraction Selectivity

The picrate extraction method introduced by Pedersen places the MR reagent in an organic solvent, such as dichloromethane, in contact with an aqueous metal picrate solution. If complexation occurs the metal ion extraction is monitored by spectrophotometry, following the co-extraction of the highly coloured picrate counterion ($\lambda_{max} = 355$ nm, $\varepsilon = 14\,416$ mol$^{-1}$ L cm$^{-1}$) into the organic solvent. The percentage cation extraction is calculated as the ratio $100 \times (A_0 - A)/A_0$, where $A_0$ is the measured absorbance of an aqueous blank metal picrate solution without complexation reagent and $A$ is the absorbance recorded in the aqueous layer after equili-

bration. The method is particularly suited to the extraction of alkali and alkaline earths, but has also been applied to the assessment of the percentage extraction of transition metal ions. Thus, a liquid–liquid extraction method is used to determine the percentages of extraction for a series of metal ions and frequently tabulated or plotted against ionic radii to illustrate the ion preference of the reagent.

## Selectivity of Molecular Recognition Ligands

Among the factors that contribute to the selectivity of metal complexation by macrocyclic receptors is the complementarity in size of the host cavity to the cation diameter. In Figure 1 this is readily seen with the crown ethers, where selectivity moves from Li$^+$ to Cs$^+$ as the host varies from 12-crown-4 up to the larger 21-crown-7. This matching of cation diameter and cavity size is also evident in the illustrated cryptand and calixarene series, with, for example, in the latter, Cs$^+$ selectivity being displayed by the calix[6]arene hexaester derivative. The principle of pre-organization of the host cavity is evident in the ion complexation by macrocycles such as the cryptands and spherands, which display greater complexing power than the earlier crown ethers. The influence of the nature of the donor atom on the selectivity of complexation is also evident in the illustrated macrocyclic compounds. The hard and soft acid–base theory of Pearson is a useful guideline as to the behaviour, with oxygen donor atoms considered to be hard bases and nitrogen and sulfur soft bases. Alkali and alkaline earth metal cations, considered hard acids, are preferred by hard bases, while soft acids such as heavy metals are preferred by soft bases. Shifts towards the complexation of soft cations such as Ag$^+$, Pb$^{2+}$ and Hg$^{2+}$ are seen in the aza and thia crowns. It is also clearly seen in the versatile calixarene series, as for example among the tetrameric series where the change from ester to amide to thioamide moves the ion preference from Na$^+$, to Ca$^{2+}$ and on to Pb$^{2+}$ and Ag$^+$. Hancock has emphasized the role of chelate ring size within the macrocycle – an effect which influences the stability of complexation by such macrocycles as the azacrowns and the calixarene tetrahydroxamate, shown in Figure 1.

Of particular note is the influence of conformation on the stability and extraction of metal ions by molecular recognition compounds. This has been powerfully illustrated recently with calixcrowns, which are capable of cone, partial cone and 1,3-alternate conformations. Calixcrowns derived from calix[4]arenes, unsubstituted on the upper rim, can be fixed in the 1,3-alternate conformation, as shown in Figure 1.

**Table 3** Percentage extraction and log $\beta$ values for alkali and alkaline earth metal ion complexation by selected functionalized calixarene derivatives

| Host macrocycle | | Li | Na | K | Rb | Cs | Mg | Ca | Sr | Ba |
|---|---|---|---|---|---|---|---|---|---|---|
| p-t-Bu-Calix[4]arene-tetraethylester | %E | 15 | **94.6** | 49.1 | 23.6 | 48.9 | | | | |
| $S_{Na/K} = 400$ | Log $\beta$ | 2.6 | **5.0** | 2.4 | 3.1 | 2.7 | | | | |
| p-t-Bu-Calix[4]arene-tetraethylamide | %E | 63 | **95.5** | 74 | 24 | 12 | 9 | **98** | 86 | 74 |
| $S_{Ca/Mg} > 7.8$ log units | Log $\beta$ | 3.9 | **7.9** | 5.8 | 3.8 | 2.4 | 1.2 | >9 | >9 | 7.2 |
| 1,3-Dialkoxycalix[4]arene-crown-6 | %E | 2.5 | 2.6 | 13.8 | 41.7 | **63.5** | | | | |
| $S_{Cs/Na} > 4.9$ log units | Log $\beta$ | <1.5 | <1.5 | 4.3 | 6.0 | **6.4** | | | | |
| 1,3-Dialkoxycalix[4]arene-crown-5 | | – | – | – | – | – | | | | |
| $S_{K/Na} = 5.53$ log units | Log $\beta$ | 4.78 | **4.30** | 9.83 | 9.41 | 6.87 | | | | |
| Valinomycin | | – | – | – | – | – | | | | |
| | Log $\beta$ | 5.83 | **6.09** | 9.35 | 9.83 | 8.97 | | | | |

Data compiled from Arnaud-Neu *et al.* (1989) and (1991), Dozol *et al.* (1997) and Casnati *et al.* (1996).
Bold figures, highest selectivity and extraction.

While the extraction and complexation profiles are strongly dependent on the size of the crown, the behaviour is dependent on the conformation, with the highest extraction and complexation levels found for the 1,3-alternate ligands. Two striking selectivities have been achieved, as shown in **Table 3**. Remarkable Cs$^+$/Na$^+$ selectivity ($S_{Cs/Na} > 4.9$ log units) was obtained for the 1,3-alternate calix[4]-crowns-6, making applications to the extraction of caesium from radioactive waste possible. The conformation has a direct bearing on the relative stabilities, with cation-$\pi$ interactions possible for caesium but not for the smaller sodium in the rigid conformation. In addition, 1,3-alternate-calix[4]arenecrown-5 conformers have been shown in extraction experiments from water to chloroform to have better K$^+$/Na$^+$ selectivity ($S_{K/Na} = 5.53$ log units) than the naturally occurring ionophore, valinomycin.

Numerous functionalized calixarenes and, more recently, calixcrowns have been studied for their ability to extract metal ions and the stabilities of the metal–ligand complexes measured. In general, changes in the extraction profiles as the ion is varied mirror the changes in stability of the complexes. Four representative examples chosen to illustrate the influences of donor atoms and conformation on extraction and selectivity are provided in Table 3, alongside stability data reported for valinomycin.

## Molecular Recognition Ligands in Inorganic Extraction

### Solvent Extraction of Metal Cations

The Pedersen picrate extraction method is particularly suited to alkali and alkaline earths; using it McKervey *et al.* were able to show that p-t-butyl-calix[4]arene esters and ketones were Na$^+$-selective over other alkali metal ions and that calixarene complexation of metal ions is determined by the cavity size and the type of functionalization.

In sample preparation prior to trace metal analysis, a well-established procedure is the extraction of transition and heavy metal ions such as Cu$^{2+}$ and Pb$^{2+}$ into methylisobutylketone using extraction reagents such as sodium diethyldithiocarbamate and dithizone. Extraction can be monitored by spectrophotometry, as many of the complexes are coloured, or by atomic absorption spectroscopy. While such liquid–liquid extraction methods using conventional extraction reagents achieve preconcentration and sample clean-up, today the challenge is for greater selectivity of extraction, even as far as the extraction of specific metal ions from complex matrices. This can be achieved by molecular recognition technology. As far back as 1983, Izatt *et al.* demonstrated selective extraction of Cs$^+$ through organic liquid membranes by macrocyclic p-tert-calix[n]arenes through the formation of neutral complexes following proton loss. Another illustrative example is the challenge of selective extraction of uranium from sea water, where lipophilic calix[6]arene carboxylates and phosphonates have been reported as uranophiles, capable of selective transport of UO$_2^{2+}$ ions from water into organic media. In this manner, calixarenes functionalized with ionizable or chelating groups have been extensively studied over a range of pH for the extraction of transition and heavy metal ions from aqueous to organic solution. In particular, Shinkai has studied calix[5]- and calix[6]arenes with sulfonate, phosphonate, carboxylate and hydroxamate groups as uranophiles in extraction and transport experiments. The great strength of calixarenes as selective metal extraction reagents is that any chelating moiety can be attached to the macrocycle at the upper or lower rim for targeted metal ion extraction.

## Extraction of Metal Ions using Molecular Recognition Solid Phases

The immobilization of extractants onto supports for solid-phase extraction has many advantages over liquid–liquid extraction, including minimization of the use of organic solvents and amenability to automation. Important approaches taken to immobilize extractants include dissolution in a support-held organic liquid and chemically bonded solid phases.

**Supported liquid membrane enrichment technique**    The supported liquid membrane (SLM) enrichment technique involves using a solid membrane, such as porous polytetrafluoroethylene, which is impregnated with an organic solvent which acts as a stationary liquid positioned between two aqueous solutions (**Figure 2A**). When the membrane separator is configured to allow the use of flowing aqueous solutions, the technique can combine the selectivity and enrichment capabilities of liquid–liquid extraction with efficient matrix constituent removal for automated sample preparation. While the use of liquid membrane technology for extraction in industrial processes is well established, the use of SLM technology for analytical applications began in the mid 1980s for trace organic extraction. For trace metal extraction, recent examples have focused on the use of extractants such as 8-hydroxyquinoline and organophosphates. The use of a lipophilic diazo-18-crown-6 in an SLM for the separation of $Cu^{2+}$ from natural water samples has been studied. SLM technology has in particular been used to complement liquid–liquid extraction studies in determining the selectivities of some of the more recently synthesized calixarene ligands. Further studies on the use of molecular recognition reagents in SLMs are likely to lead to further useful demonstrations of enhanced selectivities of extraction using molecular recognition ligands.

**Solid-phase extraction technology**    This approach to the extraction of metal ions involves the partitioning or complexation of the metal species from the aqueous phase onto a solid phase, usually a polymer or silica. From literature reviews of solid-phase extraction, it is clear that chemically bonded phases derived from liquid chromatography are generally used for organic solute extraction. For metal ions, ion exchange and chelating phases have been widely used and have clear advantages over liquid sorbent-based approaches. How molecular recognition has impacted and can further impact on this approach to extraction will be seen from the next illustrative examples.

*Silica-bond thiacrown macrocycles*    Considerable work has been done on the chemical attachment of crown ethers to silica and polymeric supports for chromatographic separation of ions and chiral solutes. Separations of alkali and alkaline earths using water-based eluents have been demonstrated and ion-modulated separations carried out. For inorganic extraction, by far the greatest impact has been made with silica-bonded thiacrown phases. Highly selective silica-bonded, sulfur-containing crown phases are commercially marketed in packed beds or columns as SuperLig™ and AnaLig™ (for industrial and analytical separations, respectively) by IBC Advanced Technologies (Provo, UT, USA). These molecular recognition materials have resulted from the research work of Izatt *et al.* at Brigham Young University, and have been applied to the extraction of targeted metals in complex matrices, such as the large scale removal of $Pd^{2+}$ from $AgNO_3$ streams, the removal of $Cs^+$, $Sr^+$ and $Pb^{2+}$ from nuclear waste streams, and for analytical extractions, to the concentration and analysis of low level $Hg^{2+}$. The selectivity of silica-bound 1,4,7,10-tetrathia-18-crown-6 for $Hg^{2+}$ over $Ag^+$ is reported as $10^6$; in comparison, the selectivity of a typical ion exchange resin for $Ag^+$ over $Hg^{2+}$ is 1.06.

*Particle-loaded membranes*    The incorporation of solid-phase extraction particles into a web of microfibrils to form an extraction membrane with fast mass transfer kinetics has been described (Figure 2B). Highly selective and efficient removal of metal ions from solution is possible; for example, rapid sample processing and determination of radioactive strontium, counted from the surface of the membrane disc, can be achieved using Empore™ Strontium Rad Disks, containing AnaLig™ molecular recognition technology.

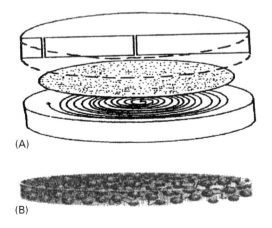

(A)

(B)

**Figure 2**    (A) Supported liquid membrane and (B) particle-loaded membrane.

*Silica-bound molecular baskets for extraction and chromatography* Molecular baskets, as calixarenes have been described, can be immobilized onto solid supports to yield new molecular recognition materials for extraction and chromatography. The obvious supports are polymer and silica phases. The nature of the surface silica-bound species in chemically bonded molecular baskets on silica has been elucidated using solid-state NMR spectroscopy. When packed as a chromatography column, a silica-bonded calix[4]arene tetraamide phase displayed significant retention of $Ca^{2+}$ over $Mg^{2+}$, a result in keeping with the reported high complexation selectivity of the calixarene for $Ca^{2+}$ over $Mg^{2+}$ ions. For transition metal ion extraction, a macrocycle that combines the sequestering ability of siderophores with the ionophoric properties of calixarenes is another interesting example of these new phases. Calix[4]arene tetrahydroxamic acid can be chemically bonded to silica particles or partitioned onto a solid support. Metal uptake profiles as a function of pH have been determined using solid-phase extraction cartridges filled with these new molecular recognition phases and the phases have been characterized by diffuse reflectance infrared Fourier transform (DRIFT) and solid-state NMR spectroscopy. The molecular recognition phase is capable of selective removal of $Pb^{2+}$ and $Fe^{3+}$ from acidic aqueous solution with $Ni^{2+}$,

$Zn^{2+}$, $Co^{2+}$, $Mn^{2+}$ and $Cd^{2+}$. The complexation of $Pb^{2+}$ takes place at a more acidic pH than is achievable with linear hydroxamic acids, as a result of host–guest complexation in the basket cavity. The structure of the silica-bonded molecular basket is schematically given in **Figure 3**, with coupling through two of the three available ethoxy groups on the derivatized upper rim. The sites for complexation of metal ions are shown on the lower rim. (It is acknowledged that there is a redundant hydroxamate group in the tetrahydroxamate when metal ions are complexed in an octahedral fashion.)

## Supercritical Fluid Molecular Recognition Technology for the Extraction of Metal Ions

Instead of carrying out liquid–liquid extractions or solid-phase extractions for the pretreatment and preconcentration of metal ions from solid and liquid matrices, it is possible to send in a selective courier molecule to permeate through the sample, grab on to the targeted metal and deposit it in concentrated form into a collection vessel for further analysis. Such a technology utilizes the solvating power of supercritical $CO_2$ and the metal–ion complexing power and selectivity of organic ligands, and is currently the focus of considerable research attention. Several studies have been reported on the supercritical fluid extraction (SFE) of metal ions via the formation of

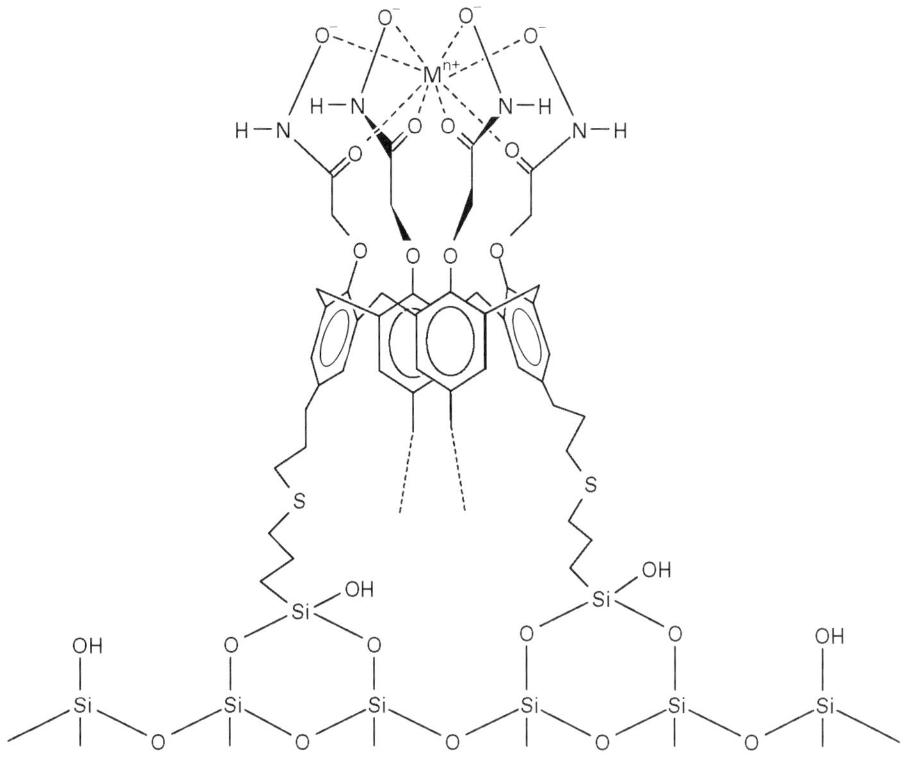

**Figure 3** Proposed structure of silica-bonded calix[4]arene tetrahydroxamate phase.

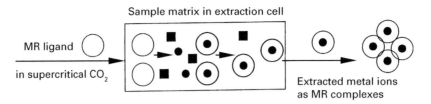

**Figure 4**   Schematic diagram of selective SFE of targeted metal ions using macrocyclic extractants. Filled circles, targeted metal ions; filled squares, diverse metal ions.

neutral metal–ligand complexes resulting in useful solubilities in supercritical $CO_2$. These solubilities can be improved by several orders of magnitude by substituting fluorine for hydrogen in the chelating ligand. For example, $\beta$-diketones and dithiocarbamates have been fluorinated and successfully used as metal extraction reagents in SFE.

More advanced complexation in SFE is achievable using molecular recognition ligands, where enhanced selectivity of extraction is achievable by careful choice of donor atoms and host cavity size in a macrocyclic reagent, as illustrated in **Figure 4**.

Recently, such selective extractions using macrocyclic reagents in SFE has been demonstrated. The selective extraction of mercury from sand, and cellulose filter paper using ionizable dibenzobistriazolo crown ether in methanol-modified supercritical $CO_2$ has been reported. The synthesis and use of fluorinated molecular baskets for metal extraction in unmodified supercritical $CO_2$ has been described. Fluorinated calixarenes are the templates on which carefully selected chelating groups can be incorporated around the cavity to yield selective extractants for targeted metals. The ability of one such molecular basket, a fluorinated calix[4]arene tetrahydroxamate, selectivity to extract $Fe^{3+}$ from metal mixtures on

cellulose paper has been monitored using atomic absorption analysis (**Figure 5**). Further examples of where targeted metal ions in a matrix, present as unwanted contaminants or as valuable metals to be recovered, are selectively complexed and removed by such macrocyclic extractants dissolved in supercritical $CO_2$ are likely in the future.

## Future Developments

Molecular recognition ligands for metal extraction are likely in the future to be cage-like and pre-organized to provide the preferred symmetry of the targeted metal ion. Higher selectivities of extraction for cations and anions can be expected, by reagents which subsequently release the guest under an applied stimulus. Molecular recognition speciation, allowing preferential extraction of individual oxidation and chemical species, will receive more attention. In chemical analysis, the incorporation of molecular recognition into miniaturized extraction devices or layers followed by detection will be further examples of the powerful role of designed molecular recognition reagents in inorganic extraction.

*See also:* **II / Affinity Separation:** Immobilised Metal Ion Chromatography: **Extraction:** Analytical Inorganic Extractions; Solid-Phase Extraction; Supercritical Fluid Extraction. **III / Ion Analysis:** Liquid Chromatography; **Metal Complexes:** Ion Chromatography; **Solid-Phase Extraction With Disks.**

## Further Reading

Arnaud-Neu F, Collins EM, Deasy M *et al.* (1989) Synthesis, X-ray crystal structures, and cation-binding properties of alkyl calixaryl esters and ketones, a new family of macrocyclic molecular receptors. *Journal of the American Chemical Society* 111: 8681–8691.

Arnaud-Neu F, Schwing-Weill M-J, Ziat K *et al.* (1991) Selective alkali and alkaline earth complexation by calixarene amides. *New Journal of Chemistry* 15: 33–37.

Böhmer V (1995) Calixarenes, macrocycles with (almost) unlimited possibilities. *Angew. Chem. Int. Ed. Engl.* 34: 713–745.

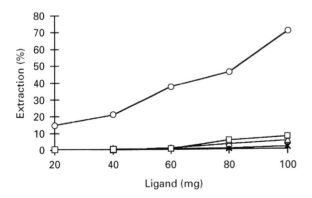

**Figure 5**   Percentage extraction of metal ions, as determined by atomic absorption analysis, versus mass of fluorinated calixarene tetrahydroxamate reagent used in the SFE of p.p.m. levels of $Fe^{3+}$ (circles), $Cu^{2+}$ (squares), $Ni^{2+}$ (triangles) and $Mn^{2+}$ (crosses) from spiked cellulose paper. 40 µL of water, 60°C, 350 atm, 30 min static and 15 min dynamic.

Casnati A, Pochini A, Ungaro R *et al.* (1996) 1,3-Alternate calix[4]arenecrown-5 conformers: new synthetic ionophores with better K$^+$/Na$^+$ selectivity than valinomycin. *Chemistry-A European Journal* 2(4): 436–445.

Cooper SR (1988) Crown thioether chemistry. *Accounts of Chemical Research* 21: 141–146.

Dozol JF, Bohmer V, McKervey MA *et al.* (1997) *EUR 17615 – New Macrocyclic Extractants for Radioactive Waste Treatment: Ionizable Crown Ethers and Functionalised Calixarenes.* Luxembourg: Office for Official Publications of the European Communities 1997 – XII.

Glennon JD, Hutchinson S, Harris SJ *et al.* (1997) Molecular baskets in supercritical CO$_2$. *Analytical Chemistry* 69: 2207–2212.

Hancock RK (1992) Chelate ring size and metal ion selection – the basis of selectivity for metal ions in open chain ligands and macrocycles. *Journal of Chemical Education* 69(8): 615–621.

Ikeda A and Shinkai S (1997) Novel cavity design using calix[*n*]arene skeletons: towards molecular recognition and metal binding. *Chemical Reviews* 97: 1713–1734.

Izatt RM (1997) Review of selective ion separations at BYU using liquid membrane and solid phase extraction procedures. *Journal of Inclusion Phenomena and Molecular Recognition in Chemistry* 29: 197–220.

Jönsson JA and Mathiasson L (1992) Supported liquid membrane techniques for sample preparation and enrichment in environmental and biological analysis. *Trends in Analytical Chemistry* 11(3): 106–114.

Kolthoff IM (1979) Applications of macrocyclic compounds in chemical analysis. *Analytical Chemistry* 51(5): 1R–22R.

Krasnushkina EA and Zolotov YuA (1983) Macrocyclic extractants. *Trends in Analytical Chemistry* 2(7): 158–162.

McKervey MA, Schwing MJ and Arnaud-Neu F (1996) Cation binding by calixarenes. *Comprehensive Supramolecular Chemistry* 1: 537–603.

Pedersen CJ (1967) Cyclic polyethers and their complexes with metal salts. *Journal of American Chemical Society* 89: 7017.

Wai CM and Wang S (1997) Supercritical fluid extraction: metals as complexes. *Journal of Chromatography A* 785: 369.

Weber E and Bartsch RA (1989) *Crown Ethers and Analogs.* Chichester: John Wiley.

# MULTIRESIDUE METHODS: EXTRACTION

**S. J. Lehotay**, Agricultural Research Service, Eastern Regional Research Center, Wyndmoor, PA, USA
**F. J. Schenck**, US Food and Drug Administration, Southeast Regional Laboratory, Atlanta, GA, USA

## Introduction

'Killing two birds with one stone' is a common expression that captures the essence of multiresidue methods of analysis. Multiresidue methods are almost always more efficient than separate single analyte methods for multiple analytes. However, a possible drawback of multiresidue methods that cover a wide polarity range or diversity of analytes is a potential loss of selectivity for individual analytes. The use of high efficiency analytical separation techniques and/or very selective detectors can compensate for a lack of selectivity in preceding steps, but as a general rule, a greater degree of selectivity leads to higher quality results. Multiresidue methods often involve a balancing act between the analytical scope of the method and the quality of the results for all analytes. It is sometimes difficult 'to have your cake and eat it, too'.

## Residues

In general, residues consist of synthetically derived chemicals that are not intended to occur in the sample, but may be present at trace concentrations as a by-product of a preliminary process related to the sample, or as a separate process altogether. Residues may be inorganic or organic, but inorganic compounds are generally analysed separately from organics. Multielemental analysis measures the natural occurrence of elements as well as any residues that may occur in the sample. Microorganisms and dirt may also be considered residues according to some definitions, but their analysis requires different techniques from organic compounds and they will not be considered further in this discussion. In the case of organic chemicals, many natural components are capable of being analysed in the same approach as the residue method, but these compounds are usually termed interferences, and great effort is often spent trying selectively to remove or avoid them (however, other chemists may be very interested in these matrix interferants).

The most common type of multiresidue application is the analysis of organic chemical contaminants in food and environmental samples. There are instances when a residue is intended, such as a fungicide designed to extend product life, but for the large majority of situations, residues are not desired in the sample. Residues may consist of pesticides, drugs, industrial by-products and/or other pollutants. Within each of these categories are subcategories known as classes of analyte. For example, classes of insecticides include organophosphorus (OP), organochlorine (OC), carbamate, pyrethroid, and others; classes of pollutants include volatile organic compounds (VOC), polycyclic aromatic hydrocarbons (PAH), polychlorinated biphenyls (PCB) alkylphenol ethoxylates (APE) and others; and veterinary drugs include antibiotics ($\beta$-lactams, aminoglycosides and tetracyclines), antibacterials (nitrofurans, fluoroquinolones and sulfonamides), and anthelminthics (benzimidazoles, avermectins and milbemycins). Thus, multiresidue methods may be single-class or multiclass depending on the number of analytes and classes covered by the analytical scope of the method.

A few existing multiresidue methods may be used to analyse more than one type of residue (e.g. pesticides and industrial pollutants), but most applications generally require the analysis of a single residue category. Analytical needs do exist that would entail the monitoring of all types of residues in some sample matrices (e.g. certain foods), but different methods are usually conducted for the separate analysis of the different residue types. It is difficult enough to develop and perform multiclass, multiresidue methods, and very few multi-type, multiclass, multiresidue methods have been attempted. However, some overall analytical schemes may include a wide array of analyte types in the extraction procedure and then divide the extract into separate aliquots for different clean-up and analytical steps.

## The Analytical Process

The analytical process goes through a series of steps which leads to the analytical results. Much like a chain that is only as strong as its weakest link, an analytical method is only as good as its weakest step. These step consist of: (i) sample collection and handling; (ii) sample preparation (extraction, clean-up); (iii) analyte determination (analytical separation, detection); and (iv) reporting of results.

The primary consideration that must be addressed independently of the analytical steps, however, is the need for the data. The analytical method should be tailored to meet the minimum needs in terms of the scope of analytes to be detected, desired limits of detection (LOD) and acceptable precision and accuracy of the results. In many cases, the data needs will

require the best possible results for as many analytes as possible. However, no analytical method can detect all possible analytes in the same procedure, and all laboratories are constrained by available personnel, space, instrumentation and other resources. Thus, the analytes must be prioritized according to importance and weighed against the cost and availability of analytical methods for their detection. The analytical chemist considers the most efficient overall approach to determine the analytes of interest that meets the acceptable data quality requirements and fits within the laboratory budget. This process may involve trade-offs between the quality of results for a particular analyte balanced against the quality of results for one or more other analytes.

The first step in the analytical process involves sample collection and handling. Unless representative samples are collected of the appropriate matrix and the samples are treated properly to avoid losses of analytes or potential contamination, then the results may not provide the information necessary to meet the needs for the analysis. In fact, a false sense of security in misleading results is often the outcome unless each step in the analytical process is carefully considered and controlled. For example, excellent recoveries and reproducibilities may be achieved for the analysis of several OP insecticides in liver tissue, but nearly all OPs partition into fat tissue in animals, and very few appear in the liver. If the purpose of the analysis is to determine animal exposure to OPs and assess risk to humans, then fat tissue should be sampled, not liver. Otherwise, the analysis may accomplish nothing of real significance.

## Extraction

Once the appropriate sample has been collected and handled properly, the next sequential step in the analytical process is the sample preparation procedure which is the subject of this article – *extraction*. Extraction is the separation of the analyte from the matrix. A few techniques, such as direct analysis of chemicals in liquids, may avoid the extraction step, but in most cases, the analytes must also be concentrated prior to analysis, which often necessitates an extraction process. The post-extraction steps in the analytical process may include clean-up of extracts and an analytical separation, but some sample preparation techniques incorporate clean-up and analyte separation into the extraction procedure. In other cases, the final analyte detection step may not require extra clean-up or analytical separation steps, but usually the price to pay for an approach that avoids clean-up steps is reduced ruggedness and higher instrument maintenance.

The type of extraction step that is used for a particular matrix depends on the nature of the matrix and analytes. Fundamentally, typical samples consist of solids, liquids and gases, or combinations thereof. Solids are usually not reasonably extracted with solids, and gases may not be extracted with gases (unless a solid membrane is placed between them), but all other combinations of liquid–solid, solid–gas, liquid–liquid and liquid–gas extractions are common in extraction techniques; supercritical fluids have also become a useful medium for extraction processes. With modern extraction techniques and the number of different solvents available, the control of pressure and temperature in the separation process has provided an essentially limitless number of possible separation conditions for the chemist to employ.

## Extractions Involving Gases

For gaseous samples, the extraction process nearly always involves passing large volumes of sample through a solid phase or liquid trap. In solid-phase systems, the analytes are adsorbed on to the particle surfaces, and in liquids the analytes are partitioned into the liquid. In the case of low temperature trapping, the analytes may simply condense on to the cold surfaces. For the most volatile analytes, a combination of condensation and adsorption may be conducted by maintaining low temperature of an adsorptive surface. The analytes are concentrated in the trapping medium and may undergo clean-up or be directly analysed in a chromatographic system. The most common approach is probably to use active materials, such as Tenax, polyurethane foam, octadecylsilyl-derivatized silica, polymer resins, or a variety of other materials to adsorb the analytes.

The use of liquid trapping by bubbling the gas through a liquid is another easy approach to extracting air-borne substances, but evaporative losses of the liquid, greater temperature limitations and less convenience generally make large volume liquid trapping a less common approach. However, liquid coatings that are useful in gas chromatography (GC) are also useful at trapping analytes from gaseous samples. It is not uncommon to perform direct collection of gases from the atmosphere at the head of a GC column kept at relatively low temperature (typically $-50°C$ to $50°C$). The chemist must be careful, however, not to expose the column to temperatures outside its range of operation. Once the sample has been extracted/collected on the column, injection occurs by simply beginning the oven temperature programme to perform separation and analysis. Purge-and-trap techniques are another way of accomplishing this without introducing air into the GC system.

### Thermal Desorption

Thermal desorption is an extraction technique which utilizes a flowing gas to extract a small heated solid or liquid sample. This process occurs during injection in GC systems and, in some cases, the approach is useful to separate thermally the analyte from the matrix for direct analysis in a flowing gas stream. The approach is not used widely in direct sample analysis, even for stable volatile and semivolatile analytes for a variety of practical reasons. For example, analyte–matrix interactions may be too strong in some cases, or matrix interferences may be too great.

Certain analytes are more prone to degradation during thermal desorption and, in multiresidue methods, these analytes may be deemed too important to sacrifice. Another potential pitfall may be that the sample size is too small to achieve the desired LOD. Thus, a liquid concentration step may be needed to increase the injected relative sample size, but then clean-up is usually required. Otherwise, thermal desorption can lead to a very rugged approach for sample introduction in GC analysis because the selectivity of the extraction matches the selectivity of the analysis. Unlike liquid sample introduction, nonvolatile components in the extract are not introduced into the GC column and the life of the chromatographic system is extended.

### Solid-Phase Microextraction (SPME)

In the late 1980s, Pawliszyn and his group at the University of Waterloo in Canada invented a technique dubbed SPME, which conveniently takes advantage of the absorption and desorption processes between gases, liquids and solids. An SPME device is a small fibre rod that has been coated with a solid or liquid phase and which is contained in a pen-like sleeve. The coated fibre is exposed to the gaseous or liquid sample and then retracted into the sleeve for brief storage. Analysis may involve thermal desorption of the fibre in a GC injection port or solvent elution of the analytes from the fibre coating into a liquid chromatographic (LC) system.

SPME is applicable to extraction of liquid samples, but it has been most noteworthy for its effectiveness in the extraction of gases. In fact, one of its modes of operation for sampling liquids is to place the fibre in the headspace above the liquid sample in an enclosed volume. The analyte will eventually partition into the headspace and then into the coating on the fibre which can be desorbed into a chromatographic system for analysis.

The major advantage of SPME is ease of use. Other advantages include low cost and avoidance of hazardous solvents. At this time, only a few fibre coatings are available, and this has limited the selectivity of SPME, but more coatings are expected to become available. The most common coating is polydimethylsiloxane (PDMS), which is also a common phase in GC columns.

## Extractions Involving Only Liquids

### Liquid–Liquid Extraction (LLE)

Water extraction is a common application for multiresidue methods. Water generally contains fewer matrix interferences than solid matrices and large volumes may often be extracted to decrease LOD. Before the widespread introduction of convenient solid-phase extraction (SPE) cartridges, LLE was the method of choice for extraction of water pollutants. Dichloromethane (DCM) is the most common solvent for extraction in LLE of water because: (i) it is only slightly miscible with water, (ii) it extracts an acceptably wide range of nonpolar analytes; (iii) it possesses a low boiling point to speed evaporation/concentration steps; and (iv) it is heavier than water and thus exists as the lower phase during partitioning in a separatory funnel or LLE glassware. Of course, other immiscible solvents are also used in LLE.

Traditionally, continuous LLE is conducted using specialized glassware which passes redistilled DCM through the water for an extended period of time (16–24 h). The extract collects with the DCM in a boiling flask where the DCM is removed by distillation. Otherwise, either manual or mechanical shaking is used to speed the extraction process (at the cost of more solvent volume and effort). In most cases, sample volume is limited to 1 L by the practical nature of the extraction process and size of the glassware. SPE has virtually displaced LLE in water methods due to its greater versatility, convenience, solvent reduction and sample volume capacity.

## Extractions Involving Liquids and Solids

The most common application in chemical residue analysis concerns the extraction of a solid sample using a liquid. A variety of liquid solvents are readily available to provide a medium for easy homogenization in a blending device. **Table 1** lists key properties of common liquids used in multiresidue methods. These parameters indicate the relative polarity of the solvent, volatility and miscibility with other liquids. In extraction processes, the tenet that like dissolves like (and conversely, opposites do *not* attract) is the primary consideration in choosing the extraction solvent. For example, hexane often provides a selective extraction for nonpolar analytes, and toluene may provide more selectivity for aromatic analytes. Practical considerations involving ease of evaporation, cost, safety and hazardous waste disposal also play a role in the selection of the extraction solvent. In situations involving acidic/basic analytes, pH is often the most critical property in the extraction and buffered aqueous solvents are often necessary. Another important consideration is the stability of the analytes in the extraction medium.

### Soxhlet Extraction

In this technique, the sample is mixed with a dispersant and/or drying agent and placed in a permeable paper thimble. The extraction thimble is placed in a glass apparatus which is exposed to the extraction solvent. Fresh solvent enters the extraction section

**Table 1**   Selected properties of common solvents used in extractions

| Solvent | Polarity index | Dielectric constant[a] | Boiling point (°C) | Viscosity (mN s$^{-1}$ m$^{-2}$) | Density[b] (g mL$^{-1}$) | Solubility in water (%w/w) |
|---|---|---|---|---|---|---|
| Acetone | 5.1 | 20.7 | 56.2 | 0.337 | 0.791 | 100 |
| Acetonitrile | 5.8 | 37.5 | 81.6 | 0.375 | 0.786 | 100 |
| Cyclohexane | 0.2 | 2.02 | 80.7 | 0.980 | 0.779 | 0.01 |
| Dichloromethane | 3.1 | 9.08 | 40.7 | 0.449 | 1.326 | 1.6 |
| Diethyl ether | 2.8 | 4.34 | 34.6 | 0.245 | 0.713 | 6.89 |
| Ethanol | 5.2 | 24.55 | 78.4 | 1.08 | 0.789 | 100 |
| Ethyl acetate | 4.4 | 6.02 | 77.2 | 0.455 | 0.901 | 8.7 |
| Hexane | 0.0 | 1.89 | 69.0 | 0.313 | 0.659 | 0.001 |
| Iso-octane | — | 1.94 | 99.2 | 0.504 | 0.692 | — |
| Methanol | 5.1 | 32.70 | 64.6 | 0.544 | 0.791 | 100 |
| Toulene | 2.4 | 2.57 | 110.8 | 0.587 | 0.866 | 0.051 |
| Water | 9.0 | 78.30 | 100.0 | 0.890 | 0.998 | — |

[a]At 25°C.
[b]At 20°C.

from the distillation section of the apparatus. When the solvent reaches a certain level, the extract siphons into a boiling flask where the extracted components are concentrated. The solvent is boiled and redistilled to fall back into the region where the sample is contained. In this way, the Soxhlet glassware is designed to repetitively conduct a number of extractions of the matrix with fresh (redistilled) solvent each time. This process is rather time-consuming (up to 16–24 h) to achieve adequate extraction efficiencies and takes up a great deal of glassware and laboratory space. Automated Soxhlet instruments have been introduced, but the Soxhlet approach is frequently regarded in modern laboratories as archaic.

## Blending and Sonication

Blending the sample with the solvent is also an old-fashioned approach, but it is very fast, convenient and inexpensive. Thus, the use of blenders, choppers, shakers, probes and other mixing devices is not likely to disappear even as newer instrumental techniques are being introduced. In the case of matrices such as clay soils that tightly retain certain analytes, sonication using a high energy probe is an alternative method that can break matrix–analyte interactions. However, due to the higher energy input involved, sonication has a greater potential for degrading analytes than simple blending, but the approach can be useful for stable analytes.

## Microwave-assisted Extraction (MAE)

MAE is a technique used in the 1990s for the extraction of organic residues in solid samples (microwave digestion has been used in the analysis of metals for several years). The approach simply involves placing the sample with the solvent in specialized containers and heating the solvent using microwave energy. The extraction process is more rapid than Soxhlet, and reduces solvent consumption, but it is more complicated and time-consuming than blending. As in the case of sonication, MAE may overcome retention of the analyte by the matrix, but analyte degradation can be a problem at higher temperatures in certain applications.

The selection of solvent, microwave energy applied and extraction time are the main parameters controlled in MAE. The user should use proper extraction vessels and equipment in MAE, because very high pressures can be generated and explosions may result if appropriate precautions are not taken. MAE instruments are available that conduct batch extractions to increase sample throughput, which is an advantage over automated instruments in other techniques that perform sequential extractions.

## Pressurized Liquid Extraction (PLE)

PLE is another time-saving and solvent-reducing approach that was developed in the mid-1990s. The instrumental approach generally involves first dispersing the sample with an inert material (e.g. drying agent or sand) and placing the mixed sample in an extraction vessel. The general approach consists of introducing the solvent into the vessel followed by heating the vessel and a static extraction step (no flow). After this 0.5–20 min step, flow is initiated (dynamic extraction step) and the extract is collected in a vial. The process may be repeated if necessary to increase analyte recoveries. Although increased temperature is not a necessity in PLE, higher temperature is usually used to speed the extraction and break analyte–matrix interactions.

The order of importance of parameters for an application in PLE (and extraction in general) is typically: (1) solvent; (2) temperature; (3) time; (4) repetitions; (5) pressure. The same types of solvents can be used in PLE as in traditional approaches, but relatively viscous solvents, such as ethanol and water, can be difficult to permeate through the sample even at high pressures. Also, highly acidic and basic conditions can be damaging to instrument components, which limits the use of PLE in certain applications. The properties of solvents can change dramatically at different temperatures and pressures (the boiling point at room temperature is commonly exceeded in PLE and MAE), thus it may be possible to replace potentially more hazardous solvents with more benign solvents. Unfortunately, physicochemical properties of many common solvents are not yet known at the elevated temperatures and pressures possible in PLE and MAE.

## Solid-Phase Extraction

SPE, sometimes referred to as liquid–solid extraction, is a popular technique for the isolation and separation of analytes from a liquid matrix. SPE columns, packed with small quantities of various chromatographic sorbents, are commercially available. SPE columns may contain polar sorbents such as silica, Florisil or alumina for normal-phase separations, and nonpolar bonded silica phases or polymers for reversed-phase separations. One of the most widely used types of SPE columns is packed with nonpolar octadecylsilyl-derivatized silica, or $C_{18}$. When liquids such as water, plasma and in some cases, milk, are eluted through $C_{18}$ SPE columns, nonpolar organic compounds such as certain pesticides, drugs or industrial pollutants will be adsorbed onto the column. These adsorbed analytes can be later eluted from the column with a relatively small amount of solvent.

SPE discs containing $C_{18}$ are widely used for the isolation of contaminants from water. Large volumes (litres) of water can be fairly rapidly eluted through the discs, and organic compounds, such as OC pesticides, PCBs and PAHs will be retained. These trapped organic compounds can then be eluted from the SPE discs with organic solvents. In many cases, the choice of solvent, pH and SPE phase provides clean-up of matrix components during the extraction process.

### Matrix Solid-phase Dispersion (MSPD)

MSPD is an extraction technique that entails mixing a sample of tissue or milk with an SPE sorbent. Stephen Barker and his group at Louisiana State University first developed the concept in the late 1980s. Typically, a small quantity (0.5 g) of liver, muscle, fat or milk is homogeneously dispersed with 2 g of $C_{18}$ silica in a mortar and pestle. The $C_{18}$ silica will disrupt the cells and disperse the contents over a large surface area, thereby exposing the entire sample to the extraction process. This homogeneous dispersion is then placed in a column, and the various components of the dispersion can be eluted from the column with a range of solvents. For example, lipids or fats can first be eluted from the column with hexane, while drugs, which are more polar, can be eluted with more polar solvents such as ethyl acetate and/or methanol. Thus, extraction and clean-up can be performed in a convenient procedure. Disadvantages of MSPD include the small sample size and potentially high cost of the solid-phase material.

# Extractions Involving Supercritical Fluids

### Supercritical Fluid Extraction (SFE)

SFE is an instrumental approach not unlike PLE, except a supercritical fluid is used as the extraction solvent rather than a liquid. SFE and PLE employ the same procedures for preparing samples and loading extraction vessels, and the same concepts of static and dynamic extractions are also pertinent. SFE typically requires higher pressure than PLE to maintain supercritical conditions and, for this reason, SFE usually requires a restrictor to better control flow and pressure of the extraction fluid. $CO_2$ is by far the most common solvent used in SFE due to its relatively low critical point (73 atm and 31°C), extraction properties, availability, gaseous natural state and safety.

A major advantage of SFE over liquid-based methods is that the extraction solvent becomes a gas after extraction and the analytes are conveniently concentrated in the collecting medium (solid-phase trap or liquid). Liquid extraction methods nearly always require a concentration step after extraction. Another key advantage of SFE is that the density of the supercritical fluid and other physicochemical properties can be dramatically altered through control of temperature and pressure. This permits a somewhat higher degree of selectivity and versatility in the extraction process without having to use different solvents. In some cases, SFE can eliminate post-extraction clean-up steps, or at least make clean-up using SPE exceptionally convenient by using the SPE sorbent as a trapping medium in SFE. Due to its many practical advantages, SFE may be considered the first choice for extraction if it is able to meet the needs of the application.

However, SFE also has several disadvantages which have delayed the widespread implementation of the approach. The higher selectivity of SFE limits the range of analytes that can be extracted under the same conditions. Furthermore, SFE can have difficulty in overcoming analyte–matrix interactions in certain applications (soils in particular). Organic solvents (and water), often called modifiers in SFE, are sometimes added to the supercritical fluid to increase the polarity range of the extraction process and to help overcome analyte retention in the matrix. Other problems with SFE include the high cost of automated instruments, relatively small sample sizes and more involved method development process. SFE has been demonstrated to be effective in the extraction of a variety of residues from a variety of matrices, but it remains to be seen if the technique can overcome its drawbacks and become more widely implemented.

# Conclusions

The analytical range of the overall analysis cannot exceed the analytical range of the extraction process, and in the case of multiresidue methods it is not uncommon to have a wide polarity range of analytes. For this reason, the use of rather exhaustive extraction conditions has been the traditional approach in multiresidue methods. The cost of a wide scope of analytes is often reduced selectivity, solvent-consuming and longer extractions, and additional clean-up of extracts. Ideally, however, the selectivity of the extraction process should match the polarity range of the targeted analytes, and no further clean-up would be required prior to analysis. Modern techniques may permit the realization of the ideal extraction process which is fast, automated, precise, efficient and safe. Furthermore, the variety of solvents and solid-phase sorbents, in combination with temperature and pressure control of modern instruments, can give the chemist the ability to achieve the desired selectivity in a single, convenient extraction procedure.

Due to the additional parameters of temperature and pressure that modern instrumental techniques provide, it has become more difficult to compartmentalize extraction techniques based on whether the extraction fluid is a dense gas, liquid, supercritical fluid, or combination thereof. Strictly speaking, pressurized fluid extraction PFE includes all types of pressurized extractions independent of the solvent's state of matter. Subcategories of PFE include SFE and PLE, but instrument companies have confused the terminology by marketing PLE as accelerated solvent extraction (ASE™) and enhanced solvent extraction (ESE). Other scientists have developed other terms to describe extraction techniques, such as enhanced fluidity extraction, which connotes a mixture of gas, liquid and/or supercritical fluid, and subcritical water extraction, which is meant to represent PLE using water at high temperatures. However, the unifying principles of extraction are the same no matter what instrument-makers or scientists may call a particular approach.

*See also:* **II/Chromatography: Gas:** Headspace Gas Chromatography. **Extraction:** Solid-Phase Extraction; Solid-Phase Microextraction; Solvent Based Separation; Supercritical Fluid Extraction. **III/Environmental Applications:** Soxhlet Extraction; **Microwave-Assisted Extraction: Environmental Applications. Solid-Phase Matrix Dispersion: Extraction.**

## Further Reading

Cairns T and Sherma J (eds) (1992) *Emerging Strategies for Pesticide Residue Analysis.* Boca Raton, FL: CRC Press.

Environmental Protection Agency (1998) *Handbook of Environmental Methods,* 3rd edn. Schenectady, NY: Genium.

Font G, Manes J, Molto JC and Pico Y (1993) Solid phase extraction multi-residue pesticide analysis of water. *Journal of Chromatography* 642: 135–161.

Food and Drug Administration (1994) *Pesticide Analytical Manual,* vol. I: *Multiresidue Methods,* 3rd edn. Washington, DC: US Dept. of Health and Human Services.

Lehotay SJ (1997) Supercritical fluid extraction of pesticides in foods. *Journal of Chromatography A* 785: 289–312.

Lopez-Avila V, Young R, Benedicto J *et al.* (1995) Extraction of organic pollutants from solid samples using microwave energy. *Analytical Chemistry* 67: 2096–2102.

Moats WA and Medina MB (eds) (1996) *Veterinary Drug Residues: Food Safety.* ACS Symposium Series 636. Washington, DC: American Chemical Society.

Pawliszyn J (1997) *Solid Phase Microextraction: Theory and Practice.* New York: Wiley-VCH.

Richter BE, Jones BA, Ezzell JL *et al.* (1996) Accelerated solvent extraction: a technique for sample preparation. *Analytical Chemistry* 68: 1033–1039.

Walker CC, Lott HM and Barker SA (1993) Matrix solid-phase dispersion extraction and the analysis of drugs and environmental pollutants in aquatic species. *Journal of Chromatography* 642: 225–242.

# NATURAL PRODUCTS

# High-Speed Countercurrent Chromatography

**A. Marston and K. Hostettmann**, Institut de Pharmacognosie et Phytochimie, Lausanne University, Switzerland

In high speed countercurrent chromatography (HSCCC) a sample is partitioned between two non-miscible liquid phases. One phase is held stationary by a centrifugal force (applied by spinning the separation element at high speed), while the second phase is pumped through the apparatus, hence the alternative term for the process centrifugal partition chromatography (CPC).

Unlike high performance liquid chromatography (HPLC), in which the stationary phase occupies 5–7% and the mobile phase about 75% of the column, the relative proportions in HSCCC are 50–75% for the stationary phase and 20–50% for the mobile phase. As a consequence, large sample loads are possible with HSCCC. Another important advantage of the absence of a solid support is that irreversible adsorption is avoided. There is total recovery of the injected sample and tailing is minimized. HSCCC is thus of special importance for the separation of sensitive and easily degraded samples. Although the efficiency of HSCCC separations is lower than that encountered in HPLC, the optimization of selectivity is the great advantage offered by the former technique.

The potential of HSCCC is further shown by the possibility of applying gradients for separations.

Solvent proportions can be changed during a chromatographic run. Furthermore, solvent elution can be reversed in the course of a separation by changing over stationary and mobile phases. Consequently, one of the characteristics of HSCCC is its extreme flexibility.

Most applications of HSCCC have been performed on two types of instrument. The first category, namely rotating coil instruments, consists of a polytetrafluoroethylene tube (column) wrapped around a spool. The spool is rotated around a central axis in such a way that it describes a planetary motion. Alternatively, hydrostatic equilibrium instruments can be employed. These have either cartridges or discs arranged around a central axis. The separation column, in effect, consists of a series of cells in the cartridges or discs.

## Preparative Applications

HSCCC is becoming a routine preparative technique in both industrial and university laboratories. Sample sizes ranging from milligrams to grams (and even kilograms in specialized instruments) can be successfully chromatographed. An extensive range of separations by HSCCC has been published (see Further Reading).

Aqueous and nonaqueous solvent systems are used and the separation of compounds with a wide range of polarities is possible. Two-phase solvents are chosen according to the hydrophobicity of the sample. For polar compounds, *n*-butanol can be employed (e.g. ethyl acetate–butanol–water or chloroform–butanol–water systems), for moderately hydrophobic compounds chloroform (e.g. chloroform–methanol–water) solvent systems, and for more hydrophobic compounds, *n*-hexane (e.g. *n*-hexane–ethyl acetate–methanol–water) solvent systems. Some of the more frequently used solvent systems are shown in **Table 1**.

A number of representative separations by HSCCC are presented here, in order to give an idea of the possibilities available, together with the chromatographic conditions employed.

### Plant-derived Natural Products

**Flavonoids** Many HSCCC separations of natural products involve polyphenols. The reason for this fact is that there is a tendency to 'tail' or even to adsorb irreversibly on conventional chromatographic supports (silica gel, polyamide etc.). This problem does not occur with all-liquid separation techniques and quantitative recovery of injected sample is achieved.

Separations of flavonoid aglycones and flavonoid glycosides have been performed mainly with chloro-

**Table 1** Frequently used solvent systems for HSCCC

| Sample | Solvent system |
|---|---|
| Polar | CHCl$_3$–MeOH–*n*-BuOH–H$_2$O 7:6:3:4 |
| | *n*-BuOH–*n*-PrOH–H$_2$O 4:1:5 |
| | *n*-BuOH–*n*-PrOH–H$_2$O 2:1:3 |
| | EtOAc–*n*-BuOH–H$_2$O 3:2:5 |
| Moderately polar | CHCl$_3$–MeOH–H$_2$O 4:3:2 |
| | CHCl$_3$–MeOH–H$_2$O 5:6:4 |
| | CHCl$_3$–MeOH–H$_2$O 10:10:6 |
| | CHCl$_3$–MeOH–H$_2$O 13:7:8 |
| | CHCl$_3$–MeOH–H$_2$O 7:13:8 |
| | *n*-Hexane–EtOAc–MeOH–H$_2$O 1:1:1:1 |
| Apolar | *n*-Hexane–EtOAc–MeOH–H$_2$O 10:5:5:1 |
| | *n*-Hexane–CH$_3$CN–MeOH 8:5:2 |
| | *n*-Hexane–CH$_3$CN–CH$_2$Cl$_2$ 10:7:3 |

form–methanol–water or chloroform–methanol–propanol–water systems.

The time required for a particular separation can be reduced by incorporating reversed-phase (RP) operation into the procedure. **Figure 1** shows the situation for the flavanone hesperitin (**1**) and the flavonols kaempferol (**2**) and quercetin (**3**). When eluting with the upper phase (mobile phase), the most polar component eluted first and the time for complete separation is over 8 h. Reversing the elution mode and changing to the lower phase as mobile phase at 70 min leads to completion of the separation in just over 2 h.

Alternatively, gradient operation can be used to speed up HSCCC. By pumping simultaneously the upper and lower phases of the two-phase solvent system by separate pumps, the proportions of the phases are changed in the coil during separation. In the example shown in **Figure 2**, the coil (capacity 360 mL) of the chromatograph was initially filled with equivalent amounts of upper and lower phases of the solvent used in Figure 1. The flavonoid mixture was then injected. By pumping upper phase at 4 mL min$^{-1}$ and lower phase at 1 mL min$^{-1}$ through the apparatus, the content of the lower phase in the coil was increased from 180 to 340 mL over 3 h. The separation time of the three flavonoids was thus reduced by 5 h, when compared with normal operation.

**Xanthones** This is another class of polyphenols to which liquid–liquid chromatography has successfully been applied. Extensive use of HSCCC was made for the separation of xanthones from a plant, *Hypericum roeperanum* (Guttiferae), from Zimbabwe. A dichloromethane extract of the roots gave a fraction (363 mg) rich in xanthones after initial chromatography. Separation of the individual xanthones from this

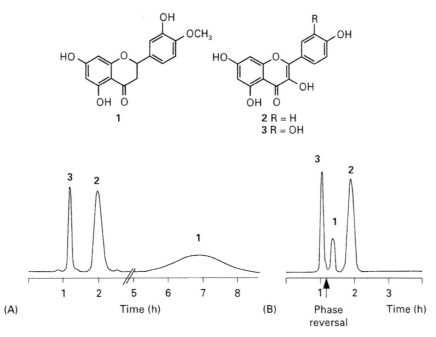

**Figure 1** HSCCC separation of hesperetin (**1**), kaempferol (**2**) and quercetin (**3**) by normal and reversed-phase solvent elution. Instrument: multilayer countercurrent chromatograph (PC Inc.). Solvent system: $CHCl_3$–MeOH–$H_2O$ 5 : 6 : 4. Detection: 254 nm. (A) Upper phase as mobile phase; flow rate 3 mL min$^{-1}$; rotational speed 700 rpm. (B) Upper phase as mobile phase to 70 min, then lower phase as mobile phase; flow rate 3 mL min$^{-1}$; rotational speed 700 rpm. (Reproduced with permission from Marston A, Slacanin I and Hostettmann K (1990) Centrifugal partition chromatography in the separation of natural products. *Phytochemical Analysis* 1: 3–17. Copyright 1990, John Wiley and Sons Limited.)

fraction by other methods, including semipreparative HPLC, proved very difficult. However, HSCCC gave six fractions which, after final purification steps

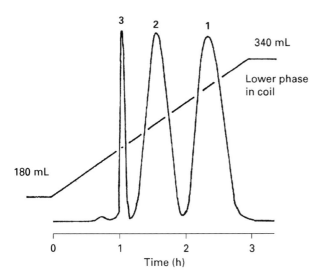

**Figure 2** Gradient elution for the separation of flavonoids **1–3**. Instrument: multilayer countercurrent chromatograph (PC Inc.). Solvent system: $CHCl_3$–MeOH–$H_2O$ 5 : 6 : 4. Composition of mobile phase: upper phase at 4 mL min$^{-1}$, lower phase at 1 mL min$^{-1}$. (Reproduced with permission from Slacanin I, Marston A and Hostettmann K. (1989) Modifications to a high-speed countercurrent chromatograph for improved separation capability. *Journal of Chromatography* 482: 234–239. Copyright 1989, Elsevier Science.)

provided eight xanthones. For the HSCCC separation, the solvent hexane–EtOAc–MeOH–$H_2O$ 1 : 1 : 1 : 1 was employed, with the upper phase as the mobile phase. One of the new xanthones isolated, 5-O-dimethyl-paxanthonin (**4**), could only be satisfactorily separated by HSCCC.

**Chalcone derivatives** A further example of the separation of polyphenols by HSCCC is provided by the isolation of chalcone derivatives **5–9** from *Brackenridgea zanguebarica*, a tree found in central Africa and belonging to the Ochnaceae family. This application of HSCCC also illustrates that large amounts of crude plant extract can potentially be handled.

Direct fractionation of 20 g of a methanol extract of the yellow bark layer from the tree is possible with the solvent system cyclohexane–ethyl acetate–methanol–water 8 : 8 : 6 : 6 (upper layer as mobile phase). The extract was dissolved in 20 mL of each phase and

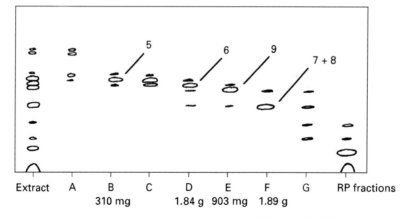

introduced via a 60 mL sample loop. The separation column (660 mL) was first filled with 50% of each phase and then eluted with mobile phase. When steady run conditions had been achieved, the sample was injected. Seven fractions were obtained (A–G; **Figure 3**), three of these giving the antifungal polyphenols 5–8. Compound 5 crystallized from fraction

B, while compounds 6–8 were purified by a final low pressure liquid chromatography step on a $C_{18}$ support. A novel, inactive, spiro derivative (9) was isolated from fraction E after low pressure liquid chromatography on a $C_{18}$ stationary phase.

As can be seen, no more than two separation steps were required for each polyphenol isolated.

**Figure 3** Thin-layer chromatography monitoring (solvent: lower phase of $CHCl_3$–MeOH–$H_2O$ 2 : 2 : 1) of fractions from the CCC-1000 HSCCC separation of *Brackenridgea zanguebarica* stem bark methanol extract. Solvent system: cyclohexane–EtOAc–MeOH–$H_2O$ 8 : 8 : 6 : 6 (upper phase as mobile phase). Flow rate: 3 mL min$^{-1}$. Rotational speed: 1000 rpm. Detection: 254 nm. Sample size: 20 g.

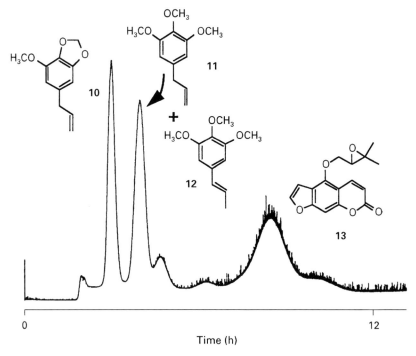

**9**

**Phenylpropanoids and coumarins**  It is possible to isolate natural products by HSCCC alone, as exemplified by the purification of phenylpropanoids and a furanocoumarin from a dichloromethane extract of the leaves of *Diplolophium buchanani* (Apiaceae). Initial fractionation gave semipure products (**Figure 4**). A subsequent liquid–liquid step, using a different nonaqueous solvent system in each case, gave pure myristicin (**10**, using hexane–*t*-butyl methyl ether–acetonitrile 5 : 1 : 5; upper layer as mobile phase) and a mixture of elemicin (**11**) and *trans*-isoelemicin (**12**) (using heptane–acetonitrile–methanol 6 : 3 : 1; upper layer as mobile phase). The furanocoumarin oxypeucedanin (**13**) was obtained by simple crystallization of the corresponding HSCCC

fraction. All four isolated compounds had both antifungal and larvicidal activities.

**Lignans**  Initial fractionation of insecticidal neolignans from *Magnolia virginiana* (Magnoliaceae) by HSCCC was better, less expensive and more efficient than traditional open-column or more recent flash chromatographic methods. A hexane extract of the leaves was chromatographed with the lower layer of the solvent system hexane–acetonitrile–ethyl acetate–water 8 : 7 : 5 : 1 as mobile phase. This solvent contained only a small proportion of water to provide compatibility with the very lipophilic extract. Subsequent purification of the fractions provided a biphenyl ether (**14**) and two biphenyls (**15, 16**), which were not only insecticidal to *Aedes aegypti* (the vector of yellow fever) but also fungicidal, bactericidal and toxic to brine shrimp.

During the investigation of anti-human immunodeficiency virus type 1 lignans from creosote bush, *Larrea tridentata* (Zygophyllaceae), a cross-axis coil planet centrifuge was employed with aqueous phases containing sodium chloride, e.g. hexane–ethyl acetate–methanol–0.5% NaCl 6 : 4 : 5 : 5. The presence of salt reduced emulsification and gave improved stationary-phase retention.

**Tannins**  Tannins include a wide variety of phenolic compounds, ranging from single glycosides of gallic

**Figure 4**  HSCCC initial fractionation on a CCC-1000 instrument of a dichloromethane extract of *Diplolophium buchanani* leaves. Solvent system: hexane–EtOAc–MeOH–H$_2$O 10 : 5 : 5 : 1 (upper phase as mobile phase). Flow rate 3 mL min$^{-1}$. Rotational speed: 1000 rpm. Detection: 254 nm. Sample size: 1.7 g.

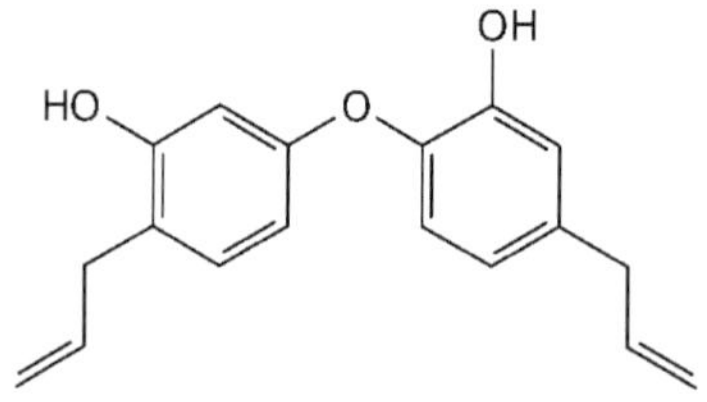

**14**

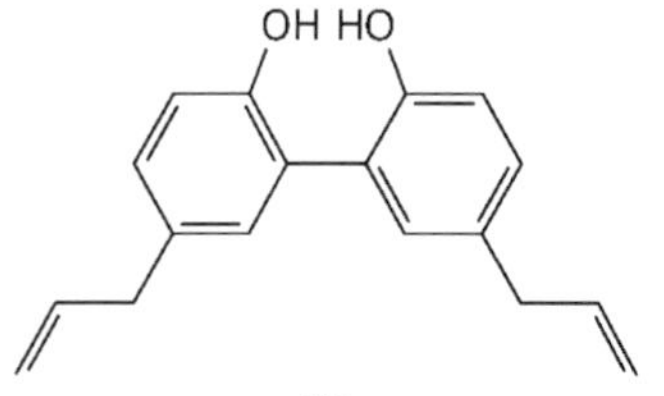

**15**

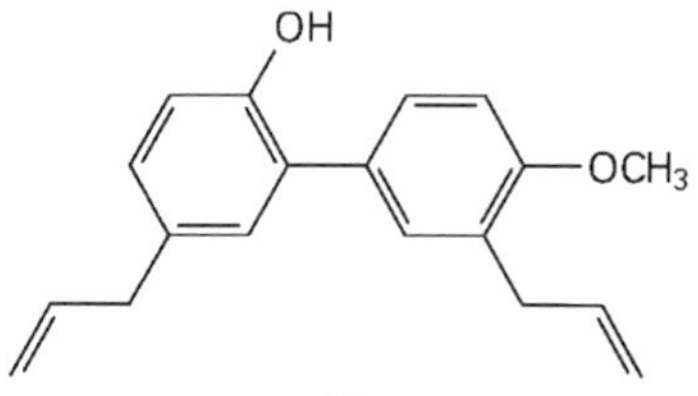

**16**

acid to complex condensed and polymerized derivatives of catechin, epicatechin and related compounds. Their separation poses special problems since there is often irreversible adsorption and even hydrolysis on solid supports. Preparative HPLC is accompanied by sample loss and deterioration or contamination of the column. HSCCC has proved to be an ideal technique for the resolution of these particular problems.

Among the examples of successful separations performed on cartridge instruments is the resolution of two diastereomers, castalgin (**17**) and vescalagin (**18**), which differ only in the configuration of a single hydroxyl group. They were extacted from *Lythrum anceps* (Lythraceae) leaves and chromatographed with the solvent system *n*-butanol–*n*-propanol–water (4 : 1 : 5), using the upper phase as the mobile phase.

The same solvent system was used to obtain a trimeric (nobotanin J; molecular weight 2764) and a tetrameric (nobotanin K; molecular weight 3742) hydrolysable tannin from the leaves of *Heterocentron roseum* (Melastomataceae). The isolation procedure involved chromatography on Toyopearl HW-40 (methanol–acetone–water gradient). Final purification of the two tannins was on an instrument with 12 cartridges (total capacity 240 mL; 700 rpm; flow rate 3 mL min$^{-1}$).

**Monoterpene glycosides**   Among the examples of monoterpene glycosides that have been separated by HSCCC are two secoiridoid glycosides from the South American plant *Halenia campanulata* (Gentianaceae). Size exclusion chromatography of a crude methanol extract of the plant on Sephadex LH-20, followed by separation by HSCCC (1000 rpm) with the solvent system CHCl$_3$–MeOH–H$_2$O  9 : 12 : 8 (lower phase as mobile phase; flow rate 3 mL min$^{-1}$) gave the two glycosides (**19, 20**). Their separation is possible by HSCCC, despite the fact that they only differ in their configuration at C$_7$.

**Triterpene glycosides**   Liquid–liquid chromatography is particularly suitable for the separation of highly polar compounds. Triterpene glycosides enter into this category and many successful isolations have

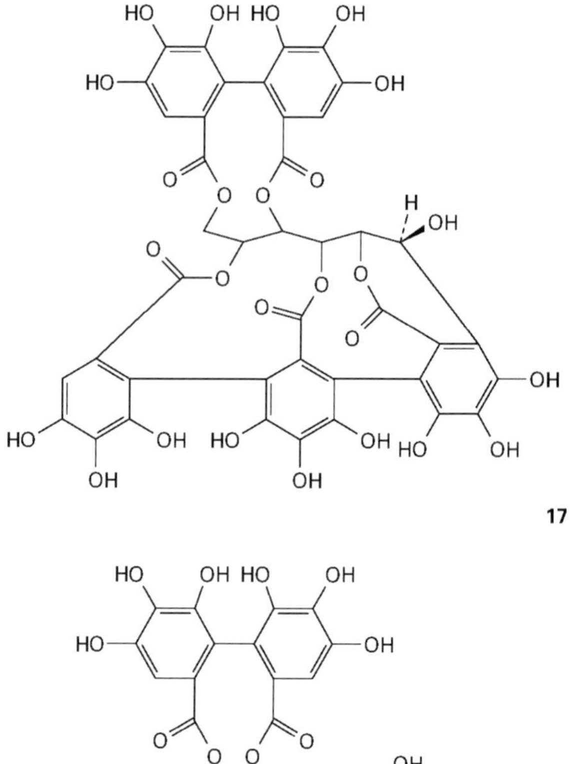

**17**

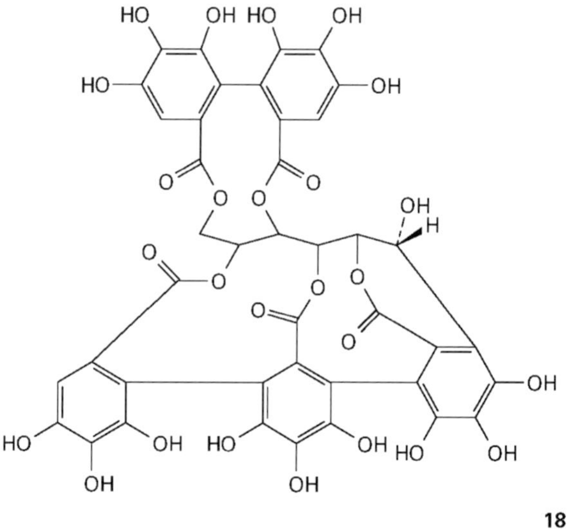

**18**

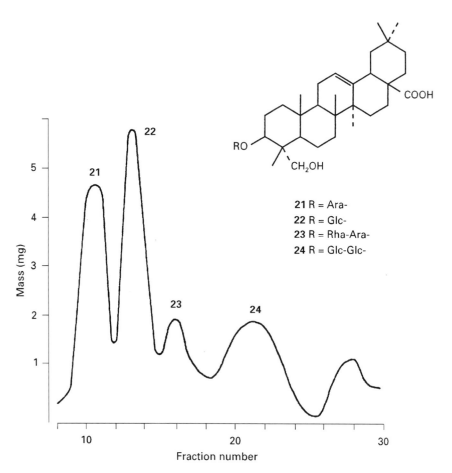

**19** R$_1$ = OCH$_3$, R$_2$ = H
**20** R$_1$ = H, R$_2$ = OCH$_3$

been performed using HSCCC either alone or in combination with other chromatographic methods. The direct separation of pure saponins from the crude methanol extract of *Hedera helix* (Araliaceae) berries has proved possible by this method. The extract was first partitioned between *n*-butanol and water. The *n*-butanol fraction was subjected to HSCCC, eluting with the lower layer of the solvent system CHCl$_3$–MeOH–H$_2$O    7 : 13 : 8.    Elution    was monitored    by    thin-layer    chromatography    and gravimetry. Molluscicidal saponins **21–24**, with hederagenin as aglycone, were separated within 2 h by this technique (**Figure 5**).

**Alkaloids**    The utility of CPC in separating alkaloids from gummy or tarry matrices has been shown in the preparative separation of pyrrolizidine alkaloids from different    plant    sources.    After    classical    alkaloid extraction procedures, batches of up to 800 mg of extract could be chromatographed on a rotating coil apparatus (380 mL capacity). Potassium phosphate buffer    (0.2 mol L$^{-1}$)    at    an    appropriate    pH    was used as stationary phase, while the mobile phase was chloroform. With the buffered stationary phase, good solute resolution was achieved since structurally similar pyrrolizidine alkaloids differ in their pKa values.

HSCCC has been employed for the separation of the antitumour drug, camptothecin (**25**). From sources    such    as    *Nothapodytes    foetida*    (Icacinaceae), camptothecin is often mixed with 9-methoxy-camptothecin (**26**). These can be easily separated with the    solvent    systems    CHCl$_3$–CCl$_4$–MeOH–H$_2$O 2 : 2 : 3 : 1 or CH$_2$Cl$_2$–MeOH–H$_2$O 5 : 3 : 1. The low solubility of the samples poses a problem with traditional chromatographic techniques. However, for the HSCCC separation, 500 mg of sample can be dissolved

**21** R = Ara-
**22** R = Glc-
**23** R = Rha-Ara-
**24** R = Glc-Glc-

**Figure 5**    Separation of a methanol extract of *Hedera helix* berries on a Sanki CPC instrument. Solvent system: CHCl$_3$–MeOH–H$_2$O 7 : 13 : 8 (lower phase as mobile phase). Flow rate 1.5 mL min$^{-1}$. Rotational speed: 700 rpm. Sample size: 100 mg.

in 70 mL lower phase and 10 mL upper phase (i.e. a large volume) for injection via a sample loop.

**25** R = H (Camptothecin)
**26** R = OCH₃ (9-Methoxycamptothecin)

Repetitive sample injections are possible for the separation of close-running compounds on rotating coil instruments. This has been shown for the separation of vincamine (**27**) and vincine (**28**) from *Vinca minor* (Apocynaceae). After 20 successive injections (at 42 min intervals), each of 1.7 mg sample mixture, 16.5 mg of **27** and 14 mg of **28** were obtained on a 230 mL rotating coil instrument. The solvent system was *n*-hexane–ethanol–water (6 : 5 : 5; lower phase as mobile phase). The resolution of the HSCCC system was not changed when the instrument was shut down overnight and restarted the next day with the same stationary phase in the column.

**27** R = H (Vincamine)
**28** R = OCH₃ (Vincine)

## Marine Natural Products

The mild conditions achieved with HSCCC and the rapidity of the method are ideal for the separation of delicate marine natural products. Attempts at purification of antitumour ecteinascidins from the tunicate *Ecteinascidia turbinata*, for example, by normal-phase or RP chromatography leads to extensive loss of activity. HSCCC, however, proved to be an effective means of separating these light- and acid-sensitive alkaloids.

The pyrroloquinoline alkaloids isobatzellines A–C have been isolated from a *Batzella* sponge. They exhibited *in vitro* cytotoxicity against the P-388 leukaemia cell line and antifungal activity against *Candida albicans*. HSCCC with a rotating coil apparatus was used in their purification, following extraction and solvent partitioning. Elution was performed with the upper phase of the solvent system heptane–chloroform–methanol–water (2 : 7 : 6 : 3).

The nonaqueous solvent system heptane (or hexane)–dichloromethane–acetonitrile (10 : 3 : 7) has been applied to the separation of a variety of marine natural products, including a long chain methoxylamine pyridine, xestamine A (**29**), isolated from the sponge *Xestospongia wiedenmayeri*.

## Antibiotics

Since its inception, HSCCC has been associated with the field of antibiotics. Liquid–liquid partition techniques are particularly suitable for their separation because bioactive metabolites are often produced in small amounts and have to be removed from other secondary metabolites and nonmetabolized media ingredients. Antibiotics are normally biosynthesized as mixtures of closely related congeners and many are labile molecules, thus requiring mild separation techniques with a high resolution.

To illustrate the applications of HSCCC to the separation of antibiotics, one example is provided by the sporavidins, which are water-soluble basic glycoside antibiotics with complex structures. These are unstable under basic conditions and exist as mixtures of closely related compounds. A sample comprising six sporavidins was resolved on a rotating coil countercurrent instrument (total capacity 325 mL; 800 rpm). Selection of the solvent system was based on partition coefficient data from chloroform–methanol–water, chloroform–ethanol–methanol–water and *n*-butanol–diethyl ether–water mixtures. After HPLC analysis, the final system adopted was *n*-butanol–diethyl ether–water (5 : 2 : 6). Sample was introduced by the so-called sandwich technique, in which sample was injected after filling with stationary phase and before mobile-phase elution. The six components were separated within 3.5 h, employing a total elution volume of 500 mL.

29

*Streptomyces lusitanus* produces the hydroquinone derivative **30** (cyanocycline C), which is extremely unstable and cannot be purified by semipreparative HPLC. However, isolation of this antibacterial compound was successfully performed on a rotating coil apparatus with the solvent system CHCl$_3$–MeOH–iPrOH–H$_2$O 3 : 10 : 10 : 10.

**30**

## Analytical Applications

Reducing the radius of revolution of a rotating coil and increasing the speed of rotation have produced highly efficient HSCCC systems with an effective analytical capability. Operational speeds lie between 2000 and 4000 rpm and the tube bore is less than 1 mm i.d. Efficient separations of micro-quantities of samples in a short time are thus possible. In general, the observed resolution and speed are comparable to those of HPLC. However, it is unlikely that HSCCC will displace HPLC for purely analytical applications. By comparison with analytical HPLC, the major advantage of HSCCC is that it is always possible to reverse the elution mode – this is useful for complex samples with a wide range of polarity. The other important use of analytical HSCCC is in the rapid selection of suitable solvent systems for scale-up to preparative applications.

The instruments available typically have a single rotating coil, with a total capacity of 5–40 mL. Flow rates are directly related to the selected solvent system and must be adapted in order to minimize back-pressure and leakage of the stationary phase. Although instruments are commercially available, many applications have been performed on prototype equipment.

While applications of analytical HSCCC to natural products are not numerous, many different classes of compounds have been separated: alkaloids, anthraquinones, coumarins, flavonoids, lignans, terpenoids and macrolides.

For example, the separation of a mixture of vin-camine (**27**), the major alkaloid of *Vinca minor* (Apocynaceae) and vincine (11-methoxyvincamine) (**28**), has been performed by analytical HSCCC in hexane–ethanol–water (6 : 5 : 5) with a 0.85 mm i.d.

multilayer coil. Comparison of the separation with results obtained from analytical RP-HPLC showed that both methods gave baseline resolution, but it was possible to observe a small peak just preceding the vincine peak in analytical HSCCC which was not resolved by RP-HPLC. Analysis of the sample by HSCCC-mass spectrometry showed that the minor compound was probably an isomer of vincine.

A crude flavonoid mixture obtained from the ethanolic extract of the fruits of sea buckthorn (*Hippophae rhamnoides*, Elaeaginaceae) has been successfully separated with an analytical HSCCC instrument. The separation of a 3 mg mixture with chloroform–ethanol–water (4 : 3 : 2) was complete within 15 min when the mobile lower phase was pumped at a flow rate of 5 mL min$^{-1}$ (rotation speed 1800 rpm). The five main components of the fruit extract were resolved (including isorhamnetin, the principal flavonoid) within 8 min, a time scale wholly comparable to analytical HPLC separations. The instrument used has a 0.85 mm i.d. coil, with a 2.5 cm radius of revolution and a capacity of 8 mL.

## Conclusion

The advantages of HSCCC favour a much wider use of this technique in the future: the absence of a solid chromatographic support avoids irreversible adsorption of samples; economics are favourable, as the consumption of solvent is up to 10 times less than conventional chromatographic techniques. Other features, such as the use of step and continuous gradients and the possibility of RP operation, add to the flexibility of the method. Very high sample loading is possible and large quantities of crude plant extracts can be injected without causing contamination problems. HSCCC, however, cannot be called a competitor to preparative-scale HPLC but it is rather a complementary technique. This complementarity can also be exploited when other chromatographic techniques fail to separate a given sample: HSCCC can be tried since its selectivity may be different.

It is still necessary to make improvements on the mechanical side. A variety of instruments are available but some are not yet reliable enough. Once larger scale production is underway, this sort of problem will be eliminated and the full potential of HSCCC will be realized.

*See also:* **II / Chromatography: Liquid:** Countercurrent Liquid. **III / Alkaloids:** Gas Chromatography; Liquid Chromatography; Thin Layer (Planar) Chromatography. **Antibiotics:** High-Speed Countercurrent Chromatography; Supercritical Fluid Chromatography. **Essential Oils:** Thin-Layer (Planar) Chromatography. **Natural Products:** High-Speed Countercurrent Chromatography;

Liquid Chromatography; Liquid Chromatography-Nuclear Magnetic Resonance; Supercritical Fluid Extraction; Thin Layer (Planar) Chromatography. **Terpenoids: Liquid Chromatography.**

## Further Reading

Conway WD (1990) *Countercurrent Chromatography: Apparatus, Theory and Applications.* New York: VCH Publishers.

Foucault AP (ed.) (1995) *Centrifugal Partition Chromatography.* New York: Marcel Dekker.

Hostettmann K, Marston A and Hostettmann M (1998) *Preparative Chromatography Techniques: Applications in Natural Product Isolation*, 2nd edn. Berlin: Springer.

Ito Y and Conway WD (eds) (1996) *High-speed Countercurrent Chromatography.* New York: John Wiley.

McAlpine JB and Hochlowski JE (1989) Countercurrent chromatography. In: Wagman GH and Cooper R (eds) *Natural Products Isolation – Separation Methods for Antimicrobials, Antivirals and Enzyme Inhibitors*, *Journal of Chromatography Library*, vol. 43, pp. 1–53. Amsterdam: Elsevier.

Mandava NB and Ito Y (eds) (1988) *Countercurrent Chromatography: Theory and Practice.* New York: Marcel Dekker.

Marston A and Hostettmann K (1994) Countercurrent chromatography as a preparative tool – applications and perspectives. *Journal of Chromatography A* 658: 315.

Oka H, Harada K and Ito Y (1998) Separation of antibiotics by counter-current chromatography. *Journal of Chromatography A* 812: 35.

Slacanin I, Marston A and Hostettmann K (1989) Modifications to a high-speed countercurrent chromatograph for improved separation capability. *Journal of Chromatography* 482: 234.

# Liquid Chromatography

**K. Hostettmann and J. L. Wolfender**, Institut de Pharmacognosie et Phytochimie, Université de Lausanne, Lausanne, Switzerland

Chemical investigation of plant constituents is strongly linked to the use of liquid chromatography (LC) at both the analytical and preparative level. Indeed, most of the secondary metabolites can be efficiently separated or isolated by different liquid chromatographic techniques. The plant extracts are generally screened by different bioassays and submitted to fractionation by chromatography. The fractions obtained are further tested for their biological activities. This process is repeated until the isolation of a pure active constituent, which is finally identified by spectroscopic methods (bioactivity guided isolation; **Figure 1**).

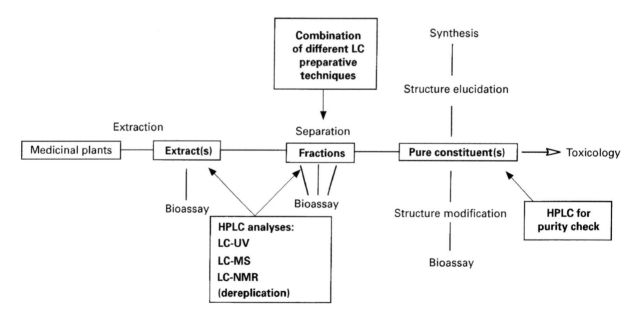

**Figure 1** Procedure for obtaining the active principles from plants. LC techniques are used at the analytical and preparative level during the whole isolation procedure. HPLC hyphenated techniques play an important role in the early recognition of well-known compounds in the extract prior to isolation.

Generally the plant material is extracted by solvents of increasing polarity. This extraction step is very important because it allows a first rough fractionation of the plant constituents. Initial extraction with low polarity solvents yields the more lipophilic components, while alcoholic solvents give a larger spectrum of apolar and polar material. The plant extracts are usually very complex mixtures which contain hundreds or thousands of different constituents. Separation of all these constituents with a single chromatographic technique is often difficult to achieve. Thin-layer chromatography (TLC) and high performance liquid chromatography (HPLC) are the most commonly used techniques for a rapid check of the chemical composition of these extracts. If the isolation of a given constituent is required, a scale-up of the analytical separation conditions to preparative chromatography techniques is needed. These preparative techniques are generally open-column chromatography, low pressure LC (LPLC), medium pressure LC (MPLC) and semi-preparative HPLC. When irreversible adsorption problems or denaturation of natural products occur, countercurrent chromatography (CCC) techniques such as centrifugal partition chromatography (CPC) are preferred.

## Analytical Techniques

### High Performance Liquid Chromatography

Of all the LC chromatography techniques used in the analysis of plants, HPLC is probably the most useful and versatile. HPLC is used routinely in phytochemistry to pilot the preparative isolation of natural products (optimization of the experimental conditions, checking of the different fractions throughout the separation) and to control the final purity of the isolated compounds. For chemotaxonomic purposes, the botanical relationships between different species can be shown by chromatographic comparison of their chemical composition. Comparison of chromatograms, used as fingerprints, between authentic samples and unknowns permits identification of the unknown material and/or the search for adulteration.

HPLC of complex mixtures such as crude extracts is usually carried out on reversed-phase columns. For the separation of the different constituents, acetonitrile–water or methanol–water gradients are applied. Linear gradients are generally used but in the case of a very complex separation, a succession of isocratic and gradient steps may be necessary. For the suppression of tailing due to polyphenolic constituents, modifiers such as trifluoroacetic or acetic

acid are added, while for the separation of basic products such as alkaloids, amines or $NH_4OH$ can be used. Crude extracts or fractions are usually injected without any sample preparation, but solid-phase extraction (SPE) prepurification or derivatization (usually to enhance the UV chromophore) can be performed in certain cases. The sample is preferably dissolved in the eluent but when solubility problems occur, as is often the case with plant extracts, stronger solvents such as tetrahydrofuran are used. When working with 4 mm i.d. columns, 50–200 µg of extract can be injected. In most cases, the separation is usually followed by UV detection because a majority of natural products possess UV chromophores. However, for the detection of constituents without a UV chromophore such as sugars, other techniques such as refractive index (RI), light scattering (LS) or electrochemical detection (ECD) are necessary.

A typical HPLC-UV chromatogram of the crude methanol extract of the roots of the African plant *Chironia krebsii* (Gentianaceae) is shown in **Figure 2**. The separation was achieved by applying a linear acetonitrile–water gradient on an RP-18 column. A good separation of the three main classes of constituents found in the plant was achieved.

### HPLC Hyphenated Techniques

In many applications, it may be necessary not only to detect but also to identify compounds in extracts. With conventional detection methods such as UV, the identity of peaks can be confirmed only from their retention times, by comparison with authentic samples. In order to get more information on the metabolites of interest, there is a need for a multiple detection system offering the possibility of taking advantage of both chromatography as a separation method and spectroscopy techniques as detection and identification methods. HPLC has thus been coupled to various sophisticated detectors such as UV photodiode array detectors (LC-UV-DAD), mass spectrometers (LC-MS) or, more recently, to nuclear magnetic resonance instruments (LC-NMR).

When searching plant extracts for compounds with interesting properties, a multidimensional approach to their chromatographic analysis is of great significance. By combining HPLC online with UV, MS and NMR, a large amount of preliminary information can be obtained about the constituents of an extract before their isolation (Figure 1). In the case of polyphenols, for example, UV spectra recorded online give useful complementary information (type of chromophore or pattern of substitution) to that obtained with LC-MS and already provide a precise assignment of the peak of interest. When these data are not sufficient, LC-NMR can give a useful

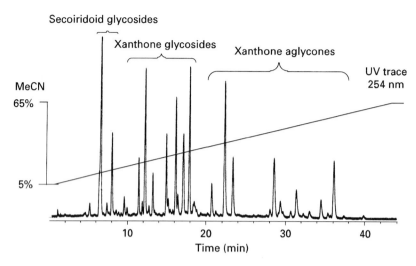

**Figure 2**   HPLC-UV analysis of the methanolic extract of the root of *Chironia krebsii* (Gentianaceae). HPLC: column, RP-18 NovaPak (4 μm, 150 × 3.9 mm i.d.); gradient, MeCN–H$_2$O (0.1% TFA) 5 : 95 → 65 : 35 in 50 min (1 mL min⁻¹).

complement for a full structural identification online. This preliminary LC chemical screening avoids the useless isolation of known constituents and concentrates the search only on the compounds of potential further interest.

This approach is illustrated by the following example: a Gentianaceous plant, *Swertia calycina*, presenting interesting antifungal activities, was studied by LC-UV, LC-MS and LC-NMR. The LC-UV analysis of the dichloromethane extract was quite simple and exhibited three main peaks. The LC-UV spectra of these peaks permitted a first rapid online recognition of the various types of constituents of the extract. The LC-MS provided molecular weight information for each of these compounds (**Figure 3**).

These online data, together with chemotaxonomical considerations, allowed the identification of a xanthone (**3**) and a secoiridoid (**1**) in this plant. However, the structure of compound (**2**) could not be completely ascertained by LC-UV-MS alone, because its UV spectrum was characteristic for a quinonic chromophore and no compound of this type has been reported in the Gentianaceae family. The LC-MS of (**2**) exhibited a protonated [M + H]⁺ ion at *m/z* 189, indicating a molecular weight of 188 amu. The LC-NMR analysis of the extract gave well resolved ¹H-NMR online spectra for the three major compounds. For the unknown (**2**), the online ¹H-NMR spectrum revealed the presence of a methoxyl group together with a quinonic ring proton while four other aromatic protons appeared at lower field (**Figure 4**). These online data were in good agreement with those of O-methyl lawsone, a known naphthoquinone. This type of compound has, however, never been reported in the Gentianaceae family.

## Preparative Techniques

Preparative-scale separation is one of the most important operations carried out in a natural products laboratory. It is often tedious and time-consuming, especially when the mixture to be separated is complex. The nature of the separation problem varies considerably from the isolation of small quantities (mg or less), for structure determination purposes, to the isolation of very much larger amounts (hundreds of mg to kg quantities), for comprehensive biological testing or even for production of therapeutic agents.

The most important preparative techniques which have found application in the isolation and purification of natural products are listed in **Table 1**. A distinction has to be made between techniques using a solid stationary phase or a liquid fixed on an inert solid support and all-liquid partition techniques. For all these techniques, the use of solvents such as acetone or in some cases chloroform is inadvisable because in their presence natural products may undergo transformations.

### Flash Chromatography/Open-column Chromatography

Open-column chromatography and flash chromatography are the most popular techniques for natural product isolation. Flash chromatography is a preparative air-driven liquid chromatographic technique with moderate resolution. It has the advantage over conventional open-column chromatography of minimizing sample loss and the risk of decomposition of natural products, due to the fast elution of the sample. These two techniques allow the use of various column sizes and permit the introduction of

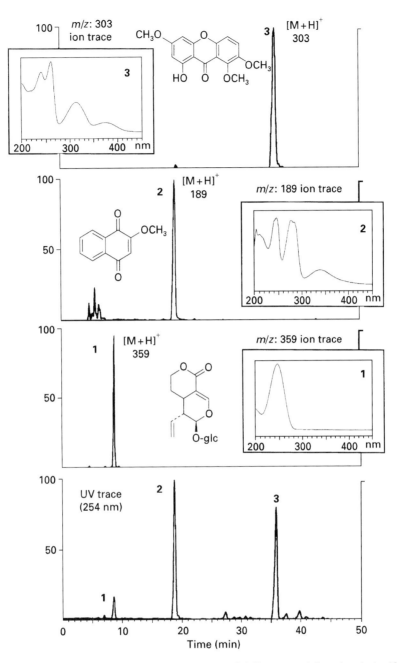

**Figure 3**  LC-UV and LC-thermospray (TSP)-MS analyses of the crude $CH_2Cl_2$ extract of *Swertia calycina* (Gentianaceae). For each major peak, the single ion LC-MS traces of the protonated molecular ions $[M + H]^+$ are displayed, together with the UV spectra obtained online.

mg to g quantities of sample. They are generally used with silica gel supports. As a rule of thumb, for natural product applications, around 30 mg sample loading per g of 50–200 μm support is feasible. When used only for filtration (first fractionation step of a crude plant extract), silica gel chromatography can be performed under overloaded conditions and quantities up to 100 mg g$^{-1}$ of support can be loaded.

These techniques can be used at different steps in the isolation process of natural products. They are mostly well adapted to the separation of nonpolar to medium polarity plant extracts. In order to find the appropriate solvent system, preliminary TLC analyses are performed and the plate is sprayed with different reagents in order to detect the natural products of interest. For a given product, solvent systems producing $R_F$ values between 0.2 and 0.3 are

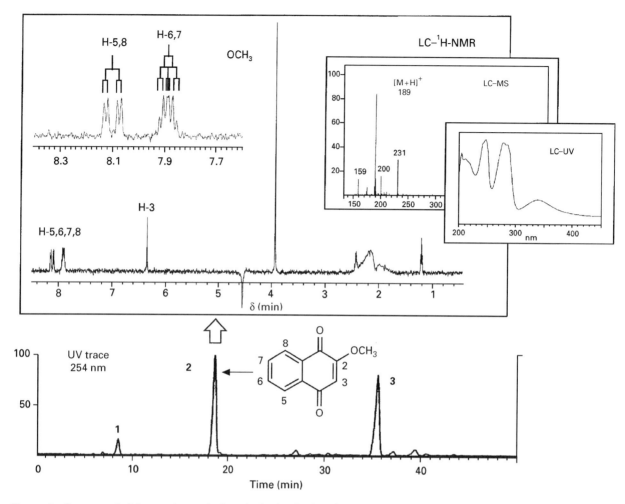

**Figure 4** Summary of all the spectroscopic data obtained online by LC-UV, LC-MS and LC-NMR for the naphthoquinone (**2**) in the dichloromethane extract of *Swertia calycina* (Gentianaceae).

usually optimum for scale-up. When flash and open-column chromatography are used as first fractionation steps of crude plant extracts, solvent systems of increasing polarity are often employed. The fractions are checked by TLC at the outlet of the column and when compounds of a given polarity are eluted, the solvent system is changed for a more polar one. This allows compounds of different polarities to be efficiently separated.

A typical step gradient solvent system for the separation of a dichloromethane extract by flash chromatography on silica gel is: chloroform–

**Table 1** Preparative separation methods for plant constituents

| Type | Technique | Abbreviation |
|---|---|---|
| Preparative thin-layer chromatography | Centrifugal TLC | CTLC |
| Open-column chromatography | | CC |
| Vacuum liquid chromatography | | VLC |
| Pressure liquid chromatography | Flash | |
| | Low pressure LC | LPLC |
| | Medium pressure LC | MPLC |
| | High performance LC | HPLC |
| Liquid–liquid chromatography | Droplet countercurrent chromatography | DCCC |
| | Centrifugal partition chromatography | CPC |

petroleum spirit (80 : 20), then pure chloroform, chloroform–methanol (80 : 20) and finally pure methanol for eluting the polar constituents.

## Preparative Liquid Chromatography

Because of the complexity of the chemical composition of crude plant extracts, open-column chromatography or flash chromatography techniques alone often do not provide sufficient resolution for the isolation of pure natural products. The introduction of pressurized liquid chromatography in natural product chemistry has allowed the use of silica gel of smaller particle size and of bonded phases such as RP-8, RP-18 and diol, thus giving more versatility and better resolution.

Depending on the sample size and the resolution to be achieved, three techniques are generally used for phytochemical applications: LPLC, MPLC and semipreparative HPLC. Polar natural products such as glycosidic derivatives can mainly be separated on RP-8 or RP-18 bonded phases while compounds of intermediate polarity such as polyphenol aglycones, polyacetylenes and sesquiterpenes can be resolved on silica gel supports (**Table 2**).

## Low and Medium Pressure Chromatography

LPLC makes use of columns containing packings of c. 40–60 μm, allowing high flow rates up to a pressure of 10 bar. Prepacked columns with various sizes and supports are available. MPLC accommodates much larger sample loads (100 mg–100 g) than LPLC (10 mg–1 g) and is designed to be operated at higher pressure (c. 5–40 bar). MPLC uses refillable columns of various sizes. Supports with particle sizes ranging from 15 to 200 μm can be used. MPLC is more versatile than LPLC but the basic operation of the two techniques is similar.

The chromatographic conditions for these techniques are selected by HPLC or by TLC. The selectivity of the eluent is first optimized and then the composition of the mobile phase is adjusted to suit the preparative conditions. Retention factors ($k$) between 1 and 5 or $R_F$ values $< 0.4$ are appropriate. Different representative solvent systems used for the separation of various classes of natural products are given in Table 2.

In some cases, similar resolution can be obtained by both LPLC or MPLC and analytical HPLC. This is of importance for the transposition of conditions from analytical to preparative level. It is illustrated here by the separation of iridoids and a phenylpropane glycoside from the root of *Sesamum angolense* (Pedaliaceae), a plant reputed to have antihaemorrhagic properties in African traditional medicine. Analytical HPLC of the relevant fractions is shown in **Figure 5A** and 5B. Baseline separation of phlomiol (**4**) and puchelloside-I (**5**) was possible with methanol–water (10 : 90). Verbascoside (**6**) was, however, only eluted with 32% methanol. These analytical conditions could be applied directly to a preparative separation on a Lobar LPLC column by performing a step-gradient elution from 10% to 32% methanol.

Following a similar procedure, several xanthone glycosides (**7–12**) from *Halenia corniculata* (Gentianaceae) were successfully isolated from an enriched fraction by MPLC in a single isocratic run (**Figure 6**).

## Semipreparative HPLC

While analytical HPLC is useful for obtaining information about sample mixtures and does not rely on their recovery, the aim of semipreparative HPLC is to isolate pure substances. Compared to the other column LC techniques, HPLC can handle very small particle sizes (5–30 μm) which provide a large gain in separation efficiency. In phytochemical investigations, this technique is often used for final purification steps for a small number of compounds which are

**Table 2** Representative solvent systems used for LPLC or MPLC

| Substance class | Support | Eluent |
| --- | --- | --- |
| Polyacetylenes | SiO$_2$ | Toluene–EtOAc (85 : 15) |
| Flavonoid aglycones | SiO$_2$ | C$_6$H$_{14}$–EtOAc, CHCl$_3$–MeOH |
|  | DIOL | CHCl$_3$–MeOH–HOAc (950 : 50 : 1) |
| Flavonoid glycosides | RP-8 | MeOH–H$_2$O gradient |
| Phenylpropanoid glycosides | RP-18 | MeOH–H$_2$O gradient |
| Chromenes | SiO$_2$ | CHCl$_3$–C$_6$H$_{14}$–MeOH |
| Sesquiterpenes | SiO$_2$ | C$_6$H$_{14}$–EtOAc |
| Diterpenes | RP-18 | MeOH–H$_2$O (1 : 1) |
| Triterpenes | RP-8 | MeOH–H$_2$O (4 : 1) |
| Iridoid glycosides | RP-18 | MeOH–H$_2$O gradient |
| Alkaloids | RP-8 | MeOH–0.02 mol L$^{-1}$ NH$_4$OAc (3 : 2) |

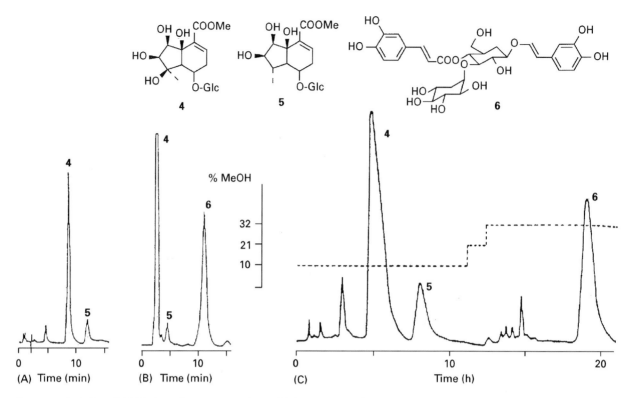

**Figure 5** Isolation of iridoids from *Sesamum angolense*. (A) Analytical HPLC: column, LiChrosorb RP-8; eluent, MeOH–H₂O 10 : 90; detection, 254 nm. (B) Analytical HPLC: column, LiChrosorb RP-8 eluent, MeOH–H₂O 32 : 68; detection, 254 nm. (C) Preparative separation: column, Lobar RP-8 (310 × 25 mm i.d.); eluent, step gradient MeOH–H₂O 10 : 90 and 32 : 68; sample, 130 mg; detection, 254 nm.

difficult to separate. The loading capacity of the column is usually rather small (1–100 mg) and the isolation of a pure natural product often requires multiple injections of a given prepurified fraction. The number of applications of this technique to natural product isolation is very large.

**Figure 7** shows the separation of closely related antifungal chromenes from *Hypericum revolutum* (Guttiferae) by semipreparative HPLC. The petroleum ether extract of the leaves and twigs after flash chromatography and LPLC on silica gel, gave an active fraction which appeared homogeneous by TLC. However, analytical HPLC on an RP-18 column showed the presence of two homologues (**13** and **14**). Since chromenes degrade in the presence of acid, forming the corresponding dichromenes, no acid was used in eluent for semipreparative separation. A total of 120 mg was separated by repetitive injections.

### Centrifugal Partition Chromatography

The use of CCC represents an interesting complementary approach to LC on solid supports. This technique is an all-liquid method without the presence of a solid support and thus has the advantage over other liquid chromatographic methods of avoid-

ing irreversible adsorption of sample, allowing a quantitative recovery and minimizing the risk of sample denaturation. These points are extremely important when labile or polar natural products have to be isolated. Among the different countercurrent liquid chromatography techniques, CPC is probably the most popular technique because of its speed and ease of use. The choice of the solvent system can be guided by TLC and for rotating coils the best $R_F$ range is 0.1–0.4. According to the type of instrument used – single coil, multiple coil or cartridge – the sample size ranges from 10 mg to 10 g. Selected examples of solvent systems used for the separation of various classes of natural products are given in **Table 3**.

In order to illustrate this approach, the separation of the active anthranoid pigments of *Psorospermum febrifugum* (Guttiferae), an African medicinal plant that exhibits strong growth inhibition of cancer cells and antimalarial activity, is shown in **Figure 8**. The lipophilic root bark extract of this plant was first separated by flash chromatography and LPLC but resulted in considerable material losses, owing to irreversible adsorption on the supports. However, in a single CPC step, three pure compounds (**15**, **16** and **17**) and a mixture of a fourth anthranoid pigment

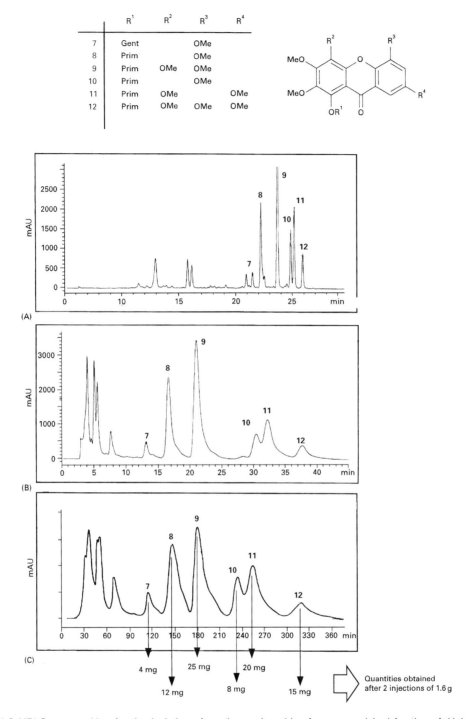

**Figure 6** HPLC-MPLC transposition for the isolation of xanthone glycosides from an enriched fraction of *Halenia corniculata* (Gentianaceae) after separation by size exclusion chromatography on Sephadex LH-20 of the methanolic extract. (A) HPLC-gradient. Conditions: column: NovaPak RP-18 (3.9 × 150 mm); H₂O–MeCN; 5–65% MeCN in 50 min; 1 mL min⁻¹. (B) HPLC-isocratic. Conditions: column: LiChroCart; Lichrosorb RP-18 (7 μm); 4 × 250 mm; H₂O–MeOH (60 : 40); 1 mL min⁻¹. (C) MPLC-isocratic. Conditions: column: MPLC home-packed Lichrosorb Rp-18 (15–25 μm); 12 × 460 mm; H₂O–MeOH (60 : 40); 3 mL min⁻¹; 12 bar. Sample size 300 mg.

(18) with an unidentified constituent (19) were obtained without loss of product. A nonaqueous solvent system was used for the separation, which could be scaled up to a 500 mg sample size.

**Combination of Methods**

No single liquid chromatographic separation method is able to solve all separation problems and moreover

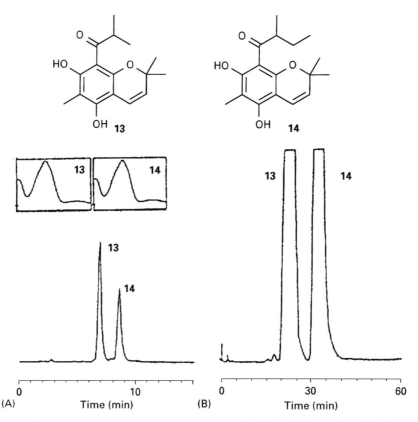

**Figure 7** Separation of antifungal chromenes from *Hypericum revolutum* (Guttiferae). (A) Analytical HPLC: column, LiChrosorb RP-18 (250 × 4.6 mm i.d.); solvent, MeOH–H$_2$O 80 : 20; flow, 1.5 mL min$^{-1}$; detection, 254 nm. (B) Semipreparative HPLC: column, μ-Bondapak RP-18 (300 × 7.8 mm i.d.); solvent MeOH–H$_2$O 67 : 37; flow, 5 mL min$^{-1}$; detection, 254 nm.

it is very common to find multistep chromatographic operations for the isolation of pure natural products. Although it is possible to obtain a pure compound by a one- or two-step procedure, a combination of techniques is normally required. Of course there are many different ways of putting together all the possible separation techniques, but in reality the choice of strategy is limited

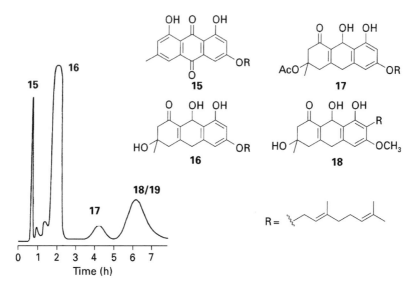

**Figure 8** CPC separation of a light petroleum ether extract of *Psorospermum febrifugum* (Guttiferre) root bark. Solvent, *n*-C$_6$H$_{14}$–MeCN–MeOH (40 : 25 : 10, mobile phase = upper phase); flow, 5.5 mL min$^{-1}$; rotational speed, 1500 rpm; sample, 100 mg; detection, 254 nm.

**Table 3** Representative solvent systems used for CPC separations

| Substance class | Eluent |
|---|---|
| Flavonoids | $CHCl_3$–MeOH–$H_2O$ 4 : 3 : 2 |
| Xanthones | Petrol ether–EtOAc–MeOH–$H_2O$ 1 : 1 : 1 : 1 |
| Tannins | nBuOH–nPrOH–$H_2O$ 4 : 1 : 5 |
| | nBuOH–nPrOH–$H_2O$ 2 : 1 : 3 |
| Saponins | $CHCl_3$–MeOH–nPrOH–$H_2O$ 5 : 6 : 1 : 4 |
| | $CHCl_3$–MeOH–iBuOH–$H_2O$ 7 : 6 : 3 : 4 |
| | $CHCl_3$–MeOH–$H_2O$ 7 : 13 : 8 |
| Polyacetylenes | Hexane–MeCN–TBME 10 : 10 : 1 |
| Aporphine alkaloids | $CHCl_3$–MeOH–0.5% HOAc 5 : 5 : 3 |

by a number of constraints: the extraction method, the complexity of the extract, the sample preparation, the polarity, the stability, the solubility, the sample size and the complementarity of the separation techniques.

When choosing a separation strategy, it is often useful to pick steps which differ as much as possible in selectivity. During an isolation procedure, the scale of the operation decreases: as the purity of the product increases, there is a corresponding diminution of sample quantity. This implies that the initial fractionation steps are those which can separate large

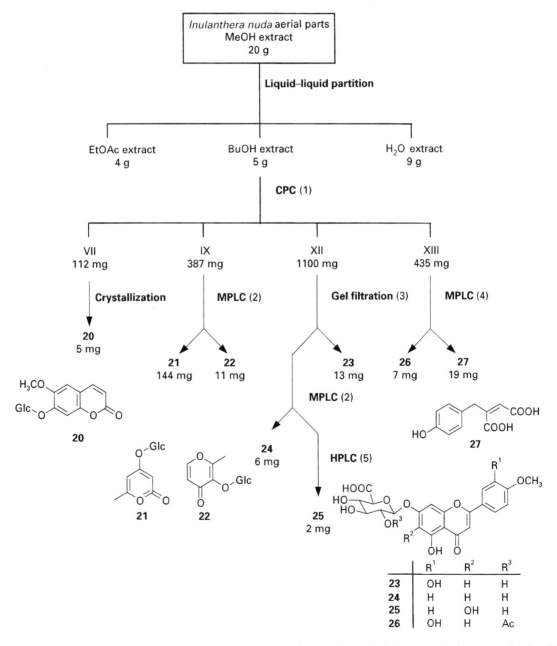

**Figure 9** Strategy for the isolation of various aromatic constituents from *Inulanthera nuda* (Asteraceae). Conditions: 1, $CHCl_3$–MeOH–$H_2O$-*i*-PrOH; 2, RP-18, MeOH–$H_2O$ (1 : 9); 3, Sephadex LH-20, MeOH–$H_2O$ (1 : 1); 4, RP-18, MeOH–$H_2O$ (1 : 4); 5, RP-18, MeCN–$H_2O$ (7 : 43).

amounts of material, e.g. column chromatography using relatively cheap stationary phases (silica, alumina, polyamide or XAD ion exchange resins), flash chromatography or CCC. Size exclusion chromatography (SEC) is also becoming increasingly popular as a first purification step. Subsequent chromatographic steps on smaller quantities can be performed with more expensive column packings and equipment. Semipreparative HPLC is often reserved for final purification.

The combination of different preparative chromatography techniques for the isolation of various aromatic compounds from *Inulanthera nuda* (Asteraceae) is presented in **Figure 9**. A first liquid–liquid partition of the methanolic extract, dissolved in water, gave an enriched butanol extract which was further separated by CPC, affording 13 fractions. One of these fractions yielded (**20**) after recrystallization while the other constituents were further separated by MPLC or SEC. The final purification of flavonoid glycoside (**25**), for example, required a combination of SEC on Sephadex LH-20, followed by MPLC and semipreparative HPLC.

## Conclusions

The introduction of modern liquid chromatographic methods has revolutionized the science of separation of natural products of plant origin. These new methods allow faster separations and facilitate the resolution of complex mixtures. As shown, techniques such as HPLC can be used both at the analytical and preparative level. At the analytical level and in combination with sophisticated detectors, structural information can be obtained online, while on the preparative scale closely related compounds can be successfully isolated.

The actual separation method or methods depend(s) on a number of factors relevant to the separation problem, but a judicial choice of strategy enables most targets to be reached. New methods and improvements are continually being introduced, with the result that the number of combinations available is steadily expanding, hopefully leading to a progressive simplification of the ever more complex separation problems that are being undertaken.

*See also:* **II/Chromatography: Liquid:** Detectors: Mass Spectrometry; Large-Scale Liquid Chromatography. **Extraction:** Solid-Phase Extraction. **III/Flash Chromatography. Medium-Pressure Liquid Chromatography. Natural Products:** High-Speed Countercurrent Chromatography; Thin-Layer (Planar) Chromatography. **Pigments:** Thin-Layer (Planar) Chromatography.

## Further Reading

Bidlingmeyer BA (1987) *Preparative Liquid Chromatography*. Amsterdam: Elsevier.

Hostettmann K, Wolfender J-L and Rodriguez S (1997) Rapid detection and subsequent isolation of bioactive constituents of crude plant extracts. *Planta Medica* 63: 2.

Hostettmann K, Marston A and Hostettmann M (1998) *Preparative Chromatography Techniques, Applications in Natural Product Isolation*. Berlin: Springer.

Kingston DGI (1979) High performance liquid chromatography of natural products. *Journal of Natural Products* 42: 237.

Marston A and Hostettmann K (1991) Modern separation methods. *Natural Product Reports* 8: 391.

Marston A and Hostettmann K (1994) Counter-current chromatography as a preparative tool – application and perspectives. *Journal of Chromatography A* 658: 315.

Wolfender J-L, Maillard M and Hostettmann K (1994). Thermospray liquid chromatography–mass spectrometry in phytochemical analysis. *Phytochemical Analysis* 5: 153.

# Liquid Chromatography–Nuclear Magnetic Resonance

**B. Schneider**, Max Planck Institute for Chemical Ecology, Jena, Germany

Hyphenation of chromatographic and spectroscopic methods is important in analytical chemistry and is of great value in modern natural product analysis. Gas chromatography–mass spectrometry (GC-MS) has been used for many years to analyse volatile compounds and derivatives of nonvolatile natural products. The development of liquid chromatography–mass spectrometry (LC-MS) extended the scope of MS coupling techniques to allow analysis of nonvolatile compounds without derivatization. Nuclear magnetic resonance (NMR) is less sensitive than MS but represents the most informative and most universal analytical technique for natural products. Thus, using a NMR spectrometer in coupling methods does not simply mean adding another detector but represents a new dimension in analytical natural product chemistry.

The combination of NMR and chromatographic, or electrophoretic, separation methods was made possible by the introduction of high field spectrometers, with an increased dynamic receiver range, the development of suitable continuous-flow cell probes and solvent suppression techniques. High performance liquid chromatography (HPLC)-NMR was first described by Watanabe in 1979 and it is now an established method in the field of natural product research. A number of examples, mainly of plant natural products, are reviewed in this article, demonstrating the advantages and limitation of HPLC-NMR.

## Methodology

### Sample Preparation

As sensitivity of HPLC-NMR is currently in the microgram or even nanogram scale, the amount of tissue investigated can be dramatically reduced as compared with that required for conventional isolation of natural products. Small scale extractions using 0.5–1 g of dried plant tissue have been described as being sufficient to record HPLC-$^1$H NMR spectra with an excellent signal-to-noise ratio. Natural products which are present in living tissue in only trace amounts (pg g$^{-1}$ tissue) are now amenable to NMR analysis without isolation, either directly in the crude extract, or after employing simple work-up steps.

The procedures for sample preparation are essentially the same as those for normal analytical HPLC. Since extracts of biological tissues are normally complex mixtures of various substances covering a broad range of polarity, including both lipophilic and hydrophilic components, pre-purification or fractionation of the crude extract can often improve the chromatographic resolution. Enrichment of the desired natural product prevents overloading of the column by unwanted components and enhances the concentration of analytes above the detection limit. Due to the use of a NMR spectrometer as detector, deuterated solvents are strongly recommended for injecting the analyte into the chromatographic system.

### HPLC

There are only a few special requirements for HPLC combined online with NMR. A pulsation-free HPLC pump to provide proper gradient formation and efficient solvent mixing should be used. The first detector cell (usually UV), which in HPLC-NMR is no longer at the end of the process, should be as small as possible to reduce peak broadening to a minimum. In general, reversed-phase chromatography is used for most HPLC-NMR applications in natural product

chemistry. Water and protonated organic solvents cause resonances in the NMR spectrum. These eluent signals might overlap with those of the analyte and thus prevent adequate spectrum evaluation. To minimize the intensity of solvent resonances, and to improve the detection limit, deuterated solvents are utilized. In practice, fully deuterated water (99% D$_2$O) is used in combination with nondeuterated HPLC-NMR-grade acetonitrile or methanol. The phenomenon of peak broadening, often occurring in longer isocratic HPLC runs, reduces the fraction of the peak transferred to the flow cell. To compensate for this broadening, solvent gradients are recommended for elution. Addition of trifluoroacetic acid or phosphoric acid also contributes to peak focusing and does not cause additional signals in the $^1$H NMR spectrum. Due to the implicit requirements of NMR methodology (solvent suppression, lock solvent), reversed-phase gradients cannot begin below a minimum concentration of 1% of the organic component and are not useful when exceeding about 95%. Flow rates between 0.6 and 1.0 mL min$^{-1}$, usually employed in analytical HPLC, are also convenient under HPLC-NMR conditions. However, adaptation of flow rate to the particular HPLC-NMR mode (continuous-flow, stopped flow) is required. Recent developments, allowing the use of microbore and capillary columns, which require lower flow rates and consume smaller amounts of solvents, permit the economical use of completely deuterated eluents. In general, the highest sample amount possible should be injected to reduce measuring time. Even column overloading, and partial peak overlap in the UV trace, may be acceptable to some extent because only a fraction of the desired peak is located in the active volume of the flow cell during spectrum acquisition. It is important to note that the quality of chromatographic separation determines the success of the NMR measurement and, thus, should be executed as carefully as possible.

### NMR

HPLC-NMR probes do not make use of conventional removable NMR tubes but contain a continuous-flow cell, fitted to the HPLC via a polyetheretherketone (PEEK) transfer capillary. The capillary connection should be as short as possible, otherwise the stray field of the NMR magnet has to be considered. As a compromise, the HPLC is usually positioned slightly outside the 5 mT line (corresponding to about 1.5 m for a 500 MHz magnet) of the stray field. A valve interface between the HPLC detector and the NMR probe allows selection of different modes, like continuous-flow, stopped-flow and storage mode. The active volume of the flow cells is between 40 and 240 μL in size. In continuous-flow mode, the detector volume and the flow rate determine the residence time of the sample in the flow cell and thus have a significant impact on sensitivity. Capillary flow cells with a detection volume in the order of 50–900 nL have been developed for microbore and capillary HPLC. Since there is no sample rotation it is possible to fit the radiofrequency coils directly on the glass body of the flow cell. This arrangement affords an optimal filling factor and, consequently, results in extraordinarily high sensitivity of the HPLC-NMR probes.

Commercially available HPLC-NMR probes are designed as inverse detection probes and, therefore, are most efficient for acquiring $^1$H NMR spectra. **Figure 1** shows $^1$H NMR spectra recorded in MeCN-D$_2$O with and without solvent suppression. In the nonsuppressed spectrum only the eluent signals are visible. To visualize the resonances of the analyte, suppression of the solvent signals is necessary. This can be accomplished by presaturation or by the 'water suppression enhanced through T1 effects' (WET) sequence. WET is more efficient in con-

tinuous-flow measurement because it requires shorter delays in comparison with the presaturation technique. However, even most efficiently suppressed solvent signals cover a certain part of the spectrum and may overlay some of the resonances of the analyte. This general drawback of HPLC-NMR can be reduced by running the same sample again in another eluent system having different chemical shift values, e.g. using acetonitrile ($\delta$ 2.0)–D$_2$O in the first and methanol ($\delta$ 3.2)–D$_2$O in a second run. The sensitivity of HPLC-NMR significantly depends on the operation mode and a number of further factors discussed above and by other authors. For the stopped-flow technique, which is the most sensitive mode, the detection limit in routine analysis (500 MHz; 120 μL flow cell) is below 1 μg in reasonable times. Using a capillary column and a nanolitre flow cell, the detection limit is now in the nanogram range. The stopped-flow technique is also suitable for acquiring homonuclear correlation spectra (COSY, TOCSY, NOESY and ROESY) of samples below 10 μg. Moreover, gradient-assisted inverse-detected heteronuclear correlation spectroscopy (GHSQC and GHMBC)

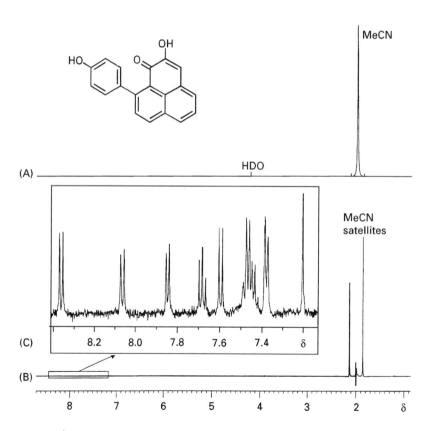

**Figure 1** Stopped-flow HPLC-$^1$H NMR spectra of an aromatic natural product measured in MeCN-D$_2$O. Spectrum (A), which was acquired without solvent suppression, exhibits the large signal of nondeuterated MeCN, small satellites of MeCN and the signal of HDO. Resonances of the analyte are not visible. Spectrum (B) was acquired using solvent suppression by double pre-saturation of MeCN and HDO. Spectrum (C) is an enlargement of (B), showing the well-resolved resonances in the aromatic part of the spectrum.

is also possible under HPLC-NMR conditions. The measurement in stopped-flow mode requires an accurate determination of the transfer time between the first detector, usually an UV or diode array detector, and the active volume of the flow cell. This is to ensure that the HPLC pump stops just at the moment when the top of the peak is in the magnet. Synchronization of HPLC and NMR is also required in fully automated mode that enables measurement of several peaks without further interaction of the operator.

The continuous-flow mode is much less sensitive than the stopped-flow technique but provides the opportunity to scan an extract rapidly for interesting natural products. Since the NMR resonances of the solvent depend on the composition ratio of the eluent, gradient elution requires continuous adaptation of the solvent suppression frequency to the moving signal. This suppression frequency is determined for each increment by the so-called scout scan prior to the WET sequence. In continuous-flow spectra the retention times ($y$ axis) are plotted versus the chemical shifts ($x$ axis). The extraction of traces from these unusual two-dimensional spectra yield one-dimensional $^1$H spectra of the desired increment that can collectively be outlined as stacked plots.

## Applications

### Identification of Natural Products

An increasing number of applications of HPLC-NMR are devoted to the identification and structure elucidation of natural products. The main part of these investigations covers plant natural products as novel biologically active components for pharmacological and agricultural preparations. The aims and methodology of a number of these investigations are discussed in the following paragraphs. **Table 1** summarizes the classes of plant natural products and the plant families that have been analysed by hyphenated HPLC-NMR techniques.

Extracts equivalent to 250 mg of dried leaves of *Zaluzania grayana* (Asteraceae) were used for HPLC separation with direct measurement of $^1$H NMR spectra in the online mode. Overlapping peaks could be separated more efficiently by the stopped-flow mode, collected in a sampling unit and analysed later by $^1$H and 2D COSY measurements. These investigations provided information on the structure of two known and a novel sesquiterpene lactone of taxonomic relevance. Using a microsampling technique, glandular trichomes from the leaf surface of *Scalesia* species (Asteraceae) were collected. A sample combined from several species of the genus was used for

online HPLC-NMR analysis. Flavones and sesquiterpene lactones were identified by comparison with spectra of authentic reference compounds or literature data.

Unstable and structurally closely related bitter acids from dried female flowers of *Humulus lupulus* were extracted by supercritical carbon dioxide and analysed by HPLC-NMR in the stopped-flow mode without any degradation. Using an acetonitrile–D$_2$O eluent containing H$_3$PO$_4$, well-resolved spectra of $\alpha$- and $\beta$-hop acids were obtained despite column overloading (2.5 mg of extract was loaded on to an analytical reversed-phase column).

The online analysis of a CH$_2$Cl$_2$ extract of *Swertia calycina* (Gentianaceae) provided $^1$H NMR spectra of all major constituents. Extraction of single traces from the 2D plot allowed a precise assignment of their specific resonances. Approximately 0.05 µmol per peak was needed to obtain a $^1$H NMR spectrum in the online mode using a 500 MHz NMR instrument. To improve the quality of the $^1$H spectra and to measure a 2D COSY spectrum of one of the components, sweroside, the same extract was investigated in the stopped-flow mode. The detection limit for a $^1$H NMR spectrum could be lowered by a factor of about 100 under stopped-flow conditions but longer acquisition times were required in comparison with the continuous-flow mode. In the case of the more complex methanol extract of another Gentianaceae species, *Gentiana ottonis*, clear $^1$H NMR spectra of secoiridoids, flavones and xanthones were only obtained in the stopped-flow mode. LC-UV and LC-MS data were also needed for full identification of compounds from both species.

Complementary HPLC-NMR and HPLC-MS studies were also performed on crude extracts and bioactive fractions from *Monetes engleri*. On-flow experiments indicated two major prenylated flavanone components in the CH$_2$Cl$_2$ extract of this plant but did not monitor minor components. A bioactive fraction obtained by medium pressure liquid chromatography (MPLC), containing these components, was subjected to stopped-flow HPLC-NMR analysis. $^1$H NMR, 1D TOCSY, 2D NOESY and gradient-enhanced inverse $^1$H, $^{13}$C correlation experiments (GHSQC, GHMBC; **Figure 2**) were recorded from the enriched sample in a total acquisition time of 9.6 h. The WET sequence was used to suppress the eluent signals of the residual HDO resonance, the resonances of MeCN and its two $^{13}$C satellites, and those of the propionitrile impurities of MeCN. The constitution of monotesone A, a new prenylated flavanone, was elucidated online using the strategy described. However, determination of the absolute configuration was only possible after isolation.

**Table 1** Identification of plant natural products by HPLC-NMR

| Species | Family | Natural products | HPLC | NMR | Reference |
|---|---|---|---|---|---|
| Zaluzania grayana | Asteraceae | Sesquiterpene lactones | RP-18, MeCN-$D_2O$, UV | 500 MHz; online: ¹H stopped-flow (peak sampling): ¹H, 2D-COSY | Spring et al. (1995) Phytochemistry 39: 609 |
| Scalesia species[a] | Asteraceae | Flavanones, sesquiterpene lactones | RP-18, MeCN-$D_2O$, MeOH-$D_2O$, UV | 500 MHz; online: ¹H | Spring et al. (1997) Phytochemistry 46: 1369 |
| Humulus lupulus | Moraceae | Lupulones | RP-18, MeCN-$D_2O$, UV | 400 MHz; stopped-flow: ¹H | Hötzel et al. (1996) Chromatographia 42: 499 |
| Swertia calycina | Gentianaceae | Naphthoquinones, secoiridoids, xanthones | RP-18, MeCN-$D_2O$, UV | 500 MHz; online: ¹H; stopped-flow: ¹H, 2D-COSY | Wolfender et al. (1997) Phytochem. Anal. 8: 97 |
| Gentiana ottonis | Gentianaceae | Flavones, secoiridoids, xanthones | RP-18, MeCN-$D_2O$, UV | 500 MHz; stopped-flow: ¹H | |
| Monotes engleri | Dipterocarpaceae | Flavanones | RP-18, MeCN-$D_2O$, UV | 500 MHz; online: ¹H; stopped-flow: ¹H, 1D TOCSY, 2D NOESY, GHSQC, GHMBC | Garo et al. (1998) Helv. Chim. Acta 81: 754 |
| Lisianthius seemannii | Gentianaceae | Secoiridoid dimer glycosides | RP-18, MeCN-$D_2O$, UV-DAD | 500 MHz; stopped-flow: ¹H | Rodriguez et al. (1998) Helv. Chim. Acta 81: 1393 |
| Vernonia fastigiata | Asteraceae | Sesquiterpene lactones | RP-18, MeCN-$D_2O$ MeOH-$D_2O$, UV | 500 MHz; online: ¹H; stopped-flow: ¹H, 1D selective NOESY, 2D COSY, 2D NOESY | Vogler et al. (1998) J. Natl. Prod. 61: 175 |
| Terminalia macroptera | Combretaceae | Sapogenines | RP-18, MeCN-$D_2O$, MeOH-$D_2O$, UV | 500 MHz; online: ¹H; stopped-flow: ¹H, 1D selective NOESY, 2D COSY, 2D TOCSY, 2D NOESY | Vogler et al. (1998) Natural Product Analysis, Braunschweig/Wiesbaden: Vieweg, p. 143 |
| Rubia tinctorum | Rubiaceae | Anthraquinones | RP-18, MeCN-$D_2O$, UV | 500 MHz; stopped-flow: ¹H | Schneider et al. (1998) Natural Product Analysis, Braunschweig/Wiesbaden: Vieweg, p. 137 |
| Taxus baccata | Taxaceae | Taxanes | RP-18, MeCN-$D_2O$, UV | 500 MHz; stopped-flow: ¹H | |
| Taxus canadensis Taxus chinensis var. mairei Taxus × media | Taxaceae | Taxanes | RP-18, MeCN-$D_2O$, UV | 500 MHz; stopped-flow: ¹H, 2D TOCSY | Schneider et al. (1998) Phytochem. Anal. 9: 237 |
| Ancistrocladus guinensis | Ancistrocladaceae | Naphthylisoquinolines | RP-18, MeCN-$D_2O$, UV | 600 MHz; online: ¹H; stopped-flow: ¹H, 2D-TOCSY, 2D ROESY | Bringmann et al. (1998) Anal. Chem. 70: 2805 |

| Species | Family | Compound class | Conditions | NMR | Reference |
|---|---|---|---|---|---|
| Triphyophyllum peltatum | Dioncophyllaceae | Naphthylisoquinolines | RP-18, MeCN-D$_2$O, UV | 600 MHz; online: $^1$H; stopped-flow: 2D TOCSY, 2D ROESY | Bringmann et al. (1998) Natural Product Analysis, Braunschweig/Wiesbaden: Vieweg, p. 147 |
| Ancistrocladus likoko | Ancistrocladaceae | Naphthylisoquinolines | RP-18, MeCN-D$_2$O, UV | 600 MHz; stopped-flow: $^1$H, 2D TOCSY, 2D ROESY | Bringmann et al. (1999) Magn. Res. Chem. 37: 98 |
| Dioncophyllum thollonii | Dioncophyllaceae | Naphthylisoquinolines, tetralones | RP-18, MeCN-D$_2$O, UV | 600 MHz; online: $^1$H; $^1$H, 2D ROESY, $^1$H time slice | Bringmann et al. (1999) J. Chromatogr. A 837: 267 |
| Habropetalum dawei | Dioncophyllaceae | Naphthylisoquinolines, isoquinolines | RP-18, MeCN-D$_2$O, UV, CD | 600 MHz; online: $^1$H; stopped-flow: $^1$H, 2D TOCSY, 2D ROESY | Bringmann et al. (1999) Anal. Chem. 71: 2678 |
| Orophea enneandra | Annonaceae | Lignans, tocopherols, polyacetylene | RP-18, MeCN-D$_2$O, UV | 500 MHz; online: $^1$H | Cavin et al. (1998) J. Natl. Prod. 61: 1497 |
| Torreya jackii | Taxaceae | Lignans | RP-18, MeCN-D$_2$O, UV | 500 MHz; stopped-flow: $^1$H | Zhao et al. (1999) J. Chromatogr. A 837:83 |
| Senecio vulgaris[b] | Asteraceae | Pyrrolizidines | RP-18, MeCN-D$_2$O, UV | 500 MHz; online: $^1$H; stopped-flow: $^1$H, 2D COSY | Wolfender et al. (1998) Current Organic Chemistry 1: 575 |
| Cordia linnaei | Boraginaceae | Meroterpenoid, naphthoquinones | RP-18, MeCN-D$_2$O, UV | 500 MHz; online: $^1$H stopped-flow: $^1$H | Ioset et al. (1999) Phytochem. Anal. 10: 137 |
| Anigozanthos flavidus | Haemodoraceae | Phenylphenalenones, stilbenes | RP-18, MeCN-D$_2$O, UV | 500 MHz; stopped-flow: $^1$H, 2D TOCSY | Schneider et al. (1998) Natural Product Analysis, Braunschweig/Wiesbaden: Vieweg, p. 137; Hölscher and Schneider (1999) Phytochemistry 50: 155 |

[a] Combined sample of several species of the genus; [b] and other Senecio species.

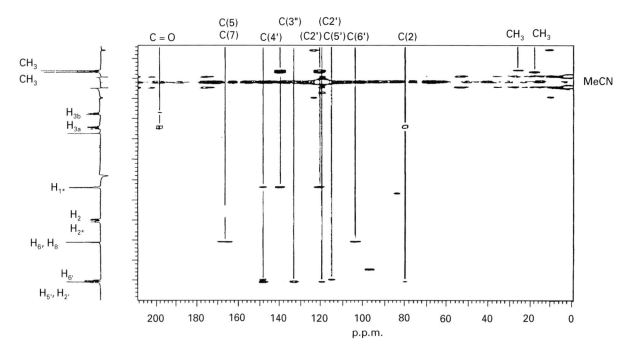

**Figure 2**   GHMBC spectrum of a prenylated flavanone acquired under HPLC-NMR conditions in 6 h 20 min. 1 mg of fraction injected. Reprinted with permission from Garo E, Wolfender JL, Hostettman K *et al.* (1998) Prenylated flavonones from *Monofes egleri*: on-line structure elucidation by LC/UV/NMR. *Helvetica Chimica Acta* 81: 754.

HPLC-NMR was shown as being the method of choice to assign the structure of a rapidly isomerizing dimeric secoiridoid glucoside carrying a (*Z*)-*p*-coumaroyl unit found in aerial parts of *Lisianth-ium seemannii* (Gentianaceae). The stopped-flow ¹H NMR spectrum of this unstable isomer was very similar to that measured for the isomerization product, as far as the monoterpenoic and glycosidic parts of the molecule were concerned. However, the resonances corresponding to the coumaroyl moieties exhibited signals of an (*E*)-double bond for the stable isomer and (*Z*)-double bond for the unstable one.

Less polar fractions of *Vernonia fastigiata* (Asteraceae) were investigated under continuous-flow conditions. In order to obtain information about signals hidden by suppressed peaks of the MeCN-D₂O eluent, HPLC-NMR spectra were measured a second time in MeOH-D₂O. The combination of both complementary spectra allowed the assignment of all proton resonances of the corresponding sesquiterpene lactones in just two HPLC-NMR runs (**Figure 3**). 2D COSY spectra of selected compounds from the more polar fractions were measured in the stopped-flow mode. 2D NOESY and a 1D selective NOESY have been employed to clarify stereochemical features. Due to the selective excitation technique in the 1D NOESY, no solvent suppression was required. A similar array of methods was reported for HPLC-NMR

investigations on active triterpenoid sapogenines from *Terminalia macroptera* (Combretaceae).

An example of how natural products were identified by simply using HPLC-¹H NMR was described for anthraquinones from hairy root cultures of *Rubia tinctorum* (Rubiaceae). First, the HPLC-¹H NMR spectrum of a known anthraquinone, lucidine, was identified by comparison with an authentical standard and was then used to assign the structures of lucidin glycosides for which no reference compounds were available.

Partially purified extracts of *Taxus baccata* (Taxaceae) leaves were used in another example. Despite incomplete chromatographic separation, identification of two isomeric taxanes was clearly possible. Pre-purified extracts of only 0.5 mg air-dried needles of further *Taxus* species, *T. canadensis*, *T. chinensis* var. *mairei*, and *T.×media* cv. Hicksii, were subjected to stopped-flow HPLC-NMR by ¹H and 2D TOCSY analysis. Taxol® and several other neutral and basic taxanes were identified by means of comparison with spectra of reference compounds or were deduced from related compounds. Due to the use of different solvents in HPLC-NMR and conventional NMR spectroscopy, differences of chemical shifts have to be considered. Comparing spectra measured by HPLC-NMR in MeCN-D₂O with that of the same compound measured under conventional conditions in deuterochloroform indicated that chem-

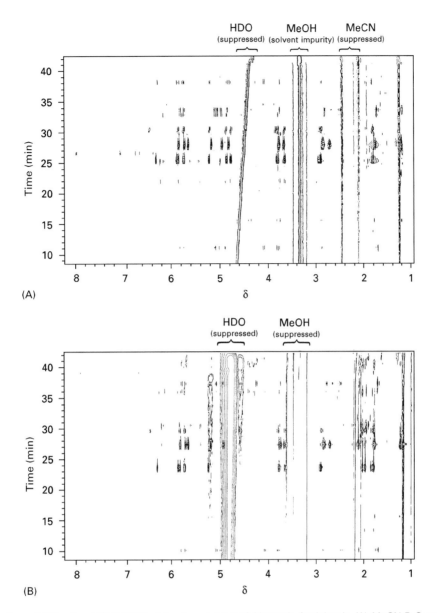

**Figure 3** Comparison of 2D online HPLC-NMR plots of an extract of *Vernonia fastigiata* in (A) MeCN-$D_2O$ and (B) MeOH-$D_2O$. Complementary HPLC-NMR eluents were used in order to provide information on signals hidden by suppressed resonances of each solvent. Adapted with permission from Vogler B, Conrad J, Hiller W, Klaiber I, Roos G and Sandor P (1998) Can LC-NMR serve as a tool for natural products elucidation? In: Schreier P, Herderich M, Humpf HU and Schwab W (eds) *Natural Product Analysis*, p. 143. Braunschweig/Wiesbaden: Vieweg.

ical shift differences did not exceed 0.2 p.p.m. However, due to the fact that some resonances were shifted upfield and others downfield, some signals appeared interchanged in sequence.

A series of HPLC-NMR analyses have been carried out at 600 MHz on naphthylisoquinolines from two plant families, the Diocophyllaceae and the Ancistrocladaceae. $^1H$ spectra extracted from the pseudo-2D continuous-flow diagram obtained under isocratic HPLC conditions using crude leaf extracts of *Ancistrocladus guineënsis* showed the typical signal pattern of naphthylisoquinolines. An optimized nonlinear

HPLC gradient was used for stopped-flow NMR analysis of a known alkaloid and two further closely related compounds. A 2D TOCSY of the HPLC fraction containing both compounds allowed the proton assignment of these isomers in a single NMR experiment. **Figure 4** shows 2D ROESY spectra of naphthylisoquinolines, the first example of the use of HPLC-NMR hyphenation to predict relative configuration. *Triphyophyllum peltatum* was analysed similarly. The major alkaloid of that species, dioncophylline A, was identified by $^1H$, 2D TOCSY and 2D ROESY spectra. Additionally, two minor components

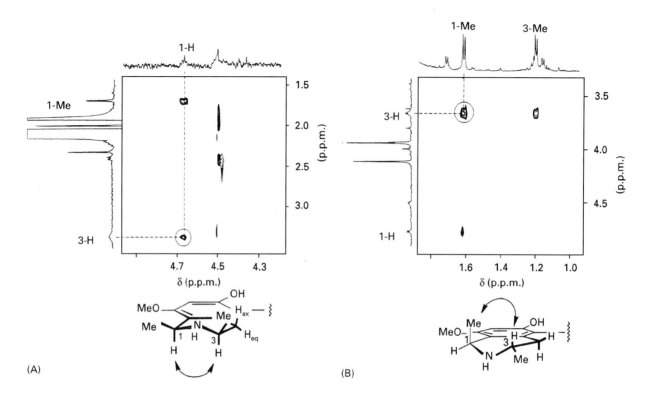

**Figure 4**  HPLC-NMR ROESY cross-peaks indicating the relative configuration at the isoquinoline moiety of diastereomeric naph-thylisoquinoline alkaloids from *Ancistrocladus guineënsis*. (A) Peak eluting at $t_R = 21.05$ min exhibits a correlation between ³H and ¹H. (B) Peak eluting at $t_R = 21.60$ min shows a correlation between 3-H and 1-Me. Reproduced with permission from Bringmann G, Günther C, Schlauer J. *et al.* (1998) HPLC-NMR on-line coupling including the ROESY technique: direct characterization of naphthylisoquinoline alkaloids in crude plant extracts. *Analytical Chemistry* 70: 2805.

were identified in the same plant using an array of analytical methods, including HPLC-NMR. The constitution and relative configuration of new naphthylisoquinoline alkaloids from *Ancistrocladus likoko*, with a 5,8′ coupling pattern, were also elucidated by application of 2D TOCSY and 2D ROESY experiments in the stopped-flow HPLC-NMR mode. Naphthylisoquinolines were also detected by online and stopped-flow HPLC-NMR techniques in *Dioncophyllum thollonii* (Dioncophyllaceae). Moreover, chromatographically unresolved diastereomeric tetralones with slightly different retention times were measured in a time-slice experiment. ¹H spectra were acquired in the stopped-flow mode at different positions of the chromatographic peak, and the diastereomers were distinguished by comparison of ¹H spectra of slices of pure and mixed components. The constitution and relative configuration of an isoquinoline and a naphthylisoquinoline from crude extracts of *Habropetalum dawei* (Dioncophyllaceae) were established by combined application of HPLC-NMR and HPLC-electrospray ionization (ESI)-MS-MS. Additional combinations with subsequent stopped-flow HPLC-circular dichroism (CD) experiments allowed deduction of the absolute configuration of these new metabolites.

While most HPLC-NMR studies make use of the stopped-flow option, either alone or after preliminary continuous-flow experiments, a variety of natural products from *Orophea enneandra* (Annonaceae) have been tentatively characterized by means of the continuous-flow technique without a subsequent stopped-flow run. Column overloading (2 mg) did not prevent proper separation of the components. The structures of three lignanes were identified by reference to literature data. In the cases of a tocopherol derivative and an unstable polyacetylene, targeted isolation and structure elucidation by complementary coupling techniques and conventional analytical methods were necessary.

A variety of lignanes were also identified from extracts of *Torreya jackii* (Taxaceae). Some of them were completely characterized by stopped-flow HPLC-¹H NMR while in other examples isolation was required to confirm the structures by conventional NMR spectroscopy. After HPLC-NMR measurements, individual lignanes were collected and subjected to MS, which was considered to be an indispensable tool for complete structure assignment.

A study on *Senecio vulgaris* (Asteraceae) has allowed identification of a variety of pyrrolizidine alkaloids and differentiation of certain isomeric macrocyclic diesters of that type. These compounds adopt *cis-trans* configurations and are not distinguishable by LC-MS. This example demonstrated again that complementary measurements in MeCN-D$_2$O and MeOH-D$_2$O are necessary to observe all resonances. Information obtained from the continuous-flow $^1$H spectrum (24 scans per increment; column overloading by 3 mg of extract) were shown to be comparable to those from a corresponding stopped-flow spectrum.

Online and stopped-flow HPLC-NMR analysis of two minor isomeric meroterpenoid napthoquinones from *Cordia linnaei* (Boraginaceae) yielded $^1$H NMR spectra exhibiting identical signals in the aromatic region. Differences were only found in methyl signals when MeCN-D$_2$O and MeOH-D$_2$O were used as complementary HPLC-NMR eluents. One of the isomers, cordiaquinone C, carried a senecioic acid moiety while the other, being a new compound, was found to contain a tigloyl substituent instead.

Extracts of *Anigozanthos flavidus* (Haemodoraceae), a plant family accumulating phenylphenalenones and stilbenes in the roots, were investigated using stopped-flow HPLC-NMR. Comparison with spectra of references and literature data was used to differentiate between known and novel compounds. A number of known compounds were identified without isolation, while others had to be isolated, especially when possessing regions poor in hydrogen atoms.

## Biosynthetic Applications

HPLC-NMR has been used in biosynthetic and enzymatic investigations of secondary plant products. Michellamines, representing dimeric naphthylisoquinoline alkaloids highly active against HIV, were formed biosynthetically by oxidative coupling of their inactive korupensamine monomers. This dimerization was catalysed by peroxidase preparations from three *Ancistrocladus* species (Ancistrocladaceae) and from *Triphyophyllum peltatum* (Dioncophyllaceae). The peroxidase was partially purified from *Ancistrocladus heyneanus* and characterized in more detail. The exclusive formation of a michelleamine from its monomeric korupensamine precursor, shown in **Figure 5**, was confirmed by HPLC-MS and stopped-flow HPLC-$^1$H NMR experiments.

Details of the phenylpropanoid metabolism preceding later steps of the biosynthesis of phenylphenalenones from *Anigozanthos preissii* (Haemodoraceae) were elucidated by Schmitt and Schneider using the stopped-flow HPLC-NMR technique. Incorporation of dihydrophenylpropanoids into phenylpropanoids in root cultures of that plant was

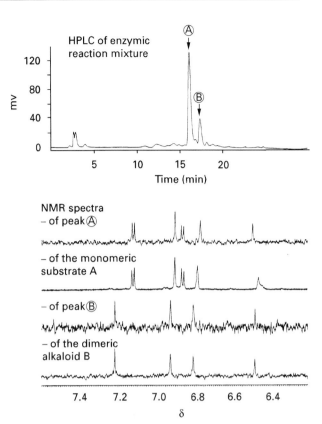

**Figure 5**  Stopped-flow HPLC-$^1$H NMR experiment confirming exclusive formation of a dimeric naphthylisoquinoline **B** from monomer **A** in the coupling reaction catalysed by peroxidase purified from *Ancistrocladus heyneanus*. Adapted with permission from Schlauer J, Rückert M, Herderich M *et al.* (1998) Characterization of enzymes from *Ancistrocladus* (Ancistrocladaceae) and *Triphyophyllum* (Dioncophyllaceae) catalyzing oxidative coupling of naphthylisoquinoline alkaloids. *Archives of Biochemistry and Biophysics* 350: 87.

proved by the coupling pattern in the HPLC-$^1$H NMR spectrum of *p*-coumaric acid biosynthesized from [2-$^{13}$C]dihydrocinnamic acid (**Figure 6**). A number of simple phenolics, which are supposed to be formed from phenylpropanoids, were also detected in these experiments by HPLC-$^1$H NMR spectroscopy in the stopped-flow mode.

## Applications Related to Natural Products

In areas related to natural product research, Careri and Mangia have reviewed the analysis of natural food components by HPLC-NMR. Lindon *et al.* have published an overview on HPLC-NMR in biomedical applications, and a small number of investigations have shown that the analysis of amino acids and peptide mixtures is also possible by HPLC-NMR.

The first HPLC-NMR analysis of biological macromolecules, published by Rückert *et al.*, utilized the combination of ion exchange chromatographic

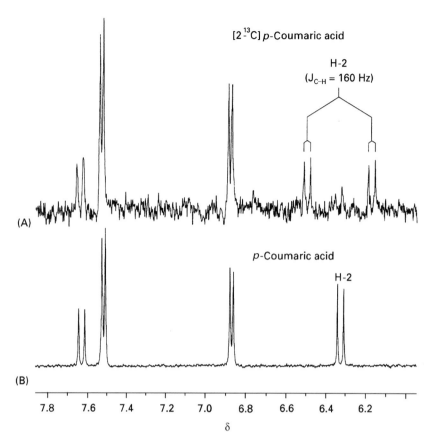

**Figure 6**   Stopped-flow HPLC-¹H NMR experiment confirming biosynthetic incorporation of ¹³C label into C-2 of *p*-coumaric acid. ¹H NMR spectra of (A) [2-¹³C]*p*-coumaric acid resulting from treatment of *Anigozanthos preissii* root cultures with [2-¹³C]dihydrocinnamic acid and (B) nonlabelled reference.

separation with ¹H, 2D TOCSY and 2D NOESY spectroscopy to characterize small proteins in mixture. The authors expect that HPLC-NMR at very high field (750 and 800 MHz) and further enhancements in sensitivity should permit online experiments and heteronuclear 2D and 3D stopped-flow experiments in the future.

## Summary and Future Developments

HPLC-NMR coupling has been developed into a valuable tool for natural product analysis. In general, the online technique is used to provide a rapid overview of the major components occurring in plants and other sources of natural products. The more sensitive stopped-flow method allows the detection and structure assignment of even minor components and enables the use of various homo- and heteronuclear correlation NMR experiments. However, unambiguous structure assignment of novel compounds of unexpected structural types requires information from other analytical methods, especially MS. Complete structure elucidation, together with stereochemical information, by multiple online com-

binations including NMR is possible but currently is rather the exception. Rapid development in analytical chemistry is expected to overcome present limitations of HPLC-NMR. The future scenario in a natural product laboratory could be an automated characterization of sources of natural products, starting with extraction and separation, followed by hyphenated instrumental analysis and finally computational structure elucidation. Additional combination with biological screening could avoid isolation of inactive compounds.

HPLC-NMR is an excellent approach to search for novel biologically active structures to be tested as new medicinal and agricultural agents, to identify known compounds without isolation, and to avoid unwanted re-isolation of known constituents from living organisms. Due to the large amount of structural information provided by NMR spectroscopy, its combination with HPLC and further spectroscopic techniques is also suitable when searching for new sources of rare natural products, for clarification of uncertain chemotaxonomic relationships and distribution of secondary compounds in various tissues. The introduction and routine application of capillary HPLC and innovative fused capillary nanolitre flow cells in NMR probes, and further development in cryoprobe technology along with the use of improved processing procedures, will continue to enhance the sensitivity of HPLC-NMR coupling. As a microanalytical method, HPLC-NMR allows the detection of various groups of natural compounds and other biomolecules in the nanogram or even picogram range and, therefore, can contribute to the solution of problems of biochemical, physiological and chemoecological research.

*See also:* **II/Chromatography: Liquid:** Mechanisms: Reversed Phases; Nuclear Magnetic Resonance Detectors. **III/Medium-Pressure Liquid Chromatography. Natural Products:** Liquid Chromatography. **Terpenoids: Liquid Chromatography.**

## Further Reading

Albert K (1995) On-line use of NMR detection in separation chemistry. *Journal of Chromatography A* 703: 123.

Albert K (1997) Supercritical fluid chromatography–proton nuclear magnetic resonance spectroscopy coupling. *Journal of Chromatography A* 785: 65.

Behnke B, Schlotterbeck G, Tallerek U *et al.* (1996) Capillary HPLC-NMR coupling: high resolution ¹H NMR spectroscopy in the nanoliter scale. *Analytical Chemistry* 68: 1110.

Careri M and Mangia A (1996) Multidimensional detection methods for separations and their application in food analysis. *Trends in Analytical Chemistry* 15: 538.

Lindon JC, Nicholson JK, Sidelmann UG and Wilson ID (1997) Directly coupled HPLC-NMR and its application to drug metabolism. *Drug Metabolism Reviews* 29: 705.

Pusecker K, Schewitz J, Gfrörer P *et al.* (1998) On line coupling of capillary electrochromatography, capillary electrophoresis, and capillary HPLC with nuclear magnetic resonance spectroscopy. *Analytical Chemistry* 70: 3280.

Rückert M, Wohlfarth M and Bringmann G (1999) Characterization of protein mixtures by ion-exchange liquid chromatography coupled on-line to NMR spectroscopy. *Journal of Chromatography A* 840: 131.

Seddon MJ, Spraul M, Wilson ID *et al.* (1994) Improvement in the characterization of minor drug metabolites from HPLC-NMR studies through the use of quantified maximum entropy processing of NMR spectra. *Journal of Pharmaceutical and Biomedical Analysis* 12: 419.

Smallcombe SH, Patt SL and Keiffer PA (1995) Wet solvent suppression and its application to LC NMR and high resolution NMR spectroscopy. *Journal of Magnetic Resonance Series A* 117: 295.

Schmitt B and Schneider B (1999) Dihydrocinnamic acids are involved in the biosynthesis of phenylphenalenones. *Phytochemistry* 52: 45.

Schreier P, Herderich M, Humpf HU and Schwab W (eds) (1998) *Natural Product Analysis.* Braunschweig/Wiesbaden: Vieweg.

Watanabe N, Niki E and Shimizu S (1979) An experiment on direct combination of high performance liquid chromatography with FT-NMR (LC-NMR). *Jeol News* 15A: 2.

Wolfender JL, Ndjoko K and Hostettmann K (1998) LC/NMR in natural products chemistry. *Current Organic Chemistry* 1: 575.

Wu N, Webb L, Peck TL and Sweedler JV (1995) On-line NMR detection of amino acids and peptides in microbore LC source. *Analytical Chemistry* 67: 3101.

# Supercritical Fluid Chromatography

**E. D. Morgan**, Keele University, Staffordshire, UK

The mild elution temperatures and the wide range of molecular masses it can accommodate makes supercritical fluid chromatography (SFC) particularly applicable to natural products. It is becoming the preferred method for the separation of enantiomers, and is especially useful for combined or hyphenated techniques. It forms a link between liquid chromatography (LC) and chromatography (GC), it has capabilities between the two and shares the instrumental set-ups of both, so both capillary column and packed column applications are recorded here.

The advantages and disadvantages of the method are debated elsewhere, but some of its strong points are indicated here. In all, 99% of supercritical fluid applications use supercritical carbon dioxide, since it has the great advantage that it is a nontoxic, nonflammable, pure, cheap mobile phase that presents no disposal problems. The greatest usefulness of SFC comes in connection with supercritical fluid extraction, which has received much attention for the isolation of natural products. If a substance can be extracted from plant or animal material with a supercritical fluid and some of the extract can be diverted to an online SFC, the course of the extraction can be followed very easily. Many of the applications of SFC recorded for natural products are of this type. The greatest disadvantage of supercritical carbon dioxide is its relatively nonpolar nature. In its solvent powers, it resembles hexane at lower pressures, becoming slightly more polar at higher pressure. The polarity of the fluid can be increased by the addition of a small proportion of a highly polar organic solvent, miscible with the supercritical $CO_2$. This is most commonly methanol. The solubility of water in supercritical $CO_2$ is too low to be of much use to increase the polarity (but see below). The proportion of the so-called modifier solvent can range from 1 to 25%, but the critical point of the mixed fluid increases with the proportion of organic liquid and the advantages mentioned in this paragraph are steadily eroded with increasing proportion of modifier. In the chromatography of natural products, therefore, SFC is most useful for products of low polarity, such as terpenes, lipids and essential oils.

Free fatty acids, methyl esters, mono-, di-, and triglycerides can all be separated by SFC methods. The technique is particularly useful for triglycerides. Because triglycerides lack a useful UV absorption for high performance liquid chromatography (HPLC) and are at the limit of volatility for GC, they are not easy to separate and quantify without conversion to methyl esters. Using SFC with capillary columns and a flame ionization detector, they can be analysed directly. For capillary column determination of free fatty acids in an ethanol extract of *Sabal serrulata* two alternatives have been proposed: derivatization of the carboxyl group and saturating the $CO_2$ with water. Both methods produce a drastic improvement in resolution. Hydroxy acids are too polar for direct determination, and the separation of triglycerides is improved by the addition of a little methanol or acetonitrile, which then is detrimental to the use of a flame detector. It is possible to convert free carboxylic acids to their methyl esters in a flow-through system with $CO_2$ containing 10 mol% methanol at 80°C over a cation exchange resin in the H form. Capillary columns with a flame ionization detector have been used to separate the fatty acids and alcohols from hydrolysis of jojoba oil. The unhydrolysed portion of the wax esters could also be seen in the chromatogram. Wool wax alcohols from hydrogenated lanolin have been examined similarly. Using a microextractor, the triglycerides of a single cotton seed kernel were extracted by supercritical fluid extraction (SFE) and linked directly to SFC for analysis. In a rare example of application to insects, the same microcell (at 45°C and 20.2 MPa or 200 atm) has been used to extract the cuticular hydrocarbons and waxes from the cuticle of a dried fruit beetle *Carcophilus hemipterus*, which were then separated by capillary SFC. The results, probably not optimized, were not as good as normally obtained by GC. Prostaglandins have been separated on a capillary column at 100°C with a $CO_2$ density gradient.

Staby and Mollerup have produced a comprehensive review of the separation and chromatography of fish oil constituents with supercritical fluids. These include triglycerides, free fatty acids, methyl and ethyl esters, cholesterol, $\alpha$-tocopherol, phospholipids and squalene. The review is particularly directed towards pilot plant separations. Another review by Borch-Jensen and Mollerup in 1997 compared chromatographic systems for natural products like fats, seeds, oils and tissues.

The less polar steroids are usefully separated by SFC. Eleven steroids, including testosterone, oestrone, oestradiol, oestriol, cortisone and hydrocortisone, can be separated in less than 2 min with 6.1% methanol in $CO_2$ on a cyanopropyl HPLC column. Even bile acid conjugates (e.g. glycocholic acid and taurocholic acid) can be subjected to SFC. The ecdysteroids (insect moulting hormones, with a polyhydroxycholesterol structure) have been separated on packed columns under very similar conditions. Boronic ester derivatives of a diol functional group in some ecdysteroids improve selectivity for that group. The ecdysteroids, which have a strong UV chromophore, are also found in many plants of diverse type, and extracts of plants can be very quickly scanned for the presence of so-called phytoecdysteroids. With the coupling of a mass spectrometer to the SFC column, separation and identification of phytoecdysteroids can be performed in a matter of minutes.

Capillary SFC has been used to analyse the terpenes of some aromatic plants and the results compared with those obtainable by GC. Thyme (*Thymus vulgaris*) gave the same information as GC; for peppermint (*Mentha × piperita*) and basil (*Ocimum basilicum*) the separation is much better by GC but SFC quantification is more reliable. The monoterpenes of lemon peel oil have been examined on a packed SFC column. Using two different silica columns (Nucleosil 100 and Spherisorb Si) linked in series with SFC Fourier transform infrared (FTIR; see later) eight sesquiterpenes (longicyclene, longifolene, aromadendrene, ledene, valencene, *cis*- and *trans*-calamenene and humulene) have been separated and identified, as have five sesquiterpenes from copaiba balsam and ylang-ylang oil under similar conditions.

Carotenes, with their strong visible and UV chromospheres, are ideal subjects for SFC. In spite of their low polarity, there are frequent reports that addition of a very small amount of an organic modifier improves selectivity. Vitamin A can exist in five pairs of *cis-trans* isomers which can be separated at temperatures below 50°C, so avoiding any fear of isomerization.

## More Polar Compounds

As already indicated, many examples of SFC of natural products are found where extraction with a supercritical fluid (SFE) and chromatography are coupled. A classic SFC separation, performed in 1982, was the separation of caffeine, theophylline and theobromine with $CO_2$–methanol. The pungent phenolic oil of ginger (*Zingiber rhyzoma*) has been analysed for [6]-gingerol (see **structure**), [8]-gingerol and [10]-gingerol. SFE-SFC has been used in the extraction of ginkgolides (oxidized diterpenes from the leaves of *Ginkgo biloba*), paclitaxel (the antitumour agent Taxol) from *Taxus brevifolia*, and the antimalarial artemisinin and its precursor artemisinic acid from *Artemisia annua*. The chromatography of artemisinin was carried out on both a capillary column with $CO_2$ (and 3% methanol) and a flame detector and on an aminopropyl packed column with a $CO_2$–methanol gradient (17.18 MPa and 40°C) and an evaporative light-scattering detector. The packed column method was faster: both compounds were eluted in 7 min as against 25 min by the capillary method. Paclitaxel has a strong UV absorption, but the ginkgolides do not, so the evaporative light-scattering detector is very helpful. Ginkgolide A has a molecular formula $C_{20}H_{24}O_9$, ginkgolide B $C_{20}H_{24}O_{10}$, paclitaxel $C_{47}H_{51}NO_{14}$, and artemisinin $C_{15}H_{22}O_5$, which shows that highly functionalized compounds can be suitable for SFC.

Limonoids from hexane extracts of the plants *Aphanamixis polystacha*, *Harpephyllum caffrum* and *Entandrophragma delevoyii* have been ana-

lysed by capillary SFC. A liquid crystal-modified polysiloxane phase that discriminated molecular shape was used in a capillary column to separate triterpene acids, including geometric isomers, from *Disoxylum pettigrewianum*. The natural insecticide azadirachtin (see **structure**) is usually analysed by reversed-phase HPLC, but because the extract obtained from Neem seeds usually contains a lot of less polar contaminants, including triglyceride oil, the column has to be flushed with pure methanol or acetonitrile at the end of each run, and total cycling time can be of the order of 1 h. With packed column SFC, using an aminopropyl silica or cyanopropyl silica column and $CO_2$–methanol (24 : 1, v/v) mobile phase at 20.6 MPa (3000 psi or 207 bar) at 55°C and at a flow rate of 2 mL min$^{-1}$, the oily impurities are eluted in the solvent front, good resolution and good peak shape are obtained, and the cycle time is about 20 min (**Figure 1**). The example illustrates how SFC in the right situation can have marked advantages over other chromatographic methods.

The cannabinoids and their metabolic products have been examined on capillary columns. SFC is perhaps not the most useful way to examine alkaloids, but a number of separations have been made. Six opium alkaloids (narcotine, papaverine, thebaine, ethylmorphine, codeine and morphine) have been separated from poppy straw on an aminopropyl column with $CO_2$–methanol–methylamine–water. Seven ergot alkaloids were separated from *Claviceps purpurea* and eight pyrrolizidine alkaloids from *Senecio anonymus*.

Flavonoids from citrus fruits that contain several methoxy groups (hexa- and heptamethoxyflavone, tangeretin, nobeletin, sinensetin and others) were separated on a silica column with $CO_2$–methanol.

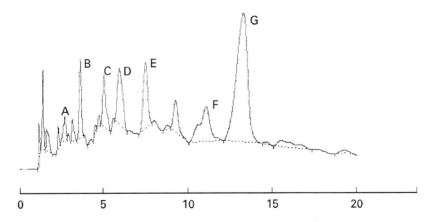

**Figure 1** SFC chromatogram of an extract of triterpenoids from the seeds of *Azadirachta indica*, on a Spherisorb cyanopropyl column (150 × 4.6 mm i.d.) of 5 μm particle size, flow rate 2 cm³ min$^{-1}$ of $CO_2$–methanol (94 : 6) at 3000 psi (20.7 MPa) and 50°C with UV detection at 217 nm. Compounds are: A, nimbin; B, salannin; C, 6-desacetylsalannin; D, 3-desacetylnimbin; E, 3-tigloylazadirachtol; F, 3-acetyl-1-tigloylazadirachtinin; G, azadirachtin. Unpublished results of A.P. Jarvis and E.D. Morgan.

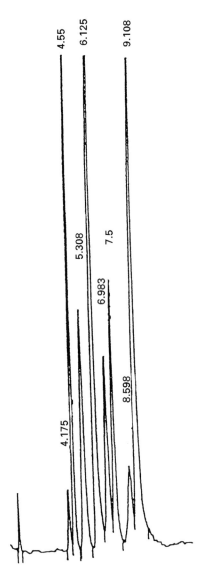

**Figure 2** SFC of monosaccharides on Zorbax TMS column (250 × 4.6 mm i.d.), flow rate 5 cm³ min⁻¹ of CO₂-modifier (80:20, v/v) at 200 bar (20 MPa) and 60°C. The modifier was methanol–water–triethylamine (91.5:8.0:0.5, v/v). Compounds in order of elution are D-ribose, m-erythrose, D-xylose, xylitol, L-sorbose, D-mannose, D-glucose and mannitol. (Reproduced with permission from Salvador A, Herbreteau B, Lafoose M and Dreux M (1997) Subcritical fluid chromatography of monosaccharides and polyols using silica and trimethylsilyl columns. *Journal of Chromatography A* 785: 195.)

A number of monosaccharides, polyols and glycolipids have been separated on silica and trimethylsilyl-bonded silica under subcritical conditions (41°C, 200 bar, and CO₂–methanol 80:20, or a modifier of methanol with 4% or 8% water; Figure 2).

An extensive review (in French) by Lubke cites 207 references on the advantages of SFC for the analysis of natural products.

At the time of writing there is increasing interest in the use of subcritical water at 100–200°C as a mobile phase with divinylbenzene polymer columns for the chromatography of highly polar compounds like alcohols, phenols, amino acids and flavones. Clearly any compounds to be analysed must be stable to hydrolysis, as must the stationary phase. As temperature and pressure increase, water becomes less polar – at 200°C it is rather nonpolar – so by controlling these conditions a fluid phase of intermediate polarity similar to the water–methanol mixtures used in HPLC can be achieved.

## Chiral Separations

Another advantage of SFC is in the separation of enantiomers. Chiral separations can be carried out at lower temperatures (often with greater efficiency than by GC on the same phases), and on larger molecules than with GC and with improved resolution over HPLC. The subject is still at a very early stage of development, with new discriminator phases appearing rapidly. If insufficient resolution is obtained under supercritical conditions, some investigators recommend subambient (and consequently subcritical) conditions. Separations have included the plant hormone abscisic acid (see **structure**), benzoin, ephedrine, mandelic acid, tropic acid (**Figure 3**), underivatized amino acids and their derivatives. For the separations shown in Figure 3, the composition of the mobile phase had the largest effect on retention, peak shape and enantioselectivity, with temperature the second most important influence. Tyrosine and tryptophane enantiomers are separated under subcritical conditions on a Chirobiotic T phase at 30°C and 200 bar isocratically with 40% modifier (methanol–water–glycerol, 92.8:7.0:0.2 v/v) in CO₂ with 0.1% triethylamine and 0.1% trifluoroacetic acid. Enantioseparations of five lignans were carried out on polyWhelk-O between 0°C and −42°C with 5–15% methanol in CO₂ with α values from 1.28 to 1.44. The potential of chiral separations of less volatile pheromones unsuitable for GC has not yet been explored.

## Hyphenated Methods

All the predictable couplings of SFC to spectroscopic methods have been tried. SFC-FTIR has been used to look for isomerized fatty acids in the triglycerides of partially hydrogenated soya bean oil and in the free fatty acids after hydrolysis using a flow-through FTIR cell. SFC is more easily coupled to a mass spectrometer than HPLC, therefore many examples of SFC-mass spectrometry (MS) can be

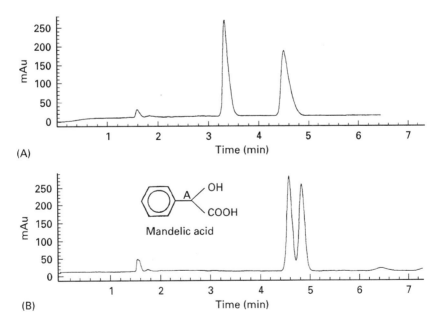

**Abscisic acid**

**[6]-Gingerol**

**Azadirachtin**

found. The spectra resemble chemical ionization spectra, with prominent M+ or M + 1 ions. Alkaloids of *Securidaca longipendunculata* have been separated by SFC-MS and SFC-MS-MS with a moving belt. Thermospray in the electron ionization mode with an SFC gradient was used for the indole alkaloids of *Catharanthus roseus* to identify 60 compounds. By SFC-UV 10 major alkaloids and 30–40 minor compounds were detected in the extract. A few antibiotics, including penicillin, cyclosporin, tetracyclin, oxytetracyclin and mitomycin C have been determined, in some cases after extraction from blood. Ecdysteroid spectra can be varied by the operating conditions to give additional ions for M-$H_2O$, M-$2H_2O$, etc.

There are a number of examples of direct coupling between SFC and proton nuclear magnetic resonance (NMR) spectrometers. $CO_2$ has the great advantage of being transparent and only a small proportion of $CD_3OD$ may be required. For polar compounds, subcritical $D_2O$ can be used (see discussion above); it is both transparent and does not generate a very high pressure. The vapour pressure of water at 200°C is only 1.55 MPa (225 psi or 15.5 bar). Resolution of

**Figure 3** Separation of the enantiomers of mandelic acid on (A) Chiralpak OD (3,5-dimethylphenylcarbamate derivative of cellulose coated on 10 μm silica gel) and (B) Chiralpak AD (3,5-dimethylphenylcarbamate derivative of amylose coated on 10 μm silica gel). Both columns were 250 × 4.6 mm i.d., at flow rate 2 cm³ min⁻¹ of $CO_2$–methanol containing 0.1% triethylamine and 0.1% trifluoroacetic acid and programmed from 5% modifier (5 min) to 30% at 5% min⁻¹ at 200 bar (20 MPa) and 30°C. (Reproduced with permission from Medvedovici A, Sandra P, Toribio L and David F (1997) Chiral packed column subcritical fluid chromatography on polysaccharide and macrocyclic antibiotic chiral stationary phases. *Journal of Chromatography A* 785: 159.)

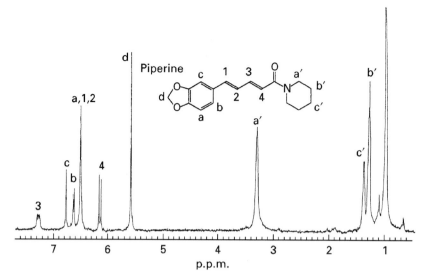

**Figure 4** A stopped-flow $^1$H NMR spectrum at 400 MHz of piperine extracted from pepper with supercritical $CO_2$ at 0.5 cm$^3$ min$^{-1}$. Pressure was 294 bar (29.4 MPa), and temperature 44°C. (Reproduced with permission from Albert, 1997.)

the NMR spectra under continuous flow approaches the quality of conventional spectra. SFC-NMR has been applied to vitamins and a range of natural products, including extracts of coffee, hops and pepper (**Figure 4**). Two problems are, firstly, the dependence of NMR signals on pressure, and secondly, the increased spin-lattice relaxation times in a supercritical fluid.

Mycotoxins from *Fusarium roseum* culture extracts have been studied with a combination of SFC-UV and SFC-MS. Some experiments have been conducted with SFE-SFC-NMR-MS and we can expect to see more of such techniques and the hyphens extended.

## Future Directions

SFC will not replace GC or HPLC, but predictions are dangerous. We already see new ideas coming forward, such as subcritical cryoseparations and use of superheated water as the mobile phase, that will extend its potential. In the field of polymers, surface active compounds are being used to render growing polymers soluble in supercritical $CO_2$. The advantages of SFC are such that we can expect every opportunity to be seized to extend its possibilities, particularly in the areas of chiral separations and hyphenated methods.

## Further Reading

Albert K (1997) Supercritical fluid chromatography–proton nuclear magnetic resonance spectroscopy coupling. *Journal of Chromatography A* 785: 65.

Anton K and Berger C (1998) *Supercritical Fluid Chromatography with Packed Columns. Techniques and Applications.* New York: Marcel Dekker.

Arpino PJ and Haas P (1995) Recent developments in supercritical fluid chromatography–mass spectrometry coupling. *Journal of Chromatography A* 703: 479.

Berger TA (1995) *Packed Column SFC.* RSC Monographs. London, UK: Royal Society of Chemistry.

Berger TA (1997) Separation of polar solutes by packed column supercritical fluid chromatography. *Journal of Chromatography A* 785: 3.

Bevan CD and Marshall PS (1994) The use of supercritical fluids in the isolation of natural products. *Natural Product Reports* 11: 451.

Borch-Jensen C and Mollerup J (1997) Phase equilibria of fish oil in sub- and super-critical carbon dioxide. *Fluid Phase Equilibria* 138: 179.

Charpentier BA and Sevenants MR (eds) (1988) *Supercritical Fluid Extraction and Chromatography. Techniques and Applications.* ACS Symposium Series 366. Washington: American Chemical Society.

Chester TL, Pinkston JD and Raynie DE (1996) Supercritical fluid chromatography and extraction. *Analytical Chemistry* 68: 487R.

Combs MT, Ashraf-Khorassani M and Taylor LT (1997) Packed column supercritical fluid chromatography-mass spectroscopy: a review. *Journal of Chromatography A* 785: 85.

Greibrokk T (1995) Application of supercritical fluid extraction in multidimensional systems. *Journal of Chromatography A* 703: 523.

Lubke M (1991) The advantages of supercritical fluid chromatography for analysing natural products. *Analysis* 19: 323.

Staby A and Mollerup J (1993) Separation of constituents of fish oil using supercritical fluids – a review of experimental solubility, extraction, and chromatographic data. *Fluid Phase Equilibria* 91: 349.

Williams KL and Sander LC (1997) Enantiomer separations on chiral stationary phases in supercritical fluid chromatography. *Journal of Chromatography A* 785: 149.

# Supercritical Fluid Extraction

**E. D. Morgan**, University of Keele,
Staffordshire, UK

Copyright © 2000 Academic Press

There are two interesting points of the phase diagram of a pure substance: the triple point, where solid, liquid and gas are all in equilibrium, and the critical point, at which liquid and gas phases cease to have separate existence. At temperatures and pressures beyond the critical point there exists only the supercritical phase, with properties between those of a gas and a liquid, varying with conditions. For example, at high pressures a supercritical fluid has solubility and density properties close to those of the liquid, but with greater diffusibility and lower viscosity. High diffusibility and low viscosity improve mass transfer and so help to decrease extraction time. All this would be of only academic or research interest, were it not that one substance, carbon dioxide, which is cheap, readily available in a pure state, and non-toxic, has a readily accessible critical point (31.1°C, and 72.9 atm, 73.8 bar, 1071 psi, or 7.38 MPa). The density of supercritical carbon dioxide at various values of temperature and pressure is given in **Figure 1**. Its vaporization on release of pressure avoids the step of concentrating a liquid solution after extraction.

The possibilities of using supercritical carbon dioxide as an extraction fluid were recognized first in industry in the 1950s and 1960s. It entered the laboratory in the 1980s at the same time pumps and control equipment were developed for supercritical fluid chromatography (SFC). Since then supercritical fluid extraction (SFE) has been explored with a wide variety of materials, and is generally recognized as a possible alternative to chlorinated or other toxic solvents for extraction of organic substances. There is no close competitor for carbon dioxide as a supercritical fluid for extraction. Other substances with easily accessible critical points are either too expensive (xenon), toxic (ammonia, nitrous oxide), flammable (ethane, pentane) or corrosive (ammonia). Extraction may be described as static (under pressure without flow of the supercritical fluid) or dynamic (the supercritical fluid flowing through the material to be extracted). Dynamic extraction is more common, but it can be preceded by a period of static extraction.

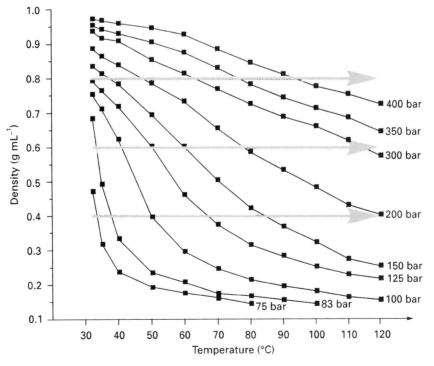

**Figure 1** A plot of carbon dioxide density against pressure and temperature above the critical point. 10 bar = 1 MPa. Horizontal arrows represent constant densities of 0.4, 0.6 and 0.8. Fatty acids are extracted only above 0.4 g mL$^{-1}$, triglycerides above 0.6 g mL$^{-1}$. Reproduced with permission from Gere DR and Derrico EM (1994) *LC.GC International* 7: 325.

The subject roughly divides itself into two aspects: commercial-scale separation of valuable products (e.g. vitamins, drugs, flavours, fragrances, pigments) and laboratory-scale extraction for research or analysis of components. Bevan and Marshall have considered the design of large scale extractors. There are a number of commercial extractors available for small scale work. They usually have six or more chambers so that extractions can be carried out on several samples simultaneously or consecutively. Pressure can be maintained and controlled by an electronic valve or by use of a fixed restrictor, such as a length of silica capillary. After release of the pressure the extract can be collected on a solid adsorbant or the $CO_2$ bubbled through a solvent. Precautions have to be taken to prevent loss of material as an aerosol.

It is possible to build an extractor at modest cost. The chief needs are high pressure pumps (one for the $CO_2$, another for the modifier (if used, see below), a cooler for the $CO_2$, since it is pumped as a liquid, an old gas chromatography (GC) oven to maintain the desired temperature of the extraction, some empty high performance liquid chromatography (HPLC) columns for extraction chambers, a pressure gauge and some means of controlling and releasing the pressure. The simplest solution for this is a length of about 30–50 cm of silica capillary. Whatever the purpose of the extraction, it can be convenient to couple the equipment to a chromatographic system to monitor the course of extraction. Most convenient is on-line SFC. The composition of the extract can be sampled periodically by inserting a switching valve leading to the chromatograph. GC linked to the extraction is also much used, particularly for the extraction of essential oils and fragrances. The chromatograph can in turn be linked to a mass spectrometer. Now there is no end to the complexity of equipment that can be linked to the extraction. Albert has particularly explored the linking of SFE to nuclear magnetic resonance (NMR) spectroscopy, and gives examples that include natural products.

The solubility properties of supercritical $CO_2$ are close to those of hexane (polarity increases slightly with pressure), so its chief disadvantage is poor solubility for more polar substances. To extract lipids, the density needs to be above $0.4 \, \text{g mL}^{-1}$ (see Figure 1),

for triglycerides it should be above $0.6 \, \text{g mL}^{-1}$. The best way to improve solubility (if increasing pressure is not convenient) is to add a small proportion (5–10%) of a polar organic solvent, such as methanol. The organic solvent in this context is commonly called a modifier. The plant, animal or other material from which substances are being extracted is often called the matrix. The solubility of water in supercritical $CO_2$ is very low (0.4 mol% at 8.0 MPa, rising to 0.8 mol% of water at 20.0 MPa), but there are a number of examples where water has been used as a modifier for extraction of more polar compounds. The polarity of water drops remarkably as it approaches its critical point.

An early landmark in SFE was the commercial extraction of caffeine from coffee with $CO_2$. Caffeine is a relatively polar compound, so its extraction with the very apolar $CO_2$ is surprising. Hops and spices followed closely.

## Lipids

Lipids from plant and animal sources are obvious targets for SFE, both on a commercial scale and as an analytical method in the laboratory. There are many reports of its use to extract fish oils (particularly to obtain concentrates of polyunsaturated acid glycerides), dairy products and seed oils. In many cases attempts are recorded to make selective extraction of valuable minor substances accompanying the oil, such as tocopherols, carotenes or sterols. The content of tocopherols is said to be significantly higher from rapeseed and soybeans by SFE. One can find examples of SFE of natural products in the reviews by Bevan and Marshall and by Jarvis and Morgan. The lipids of rapeseed and soybeans were extracted with $CO_2$ alone and with added propane or nitrous oxide. In another case, oil from a fungus *Mortierella ramanniane* was extracted with $CO_2$, $N_2O$, $CHF_3$ and $SF_6$. Extraction was best at 60°C and 157–295 bar with $N_2O$, followed by $CO_2$, $CHF_3$ and $SF_6$. Addition of 20% ethanol greatly increased the solubility of the oil and decreased its acidity. Some examples are listed in **Table 1** with some of the conditions used for the extraction, although in most cases a range of differing conditions were explored. Unfortunately, there is a scarcity of information directly comparing efficiency of SFE with solvent extraction.

**1** Bixin

**Table 1** Examples of extraction of lipids with conditions used

| Extracted | Matrix | Temperature (°C) | Pressure (MPa) | Additional information[a] |
|---|---|---|---|---|
| Bixin (**1**) | *Bixa orellana* (annatto) | 40 | 60.62 | 4% Acetonitrile with 0.05% trifluoroacetic acid, yield 0.27% |
| Carotenes | *Daucus carota* (carrots) | 40 | 60.6 | 5% CHCl₃ 92.7%, 1 h |
| Carotenes | *Mauritia flexuosa* (buriti fruit) | 40–55 | 30 | 80% |
| Carotenes and lutein | Leaf protein concentrate | 40 | 30 | |
| Fatty acid methyl esters | Fish oil | 40 | 8.0 | 20 min |
| $\gamma$-Linolenic acid | *Oenothera paradoxa* (evening primrose) seed | 40–60 | 20–70 | 30 min, 95% |
| Phospholipids | Rape seed (Canola) | 70 | 55.2 | 10% EtOH |
| Phytosterols | Seed, corn oil, margarine | | | |
| Sterols | Egg yolk | 45 | 17.7 | **a**[b], 1 h |
| Tocopherols | *Hordeum vulgare* (barley) | 40 | 23.69 | 1 h, 4.38%, density 0.921 |
| Triglycerides | Rapeseed, soybeans | 20–40 | 25 | **a** |
| Triglycerides | Ground rapeseed, linseed meal | 90 | 34.3 | |
| Triglycerides | Sunflower | 42–80 | 15.2–35.4 | 20% EtOH |

[a]High percentages refer to total content, usually compared to that obtainable by solvent extraction; low percentages refer to total extracted as a percentage of total mass.
[b]Yield comparable to solvent extraction.

Glycolipids and phospholipids are less easily extracted and usually require the addition of methanol as modifier. Sterols and sterol esters are more difficult to extract than triglycerides, but have also been explored. By stepwise increases of pressure, the concentration of phytosterols in soybean extract increased 30 times, from corn fibre by 12 times and from corn bran by 37 times. Methods are available for the analytical determination of sterols in animal skin, meat and fat. SFE is particularly useful for unstable compounds such as $\gamma$-linolenic acid and carotenes. The latter have been extracted from leaf protein concentrates and carrots (see Table 1). **Figure 2** shows a chromatogram of a carotene extract obtained from carrots. The content of tocopherols is greater in the oil of rapeseed and soybeans extracted by SFE than by solvent extraction. Fluorometric and electrochemical detectors with HPLC have been used to follow the extraction.

Lanolin has been extracted from wool using 20% acetone as modifier, giving comparable results to dichloromethane Soxhlet extraction, but it gave a cleaner product (less mineral salts and protein).

## Essential Oils

Essential oils are an obvious target for SFE (**Table 2**), and this is probably the area that has received most attention in recent years, judging by the number of publications. Many compounds in essential oils are either highly unsaturated or subject to thermal or oxidative degradation, so that SFE offers a clear

advantage for them. One complication is that many plant oils have been obtained by steam distillation, which extracts the monoterpenes but leaves most of the sesquiterpenes behind. SFE removes monoterpenes and sesquiterpenes together, so that the composition and odour of the SFE product may be distinctly different. When the price paid for a

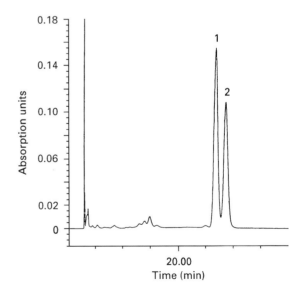

**Figure 2** HPLC chromatogram of the SFE extract of carrots collected at 40°C and 50.5 MPa for 1 h with a flow rate of 600–750 mL min⁻¹ of CO₂. 1, $\alpha$-carotene; 2, $\beta$-carotene. (Reproduced with permission from Chandra A and Nair MG (1997) *Phytochemical Analysis* 8: 244. Copyright John Wiley & Sons Ltd.)

**Table 2** Examples of extraction of essential oils, flavours and fragrances with conditions

| Extracted | Matrix | Temperature (°C) | Pressure (MPa) | Additional information |
|---|---|---|---|---|
| Onion flavour | Allium cepa (onion) | 37 | 24.5 | Flow 0.5 L min$^{-1}$, yield improved by EtOH modifier |
| Organo–sulfur compounds, cepaenes, allicin | Allium tricoccum (ramp) | | | |
| Root oil (118 compounds identified) | Angelica archangelica | 40 | 12.0 | 1 h static, 2 h dynamic |
| Alkylpyrazines | Arachis hypogaea (roasted peanuts) | 50 | 9.6 | Density 0.35, lipids not extracted |
| Carvone, limonene | Carum carvi (caraway seed) | 32 | 12.5 | Time and flow rate affected yield |
| Capsacinoids | Capsicum frutescens (chili), Capsicum annum (paprika) | 80 | | Density 0.75, H$_2$O, yield lower or equal to solvent extraction |
| Essential oil | Cuminum cyminum (cumin seed) | 40 | 10.0 | |
| Curcumin (**3**) | Curcuma longa (turmeric) | 60 | 28.0 | 20% MeOH, 2 mL min$^{-1}$ |
| Neral, geranial, geraniol, nerolic and geranic acids | Cymbopogon citratus (lemongrass) | | | |
| $\beta$-Phellandrene, p-cymene, cryptone, spathulenol and 86 others | Eucalyptus camaldulensis | | | |
| Limonene, fenchone, methylchavicol, anethole | Foeniculum vulgare (fennel) | 31–35 | 8.0–8.4 | 10.0% |
| Anethole | Illicium verum (star anis) | 80 | | Density 0.35, 90% pure |
| Olive oil aroma (hexanal, 2-hexenal, hexanol, 3-hexenol and others) | Olive oil and olives | 40–45 | 7.7–11.5 | Static 1–5 min, dynamic 30 min |
| Essential oil (limonene) | Orange peel | 20–50 | 8–28 | Optimum for limonene (99.5%) 35°, 12.5 MPa; optimum for linalool, 35°, 8.0 MPa |
| Kavain, yangonin, methysticin and derivatives | Piper methysticum (kava) | | | |
| Essential oil | Piper nigrum (black pepper) | 30–50 | 15–30 | |
| Carnosic acid | Rosmarinus officinalis (rosemary) | 37–47 | 10–16 | |
| Eugenol, eugenol acetate, $\alpha$- and $\beta$-caryophyllenes | Syzygium aromaticum (clove buds) | 50 | 24 | |
| Tanshinone IIA | Salvia militiorrhiza (bunge) | 60 | 24.5 | 0–10% MeOH |
| Pyrazines | Theobromo cacao (cocoa beans) | 60 | 20.0 | 2% MeOH, 20 min |

product depends upon odour, or content, this is important. Cedarwood oil obtained by SFE is closer in aroma to the original wood than steam-distilled oil. A disadvantage of SFE is that, at higher pressure, leaf waxes, which are not extracted by steam, are extracted as well. Freshly cut peppermint and spearmint plants extracted with supercritical and subcritical CO$_2$ (temperature 24–43°C) gave oils similar to that from steam distillation. A comparative study of essential oils and wax from lavender showed that the SFE extract contained three times as much linalyl acetate as the steam distillate; presumably the lower content was caused by hydrolysis. In the citrus industry it is reported that removing terpenes from citrus oil avoids oxidation to undesired products, while an

SFE product can be used to re-blend to give new flavour. SFE of rosemary leaves for 10 min gave similar yield to that of 4 h sonication with CH$_2$Cl$_2$ (Table 2). There is interest in the antioxidants obtainable from rosemary and sage, obtainable by a two-stage SFE extraction of the essential oil followed by the antioxidants. In a study of extraction of orange peel, the dried peel should be reduced to 2 mm particles for rapid extraction. For particles of 0.3 mm, 75% of the total oil was extracted with a ratio of 6 kg of CO$_2$ per kg of orange peel. In a study of effect of different modifiers on the SFE of lemongrass oil, GC–mass spectrometry (GC-MS) indicated a different profile of monoterpenes depending upon the modifier.

## Flavours and Fragrances

The subject of flavours and fragrances overlaps with essential oils (Table 2). The mild conditions used for SFE with $CO_2$ can provide an accurate representation of the taste, colour and odour of natural substances found in herbs, spices, beverages and foods. Many studies have been of an analytical nature, to compare products obtained by different processes, to compare plant materials for quality, or to find the essential source of the desirable odour, as for example in the cases of coffee and olive oil. A brewed coffee aroma as similar as possible to the original brewed coffee has been obtained by SFE, monitored by smelling the product. Attempts have been made also to extract the aroma of virgin olive oil, from the oil and the olives, trap it on Tenax and analyse the product by GC-MS. SFE of cocoa beans with GC-MS analysis of the extract was used to assess the effect of storage on quality of the beans. Some alkylpyrazines were reduced after storage. There have been a number of studies of clove oil, none of which have reported a distinctly different yield between SFE and distillation methods. In many reports the yield is slightly lower by SFE. Hops have been the subject of study to produce bitter extracts for the brewing industry. The optimum conditions for an extract for the beer bitterness and aroma have been developed. Extraction of the leaves of hops gave no bitter extract. By adjusting the conditions the flavour of roasted peanuts could be collected without extracting the oil (Table 2).

Extraction of fennel seeds gave a higher yield by SFE (10.0%) than steam distillation (3.0%), about the same as extraction with hexane (10.6%), and less than ethanol extraction (15.4%), but the SFE and distilled products had a much more intense odour and taste than the solvent extracts. Cumin seed oil by SFE contained valuable components which would be thermally degraded by steam distillation. In the case of onion flavour oil SFE and liquid $CO_2$ extracts had the flavour of fresh onions, while the steam distillation solvent-extracted oil had a cooked onion flavour. The flavour of Emmentaler cheese during ripening has been followed by SFE and GC-MS, but further fractionation was needed because of the dominance of fatty acids, in order to analyse the less abundant alcohol, carbonyl and lactone aroma compounds. Guaca or quemadora (*Spilanthes americana*), with a slightly burning and numbing taste, is used in South American cooking. Comparison of steam-istilled and SFE extracts of leaves, stem and flowers showed significant differences (**Figure 3**). Eighty-eight compounds were identified in the extracts.

**2** Safranal

**3** Curcumin

Saffron is such as expensive spice, that it is very liable to fraudulent imitation. The most important aroma compound, safranal (**2**), has been studied by isotope analysis. Synthetic safranal can easily be distinguished from the natural by [13]C-isotope content. SFE gave a cleaner and faster method of obtaining an extract of volatiles than solvent extraction, but there was some isotopic fractionation depending upon extraction yield.

## Medicinal Compounds and Alkaloids

There are many medicinal compounds in plants that are targets for SFE, but many of these are more polar substances and are therefore more difficult to extract efficiently. The conditions must be explored for each example at our present state of knowledge (**Table 3**). Feverfew (*Tanacetum parthenium*) is much in demand as a herbal remedy and other worthless dried plants of similar appearance are frequently sold as feverfew. The value of the plant can be checked by SFE with $CO_2$ for the content of parthenolide (**4**), the active ingredient. Artemisinin (**5**) is an antimalarial present in *Artemisia annua*. Texanes and baccatins have been extracted from ground needles and seeds of *Taxus* spp. with 3% ethanol modifier; ethyl acetate, methanol, dichloromethane and diethyl ether have all been used as modifier for this purpose. In all cases waxy materials are co-extracted, so hexane solvent extraction was used first

**4** Parthenolide

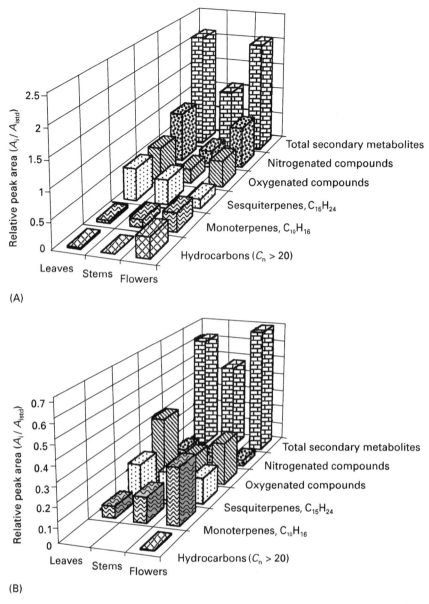

**Figure 3** The composition of extracts obtained from different parts of the plant *Spilanthes americana* by (A) SFE and (B) simultaneous steam distillation–extraction. (Reproduced from Stashenko EE, Puertas MA and Combariza MY (1996) *Journal of Chromatography A* 752: 223 with permission from Elsevier Science.)

**5** Artemisinin

to remove the waxes and then SFE was applied for the precursors of paclitaxel (taxol). Ginkgolides and bilobalides are extractable from *Ginkgo biloba* with 10% methanol added. There is a report of SFE of medicinal plants being directly coupled to a uterotonic bioassay on abdominal muscle to discover possible substances to induce uterine contraction.

The bark of *Magnolia officinalis* is used in Chinese medicine for a number of purposes. A major active compound is magnolol (**6**), a neolignin. SFE was compared with solvent extraction with phytosols, a series of new nonchlorinated fluorocarbon

**Table 3** Examples of extraction of medicinal compounds, alkaloids and polar compounds with conditions used

| Extracted | Matrix | Temperature (°C) | Pressure (MPa) | Additional information |
|---|---|---|---|---|
| Atractylon (**8**) | *Atractyloides* rhizomes | 40 | 10 | 2 mL min$^{-1}$, 20 s |
| Bile acids | Bovine bile | 70 | 22 | 15% MeOH, 20 min, 88% recovery |
| Cedrelone (**9**) | *Cedrela toona* | 40 | 40.0 | 30 min static, 40 min dynamic, 0.6 g sample wetted with 40 μL MeOH |
| Pyrethrins | *Chrysanthemum cinerariaefolium* | 40 | 8.3 | Most extracted in 3 h |
| Uterine contractants | *Clivia miniata, Ekebergia capensis, Grewia occidentalis, Asclepias fruticosa* | | 20–40 | |
| Podophyllotoxin (**6**) | *Dysosma pleiantha* roots | 40–80 | 13.6–34.0 | With MeOH added, yield 95% |
| Phenols | Olive leaves | 100 | 33.4 | $CO_2$ density 0.70, 10% MeOH, total 2 mL min$^{-1}$, 140 min |
| Glycosylated flavonoids | *Passiflora edulis* (passion fruit leaves) | 75 | 10.1 | 15% MeOH, 5 min, 1.75% |
| Schisandrols, schisandrins (lignans) | *Schisandra chinensis* | 40–80 | 13.6–34.0 | 80% of that from MeOH extraction |
| Flavonoids | *Scutellariae radix* | 50 | 20.0 | $CO_2$–MeOH–$H_2O$ 20:2:0.9 |
| Isoflavones | Soybean products | 50 | 60 | 20% EtOH, 1 h, 93% |
| Theobromine, caffeine, cocoa butter | *Theobromo cacao* (cocoa) | 40–90 | 8.0–30 | EtOH modifier |

solvents of varying polarity. SFE with 10% added methanol gave the highest yield of magnolol (1.86%), and phytosol A gave the lowest (0.78%). Digoxin can be obtained from *Digitalis lanata* leaves by SFE, but the process is not very selective and various strategies have been tried to improve selectivity, including use of trifluoromethane and tetrafluoroethane as extractives, but no one alternative had a clear advantage.

The antiviral compound podophyllotoxin (**7**) from *Dysosma pleiantha* and atractylon (**8**) from the oriental drug *Atractyloides* rhizome illustrate further the types of separations that have achieved. Atractylon, an oxidatively unstable compound, was extracted analytically in 20 s with a recovery 30% higher than by solvent extraction. The phototoxic furocoumarins (psoralen and derivatives) were extracted analytically from the vegetable celariac (*Apium graveolens*) by SFE, Soxhlet extraction with ethanol and sonication

with chloroform. SFE gave higher extractions (**Figure 4**).

## Polar Compounds

The wide range of polar compounds of interest as natural products is a greater challenge to the power of

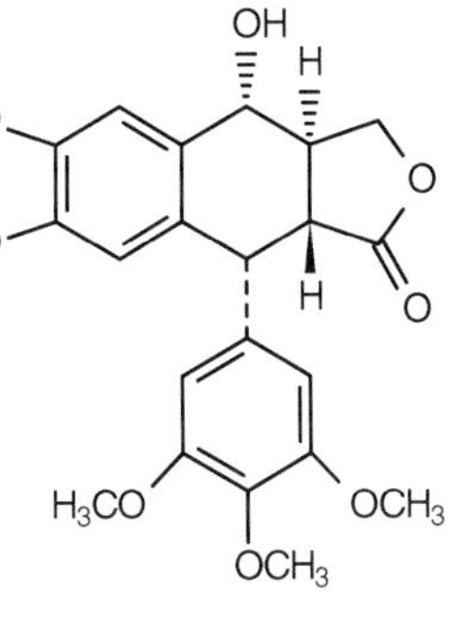

**7** Podophyllotoxin

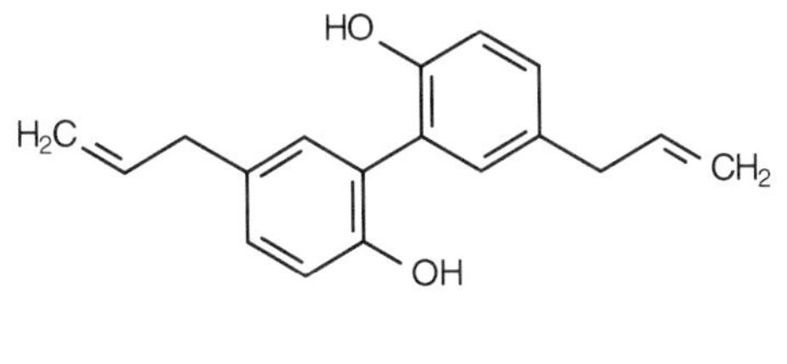

**6** Magnolol

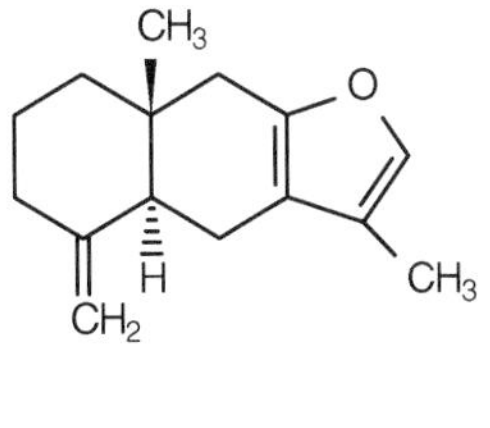

**8** Atractylon

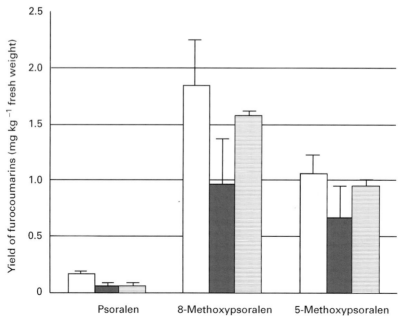

**9** Cedrolone

SFE. The success obtained varies considerably, and our understanding of the physical process of diffusion, cell wall penetration, rate of dissolution and solubility are too rudimentary to make predictions. Attempts to extract the tetranortrieterpenoid azadirachtin ($C_{35}H_{44}O_{16}$) from neem (*Azadirachta indica*) seeds with or without added methanol did not give as high a yield as methanol solvent extraction. Attempts to remove triglyceride oil from the seeds first by selecting extraction conditions were not successful. Trichothecene mycotoxins at p.p.m. levels in wheat have been extracted for analytical purposes and determined online by chemical ionization–mass spectrometry (CI-MS).

McHugh and Krukonis have discussed the decaffeination of coffee extracts. The removal of nicotine from tobacco and snuff has also been achieved, chiefly for analysis. Some lignans and coumarins with lower numbers of hydroxyl groups can be extracted. Hydroxypinoresorcinol was extracted from *Fraxinus japonica* and other *Fraxinus* species using $CO_2$ with water modifier. The extraction of lignans by SFE from *Forsythia* species was as good as solvent extraction with refluxing hexane or ethanol. SFE gave better extraction of flavanones and xanthones from the root bark of the Osage orange (*Maclura pomifera*), provided 20% methanol was used as modifier, than liquid extraction, and in much shorter time. An unusual example is the determination of the alkaloids berberine and palmatine in *Phellodendri cortex* using ion pair SFE. The ion-pairing agent was dioctyl sodium sulfosuccinate, with 10% methanol as the modifier. The extraction required only 10 min. Microcystins, toxic peptides, have been extracted from cyanobacteria with a ternary mixture of 90% $CO_2$, 9.5% acetic acid and 0.5% water. Even the particulate material from hardwood smoke has been examined (the principal substances identified by GC-MS were guiacol and syringol derivatives).

Where cost is mentioned, SFE is in many cases admitted to cost more at present than Soxhlet extraction, but some studies suggest that costs can be reduced with proper experiment design, even for poorly soluble natural products.

## Looking Ahead

SFE is firmly established as an industrial process for the isolation of a small number of natural products. Patents exist for a larger number of examples. As new industrial plant is bought into operation, and old

**Figure 4** Comparison of extraction of three phototoxic furocoumarins from celeriac (*Apium graveolens*) by SFE (open columns), Soxhlet extraction (ethanol: filled columns) and sonication with chloroform (hatched columns). (Reproduced with permission from Järvenpää EP, Jestoi MN and Huopalahti R (1997) *Phytochemical Analysis* 8: 250. Copyright John Wiley & Sons Ltd.)

equipment removed, the number of examples will increase. Essential oils are going that way quickly. For polar substances, the published work shows that each material has to be examined to find the best conditions. It is difficult to assess how many natural products are being routinely extracted in this way, but the steady output of papers indicates continuing interest. As the chlorinated solvents are withdrawn, we can expect to see the use of SFE increase strongly. There is a ready market for some cheaper and less elaborate extraction equipment to meet this growing demand.

*See also:* **II / Chromatography: Supercritical Fluid: Theory of Supercritical Fluid Chromatography. Extraction:** Supercritical Fluid Extraction. **III / Supercritical Fluid Extraction-Supercritical Fluid Chromatography.**

## Further Reading

Albert K (1997) Supercritical fluid chromatography-proton nuclear magnetic resonance spectroscopy coupling. *Journal of Chromatography A* 785: 65.

Bevan CD and Marshall PS (1994) The use of supercritical fluids in the isolation of natural products. *Natural Products Reports* 11: 451.

Jarvis AP and Morgan ED (1997) Isolation of plant products by supercritical-fluid extraction. *Phytochemical Analysis* 8: 217. [The whole of *Phytochemical Analysis* 1997; 8(5) is devoted to papers on supercritical fluid extraction.]

Kalampoukas G and Dervakos GA (1996) Process optimization for clean manufacturing: supercritical fluid extraction for β-carotene. *Computers and Chemical Engineering* 20: S1383.

McHugh MA and Krukonis VJ (1986) *Supercritical Fluid Extraction.* London: Butterworth.

Modey WK, Mulholland DA and Raynor MW (1996) Analytical supercritical fluid extraction of natural products. *Phytochemical Analysis* 7: 1.

Reverchon E (1997) Supercritical fluid extraction and fractionation of essential oils and related products. *Journal of Supercritical Fluids* 10: 1.

Smith RM and Hawthorne SB (1997) Supercritical fluids in chromatography and extraction. Special volume of *Journal of Chromatography A* 785.

# Thin-Layer (Planar) Chromatography

**J. Pothier**, University of Tours, Tours, France

## Introduction

The use of thin-layer chromatography (TLC) for the analysis of plant extracts began in the 1960s, with the work of Stahl and Randerath. The continuing use, and further development of TLC in plant analysis is justified because of its rapidity and because of the availability of a number of different sorbents. TLC is also useful in plant analysis because it is possible to work on crude extracts, which is not the case with other analytical methods. In plant analysis, derivatization methods serve to increase sensitivity and selectivity in addition to providing evidence concerning the quality of the separation. A selection of the most important derivatization/detection methods are given in **Table 1**. As well as these advantages, TLC is very economical and can be employed for routine use because the consumption of solvent is very low, and it is possible to analyse numerous samples on the same plate. The literature on plant analysis by TLC is very extensive, with the most important contributions being those of Stahl, Randerath, Fried and Sherma, Harbone, and Wagner, who is our main reference on this topic. Most TLC studies are listed in the different pharmacopoeias of the world, as reported by Wagner. The classes of plant compounds separated by TLC covered in this article are as follows:

- Alkaloids
- Glycosides: flavonoids, coumarins, anthocyanins, ginkgolides, anthraquinone glycosides, cardiac glycosides
- Saponins
- Essential oils
- Cannabinoids
- Valepotriates
- Bitter principles

## Alkaloids

Most plant alkaloids are tertiary amines; others contain primary, secondary, or quaternary nitrogen (**Figures 1–4**). The basicity of individual alkaloids varies considerably; depending on which of the four types is represented. The $pK_b$ values lie in a range 10–12 for weak bases like purines to 3–7 for stronger bases like opium alkaloids. These factors must be borne in mind for extraction, and also for derivatization. The sample sizes applied to the TLC plate must be calculated, according to the average alkaloid content of the specific extracts. The majority of workers

**Table 1** Detection methods and spray reagents

| | | |
|---|---|---|
| Acetic anhydride | 10 mL acetic anhydride, heated at 150°C for about 30 min and then inspected in UV light (365 nm) | Ginkgolides |
| Anisaldehyde–sulfuric acid | 0.5 mL anisaldehyde with 10 mL glacial acetic acid, 85 mL methanol and 5 mL concentrated sulfuric acid, in that order. Spray and heat at 100°C for 5–10 min. Then evaluated in visible light or UV 365 nm; conservation limited, not usable when the reagent is red–violet | Terpenoids, propyl-propanoids, saponins, anthocyanins |
| Antimony(III) chloride (SbCl₃) | 20% solution of antimony chloride in chloroform, or ethanol sprayed and then heated for 5–6 min at 110°C | Cardiac glycosides, saponins |
| Chloramine–trichloroacetic acid | 10 mL freshly prepared 3% aqueous chloramine T solution with 40 mL 25% ethanolic trichloroacetic acid, sprayed, then heated at 100°C for 5–10 min then evaluated in UV light (365 nm) | Cardiac glycosides |
| Dinitrophenylhydrazine | 0.1 g, 2,4-dinitrophenylhydrazine in 100 mL methanol, followed by addition of 1 mL 36% hydrochloric acid; evaluation immediately in visible light | Ketones, aldehydes |
| Dragendorff (Munier–Macheboeuf) reagent | Solution A: 0.85 g basic bismuth nitrate in 10 mL glacial acetic acid and 40 mL water under heating. Solution B: 8 g potassium iodide in 30 mL water. Stock solutions A + B are mixed | Alkaloids |
| Dragendorff, followed by sodium nitrite | After spraying with Dragendorff, the plate is sprayed with 10% aqueous sodium nitrite. The coloured zones are brown | Alkaloids |
| Dragendorff (with hydrochloric acid) | 5 g bismuth carbonate in 50 mL H₂O, then 10 mL hydrochloric acid and add 25 g potassium iodide, complete with water to 100 mL. The spray reagent is obtained by dilution of 1 mL in 25 mL HCN | Alkaloids, purines (caffeine, theobromine, theophylline) |
| Fast blue | Fast blue salt B in 100 mL. Spray then look in visible light. A second solution can be sprayed using 10% ethanolic acid, followed by inspection in visible light | Cannabinoids |
| Iodine | About 10 g solid iodine is spread in a chromatographic tank; the plate is placed into the tank and exposed to iodine vapour, yellow–brown zones are detected in visible light | Compounds with conjugated double bonds |
| Iodine–chloroform | 0.5 g iodine in 100 mL chloroform; after spraying, the plate is warmed at 70°C during 5 min, the plate is evaluated after 20 min in visible or UV light (365 nm) | Ipecacuanha alkaloids |
| Iodine–hydrochloric acid | Solution A: 1 g potassium iodide and 1 g iodine in 100 mL ethanol. Solution B: 25 mL 25% HCl with 25 mL ethanol. Spray the plate with 5 mL of A followed by 5 mL of B | Purines |
| Iodoplatinate | 0.3 g hydrogen hexachloroplatinate hydrate in 100 mL water with 100 mL 6% potassium iodide solution | Alkaloids (blue–violet) |
| Iron(III) chloride (FeCl₃) | 10% in aqueous solution evaluation in visible light | Polyphenols |
| Kedde reagent | 5 mL freshly prepared 3% ethanolic 3,5-dinitrobenzoic acid with 5 mL 2 mol L⁻¹ NaOH | Cardenolides |
| Liebermann reagent | 5 mL acetic anhydride and 5 mL concentrated sulfuric acid is added carefully to 50 mL absolute ethanol, while cooling in ice. This agent must be freshly prepared. The plate is warmed at 100°C, 5–10 min and then inspected in UV light (365 nm) | Triterpenes, steroids (saponins) |
| Marquis reagent | 3 mL formaldehyde in 100 mL concentrated sulfuric acid; evaluation in visible light | Morphine, codeine, thebaine |
| Neu (NP/PEG) | 1% methanolic diphenylboric acid β-ethylaminoester (diphenylboryloxyethylamine, NP) followed by 5% ethanolic polyethylene glycol-4000 (PEG) | Flavonoids, anthocyanins |

**Table 1** *Continued*

| | | |
|---|---|---|
| Nitric acid (HNO₃ concentrate) for alkaloids | After spraying the plate is heated 15 min at 120°C | Ajmaline, brucine |
| Nitric acid (HNO₃) + KOH | After spraying, the plate is heated 15 min at 120°C. Then sprayed with 10% ethanolic KOH reagent. Red brown in visible light yellow–brown, brown fluorescence with UV light (365 nm) | Anthracenosides, sennosides |
| Phosphomolybdic acid reagent | 20% ethanolic solution of phosphomolybdic acid spraying then heating at 100°C for 5 min | Essential oils |
| Potassium hydroxide (KOH) (Borntraeger reagent) | 5% or 10% ethanolic potassium hydroxide. In visible light, anthraquinones coloured red; anthrones yellow in UV light (365 nm). Coumarins blue in UV light (365 nm) | Anthracenosides, coumarins |
| Vanillin–hydrochloric acid | 1% in ethanol followed by 3 mL concentrated HCl. In visible light, heating 5 min at 100°C intensifies colours | Essential oils |
| Vanillin–phosphoric acid | 1 g vanillin in 100 mL of 50% phosphoric acid. Heat 10–20 min at 120°C | Essential oils |
| Vanillin–sulfuric acid | 1 g vanillin in 100 mL ethanol then add 2 mL of concentrated $H_2SO_4$. The plate is sprayed and heated at 100°C (10 min) | Essential oils |
| Van Urk reagent | 0.2 g 4-dimethylaminobenzaldehyde in a cooled mixture of 35 mL water and 65 mL concentrated sulfuric acid. Then add 0.15 mL of a 10% aqueous iron(III) chloride solution | Indolic alkaloids, ergot alkaloids |

use silica gel 60 F254 precoated TLC plates but aluminium oxide is also suitable. Many mobile phase systems contain chloroform, the eluting power of which may be decreased by addition of cyclohexane or increased by acetone, ethanol or methanol. The mobile phase is often made alkaline by the addition of ammonia or diethylamine to the less polar solvents but diethylamine is not easy to remove before spraying. The most commonly employed eluents are chloroform–methanol (90 : 10) and chloroform–diethylamine (90 : 10). It is also possible to use a screening system, suitable for the major alkaloids of most drugs employing a solvent mixture of toluene–ethyl acetate–diethylamine (70 : 20 : 10).

The detection of alkaloids is possible by quenching UV light at 254 nm and at 365 nm and by two main spray reagents Dragendorff and iodoplatinate (see Table 1). The chromatographic systems used for plant alkaloids and also the derivatization techniques used for identification are given in **Table 2**.

## Glycosides

Glycosides are compounds that yield one or more sugars upon hydrolysis. Among the products of hydrolysis, the non-sugar components of the glycosides are known as aglycones. The classification of glycosides is a difficult matter and here the therapeutic use has been chosen.

### Flavonoids

The flavonol glycosides and their aglycones are generally termed flavonoids. A large number of differ-

ent flavonoids are known to occur in nature, and these yellow pigments are widely distributed throughout the higher plants. The main constituents of flavonoid drugs are 2-phenyl-γ-benzopyrones. The various structural types of flavonoids differ in the degree of oxidation of the C ring. Most of these compounds are present in drugs as monoglycosides or diglycosides. It is possible to classify flavonoids into flavonols, flavones, flavanons, flavanols and flavanolignans in relation to substituents and the presence of double bonds (**Figure 5**).

There are numerous plants containing flavonoids and there are also numerous flavonoids and so here only the main plants and compounds are cited in **Table 3**.

Before TLC is possible, extracts of the plant must be made, and a general method for the extraction of flavonoids is as follows: 1 g of the powered plant material is extracted with 10 mL methanol for 5 min at 60°C and then filtered. The methanolic extract is then analysed by TLC. When the plant contains lipids it is often necessary to use hexane to defat the powder prior to methanolic extraction and TLC. It is also worth noting that when analysing flavonoids, it is often better to examine the aglycones present in hydrolysed plant extracts. Having obtained a suitable extract, it is then possible to separate the various components using TLC and a variety of chromatographic systems have been devised for this purpose.

It is possible to screen flavonoids on silica gel TLC plates with the following solvent: ethyl acetate–formic acid–glacial acetic acid–water (100 : 11 : 11 : 26).

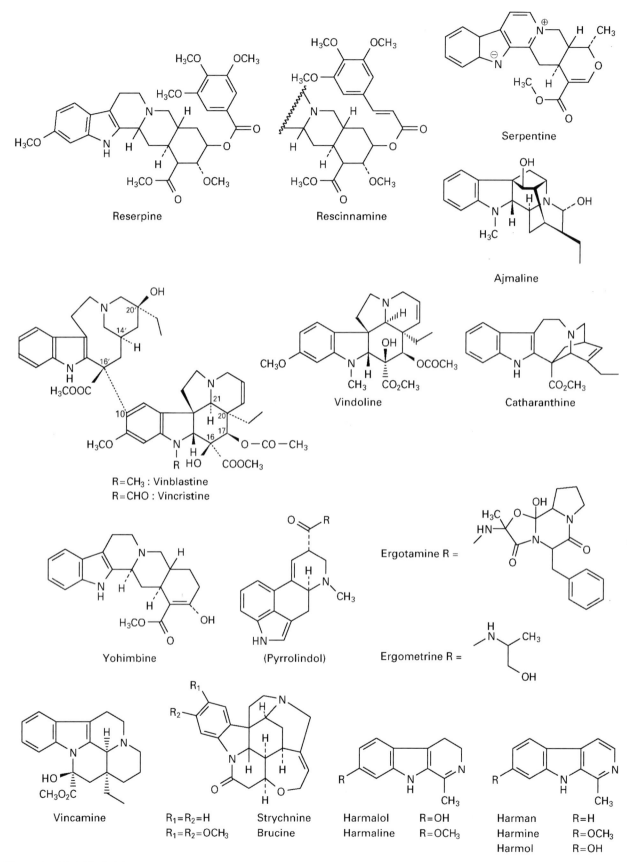

**Figure 1**    Alkaloids: indoles.

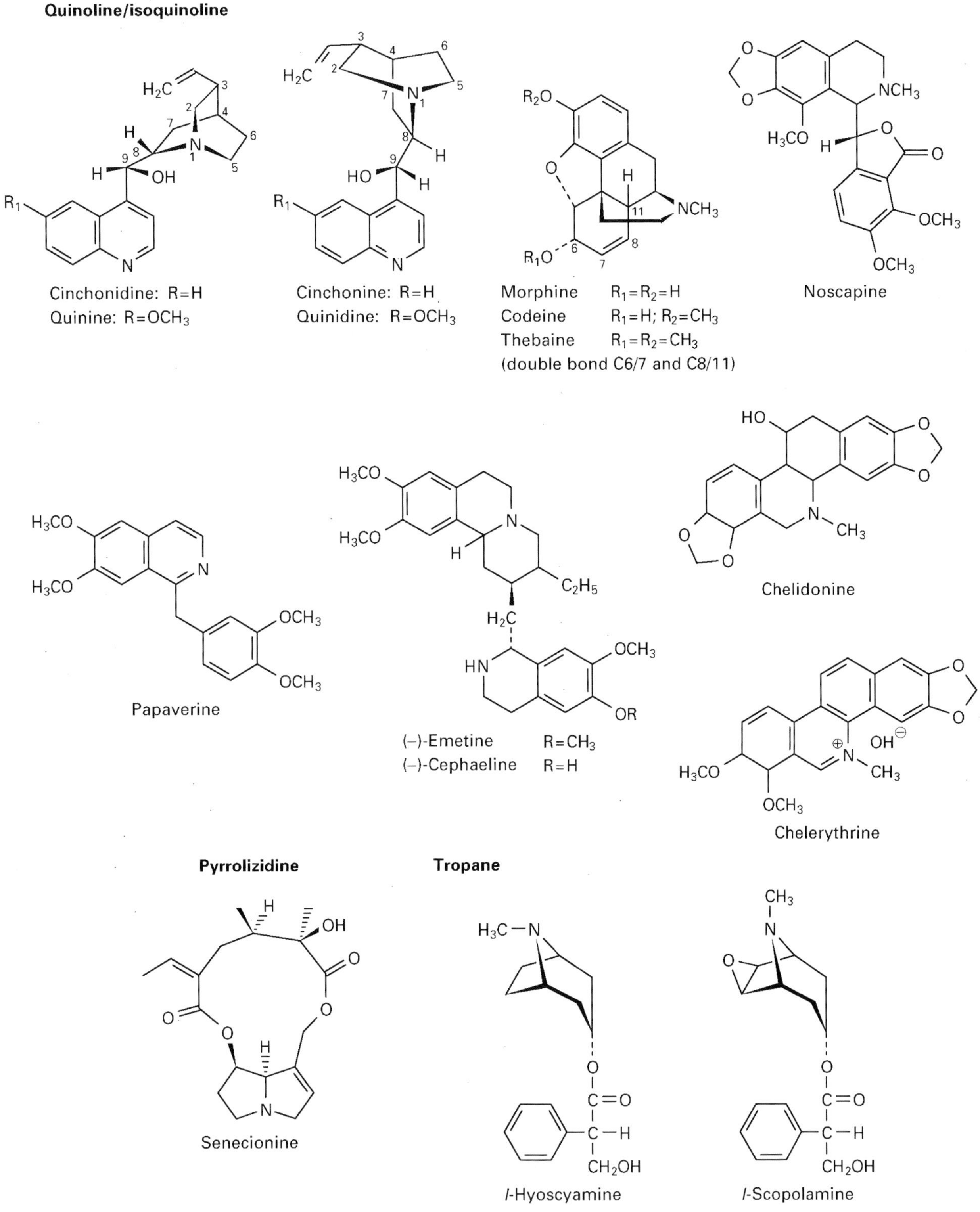

## Quinoline/isoquinoline

Cinchonidine: R=H
Quinine: R=OCH₃

Cinchonine: R=H
Quinidine: R=OCH₃

Morphine    $R_1=R_2=H$
Codeine     $R_1=H; R_2=CH_3$
Thebaine    $R_1=R_2=CH_3$
(double bond C6/7 and C8/11)

Noscapine

Papaverine

(−)-Emetine      R=CH₃
(−)-Cephaeline   R=H

Chelidonine

Chelerythrine

**Pyrrolizidine**

**Tropane**

Senecionine

*l*-Hyoscyamine

*l*-Scopolamine

**Figure 2**   Alkaloids: quinoline/isoquinoline, pyrrolizidine and tropane.

With addition of methyl ethyl ketone (MEK) to give the solvent ethyl acetate–formic acid–glacial acetic acid–MEK–water (50 : 7 : 3 : 30 : 10) it is possible to separate rutin and vitexine-2-O-rhamnoside. Numer-

ous other eluents are also used, such as chloroform–acetone–formic acid (75 : 16.5 : 8.5) for flavanolignans of milk thistle (*Silybum marianum*) and amentoflavone from black haw (*Viburnum*

**Tropolone**

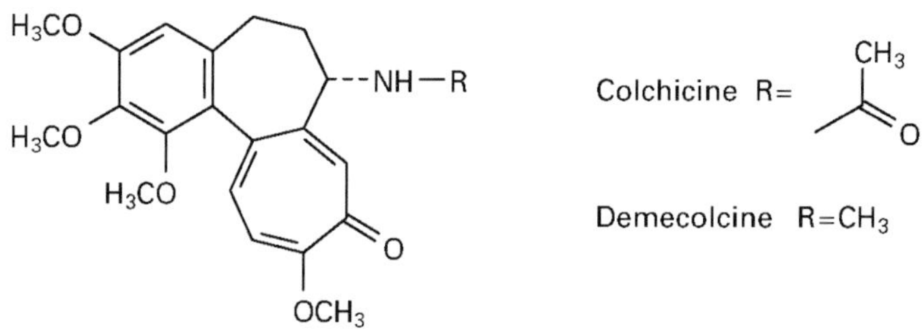
H3CO
H3CO
H3CO
NH—R
O
OCH3
Colchicine R=
CH3
O
Demecolcine R=CH3

**Quinolizidine**

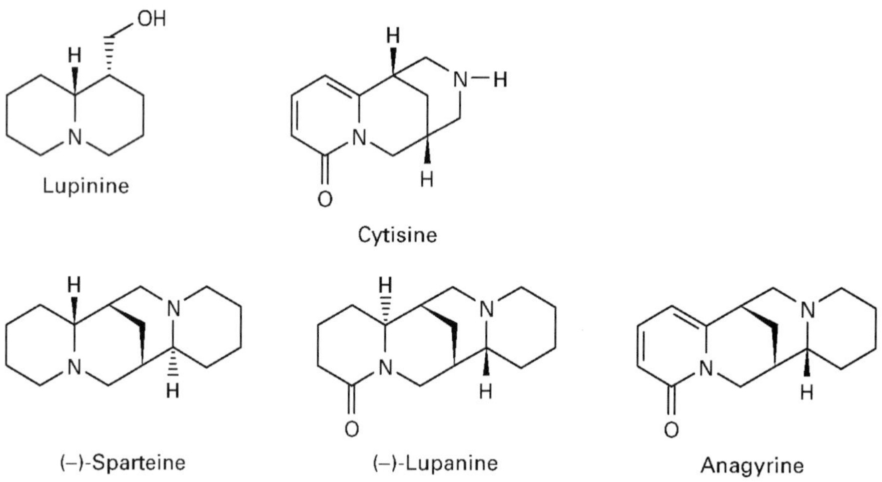
OH
H
N
Lupinine
H
N—H
N
H
O
Cytisine
(−)-Sparteine
(−)-Lupanine
Anagyrine

**Purines**

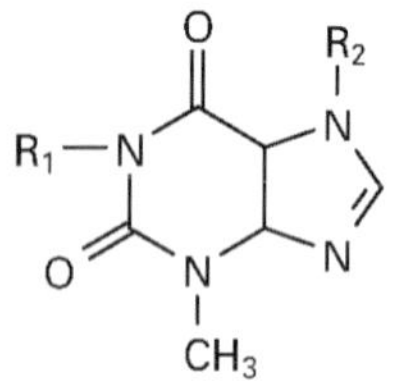
O
R2
R1—N
N
O
N
N
CH3

| | $R_1$ | $R_2$ |
|---|---|---|
| Caffeine | $CH_3$ | $CH_3$ |
| Theobromine | H | $CH_3$ |
| Theophylline | $CH_3$ | H |

**Figure 3** Alkaloids: tropolone, quinolizidine and purines.

*prunifolium*). Chloroform–ethyl acetate (60 : 40) has been used for the flavonoid aglycones of orthosiphon (*Orthosiphon aristatus*).

**Flavonoid aglycones** In addition, the following eluents can be used to separate the aglycones of flavonoids: benzene–pyridine–formic acid (72 : 18 : 10) and toluene–ethyl formate–formic acid (50 : 40 : 10). Toluene–dioxane–glacial acetic acid (90 : 25 : 4) and a further list of suitable sorbents and solvent systems for flavonoids and their aglycones is given in **Table 4**.

Flavonoids can be detected on TLC plates containing a fluorescent indicator because they cause fluorescence quenching when irradiated with UV light at 254 nm, or 365 nm depending on the structural type. Flavonoids also show dark yellow, green or blue fluorescence, which is intensified and changed by the use of various spray reagents. With the spray reagent diphenylboryloxyethanolamine/polyethylene glycol (NP/PEG), flavonoids and biflavonoids give yellow–orange and green fluorescence when irradiated at 365 nm. Acetic acid reagent gives various blue fluorescent zones after heating.

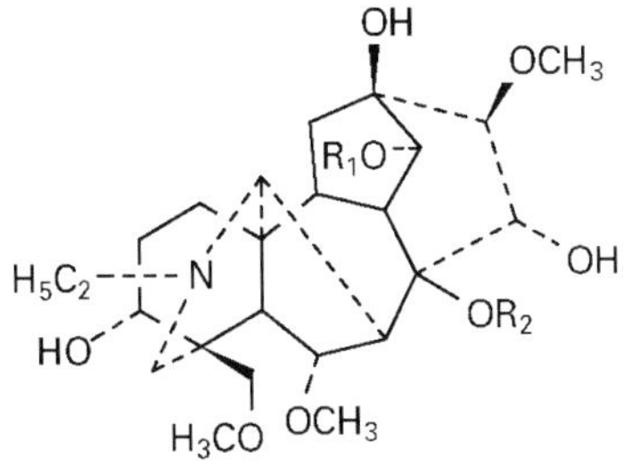

| R₁ | R₂ | |
|---|---|---|
| COC₆H₅ | COCH₃ | Aconitine |
| COC₆H₅ | H | Benzoylaconin |
| H | H | Aconin |

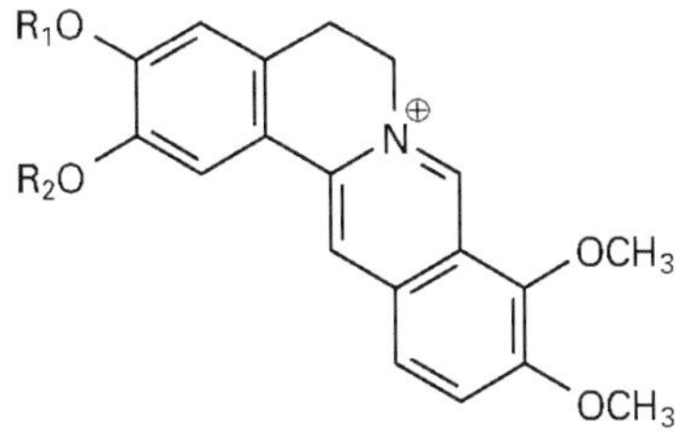

| R₁ | R₂ | |
|---|---|---|
| H | CH₃ | Jatrorrhizine |
| CH₃ | H | Columbamine |
| CH₃ | CH₃ | Palmatine |
| —CH₂— | | Berberine |

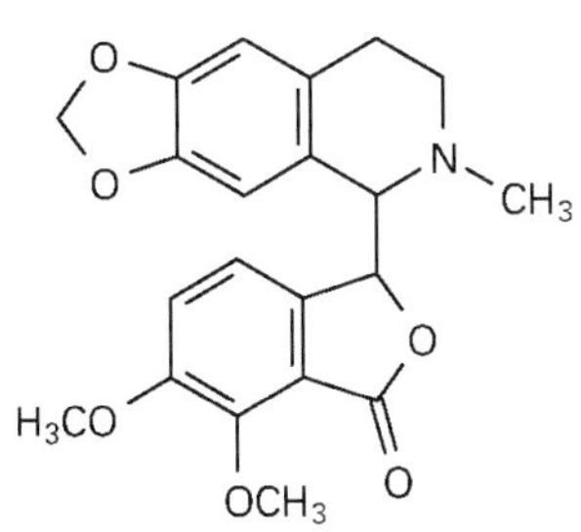

Hydrastine

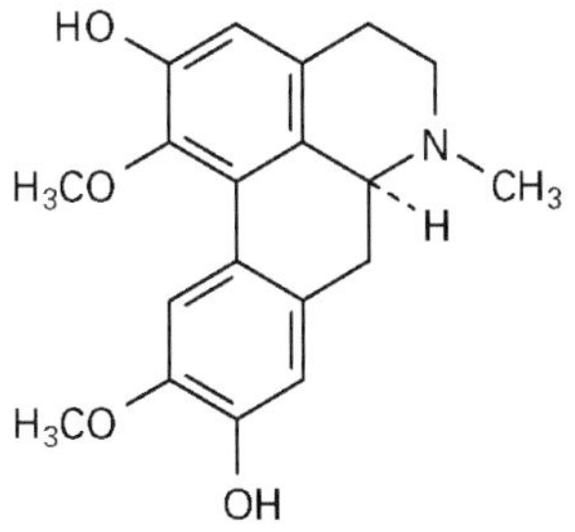

Boldine

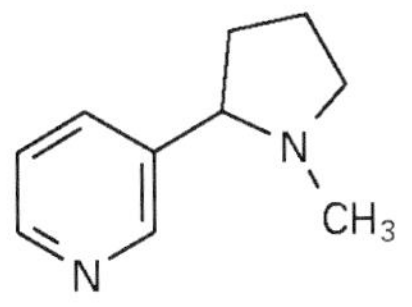

Nicotine

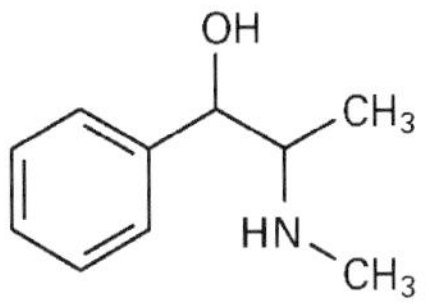

Ephedrine

**Figure 4**  Miscellaneous alkaloids.

## Coumarins

Coumarins are derivatives of benzo-α-pyrones (**Figure 6**).

- *Simple coumarins* consist of coumarin and compounds substituted with OH in umbelliferon, or OCH₃ in scopoletin, and both present in tonka beans (*Coumarouna odorata*), woodruff (*Asperula odorata*) and melilot (*Melilotus officinalis*) or in position C6- and C7-scopoletin which is common in Solanaceae C5 and C8 like fraxin, isofraxidin, fraxetin three compounds of ash bark (*Fraxinus excelsior*).

- *Complex coumarins* belong to two families of plants, the Solanaceae and especially the Apiaceae. C-prenylated coumarins like umbelliprenin are found in angelica root (*Angelica archangelica*). The furanocoumarins possess a furan ring fused at C6 and C7 like psoralen from rue (*Ruta graveolens*), imperatorin, bergapten in ammi (*Ammi majus*), angelica and burnet root (*Pimpinella major*), or a furan ring fused at C7–C8 like angelicin in angelica. Pyranocoumarins have an additional pyran ring at C7–C8, for example visnadin, samidin from ammi (Figure 6).

**Table 2**   Alkaloids

| Group | Eluents | Reagents | Compounds |
|---|---|---|---|
| **Quinoline/isoquinoline** | | | |
| Cinchona (bark): *Cinchona* sp., *Cinchona succirubra*, *Cinchona ledgeriana* | Chloroform–acetone–methanol–ammonia (60 : 20 : 20 : 10) Chloroform diethylamine (90 : 10) Toluene–ethyl acetate–diethylamine (70 : 20 : 10) | 10% ethanol $H_2SO_4$, then UV 365 nm 10% $H_2SO_4$ or 10% HCOOH, then iodoplatinate | Quinine, quinidine, strong fluorescence Quinine, quinidine cinchonine, cinchonidine main alkaloids, dihydrocompounds and epiquinine basis give coloration; pink to blue with iodoplatinate |
| Opium (*Opium*) | Toluene–acetone–ethanol–ammonia (45 : 45 : 7 : 3) Cyclohexane–ethylenediamine (80 : 20) Chloroform–acetone–diethylamine (50 : 40 : 10) Chloroform–methanol (90 : 10) Toluene–ethyl acetate–diethylamine (70 : 20 : 10) | Dragendorff and $NaNO_2$ Iodoplatinate Marquis reagent NP/PEG, then UV 365 nm | Morphine, codeine, noscapine, papaverine thebaine; all major alkaloids give orange–brown coloration Pink coloration with papaverine, noscapine, thebaine Blue with morphine and codeine Violet for codeine and morphine Except codeine, the main alkaloids give a blue fluorescence |
| Ipeac (*Cephaelis ipecacuanha*) | Toluene–ethyl acetate–diethylamine (70 : 20 : 10) | UV 365 nm Iodine reagent Dragendorff | Cepheline, emetine, fluorescence light blue Cepheline, bright blue emetine yellow–white The major alkaloids give orange–brown coloration |
| Celandine (*Chelidonium majus*) | Propanol–water–formic acid (90 : 9 : 1) | UV 365 nm Dragendorff | Bright yellow fluorescence of coptisine, sanguinarine; weak yellow–green for chelidonine, chelerytrine Brown. Not stable with main alkaloids |
| **Pyrrolizidine** | | | |
| Golden senecio (*Senecio vulgaris*) | Chloroform–methanol–ammonia–pentane (82 : 14 : 2.6 : 20) Acetone–methanol–ammonia (40 : 30 : 20) | UV 254 nm | Senecionine, Senecionine N-oxide, agmatine |
| **Tropane** | | | |
| Belladonna (*Atropa belladonna*), Thorn apple (*Datura stramonium*), Henbane (*Hyoscyamus niger*) | Toluene–ethyl acetate–diethylamine (70 : 20 : 10) Acetone–water–ammonia (90 : 7 : 3) | Dragendorff Iodoplatinate | Scopolamine, ( − )-hyoscyamine or atropine |
| **Tropolone** | | | |
| Meadow saffron (*Colchicum autumnale*) | Chloroform–methanol (95 : 5) Benzene–ethyl acetate–diethylamine–methanol–water (15 : 12 : 3 : 6 : 12) Chloroform–acetone–diethylamine (80 : 10 : 10) | UV light (254 nm) Dragendorff, 10% ethanol Hydrochloric acid gives a yellow coloration | Colchicine, demecolcine, 3-demethylcolchicine |

**Table 2** *Continued*

| Group | Eluents | Reagents | Compounds |
|---|---|---|---|
| **Indole alkaloids** | | | |
| Rauwolfia: *Rauwolfia* sp., *Rauwolfia vomitoria*, *Rauwolfia serpentina* | Toluene–ethyl acetate–diethylamine (70 : 20 : 10) | UV 254 nm | Ajmaline: prominent quenching |
| | Heptane–methyl ethyl ketone–methanol (53 : 34 : 8) | Dragendorff (orange–brown) | All alkaloids, ajmaline, serpentine, rescinnamine, rauwolcine give orange colours |
| | Cyclohexane–diethyl ether (60 : 40) | Nitric acid | Ajmaline give red colour |
| Catharanthus leaves (*Catharanthus* sp.) | Ethyl acetate–ethanol–benzene–ammonia (100 : 5 : 5 : 3) Chloroform–methanol (90 : 10) two dimensional: direction 1, ethyl acetate–methanol (80 : 20); direction 2, dichloromethane–methanol (12 : 1) | Dragendorff | Vinblastine, vincristine, vindoline Catharanthine, and other minor alkaloids give a brown coloration |
| Yohimbe bark (*Pausinystalia yohimbe*) | Toluene–ethyl acetate–diethylamine (70 : 20 : 10) | UV 365 nm | Yohimbine (blue fluorescence) |
| | | Dragendorff | Yohimbine, pseudoyohimbine, coryantheine (orange zones) |
| Ergot (*Claviceps purpurea*) | Toluene–ethyl acetate–diethylamine (70 : 20 : 10) Toluene–chloroform–ethanol (28.5 : 57 : 14.5) | Van Urk | Ergocristine, ergotamine, ergometrine give blue zone |
| Common periwinkle leaves (*Vinca minor*) | Ethyl acetate–methanol (90 : 10) | UV 254 nm | Vincamine, vincaminine, vincamajine, vincine give blue–green fluorescence |
| | | Dragendorff | Weak brown, with major alkaloids |
| Nux vomica (*Strychnos nux vomica*) Ignatius beans (*Strychnos ignatii*) | Toluene–ethyl acetate–diethylamine (70 : 20 : 10) | UV 254 nm | Strychnine, brucine give strong blue fluorescence |
| | | Dragendorff | Brown for brucine, strychnine and minor orange–brown zones for pseudostrychnine, and $\alpha,\beta$-colubrines for nux vomica |
| | | Iodoplatinate | Blue zones with brucine and strychnine |
| | | Nitric acid | Brucine give a red colour |
| Syrian rue (*Peganum harmana*) | Chloroform–acetone–diethylamine (50 : 40 : 10) Chloroform–methanol–ammonia 10% (80 : 40 : 1.5) | UV 365 nm | Harmanol, harmaline, harmine, harmane, harmon, give a strong blue fluorescence |
| **Miscellaneous alkaloids** | | | |
| Aconite (*Aconitum napellus*) | Ether–chloroform–ammonia (25 : 10 : 1) Cyclohexane–ethyl acetate–ethylenediamine (80 : 10 : 10) Hexane–chloroform (60 : 40) Chloroform–methanol (80 : 20) | UV 254 nm Dragendorff NaNO$_2$ | Aconitine, mesaconitine, hypoaconitine, give orange colours |

**Table 2** *Continued*

| Group | Eluents | Reagents | Compounds |
|---|---|---|---|
| Barberry (*Berberis vulgaris*) | n-Butanol–formic acid–water (90 : 1 : 9) | Dragendorff | Berberine, protoberberine, jateorrhizine, palmitine give orange colours |
| | n-Butanol–ethyl acetate–formic acid–water (30 : 50 : 10 : 10) | Without treatment | Berberine, yellow in visible light |
| Hydrastis (*Hydrastis canadensis*) | (See barberry, *Berberis vulgaris*) | UV: hydrastine, blue–white fluorescence with Dragendorff | Berberine, hydrastine |
| Boldo (*Peumus boldus*) | Toluene–ethyl acetate (93 : 7) | Dragendorff | Aporphinic alkaloids, boldine |
| Tobacco (*Nicotiana tabacum*) | Toluene–ethyl acetate–diethylamine (70 : 20 : 10) | UV for nicotine, Dragendorff | Nicotine, nornicotine, anabasine, give red–orange colours |
| Desert tea (ma huang) (*Ephedra* sp.) | Toluene–chloroform–ethanol (28.5 : 47 : 14.5) | Ninhydrin (violet–red for ephedrine) | Ephedrine, norephedrine, pseudoephedrine give orange colours |
| **Quinolizidine** | | | |
| Lupines (*Lupinus* sp.) Broom tops (*Sarothamnus* sp.) | Chloroform–methanol ammonia (85 : 14 : 7) Chloroform–methanol (80 : 20) Cyclohexane–diethylamine (70 : 30) Toluene–acetone–ethanol–ammonia (30 : 40 : 12 : 4) Cyclohexane–dichloromethane–diethylamine (40 : 40 : 20) | UV and iodine vapours, Dragendorff Iodoplatinate Heating the plate to 100°C, then UV 254 nm Note: quantification by densitometry (565 nm) after derivatization by Dragendorff | Lupanine, sparteine, cytisine, *N*-methylcytisine, hydroxylupanine, matrine Fluorescence blue |
| **Purines** | | | |
| Coffee (*Coffea* sp.) Thea (*Thea sinensis*) Cocoa (*Theobroma cacao*) Kola nuts (*Kola nitida*) Mate (*Ilex paraguanensis*) Guarana (*Paullinia cupana*) | Ethyl acetate–methanol–water (100 : 13.5 : 10) Ethyl acetate–formic acid–glacial acetic acid–water (100 : 11 : 11 : 26) | UV light (254 nm) Dragendorff, acidic Iodine–hydrochloric acid reagent | Fluorescence quenching for caffeine, theobromine, theophylline range coloration Dark brown coloration |

For analysis by TLC, it is necessary to first prepare an extract and for this 1 g is extracted with 10 mL methanol for 30 min under reflux on a water bath. After filtration the solution is evaporated to about 1 mL before application to the plate. For separation on silica gel TLC plates, the following eluents are used: toluene–ether (10 : 10; saturated with 10% acetic acid) – this eluent is used for coumarin/aglycones; and ethyl acetate–formic acid–glacial acetic acid–water (10 : 11 : 11 : 26) for glycosides. Following chromatography, the coumarins can be detected by irradiation with UV light as there is distinct fluorescence quenching for all coumarins at 254 nm and 365 nm. Simple coumarins give blue or blue green fluorescence, and furano and pyrano-coumarins yellow, brown, blue, or blue–green fluorescence. The non-substituted chromones show less intense fluorescence: visnagin (pale blue);

khellin (yellow brown). They can also be detected by using spray reagents and these include NP/PEG and KOH.

## Anthocyanins

Anthocyanins are the most significant group of coloured substances in plants; they are responsible for the pink, mauve, red, violet, and blue colours of flowers and other plant parts. They are present in plants as glycosides of flavylium salts in petals and leaves. In the fruits of higher plants, they are mostly present as glycosides of hydroxylated 2-phenyl-benzopyrylium (**Figure 7**). Anthocyanins found in numerous plants used therapeutically include the following: hibiscus (*Hibiscus sabdariffa*) (hibiscin); corn flowers (*Centaurea cyanus*) (cyanin; pelargonin); common mallow (*Malva sylvestris*) (malvin

**Flavonols**

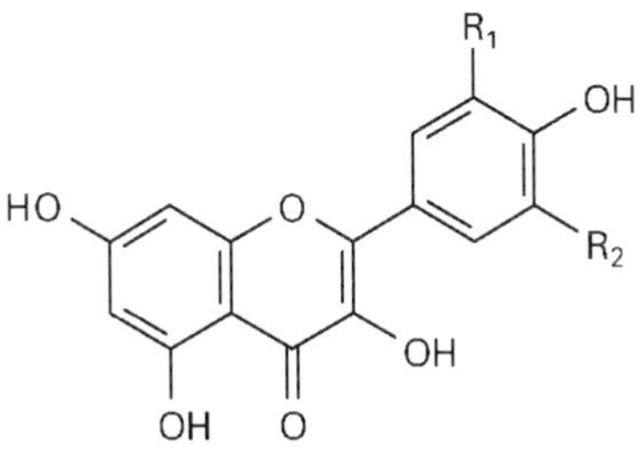

| | R₁ | R₂ | Aglycone | Quercetin |
|---|---|---|---|---|
| | OH | H | Quercetin | Q-3-O-glucoside |
| | H | H | Kaempferol | (isoquercitrin) |
| | OH | OH | Myricetin | Q-3-O-rhamnoside |
| | OCH₃ | H | Isorhamnetin | (quercitrin) |
| | | | | Q-3-O-galactoside |
| | | | | (hyperoside) |
| | | | | Q-3-O-rutinoside |
| | | | | (rutin) |
| | | | | Q-4'-O-glucoside |
| | | | | (spiraeoside) |

**Flavones**

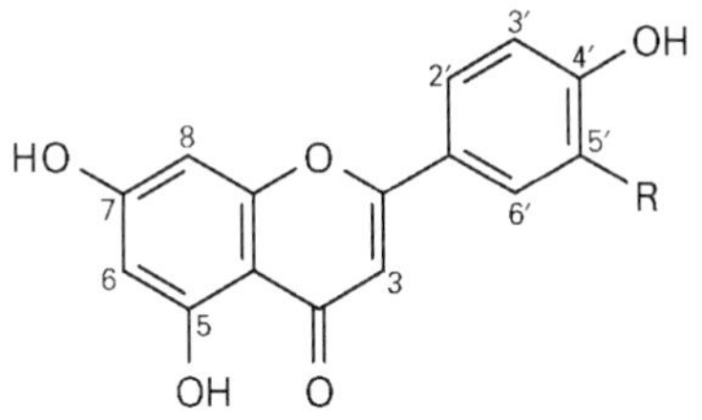

| Aglycone | Glycoside |
|---|---|
| Apigenin | A-8-C-glucoside (vitexin) |
| R=H | A-6-C-glucoside (isovitexin) |
| | A-7-O-apiosyl-glucoside (apiin) |
| | A-6-α-L-arabinopyranoside-8- |
| | C-glucoside (schaftoside) |
| Luteolin | L-5-O-glucoside (galuteolin) |
| R=OH | L-8-C-glucoside (orientin) |
| | L-6-C-glucoside (iso-orientin) |

**Flavanones**

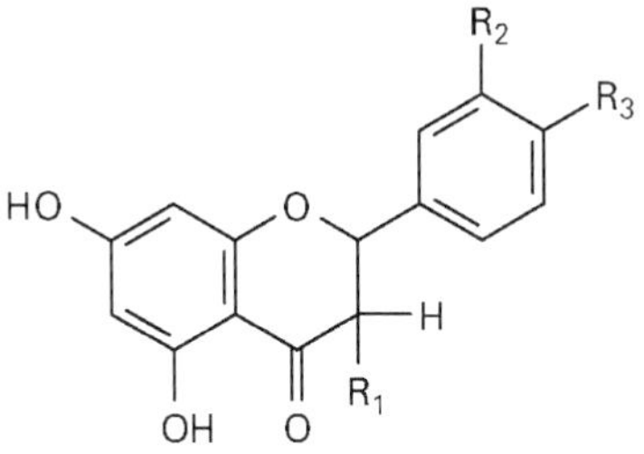

| R₁ | R₂ | R₃ | |
|---|---|---|---|
| H | H | OH | Naringenin |
| | | | Naringin |
| H | OH | OH | Eriodyctiol |
| | | | Eriocitrin |
| H | OCH₃ | OH | Homocriodyctiol |
| H | OH | OCH₃ | Hesperetin |
| | | | Neohesperidin |
| | | | Hesperidin |

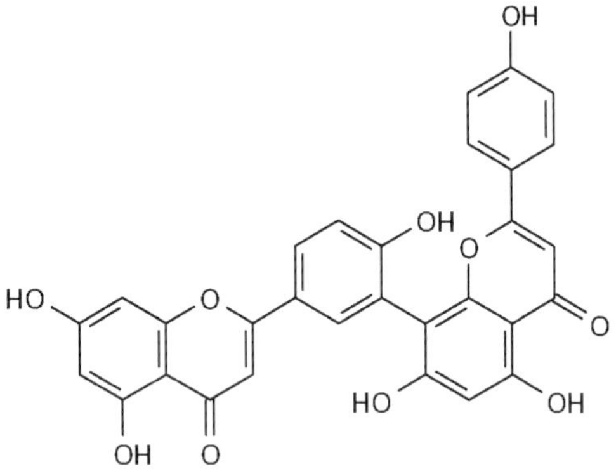

Amentoflavone

**Figure 5** Flavonoids.

**Table 3** Plant flavonoids: eluents

| Plant | Eluents | Compounds |
|---|---|---|
| Arnica (*Arnica montana*) | Ethyl acetate–glacial acetic–formic acid–water (100 : 11 : 11 : 26) | Quercetin-3-O-glucoside and 3-O-glucogalacturonide, luteonin-7-O-glucoside, kaempferol-3-O-glucoside |
| Ginkgo leaves (*Ginkgo biloba*) | Ethyl acetate–glacial acetic–formic acid–water (100 : 11 : 11 : 26)<br>Chloroform–acetone–formic acid (75 : 16.5 : 8.5)<br>Toluene–acetone (70 : 30) | Quercetin, kaempferol and isorhamnetin glycosides: flavonol acylglycosides<br>Biflavonoids: amentoflavone, bilobetin, ginkgogetin, isoginkgogetin<br>Ginkgolides a, b, c, catechin and epicatechin |
| Acacia flowers (*Robinia pseudoacacia*) | Ethyl acetate–formic acid–glacial acetic acid–water (100 : 11 : 11 : 26) | Kaempferol-3-O-rhamnosylgalactosyl-7-rhamnoside (robinin)<br>Acacetin-7-O-rutinoside, acaciin |
| Roman camomile (*Chamaemelum nobile*) | Ethyl acetate–formic acid–glacial acetic acid–water (100 : 11 : 11 : 26) | Apigenin-7-O-glycoside, 7-apiosil glucoside (apiin)<br>Quercitrin |
| Marigold flowers (*Calendula officinalis*) | Ethyl acetate–formic acid–glacial acetic acid–water (100 : 11 : 11 : 26) | Isorhamnetin glycosides<br>Isorhamnetin-3-O-glucoside (narcissin) |
| Hawthorn flowers, leaves (*Crataegus* sp.) | Ethyl acetate–formic acid–glacial acetic acid–water (100 : 11 : 11 : 26) | Quercetin glycosides: rutin, hyperoside spiraeoside<br>Flavon-C-glycosides: vitexin, isovitexin rhamnoside |
| Coltsfoot (*Tussilago farfara*) | Ethyl acetate–formic acid–glacial acetic acid–water (100 : 11 : 11 : 26) | Quercetin glucosides, rutin, hyperoside isoquercetin |
| German chamomile flowers (*Matricaria chamomilla*) | Ethyl acetate–formic acid–glacial acetic acid–water (100 : 11 : 11 : 26) | Flavonoid aglycones, apigenin-7-O-glucoside, luteolin-7-O-glucoside |
| Meadow-sweet (*Filipendula ulmaria*) | Ethyl acetate–formic acid–glacial acetic acid–water (100 : 11 : 11 : 26) | Quercetin-4′-O-glucoside (spiraeoside)<br>Hyperoside<br>Kaempferol glycosides |
| Lime flowers (*Tilia* sp.) | Ethyl acetate–formic acid–glacial acetic acid–water (100 : 11 : 11 : 26) | Quercetin glycosides: quercitrin, isoquercitrin<br>Kaempferol glycosides |
| Mullein flowers (*Verbascum album*) | Ethyl acetate–formic acid–glacial acetic acid–water (100 : 11 : 11 : 26) | Kaempferol, rutine, hesperidin, apigenin |
| Blackcurrant (*Ribes nigrum*) | Ethyl acetate–formic acid–glacial acetic acid–water (100 : 11 : 11 : 26) | Quercetin, kaempferol, myricetin and isorhamnetin glycosides |
| Round-headed bush clover (*Lespedeza capitata*) | Ethyl acetate–formic acid–glacial acetic acid–water (100 : 11 : 11 : 26) | Flavon-C-glycosides: orientin, iso-orientin, vitexin, isovitexin |
| Passion flower (*Passiflora incarnata*) | Ethyl acetate–formic acid–glacial acetic acid–water (100 : 11 : 11 : 26) | Flavon-C-glycosides: isovitexin, vitexin, orientin, iso-orientin<br>Flavon-O-glycosides: rutin, hyperoside, isoquercitrin |
| Lemon and other Aurantiaceae (*Citrus* sp.) | Ethyl acetate–formic acid–glacial acetic acid–water (100 : 11 : 11 : 26) | Flavanon glycosides, eriocitrin, naringin, hesperidin |
| Sophora buds (*Sorphora japonica*) | Ethyl acetate–formic acid–glacial acetic acid–water (100 : 11 : 11 : 26) | Flavonol glycosides: rutin |

**Table 4**  Plant flavonoids: sorbents/eluents

| Plant/sorbent | Eluents[a] | Compounds |
|---|---|---|
| **Silica gel** | | |
| Elm (*Ulmus* sp.) | Ethyl acetate–formic acid–acetic acid–water (100 : 11 : 11 : 27) | Quercetin glycoside, kaempferol glycoside |
| *Lipocedrus* | Chloroform–methanol–formic acid (90 : 10 : 1) | Flavonol glycoside |
| *Calendula officinalis* (flowers) | Benzene–methanol–acetic acid (90 : 16 : 8) | Flavonol glycoside |
| Henry anisetree (*Illicium henryii*) (root cortex) | Butanol–acetic acid–water–methanol (40 : 20: 10 : 50) | Flavonoids |
| *Sedum sediform* | Toluene–acetone–formic acid (60 : 60 : 12) | Phloroglucinol glycoside |
| *Olea europea* | Ethyl acetate–formic acid–water (60 : 10 : 10) | Flavonoids |
| **Polyamide (aglycones)** | | |
| *Alnus glutinosa* | Toluene–petroleum ether–MEK–methanol (50 : 25 : 11 : 13) | Flavonoid aglycones |
| *Keckiella* | Benzene–MEK–methanol (80 : 13 : 7) | Flavonoid aglycones |
| *Viguiera* sp. | Toluene–MEK–methanol (60 : 25 : 15) | Flavonoid aglycones |
| **Polyamide (flavonoids)** | | |
| *Lastenia californica* | Water–butanol–acetone–dioxane (70 : 15 : 10 : 5) | Flavonoids |
| *Rosa* cultivars | Methanol–acetic acid–water (90 : 5 : 5) | Flavonoids |
| *Cleome* sp. | Methylene chloride–benzene–methanol (75 : 5 : 5) Benzene–petroleum ether–MEK–methanol (60 : 60 : 7 : 7 or 60 : 30 : 7 : 7) Benzene–MEK–methanol (40 : 30 : 30) | Flavonoids |
| *Illicium henryii* | Butanol–acetic acid–methanol–water (40 : 10 : 20 : 50) | Flavonoids |

[a]MEK = methyl ethyl ketone.

and delphinidin glycosides); hollyhock (*Althea rosea*) (delphinidin-3-glycoside, malvidin-3-glycoside); bilberry (*Vaccinium myrtillus*) (delphinidin-3-glycoside = myrtillin A).

For analysis, anthocyanins must be extracted from plants with solvents containing acetic or hydrochloric acid. Plants are extracted for 15 min with methanol–HCl 25% (9 : 1) and the filtrates are subsequently used for chromatography. For TLC on silica gel, the following eluents are commonly used: ethyl acetate–glacial acetic acid–formic acid–water (100 : 11 : 11 : 26); and n-butanol–glacial acetic acid–water (40 : 10 : 20 or 40 : 10 : 50). Because these compounds are coloured, detection is possible visually without the need for chemical treatment or the developed TLC plates can be sprayed with anisaldehyde–H$_2$SO$_4$ reagent.

### Anthraquinone glycosides

A number of glycosides with aglycones related to anthracene are present in such drugs as cascara (*Cascara sagrada*), aloes (*Aloe* sp.), alder buckhorn (*Rhamnus frangula*), rhubarb (*Rheum officinalis*) and senna (*Cassia senna*). These drugs are employed as cathartics. On hydrolysis, the glycosides yield aglycones which are di-, tri-, or tetrahydroxy-anthraquinones or derivatives of these compounds (**Figure 8**). The anthraquinones possess phenolic

groups on C1 and C8 and keto groups on C9 and C10; in the anthrones and anthranol, only C9 carries an oxygen function. Most compounds in this group are present in the plant as O-glycosides. In the O- and C-glycosides, the only sugars found are glucose, rhamnose and apiose.

Prior to TLC, the powdered plant material is extracted for 5 min with methanol (1 g of plant in 100 mL) then filtered. It is necessary to hydrolyse the extract to characterize the aglycones and for this 1 g of powder plant is heated under reflux with X mL 7.5% hydrochloric acid for 15 min. After cooling, the mixture is extracted by shaking with X mL of chloroform or ether. The organic phase is then taken and concentrated to about 1 mL, and then used for TLC. Chromatography is performed on silica gel precoated plates with light petroleum–ethyl acetate–formic acid (75 : 25 : 1) or ethyl acetate–methanol–water (100 : 13.5 : 10) for all anthracene drug extracts except for senna. In this case, n-propanol–ethyl acetate–water–glacial acetic acid (40 : 40 : 29 : 1) is used.

For the non-laxative dehydrodianthrones of St John's wort (*Hypericum perforatum*) (Figure 8), TLC is performed with the eluent toluene–ethyl formate–formic acid (50 : 40 : 10).

Following TLC, all anthracene derivatives can be readily detected because they quench fluorescence

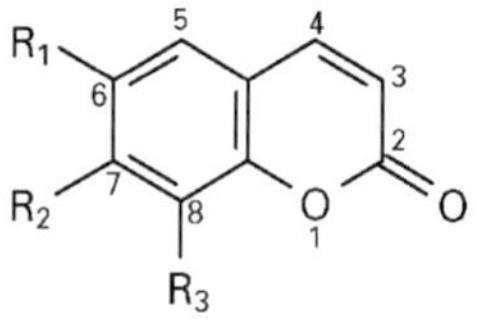

|  | R₁ | R₂ | R₃ |  |
|---|---|---|---|---|
| | H | H | H | Coumarin |
| | H | OH | H | Umbelliferone |
| | OH | OH | H | Aesculetin |
| | OCH₃ | OH | H | Scopoletin |
| | OCH₃ | OH | OH | Fraxetin |
| | OCH₃ | OH | OCH₃ | Isofraxidin |
| | OCH₃ | OH | O-gluc | Fraxin |

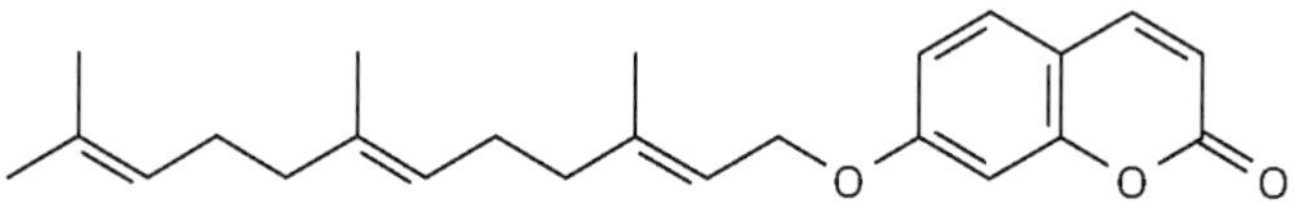

Umbelliprenin

**7,6-Furanocoumarins**

|  | R₁ | R₂ |  |
|---|---|---|---|
| 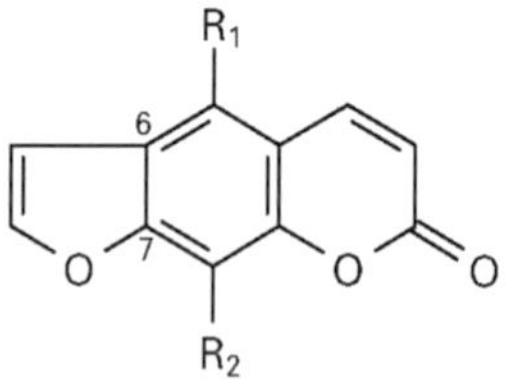 | H | H | Psoralen |
| | H | OCH₃ | Xanthotoxin |
| | H | OH | Xanthotoxol |
| | OCH₃ | H | Bergapten |

**7,8-Furanocoumarins**

|  | R₁ | R₂ |  |
|---|---|---|---|
| 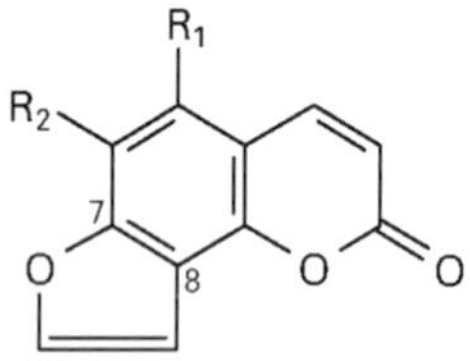 | H | H | Angelicin |
| | OCH₃ | H | Isobergapten |

**Pyranocoumarins**

| | R | |
|---|---|---|
| | —CO—CH=C(CH₃)₂ | Samidin |
| | —CO—CH₂—CH(CH₃)₂ | Dihydrosamidin |
| | —CO—CH—C₂H₅ | Visnadin |
| | CH₃ | |

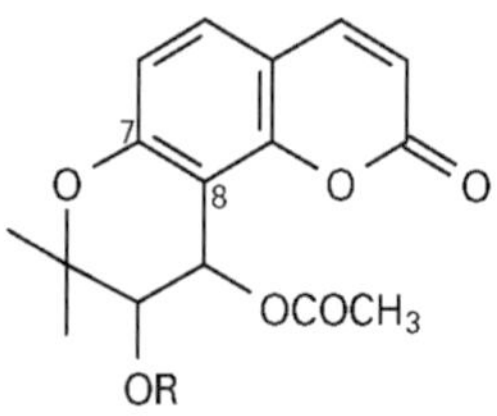

**Figure 6** Coumarins.

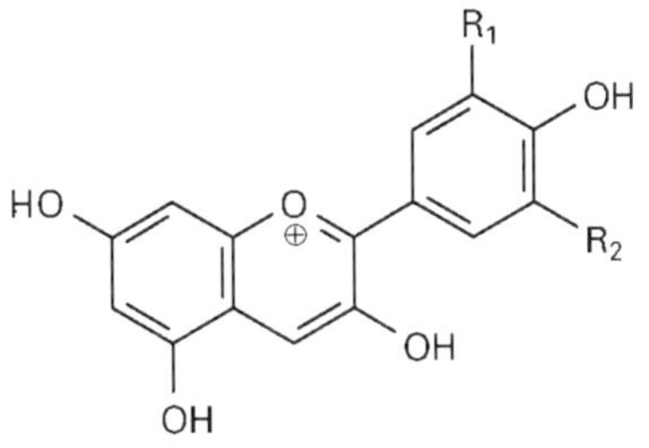


| | $R_1$ | $R_2$ |
|---|---|---|
| Pelargonidin | H | H |
| Cyanidin | H | OH |
| Malvidin | $OCH_3$ | $OCH_3$ |
| Petunidin | OH | $OCH_3$ |
| Delphinidin | OH | OH |

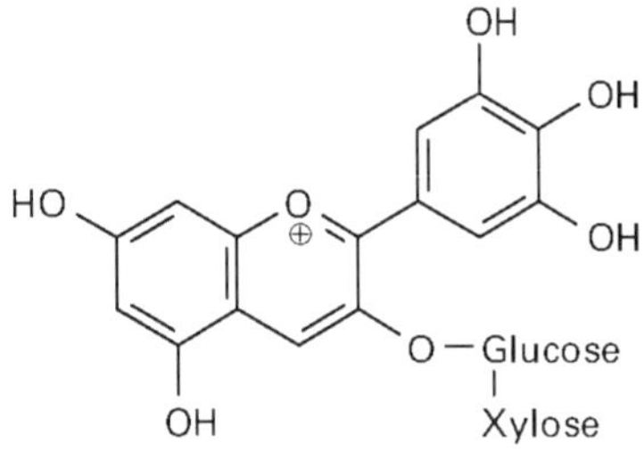


Hibiscin

**Figure 7** Anthocyanins.

when irradiated at UV 254 nm and give yellow or red–brown fluorescence. Different specific reagents are also used for detection (see the Appendix). Thus anthraquinone appears red in the visible after spraying with KOH. With NP/PEG, anthrones and anthranones give intense yellow fluorescence when irradiated at 365 nm. For the characterization of sennosides, the TLC plate is sprayed with $HNO_3$ and then heated for 10 min at 120°C. Before spraying with ethanolic KOH, these appear brown–red in UV 365 nm and brown in visible light. Hypericin gives red fluorescence when irradiated at 365 nm.

### Cardiac glycosides

There are some steroids present in nature, known as the cardiac glycosides, which are characterized by the highly specific and powerful action that they exert upon cardiac muscle. These steroids occur as glycosides with sugars in the 3-position of the steroid nucleus. The steroid aglycones or genins are of two types, either a cardenolide or a bufadienolide.

The steroids are structurally derived from the tetracyclic 10,13-dimethylcyclopentanoperhydrophenanthrene ring system. They possess a $\gamma$-lactone ring for the cardenolide or a $\delta$-lactone ring for the bufadienolide attached in the position at C17. The sugar residues are derived from deoxy- and/or C3-O-methylated sugars, and they are linked glycosidically by the C3-OH groups of the steroid aglycones (**Figure 9**). The main plants containing cardenolides are white foxglove (*Digitalis lanata*), red foxglove (*Digitalis purpurea*), oleander (*Nerium oleander*), strophanthus (*Strophantus gratus*; *Strophantus kombe*), adonis (*Adonis vernalis*) and lily of the valley (*Convallaria majalis*). The main plants containing bufadienolides are hellebores (*Helleborus* sp.) and squill (*Urginea maritima*).

For analysis, 1–10 g of powered plant is extracted by heating for 15 min under reflux with 20 mL 50% ethanol, with the addition of 10 mL 10% lead(II) acetate solution. After cooling and filtration, the solution is extracted twice with 15 mL dichloromethane. The combined lower organic phases are then filtered over anhydrous sodium sulfate and evaporated to dryness. The residue to dissolved in 1 mL of dichloromethane–ethanol (1 : 1) and the solution obtained is used for chromatography.

The TLC of the cardiac glycosides is accomplished on silica gel with the following solvents: ethyl acetate–methanol–water (100 : 13.5 : 10) or (81 : 11 : 8) and ethyl acetate–methanol–ethanol–water (81 : 11 : 4 : 8); and the lower phase of chloroform–methanol–water (35 : 25 : 10) for Hellebore bufadienolides.

The separated analytes can be detected under UV light at 254 nm as there is weak fluorescence quenching for cardenolides which is more distinct for bufadienolides. Spray reagents for detection include antimony chloride in chloroform and heating at 100°C, chloramine T, sulfuric acid and Kedde reagent (see the Appendix).

## Saponins

The formation of persistent foams during the extraction or concentration of plant extracts indicates the presence of saponins. The saponins are mainly triterpene derivatives, with similar amounts of steroid present. The most important plants are described in **Table 5**. Ginseng roots (*Panax ginseng*) contain triterpene glycosides: the ginsenosides a, b, c, d, e, f,

**Anthraquinone glycosides**

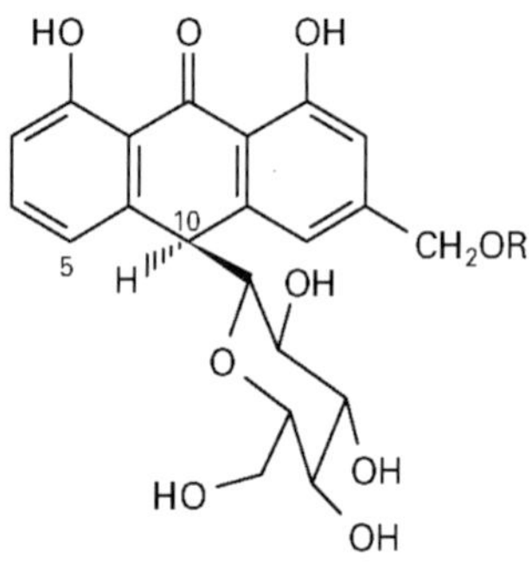

R=H, aloine A
R= α-L-Rha, aloinoside A

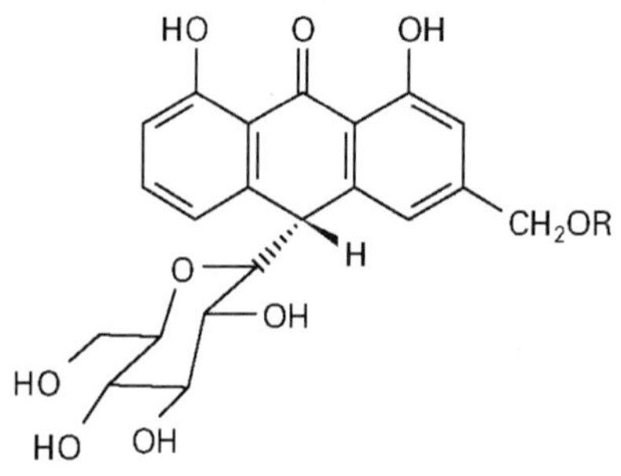

R=H, aloine B
R= α-L-Rha, aloinoside B

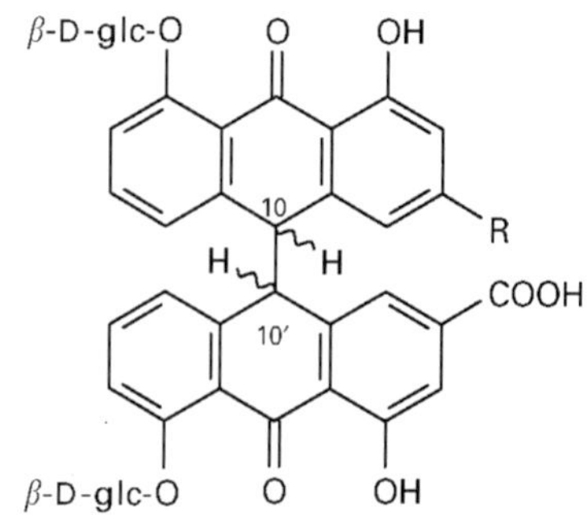

| R | C-10 | C-10′ | |
|---|---|---|---|
| COOH | R | R | sennoside A |
| COOH | R | S | sennoside B |
| CH$_2$OH | R | R | sennoside C |
| CH$_2$OH | R | S | sennoside D |

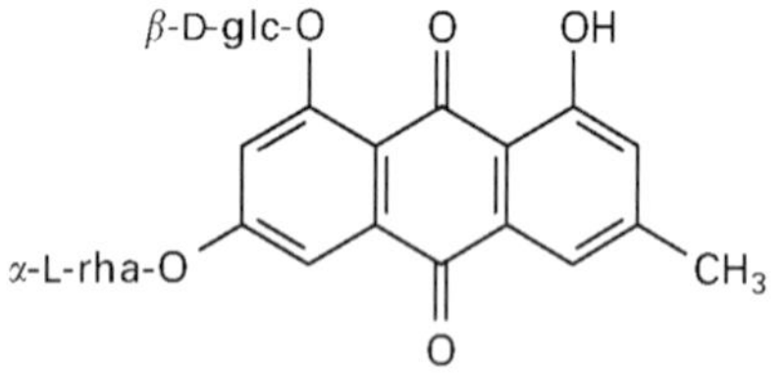

Glucofranguloside A

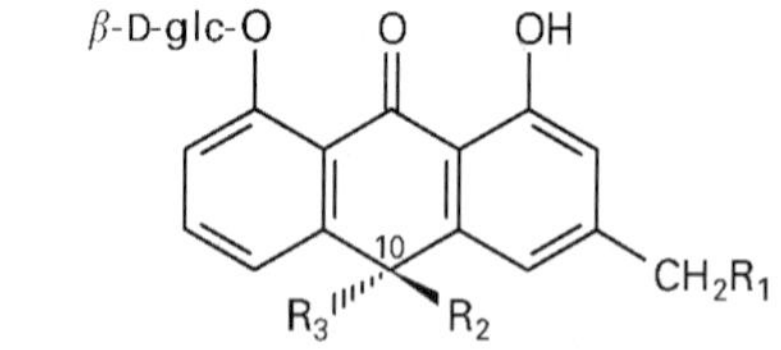

R$_1$=OH, R$_2$=β–D-glc, R$_3$ =H,  cascaroside A
R$_1$=OH, R$_2$=H, R$_3$=β–D-glc,  cascaroside B
R$_1$=H, R$_2$=β–D-glc, R$_3$ =H,   cascaroside C
R$_1$=H, R$_2$=H, R$_3$=β–D-glc,   cascaroside D

**Dehydrodianthrones**

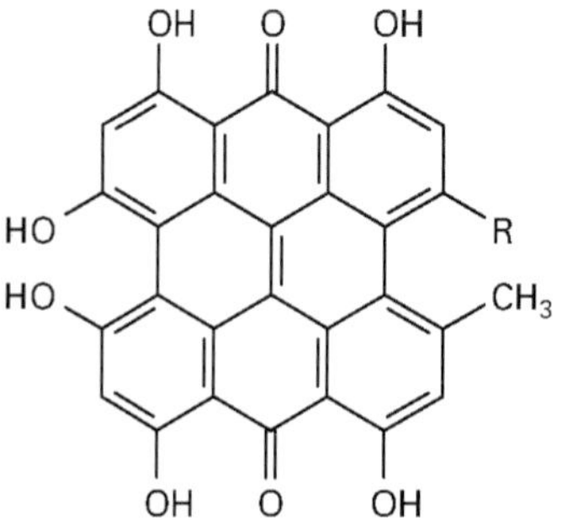

R=CH$_3$,    hypericine
R=CH$_2$OH,  pseudohypericine

**Figure 8**  Anthraquinones.

g, h. Horse chestnuts (*Aesculus hippocastanum*) contain the pentacyclic triterpenes glycosides aescine and aescinol, liquorice roots (*Glycyrrhiza glabra*) contain saponins aglycone from the glycyrrhetic acid, milkort root (*Polygala senega*) triterpene ester saponins 'senegenines', red soapwood root (*Saponaria officinalis*) triterpene saponins, Indian pennywort (*Centella asiatica*) asiaticoside A, B, 'madecassoside'

## Cardenolides

| Digitalis lanatae and Digitalis purpureae | | $R_1$ | $R_2$ | $R_3$ |
|---|---|---|---|---|
| Cardenolide aglycones | Digitoxigenin | H | H | H |
| | Gitoxigenin | H | OH | H |
| | Digoxigenin | H | H | OH |
| | Diginatigenin | H | OH | OH |
| | Gitaloxigenin | H | O—CHO | H |

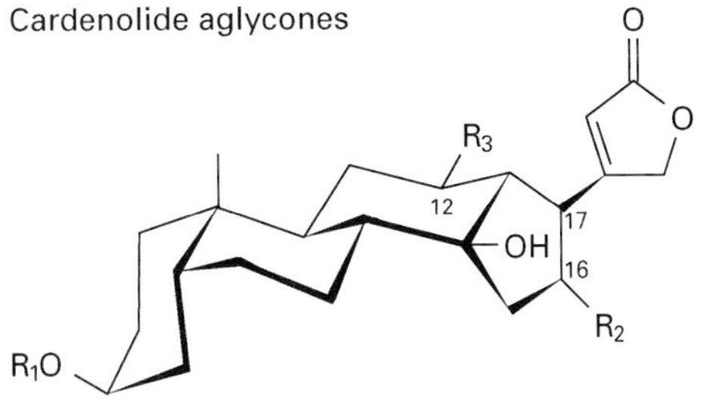

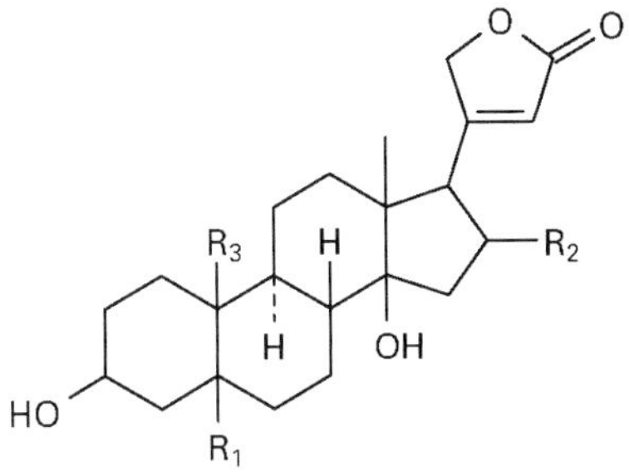

| | | $R_1$ | $R_2$ | $R_3$ | |
|---|---|---|---|---|---|
| Adonis | | OH | H | CHO | |
| | K-Strophanthidin (S) | | | | Cymarin (S-cymaroside) desglucocheirotoxin (S-gulomethyloside), k-Strophanthidin-β, k-strophanthoside |
| | | H | OH | CHO | |
| | Adonitoxigenin (A) | | | | Adonitoxin (A-rhamnoside), A-2-O-acetyl-rhamnoside, A-3-O-acetylrhamnoside, and glucosides and xylosides |
| | | H | OH | CH₂OH | Adonitoxigenol (-rhamnoside) |
| | | OH | OH | CHO | Strophadogenin (-diginoside) |
| Strophanthus | | OH | H | CHO | Cymarin (S-cymaroside), helveticoside (S-β-D-digitoxide) |
| | k-Strophanthidin (S) | | | | Erysimoside (S-digitoxoside-glucoside), k-strophanthin-β, k-strophanthoside |

## Bufadienolides

Helleborus

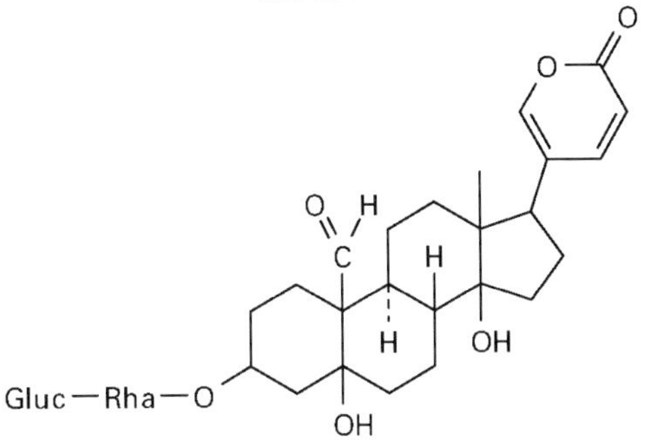

Gluc—Rha—O

Hellebrin

Scilla

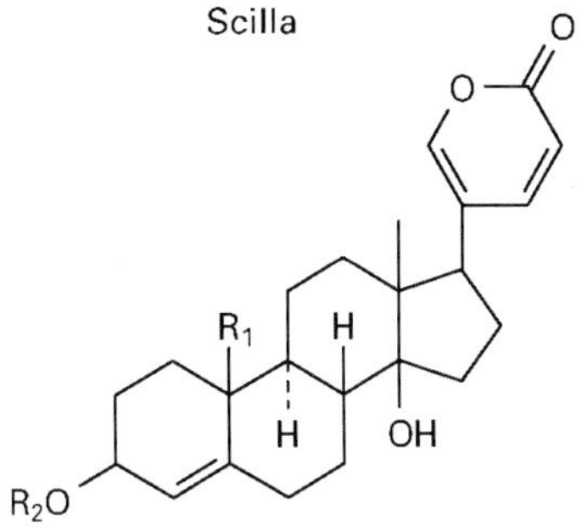

| | $R_1$ | $R_2$ |
|---|---|---|
| Scillarenin | CH₃ | H (Aglycon) |
| Proscillaridin A | CH₃ | Rham |
| Scilliphaeoside | H | Rham |
| Scillaren A | CH₃ | Gluc-Rham |
| Glucoscillaren A | CH₃ | Gluc-Gluc-Rham |

**Figure 9** Cardiac glycosides.

**Table 5**  Saponosids

| Plants | Eluents | Reagents | Compounds |
|---|---|---|---|
| Ginseng roots (*Panax ginseng*) | Chloroform–methanol–water (70 : 30 : 40)<br><br>Butanol–ethyl acetate–water (40 : 10 : 10) | Vanillin–phosphoric acid gives red zones in the visible and red; fluorescence in UV 365 nm Sulfuric acid then heated 110°C, 7 min | Triterpene glycosides, ginsenosides R$x$ ($x$ = a, b$_1$, b$_2$, d, e, f, g$_1$, h) Derived from dammaranne (protopanaxtriol, panaxadiol) |
| Eleutherocoque roots (*Eleutherococus senticosus*) | 1,2-Dichloroethane–ethanol–methanol–water (65 : 22 : 22 : 7) | Vanillin–sulfuric acid detection at UV 285 nm | Triterpenes: eleutherosides |
| Liquorice roots (*Glycyrrhiza glabra*) | Chloroform–glacial acetic acid–methanol–water (60 : 32 : 12 : 8) Ethyl acetate–ethanol–water–ammonia (65 : 25 : 9 : 1) | Anisaldehyde–sulfuric acid | Saponosids: glycyrrhizin, glycyrrhizic acid<br><br>Aglycone: glycyrrhetic acid |
| Milkwort roots (*Polygala senega*) | Chloroform–glacial acetic acid–methanol–water (60 : 32 : 12 : 8) | Anisaldehyde–sulfuric acid: five red saponin zones | Triterpene ester saponins = senegenins |
| Red soapwood (*Saponaria officinalis*) | Chloroform–glacial acetic acid–methanol–water (60 : 32 : 12 : 8) | Anisaldehyde–sulfuric acid give six violet zones and one brown band | Triterpene saponins derived from gypsogenin (quillaic acid) |
| Butcher's broom (*Ruscus aculeatus*) | Chloroform–glacial acetic acid–methanol–water (60 : 32 : 12 : 8) | Anisaldehyde–sulfuric acid: six to eight yellow or green bands | Steroid saponins = neoruscogenin glycosides Aglycones: ruscogenin and neoruscogenin |
| Sarsapilla (*Smilax* sp.) | Chloroform–glacial acetic acid–methanol–water (60 : 32 : 12 : 8) | Anisaldehyde–sulfuric acid: six yellow–brown saponins | Steroid saponins: smilax saponin, spirostanol-saponin Aglycones: sarsapogenin and its isomer smilagenin |
| Indian pennywort (*Centella asiatica*) | Chloroform–glacial acetic acid–methanol–water (60 : 32 : 12 : 8) | Anisaldehyde–sulfuric acid: violet–blue in fluorescence. Brown–violet zone | Esters saponins Madecassoside, a mixture of asiaticoside A and B |
| Soap bark (*Quillaja saponaria*) | Chloroform–glacial acetic acid–methanol–water (60 : 32 : 12 : 8) | Anisaldehyde–sulfuric acid: brown-to-violet zones | Quillaja saponins constitute a mixture of hydroxy gypsogenins |
| Horse chestnut seeds (*Aesculus hippocastanum*) | Chloroform–glacial acetic acid–methanol–water (60 : 32 : 12 : 8) Propanol–ethyl acetate–water (40 : 30 : 30) | Anisaldehyde–sulfuric acid gives main blue–violet–black of aescins Iron(III) chloride at UV 540 nm | Pentacyclic triterpene glycosides, aescine, aescinol<br><br>$\beta$-Aescine |

and soap bark (*Quillaja saponaria*) quillaja saponins (**Figure 10**).

Steroid saponins are present in Butcher's broom (*Ruscus aculeatus*) like ruscogenin and *Sarsaparilla smilax* sp. smilax saponins.

In order to obtain a suitable sample for TLC the plant powder is extracted by heating for 10 min under reflux with 10 mL of 70% ethanol. After filtration and evaporation, this solution is used for TLC. Ginseng radix is extracted with 90% ethanol under the same conditions.

Chromatographic solvents for the separation of these compounds on silica gel include chloroform–glacial acetic acid–methanol–water (64 : 32 : 12 : 8) which is suitable for separation of numerous saponin mixtures. For ginsenosides chloroform–methanol–water (70 : 34 : 4) is used whilst ethyl acetate–ethanol–water–ammonia (65 : 25 : 9 : 1) is useful for glycyrrhetic acid.

Once separated, the various analytes can be seen by inspection under UV light at 254 or 365 nm for glycyrrhizin and glycyrrhetic acid. Spraying with va-nillin–sulfuric acid reagent gives a range of colours for the saponins in the visible spectrum mainly blue, blue–violet and sometimes red and yellow–brown zones. The anisaldehyde–sulfuric acid reagent gives the same colours as vanillin. Vanillin–phosphoric acid reagent with ginsenosides gives red–violet colours in the visible spectrum, and reddish or blue fluorescence when viewed under UV light at 365 nm.

## Essential Oils

Essential oils are the odorous principles found in various plant parts (**Table 6**) because they evaporate when exposed to the air at room temperature, they are called 'volatile oils', 'ethereal oils' or 'essential oils'; the last term is applied since volatile oils represent the 'essences' or odoriferous constituent of the plants. Odorous principles consist either of (a) terpenes, i.e. alcohols (borneol, geraniol, linalool, menthol), aldehydes (anisaldehyde, citral), ketones (carvone fenchone, menthone, thujone), esters (bornyl acetate, linalyl acetate, menthyl acetate oxides, 1,8-cineole) or (b) phenylpropane derivatives i.e. anethole apiole, eugenol and safrole (**Figure 11**). Es-

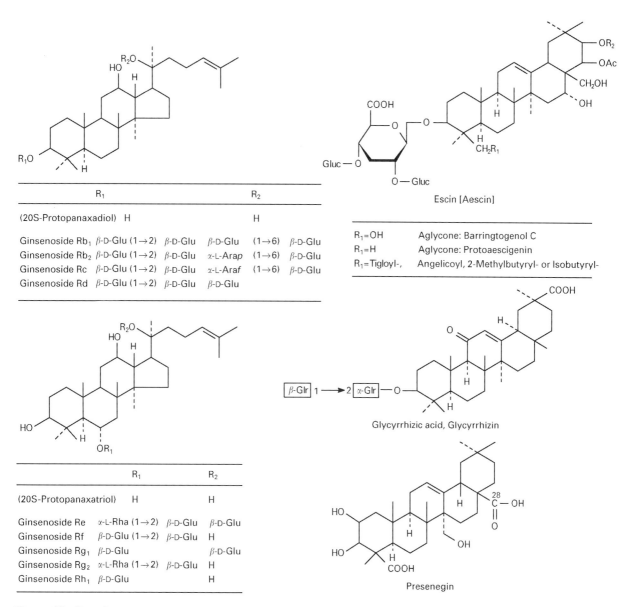

**Figure 10**  Saponins.

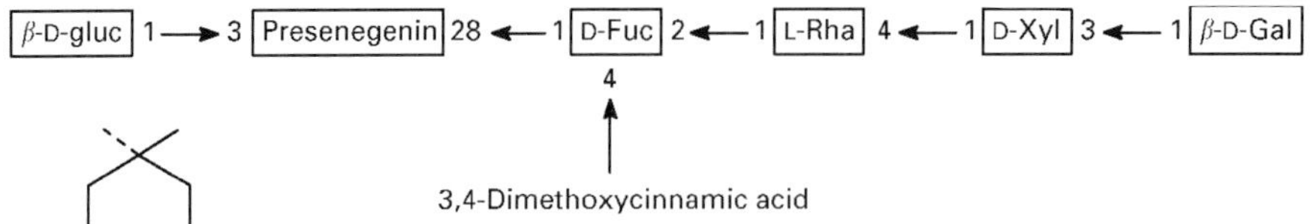

$\beta$-D-gluc 1 ⟶ 3 Presenegenin 28 ⟵ 1 D-Fuc 2 ⟵ 1 L-Rha 4 ⟵ 1 D-Xyl 3 ⟵ 1 $\beta$-D-Gal

4

↑

3,4-Dimethoxycinnamic acid

Senegin II

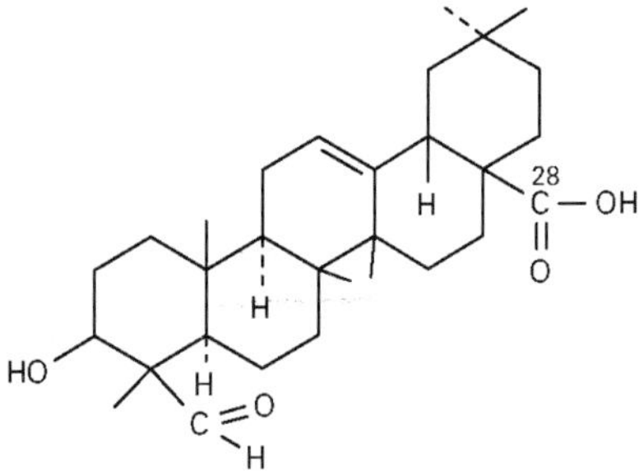

Gypsogenin

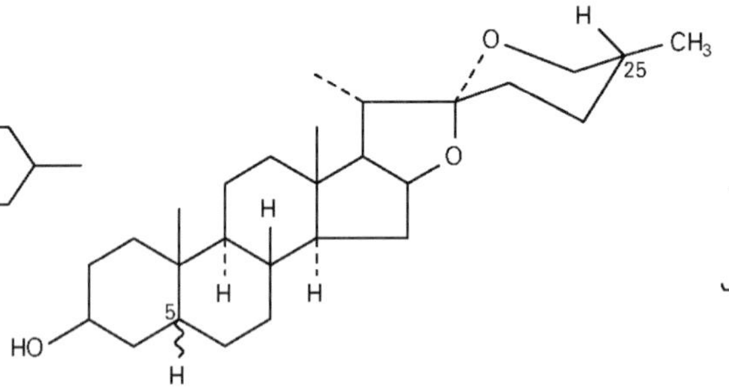

Ruscogenin

Smilagenin (5$\beta$, 25$\alpha$)

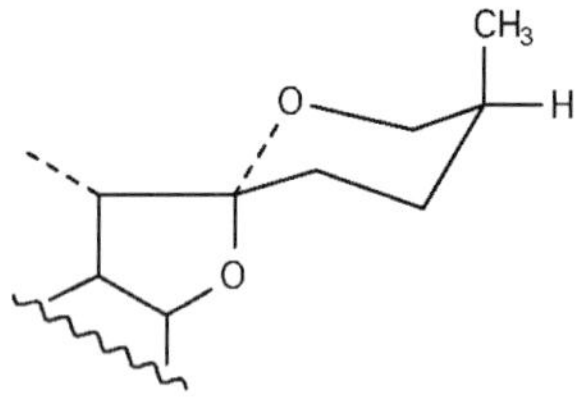

Sarsapogenin (5$\beta$, 25$\beta$)

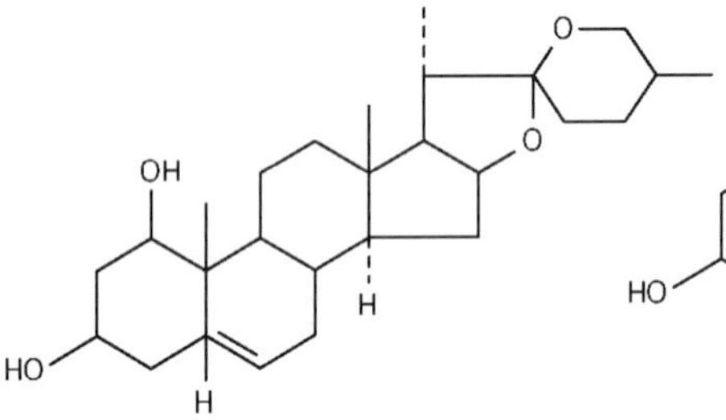

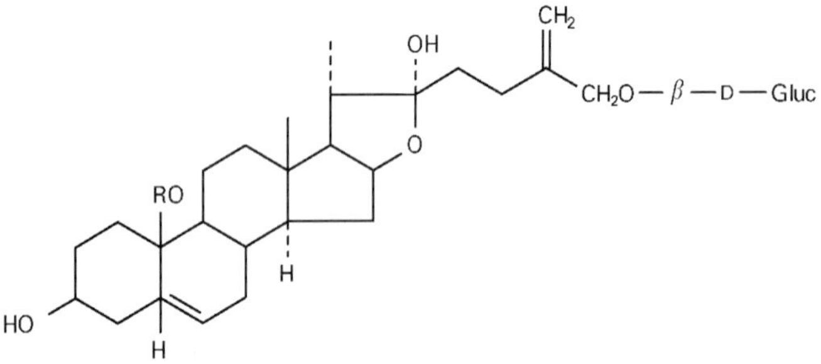

Ruscoside   R=O-$\beta$-D-Gluc-(1→3)-O-$\alpha$-L-Rha-(1→2)-O-$\alpha$-L-Ara(1→)

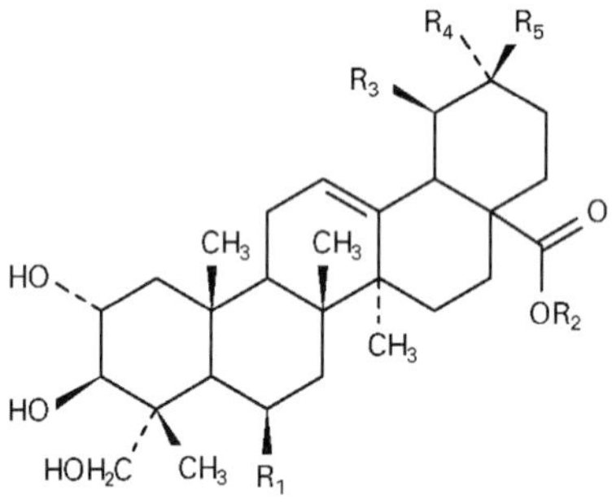

| | $R_1$ | $R_2$ | $R_3$ | $R_4$ | $R_5$ |
|---|---|---|---|---|---|
| Asiaticoside | –H | →1)-$\beta$-Gluc-(6→1)-$\beta$-D-Gluc-(4→1)-$\alpha$-L-Rha | –CH$_3$ | –CH$_3$ | –H |
| Asiaticoside A | –OH | →1)-$\beta$-Gluc-(6→1)-$\beta$-D-Gluc-(4→1)-$\alpha$-L-Rha | –CH$_3$ | –CH$_3$ | –H |
| (Asiaticoside B | –OH | →1)-$\beta$-Gluc-(6→1)-$\beta$-D-Gluc-(4→1)-$\alpha$-L-Rha | –H | –CH$_3$ | –CH$_3$) |

**Figure 10** *Continued.*

sential oils are soluble in ethanol and toluene and are mostly obtained by steam distillation of plant material.

For the preparation of extracts, a micro-steam distillation method is used to obtain the essential oil; a standard method is described in some pharmacopoeias. The essential oil is recovered in toluene or xylene and constitutes the sample to be analysed by TLC, but it is also possible to isolate it with hexane, ether or acetone.

Silica gel is the most widely used sorbent for the essential oils, with solvents such as benzene or toluene, chloroform, methylene chloride ethyl acetate for development. Because of its toxicity, however, benzene can no longer be recommended as a solvent for TLC. The eluent toluene–ethyl acetate (93 : 7) is suitable for the analysis and comparison of all of the important essential oils. Different eluents can be employed in special cases, e.g. toluene (*Pimpinella*

**Table 6** Essential oils

| Plant family | Plants | Main compounds | Colour with vanillin–sulfuric acid |
|---|---|---|---|
| Apiaceae | Anise (*Pimpinella anisum*) | *Trans*-anethole | Red–brown |
| | Fennel seed (*Foeniculum vulgare*) | *Trans*-anethole | Red–brown |
| | Parsley fruits (*Petroselinum crispum*) | Apiol | Violet–brown |
| | | Myristin | Violet–brown |
| | Caraway fruits (*Carvum carvi*) | Carvone | Red–violet |
| | Coriander fruits (*Coriandrum sativum*) | Linalool | Blue |
| Asteraceae | Camomile flowers (*Chamomilla reticula*) | Chamazulene | Red–violet |
| | | Bisabolol | Violet |
| | Roman camomile (*Chamaemelum nobile*) | Easters of angelicin | Grey–violet |
| | Worm seed (*Artemisia cina*) | 1,8-cineole, thujone | Blue |
| Lamiaceae | Peppermint leaves (*Mentha* sp.) | Menthol | Blue |
| | | Menthone | |
| | Rosemary leaves (*Rosmarinus officinalis*) | 1,8-cineole | Green |
| | | Borneol, pinene, camphene | Blue |
| | Lemon balm (*Melissa officinalis*) | Citronellal | Blue–violet |
| | | Citral | Black–blue |
| | | Citronellol | Violet–blue |
| | | | Black–blue |
| | Lavander flowers (*Lavandula officinalis*) | Linalool | Blue |
| | | Nerol | Blue |
| | | Borneol | Blue–violet |
| | Basil (*Ocinum basilicum*) | Methyl chavicol | Red |
| | Thyme (*Thymus vulgaris*) | Thymol | Red–violet |
| | | Carvacrol | Red |
| | | Linalool | Blue |
| | Sage leaves (*Salvia officinalis*) | Thujone | Pink–violet |
| | | 1,8-Cineole | Blue |
| | | Borneol | Blue–violet |
| | Greek sage (*Salvia triloba*) | 1,8-Cineole | Blue |
| | | Thujone | Pink–violet |
| Lauraceae | Cinnamon bark (*Cinnamomum zeylanicum*) | Cinnamaldehyde | Grey–blue |
| | Chinese cinnamon (*Cinnamomum aromaticum*) | Cinnamaldehyde | Grey–blue |
| Myrtaceae | Cloves (*Syzygium aromaticum*) | Eugenol | Yellow–brown |
| | Blue gum leaves (*Eucalyptus globulus*) | 1,8-Cineole = eucalyptol | Blue |
| Rutaceae | Bitter orange peel (*Citrus aurantium* sp.) | Limonene | Grey–violet |
| | | Citral | Blue–violet |
| | Bergamot (*Citrus aurantium* var. *amara*) | Limonene | Grey–violet |
| | | Citral | Blue–violet |
| | Orange flowers (*Citrus sinensis*) | Linalyl acetate | Blue |
| | | Linalool | Blue |
| | Lemon peel (*Citrus limon*) | Limonen | Grey–violet |
| | | Citral | Violet–blue |

*anisum*), chloroform (*Melissa officinalis*), methylene chloride (*Pimpinella, Juniperus, Lavandula, Rosmarinus,* and *Salvia*), toluene–ethyl acetate–(*Eucalyptus, Mentha*), chloroform–toluene (75 : 25) for *Thymus vulgaris* and *Chamomilla recutica* (Table 2).

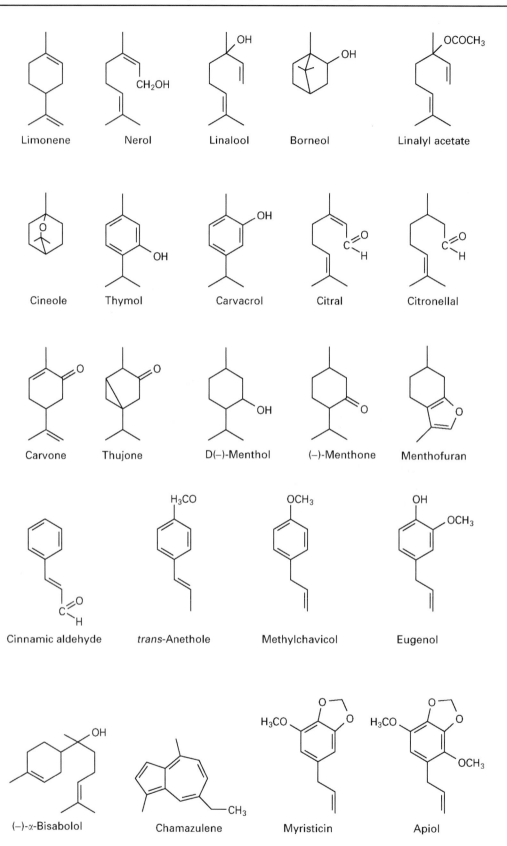

**Figure 11**    Essential oils.

Under UV light at 254 nm, compounds containing at least two conjugated double bonds quench fluorescence and appear as dark zones against the light-green fluorescent background of the TLC plate. This is the case for the derivatives of phenylpropane (anethole, safrole, apiol, myristicin, eugenol) and compounds such as thymol. The spraying reagents that can be used for the essential oils are (see Appendix) anisaldehyde–sulfuric acid, which gives blue, green, red and brown coloration, and phosphomolybdic acid, which gives uniform blue zones on a yellow background. However, the reagent most widely used for these compounds is vanillin–sulfuric acid which gives a range of different colours (Table 6).

## Cannabinoids

The cannabinoids are found in Indian hemp (marihuana; *Cannabis sativa* var. *indica*) (**Figure 12**). The cannabinoids are benzopyran derivatives but only $\Delta^{9,10}$-tetrahydro cannabinol (THC) shows hallucinogenic activity. The type and quantity of the constituents present in the plant depends on the geographical origin and climatic conditions. Marihuana is the flowering or seed-carrying, dried branch tips of the female plant. Hashish is the resin from the leaves and flower of the female plant. The most important cannabinoids are cannabidiol, cannabidiol acid, cannabinol, and $\Delta^9$-THC.

For chromatography, the powdered plant material is extracted with chloroform or hexane and separation can be performed on silica gel TLC plates with hexane–diethyl ether (80 : 20) or hexane–dioxane (90 : 10).

The cannabinoids can be detected by irradiation under UV (254-nm) light as they show fluorescence quenching. With the Fast blue reagent the cannabinoids form violet–red, orange–red or carmine zones; standard thymol gives an orange zone.

## Valepotriates

The main active constituents of these drugs, the valepotriates are triesters of a terpenoid, trihydric alcohol. This alcohol has the structure of an iridoid cyclopentanopyran with an attached epoxide ring. Valepotriates are present in valerian rhizome (*Valeriana officinalis*). The drug is extracted with dichloromethane at 60°C then filtered and evaporated to dryness. The chromatographic system is silica gel with toluene–ethyl acetate (75 : 25) or n-hexane–methyl ethyl ketone (80 : 20) as eluents. Under UV light (254 nm) they give a yellow fluorescence and in the visible region with the dinitrophenylhydrazine reagent, after heating, green–grey or blue zones appear. The valepotriates characterized in this way are valtrate, isovaltrate, and acevaltrate.

## Bitter Principles

The bitter principles are other compounds in plants which can be characterized by TLC. Plants with bitter principles include gentian, hops, condurango, artichoke and bryony root. Most of them possess a terpenoid structure and can be characterized in TLC with ethyl acetate–methanol–water (77 : 15 : 8) and then derivatization with vanillin-sulfuric reagent. However, these compounds are less important

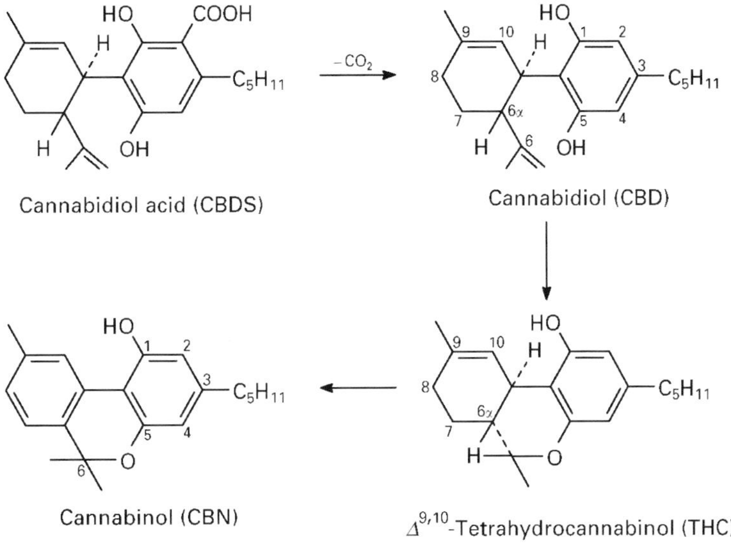

**Figure 12**  Cannabinoids.

than the compounds mentioned elsewhere in this chapter.

## Conclusion

TLC has many advantages for the analysis of herbal products, especially phytopharmaceuticals, for the identification of plants and the quantification of certain marker substances. Planar chromatography has advantages because it allows a parallel evaluation and comparison of multiple samples. In addition, various chromatographic separation systems can be combined with a multitude of specific and non-specific derivatizing agents. Even in samples having complex matrixes such as the pharmaceutical preparations of extracts of plants, sample preparation can be kept simple because of the use of the stationary phase for only one analysis. Unlike column chromatography, contamination of the chromatographic system by carryover cannot occur. In many instances the chemical composition of the herb is not completely known and for many plants, there are often no established methods of analysis available so that a rapid screening technique like TLC is very valuable. Constituents of herbals that belong to very different classes of chemical compounds can often create difficulties in detection, but with this in mind, TLC can offer many advantages.

**See Colour Plates 105, 106.**

*See also:* **III/Alkaloids:** High Speed Counter Current Chromatography; Liquid Chromatography; Thin-Layer (Planar) Chromatography. **Citrus Oils: Liquid Chromatography. Essential Oils:** Distillation; Gas Chromatography; Thin-Layer (Planar) Chromatography. **Pigments:** Liquid Chromatography; Thin-Layer (Planar) Chromatography. **Terpenoids: Liquid Chromatography. Appendix 17/Thin-Layer (Planar) Chromatography: Detection.**

## Further Reading

Anonymous (1998) Camag bibliography service. CD ROM. Muttenz, Switzerland: Camag.

Baerheim Svendsen A and Verpoorte R (1983) *Chromatography of Alkaloids. Part A: Thin Layer Chromatography.* Elsevier: Amsterdam.

Bruneton J (1999) *Pharmacognosy, (Phytochemistry Medicinal Plants),* 2nd edn. Paris: Tec and Doc.

Fried B and Sherma J (1994) *Thin Layer Chromatography: Technique and Applications.* New York: Marcel Dekker.

Fried B and Sherma J (1996) *Practical Thin Layer Chromatography,* pp. 31–47. Boca Raton: CRC Press.

Geiss F (1987) *Fundamentals of Thin Layer Chromatography.* Heidelberg: Hüthig.

Harbone JB (1973) *Phytochemical Methods. A Guide to Modern Techniques of Plant Analysis.* London: Chapman and Hall.

Jork H, Funk W, Fisher W and Wimmer H (1990) *Thin Layer Chromatography Reagents and Detection Methods,* Vol 1a. Weinheim: VCH.

Randerath K (1967) *Dünnschicht Chromatographie,* 2nd edn. Berlin: Springer Verlag.

Stahl E (1967) *Dünnschicht Chromatographie,* 2nd edn. Berlin: Springer Verlag.

Touchstone J (1992) *Practical Thin Layer Chromatography,* 3rd edn. London: John Wiley.

Wagner K, Bladt S and Zgainski EM (1966) *Plant Drug Analysis.* Berlin: Springer Verlag.

Wichtl M (1994) *Herbal Drugs and Pharmaceuticals.* Boca Raton: CRC Press.

# NEUROTOXINS: CHROMATOGRAPHY

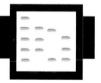

**K. J. James and A. Furey**, Cork Institute of Technology, Cork, Ireland

Chromatography has had a major impact on the discovery and detection of potent, naturally occurring neurotoxins. The neurotoxins discussed in this article were selected because they significantly impact on human health as a result of intoxications from bites and stings or the consumption of contaminated food and water. Many of these toxins target receptors that have implications for the development of potential therapeutic agents. In the neurotoxin topics that have been highlighted here, the role of chromatography in toxin discovery, purification and analysis is emphasized.

## Neurotoxins from Marine and Freshwater Algae

It was only in the latter part of the 20th century that scientists appreciated that certain species of microalgae can cause sporadic toxic events that can lead to serious illness, with occasional deaths, in humans as well as farmed and domestic animals. When high populations of toxin-producing microalgae occur

**Table 1**  Representative neurotoxins that are found in microalgae and marine food

| Toxin | Toxin potency, $LD_{50}\,\mu g\,kg^{-1}$ | Poisoning syndrome | Food type | Typical LC method |
|---|---|---|---|---|
| Saxitoxin | 3 | Paralytic shellfish poisoning (PSP) | Shellfish, freshwaters | LC-FL |
| Azaspiracids | 140–200 | Azaspiracid poisoning (AZP) | Shellfish | LC-MS |
| Anatoxin-a | 200 | 'Very fast death factor' | Freshwaters | LC-UV, LC-FL |
| Anatoxin-a(s) | 20 | – | Freshwaters | LC-MS |
| Tetrodotoxin | 8 | Puffer fish poisoning | Puffer fish, crabs | LC-FL |
| Brevetoxin-a | 95 | Neurological shellfish poisoning (NSP) | Shellfish | LC-UV |
| Ciguatoxin | 0.45 | Ciguatera | Finfish | LC-MS |

$LD_{50}$, lethal dose at 50% mortality, for 14–20 g mouse, when delivered intraperitoneally (IP).

they are termed harmful algal blooms (HABs) and there is compelling evidence to suggest that there is a global increase in the frequency of such events. Molluscs are the marine animals most susceptible to this toxicity, especially the filter-feeding varieties such as mussels, clams, scallops and oysters, which accumulate these toxins to hazardous concentrations. However, potent neurotoxins can also occur in finfish, toxins can accumulate in food plants, while skin absorption of toxins from bacteria is also possible. The comparative potencies of the main neurotoxins that are found in microalgae and marine food are shown in **Table 1**. Their high toxicities, together with the worldwide occurrence of these toxins, have led to a requirement for sensitive analytical methods for their determination. Ciguatoxin is a lipid-soluble toxin and occurs in the flesh of finfish in tropical and subtropical waters. It exerts its effect by activating voltage-dependent sodium channels, producing both gastrointestinal and neurological symptoms with occasional fatalities. Other neurotoxins that act on sodium channels include (1) tetrodotoxin, first discovered in puffer fish, (2) saxitoxin, responsible for paralytic shellfish poisoning (PSP) and (3) brevetoxins, responsible for neurological shellfish poisoning (NSP). A new human toxic syndrome, azaspiracid shellfish poisoning (AZP), which is caused by the consumption of mussels, has recently been identified in Europe. Cyanobacteria (blue-green algae) can produce both hepatotoxins and neurotoxins and represent a serious hazard because of their contamination of freshwater lakes and reservoirs used by animals and for human consumption. Anatoxin-a is the most common neurotoxin in freshwaters but PSP toxins have also been reported. Chromatography continues to play an important role, not only in the discovery of neurotoxins, but also in the development of analytical methods that are used for their quantitative determination for regulatory control and for forensic investigation of intoxications. Highly sensi-

tive chromatographic detection methods, especially using fluorescence (liquid chromatography–fluorescence, LC-FL) and mass spectrometry (liquid chromatography–mass spectrometry, LC-MS), have been particularly important and have been critical for the identification of the microalgae responsible for producing specific neurotoxins.

## Neurotoxins from Cyanobacteria (Blue-Green Algae)

There have been reports since the 19th century of animal mortalities associated with drinking water contaminated by cyanobacteria. These phenomena have occurred throughout the world and have been attributed to both hepatotoxic microcystins and to neurotoxins which are produced by some species of cyanobacteria. There are three classes of neurotoxins that are commonly found in cyanobacteria: (1) saxitoxin and analogues, (2) anatoxin-a(s), (3) anatoxin-a and analogues. These compounds are shown in **Figures 1–3**.

**Saxitoxin and analogues**  The most spectacular neurotoxic event due to cyanobacteria occurred in Australia in 1991 when a toxic bloom in the Darling river resulted in the deaths of 1600 sheep and other animals. These intoxications were attributed mainly to neurotoxins belonging to the saxitoxin group, which were previously identified as PSP toxins. These toxins are discussed elsewhere. The investigation of

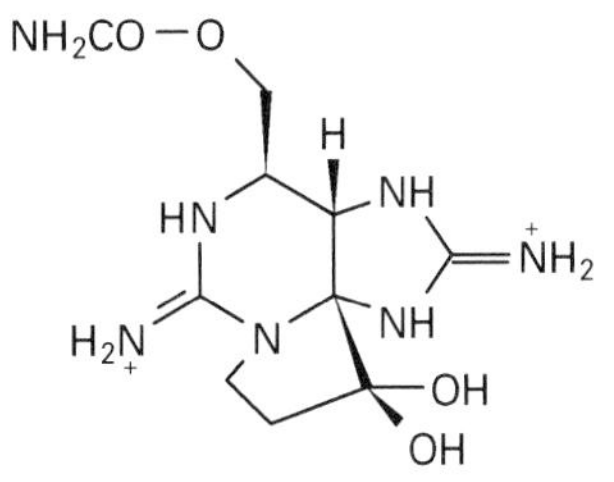

**Figure 1**  Saxitoxin.

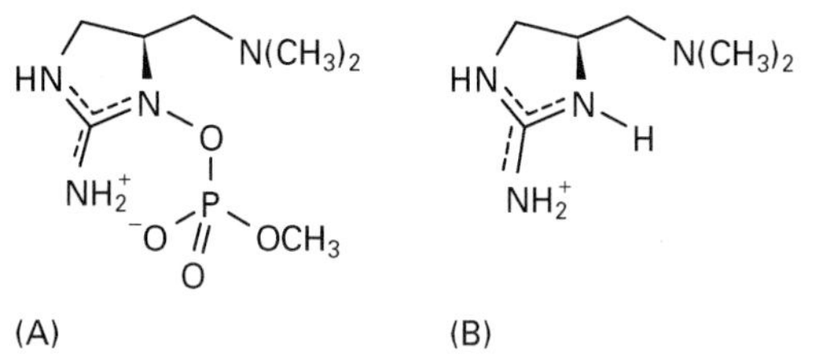

**Figure 2** (A) Anatoxin-a(s), $m/z$ 253.1; (B) fragment ion, $m/z$ 143.1.

these events relied heavily on the application of the fluorimetric high performance liquid chromatography (HPLC) method (LC-FL) for saxitoxins, developed by Oshima. This method uses paired ion reagents with three sets of isocratic reversed-phase chromatographic conditions to separate 18 analogues, which are detected by post-column oxidation and form highly fluorescent products. In addition to the PSP toxins that have previously been identified in shellfish, a number of new PSP analogues have recently been isolated from cyanobacteria.

**Anatoxin-a(s)**   Anatoxin-a(s) is a unique organophosphate toxin and it is a potent cholinesterase inhibitor. This neurotoxin has been identified in cyanobacteria in North America and in Europe where it has been fatal to dogs and birds. However, this toxin is difficult to detect as the chromatographic sensitivity using ultraviolet light (LC-UV) is very poor and it is probably more widespread in nature than has so far been discovered. Anatoxin-a(s) is unstable, particularly at slightly basic pH, but has been determined using fast atom bombardment–mass spectrometry (FAB-MS). However, even a 'soft' ionization technique such as electrospray LC-MS produces mainly the fragment ion at $m/z$ 143.1 (Figure 2B), due to loss of the phosphate moiety. Fortunately, this ion is sufficiently characteristic to allow the screening of water and algae samples for the presence of anatoxin-a(s).

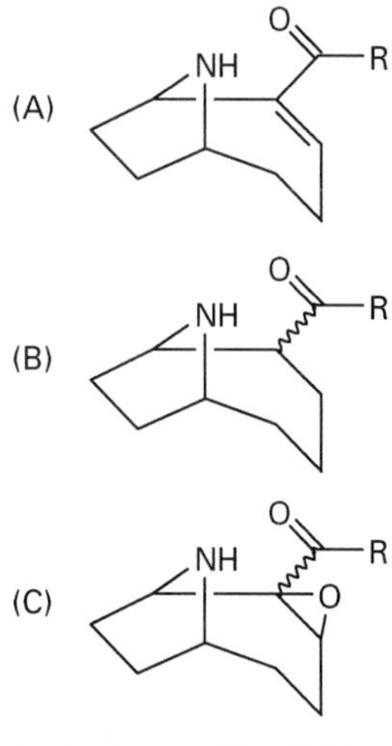

**Anatoxin-a and analogues**   Anatoxin-a (Figure 3A, $R = CH_3$) was the first cyanobacterial toxin to be structurally elucidated and it is a potent nicotinic agonist which acts as a depolarizing neuromuscular blocking agent. Typical symptoms in animals include muscle fasciculations, gasping and convulsions, with death due to respiratory arrest within minutes after drinking contaminated water. In fact, fatalities to animals, including cattle and dogs, were so rapid that before the identification of anatoxin-a this toxin was referred to as 'very fast death factor'. A related toxin, homoanatoxin-a (Figure 3A, $R = C_2H_5$), was isolated recently in Norway.

Several chromatographic methods are available for the analysis of anatoxin-a in cyanobacterial bloom material, including LC-UV and LC-MS using electrospray ionization. Derivatizations of anatoxins, followed by gas chromatography (GC) with electron capture or MS detection have also been successful. A highly sensitive LC-FL method has been developed by the authors for the determination of anatoxin-a, using derivatization with 4-fluoro-7-nitro-2,1,3-benzoxadiazole (NBD-F). This has been applied to the analysis of the anatoxins and their degradation products, the dihydro (Figure 3B) and epoxy (Figure 3C) analogues, which result from the reduction or oxidation of the alkene moiety. These degradation products are not detected by the commonly used LC-UV method as they do not have the $\alpha,\beta$-unsaturated ketone that is present in the parent toxins.

The determination of anatoxin-a in raw waters poses greater analytical problems mainly due to the low natural concentration of this toxin. However, Harada has developed an efficient solid-phase extraction (SPE) procedure in which anatoxin-a and analogues are efficiently extracted from water using a weak cation exchange phase. After trapping the anatoxins on the SPE cartridge, which is washed with methanol–water, they are readily eluted using acidic methanol since they are basic compounds. All of these anatoxins can be derivatized by reaction with NBD-F (**Figure 4**), at room temperature for several minutes, to produce highly fluorescent products that are readily separated using isocratic reversed-phase LC-FL. A typical chromatogram showing the separation of the NBD derivatives of anatoxin-a, homoanatoxin-a and their degradation products is shown in **Figure 5A**. The detection limit for anatoxin-a

**Figure 3** (A) Anatoxin-a ($R = CH_3$), homoanatoxin-a ($R = CH_2CH_3$); (B) dihydroanatoxin-a ($R = CH_3$), dihydrohomoanatoxin-a ($R = CH_2CH_3$); (C) epoxyanatoxin-a ($R = CH_3$), epoxyhomoanatoxin-a ($R = CH_2CH_3$).

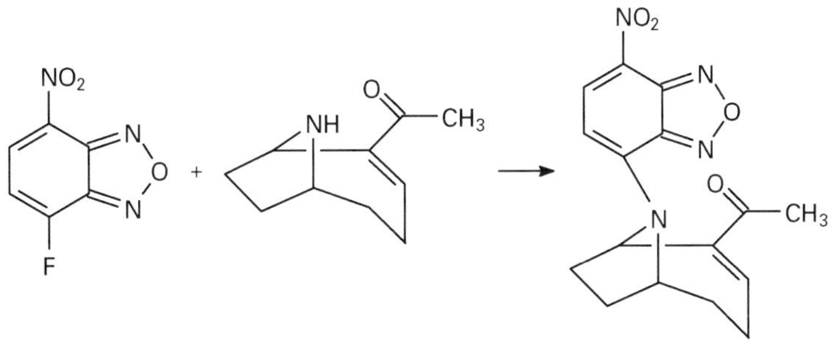

**Figure 4**    Reaction of anatoxin-a with NBD-F to produce a fluorescent product.

is $0.02\ ng\ mL^{-1}$, which allows this method to be applied to the routine monitoring of water supplies as well as for the forensic investigation of toxic incidents. This method was used to investigate the deaths of two dogs near Lough Derg, Ireland, two weeks after the event. Figure 5B shows a chromatogram from the investigation in which the dihydroanatoxin-a isomers are separated and are present in higher levels than anatoxin-a. Frequently, the detection of dihydroanatoxins in old samples may be the only evidence to implicate anatoxins in an intoxication event as anatoxin-a is unstable and readily decomposes at basic pH and in light.

### Azaspiracid Poisoning (AZP) – A New Human Toxic Syndrome

Azaspiracid (formerly KT-3) is a marine toxin responsible for a new toxic syndrome, AZP. The first confirmed occurrence of this toxicity was in November 1995 in the Netherlands when at least eight people reported severe illness after the consumption of cultured mussels from the west coast of Ireland (Killary Harbour). Although human symptoms, which included vomiting, severe diarrhoea and stomach cramps, were similar to diarrhoetic shellfish poisoning (DSP), only insignificant levels of DSP toxins were detected using LC-FL. The isolation of the major toxin was difficult, which is typical when dealing with a complex matrix such as shellfish tissue, and relied on the use of a variety of preparative chromatographic phases, as described by Satake and co-workers. Starting with 20 kg of mussel meat, the extract from several solvent extraction procedures was first subjected to adsorption chromatography using silica, followed by gel permeation chromatography. Next, carboxymethyl (CM) and diethylaminoethyl (DEAE) weak ion exchange phases were used and the final purification again used gel permeation to give 2 mg of azaspiracid.

Azaspiracid is characterized by a trispiro assembly and an azaspiro ring moiety that is unique in nature (**Figure 6**). In 1997 azaspiracid was again responsible for a toxic incident, which occurred in Arranmore Island, Ireland, with more than 12 local human intoxications. There have been several other reported incidents, in Italy and France, and two analogues of azaspiracid have also been isolated, namely methylazaspiracid (AZ-2) and demethylazaspiracid (AZ-3). There have only been limited toxicological studies of azaspiracids. Mice administered high doses of azaspiracid died after short periods, showing neurotoxic symptoms, while morphopathological studies showed that the target organs are the liver, spleen and digestive tract.

Azaspiracids, like most other shellfish toxins, are produced by dinoflagellates but the causative organism is as yet unknown. However, unlike other types of shellfish toxicity, natural depuration of azaspiracids is very slow and toxins can persist in shellfish for as long as eight months. The development of analytical methods to determine azaspiracids in seafood is therefore a priority research topic and LC-MS has proved invaluable for the monitoring and management of toxic outbreaks. **Figure 7** shows the chromatograms obtained from a crude extract from mussels using electrospray ion-trap mass spectrometry, without using any clean-up procedure. To achieve further confirmation of toxin identity, liquid chromatography–collision-induced dissociation–mass spectrometry (LC-CID-MS), with a collision energy of 40%, gave a characteristic fragmentation due to sequential loss of water molecules, as shown in **Figure 8**.

## Venoms from Snakes and Spiders

### Neurotoxic Peptides

More than 90% of the snake venoms produced by mambas, cobras and tiger snakes (which all belong to the family *Elapidae*), contain small protein molecules that are responsible for a wide range of toxicological and pharmacological activities. The complex mixtures of polypeptides that are present in most snake

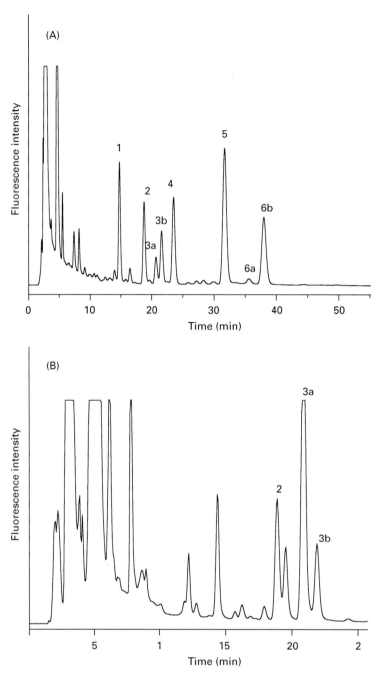

**Figure 5** (A) Chromatogram from the fluorimetric HPLC analysis of anatoxin standards following derivatization with NBD-F. 1, NBD-anatoxin-a epoxide (14.7 min, 4.1 ng); 2, NBD-anatoxin-a (18.7 min, 3.5 ng); 3a, NDB-dihydroanatoxin-a isomer 1 (20.7 min, 1.2 ng); 3b, NDB-dihydroanatoxin-a isomer 2 (21.6 min, 2.5 ng); 4, NBD-homoanatoxin-a epoxide (23.5 min, 4.3 ng); 5, NBD-homoanatoxin-a (31.6 min, 8.8 ng); 6a, NBD-dihydrohomoanatoxin-a isomer 1 (35.5 min, 0.4 ng); 6b, NBD-dihydrohomoanatoxin-a isomer 2 (37.9 min, 5.2 ng). Reproduced with permission from James KJ *et al.* (1998) *Journal of Chromatography* 798: 147–157. (B) Chromatogram obtained using a water sample from Lough Derg, Ireland, showing the presence of NBD derivatized anatoxins. This water contained anatoxin-a (2.1 $\mu g\,L^{-1}$), dihydroanatoxin-a isomer 1 (49 $\mu g\,L^{-1}$) and dihydroanatoxin-a isomer 2 (1.6 $\mu g\,L^{-1}$). HPLC conditions: 5 $\mu$m Prodigy $C_{18}$ column (250 × 3.2 mm); temperature, 35°C; mobile phase, acetonitrile–water (45 : 55, v/v); flow rate, 0.5 mL min$^{-1}$; fluorescence detection, $\lambda_{ex}$ 470 nm, $\lambda_{em}$ 530 nm.

venoms can be generally divided into neurotoxins and cardiotoxins (6000–10 000 amu), together with phospholipases $A_2$ (*c.* 13 000 amu) and larger enzymes. Often, the toxic effects of these peptides result from a synergistic effect of the polypeptides rather than from a high intrinsic toxicity of individual compounds, and they have attracted interest for their potential therapeutic applications.

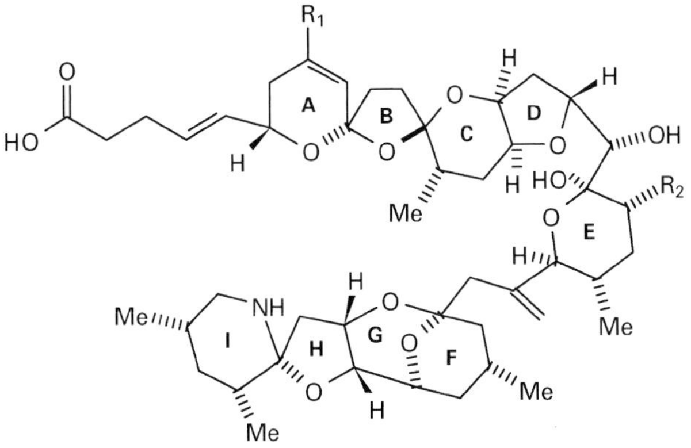

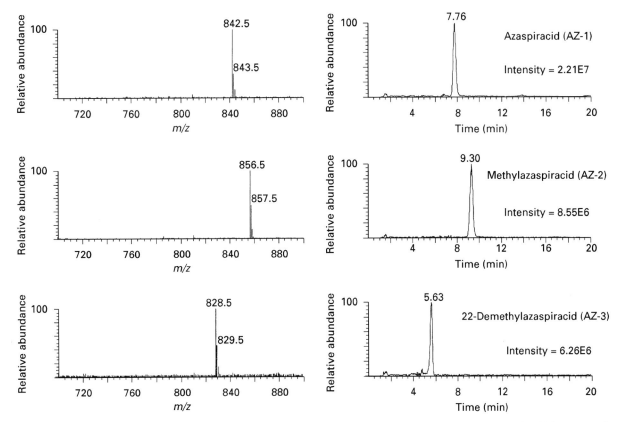

**Figure 6** Azaspiracid (AZ-1, $R_1 = H$, $R_2 = Me$), methylazaspiracid (AZ-2, $R_1 = R_2 = Me$), demethylazaspiracid (AZ-3, $R_1 = R_2 = H$).

The analysis of snake venoms has traditionally relied on chromatographic separations using gel filtration and ion exchange phases, with protein size determined using methods such as sodium dodecyl sulfate–polyacrylamide gel electrophoresis (SDS-PAGE). Subsequent sequencing of the amino acids in the isolated proteins can be carried out by automated Edman degradation with a gas phase microsequencer.

Using these techniques, the isolation of 28 peptides from the venom of the black mamba snake has been reported, most of which are structurally related cationic peptides, called dendrodotoxins, with similar activities. However, the application of capillary electrophoresis–electrospray ionization–mass spectrometry (CE-ESI-MS) has been shown to be a particularly effective technique for the separation and analysis of such complex mixtures of small proteins as are found in these venoms. A problem that is often encountered when separating basic peptides is the retention of a nett positive charge that leads to peak broadening due to the sorption of analytes to the negatively charged column wall. Several column wall derivatizing reagents have been reported to minimize this problem and it has been shown by Tomer and co-workers that when CE is carried out using a fused silica column derivatized with 3-aminopropyl-trimethoxysilane (APS), the charge on the capillary wall becomes positive. Excess negative ions in solution drive the electroosmotic flow from a high negative potential to ground, and positively charged analyte ions are repelled by the column wall. ESI-MS is particularly useful for the analysis of large protein analytes, as these become multiply charged. The $m/z$

**Figure 7** Electrospray LC-MS analysis of a toxic shellfish sample implicated in human intoxication. Azaspiracid (14.7 $\mu g\,g^{-1}$), methylazaspiracid (13 $\mu g\,g^{-1}$), 22-demethylazaspiracid (8 $\mu g\,g^{-1}$). Chromatographic conditions: $C_{18}$ Luna column (5 $\mu m$, 250 × 3.2 mm, Phenomenex); 25°C; acetonitrile–water (70 : 30) containing 0.50% TFA; Flow rate 0.2 mL min$^{-1}$.

T: + c sid full ms [50.00 – 1300.00]

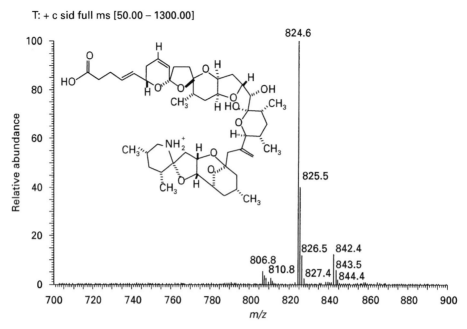

**Figure 8**  Mass spectrum obtained by LC-CID-MS of azaspiracid at 40% collision energy showing fragment ions for [M + H–H₂O]⁺ and [M + H–2H₂O]⁺ at m/z 824.5 and 806.4, respectively.

ratio (determined by the mass spectrometer) is consequently reduced, which permits the determination of higher charged large molecules using spectrometers with significantly lower mass ranges. For example, the CE-ESI-MS analysis of the black mamba snake venom showed two dominant ions in the spectrum at $m/z$ 1020 and 1090 which were related to the $[M + 7H]^{7+}$ and $[M + 6H]^{6+}$ ions of toxin 1 (7133.5 amu), previously shown to be the predominant dendrodotoxin.

### Acylpolyamine Neurotoxins from Spider Venoms

The tendency among humans to avoid contact with spiders is attributed, in part, to the ability of some species to deliver potent venoms. This venomous capability is used largely to paralyse or kill prey, particularly insects, but also affects a wide range of invertebrate and vertebrate animals. Neurotoxins are important as tools in neurochemical research and to investigate the functioning of neural receptors and ion channel modulators. Chemical studies on spider venoms have been hampered by the fact that many proteins, polypeptides and polyamines are often present in samples that are difficult to acquire in sufficient quantities. Special interest has been shown in acylpolyamines, low molecular weight toxins from spiders that antagonize specific glutamate receptors, since there are few other examples of this activity. Neural functions affected include memory and motor control and these toxins are important in studies to design potential therapeutic agents.

Acylpolyamines were first discovered in the 1980s and some examples are shown in **Figure 9**. They contain an aromatic ring connected to chains containing amide and amine moieties with some also incorporating amino acids, particularly arginine. The lengths of the polyamine chains vary considerably, as they can contain from 7 to 43 atoms, and they are linked to various phenol or indole rings. Examples of toxins with a dihydroxybenzene ring include NSTX-3 (Figure 9A-2, X = C) and JSTX-3 (Figure 9A-3, X = C).

Several chromatographic techniques have been applied to separate these complex mixtures, which can contain as many as 50 acylpolyamines in a single spider venom sample. Preparative HPLC can be used directly to separate toxins in an aqueous extract of spider venom using photodiode-array UV detection. However, tandem LC, combining UV detection with on-line fluorescence detection, following reaction with o-phthalaldehyde (OPA), has proved valuable for the detection of minor components. For these separations, a linear gradient of water (containing 0.1% trifluoroacetic acid), for example 5–60% acetonitrile, can be used for reversed-phase chromatography. Fractions containing active constituents typically require up to three further preparative LC steps, with both ion exchange and octadecyl-silica (ODS) columns, to purify toxins to homogeneity. Bioassays of fractions to detect toxin activity are also frequently used in these studies and an example is the assay of histamine release from rat peritoneal

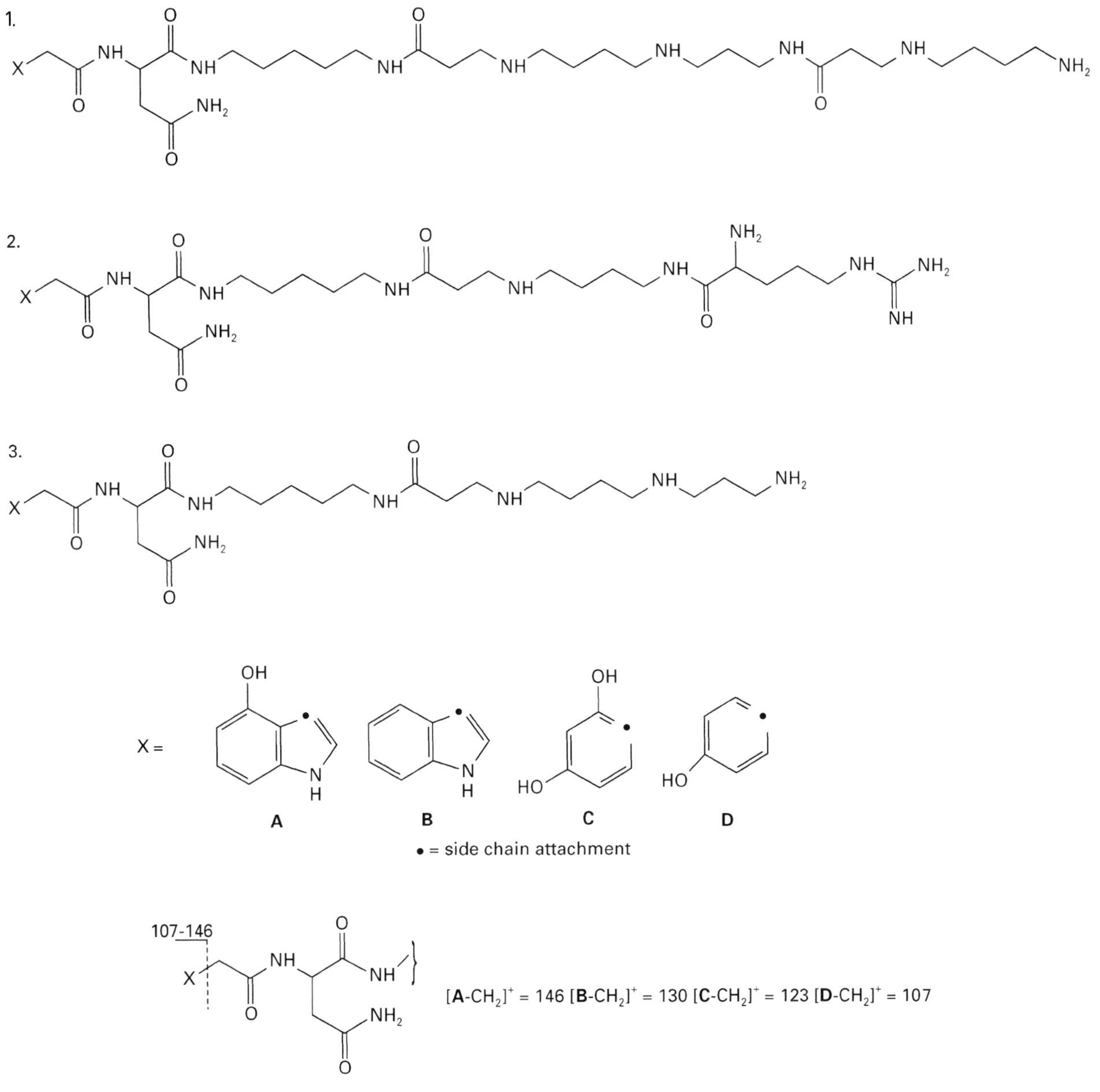

**Figure 9**   Structures of acylpolyamine toxins from spiders.

mast cells. Toxin identification requires NMR and/or hydrolysis of toxins to amino acids and amines, but for structural confirmation of very small quantities of toxins synthesis has also been employed.

In recent years, major advances have been made possible by the use of μ-column LC with FAB-MS. **Figure 10** shows the two-dimensional MS chromatogram from a spider venom extract obtained using on-line μ-column LC-FAB-MS. The acylpolyamines in this venom have an arginine terminal group with structures similar to those in Figure 9A–2A, –2B and –2C. Further structural information can be obtained

using collision-induced dissociation (CID) tandem mass spectrometry (MS-MS) which is often sufficient for the full structural elucidation of these toxins. The aromatic ring in each toxin is readily identified from the strong MS signal because fragmentation at the carbon adjacent to the ring produces ions of $m/z$ 107 (phenol), 123 (dihydroxybenzene), 130 (indole) or 146 (hydroxyindole), as illustrated in Figure 9B. From such studies, over 40 acylpolyamines have been identified in a crude spider venom extract. A combination of LC-MS and matrix-assisted laser desorption/ionization (MALDI) MS has emerged as

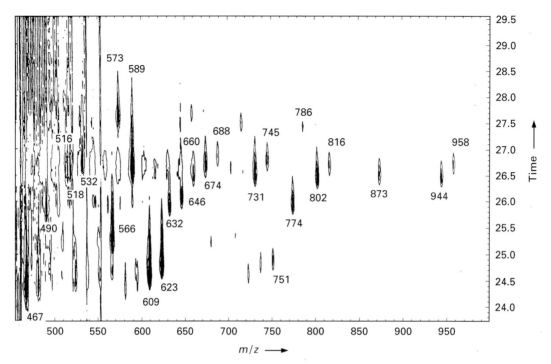

**Figure 10** Two-dimensional MS chromatogram display of protonated molecular ions [M + H]⁺ obtained from spider venom (*Nephilengys cruenta*) extracts using on-line µ-column LC-FAB–MS. (Reproduced with permission from Palma MS *et al.* (1997) *Natural Toxins* 5: 47.)

a powerful technique for the analysis of small samples of venom containing complex mixtures of these toxins.

**See Colour Plate 107.**

*See also:* **II/Chromatography: Gas:** Derivatization; Detectors: Mass Spectrometry. **Chromatography: Liquid:** Detectors: Fluorescence Detection; Detectors: Mass Spectrometry. **Electrophoresis:** Capillary Electrophoresis-Mass Spectrometry; One-dimensional Sodium Dodecyl Sulphate Polyacrylamide Gel Electrophoresis. **III/Toxins: Chromatography. Venoms: Chromatography.**

## Further Reading

Botana LM, Rodriguez-Vieytes M, Alfonso A and Louzao MC (1996) Phycotoxins: paralytic shellfish poisoning and diarrhetic shellfish poisoning. In: Nollet LML (ed.) *Handbook of Food Analysis*, vol. 2, pp 1147–1169. New York: Marcel Dekker.

Chorus I and Bartram J, eds (1999) *Toxic Cyanobacteria in Water*. World Health Organisation. London: E&FN Spon.

Codd GA, Jefferies TM, Keevil CW and Potter E, eds (1994) *Detection Methods for Cyanobacterial Toxins*. Cambridge: Royal Society of Chemistry.

Falconer I, ed. (1993) *Algal Toxins in Seafood and Drinking Water*. London: Academic Press.

Hu AT, ed. (1988) *Handbook of Natural Toxins*, vol. 3 *Marine Toxins and Venoms*. New York: Marcel Dekker.

James KJ, Furey A, Sherlock IR, *et al.* (1998) Sensitive determination of anatoxin-a, homoanatoxin-a and their degradation products by liquid chromatography with fluorimetric detection. *Journal of Chromatography* 789: 147–157.

McCormick KD and Meinwald J (1993) Neurotoxic acylpolyamines from spider venoms. *Journal of Chemical Ecology* 19: 2411–2413.

Oshima Y (1995) Postcolumn derivatisation liquid chromatographic method for paralytic shellfish toxins. *J AOAC Int.* 78: 528–532.

Palma MS, Itagaki, Y, Fujita T, Naoki H and Nakajima T (1997) Mass spectrometric structure determination of spider toxins: arginine-containing acylpolyamines from venoms of Brazilian garden spider. *Nephilengys cruentata*. *Natural Toxins* 5: 47–57.

Perkins JR, Parker CE and Tomer KB (1993) The characterisation of snake venoms using capillary electrophoresis in conjunction with electrospray mass spectrometry: Black Mambas. *Electrophoresis* 14: 458.

Reguera B, Blanco J, Fernández, ML and Wyatt T, eds (1998) *Harmful Algae*. Santiago de Compestela: Xunta de Galicia and Intergovernmental Oceanographic Commission of UNESCO.

Satake M, Ofuji K, Naoki H, *et al.* (1998) Azaspiracid, a new marine toxin having unique spiro ring assemblies, isolated from Irish mussels, *Mytilus edulis*. *Journal of the American Chemical Society* 120: 9967–9968.

Schäfer A, Benz H, Fieler W, *et al.* (1994) Polyamine toxins from spiders and wasps. In: Cordell GA and Brossi A (eds) *The Alkaloids: Chemistry and Pharmacology*, vol. 45, pp 1–125. San Diego: Academic Press.

ISBN 0-12-226770-2